CONTENTS

The Climate Near the Ground

The Climate Near the Ground

Sixth Edition

RUDOLF GEIGER, ROBERT H. ARON,
AND PAUL TODHUNTER

ROWMAN & LITTLEFIELD PUBLISHERS, INC.
Lanham • Boulder • New York • Oxford

ROWMAN & LITTLEFIELD PUBLISHERS, INC.

Published in the United States of America
by Rowman & Littlefield Publishers, Inc.
A Member of the Rowman & Littlefield Publishing Group
4501 Forbes Boulevard, Suite 200, Lanham, Maryland 20706
www.rowmanlittlefield.com

P.O. Box 317, Oxford OX2 9RU, United Kingdom

Fourth edition published by Harvard University Press in 1950.
Fifth edition published by Vieweg (a subsidiary of Bertelsmann Professional Information) in 1995.

British Library Cataloguing in Publication Information Available

Library of Congress Cataloging-in-Publication Data Available

ISBN 0-7425-1857-4 (cloth : alk. paper)

Printed in the United States of America

™ The paper used in this publication meets the minimum requirements of American National Standard for Information Sciences—Permanence of Paper for Printed Library Materials, ANSI/NISO Z39.48-1992.

IN COMMEMORATION OF RUDOLF GEIGER'S ONE HUNDREDTH ANNIVERSARY

Rudolf Geiger was born on 24 August 1894 and died 22 January 1981. He was born into a family of scholars. His father Wilhelm Geiger, 1856-1943, was Professor of Iranian and Indian Studies in Erlangen and in Munich. His brother Hans Geiger, 1882-1945, became famous in the field of nuclear physics and for assisting in developing the Geiger-Müller counter (Geiger counter) in 1928, in collaboration with E. W. Müller.

Rudolf Geiger attended a humanitarian gymnasium (secondary school) in Erlangen and later studied mathematics. His dissertation was entitled: *Geodesy of India in Antiquity and in the Middle Ages*. This was the same general area of study as his father. His father's research concentrated upon the arts and humanities, while Rudolf Geiger's concentrated on the mathematical and natural science aspects of this field.

In 1920 R. Geiger became an x-ray physicist at the Women's Clinic, University of Erlangen. He was, however, forced to abandon this line of work because of anemia, resulting from his exposure to x-rays. From 1920 until 1923, he worked as an assistant at the Institute of Physics, Technological University of Darmstadt, where he became interested in meteorology. In 1923 he joined the recently established Meteorological Department at the Forestry Experimental Station in Munich, which was attached to the Bavarian Weather Service. At the experimental station, he wrote his initial papers on the climate of plants and the nature of the air layers near the surface. In 1927 he was invited by the faculty to become a member of the academic community in meteorology and climatology. In 1937 he became Professor (ordinarius) of Meteorology, Mathematics and Physics, and Director of the Institute of Meteorology and Physics at the University of Forestry in Eberswalde.

In 1948 he became Professor (ordinarius) and Chairman of the Institute of Meteorology and Forestry Research Station at the University of Munich. In 1958 he attained the rank of Professor Emeritus.

Rudolf Geiger conducted research in the microclimatology of the air layers near the surface. He showed that temperature and moisture of air, soil, and vegetation are all interdependent.

The Bavarian network of weather stations made available abundant observational data for the exploration of the air layers near the surface. The investigations were performed in wooded patches of the Ebersberger Forest and although the research was very basic, it soon proved to have practical application in the afforestation of stands which had been damaged by localized late-frosts.

From the beginning, Geiger's research focused on the problems, consequences, and prevention of frost damage, both in forests, and gardens. As a result of this research, he was able to offer practical advice dealing with the prevention and restoration of frost damage.

Within forests his analysis of the variation of meteorological variables and the difference between stands with uniform crowns as compared to those with irregular heights, paved the way to his classical work on the climate of forests. This research concentrated upon the difference in the climate of the crown and trunk area in both coniferous and deciduous forests. He found that while the climate in the trunk zone is more uniform, the crown layer is the active zone in which the energy exchange between the surface below and the atmosphere above takes place.

Geiger's measurements in the Hokenkarfen Mountains in Wurtemberg contributed to the study and understanding of the effects of hills and slopes upon the local climate (topoclimatology). Rudolf Geiger also studied the variations of solar radiation, wind, and precipitation in different settings.

This research was further expanded to include microclimatic investigations in forested mountain regions (Grosser Arber) and in the beech forests which cover the thermal belts on many hills. Today this research constitutes the basis for further research into forests. Geiger not only actively took part in the development of scientific knowledge throughout the world, but he was also able to influence many other researchers.

In 1930, Geiger took part in an exploratory voyage to West Africa to measure high altitude winds in preparation for air traffic by blimps (Zeppelin air-ships). He became very active in international meteorological organizations. His appointment in 1937 as Secretary to the Commission for Agricultural Meteorology increased his responsibilities and widened his sphere of influence.

In 1950, as a result of his international reputation and connections, Rudolf Geiger was invited by C. W. Thornthwaite to visit his climatology laboratory at Johns Hopkins University in Seabrook, New Jersey. During this stay, he participated in several research projects and was a guest lecturer. C. W. Thornthwaite later returned this visit and collaborated with Geiger in Munich.

Rufolf Geiger became a member of the Bavarian Academy of Science. In 1956, through the mediation of UNESCO, he was invited as a specialist in microclimatology to participate in a symposium in Canberra, Australia on the climatology of arid regions. For the benefit of his students, he took numerous pictures of typical cloud formations over the archipelago of Indonesia while flying to Australia.

Rudolf Geiger's research is characterized by the organization and integration of countless observations into a unified explanation of the workings and interrelationships of the various elements of the local climate. This fundamental research has many practical applications and is in use in commercial aspects of forest climatology. As suggested by A. Baumgartner in *Forstarchiv* No. 35, 1965, the most rational form of fundamental research is when it has applications in the real world.

Geiger always presented the results of his research in a clear, precise and perceptual way. He was also an enthusiastic, inspiring, and exemplary teacher. He was a congenial person, had a calm disposition, and his personality exuded warm humanity. His considered, profound, and well thought out opinions fascinated not only his students, but everyone who had the good fortune to come into contact with him.

In 1952 R. Geiger became a full member of the Bavarian Academy of Science. From 1958 to 1961 he functioned as Class Secretary and during the period from 1963-1965 as Secretary of the Commission for Glaciology. Rudolf Geiger received the following honors: in 1955, he became a member of the Leopoldina Academy in Halle; in 1959, he received the Bavarian Order of Merit; in 1964, he became an honorary member of the Agrometeorological Society of Japan and the Munich Meteorological Society; in 1968, he received an Honorary Doctorate in Natural Sciences from the University of Hohenheim; and in 1977, he receive the Peter Lenné Gold Medal from the Goethe Foundation in Basle.

PREFACE SIXTH EDITION

When Rudolf Geiger completed his fourth edition of The Climate Near the Ground in 1961 it provided a complete, balanced and coherent presentation of the field of microclimatology up to that point in time. In the intervening four decades tremendous advances have been made in micrometeorological theory, instrumentation, modeling approaches, and our understanding of the role of plant and animal physiology in controlling energy and mass fluxes. It was not been possible to systematically incorporate these developments into the fifth and sixth editions of Geiger's work, nor did we intend to do so since several excellent texts have been written with that express purpose in mind. Because the surface microclimate has such broad applications to such a wide range of disciplines, however, we felt, and still feel, that the book provides a valuable resource for those workers who desire a thorough introduction to the characteristics and causes of the surface microclimate. This work is intended for persons from related fields who desire to understand the surface microclimate but who do not have the background, inclination, time or need to consult these other works. The goals of this book remain the same as the previous editions, namely to develop a clear and vivid textbook for those who are just taking up the study of microclimatology, and to illustrate application of these principles to as wide an array of subfields as possible. The text has been revised in order to keep more abreast of developments on all topics covered, to provide more examples of work from North America, and to more clearly illustrate how surface energy fluxes are linked to the nature of the surface environment. We have added discussions dealing with the speed of sound, the use of vegetative barriers to reduce sound, physiological control of energy and moisture exchanges, and expanded our discussion of microclimatic effects on animal behavior and dwellings and human behavior and dwellings. In the sixth edition we chose to change the units used from cgs to SI, to adopt a more modern notation system, and to clarify some of the figurcs and illustrations.

We wish to thank the following people for sending or making us aware of additional important works in the field.

Dennis Bohn, Rane Corporation

Günther Flemming, Technische Universität, Dresden

Yair Goldreich, Bar-Llan University, Israel

Helmut Klug, Deutsches Windenergle-Institut

David Martsolf, University of Florida

Roddam Narasimha, Indian Institute of Science

Barbara Obrebska-Starklowa, Jagiellonian University, Poland

Stanley Ring, Iowa State University

Michelle Swearingen, Pennsylvania State University

Eugene S. Takle, Iowa State University

Dennis W. Thomson, Pennsylvania State University

Our thanks also extend to Cheryl Dusty-Delauro and Stacie Young at Central Michigan University Graphics Production, for their assistance in redrafting many of our figures.

Robert Aron, Ph.D.
Department of Geography
Central Michigan University
Mt. Pleasant, MI. U.S.A.
E-mail: Robert.H.Aron@cmich.edu

Paul Todhunter, Ph.D.
Department of Geography
University of North Dakota
Grand Forks, ND. U.S.A.
E-mail: paul_todhunter@und.nodak.edu
September, 2002

PREFACE FIFTH EDITION

Few scientific books ever written become recognized as classics within their field. Dr. Geiger's *The Climate Near the Ground* was such a book. Sometimes referred to as 'The Bible of Climatology', it trained an entire generation of climatologists, and was a standard reference work for environmental scientists concerned with the surface microclimate. It not only presented the historical development of microclimatic understanding but also attempted to lead up to the current state of knowledge.

Although the last revision of the book was completed in 1960, a reading of that edition still reveals the depth and breadth of understanding, keen insight, and clarity of expression which made it such a revered work. It is our belief that *The Climate Near the Ground* still has much to offer the current generation of environmental scientists. It is for this reason that we have undertaken the current revision.

In updating this book, our goals remain the same as those of Rudolf Geiger, namely to develop a clear and vivid textbook for those who are just taking up the study of microclimatology and, at the same time, to serve as a reference work for those already familiar with the subject. As did Rudolf Geiger, we have attempted to bring all the topics covered in this book up to the present state of knowledge and have suggested some areas where additional research might prove both beneficial and fruitful.

With the proliferation of literature since 1960 we are certain that we have missed many advances that should have been included. As we plan to continue in the updating of this book, we invite readers to send any materials that they feel relevant. We, in turn, promise to acknowledge not only all contributions used but also those who have taken the time to send us the material.

We have made a concerted effort to conform to the organization, writing style, system of notation, and symbols of the original version of the book. This was done in a desire to assist in making those familiar with previous editions of the book feel comfortable with the revision.

In 1960 the International System of Units (SI) was adopted at the International Conference of Weights and Measures, organized by the International Bureau of Weights and Measures in Paris, as the world standard. Since that time the SI system has largely supplanted the cgs system upon which the last edition of the book was based. Since both the SI and cgs systems are in such wide use today, most readers will be familiar with each. An appendix listing unit conversions has been included to assist the reader when needed. We have also, for the most part, retained the notation and terminology used by Dr. Geiger even though alternate symbols and terms may be more commonly used at present. In order to assist the reader we have included alternative terms in parentheses following the first use of a term in the text.

In comparing this new edition with the previous edition, the reader will find that many of the diagrams and substantial portions of the text remain essentially unaltered. In many cases we have purposely not replaced older figures with more recent ones which show essentially the same features in order to provide the reader with a sense of the historical development of the field of microclimatology. The organizational structure of the book remains largely unaltered and the writing style remains qualitative in nature. Quantitative developments in the field since 1960, including theoretical micrometeorology, micrometeorological instrumentation, numerical modelling, remote sensing and digital elevation models have not been pursued in depth in order to retain the qualitative approach of the original work.

Our goal was not to totally rewrite this book, but simply to update those portions that have become dated and add new findings or topics that are relevant. Despite this original goal, substantial changes have been made both in topics added or expanded and those that have been reduced or eliminated. In addition, because the original translators were very faithful to the text, we have changed the wording in many places in an attempt to make the text read smoother.

We offer this revision in the hope that this important work remains as a useful reference in the field.

We wish to acknowledge I-Ming Aron for her time and many helpful suggestions. We also wish to thank Central Michigan University for its generous support in awarding Robert Aron a sabbatical to work on this book, and for furnishing financial assistance for typing and graphic arts.

Our thanks also extend to Dennis Pompilius at Central Michigan University Graphic Arts and our typist, Martha Brian.

Robert Aron, Ph.D.
Department of Geography
Central Michigan University
Mt. Pleasant, MI, U.S.A.

Paul Todhunter, Ph.D.
Department of Geography
University of North Dakota
Grand Forks, ND, U.S.A.
September 1994

PREFACE TO THE FOURTH EDITION

Anyone who merely turns the pages of this new edition will find that 48 percent of the figures are familiar to him from the third edition. But whoever reads it will discover that no three consecutive pages of text have been transferred unaltered. The enormous development that has taken place since 1950, particularly the surprising extension in the practical applications of micrometeorology, have made it necessary to rewrite the book.

In producing this work, I had in mind two aims which were linked more closely to each other than I had at first dared to hope. The new edition was to be a clear and vivid textbook for those who were just taking up the study of microclimatology, and at the same time a reference work for those already familiar with the subject. For the first task, I had in mind the students who would recoil with horror at the insurmountable barrier of an apparently unlimited and ever-increasing pile of literature and thus were in need of assistance. In addition, I was thinking of colleagues working in related sciences, who have no time to study our literature. And finally, but not least important, I was also thinking of all who work on the land, in forests and gardens, the architects, geographers, country planners, entomologists, doctors, transportation engineers, and others who – without having studied much physics – are anxious to acquire a knowledge of the rational physical principles governing the meteorologic laws that they have to put into practice. For the benefit of all these, I have at all times made renewed efforts to state the facts in the simplest and most uncomplicated manner possible. I have also tried at all times to improve the style so that the reader would be able to go on his way lightly. The extent to which I have been successful in providing genuine help to the reader remains to be seen.

The book should, in addition, lead up to the present day status in research and thus be of assistance to those already familiar with the subject. To stay within the required limits of space permitted, only brief references are made to results that pointed the way to the future. The novice will easily pass over these; whereas to the initiated they provide access to the literature on the subject.

I owe a special debt of gratitude to my university colleague, Dr. Gustav Hofmann, who read the finished manuscript. The lively discussions that followed resulted in the introduction of many improvements in the book.

I would like, further, to thank all those who assisted me by sending reprints of their works. Much came to my knowledge only by this means. Such assistance is indispensable in a technical field that has so many contacts with neighboring sciences. It is precisely the representatives of these adjacent fields of knowledge whom I would ask most urgently to bring to my notice any of the deficiencies and gaps in treatment, which are unavoidable with the present day scope and tempo of research in microclimate.

During my work on the new edition, it became clear to me why, in spite of the ever-swelling flood of published papers, the number of usable textbooks is increasing much more slowly. In drafting almost every chapter, I could feel the spirit of a cherished colleague, at home or abroad, looking over my shoulder, and it seemed to me that he was more fitted to write just this chapter than I was. Then came the feeling of suffocation under the weight of good new literature, and the anxiety that the first chapter would be out of date before the last one could be finished. However, I was able to bear these tensions daily for three years, because without useful signposts no one can find his way in the labyrinth of science. If I have been able to point the way to a few, this will be my best reward for all the pains taken.

Munich–Pasing, November 1960 RUDOLF GEIGER
Perlschneiderstrasse, 18

PREFACE TO THE FIRST EDITION

I was introduced to microclimatology by Professor A. Schmauss. He put me in charge of the organization and direction of the Bavarian special network for investigation of the air layer near the ground. Later I also had to make two extensive open-air investigations in forest meteorology. I had a good opportunity to get in closer touch with people dealing in forestry, moor cultivation, and agriculture. Thus I became acquainted with the difficulties in the practical application of the results of climatological research. The problem of application is indeed not new. While a systematic study has not been undertaken as yet, many valuable contributions have already been made, as, I hope, this book will demonstrate. This book is intended for the practical person who has neither time nor opportunity to look for the available papers from the vast meteorological literature. When, therefore, I was invited to write "Climate near the Ground," I was glad to have the opportunity to attempt a first survey of microclimatological problems. This is the best way for me to thank people for the manifold suggestions that I have received from them, particularly those with a special interest in forestry. It is also a pleasure to express here my sincerest thanks to Professor Schmauss for his constant and unselfish furtherance of my work.

RUDOLF GEIGER

Munich, July 1927

INTRODUCTION

1 Microclimate and Research

Most people are familiar with the meteorological shelters (screens) belonging to the observing stations of various national meteorological networks. The instruments housed in these shelters are protected against radiation and precipitation, yet they are well exposed to the air by means of numerous openings in the sides at a height of about 1.5 m above the level of the ground.

This positioning level was decided upon after a long series of observations toward the end of the 19th century. At this height, the random influences of the position selected, the nature and state of the immediate ground situation, and the vegetation growing on it no longer play a significant role. The meteorological station is thus "representative" of the more general surrounding area, and the climate thus measured is that experienced by a human walking upright, or human climate, as it is sometimes termed.

The air layer below this agreed upon level of about 1.5 m will be called the air layer close to the ground. As the ground surface is approached, many atmospheric elements change rapidly. For example, the closer to the ground, the more the wind speed is reduced by friction, and the less the mixing of the air. The ground surface absorbs radiation from the sun and sends out radiation influencing the air in contact with it. The ground surface is also a source of water vapor, which escapes into the atmosphere by evapotranspiration, and particulates and gases which diffuse from the soil.

It is this layer near the ground where plants grow, where young plants, which are especially sensitive to weather fluctuations, germinate and become established, and where animals and insects that are restricted to near the surface, or that live on or below the ground, must depend for their survival. The study of the climate near the ground is, therefore, the climate of the habitat of most living organisms.

The immediate consequence of this is that the climate conditions to which young growing plants are exposed cannot be deduced directly from the climate summaries published for the network of official stations. One example will be discussed in more detail in Section 42, where only one night of frost (-1.8°C) was recorded during a particular month of May at the meteorological station in Munich. While in the same month, 23 nights of frost with temperatures as low as -14.4°C were measured in the air layer near the ground 20 km outside the city. These temperature variations are of vital importance to farmers, foresters, gardeners, engineers, and architects.

Meteorological variables are not only subject to vertical changes near the ground, but also vary horizontally within short distances. These variations are brought about by changes in the nature and moisture content of the soil, by even minute differences in surface slope, and by the type and height of vegetation growing on it. All these climates found within a small space are grouped together under the general description of microclimate.

Climatic phenomena are frequently grouped into one of four scales based upon their characteristic horizontal, vertical, and temporal dimensions. These scale ranges, given in Figure 1-1 and Table 1-1, are not precise and have been the subject of much debate largely because their values vary with topography and the nature of the underlying surface. Microclimate, which describes the climate of an individual site or station, is characterized by rapid vertical and horizontal changes due to the effects of surface frictional drag, soil type, surface slope and orientation, vegetation cover, and surface moisture content. Local climate is associated with the climate of a locality where surface conditions are clearly distinct from those of the surrounding area. M. M. Yoshino [110]* states that the formation of local climates (often synonymous with topoclimates or terrain climates), are controlled by large-scale changes in ground cover (such as forests and cities), the distribution of land and water (such as rivers, lakes, and coasts), and topographic controls (such as slope, orientation, elevation, and terrain configuration). Mesoclimate describes the climate of a region while macroclimate is concerned with climate at a continental scale as it is governed by large-scale atmospheric circulation systems. The focus in this book will be at the microclimate and local climate scales.

Table 1-1. Spatial and temporal scales of climate. (After M. M. Yoshino [110])

Scale	Horizontal range (m)	Vertical range (m)	Primary time scale (sec)
Microclimate	10^{-3} - 10^{2}	- 10 - 10^{1}	$< 10^{1}$
Local climate	10^{2} - 10^{4}	$5 \cdot 10^{0}$ - 10^{3}	10^{1} - 10^{4}
Mesoclimate	10^{3} - $2 \cdot 10^{5}$	$5 \cdot 10^{2}$ - $4 \cdot 10^{3}$	10^{4} - 10^{5}
Macroclimate	$> 2 \cdot 10^{5}$	10^{3} - 10^{4}	10^{5} - 10^{6}

*The figures in brackets refer to the bibliography at the end of this book.

"Climate" is an abstract concept. E. T. Linacre [104a] in a review of sixteen published definitions, offered the following definition: "Climate is the synthesis of atmospheric conditions characteristic of a particular place in the long term. It is expressed by means of averages of the various elements of weather, and also the probabilities of other conditions, including extreme events." The climate of a given site is comprised of the average conditions, the regular sequences of weather events, and the repeatedly observed special phenomena such as tornadoes, dust storms, and late frosts. An understanding of the microclimate or local climate is not possible without adequate consideration of their linkages to weather elements occurring at the mesoclimate and macroclimate scales. The climate of a site is the product of local-scale surface-atmosphere interactions, as well as the frequency of occurrence of particular synoptic circulation patterns.

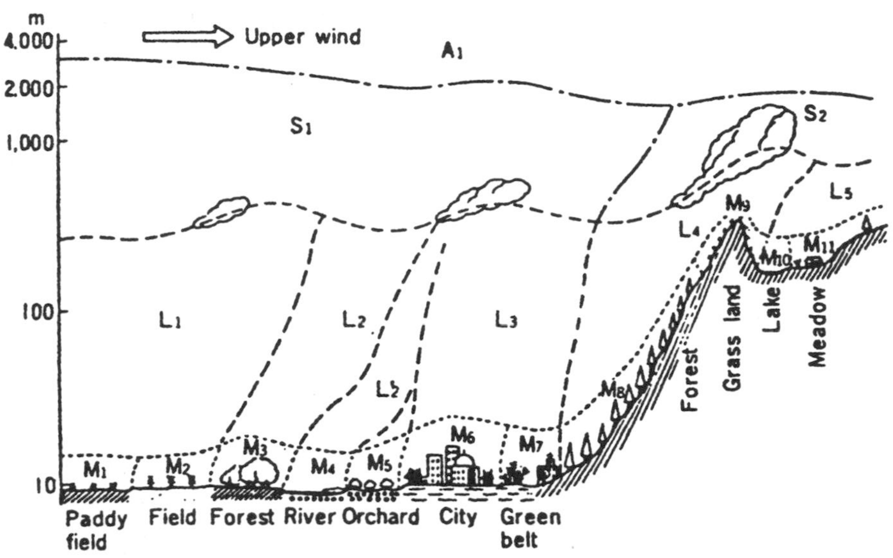

Figure 1-1. Schematic illustration of the micro- (M), local (L), meso- (S) and macro- (A) climate scales. (After M. M. Yoshino [110]. Reprinted with permission of Van Nostrand Reinhold)

Special weather phenomena can be found in the air layer near the ground, as shown by shallow ground fogs over forest meadows, or by sand sweeping above the ground over prairies. Normally, however, the special phenomena of the ground layer are not as obvious and can only be deduced through the use of instruments. Without an understanding of the processes involved, it is not possible to understand the microclimate. For this reason, a textbook on microclimatology must, of necessity, begin as a textbook on micrometeorology; thus, Chapters I to IV are devoted to basic physical principles. Those who are primarily interested in theoretical micrometeorology, which has been called the "mathematician's paradise" by O. G. Sutton [109], should consult the works of R. E. Munn [105], S. P. Arya [100], R. B. Stull [108], and R. S. Scorer [107]. H. G. Jones [102] is recommended for those who desire a quantitative approach to the interrelationship of plants and the environment. The present book is not intended for the theoretical atmospheric scientist, but will serve the ever-increasing number of people who hope to learn about the average conditions of life near the ground. These include people interested in plant life: farmers, foresters, gardeners, vine growers, and botanists; those interested in animal life: zoologists, entomologists, farmers, and breeders; those interested in the state of the ground: traffic engineers, road builders, architects, hydrologists, soil scientists, geographers, land use planners, and bioclimatologists. It is hoped that this book will heighten the reader's awareness and understanding of the processes by which microclimates develop and their interrelationship with plants, animals, and humans.

The beginnings of microclimatology date back about a century. In Finland, in August of 1893, Theodor Homén (1858-1923) – far ahead of his time – made comparative measurements of the energy budget in various types of soil. This seems to us today to be the real beginning of thinking in terms of microclimatology. Gregor Kraus (1841-1915), who published a book on soil and climate in very small places in 1911 [103], can be described as the Father of microclimatology. He became aware of the extreme local conditions affecting the chalk countryside of the Main area near Karlstadt, and investigated them out of pure love of research. The fundamentals and practical applications of micrometeorology were then developed mainly by Wilhelm Schmidt (1883-1936) in Vienna and August Schmauss (1877-1954) in Munich.

CHAPTER I

EARTH'S SURFACE ENERGY BUDGET

2 Physical Basis of Earth's Radiation Balance

Radiation is the most important of all meteorological elements. It is the source of power that drives the atmospheric circulation, the oceanic circulation, and the hydrological cycle, and is the only means of energy exchange between the earth and the rest of the universe. In spite of this paramount importance, it is only in the present century that meteorologists have seriously begun to investigate radiation. The primary reasons for this involve the technical difficulties in the design and use of instruments for measuring radiation. These difficulties are much greater than those encountered with the coarser instruments for measuring temperature, air pressure, humidity, wind, and most other climatic elements. There is widespread recognition of the importance of radiation processes in all energy exchanges in the atmosphere. These processes are of utmost importance in microclimatology. To provide the reader with a solid background for later chapters, the most important radiation laws and the basic theory of the earth's radiation balance are summarized below.

Energy can be transferred within the atmosphere by four methods: conduction, convection, evaporation, and electromagnetic radiation. The process of conduction involves molecular diffusion. If one end of an iron rod is heated, energy travels within the rod in a way not visible to the eye; the livelier motion of the heated molecules is transmitted by collision to the more slowly moving neighboring molecules.

Likewise, when water is heated in a beaker, the same kind of conduction takes place as that in the iron rod; however, the heating of the whole body of water is brought about much more effectively by the stream of heated particles rising from below. This motion can easily be observed in artificially colored water: energy is transported as the water mixes. This process of mass exchange, characteristic of both liquids and gases, is almost always present under atmospheric conditions, and plays a much more important role in minimizing temperature differences than does conduction. This motion is called convection (sensible heat flux) and will be discussed in greater detail in Sections 7 and 8.

Evaporating water requires about 2.47 megajoules (MJ) of energy for every kilogram released as vapor. This energy becomes available to the air when the water vapor condenses. In this case, energy is transported by means of the change of state of the water. Evaporation and other forms of latent heat flux are discussed in Chapter III.

Method four is the process of electromagnetic radiation which, in contrast to the previous three forms, does not require a physical medium for transport. Electromagnetic radiation relevant to Earth atmosphere is in the range from 0.15 - 100 microns (μ) (1μ = 0.001 mm = 1000 mμ = 10,000 Å). Longer wave radiation is familiar in the form of VHF and other radio waves; examples of shorter wave radiation include x-rays, and gamma rays from radioactive substances.

The human eye perceives radiation between about 0.36 and 0.76 μ as light, and is most sensitive in the range 0.55-0.56 μ. Because human visual perception is restricted to this spectral range, these wavelengths of radiation are referred to as visible radiation. Color perception is dependent on wavelength, as follows: 0.36 ← violet → 0.42 ← blue → 0.49 ← green → 0.54 ← yellow → 0.59 ← orange → 0.65 ← red → 0.76. Longer wave radiation, which is not perceptible to the eye, is called near infrared. Radiation below 0.36 μ is called ultraviolet or UV for short.

The atmosphere of Earth reflects, scatters, and absorbs part of the incoming solar radiation. Solar radiation received at the top of the atmosphere before these processes begin to attenuate it is termed extraterrestrial radiation (or potential insolation). For a mean Earth-sun distance M of 149.7×10^6 km, the intensity of solar radiation incident upon a surface perpendicular to the sun's rays, according to the best available observations, is 1367.7 W m^{-2} (± 2 percent). This quantity is called the solar constant S. It varies by about ± 3.4 percent with the changing distance of the earth from the sun during the course of the year. This potential insolation falls on an area of πR^2, where R is the Earth's radius (6371 km).

According to the Stefan-Boltzmann law of radiation, $E = \sigma T_S{}^4$, all black bodies radiate such that the flux density of emitted radiation E (summed over all wavelengths) is proportional to the fourth power of the absolute surface temperature T_S in Kelvin (K = 273.15 + °C). The constant of proportionality σ has the value 5.675 x 10^{-8}W m^{-2} K^{-4}. Therefore:

$$E = 5.675 \times 10^{-8}\, T_S{}^4 \qquad (\text{W m}^{-2}).$$

A blackbody is a substance which absorbs all radiation incident upon it and which perfectly emits radiation in a continuous spectrum according to the Stefan-Boltzmann Law. The amount and wavelength of peak emission are dependent upon its surface temperature. Although no natural substance is a perfect blackbody, many substances approach a blackbody over large portions of their spectra. The concept of a blackbody is also useful as a reference value. If the sun is considered to approximate a blackbody, its surface temperature T_S can be calculated from this law with the aid of the solar constant S. Since the flux density of solar radiation decreases with the square of the distance from the sun, and the sun's radius is s = 695,560 km, while Earth's radius R is negligible in comparison with the distance M from the sun to Earth, then $\sigma T_S{}^4/S = M^2/s^2$, from which the value of T_S is found to be 5780 K.

The Stefan-Boltzmann law also provides a solution to Earth's mean temperature, based on the assumption that Earth radiates like a blackbody. If we assume that a condition of radiative equilibrium exists between the radiation emitted by the Earth and the radiation received from the sun, then the amount radiated by the surface of a sphere of area $4\pi R^2$ must equal the quantity received by a cross-sectional area πR^2 multiplied by the solar constant S and the planetary albedo a_p. The mean blackbody surface temperature of the earth T_E, calculated from the equation:

$$\sigma T_E{}^4\, 4\pi R^2 = S\pi R^2(1\text{-}a_p)$$

with a planetary albedo of 0.30 is found to be 255 K = -18°C (Earth for humans is thus at a relatively ideal distance from the sun; Venus, which is closer, has an equilibrium temperature of 57°C, and the more distant Mars is -46°C). The surface temperature of Earth observed near the ground is higher (287 K = 14°C) because of its atmosphere (Section 4).

Even if the quantity of radiation received from the sun is equal to that radiated by the earth, the two types of radiation are fundamentally different. According to Wien's displacement law, the maximum intensity of radiation, λ_{max}, is constant for a given temperature. With T_s in Kelvin and λ_{max} in microns:

$$\lambda_{max} = 2897/T_s$$

The higher the surface temperature of a body, the farther the radiation maximum is displaced toward shorter waves. For the surface temperature of 5780 K, evaluated above for the sun, λ_{max} is about 0.50 μ; the observed maximum at 0.47 μ implies a higher solar temperature (the difference shows that the sun radiates only approximately as a blackbody). In either case, the most intense solar radiation occurs in the blue-green range of visible light. The wavelength of maximum intensity of radiation for Earth's actual surface temperature of 287 K is about 10.0 μ which is well into the infrared.

The emission spectrum of a blackbody, however, is highly asymmetric (Figure 2-1). The wavelength of peak emission given by Wien's Law does not coincide with the median wavelength λ_m that divides the emitted energy into two equal portions. The median wavelength dividing the energy in equal halves is given by:

$$\lambda_m = 4110/T_s$$

with T_s in Kelvin and λ_m in microns. For the blackbody temperatures of 5780 K and 287 K cited earlier, this yields a median wavelength λ_m of 0.71 μ and 14.3 μ for the sun and Earth, respectively (Figure 2-1).

Figure 2-1a shows the relative distribution of emitted energy for two black bodies. The curve on the left is for the sun and the curve on the right is for the earth. In Figure 2-1a the vertical scale has been normalized for comparison purposes. Figure 2-1b illustrates why in meteorology one normally distinguishes between two fundamentally different streams of radiation. About 99% of the solar radiation is in the range from 0.15 to 4.0 μ. Radiation emitted by the earth and its atmosphere lies between 3.0 and approximately 100 μ. The very small region of overlap between 3.0-4.0 μ is less than 1 percent of the total output of the earth and the sun. As a result, 4.0 μ is usually used as the division between shortwave (solar) and longwave (terrestrial) radiation.

Attention will now be turned from this planetary consideration of radiation processes to conditions prevailing at Earth's surface, and to the energy balance.

BLACKBODY (PLANCK) CURVES

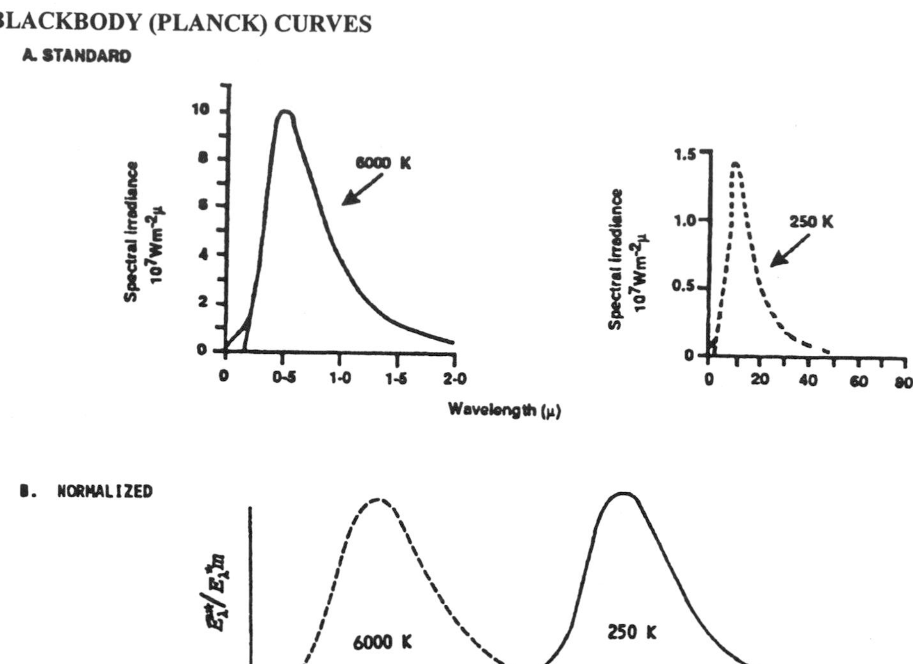

Figure 2-1. A. Standard blackbody curves; B. Normalized blackbody curves for the sun (6000 K) and Earth (250 K). (Normalized curve after R. G. Fleagle and J. A. Businger [2006])

3 Components of the Energy Balance and Their Importance

Let us first consider the ideal case in which Earth's surface is extended horizontally. In this case, the ground-atmosphere boundary is a two dimensional plane and without mass. The plane stores no energy, but under normal circumstances a considerable exchange of energy occurs across it. The quantities that determine this energy exchange will now be considered.

First, radiation is the major factor of energy exchange. Incoming radiation arrives at Earth's surface in the form of shortwave radiation coming from both the solar disk and the sky, and longwave radiation coming from the sky. Outgoing radiation is returned to space from the surface of earth in the form of reflected shortwave radiation and longwave (or terrestrial) radiation. In terms of notation, factors that add energy to the ground surface are considered positive; factors that subtract energy from it will be considered negative. The sum or net balance of incoming and outgoing radiation--termed net radiation Q^*, may be positive or negative (unit: W m^{-2}. In the fifth edition we used c.g.s. units. For those making the transition to SI: W m^{-2} = J sec^{-1} m^{-2} = 2.388 x 10^{-5} cal cm^{-2} sec^{-1}).

The second factor Q_G is determined by the flow of energy from within the ground to the surface or in the reverse direction. During a cold winter night, energy flows upward through the ground, and Q_G is therefore positive; on a summer afternoon, Q_G is negative because energy is transported downward from the surface. The laws governing the transport of energy in the ground will be dealt with in Section 6.

Third, the air above the ground plays a part in the exchange of energy Q_H. This factor may also be positive or negative. It is positive when energy (heat) flows from the air toward the ground surface, and negative when energy is transported away from the surface. Transport of energy to or from the ground depends not only on the molecular diffusion of heat from the ground to the air layer in contact with the ground, but also on mass exchange (eddy diffusion). More detailed explanations are found in Sections 7-9.

Fourth is latent heat Q_E. This is measured, like all the other energy balance factors, in Watts per square meter. The quantity of energy required to evaporate 1 kg of water is called the latent heat of vaporization r_v and varies with temperature (Table 3-1). The energy flux term Q_E can be converted into its mass flux equivalent E (kg sec^{-1} m^{-2}) by dividing it by r_v.

Table 3-1. Latent heat for changes of state.

T:	-40	-30	-20	-10	0	10	20	30	40	°C
r_v:	2.602	2.575	2.550	2.525	2.501	2.478	2.454	2.430	2.407	MJ kg^{-1} vaporization
r_f:	0.236	0.264	0.289	0.312	0.334					MJ kg^{-1} fusion
r_s:	2.839	2.839	2.838	2.837	2.835					MJ kg^{-1} sublimation

When ice "evaporates" (sublimation) in winter, the energy requirement of water evaporating is increased by the latent heat of fusion (r_f) of the ice (0.334 MJ kg^{-1} at 0 °C). The latent heat of sublimation (r_s) requires 2.835 - 2.839 MJ kg^{-1} at 0 °C to -40 °C. When water and ice evaporate, Q_E is negative. Positive values of Q_E occur when dew or frost form on the surface and latent heat of condensation or sublimation is released.

Since the earth's surface is a boundary interface and can absorb no energy, according to the law of conservation of energy, the following equation must be satisfied for all units of time:

$$Q^* + Q_G + Q_H + Q_E = 0$$

This is the fundamental equation, containing the basic factors, which governs energy exchange at Earth's surface. Over oceans, lakes, and rivers the factor Q_W, for the exchange of energy between water and its surface, is used instead of Q_G. The equation then becomes

$$Q^* + Q_W + Q_H + Q_E = 0$$

Horizontal airflow (advection) from the surrounding area may be warmer or colder, moister or drier than the area under consideration. This flow was not considered in the introductory ideal situation, but often occurs under actual conditions. This advection process has an effect on the energy budget of the area and changes the assumptions on which the previous discussion was based. Without verifying at this stage the effects it must have on the energy balance equation, we shall introduce this additional advection process, which is of considerable practical significance, into the equation as an extra factor Q_A. Advection processes will be considered in Section 27.

Precipitation may entail a gain or a loss of energy for the ground, depending on its temperature, and this is given the symbol Q_R. This process was also neglected in our initial assumption and is discussed in Section 6.

The complete equation for the energy exchange at a horizontal ground surface free of vegetation consequently becomes:

$$Q^* + Q_G + Q_H + Q_E + Q_A + Q_R = 0$$

If the ground is covered with vegetation, new factors are introduced (Chapter V).

To illustrate the significance of the principal factors of this energy exchange, Table 3-2 gives the results of an investigation by F. Albrecht [2501] at Potsdam, for selected times of day in 1903. During the day, the surface energy balance is dominated by positive net radiation Q^*; the surplus of radiant energy (Q^*) flows into the ground (Q_G), into the air (Q_H) and is used to evaporate water or melt ice (Q_E). During the night, energy is lost from the surface by outgoing radiation and by evaporation (the role of dew or frost is quantitatively insignificant); this surface energy loss is offset by a gain of heat partly from the ground and partly from the air above it.

Table 3-2. Energy budget in Potsdam (W m^{-2}). (After F. Albrecht [2501])

		Factors			
Typical Period	Mean value (1903)	Q^*	Q_G	Q_H	Q_E
Summer day	June 12-13 hr	284.0	-115.1	-65.6	-103.3
Winter day	Jan. 12-13 hr	63.5	-57.2	- 2.1	-4.2
Summer night	June 0-1 hr	-55.8	48.8	14.7	-7.7
Winter night	Jan. 0-1 hr	-45.4	14.7	41.9	-11.2

4 Radiation Balance of Earth's Surface

Of all the factors mentioned in Section 3 as taking part in the energy exchange at the surface of the earth, radiation is the most important. The symbol Q^* represents the radiation balance or net radiation. If incoming radiation is greater than outgoing radiation, the balance is positive; if it is less, the balance is negative. A negative balance is described as a net loss of radiation.

The radiation balance consists of two radiation streams of different spectral ranges, as distinguished in Figure 2-1. The first is shortwave radiation from the sun. Solar radiation reaching the surface of earth consists of direct-beam solar radiation S (sometimes called direct solar radiation), a directional component emanating from the solar disk that is not reflected by clouds, or absorbed or scattered by the atmosphere and diffuse solar radiation D (sometimes called sky radiation), which is a nondirectional component that is comprised of scattered solar radiation that has reached the ground. The value of $S + D$ reaching a horizontal surface is called global solar radiation (sometimes called solar irradiance) and is represented by $K\downarrow$. Part of this radiation is reflected by the Earth's surface. This reflected solar radiation $K\uparrow$ depends on the nature of the ground in contrast to $K\downarrow$.

Although incoming extraterrestrial longwave radiation is of no significance for the earth radiation balance discussed in Section 2, longwave radiation is of great importance for the radiation balance of Earth's surface. The atmosphere of the earth contains water vapor, water drops, carbon dioxide, particulates, ozone and other trace gases. All of these absorb and emit radiation according to Kirchhoff's law. Kirchhoff's law states that, for a given wavelength and temperature, the absorptivity of radiation is equal to the emissivity of radiation (absorptivity equals emissivity). In other words, if a substance absorbs energy well at a given wavelength, it can also emit well at that wavelength. Earth, for example, absorbs shortwave energy quite well. Thus, if for any reason its temperature were to rise to the level of the sun, it would also emit well at these wavelengths. Water vapor absorbs energy poorly at around 4 and 10 μ. Thus, regardless of its temperature, it will not be able to emit much energy at these wavelengths.

Longwave radiation emitted by the atmosphere $L\downarrow$ is termed counterradiation (sometimes called longwave irradiance or atmospheric radiation) since it counteracts the terrestrial radiation loss from the surface. It occurs both by day and by night, and increases somewhat during the day since it is dependent on atmospheric temperature, humidity, and cloud cover.

Surfaces also do not absorb all the longwave energy incident upon them. The reflected longwave energy is typically referred to as longwave reflectivity rather than longwave albedo, and is equal to $L\downarrow$ (1-ε), where ε is the surface emissivity (expressed as a fraction). The surface emissivity is defined as the ratio of radiation emitted by a body at a surface temperature to the blackbody radiation from the body at the same surface temperature. The emissivity (absorptivity is equal to emissivity) for a number of natural surfaces for the spectral region from 9 to 12 μ is given in Table 4-1. It can be seen that most natural surfaces have a relatively low reflectivity for longwave radiation.

D. M. Gates and W. Tantraporn [209] recorded longwave reflectivity for a number of plants and trees over seven bands of wavelength between 3 and 25 μ and found them to be between zero and six percent. These values were exceeded only in isolated instances (lemon-tree leaves had a longwave reflectivity of 17 percent at 10 μ). G. Falckenberg [207] recorded longwave reflectivities between 0 and 8 percent in the 10 μ region; only light sand had a value of as high as 11 percent. In general, the longwave reflectance of natural surfaces is less than 5 percent. Snow cover, which reflects so strongly within the visible spectrum that newly fallen snow produces a striking improvement in light conditions, is nearly an ideal blackbody for long waves, reflecting around 1.4 percent of the incident radiation. As G. Falckenberg so aptly puts it, "Snow can only be made more transparent to long waves by spreading soot on it."

According to the Stefan-Boltzmann law, the radiation emitted by the soil surface by day and by night would be exactly σT_S^4 (T is the surface temperature in K) if the ground were a blackbody. It has just been shown that this condition is largely fulfilled. To the extent that it is not fulfilled, the outgoing radiation will be reduced (at a given surface temperature) based upon the value of the surface emissivity. For a graybody surface the outgoing longwave radiation will be equal to $\varepsilon \sigma T_S^4$ where ε is the surface emissivity. The relationship between the true surface temperature T_S for graybody emission and the apparent surface radiative temperature T_R (sometimes called the radiant temperature or equivalent blackbody temperature) for blackbody emission may be determined by setting $\varepsilon \sigma T_S^4$ equal to σT_R^4 and solving for T_S and T_R. It can be shown that $T_R = \varepsilon^{0.25} T_S$ and $T_S = T_R / \varepsilon^{0.25}$. For $T_S = 300$ K and $\varepsilon = 0.98$, T_R is within 1.5°C of T_S. For a more extreme surface emissivity of 0.95, with T_S again set to 300 K, the departure of T_R from T_S is still less than 4°C. As the value of ε decreases below

Table 4-1. Emissivity from natural surfaces expressed as a fraction of the radiant energy of a blackbody at the same temperature for the spectral region from 9 to 12 μ. (From U. L. Gayevsky [210])

Water	0.960
Fresh snow	0.986
Coniferous needles	0.971
Leaves	
Corn, Beans	0.940
Cotton, Tobacco	0.980
Sugar Cane	0.940
Dry peat	0.970
Wet peat	0.983
Dry fine sand	0.949
Wet fine sand	0.962
Thick green grass	0.986
Thin green grass on wet clay soil	0.975
Forest, Deciduous	0.950
Forest, Coniferous	0.970
Fur, Hair	
Mouse	0.940
Squirrel	0.980
Hare, Wolf	0.990
Human Skin	0.980
Glass	0.940

1.0, however, the amount of reflected longwave radiation (which is equal to $L\downarrow$ (1-ε)), increases. The outgoing longwave radiation from the surface will be represented by the symbol $L\uparrow$.

The surface will be treated as a blackbody throughout the remainder of the book since, for the range of natural surface emissivities, the departure of T_R from T_s is small. The longwave reflectance term will also be omitted. Although these assumptions are not particularly problematic for the study of surface microclimates, they do greatly complicate the interpretation of thermal infrared images obtained from airborne or satellite remote sensing platforms and have been the subject of renewed study in recent years. A. A. Van de Griend, et al. [217] illustrates one such study. They measured the surface emissivity in the wavelength band 8-14 μ for a range of surfaces in a bush-savanna environment in Botswana and found surface emissivities ranging from 0.914 for a bare loamy sand soil, to 0.986 for a complete covering of savanna shrub (*Euclea undulata*). A large spatial variability of surface emissivity was found within the study area, although repetitive measurements within a given surface type were reproducible. Their work should be examined for a discussion of the consequences of variable surface emissivity in thermal infrared remote sensing applications.

The radiation balance Q^* is therefore given by the equation:

$$Q^* = K\downarrow + L\downarrow - L\uparrow - K\uparrow \qquad (\mathrm{W\ m^{-2}})$$

The last two factors depend on the nature of the ground surface, while the first two are independent of it.

Figure 4-1 shows the magnitude of these factors from 12:00 to 13:00 hr on a summer day (5 June 1954), and from 0:00 to 1:00 of the following night, from measurements made at the Meteorological Observatory in Hamburg and published by R. Fleischer and K. Gräfe [208]. The width of the arrow is proportional to the intensity of radiation. The remaining factors Q_H, Q_G and Q_E (Section 3) in the energy budget are included to complete the diagram; they are taken from measurements made by E. Frankenberger [2506] on clear days with light winds, near Hamburg for the same month.

It is clear from Figure 4-1 how important radiation is in the total energy budget. During the day, shortwave radiation is intense. Terrestrial radiation lasts the whole 24 hr, and, it is largely compensated for by counterradiation. At night, however, the energy balance is completely dominated by longwave radiation, as shown in Figure 4-1. This will be discussed in more detail in Section 5.

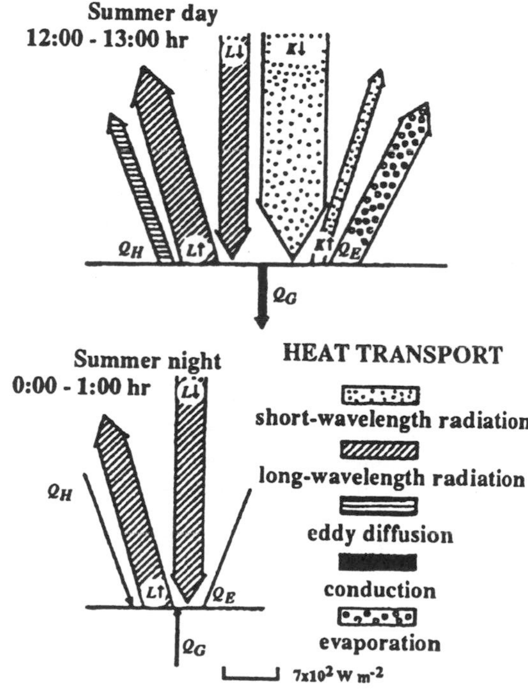

Figure 4-1. The importance of radiation as compared with the other terms in the energy budget.

W. Collmann [204] determined the following radiation balance for a year:

S = 1430.2 MJ m^{-2} yr^{-1}
D = 1819.3
$L\downarrow$ = 1,0072.6
$L\uparrow$ = -1,1257.8
$K\uparrow$ = -601.6

Radiation balance Q^* = 1462.7 MJ m^{-2} yr^{-1}

Out of a radiation balance of 1462.7 MJ m^{-2} yr^{-1}, 86 percent was used in evaporation (Q_E) and 14 percent in heating the air (Q_H); quantities of energy passing into and out of the ground (Q_G) canceled each other over the year.

It is incorrect to assume that the surface net radiation gain by day (or summer) is matched by an equal net radiation loss during the night (or winter). As A. Baumgartner [2502] pointed out, the radiation balances of day and night (or summer and winter) are not equally matched; thus, in contrast to the equilibrium of the radiation balance of the earth as a planet (Section 2), Earth's surface receives substantially more energy than it radiates. Most of this difference is used in evaporation, which shows the close relation between the energy balance and the water balance. This relation between the transfer of energy and water allows the results of energy and water balance investigations to be cross-checked; an evaporation level calculated from the water balance must reappear in the energy balance as a corresponding amount of energy used for evaporation.

In general, it is costly and time consuming to measure the individual components of short and longwave radiation separately. For most microclimatological purposes it is sufficient to determine the net radiation.

Let us now direct our attention to the reflection of solar radiation from natural surfaces. The reflection coefficient or albedo of an object is equal to the ratio of the reflected solar radiation to the incident solar radiation ($K\uparrow/K\downarrow$), and is usually expressed as a percentage (or sometimes as a fraction). The reflectivity of an object is a measure of the capacity of an object to reflect solar radiation at a specific wavelength. Thus, albedo is the integrated product of the spectral composition of the incident solar radiation and the spectral reflectivity of an object.

A distinction is made between diffuse and specular reflection. Reflection is described as diffuse when the incident rays are reflected without directional preference. This type of reflection normally occurs from the rough surfaces found in nature. Reflection is described as specular when the incident rays are reflected in a preferred direction, typically along a line tangent to the incident rays.

Table 4-2 presents albedos for the spectral region from 0.3 to 2 μ. The methods employed in making the measurements were not uniform; therefore, the accuracy may vary somewhat (after K. Ya Kondratyev [212] and R. J. List [4015]). The albedo of soil shows great variability depending upon particle size, mineral composition, moisture and organic matter content, and surface roughness. The albedo of vegetation is also quite variable. Variation can be caused by vegetation type, color, canopy geometry, moisture content, wetness, percent of the ground covered, leaf area and size, and the stage of the plant's growth.

Snow's albedo varies with its crystal size and density, the amount of dirt, soot, or dust mixed in, it's surface roughness and liquid water content, and the number of thaws. K. Ya Kondratyev [212] reported the albedo of snow decreasing from 80% before to 66% after a

Table 4-2. Albedo (percent) of various surfaces, for the spectral region from 0.3 to 2 μ. (Modified after K. Ya Kondratyev [212] and R. J. List [4015])

Surface			
Clean snow		75-98	
Melting snow		66-88	
Very dirty snow		20-30	
Ice on frozen lake		12	
Sea ice-slightly porous milky blue		36	
Gray soil, dry		25-30	
Gray soil, moist		10-12	
Black soil, dry		14	
Black soil, moist		8	
Fallow fields, dry		8-12	
Fallow fields, moist		5-7	
Fields, green		10-15	
Yellow sand		35	
White sand		34-40	
Fine light sand		37	
Light clay earth			
Leveled		30-31	
Covered with small clods		25	
Clovered with large clods		20	
Newly ploughed		17	
Summer wheat		10-25	
Cotton		20-22	
Lettuce		22	
Beet		18	
High standing grass		18-20	
Oaks		18	
Pine		14	
Fir		10	
Sea surface			
Solar angle	Rough		Calm
90	13.1		2.1
60	3.8		2.2
30	2.4		6.2
Clouds			
Less than 150 m thick		5-63	
150-300 m thick		31-75	
300-600 m thick		59-84	

thaw. He also reported a smooth snow surface albedo between 70-76% and that of an identical but knobby snow surface to be between 60 and 65%. The longwave reflectance and shortwave albedo of snow are discussed further in Section 24.

Sea surface albedo variations are caused not only by the angle of incoming solar radiation, and the turbidity and phytoplankton concentration of the water, but also by the water's roughness. For calm water, the albedo increases as the solar altitude decreases (Table 4-2). For rough water, however, the albedo increases with increasing sun angle. The albedo of water also decreases with increasing cloudiness.

The albedo of clouds increases with increasing thickness. This occurs rapidly for thin clouds but more slowly as the clouds become thicker than 200-300 m. For more information dealing with variations of cloud albedo and their thickness, type, percents of cloud cover, and variations through the year, the reader is referred to K. Ya Kondratyev [212]. For a good discussion of albedo variations with cloudiness, sun angle, and variations in plant albedo with respect to cloudiness and season, the reader is referred to S. B. Ahmad and J. G. Lockwood [200]. For a discussion of the amount of solar energy reflected from various soils and types of vegetation as a function of solar elevation and wavelength, the reader is referred to K. L. Coulson and D. W. Reynolds [205].

As indicated earlier, the albedo is influenced not only by the nature of a surface but also by its moisture content. Wet surfaces appear darker than dry surfaces. In this connection A. Ångström [201] observed that the albedo of gray sand decreased from 18 to 9 percent when it became wet; tall light-colored grass showed a decrease from 32 to 20 percent. In the Amrum Sand Dunes, K. Büttner and E. Sutter [203] noted a dry to wet change of from 37 to 24 percent. Table 4-3 shows the dependence of this reduction on wavelength.

Table 4-3. Reduction of albedo (percent) when sand becomes wet as a function of wavelength (μ). (After F. Sauberer [214])

Wavelength	0.4	0.5	0.6	0.7	0.8
Dry sand	20	23	29	30	30
Wet sand	10	12	15	16	19

A. Ångström stated that when a layer of water covers parts of the ground or plants, light rays might enter the layer from any direction. They can escape, however, only if they reach the surface of the water at an angle less than the critical angle for total reflection. Since the state of the surface undergoes daily and annual changes, there are corresponding periods of variation in albedo. Monthly values of albedo for a meadow and for the surface of the Danube near Vienna, for global solar radiation, measured by F. Sauberer [215] and I. Dirmhirn [206], are shown in Table 4-4. The albedo for the meadow without snow is lower in winter probably because the meadow is not as dense and more energy is being absorbed at the surface of the soil. This results in the incoming solar energy penetrating more deeply into the mass of the vegetation (the surface is rougher), and more energy being absorbed by the ground (which has a lower albedo (Table 4-2)). The water surface of the Danube River also has a higher albedo in winter due to the lower sun angle (Table 4-2).

Table 4-4. Monthly average albedo (percent) near Vienna.

Month	Jan	Feb	Mar	Apr	May	Jun	Jul	Aug	Sep	Oct	Nov	Dec
Meadow, depending on the development of vegetation	13	13	16	20	20	20	20	20	19	18	15	13
Meadow, taking winter snow cover into account	44	39	27	20	20	20	20	20	19	18	21	36
Water surface of the Danube River	11.2	11.4	10.7	10.0	9.0	8.6	8.6	9.7	9.5	11.7	11.8	11.9

The effect of cloudiness and horizontal shielding may be found in additional data from I. Dirmhirn [206], and the daily variation in F. Sauberer and I. Dirmhirn [216]. The annual variation is determined by the condition of the ground, which has been charted for Vienna by F. Lauscher [213] in the form of frequency isopleths of snow cover, slush, wet and dry soils, and dry frost.

Reflectivities for shortwave radiation below 0.36 μ are considerably lower than the figures given in Table 4-4 for the full spectral range of solar radiation. Only snow cover shows high reflection of ultraviolet radiation (80 to 85 percent).

Specular reflection is directional, and therefore depends upon the angle of incidence of the sun's rays. This type of reflection occurs at the surfaces of water and sand and is of importance to the climate of riverbanks and seashores (Section 22). Figure 4-2 shows the relation between albedo and the elevation of the sun. From the zenith to an elevation of 40°, there is little change from the values given in Table 4-2. The albedo then increases sharply, a familiar effect experienced when the dazzling rays of the setting sun are reflected by water. The surprising thing is that this is also true to an extent for a rough sandy surface.

The radiation not reflected by the surface is absorbed. Natural surfaces are mostly irregular in form; they are made up of particles or grains between which radiation can penetrate. As far as the surface energy budget is concerned, this is of no quantitative importance. It, however, is of great significance for the life of bacteria and algae and the germination of seeds that require the stimulation of light.

As distance below the surface increases, the longer waves of the solar spectrum are more and more strongly represented. Red light predominates below the surface because it is more strongly reflected than blue on each of the many occasions when there is reflection from a particle of soil. F. Sauberer [214] showed that the amount of reflection varied considerably with different types of soil.

A. Baumgartner [202] investigated the distribution of light in the ground by using graded quartz sands, artificial light, and selenium cells. Figure 4-3 shows the results for grains with dimensions between 0.1 and 6 mm in a completely dry state. The coarser the particles the more the radiation can penetrate. In the case of fine-grained sands, especially when mixed with a little loam, the penetrating light is reduced to one thousandth in the first millimeter. M. Köhn [211] has shown that only half the incident light penetrates to a depth of 0.015 mm in soil with texture as fine as dust.

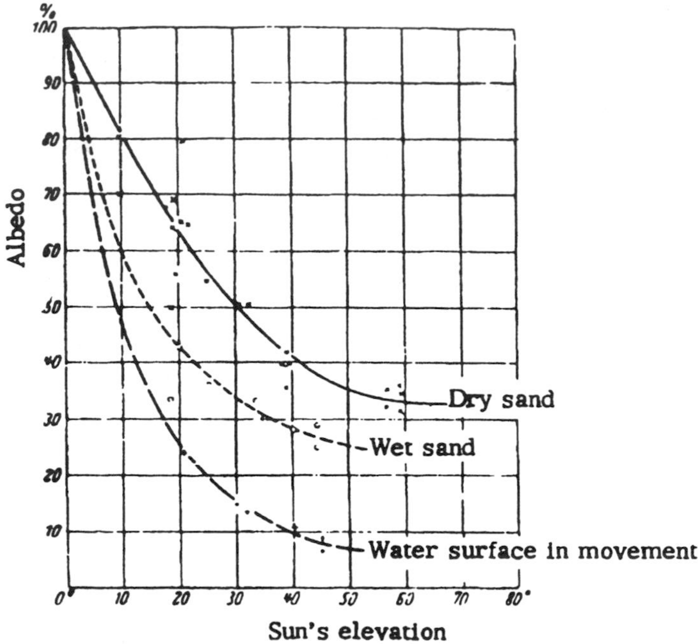

Figure 4-2. Specular reflection of sunlight from sand and water surfaces. (After K. Büttner and E. Sutter [203])

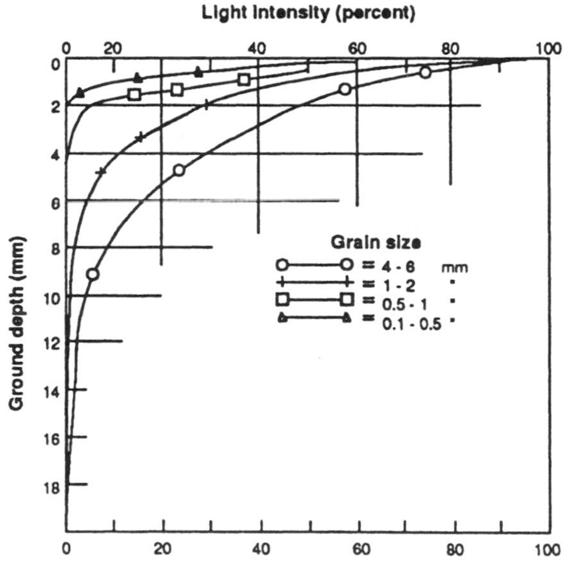

Figure 4-3. Light transmission into sand of variable grain size. (After A. Baumgartner [202])

5 Longwave Radiation at Night

At night, the terms *S*, *D*, and $K\uparrow$ in the net radiation equation are zero. The radiation balance then is $Q^* = L\downarrow - L\uparrow$. The net longwave radiation $L\downarrow - L\uparrow$ will be referred to as L^*. The nocturnal radiation balance is equal to the counterradiation from the atmosphere above the place of observation minus the Stefan-Boltzmann radiation from the ground (as a blackbody radiator). The values of Q^* are almost always negative at night. That is to say, the second term is larger than the first, except in unusual circumstances. The counterradiation comes from water vapor, cloud droplets, carbon dioxide, particulates, ozone, and other minor trace gases in proportion to the quantities present and to their temperature. Figure 5-1, after F. Schnaidt [529], illustrates the selective absorption of water vapor (*a*) and carbon dioxide (*b*). The absorption coefficients for water vapor (per 0.1 mm of precipitable liquid) and for CO_2 (per 1 m of air under standard conditions), are shown for two different wavelength λ scales. Figure 5-1 also indicates the amount of energy emitted by water vapor and carbon dioxide at various wavelengths, since, according to Kirchhoff's law, emission is equal to absorption for a given wavelength and temperature. Carbon dioxide absorbs only in a few well-marked absorption bands with centers at 2.8, 4.3, and 14.9 μ. In 2000, carbon dioxide constituted about 0.0370 percent (370 ppm) of the gases in the atmosphere on a percent by dry volume basis.

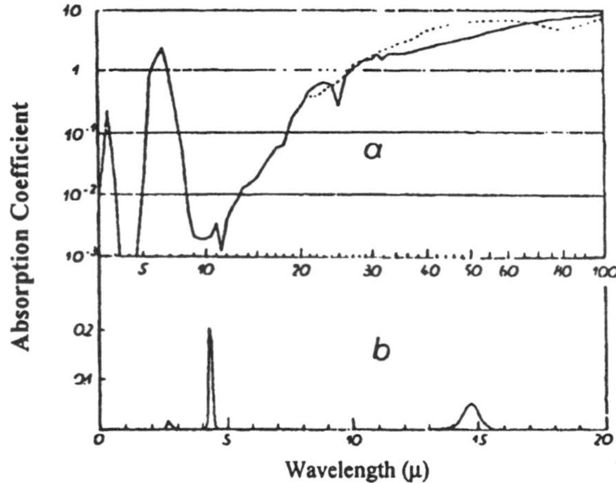

Figure 5-1. Absorption spectra for (*a*) water vapor and (*b*) carbon dioxide. (After F. Schnaidt [529])

It had been increasing (1959-2000) at Mauna Loa, Hawaii by about 0.45 percent (1.6 ppm) per year. This is thought to have been due to the burning of fossil fuels and clearing of forests. CO_2 currently accounts for about 1/6 of the total counterradiation emitted by gases. As CO_2 increases, the amount of energy absorbed in any layer of the atmosphere by CO_2 will also increase. For atmospheric levels and changes in the levels of CO_2 and other gases around the world, the reader is referred to *Trends: A Compendium of Data on Global Change* [T.A. Boden et al., 504]. The amount of counterradiation due to ozone, according to H. M. Bolz [505], is only about 2 percent of the total. It follows, therefore, that the varying quantity of water vapor and the changing temperature of the atmosphere are primarily responsible for the changing amounts of counterradiation observed. Figure 5-1 shows that water vapor has a marked absorption band centered at 2.7 μ, and a very broad band with its maximum around 6.3 μ. As the wavelength increases beyond 13 μ, the atmosphere becomes increasingly opaque until practically all radiation is absorbed. In the region between there are two spectral ranges, one at about 4 μ and the other from 8 to 13 μ, that are called "windows" of the atmosphere. In the first window (4 μ), absorption is practically nil, while in the second, it is only about 10 percent. There is very little incoming solar or outgoing longwave radiation around the 4 μ band (Section 2); hence, the second window, from 8 to 13 μ, plays the leading role in the nocturnal radiation balance. No matter how heavily the atmosphere is laden with water vapor, it absorbs little radiation at these wavelengths.

Another way of looking at the absorption of radiation by gases in the atmosphere is how effectively a gas absorbs radiation at different wavelengths irrespective of their concentration in the atmosphere (Figure 5-2). Figure 5-2 illustrates that within the 8-13 μ range, ozone does absorb strongly around 9.6 μ. The ozone concentration in the lower atmosphere is so low (normally around 20 ppb but at times well below 10 ppb (D. Kley et al. [522])) that most of the energy in this wavelength is not absorbed near the surface. In some cities, however, where concentrations have occasionally been recorded in excess of 800 ppb and quite regularly in excess of 200 ppb, ozone can have a minor effect in reducing the loss of radiation in

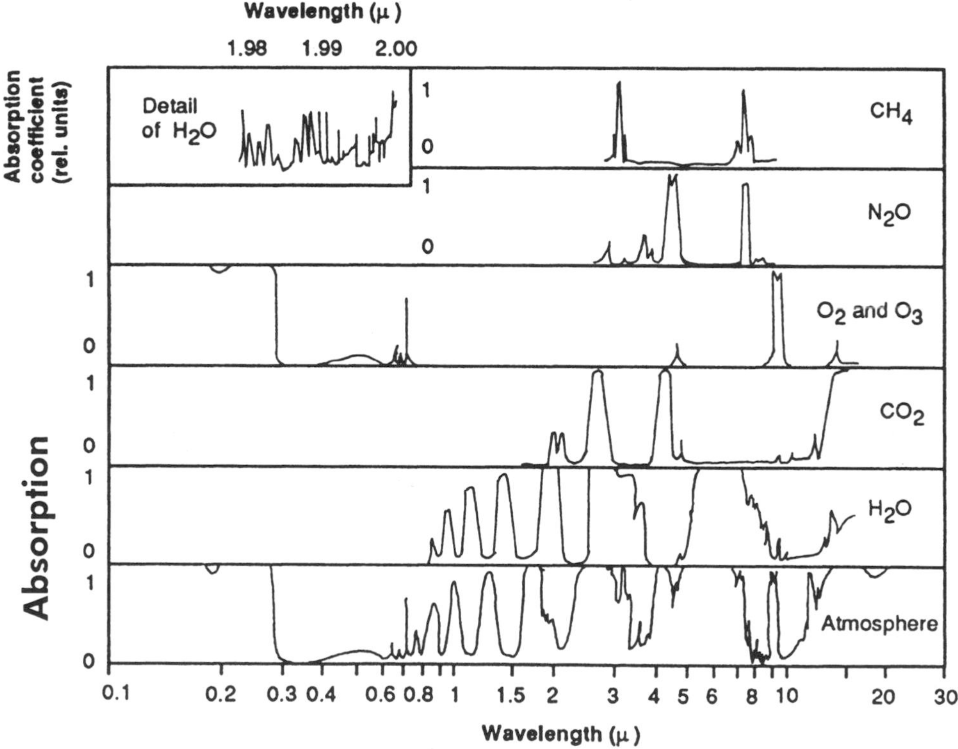

Figure 5-2. Absorption spectra for CH_4, NO_2, O_2, O_3, CO_2, and H_2O, and of the atmosphere. (From R. G. Fleagle and J. A. Businger [2006] after J. H. Howard [519] and R. M. Goody and G. D. Robinson [514])

the water vapor window. It should be noted, however, that ozone concentrations in cities typically reach a maximum during the day and decrease rapidly in the early evening, reaching a minimum around sunrise. Increased atmospheric emissions of the precursors of ozone have led to increased ozone concentrations in the lower atmosphere. J.K. Angell and J. Korshover [500] and A. Volz and D. Kley [533] have shown that ozone levels in rural areas of central Europe have approximately doubled over the past century. J. Fishman et al. [513] have calculated that increased ozone concentrations in the Northern Hemisphere have resulted in a 0.2°C increase in the mean temperature of the hemisphere. Further increases in tropospheric ozone will result in additional heating (O. Hov [518] and A. Lal et al. [523]).

Although every layer of the atmosphere takes part in counterradiation, the contributions of individual layers are vastly different. O. Czepa and H. Reuter [510] estimated the relative percentage of total counterradiation reaching the ground from a series of layers (each of which contained 0.6 mm of precipitable water), for a normal atmospheric water vapor content of 14.25 mm of precipitable water. Their computations for the six lowest layers are shown in Table 5-1. The very first layer, extending to only 87 m above the ground, contributes not less than 72 percent of the total counterradiation and the next 89 m contributes only an additional 6.4 percent. This phenomenon depends partly on the low penetrating ability of longwave radiation, a characteristic that in turn depends on the wavelength. The upper layers make only very minor contributions to the surface counterradiation, and this decreases gradually with height.

Table 5-1. Contribution of various atmospheric layers to counterradiation received at the surface.

Layer thickness (m)	87	89	93	99	102	108
Percent share of counterradiation	72.0	6.4	4.0	3.7	2.3	1.2

Counterradiation can be estimated from radiative transfer theory (K-N. Liou [526]). Radiosonde observations are, however, needed to provide data on the water vapor content and temperature for each atmospheric layer. Since such measurements are only available for a limited number of stations, a different method is used in microclimatology. Assuming that the sky hemisphere radiates as a graybody, we can write a counterradiation equation similar to the one developed for the surface of the earth (Section 4):

$$L\downarrow = \varepsilon_{sky}\sigma T_{sky}^{4}$$

where ε_{sky} and T_{sky} are the emissivity and temperature of the sky, respectively. Although this equation provides insight into the underlying theory of counterradiation, it lacks practical application because the atmosphere does not have a clearly identifiable surface for which ε_{sky} and T_{sky} can be evaluated. Consequently, climatologists have routinely estimated counterradiation based upon surface air layer characteristics:

$$L\downarrow = \varepsilon_{A}\sigma T_{A}^{4}$$

where ε_A and T_A are the atmospheric emissivity and temperature evaluated at shelter-height. From observations of $L\downarrow$ and T_A, ε_A may be empirically evaluated since $\varepsilon_A = L\downarrow/\sigma T_A^4$.

Numerous empirical equations have been developed over the years to approximate the counterradiation received by a horizontal surface from the whole sky hemisphere. A. Ångström [501] developed one of the earliest methods which used the shelter-height air temperature T_A (K) and the vapor pressure e to approximate the conditions prevailing in the atmospheric layer that dominates counterradiation. From numerous measurements, A. Ångström [501] deduced the following relation for clear sky conditions:

$$L\downarrow = \sigma T_A{}^4 (a - b \cdot 10^{-ce}) \qquad (\mathrm{W\,m^{-2}})$$

Here σ is the Stefan-Boltzmann constant and a, b, and c are empirical constants. Using comprehensive measurements, H. M. Bolz and G. Falckenberg [507] calculated these constants to be: $a = 0.820$, $b = 0.250$, and $c = 0.126$. These figures are valid primarily for the German Baltic coast where they were measured. H. Hinzpeter [517], however, pointed out that they are somewhat low in comparison with other series of readings made in Central Europe. Figure 5-3 (developed using these values) demonstrates the relation among the variables involved in determining counterradiation.

The longwave radiation balance of a body with temperature T_A is $L^* = L\downarrow - \sigma T_A{}^4$, which, with the equation given above, becomes

$$L^* = -\sigma T_A{}^4 (1 - \mathrm{a} + \mathrm{b} \cdot 10^{-ce}) \qquad (\mathrm{W\,m^{-2}})$$

The value of L^*, called "effective outgoing radiation" (or net longwave radiation), can be measured by reading the temperature and humidity of the air at shelter-height and using an instrument for measuring radiation pointed in the direction of the sky. At night, Q^* will be equivalent to L^* because of the absence of solar radiation.

Figure 5-3 compares the values of L^* (abscissa) with variations in the air temperature T_A (ordinate, °C) and humidity e at the level of the instrument shelter. Humidity may be shown by lines of equal vapor pressure e (mm-Hg) or by curves of equal relative humidity (percent). The graph is for clear night skies on the German Baltic coast. The effect of cloud cover will be discussed later.

The range of possible values for nocturnal net radiation (Figure 5-3) is restricted in three ways: in the lower right corner by the zero line of vapor pressure, on the left side by the 100 percent relative humidity curve, and at the top left where the lines of increasing vapor pressure become progressively closer together.

An increase in water vapor content of the atmosphere will only result in a limited reduction in the net loss of longwave radiation because, as previously mentioned (Figure 5-1), the two "windows" toward the sky always remain open. For example, if the air temperature is 10°C and the relative humidity is 60 percent, the value of L^* from Figure 5-3 is -85 W m^{-2}. If the ground temperature is colder than the air, which at night is usually the case, the reduction in the loss of outgoing radiation is obtained from the supplementary diagram at the top of the figure. In the above example, when the ground is 5°C colder than the air (air temperature 10°C, relative humidity 60 percent), a correction factor of 26 W m^{-2} is added, resulting in a longwave radiation balance of -59 W m^{-2} for the ground. If the ground is at a higher temperature than the air, as it may be when there is an influx of cold air, the correction is subtracted. Figure 5-3 also shows how strongly night radiation losses are reduced when the ground temperature is lower. The effective outgoing radiation, measured by a pyrgeometer directed toward the sky only, is substantially greater than that of the ground. If the pyrgeom-

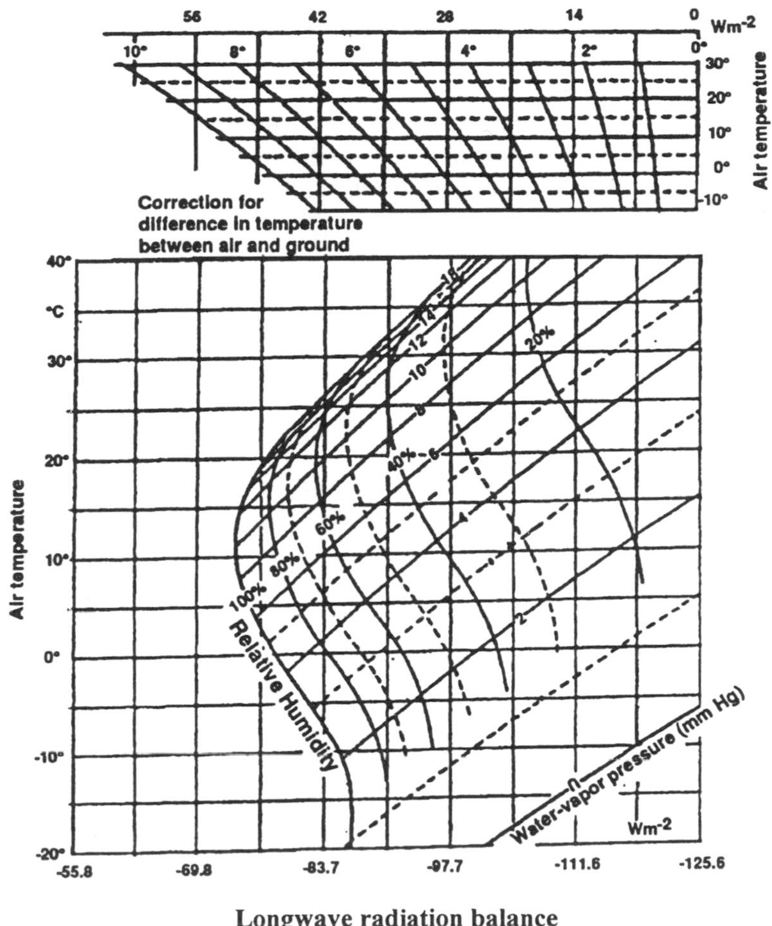

Figure 5-3. Calculation of the night radiation balance of the ground surface with clear skies (explanation in the text).

eter is also directed downward toward the ground, a radiation loss is obtained because $T_A > T_B$, this is equivalent to the correction just mentioned. This difference must be taken into account in all considerations of surface energy budgets.

Several formulas have been developed over the past few decades to estimate the full spectrum clear sky counterradiation which seem to have general application. These include formulas based upon shelter-height humidity (D. Brunt [508], W. Brutsaert [509], P. Berdahl and M. Martin [503]), shelter-height air temperature (S. B. Idso and R. D. Jackson [521], W. C. Swinbank [531]), or both shelter-height humidity and air temperature (W. Brutsaert [509], S. B. Idso [520]). Comparisons of observed versus estimated clear-sky counterradiation have demonstrated model errors within 5% of measured values (A. J. Arnfield [502], J. L. Hatfield, et al. [515], M. Sugita and W. Brutsaert [530]). These methods of determining counterradiation integrate the contributions of individual areas of the sky vault in the hemisphere above the horizontal position in question. In microclimatology it is important to note that the streams of radiation received from different parts of the sky are substantially different. The thickness of air above any point on the ground, and hence the quantity of carbon dioxide, ozone and precipitable depth of the water vapor, is least in the direction of the zenith. M. H. Unsworth and J. L. Monteith [532] examined the angular distribution of counterradiation $L\downarrow$ from the sky, by analyzing 46 observation sets over three years at the English Midlands (53°N) and The Sudan (14°N). They found that the angular contribution of $L\downarrow$ varied with the angular departure from the zenith (solar zenith angle Z), but was independent of the angular departure from true north (solar azimuth). By relating counterradiation to the shelter-height temperature, they found that, for clear skies, the apparent emissivity of the atmosphere was a linear function of ln (u sec Z) and could be described by:

$$\varepsilon(Z) = a + b \cdot \ln(u \sec Z)$$

where u is the depth of precipitable water (cm). For the English Midlands, $a = 0.70 \pm 0.05$ and $b = 0.09 \pm 0.002$; for The Sudan, $a = 0.67 \pm 0.03$ and $b = 0.085 \pm 0.002$. Scatter and systematic variations in the values for a and b were attributed to significant vertical temperature gradients, particularly inversions or strong lapse conditions, and to the radiative effects of aerosols. The angular distributions of $L\downarrow$ from overcast skies was determined to be indistinguishable from the clear sky case. The least amount of counterradiation, therefore, is received from the zenith direction. Since the outgoing radiation is emitted from the surface equally in all directions of the sky hemisphere, the net longwave radiation loss is greatest in this direction. G. Falckenberg [512] showed that counterradiation from the zenith obeys an equation similar to that given on page 22 for counterradiation from the sky hemisphere, but with different values for the constants, becoming:

$$L\downarrow_{zenith} = \sigma T_A^4 \,(0.78 - 0.30 \cdot 10^{-0.065e}) \qquad (\mathrm{W\ m^{-2}})$$

Counterradiation from the zenith is much smaller than that for the whole sky hemisphere. The optical air mass, or thickness of the atmosphere, measured directly upward from a point on the ground, is less than the distance through the atmosphere in any other direction. This thickness increases as the zenith angle increases, becoming 1.5 times the zenith thickness at 48°, twice at 60°, and three times at 71°. The effective outgoing radiation L^* decreases proportionally as the zenith angle increases toward the horizon. Table 5-2 gives relative values of L^* measured by P. Dubois [511] for a water vapor pressure of 7.2 mb taking the effective outgoing radiation in the zenith direction as 100. These values are in agreement with measurements by H. Hinzpeter [517]. Values of the optical air mass m are included for comparison purposes, and are obtained from the following equation given in J. A. Davies and D. C. McKay [510a].

$$m = 35 /[\mathrm{sqrt}\,(1 + 1224 \cdot \cos^2 Z)] \cdot P/1013$$

where Z is the solar zenith angle (degrees) and P is the atmospheric pressure (mb).

P. Dubois's [511] measurements made it possible for F. Lauscher [525] to determine effective outgoing radiation for a number of different types of topography. The results that follow are based on the Dubois [511] value of 7.2 mb vapor pressure. Five different geometric configurations are shown in the sketches in Figure 5-4. They include:

Table 5-2. Relative effective outgoing radiation for different angles from the zenith.

Departure from zenith (degrees)	0	10	20	30	40	50	60	70	80	90
Optical air mass	1.00	1.02	1.06	1.16	1.31	1.56	2.00	2.92	5.69	35.0
L^* (relative value)	100	100	98	96	93	89	81	69	51	0

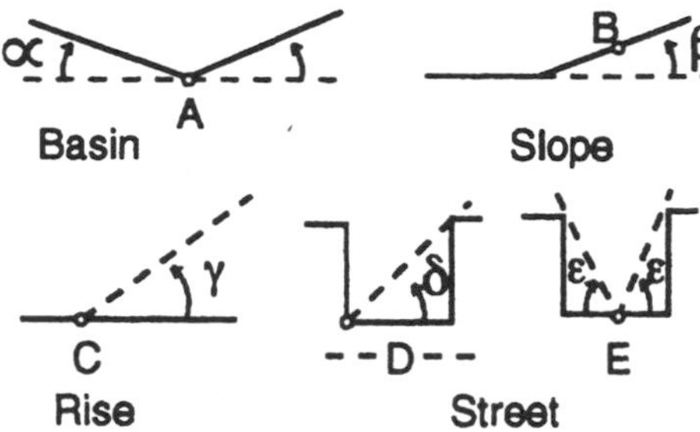

Figure 5-4. Topographic features referred to in discussion of effective outgoing radiation.

A. When the horizon surrounding a place is uniformly obscured up to an angle α, as in a hollow in the ground, in the middle of a circular clearing in a forest, or in an amphitheater, the feature is termed a basin. The figures in row A of Table 5-3 give the effective outgoing radiation in percent of the unobstructed radiation that would be emitted by a level plane, for various angles of shielding. A similar discussion of shielding angle effects on radiative exchanges based upon the more currently used concept of the 'sky view factor' is presented in T.R. Oke [3332]. It may be surprising that an angle of shielding of about 20°, which is quite noticeable on the ground, reduces the radiation loss by less than 9 percent. This is an indication of just how small the effect of obstructions are on the effective outgoing radiation close to the horizon; therefore, radiation measurements are not significantly less accurate if the horizon is not completely free of obstruction. The part of the sky above an elevation of 30° includes half of the sky, yet almost 80 percent of the radiation loss is through this half of the sky.

Table 5-3. Ratio (percent) of the effective outgoing radiation from sheltered or inclined surface to the radiation from a completely open horizontal surface. (After F. Lauscher [525])

Shielding Angle (°)		0	5	10	15	20	30	45	60	75	90
Basin (α)	A	100.0	99.6	98.2	95.5	91.5	79.3	54.9	28.2	7.9	.0
Slope (β)	B	100.0	99.6	98.6	97.0	95.1	90.0	79.6	66.7	52.8	39.6
Rise (γ)	C	100.0	99.7	99.2	98.8	97.9	95.1	87.7	77.2	63.9	50.0
Side of street (δ)	D	100.0	93.0	86.2	79.7	73.7	62.2	45.2	29.6	14.3	.0
Middle of street (ε)	E	100.0	99.3	98.4	97.6	95.8	90.2	75.4	54.4	27.9	.0

B. If a plane surface is tilted at an angle β from the horizontal, the effective outgoing radiation is reduced since no measurable quantity of radiation is directed below the horizon (The assumption of a homogeneous temperature distribution for L^* is certainly not valid for Q^*). The decrease is negligible for small angles of tilt and does not reach 10 percent for all but the steepest naturally occurring slopes. A vertical wall (β = 90°), however, experiences an effective outgoing radiation only 40 percent of that from level ground.

C. In the vicinity of a sharp rise in the ground, such as a rock face, a man-made wall, a building, a shelterbelt, a hedge, or the edge of a forested area, thought of as extending indefinitely in either direction along the ground (perpendicular to the paper in Figure 5-4), the effective outgoing radiation will be reduced on approaching the wall by an amount depending on the angle γ. This reduction is less, of course, than with the basin designated A. For γ = 90°, such as at the foot of the wall or at the edge of a stand of timber, radiation loss to the sky is 50 percent of that in open country, since half the sky is shielded. This loss is greater than the 40 percent loss incurred from the plane surface tilted to 90° because a position on the tilted surface at point B is oriented more toward the horizon where the optical thickness of the atmosphere is greater.

D. It is useful to consider a theoretical street, infinite in length, with rows of houses of equal height on either side. This model is useful in town planning, for forest clearings, and for long valleys. Line D in Table 5-3 gives the relative values for radiation as a function of the angle δ (Figure 5-4). For example, in a street with houses on both sides the same height as the width of the street, effective outgoing radiation to the sky was 45 percent of that in the open country (δ = 45°).

E. Effective outgoing radiation varies with position in the street. Areas near the houses have more protection, and net longwave radiation loss is at a maximum in the middle of the street. This maximum value is given in line E of Table 5-3, the angle ε being measured from the center of the street. It should be mentioned here that the walls of the houses bordering the street also radiate but to a markedly smaller extent than shown in line B (Table 5-3) for 90° because each side of the street presents a barrier to the radiation of the other.

When there is a cloud cover at night, there is, in addition to the atmospheric counterradiation just discussed, additional radiation from water and ice particles in the undersides of clouds. From the previous discussion it is clear that the lowest layer of air is the most important for counterradiation. It, therefore, follows that low clouds reduce effective outgoing radiation to a greater extent than do high clouds. In addition to type and height of clouds, there is also an effect from the extent of cloud cover; this is measured in tenths of the sky (0.0 to 1.0) and indicated by the symbol w. Experience shows that when w is small the few clouds present are close to the horizon and are thus in a part of the sky that is of less importance for the radiation balance than the area near the zenith. Counterradiation from clouds, therefore, does not increase linearly with w, but almost as a quadratic function. This theory was proposed in 1928 by F. Lauscher [524], and has been verified by H. M. Bolz [506].

Counterradiation from a cloudy sky $L{\downarrow}_w$ is expressed, according to Bolz [506], in terms of the value for a cloudless sky by the equation $L{\downarrow}_w = L{\downarrow}\,(1 + kw^2)$. The constant k is greater for clouds found in the lower levels of the atmosphere. Its values for different types of cloud are as follows (D.L. Morgan, et al. [527]):

Cloud type	Ci	Cs	Ac	As	Cb	Cu	Sc	St	Ns	fog
k	0.04	0.08	0.16	0.20	0.20	0.20	0.22	0.24	0.25	0.25

Thus, a cloud deck would result in an enhancement of $L\downarrow_{\mathrm{W}}$ by between 4 and 25% (cirrus to nimbostratus) in comparison to a clear sky. Comparisons of measured and estimated counter-radiation under all sky conditions made by A. J. Arnfield [502] and M. Sugita and W. Brutsaert [530] revealed only slightly larger errors than those obtained under clear sky conditions.

During the night, both air and ground temperatures decline; as a result, radiation losses also decrease, but only to a small extent (provided there is no change in cloud cover during the night). According to F. Sauberer [528], the decrease is only about 10 percent during a night with clear skies and about 15 percent on an overcast night. Energy losses through radiation from the air layer near the ground are greatest after sunset and decrease through the night (Figure 34-1). H. Kraus [2510] evaluated the energy budget of the layer of air between 50 and 600 cm above a grass covered soil on the evening of 12 October 1956. At 17:00, there was still a good deal of energy being conducted from this layer, and thus this layer made a significant contribution to the cooling process and to the formation of a radiation fog which was seen to begin forming at 17:30. The latent heat released on fog formation was found to be negligible: "the fog, which appears to the observer to be the most significant feature turns out, from the energy point of view, to be merely a side effect of cooling in the evening." However, as the nocturnal inversion strengthened, the flow of sensible heat from the air offset the loss by radiation to an increasing extent. When the fog was swept away by stronger winds (22:00) this became even greater and resulted in a temperature increase in the air layer near the ground (Section 14). However, only at 17:00 did evaporation from the grass surface transfer more water vapor into the air layer near the ground than the latter gave off upward; thus, this layer lost water in spite of the fog formation.

6 The Laws of Energy Transport in the Ground

The exchanges of radiation at the surface of the ground, discussed in Sections 4 and 5, cause periodic (daily and yearly) variations of the surface temperature. These changes affect the temperature of the soil below the surface and of the layer of air above it. In this section, the factors influencing these effects will be discussed for the ground only, and the following section will deal with their influences on the air.

We must first consider whether some other influences might be at work in the ground. The high temperature of the interior of the earth immediately comes to mind. Since the rate at which temperature increases with depth in the ground is, on the average, 1°C for every 40 m (a unit of geothermal depth), the upward flow of heat is about 0.07 W m^{-2}, which can be neglected. This geothermal energy source becomes significant only in volcanic regions, near hot springs, or where there are underground fires.

The way in which air moves into and out of the ground also needs to be considered. M. Diem [602] has shown experimentally that, with a sandy soil, the volume of air "breathed" through the surface of the ground in 1 day was equal to a column of air 22 m high over the same area. The air temperature near the ground is influenced by the nature of the soil and by this "breathing" or air-exchange process. Since the thermal capacity of soil is about 1000-3000 times that of air (depending to a large degree upon the soil's water content (Table 6-1)), the reciprocal effect of air on ground temperature is always negligible. The "breathing" of cold or dry air through the soil may result in a drying out of the soil. On the other hand, when humid air which is warmer than the soil is "breathed" through, some water vapor may condense out ("internal dew"). G. Hofmann [603] has estimated that even in the most favorable circumstances the amount of this condensation is less than 0.01 mm per hour. Nevertheless, this is equivalent to 7 W m^{-2}, which is of interest in building technology and for water conservation in time of drought. In normal situations, this can be neglected.

Table 6-1. Variation of density and thermal capacity of sandy soil with varying water content.

Water content ν_w (%)	0	10	20	30	40
Density ρ_m (10^3 kg m^{-3})	1.50	1.60	1.70	1.80	1.90
Thermal capacity $(\rho c)_m$ (10^6 J m^{-3} K^{-1})	1.25	1.67	2.09	2.51	2.93

A more important external effect is the penetration of cold or warm rain into the ground (Section 3). Figure 6-1 shows how a 20.8 mm fall of cold rain affected the ground temperature; the variation in the rainfall intensity with time is shown on the upper graph.

Nineteen minutes after the rain started, the thermometer at a depth of 1 cm recorded the arrival of the cold water. After another 19 minutes, it had penetrated to a depth of 20 cm. However, only a heavy, very cold rain on permeable ground will show such a marked influence. The soil temperature at 50 cm, for example, was unaffected by the cold rain.

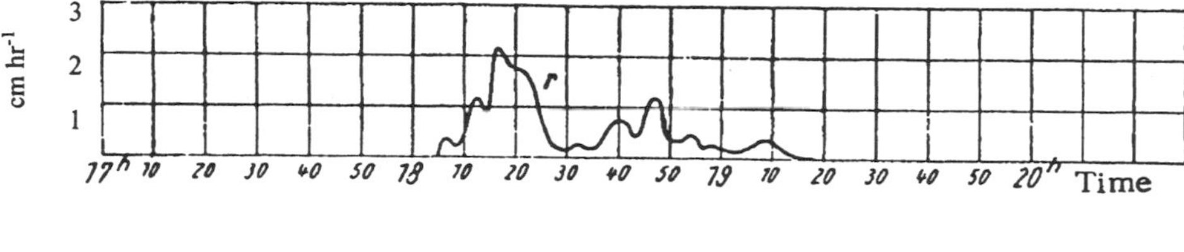

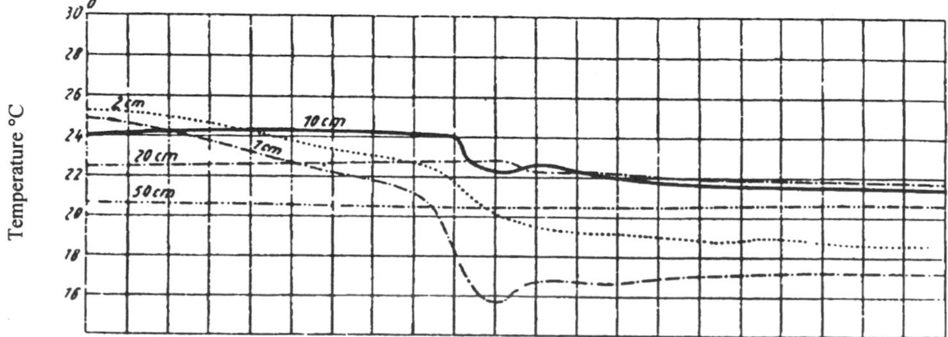

Figure 6-1. Effect of cold rain on soil temperatures. (After F. Becker [601])

All natural ground has three fundamentally different components: (l) the soil proper, consisting of organic and inorganic substances, of density ρ_s (kg m^{-3}) and specific heat c_s (J kg^{-1} K^{-1}); (2) free available water, not chemically bound to the soil; (3) the air occupying the spaces between the soil particles. The parts by volume occupied by these three elements, expressed as percentages, are v_s, v_w and v_l, totaling 100 percent. The values of these quantities, measured by S. Uhlig [2129] on 29 August 1949, in an experimental plot at the Agricultural College of Hohenheim, were: $v_s = 50$, $v_w = 16$, and $v_l = 34$ at the surface, and $v_s = 59$, $v_w = 23$, and $v_l = 18$ at a depth of 0.5 m. The density of the soil ρ_m and its specific heat can be calculated from these figures. Air can be ignored because of its low density. Therefore, taking the density of water as 1000 kg m^{-3}:

$$\rho_m = 10\,(v_s\,\rho_s + v_w) \qquad (\text{kg m}^{-3})$$

It is equally simple to evaluate the thermal capacity. The figure arrived at gives the amount of energy required to raise the temperature of 1 m^3 of the soil by 1 K. The symbol $(\rho c)_m$ is used for this quantity in subsequent discussions. It is equal to the product of ρ and c; hence, using the approximation given above, and taking the specific heat of water c_w as 4.19×10^3 J kg^{-1} K^{-1}, we obtain:

$$(\rho c)_m = 4.19 \times 10^4\,(v_s \rho_s c_s + v_w) \qquad (\text{J m}^{-3}\,\text{K}^{-1})$$

The quantity $(\rho c)_m$ is constant for a soil that is completely free of water ($v_w = 0$). Its value increases with increasing water content; for example, in a sandy soil in which 57 percent of the volume (v_s) is occupied by grains of sand having a density $\rho_s = 2630$ kg m^{-3} with a specific heat $c_s = 838$ J kg^{-1} K^{-1}, the density and thermal capacity are as shown in Table 6-1. Additional values are given in Section 21 for frozen soils.

In nature, soil characteristics are rarely uniform with depth, but show distinct vertical variation. This is illustrated in Figure 6-2 from two sets of soil profiles obtained by A. G. Price and B.O. Bauer [2122] beneath a forest soil in Chalk River, Ontario, Canada. The medium to fine sand surface soils become somewhat more coarse with depth. Soil porosity and bulk density were inversely related to one another, with porosity and bulk density decreasing and increasing with depth, respectively. Soil organic matter content was highest near the surface because of surface litter input. Soil moisture content also frequently increases with depth due to surface evaporation and the extraction of water by the shallow root system of plants.

Values are given for ρ_s and c_s in Table 6-2 for some materials of relevance to microclimatology. There is no difficulty in determining these values for a homogeneous substance such as silver because it is a true constant, which is only slightly dependent on temperature. For substances described by such general terms as "sand", "clay", or "peat", only mean values can be given. Thermal capacity is much more difficult to estimate because of its dependence on water content. To indicate the correct order of magnitude of this quantity, a distinction is drawn between "dry" and "wet" soils in Table 6-2.

To use these values for calculating energy transport within the ground, we assume that the soil is homogeneous. All points at a depth of x cm will then have the same temperature T(°C). The transfer of energy within a solid medium occurs by a process of molecular diffusion. This energy transfer process will be called conduction Q_G (sometimes called soil heat flux or substrate heat flux). By convention, it is directed upward toward the surface when the

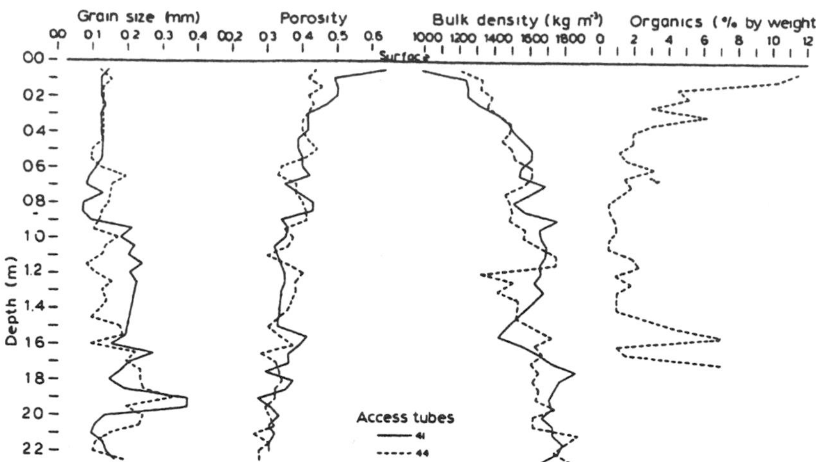

Figure 6-2. Soil profile characteristics beneath a medium- to fine-sandy forest soil at Chalk River, Ontario, Canada. (After A.G. Price and B.O. Bauer [2122])

temperature increases with depth (and downward when temperatures decrease with depth) and is proportional to the rate of change of temperature with depth:

$$Q_G = \lambda \frac{dT}{dz} \qquad (\text{W m}^{-2})$$

The constant of proportionality λ (W m^{-1} K^{-1}) which controls the flow of heat in soil is termed thermal conductivity, and is the amount of energy that will flow through 1 meter of a substance in 1 sec when the temperature difference between opposite faces is 1 K and there are no other variations in temperature.

For chemically pure substances, λ is a constant; for instance, it is 4.188 W m^{-1} K^{-1} for silver. In natural soils, however, λ varies spatially with soil composition and temporally with water content. This is illustrated in Figure 6-3, using measurements made by F. Albrecht [600] in sandy soil in Potsdam for the month of July, 1937. Precipitation is shown by the upper graph in Figure 6-3, while the lower half shows curves of thermal conductivity (λ in W m^{-1} K^{-1}) for depths of 1, 10, and 50 cm. Since soil dries from the surface downward, there is normally an increase in the amount of water present with increasing soil depth, and a corresponding increase in thermal conductivity. The conductivity curve for 1 cm depth (solid line, Figure 6-3) shows a rapid and marked response to rain. This effect is delayed and weakened as the depth increases. The curves for depths of 10 cm (broken line) and 50 cm (dotted line) frequently run in opposite directions as a result of the movement of soil water. For example, on July 20, the thermal conductivity at 10 cm is decreasing as the water moves away, that at 50 cm is increasing as the water begins to arrive. This will be referred to again in Section 21. Figure 6-3 shows that changing water content is able to change the value of λ three or fourfold. Thus, the numerical values for λ given in Table 6-2 are only approximations.

Along flat surfaces the vertical variations in soil moisture content and thermal conductivity are much greater than the horizontal variations. Only very small changes in surface microtopography, however, are able to create significant spatial variability in these two soil prop-

Table 6-2. Order of magnitude of some constants in the energy budget of the ground (arranged according to decreasing thermal conductivity).

Type of soil (or material)	Solid soil particles		Natural soil			
	Density ρ_s	Specific heat c_s	Density ρ_m	Thermal capacity $(\rho c)_m$	Thermal conductivity λ	Thermal diffusivity a
	(10^3 kg m^{-3})	(10^3 J kg^{-1} K^{-1})	(10^3 kg m^{-3})	(10^6 J m^{-3} K^{-1})	(W m^{-1}K^{-1})	(10^{-7} m^2 sec^{-1})
Silver	10.5	0.24	--	2.47	4187.6	1695
Iron	7.9	0.44	--	3.43	879.4	256
Concrete	2.2-2.5	0.88	--	2.09	46.1	22
Rock	2.5-2.9	0.71-0.84	2.5-2.9	1.80-2.43	16.8-41.9	7-23
Ice (see also Section 24)	0.92	2.12	1.7-2.3	1.93	20.9-29.3	11-15
Dry sand	2.6	0.84	1.4-1.7	0.42-1.68	1.7-2.9	1-7
Wet sand	2.6	0.84	--	0.84-2.51	8.4-25.1	3-12
Old snow (density 0.8)	--	--	0.8	1.549	12.6-20.9	8-14
Still water	1.0	4.19	--	4.19	5.4-6.3	1.3-1.5
Wet moorland	1.4-2.0	--	0.8-1.0	2.51-3.35	2.9-4.2	0.9-1.6
Dry clay	2.3-2.7	0.71-0.84	--	0.42-1.68	0.8-6.3	0.5-15
Wet clay	2.3-2.7	0.71-0.84	1.7-2.2	1.26-1.67	8.4-20.9	5-17
New snow (density 0.2)	--	--	0.2	0.38	0.8-1.3	2-3
Dry wood (wood fibers)	1.5	1.13	0.4-0.8	0.42-0.84	0.8-2.1	1-5
Dry moorland	1.4-2.0	--	0.3-0.6	0.42-0.84	0.48-1.3	0.5-3
Still air	0.0010-0.0014	1.005	--	0.0010-0.0014	0.21-0.25	147-250

erties. P. Todhunter has observed substantial changes in soil darkness, related to soil moisture variations, over very short distances within the organic soils of the Red River Valley of North Dakota, a former glacial lake bed.

The temperature of a soil depends not only on the quantity of energy transported to it, but also on its ability to absorb this energy, that is, on its thermal capacity $(\rho c)_m$. Temperature changes in a volume of soil are brought about by alterations in the rate of energy flow Q_G with depth z below the surface:

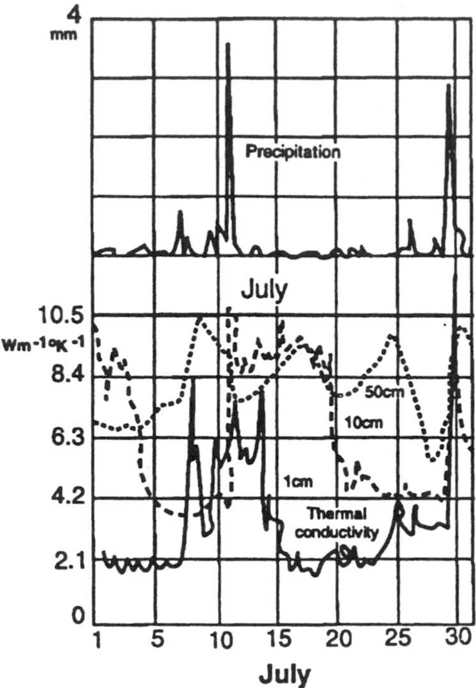

Figure 6-3. Change in thermal conductivity of soil as a result of precipitation. (After F. Albrecht [600])

$$\frac{dQ_G}{dz} = (\rho c)_m \frac{dT}{dt} \qquad (\text{W m}^{-3})$$

If the equation for Q_G from page 30 is substituted, this becomes (after a little rearrangement):

$$\frac{dT}{dt} = \frac{\lambda}{(\rho c)_m} \frac{d^2T}{dz^2} \qquad (\text{K sec}^{-1})$$

This equation gives the relation between the variations of temperature T with time t and depth z. It shows that temperature changes within a soil are controlled by the physical properties of the soil, and the rate of change of the temperature gradient with soil depth. The factor

$$a = \frac{\lambda}{(\rho c)_m} \qquad (\text{m}^2 \text{ sec}^{-1})$$

is called the thermal diffusivity. Numerical values for thermal diffusivity are given in the last column of Table 6-2. It is well known, for example, that still air is a poor conductor of heat (low thermal conductivity), but, because of its low thermal capacity, it is a good transporter of temperature (high thermal diffusivity). Thermal diffusivity is discussed in more detail in Section 21.

A rhythm of temperature variation is brought about at the surface of the earth ($z = 0$) by daily and annual variations. The oscillation period of the daily (annual) heat cycle is t = 86,400 seconds (365 days). The solution of the differential equation gives, for this case, the following relation between the daily fluctuations of temperature s_1 and s_2 at depths z_1 and z_2 for a homogeneous soil:

$$s_2 = s_1 \exp\left[(z_1 - z_2)\sqrt{\frac{\pi}{at}}\right] \qquad \text{(K)}$$

In this expression, "exp" has its usual meaning of the exponential function (e = 2.71828). Consider, for example, a day when a variation s_1 of 38°C in temperature was measured at the surface ($z_1 = 0$) of dry sandy soil ($a = 13\ \text{m}^2\ \text{sec}^{-1}$). The daily temperature variation at depth z_2 = 8 cm would work out as s_2 = 9.97°C. The interval between the times of arrival of a temperature wave at two different depths in a homogeneous soil is given by:

$$t_2 - t_1 = (z_1 - z_2)\frac{t}{2\pi}\sqrt{\frac{\pi}{at}} \qquad \text{(sec)}$$

where t_1 is the time of arrival of an extreme value (maximum or minimum) at depth z_1, and t_2 is the time of arrival at z_2. If the time of the maximum at the surface was 12:30 hours, for example, the maximum would arrive at a depth of 8 cm 18,398 seconds later, that is, at 17:37. Thus the speed at which a given temperature fluctuation at the soil surface is propagated into the soil is proportional to the square root of its thermal diffusivity.

The first of these two equations can be used to determine the depth of penetration of the daily or annual temperature fluctuation into various types of soil. Penetration depth (also called the damping depth) is defined as the depth at which the fluctuation is reduced to 0.01 of its surface value. The higher the thermal diffusivity the greater the depth to which daily and annual thermal fluctuations will occur. Table 6-3 shows the penetration depth of daily and annual temperature fluctuations for a few different types of soil. The annual fluctuation is about 19 times the daily fluctuation.

Table 6-3. Penetration depth of temperature fluctuation.

Thermal diffusivity (a) (10^{-4} m^2 sec^{-1})	0.02	0.01	0.007	0.001
Type of soil	Rock	Wet sand	Snow cover	Dry sand
Daily fluctuation (m)	1.08	0.76	0.64	0.24
Annual fluctuation (m)	20.6	14.5	12.2	4.6

When these equations are used in practice, the conditions on which the mathematical statement is based are not fulfilled since thermal conductivity and thermal capacity both vary systematically with depth and change with time (Figure 6-2). The temperature wave originating at the surface thus often departs significantly from the sine curve assumed in the evaluation. Different values of a are found when it is evaluated from the measured rate of decrease of the temperature fluctuation with depth.

These laws governing the transport of energy in the ground can be used to plot a mean curve of temperature within the ground; this will be discussed in Section 10. The influence of soil type, ground cover, cultivation, moisture content, and ground freezing will be treated in Sections 19-21.

7 Transport of Energy in the Atmosphere. Eddy Diffusion

Molecular conduction of energy also takes place in the air. Since air possesses the ability to transport heat more readily than the soil (note the high value of the thermal diffusivity of still air in Table 6-2), it follows that the air temperature changes more quickly than the ground temperature. The height to which the daily fluctuation of temperature would reach in the air through conduction is found from the equation given in Section 6 to be about 3 m. This is about three times as far as it will penetrate into the best-conducting soil. Observations show, however, that at a height of at least 1000 m above the ground, there is often a measurable difference of air temperature between day and night. Conduction, therefore, usually plays a negligible part in the actual transport of heat in the atmosphere.

The decisive factor in energy transport within the atmosphere is eddy diffusion as already mentioned in Section 3. Observations made on the flow of fluids through tubes by O. Reynolds in his famous experiment in 1883 have shown that two types of flow can be distinguished: laminar flow and turbulent flow. If a colored fluid is introduced through a thin tube into the middle of a glass tube through which another fluid is streaming, when the rate of flow is small, a well-defined trace of color can be observed and the current appears to consist of a bundle of threads running parallel to one another. As the speed of flow increases, a point is reached at which the thread of color is suddenly torn apart and, in the irregular movement that follows, the color is soon distributed uniformly throughout the tube. The first type of flow is called laminar and the second turbulent. The sudden change from one type of flow to the other depends on the value of a constant that is directly proportional to the speed of flow and the density of the fluid and inversely proportional to its viscosity (Reynolds Number).

Laminar flow is much rarer in gases (which are more easily set in motion) than in liquids. Except for a thin layer of air at the surface, motion in the air is nearly always turbulent. The depth of the layer with laminar flow is thicker at low wind speeds and becomes thinner as wind speed and turbulence increase. Turbulence may be visualized when one tries to follow the path of a snowflake in the wind, when the smoke from a locomotive or from a chimney is observed closely, or when the movement of a winged seed can be followed on a sunny day. The individual parcel of air in random motion, sometimes called the "turbulence element" or "eddy", in these examples picks up a visible suspended particle (snowflake, particulate, or seed) and deposits it in another chance position. These eddies also transport invisible properties such as their content of heat, water vapor, kinetic energy, carbon dioxide, radon, and so forth. As the parcels of air undergo haphazard movement, all of their properties move with them. This process is fundamental to the concept of eddy diffusion.

Study of fluid flow also shows that in places where the air comes into contact with a solid surface such as the ground or a wall, turbulence, and hence eddy diffusion, does not extend to the solid. A layer of air a few millimeters thick adheres with great tenacity to the wall or ground. This is termed the laminar boundary layer. The laws of eddy diffusion are not valid in this layer, but the transition from the solid surface to turbulent air is completed within it, governed only by the laws of molecular physics. In this layer, heat is transported only by conduction, and water vapor and other atmospheric elements by diffusion. This laminar boundary layer constitutes a formidable barrier to the transfer of energy, mass and momentum.

Turbulence (convection) may be thought of as a supplementary motion in all directions, superimposed on the (horizontal) wind. The horizontal components of the turbulent motion may increase or decrease the speed of the wind and cause it to deviate in direction from its mean path. This results in gustiness in wind, which can be observed from any wind speed record or swinging wind vane. As a rough approximation, the lowest wind speed is about 0.2 times and the greatest speed is about 1.9 times the mean wind speed.

Although smaller, the vertical components of the turbulent motion are of great importance since they provide a mechanism not only for the vertical transport of energy but all atmospheric properties.

The vertical transport of a characteristic s having units of P varies with height. The change of s with height can be expressed by

$$C = A\,ds/dz \qquad (\text{P m}^{-2}\ \text{sec}^{-1})$$

The coefficient of ds/dz is independent of the properties of the air mass in turbulent motion and is a measure of the vigor of the motion. The quantity A (kg m^{-1} sec^{-1}) is called the Austausch coefficient. Its importance lies in the possibility it offers of expressing the apparently irregular movements of eddy diffusion in a numerical form. In English literature, the quantity $k = A/\rho$ is used instead of the Austausch coefficient and is called the eddy diffusivity. It represents the volume transported through unit area in unit time, and has dimensions of m^2 sec^{-1}, (the same as those of thermal diffusivity). Since $A = k\rho$, the relationship between the Austausch coefficient A and the eddy diffusivity k will vary with temperature and pressure. At an atmospheric pressure of 1000 mb, over a temperature range from -40 to 40°C, ρ will vary between 1.4942 and 1.1125 kg m^{-3}.

If this general form is adapted to the treatment of the flow of sensible heat, Q_H (W m^{-2}) takes the place of C, the quantity c_pT is used for s, and we have

$$Q_H = Ac_\rho \frac{dT}{dz} \qquad (\text{W m}^{-2})$$

This equation is of the same form as that for thermal conductivity in Section 6. Here, in place of the thermal conductivity λ, the Austausch coefficient is used (multiplied by c_p). Transfer of sensible heat in the air, therefore, obeys the same type of law as heat transfer in the ground, but at different orders of magnitude because of the difference between λ and Ac_p. It should be kept in mind that the factor A *is* subject to very great variations in space and time. In the laminar boundary layer, where the principal transport mechanism is molecular diffusion, A is of the order of 10^{-5} kg m^{-1} sec^{-1}. The value of A increases with distance from the ground. Close to the ground, this rate of increase is very large. For example, at heights of 1 to 10 m's, it varies from 0.01 to 1.0 kg m^{-1} sec^{-1}. In the atmosphere as a whole, several orders of magnitude are possible. A distinction is often made between the molecular diffusivity of heat, which governs the molecular transport of heat in the laminar boundary layer, and the eddy diffusivity of heat, which controls the turbulent transport of heat in the turbulent boundary layer. Use of the Austausch coefficient is applicable to either layer.

8 Mixing Due to Friction and Convection

Eddy diffusion has two causes, frictional and convection exchange. Exchange of energy, mass, and momentum by frictional exchange may alternatively be called mechanical or dynamic mixing, that is, mixing due to shear or forced convection. It is caused by variations in the roughness of natural surfaces and by changes in wind speed and direction with height.

At night, the primary type of mixing is frictional. E. Frankenberger [2506] made detailed measurements of the energy balance of meadowland near Quickborn in Holstein from 1 September 1953 to 31 August 1954 using instruments attached to a radio mast at heights of 2, 13, 28, and 70 m. Figure 8-1 compares the 8, 15, and 30 m Austausch coefficient A, on a vertical logarithmic scale, with variations in wind speed measured at a height of 10 m on clear summer days. The lower part of the diagram shows that the nocturnal Austausch coefficient increases sharply with height and wind speed.

During the day, when the soil's surface is heated by solar input, or if for any other reason the surface is warmer than the air, convective mixing will enhance frictional mixing. When the air rises, it cools by the dry adiabatic lapse rate which is very close to 1°C/100 m (at higher elevations, the dry adiabatic lapse rate decreases slightly). When air descends, it is heated by the dry adiabatic lapse rate. If the lapse rate γ, or decrease of temperature with an increase of height, is less than 1°C/100 m, the stratification of the air is stable. This means that every vertical movement will be damped out because ascending air enters a region in which it is cooler than the surrounding air, while descending air will be warmer than its en-

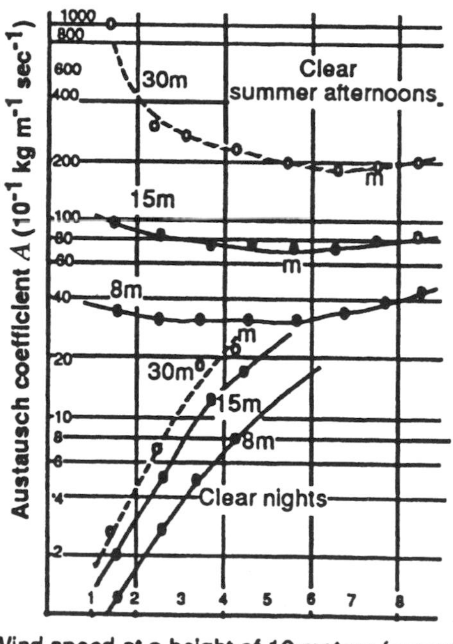

Figure 8-1. Dependence of the Austausch coefficient A on wind speed and height. (After measurements by E. Frankenberger [2506])

vironment and therefore will tend to return to its original position. Thus the average (normal) lapse rate of 0.6°C/100 m implies stable conditions. If temperature increases with height, that is, if γ is negative, stability is even more pronounced. Such a reversal of the normal vertical temperature gradient is called an inversion.

When the temperature does not change with a change in height (i.e., when two adjacent layers of the atmosphere have exactly the same temperature), this layer of air is referred to as isothermal. Isothermal conditions usually represent a transitional state in the atmosphere and typically occur in the morning shortly after sunrise as the atmosphere is changing from a stable to an unstable state, and in the late afternoon shortly before sunset when the reverse change takes place. In neutral stability, the atmospheric temperature change with height is equal to the dry adiabatic lapse rate of 1°C/100 m (in this situation, the potential temperature is constant with height).

At times of intense solar radiation, a lapse rate of more than 1°C/100 m frequently occurs near the ground. These rapid decreases of temperature with an increase of height are referred to as superadiabatic lapse rates. Air heated by the ground rises and cooler air from aloft sinks to replace it. This vertical circulation caused by surface heating is known as convection (in contrast to advection, the horizontal movement of air). Only under the most extraordinary of situations will the air be lifted enough near the ground for condensation to begin due to this lifting. Thus, a discussion of the moist adiabatic lapse rate and lifting condensation processes are beyond the scope of this book.

Figure 8-2. The visible convection process. (After L. A. Ramdas and S. L. Malurkar [704])

This convection process gives rise to irregular motions termed convective mixing (sometimes called thermal mixing or free convection). The parcels of air involved in convection are no longer randomly directed but usually tend to resolve themselves into ascending and descending currents. Figure 8-2 shows how this vertical motion was made visible by L. A. Ramdas and S. L. Malurkar [704], by spreading water over the surface of a heated plate. The lighter parts of the photograph show where the upward movement took place; the dark areas indicate descending currents. The air parcels involved in convective mixing are normally larger than the eddies created by frictional mixing.

The measurements made by E. Frankenberger [2506] for a clear summer day indicate no evidence of an increase in the coefficient *A* with increasing wind speed (shown in the upper half of Figure 8-1). On the contrary, when winds are light, its value is greater than for a strong wind. This is due to the influence of free convection. When winds are light and the ground is strongly heated, the effect of this heating and resulting convection dominates the whole mixing process. During the day, as the wind speed increases, the value *A* decreases to a minimum (*m*, Figure 8-1), from which point onward frictional mixing predominates. The values of *A* for the strongest winds of 8 to 9 m sec^{-1} are, therefore, an approximate extrapolation of the night curves when frictional mixing predominates.

Normally, one of the two forms of mixing predominates while the other plays a smaller role in the exchange process. Typically, frictional mixing predominates at night and convective mixing during the day. When both frictional and convective mixing occur in roughly equal contributions, it is often referred to as mixed convection. It has been shown by C. H. B. Priestley [703] that the time during which the two processes are roughly equal is only a transitional period of short duration (see also K. Brocks [1102]). In the early morning, mixing is, at first, almost entirely frictional. As the elevation of the sun increases there is a sudden transition to convective mixing. With increased instability this transition occurs at lower wind speeds. In addition, the greater the instability of the air and the closer the observation point is to the ground, the earlier in the morning this change is observed.

Both the surface wind speed, on which frictional mixing depends, and the intensity of surface heating, on which convective mixing depends, normally have a maximum at midday and a minimum around sunrise. Therefore the Austausch coefficient *A*, which depends on both of these factors, shows a diurnal variation. Table 8-1 gives examples taken from three different investigations. The figures are averages for 2 hr periods. The first line is from calculations by H. Lettau [701] from 5 years of measurements in the lowest 100 m of the atmosphere, made by N. K. Johnson and G. S. P. Heywood [1108]. They apply to a height of 45 m above the ground, and show considerable variation in 2 hr values between 0.4 and 12.5 kg m^{-1} sec^{-1}. The curve of diurnal variation is smooth because the results were averaged over a two hour period. A. Baumgartner [2502] measured the total energy balance for the crown area of a young pine forest in the neighborhood of Munich on six clear days in July 1952. The values of *A* derived from his results are shown on the second line. The third and fourth lines give mean 2 hr values of *A* for April and November, derived from radon measurements made by M. H. Wilkening [1734] at 0.8 m in Socorro, New Mexico, over a 6 year period. These two months correspond to the maximum and minimum of eddy diffusion over the annual cycle. Over the diurnal cycle, minimum values for both months occur in the early morning (06:00), while maximum values of *A* are found in the late afternoon (18:00 for April, 16:00 for November). Mean daily values of *A* are more than 4 times greater in April than in November due to the more active surface heating.

A fair amount of mixing occurs by night. At one time an explanation was put forward, based on research by A. Defant [700] and A. Schmauss [705], that this was attributable to convective descent or return convection, a term used to describe the vertical movement of dust particles. The particulates are carried aloft by the wind during the day and return by night to the ground, descending partly because of their weight. The small solid particulates, thousands of which are often present in a liter of air, cool the adjacent air by radiating energy and conducting heat from the air. This parcel of air, being cooled, will sink (hence the para-

Table 8-1. Diurnal variation in Austausch coefficient *A* (10^{-1} kg m^{-1} sec^{-1}).

Time of day (hour)	0	2	4	6	8	10	12	14	16	18	20	22
From temperature observations at a height of 45 m (after H. Lettau [701])	21	5	4	10	48	125	94	34	28	31	46	40
From calculations of heat budget (after A. Baumgartner [2502])	10	10	9	18	35	40	43	32	22	11	8	8
From radon measurements at 0.8 m (after M. H. Wilkening [1734])												
April	144	103	89	84	134	206	360	495	567	402	258	196
November	24	19	19	17	27	56	99	137	116	69	42	31

doxical expression "cold convection"). However correct this may be in principle, quantitatively, it is not of very great significance. At night, frictional mixing will still be present, although weak, and will occur as what H. M. Bolz [3808] called microturbulence. This results from the fact that even with the reduced nocturnal wind speeds, the physical form of the ground surface will produce a minor amount of frictional mixing.

The study of turbulent mixing has today become a special branch of science. This discussion of turbulence has purposely been confined to a descriptive level. Our present understanding of turbulence can be traced to the classic work of A. S. Monin and A. M. Obukov [702]. An introduction to the physical laws governing turbulence can be found in R. E. Munn [105], while S. P. Arya [100] and R. B. Stull [108] provide comprehensive and quantitative summaries of current theory.

9 Temperature Instability, Dissemination of Seeds, Dispersion of Air Pollutants, and Effective Stack Height as Problems of Eddy Diffusion.

All meteorological elements are inherently unsteady, a fact that may be verified by observation when using measuring instruments with a fast response time over a sufficiently long period of time. Gustiness as a disturbance of wind speed and direction has already been mentioned (Section 7). It is possible to think of temperature in a similar way. Since there is little mixing close to the ground because of reduced wind speed and friction, the individual eddies may have greatly differing characteristics. It, therefore, follows that temperature instability is particularly great in the air close to the ground.

Figure 9-1 shows the air temperature trace recorded over short grass at the Munich airport during a sunny forenoon in May 1934. Temperatures at heights of 200, 100, and 23 cm were recorded every 20 seconds. In the early hours of the day the points merge to form a

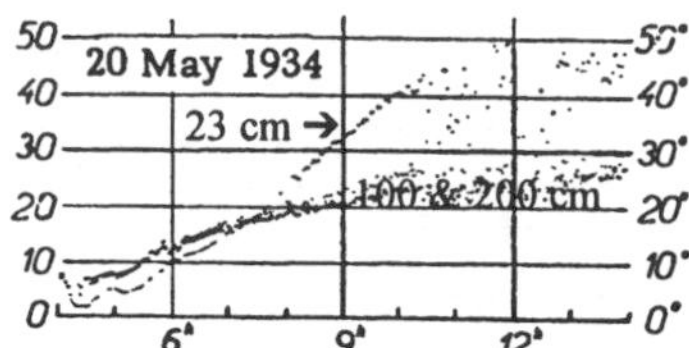

Figure 9-1. Convection, setting in at 10:00 hr, leads to great temperature instability.

line; then about 10:00 hr convection sets in so vigorously that it looks as if there is a rising cloud of dots.

Figure 9-2 shows recordings made by W. Haude [908] in 1931 over stony ground in the Gobi desert. In a 4 1/2 min interval, the air temperature at 1 mm (solid line) and at 1 cm (broken line) above the ground fluctuated up to 2 °C on either side of the mean value. The pattern of isotherms within the 9 mm layer of air is shown in the lower part of the diagram. Very close to the ground (1 mm) there is very little eddy diffusion and temperature fluctuations are suppressed since heat transfer is primarily by molecular diffusion, which is a conservative heat transfer process.

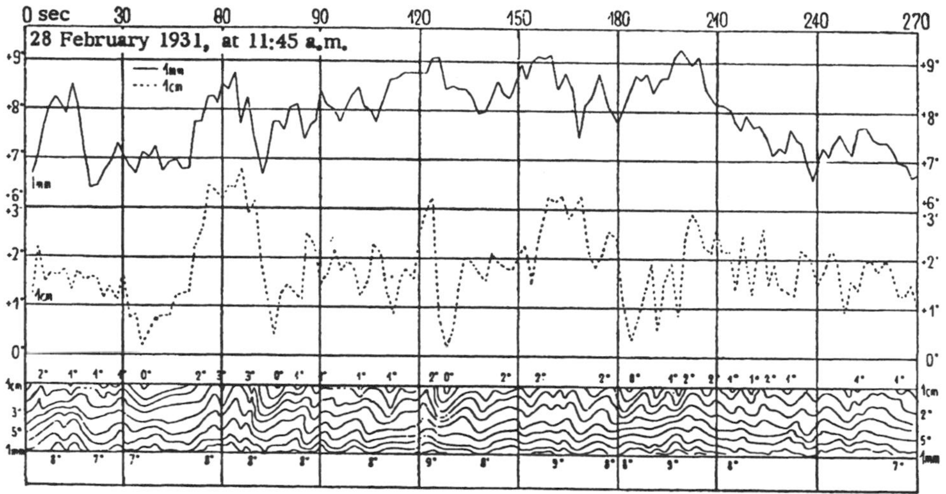

Figure 9-2. Rapid recording of air temperature over desert soil. (After W. Haude [908])

This instability characteristic shown by wind and temperature applies to other characteristics as well. Figures 9-3 and 9-4 show the instability of both temperature and vapor pressure on 3-4 August 1954. The measurements were made above a 50 cm alfalfa field in Berlin-Dahlem by U. Berger-Landefeldt, et al. [903]. Its position at a height of 55 cm was just above the mean plant height. The average temperature (Figure 9-3) and vapor pressure (Figure 9-4) for 5 minute intervals, and the highest and lowest instantaneous values within the 5 minutes were extracted. The range of temperature fluctuation above and below the mean value is shown by the hatched areas.

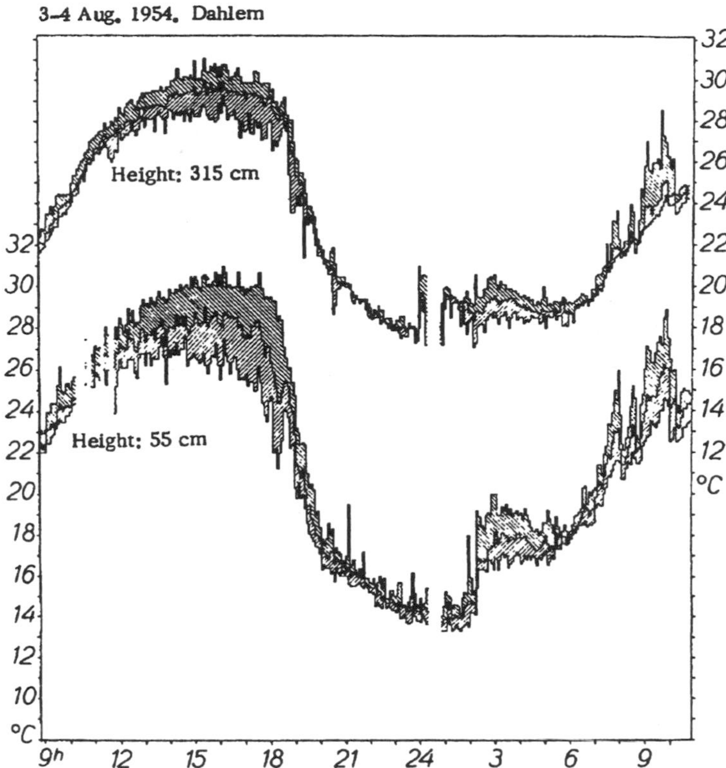

Figure 9-3. Temperature instability during the day at two heights above an alfalfa field. (After U. Berger-Landefeldt, et al. [903])

These diagrams show the increased magnitude and decreased variability of eddy diffusion with height above the ground. They also portray the marked diurnal variation, with great instability during the day and much more stable conditions at night. It is worth noting how the small increase in temperature at 02:10 hr was immediately accompanied by increased instability. The fluctuation in water vapor content (Figure 9-4) within a 5 minute interval was greater than the total diurnal variation of the mean hourly values. Thus, within the laminar boundary layer (close to the ground) during the day fluctuations in temperature and water vapor are small due to the suppression of eddy diffusion. Fluctuations are also small aloft because the air is thoroughly mixed. In the intermediate zone (55 m, Figures 9-3 and 9-4) limited eddy diffusion results in large fluctuations. As the temperature drops in the late afternoon (18:00), thermal convection and the variation in humidity is greatly reduced. Because less water vapor is mixed to higher layers, a second daily maximum of vapor pressure at 315 m is soon attained (Section 15).

Another natural occurrence, which can be explained only by eddy diffusion, is the scattering of seeds, spores, pollen, and fruits. In two pine forests in Hesse during a 7 week observation period, L. Kohlermann [913] found the scattering of seeds to be related to the relative humidity as shown in Table 9-1. The maximum fall of seeds in both cases was found to occur

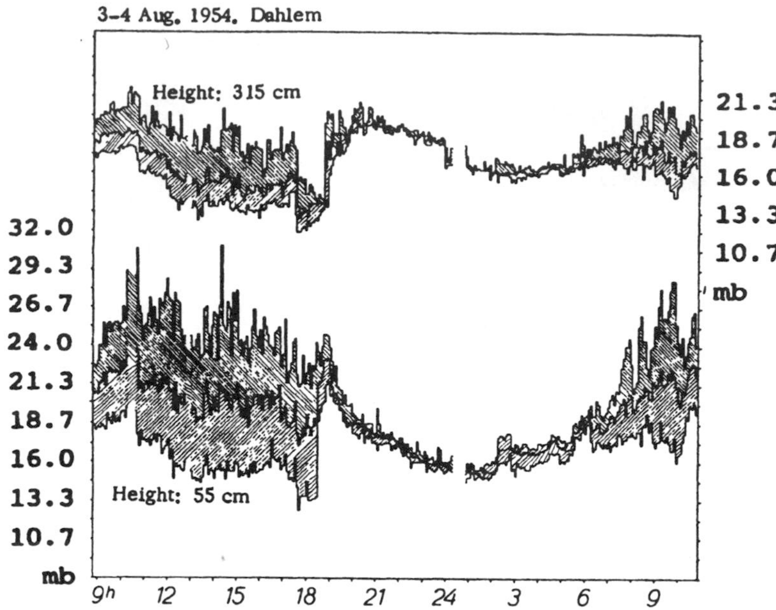

Figure 9-4. Instability of vapor pressure during the day (for the same positions and times as in Figure 9-3). (After U. Berger-Landefeldt, et al. [903])

when the relative humidity was between 55 to 65 percent. If the weather is more humid, the cones retain the seeds by swelling; drier days are rather rare.

Table 9-1. Seed dispersal and relative humidity (percent).

Relative humidity	100		75		65		55		45		35
Dudenhofen Forest Office		5		18		62		10		5	
Schlitz Forest Office		12		29		53		6		0	

If the only influences at work were the velocity of the wind u (cm sec^{-1}) and the rate of fall of the seed c (cm sec^{-1}), the trajectory would be a parabola similar to that followed by a bullet fired horizontally. However, the rate of fall varies from seed to seed. Kohlermann [913] found, for example, that the rate of fall of *Populus nigra* seeds varied from 12 to 50 cm sec^{-1} with a mean value of 26 cm sec^{-1}. This in itself provides a zone of scattering about the mean parabolic trajectory.

Under a regime of eddy diffusion, each seed has an equal chance of being released into a rising or descending eddy. This leads to a broadening of the zone of scattering. The fact that each upward movement increases the time the seed spends in the air means that, on the whole, upward movements have a greater effect on seed dispersion than downward movements. F. Firbas and H. Rempe [907] found that pollen collected during flights at 2000 m and above did not agree with expected values for size and rate of fall. It is clear that the "chimneys" of the atmosphere suck up masses of pollen of all sizes, and that the rate of fall has only a minor effect.

Modern society has produced sources of atmospheric pollution situated at some height above the ground. The most obvious of these are domestic chimneys and tall industrial stacks which pour out great masses of particulates and combustion products, sometimes resulting in pollution injurious to health.

Figure 9-5 shows the different forms that plumes from a stack will take under different atmospheric conditions. On the left of each diagram, the environmental lapse rate (actual

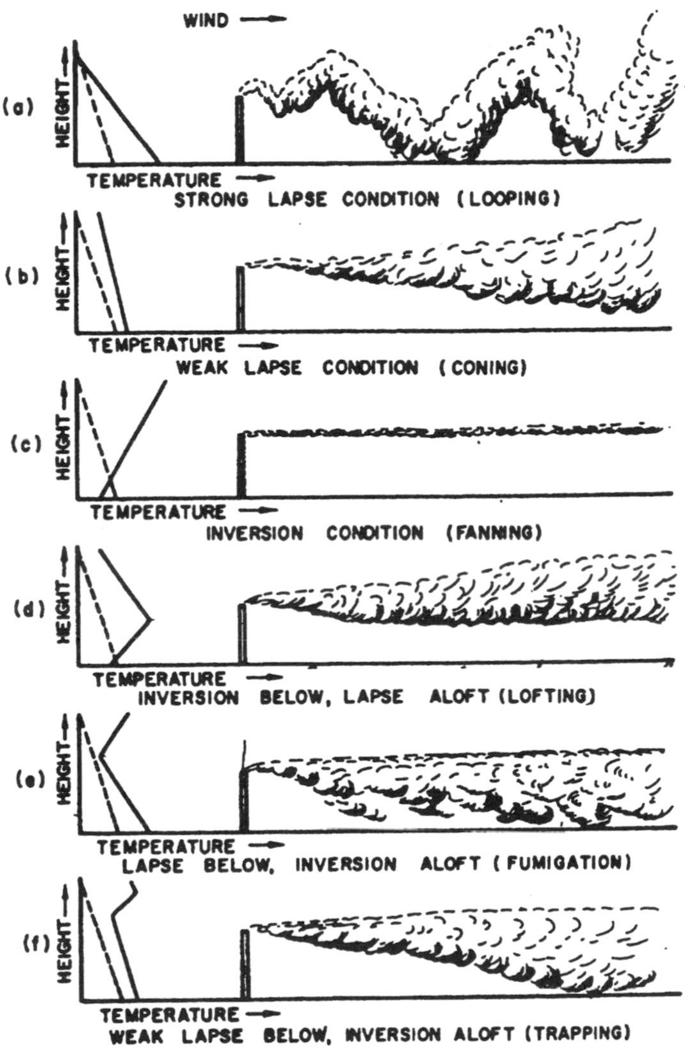

Figure 9-5. Six types of plume behavior under various atmospheric stability conditions. At the left, broken lines are the dry adiabatic lapse rate; solid lines are the environmental lapse rate. (As first proposed by P. E. Church [905] and modified by E. W. Hewson [910])

change of temperature with height, solid line) is compared with the dry adiabatic lapse rate (broken line).

Looping plumes (Figure 9-5(a)) occur when the lapse rate is superadiabatic, i.e., greater than 1°C/100 m, the air is unstable, and thermal turbulence (free convection) is highly developed. The looping occurs because large thermal eddies carry portions of the plume rapidly upward and downward. Effluents diffuse rapidly, but sporadic puffs with high concentrations are at intervals brought to the ground near the base of the stack. Looping usually occurs during clear daytime conditions in seasons when solar heating of the ground is strong and cloudiness, strong winds, or a snow cover are absent.

When the environmental lapse rate lies between the dry adiabatic lapse rate, 1°C/100 m, and the isothermal value, 0°C/100 m, the stability of the air is moderate and convective mixing in both the horizontal and vertical is present but not as intense. A plume will tend to take the form of a gradually widening cone (Figure 9-5(b)) with its vortex at the effective stack height. The distance from the stack at which the effluents first reach the surface is greater than with a looping plume since thermal eddies are greatly reduced. Coning frequently occurs on cloudy or windy days and under these conditions may even occur at night.

If the temperature increases from the surface upward, as in a radiation or advection inversion, the air is very stable and vertical turbulence and mixing are strongly suppressed; however, horizontal mixing is still present. This free horizontal mixing associated with highly suppressed vertical mixing leads to a horizontal spreading and meandering of the plume known as fanning (Figure 9-5(c)). At inland locations, fanning occurs primarily during nights with clear skies and light winds. Because of the slow rate of dispersion, fanning plumes have been observed extending great distances from their source.

Around sunset, on days with clear skies and light winds, a radiational inversion typically begins to develop from the ground up. If the top of the stack reaches above the surface inversion such that there is a shallow radiation inversion below and lapse conditions above (Figure 9-5(d)), there is a rapid upward diffusion. Rapid downward diffusion, however, extends only to the top of the surface inversion and is strongly suppressed through the inversion. This is referred to as a lofting plume. From a local air pollution point of view this represents the best time of the day for the emission of pollutants.

When the top of the stack is below the inversion such that there are adiabatic or superadiabatic conditions below and an inversion aloft (Figure 9-5(e)), this is referred to as fumigation conditions. Since there is a thorough mixing of the pollutants near the ground while upward dispersion is severely restricted, fumigation represents the worst atmospheric conditions for the release of effluents.

E. W. Hewson [910] has identified three situations that commonly lead to fumigation conditions.

Type I occurs over inland areas when solar heating in the morning destroys a nocturnal radiation inversion in which fanning of a plume has been occurring. As a superadiabatic lapse rate is established in a layer, which grows upward, thermal turbulence brings high concentrations to the ground when the top of the unstable layer reaches the fanning plume. This type was first analyzed during the trail investigation by E. W. Hewson [909].

Type II occurs in cities. Heat sources in the city maintain an unstable lapse rate up to two or three roof heights. The air above, which has come from surrounding rural areas, is stable as a result of radiational cooling. In the early evening this leads to a mild fumigation for several hours over the city until the radiational heat losses cause stability in the city as well as in adjacent rural areas.

Type III occurs when cool onshore winds result in a surface advection inversion extending above the stack height. Thus, a fanning will occur as the plume leaves the stack. But during the day the surface lake air is rapidly heated as it moves over the warm land surface and thermal turbulence grows upward to the fanning plume and brings high concentrations to the surface. W. A. Lyons and H. S. Cole [914] describe this type of fumigation condition associated with the summer lake breeze from Lake Michigan. E. W. Hewson [910] states:

> "Conversely, offshore winds on clear nights will be characterized by a surface radiation inversion extending up to and above the effective stack height where a fanning plume will develop. During autumn and winter the surface temperature of a lake large enough to remain unfrozen will be relatively high, so that an unstable layer with thermal turbulence will grow upward over the lake. When this layer reaches the fanning plume, the gases will be brought to the lake surface and lead to a fumigation. If the land is snow covered a similar fumigation will occur over the lake during the day."

E. W. Hewson [910] added a sixth plume type to the five types initially proposed by Church [905] in order to draw attention to the importance of low frontal and subsidence inversions. Trapping occurs with a weak lapse rate when, in addition, there is the base of a low inversion not far above the effective stack height, upward diffusion is greatly reduced, and surface concentrations increased. Thus, although the diffusion is good at the effective stack height and below, the plume is trapped between the ground and the low inversion base (Figure 9-5(f)). This is referred to as plume trapping. Such trapping was first observed and analyzed by E. W. Hewson and G. C. Gill [911] in a deep and narrow valley where it presented a particularly difficult problem because the physical barrier presented by the valley sides caused horizontal trapping in addition to vertical trapping.

Very tall stacks have been built at great expense because it was felt that the higher a stack, the longer it takes for the emitted pollutants to reach the surface, the greater diffusion, and the lower the surface concentration would be. While this is generally true, E. W. Hewson [910] has shown that increasing a stack's height beyond a certain level accrues little additional benefit (Figure 9-6).

The actual stack height can be affected by a number of factors. Gases released with either a high exit velocity and/or a positive buoyancy resulting from high initial temperatures will have an effective stack height, which is greater than the actual stack height. On the other hand, evaporative cooling of stack effluents containing water droplets and/or aerodynamic downwash may lead to a lower effective stack height. At times when atmospheric situations are less favorable for the dispersion of pollutants, increasing the exit velocity or temperature of the emitted gases may help to mitigate these conditions.

When stack effluents are wet washed before emission in order to remove gases such as SO_2, the air becomes saturated and, in the process, its temperature is lowered to the wet bulb temperature of the air before the washing took place. If, after leaving the wet washing stage but before release to the atmosphere, the stack gases are cooled by contact to relatively cold

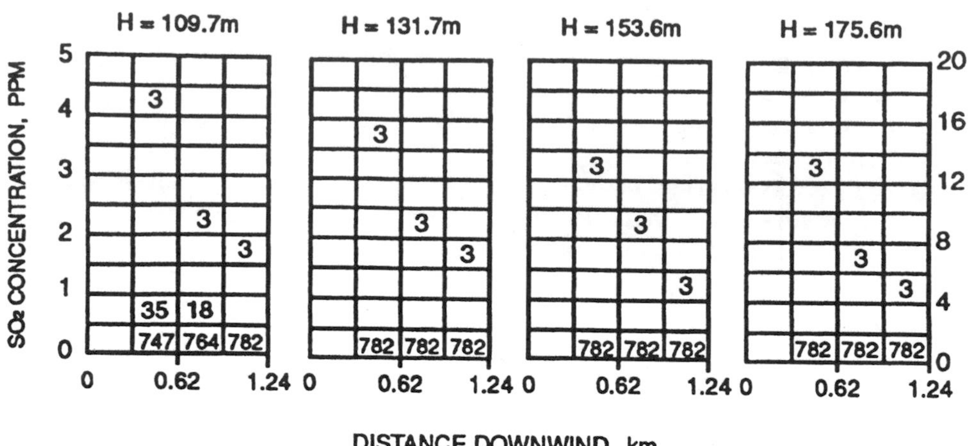

Figure 9-6. Concentrations of SO_2 in ppm by volume to be expected for four stack heights: 109.7, 131.7, 153.6, and 175.6 meters. The numbers give the hours per year when the SO_2 concentration lies within the limits indicated over the 22.5 degree segmental area to the northeast of the plant. Southwest wind; stack gas exit velocity = 36.4 m sec^{-1}. (After E. W. Hewson [910])

surfaces, condensation occurs to form small water droplets, which become part of the effluent leaving the stack. As this effluent leaves the stack and mixes with surrounding drier air, the water droplets will evaporate and the plume will cool as the required latent heat of vaporization is removed from the air. E. W. Hewson explains:

"The mass of liquid water per unit mass of dry air in the effluent as it leaves the stack is readily obtained from the upper scale of Figure 9-7, which gives the number of grams of saturated water vapor mixed with one kilogram of dry air at the temperature shown on the horizontal scale just below. Thus, if the initial wet bulb temperature of the stack gases approaching the wet washing stage is 60°C, the air leaving this stage is saturated at 60°C and contains 155 g of water vapor per kg of dry air. If subsequent cooling reduces the temperature of the gases to 30°C, they will contain 27.6 g of water vapor per kg of dry air, the difference between 155 and 27.6 g kg^{-1}, or 127.4 g kg^{-1}, having condensed into very small water droplets, most of which emerge from the stack.

The lower portion of Figure 9-7 gives approximately the evaporational cooling of the plume if all the water droplets evaporated immediately upon emerging from the stack. The ordinate scale is for the wet bulb temperature in °C of the stack gases approaching wet washing, which is the actual temperature of the gases just after wet washing; the abscissa is for the temperature in °C of the effluent just as it leaves the stack. The curved isopleths give the evaporative cooling in °C, assuming immediate and complete cooling upon emergence from the stack. Using the temperature given above, 60°C and 30°C, the instantaneous evaporative cooling is seen to be about 320°C from the graph. However, the cooling does not occur instantaneously, but slowly as the initially saturated plume mixes with surrounding drier air. If the evaporation is complete after each unit volume of the emerging plume has mixed with 100 unit volumes of surrounding air, then the evaporative cooling of each portion of the enlarged plume, consisting of 101 unit volumes, will average 3.2°C. Alternatively, we may say that after evaporative cooling has been completed the plume will behave in the same way

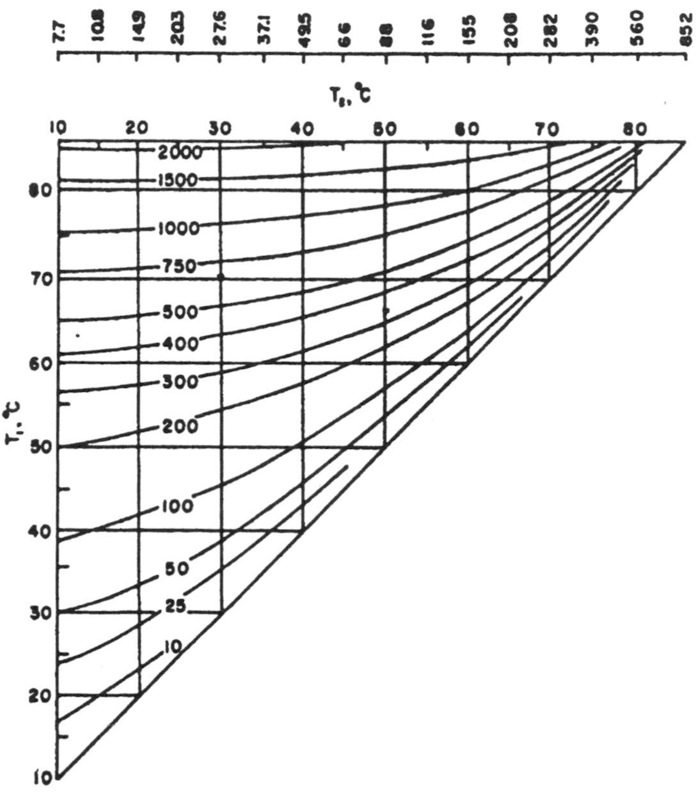

Figure 9-7. Graph for estimating the evaporative cooling of a plume containing small water droplets condensed in it after wet washing but before emission to the air. (After E. W. Hewson [910])

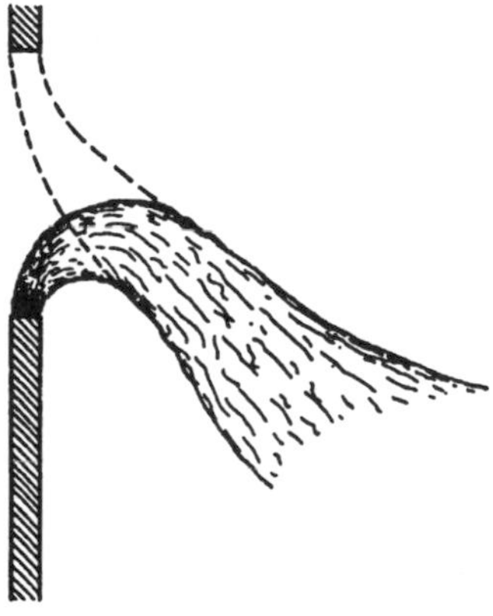

Figure 9-8. Representation of the lowered effective stack height resulting from evaporative cooling and consequent negative buoyancy of the plume. (After R. S. Scorer [915])

as a plume emerging from the stack with a temperature about 320°C less than that of the actual plume.

Thus the plume will have negative buoyancy and the effective stack height is thereby reduced. The behavior of the plume may be studied by the method of mirror images, as illustrated in Figure 9-8. The cooled plume may be thought of as emerging from the orifice of an inverted stack situated above the actual stack" (E.W. Hewson [910]).

A second factor that can lower effective stack height is aerodynamic downwash. Downwash may occur as the result of one or both of two processes. In the first, Kármán vortices may form just in the lee of the stack near its top. If the exit velocity of the plume is high, it may be drawn downward by the low pressure in these vortices. An effluent-colored stain around the upper portion of a stack suggests that downwash of this kind may be of relatively frequent occurrence. This can frequently be observed in the chimneys of residential fireplaces. The second type occurs when the stack is situated on or near a large building; the plume may descend in large eddies formed in the lee of the building as the air flows over and around it.

Aerodynamic downwash can also be accentuated or minimized by the location of the stack with respect to the building, wind direction (Figure 9-9), or even the orientation of the building with respect to the wind direction (Figure 9-10).

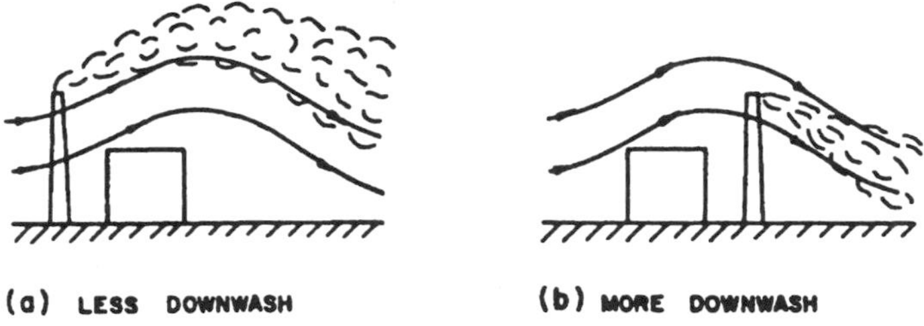

Figure 9-9. Variation of aerodynamic downwash with respect to position of the stack relative to the building. (After E. W. Hewson [910])

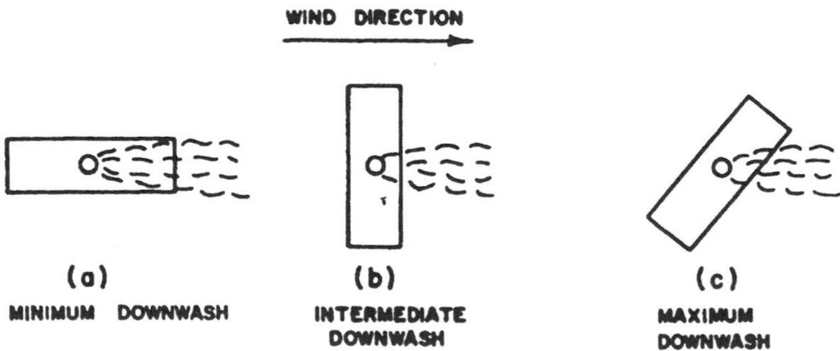

Figure 9-10. Variation of aerodynamic downwash with building orientation in relation to direction of the prevailing winds. (After E. W. Hewson [910])

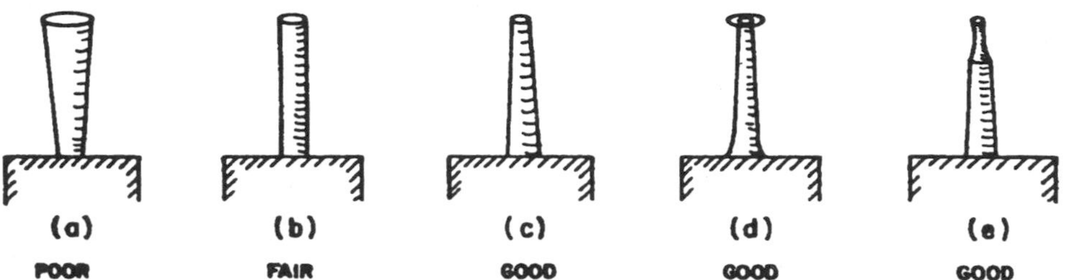

Figure 9-11. Various stack shapes and their properties with regard to aerodynamic downwash: (a) flared stack, poor design, pronounced stack downwash induced; (b) straight stack, fair design, some stack downwash; (c) tapered stack, good design, little stack downwash; (d) tapered stack with horizontal annulus at top, good design, little stack downwash; (e) nozzle at upper part of stack, good design, little stack downwash and reduced building downwash. (After E. W. Hewson [910])

The shape of a stack may also have a pronounced effect on aerodynamic downwash, especially on downwash occurring as a result of reduced pressure in Kármán or tip vortices. Several stack shapes are illustrated in 9-11, and their design characteristics with regard to aerodynamic downwash are indicated.

E. W. Hewson [910] suggests that when the prevailing wind passes over hills before reaching an industrial plant, the hills may serve either to accentuate or to alleviate the air pollution problem. If the plant is situated near the base of the hills, the plume may be carried downward by aerodynamic downwash in the lee of the hills as illustrated in Figure 9-12. Lee eddies and extreme downwash are more likely to occur when the air is relatively stable than when it is unstable. If the hills are rough and irregular in shape and spacing, there may be highly localized downwash, the general result of which is to increase the turbulence and hence the diffusion.

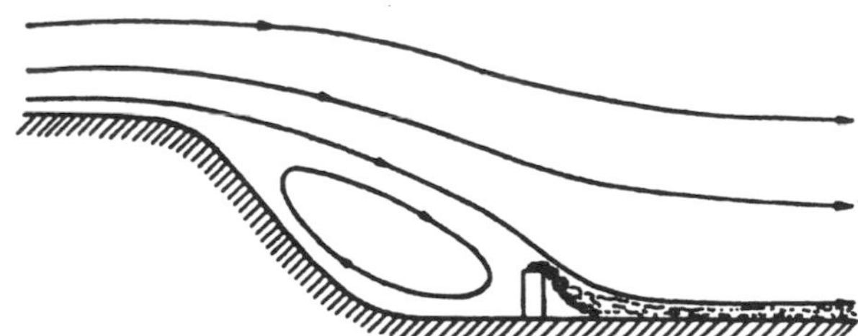

Figure 9-12. A plume carried to the ground by aerodynamic downwash if the stack is located at the point where the flow rejoins the ground at the lee side of the large eddy after separating at the hill top. (After R. S. Scorer [915])

A frequently used rule-of-thumb is that a stack must be at least 2 1/2 times as high as the tallest nearby building if aerodynamic downwash is to be avoided. This rule, however, is useful in only the most uncomplicated of situations.

In estimating the air pollution potential and concentrations for a given area, the air near the ground is often treated as a box. Three parameters are typically used to determine the air pollution concentrations within this box: the local sources of air pollution, the height through which vertical mixing takes place (mixing height), and the movement of air through the box (transport wind speed). Models employing mixing height have often yielded predictions of air pollution levels that are not as accurate as might be desirable. The most popular method for measuring the mixing height (Holzworth method, G. C. Holzworth [912]) is not consistently correlated with air pollution levels (R. H. Aron [900, 901] and R. H. Aron and I-M. Aron [902]). Efforts have to be made to develop better methods for estimating the vertical dispersion of air pollutants (J. P. Deng and R. H. Aron [906]). Additional research is required in order to establish a reliable measure of the height and degree to which air pollutants mix vertically.

This discussion of the dispersion of air pollutants is leading us away from the main focus of this book, which is the air layer near the ground. The reader is, therefore, directed to the bibliography (E. W. Hewson [910], R.W. Boubel, et al., [904], U. S. Weather Bureau [916]) for further information.

CHAPTER II

THE AIR LAYER OVER LEVEL GROUND WITHOUT VEGETATION

10 Normal Temperature Stratification in the Underlying Surface (the Ground)

The intention in Chapter I was to describe the laws that govern processes in the air layer close to the ground. In this chapter, attention is directed to these processes themselves. It is assumed, first, that the ground is completely horizontal, so that there are no topographic effects, and second, that the ground is free of vegetation, so that there are no effects from vegetation. The development of temperature, humidity, wind, and other elements will be described for these conditions on the basis of available observations.

Because the ground surface absorbs and emits radiation, evaporates and condenses water, and retards all air movement, the ground profoundly affects conditions in the overlying layer of air. The nature and state of the ground vary within wide limits from place to place and at different times. Moreover, the surface may be water, snow, or ice, instead of soil. Therefore, the ground and its influence must be discussed more broadly, as the underlying surface. The influence of the underlying surface is so great that Chapter III will be devoted exclusively to this topic.

Without some knowledge of the processes that occur within the ground, the subsequent discussion of what takes place in the air layer near the ground could not be understood. For this reason, the temperature behavior in the ground itself will be considered first. Here, too, a simplification is made by assuming a "normal ground" under a homogeneous surface, having the same composition at all depths. The following results of observation only approximately fulfill these conditions.

As was pointed out in Section 6, the temperature of the soil surface controls the temperature in the ground. The first and simplest example to be considered is a sunny summer day free of all atmospheric disturbances such as fronts, etc. Figure 10-1 shows the first classical series of ground temperature measurements made on the Wikkarais estate (60°17'N) in 1893 by T. Homén [1010], the Finnish pioneer in microclimatology.

Three quantities must always be considered in the representation of soil temperature: depth in the ground (z), temperature (T), and time (t). The variation of temperature, therefore, corresponds to a surface in space, which may be represented on a plane in three different ways. One can draw isotherms in a coordinate system with t as abscissa and z as ordinate (Figures 10-1 and 10-3). Isotherms within the ground are often called geotherms. Alternatively, one can make t the abscissa and T the ordinate and then draw the temperature pattern

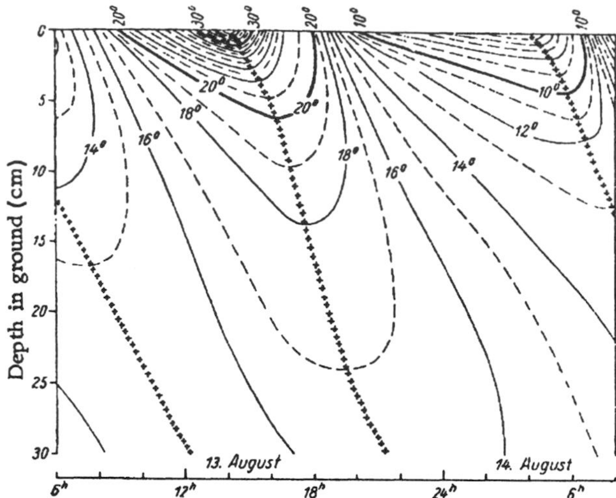

Figure 10-1. Penetration of the daily temperature wave (°C) into the ground on a clear summer day. (From observations by T. Homén [1010] in Finland)

for selected depths (Figures 10-4 through 10-6 and 10-8). Another method is to take T as abscissa and z as ordinate, using what are called tautochrones to show the variation of temperature with depth for a definite period of time (Figure 10-2). Each of these three methods of representation has advantages.

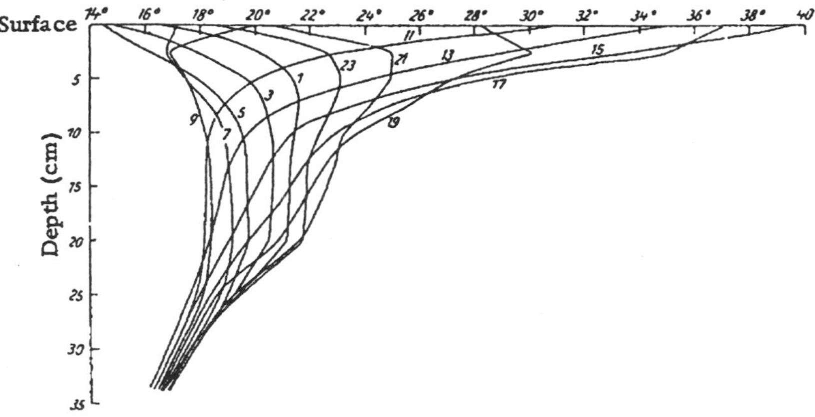

Figure 10-2. Tautochrones of soil temperature (°C) on a clear summer day. (After L. Herr [1009])

Figure 10-1 shows the typical temperature pattern on a clear summer day. In the early afternoon of 13 August 1893, surface temperatures above 34°C were recorded in the sandy heathland. Before sunrise the following day, temperatures were below 5°C. This strong daily fluctuation decreases rapidly with depth. The heating to 20°C during the day penetrates only to a depth of 6 cm, and the night cooling below 10°C only to 4 cm. The deflection of the isotherms to the right with depth shows the lag in the time of penetration of the extreme temperatures into the ground. This is shown by the lines of plus signs (+) which join the points of highest and lowest temperature at the various depths. In ideal, homogeneous ground these lines are straight. It should also be noted that with increasing depth, the maximum temperature is lower, the minimum is higher and thus, the temperature range is less.

Figure 10-2 shows the diurnal variations of soil temperature on a clear summer day in the form of tautochrones. These observations by L. Herr [1009] were taken on 10 and 11 July 1934 for ten different depths in the ground; the temperature variation with depth shown here is for the odd hours of the day. The tautochrones vary between two extremes, roughly defined by the 15 and 5 hr tautochrones. In the first case, with a strong positive radiation balance, the maximum temperature is about 40°C at the surface, and the temperature decreases at first quickly, then more slowly, with depth. In the second case, with a negative nocturnal radiation balance, the minimum is observed at the surface, and the temperature increases to a depth of 20 cm.

During the course of the day, the tautochrones move between those two extremes. Their pattern appears to be complicated by the fact that, in the intervening time, the heat at various depths in the ground may flow in different directions. For example, at 21:00 hours, the highest temperature is recorded at a depth of 5 cm. Below this point, the daytime energy still flows downward, but above it, the upward energy flow already compensates to some extent for the loss of energy by the radiating soil surface. A contrasting picture is given by the tautochrones at 09:00 hr. The decrease of daily temperature fluctuations with depth is shown by the way the curves gradually converge. The fact that the curves below 30 cm run upward from left to right shows that this particular day occurred at a time when the soil was warming up. In fall and winter when the ground is cooling, the tautochrone curves will be in the opposite direction.

It is worth mentioning that in central Europe under favorable conditions, as in poorly conducting soil and on southern slopes, a surface temperature of 60°C or more may be observed (Section 20).

The temperature of the ground surface responds more quickly than air temperature to changes in the weather. Even during the short break in solar radiation during the total eclipse of 30 June 1954, H. S. Paulsen [1016] in Norway and K. Utaaker in Kleppe (60°31'N) observed the following drops in temperature during totality: 0.9°C at 1 cm, 0.3°C at 2 cm, and 0.1°C at 5 cm below the surface. During the same eclipse B. Kullenberg [1012] in southern Sweden observed a decrease in air temperature (at a height of 10 mm above the ground) from 24.6°C (1 hour before) to 15.8°C (at totality), and a rise again to 19.6°C (1 hour later). Figure 10-3 shows the unsteady behavior of ground temperature under the influence of changing warm and cold periods for the winter of 1928-29 in Potsdam, based on a sketch by J. Bartels [3002]. The daily fluctuations penetrate, at most, to a depth of about 1 m. Below 1 m, the isotherms show minimal daily fluctuation, but temperatures gradually increase with time due to the spring warm up. Climatologically, the soil substrate consists of three layers. The first layer, extending from the surface to a depth of about 1 m, is subject to temperature variations on a daily basis. Beneath this first layer lies the second layer, extending to a depth of approximately 20 m, in which soil temperatures are subject to only annual oscillations in temperature. Below 20 m lies the third layer in which soil temperatures increase with depth due to the heat coming from the interior of Earth (Section 6). The precise depths at which these layers occur will vary, however, with the thermal diffusivity of the soil.

A 10 yr series of soil temperature observations made in Quedlinburg (51°47'N) for the period 1938-1947 has been evaluated by K. Unger [1018]. The graph of the 10 yr daily av-

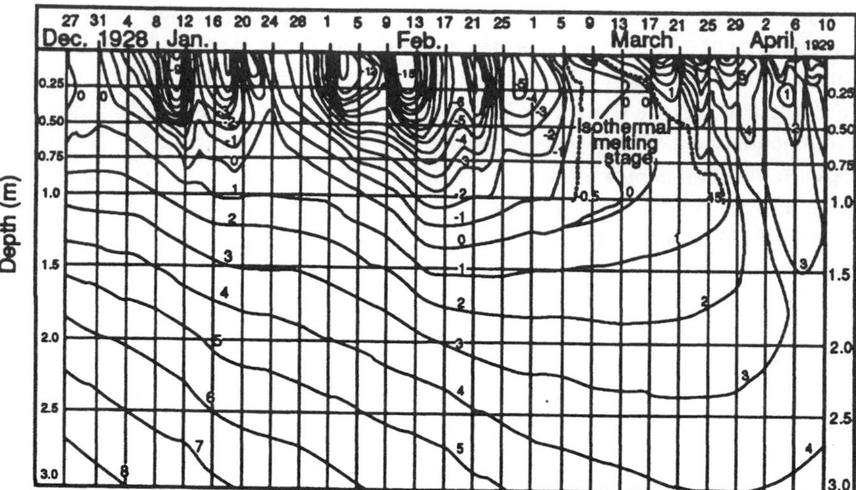

Figure 10-3. Penetration of cold and warm temperature waves (°C) into the ground in Potsdam in the winter of 1928-29.

erages for depths of 0, 20, 50, and 100 cm is shown in Figure 10-4. This figure illustrates how episodic (synoptic scale) weather events not only appear on the surface during the course of the year, but also penetrate into the ground as much as 50 cm. The cold spell that normally occurs about the middle of June, which in Germany is called the "sheep-shearing" cold spell, can be seen at a depth of 1 m.

The long series of observations made in Potsdam from 1894 to 1948 have been used by G. Hausmann [1008] to show the effects of large scale weather fluctuations and seasonal changes at greater depths below the surface. Table 10-1 shows the average yearly tempera-

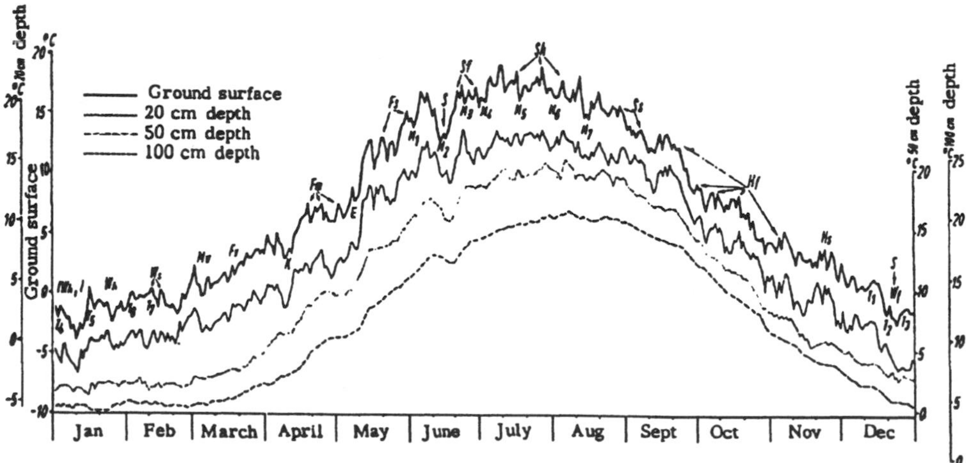

Figure 10-4. Daily average soil temperatues (°C) over a 10 yr period at Quedlinburg, illustrating the decreasing effect of changing weather events on soil temperatures with depth in the ground. (After K. Unger [1018])

ture extremes and the yearly fluctuation (all values have been rounded off to the nearest 0.1°C). The time of arrival of the extreme at these depths also shows the delay in penetrating the ground. The warmest time of the year at a depth of 12 m is the same as the coldest time of winter at a depth of 1 m. These examples illustrate the marked influence changing weather has on temperature in the ground and the relation of these temperatures with the time of year and depth in the ground.

Table 10-1. Soil temperatures (°C) at Potsdam, 1894-1948. (After G. Hausmann [1008])

Depth in ground (m)	Average annual		Average annual fluctuation	Average time of year	
	Maximum	Minimum		Maximum	Minimum
1.0	20.7	1.0	19.7	30 July	11 Feb.
2.0	17.2	3.6	13.6	15 Aug.	4 Mar.
4.0	13.7	6.3	7.4	22 Sept	3 Apr.
6.0	11.9	7.8	4.1	30 Oct.	4 May
12.0	10.0	9.3	0.7	10 Feb.	10 Aug.

Figures 10-5 and 10-6 illustrate the daily variation in two extreme months, in sandy soil in Pavlovsk (59°41'N), for 10 yr observations evaluated by E. Leyst [1013]. The air temperature maximum normally occurs between 14:00 and 15:00, while the soil temperature at 1 cm follows the radiation curve with little delay. In spring (Figure 10-5) heat is penetrating into the ground. At a depth of 80 and 160 cm soil temperatures are still cool from winter. In winter (Figure 10-6) the reverse is the case and at a depth of 1 m the ground is still unfrozen.

The daily fluctuation of temperature varies with the time of year, as is shown by a comparison of the month of May with January (Figures 10-5 and 10-6). In winter, the daily fluctuation is unusually small at the high latitudes of Pavlovsk, while in late spring it is very large. The uppermost layer of the ground shows the systematic influence of the time of year

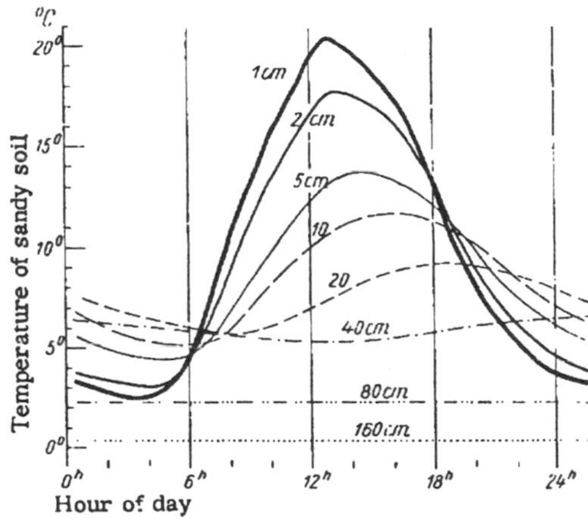

Figure 10-5. Daily sequence of temperature in sandy soil in May, from 10 yr averages at Pavlovsk. (After E. Leyst [1013])

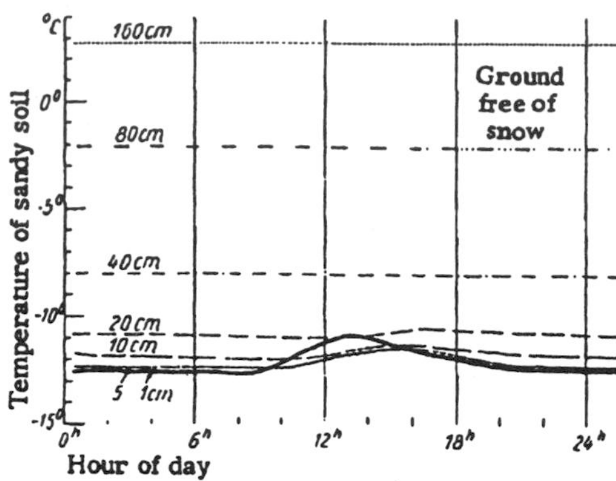

Figure 10-6. Daily sequence of temperature in sandy soil in January, from 10 yr averages at Pavlovsk. (After E. Leyst [1013])

on soil temperatures. Table 10-2 gives the average temperature for a ten year period at 2 hr intervals for every month throughout the year at a depth of 1 cm. The difference between the warmest and coldest hourly averages is called the periodic daily fluctuation of temperature and the difference between the average daily maximum and minimum is called the aperiodic daily fluctuation. The aperiodic daily fluctuation is always greater than the periodic daily fluctuation due to short-term departures from mean hourly conditions. In Table 10-2, the aperiodic daily fluctuations are approximately 2°C greater in summer, and approximately 4°C greater in winter than the periodic daily fluctuations. The daily averages shown in the last column of the table illustrate the yearly variation.

A comparison with an entirely different type of climate is given by the measurements made by W. Haude, in the sandy soil of the Gobi Desert, at Ikengung (41°54'N, elevation about 1500 m). These results were evaluated by F. Albrecht [1001]. The figures (Table 10-3) give the daily variation of temperature in the uppermost layer of the ground to a depth of 0.5 m for 12 sunny August days. The values (maximums indicated by asterisks, minimums by plus signs) show how the extremes were both delayed and became less pronounced with increasing depth. Due to the relatively high latitude and altitude, the temperature at a depth of 2 mm did not reach 50°C in midsummer.

Table 10-3. Hourly sequence of average soil temperature (°C) at Ikengung (Gobi Desert) from 1 to 12 August 1931.

Depth (cm)	0	2	4	6	8	10	12	14	16	18	20	22	24
0.2	18.5	17.4	16.9+	18.9	30.3	40.7	46.5*	45.0	37.0	28.5	23.0	20.1	18.4
0.6	18.7	17.6	17.2+	19.1	30.3	38.9	42.7	44.2*	37.0	28.9	23.5	20.4	18.6
5	21.4	20.0	19.6+	19.8	23.9	30.1	34.7	35.4*	33.6	31.0	26.9	23.7	21.3
10	24.9	23.7	22.8	22.4+	23.4	25.4	28.2	30.5	31.0*	29.9	28.4	26.3	24.8
25	26.3	25.5	25.0	24.4	24.0	23.9+	24.4	25.4	26.5	27.4	27.7*	27.1	26.2
50	24.7*	24.7*	24.7*	24.6	24.4	24.2	24.1	24.0+	24.1	24.3	24.5	24.6	24.7*

Table 10-2. Summary of temperature measurements over 10 years at Pavlovsk. (After E. Leyst [1013])

Month	Temperature (°C) at a depth of 1 cm every 2 hours for 2 hr averages												Daily fluctuation		Daily mean temper-ture (°C)
	02	04	06	08	10	12	14	16	18	20	22	24	Per-iodic	Aper-iodic	
Jan.	-12.4	-12.4	-12.4	-12.5	-12.1	-11.1	-11.0	-11.5	-12.0	-12.2	-12.3	-12.3	1.7	5.8	-12.0
Feb.	-13.2	-13.3	-13.3	-13.2	-12.0	-10.1	-9.4	-10.4	-12.0	-12.8	-13.1	-13.2	4.0	6.4	-12.2
Mar.	-12.1	-12.6	-12.9	-12.4	-9.2	-5.8	-4.0	-5.0	-8.0	-9.4	-10.3	-11.0	8.9	10.2	-9.4
April	-0.8	-1.3	-0.7	2.8	7.8	12.7	12.9	10.3	5.9	2.3	0.9	-0.1	14.3	16.0	4.4
May	2.8	2.5	4.7	10.5	15.3	19.3	19.7	17.7	13.4	8.2	5.3	3.8	17.8	20.4	10.3
June	7.7	7.6	11.2	16.6	21.6	25.6	26.6	24.8	19.9	14.3	10.8	9.1	19.2	21.6	16.3
July	11.0	10.7	13.5	18.8	23.4	25.6	26.4	24.6	21.4	16.9	14.0	12.3	15.7	17.7	18.2
Aug.	11.2	10.8	11.7	16.5	21.1	23.7	24.5	22.6	18.9	15.3	13.1	11.9	14.1	16.0	16.8
Sept.	6.8	6.4	6.2	9.0	13.6	17.0	17.2	15.0	11.3	9.0	8.0	7.1	11.3	13.8	10.6
Oct.	1.1	1.2	1.2	1.7	3.5	5.5	5.6	4.0	2.4	1.5	1.0	0.8	4.6	7.1	2.4
Nov.	-2.9	-3.0	-2.9	-3.0	-2.0	-1.2	-1.2	-1.9	-2.5	-2.5	-2.7	-2.8	1.8	3.4	-2.4
Dec.	-9.0	-9.0	-9.0	-9.1	-8.8	-8.3	-8.4	-8.8	-9.0	-9.2	-9.3	-9.4	0.9	4.8	-8.9

In addition to the seasonal influence, the daily range of temperature is also affected by cloudiness. I. Dirmhirn [1007] evaluated this in Vienna (Figure 10-7).

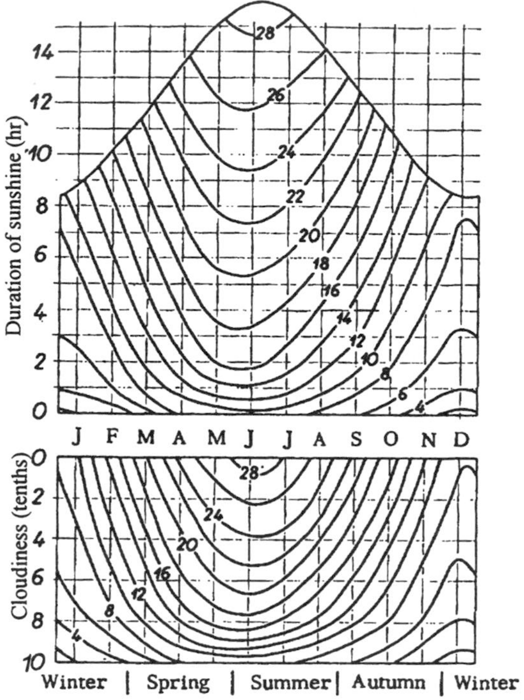

Figure 10-7. Variation of diurnal fluctuation of temperature (°C) at a depth of 1 cm in the ground with changing sunshine duration and cloudiness in Vienna. (After I. Dirmhirn [1007])

The upper diagram shows the daily variation of temperature at a depth of 1 cm throughout the year as a function of the duration of sunshine. The lower diagram shows this variation as a function of cloudiness. With changes in the declination of the sun, the duration of sunlight is a function of the season, and this forms the upper limit in Figure 10-7. Temperature fluctuations are greater in summer and on clear days. For example, on a clear day in midsummer, the temperature fluctuation exceeds 28°C, whereas on a cloudy day it is only about 9°C. In winter on a clear day, the fluctuation is also about 9°C but on a cloudy day only about 2°C. With decreasing cloudiness, the temperature fluctuation increases rapidly at first and then more slowly. This is attributable to the moderating effect of convectional mixing.

We now turn our attention from the daily temperature pattern to the variations taking place throughout a year. Figure 10-8 shows the series of observations made from 1873 to 1877 and from 1879 to 1886 at Königsberg (54°43'N) by A. Schmidt [1017] and E. Leyst [1014]. Although the curves show the damping of the annual temperature wave with depth in such a classical form that they might seem to have been constructed from theory, they are actual observations. From the positions of the curves one might conclude that the mean annual temperature scarcely alters with depth. A study of the figures, however, showed a small increase of temperature with depth, which corresponded approximately with that of the geothermal gradient (Section 6). This, however, is not always the case. The 10 yr measurements made in Pavlovsk, examples of which have already been shown (Table 10-2), indicate that from 5 to 320 cm there is an increase of the annual mean temperature from 4.3 to 6.1°C, which gives a rate of change of 1°C in less than 2 m.

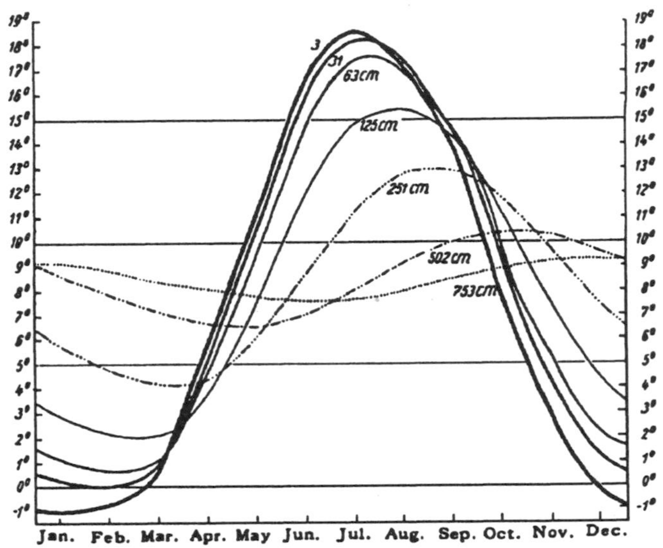

Figure 10-8. Annual sequence of soil temperature (°C) at Königsberg. (After A. Schmidt [1017] and E. Leyst [1014])

At one time, much penetrating thought was expended in the search for an explanation of this difference in the behavior of the annual mean temperature. Under stationary conditions, if one were to disregard the possibility of a systematic yearly change in the thermal conductivity of the soil, the variation in the mean annual temperature would have to be a function of the geothermal gradient. Otherwise, there would have to be a steady stream of heat downward or upward, and this cannot be reconciled with the energy budget. One might try to explain the observations by pointing out that the temperature of precipitation also influences soil temperature (Sections 3 and 6), or that thermal conductivity is a function of the season. However, G. Hausmann [1008] explained all these different possibilities in his investigations into the effect of synoptic weather fluctuations on the soil temperature in Potsdam. Over the 54 yr period the annual mean temperatures varied as follows:

Depth (m)	1	2	4	12
Temperature variation (°C)	2.6	1.9	1.5	1.3

Based on G. Hausmann's [1008] observations of the variation of annual temperature between 1896 and 1949 for depths of 1 and 12 m, an increase in the mean of the temperature of 1°C was observed for 1907-1909, and for period 1947-1949, a decrease of 1°C at 12 m. The first of these periods was a time of predominantly cold weather, especially in the cold winter of 1908-09. The years 1947 to 1949, however, were warmer with a particularly dry summer in 1947.

The soil temperature regimes of tropical environments are characterized by a greater constancy of soil temperatures and a greatly reduced annual variation in soil temperature. Important spatial and seasonal variations in mean monthly soil temperature still occur, however, due to the annual migration of the overhead sun, seasonal changes in cloud cover associated with the migration of the Inter-Tropical Convergence Zone, and variations in soil thermal diffusivity created by annual soil moisture variations. Figure 10-9 shows the mean monthly soil temperatures at 30 and 120 cm depth, along with mean monthly air temperatures, for four stations in Nigeria over the 12 yr period 1969-1980, from O. Ameyan and O. Alabi [1002]. The four sites cover a broad range of tropical environments: Port Harcourt (5°45'N, 20 m) (latitude, elevation) is located in a coastal mangrove swamp, Ibadan (7°30'N, 227 m) is within the rain forest, Yola (9°22'N, 186 m) is located in the Guinea savanna, and Katsina (13°N, 518 m) is in the drier Sudan savanna. Mean monthly soil temperatures closely follow mean monthly air temperatures, with appropriate lags. Soil temperatures are highest in March/April at the beginning of the rainy season when the noon sun is

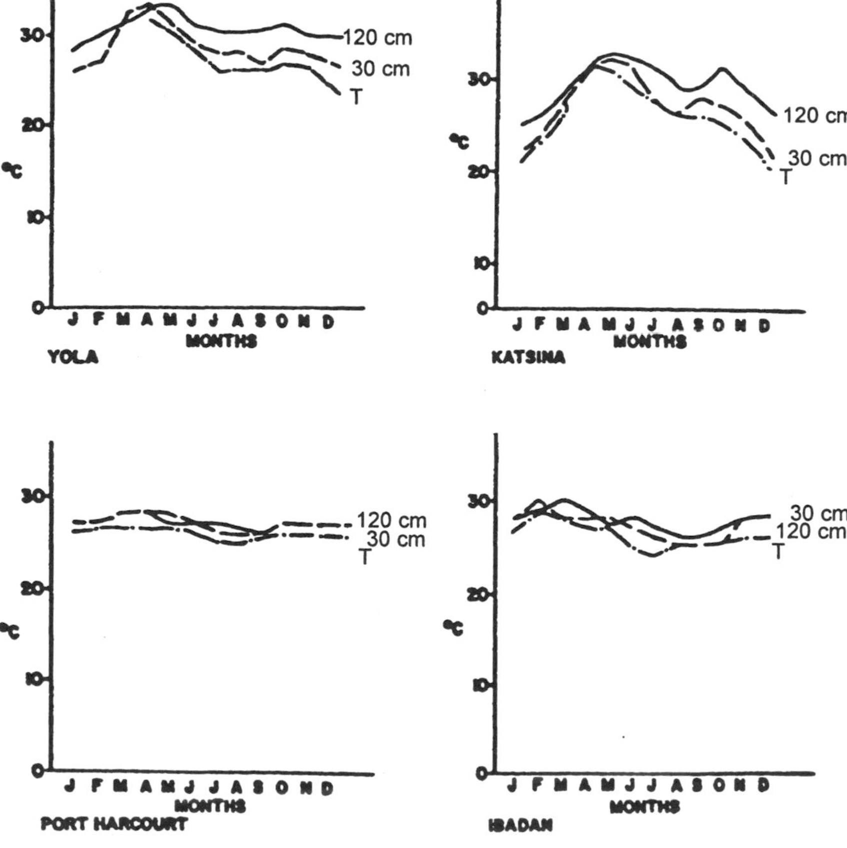

Figure 10-9. Mean monthly soil temperatures (°C) at 30 and 120 cm depth and mean monthly air temperature (T) for four stations in Nigeria. (After O. Ameyan and O. Alabi [1002])

overhead, low cloud cover allows for strong surface heating, and moist soils have a high thermal diffusivity. The northward migration of the ITCZ brings extensive cloud cover, reduced surface heating, and decreasing air temperatures which removes heat from the soil ($-Q_G$), resulting in decreasing soil temperatures from May through September. The southward movement of the ITCZ in October results in increased surface heating and soil warming. By November, the southward movement of the noon sun and the arrival of dust carried by the strong northeast Harmattan winds brings reduced surface heating, decreasing air temperatures, and lower soil temperatures, especially at Yola and Katsina in the north. A bimodal seasonal soil temperature pattern is found at Yola and Katsina due to their more variable noon sun, cloud cover, and soil moisture conditions. Port Harcourt and Ibadan show only small seasonal soil temperature variations because of their more constant environment.

J. H. Chang [1006] completed a comprehensive global evaluation of soil temperature measurements and published a world atlas of soil temperature at depths of 10, 30, and 120 cm for January, April, July, and October [1004], and the annual temperature fluctuation for those three depths [1005]. E. Batta [1003] has published measurements at Budapest for 1912 to 1941 for depths of 2 to 400 cm. P. Katíc [1011] has compiled measurements at Novi Sad in the former Yugoslavia Republic, and J. S. G. McCulloch [1015] has provided a yearly series for Muguga in Kenya (1°S).

11 Temperature in the Lowest 100 m of the Atmosphere

The temperature of the air layer near the ground is determined by surface conditions. In contrast to conditions within the ground, the role of molecular diffusion is insignificant as compared with eddy diffusion. There are other factors in addition to mixing such as longwave radiation, breathing of the ground (Section 6), the influence of the boundary layer (Section 7), and many others, which make atmospheric processes difficult to evaluate. For this reason, we will momentarily disregard the air layer very near the ground and first consider the processes taking place from 1 to around 100 m in the atmosphere.

Figure 11-1 gives the daily pattern of temperature averaged over two extreme months as observed by W. D. Flower [1105] at heights between 1 and 61 m above the ground in Egypt. The curves show how the daily fluctuation of temperature varies with distance from the ground, and how the extremes lag with height, an effect already seen in Figure 10-5. One must always bear in mind the vastly different orders of magnitude involved in the depth of temperature fluctuations in the soil, as opposed to the height of these fluctuations in the air. As in the soil, these fluctuations are greater in summer than in winter. Further from the surface the maximum temperature is lower, the minimum higher, the range less, and the times of occurrence of temperature extremes later. For example in Figure 11-1 between 1 and 61 m, the daily fluctuation of temperature in July decreases from 15.4 to 11.1°C. The time of maximum temperature displacement was delayed from 14:55 hours at 1 m to 15:33 hours at 61 m. Table 6-2 shows that the thermal diffusivity of calm air (in an atmosphere which only transports heat by molecular diffusion) is of the order of 10^{-1}, whereas that of solid earth is 10^{-3}. By using the given values for ρ and c, one can calculate the thermal conductivity λ (Section 6). From this it follows that, as a result of turbulent mixing, the thermal conductivity and diffusivity of the air have been increased 100,000

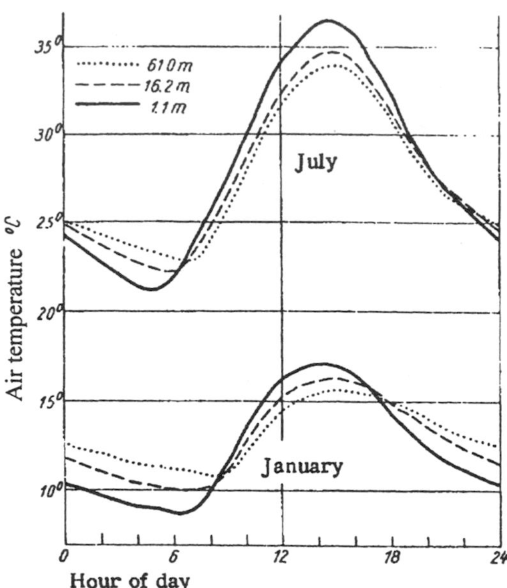

Figure 11-1. Average daily temperature sequence in an air layer over desert soil. (After W. D. Flower [1105])

fold. If air undergoing mixing is compared with dry, sandy soil, its thermal conductivity is 10^5 times and its thermal diffusivity 10^7 times greater. This example illustrates the overwhelming importance of convection in energy transfer in the lower atmosphere.

The values shown are averages for a 60 m layer of air. Within this layer, there are substantial differences, because the amount of mixing is not constant with height (Figure 8-1). The small amount of mixing that takes place close to the ground (lower Austausch coefficient) gives rise to a greater temperature gradient; while at higher levels, more lively mixing (higher Austausch coefficient) results in smaller temperature gradients. This also means that the range within which the temperature gradient can vary will become greater closer to the ground.

Figure 11-2 is based on the measurements made over 3 years in Rye, Sussex, 5 km from the south coast of England by A. C. Best, et al. [1101]. The abscissa is the temperature lapse rate γ (°C/100 m). From the ordinate, the percentage of all hours with a lapse rate above or below different levels can be evaluated. The average value for γ in the free atmosphere (normal lapse rate) is 0.6°C/100 m. Thus, in Figure 11-2, the average γ lies between the adiabatic and the isothermal lapse rates. The upper most layer of air (dotted curve) shows a superadiabatic lapse rate in only 8 percent of all cases, and these hardly ever exceed 2°C/100 m. Inversions, on the other hand, are found in 32 percent of cases.

As the ground is approached, however, there is an enormous expansion of the range within which the lapse rate may lie. In the highest layer observed, between 47 and 107 m as has just been shown, 60 percent of all gradients lie between the adiabatic and the isothermal; while between 15 and 47 m only 27 percent, and between 1 and 15 m only 9 percent are found in this range.

During the 3 yr observation period, the highest superadiabatic lapse rates occurred in the spring months of March to June. In the lowest layer (1 to 15 m), a lapse rate as high as

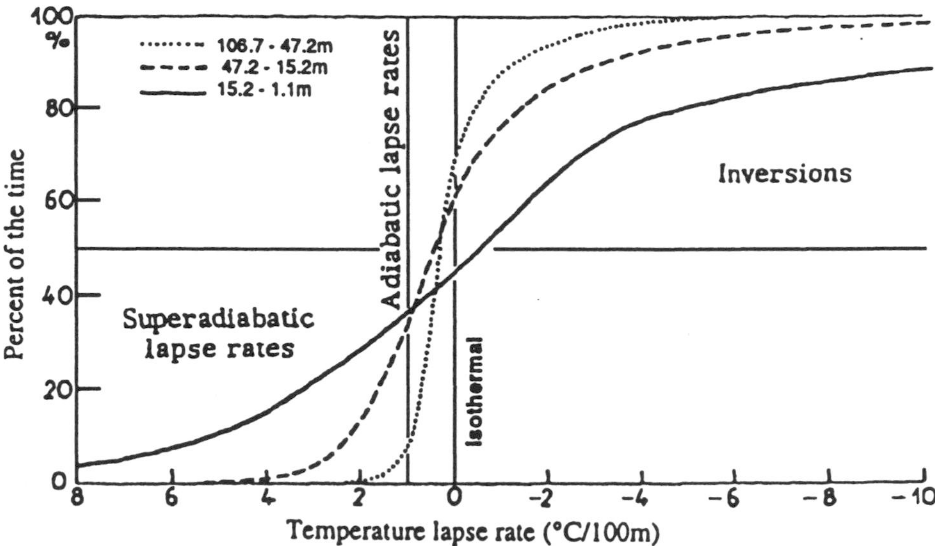

Figure 11-2. Frequency distribution of temperature change in three air layers from observations over three years in England. (After A. C. Best, et al. [1101])

20.5°C/100 m was recorded only between 11:00 and 13:00 hr on clear days. In the upper most layer (47 to 107 m), the highest value was 4.2°C/100 m. Superadiabatic lapse rates were observed in the lower, middle and upper layers approximately 38, 33 and 8 percent of the time, respectively. The frequency of inversions in the same three layers was 53, 40 and 32 percent of the time. The strongest inversions were with clear skies, sometimes in conjunction with ground fog. The maximum values in the lowest layer were -53.4°C/100 m, and in the uppermost layer, -13.6°C/100 m. As the ground is approached, the lapse rate becomes steeper due to the reduction in mixing. This reduced level of convective mixing accounts for the greater frequency of superadiabatic lapse rates in the lowest layer, which are produced by daytime solar heating, as well as the higher frequency of inversions in the lowest layer, which are largely due to nocturnal radiative cooling.

The magnitude of the temperature lapse rate increases as the surface radiation balance reaches larger positive or negative values. This effect may easily be observed from the daily and yearly variations of gradient (Tables 11-1, 11-2, 13-1, and 13-2).

Figure 11-3 shows the daily pattern of the temperature lapse rate in three layers at different heights from the observations made by W. D. Flower [1105] in Egypt. Around sunrise, which occurs shortly before 05:00 hr, the inversion is strongest. One hour after sunrise, the isothermal condition begins to break down as a result of heating from below. This occurs throughout the entire year. In the lowest layer, superadiabatic gradients begin to increase substantially under the influence of increasing radiation; this is less pronounced in the middle and upper layers because of the freer movement of the air. The closer to the ground, the greater the gradient. The initiation of convection causes the gradient curves, throughout most of the day, to be almost horizontal above 16 m (Figure 11-3). Not until 1 hour after sunset (at about 19:00 hours), or after 2 hours at higher levels, did isothermal conditions develop again. In contrast to conditions at sunrise, this time difference varied throughout the year. It occurs latest in summer and earliest in winter, often before sunset.

Table 11-1. Temperature lapse rate (°C/100 m) for the air layer from 47.2 to 106.7 m, giving the diurnal and annual variation in the form of monthly averages for the three years 1945-1948 at Rye, England. (Corresponding humidity gradients are given in Table 15-2)

Hour	2	4	6	8	10	12	14	16	18	20	22	24
Jan.	-0.26	-0.31	-0.17	-0.18	-0.03	0.30	0.38	0.19	-0.05	-0.15	-0.18	-0.21
Feb.	-0.02	-0.13	-0.02	-0.06	0.23	0.53	0.53	0.47	0.22	0.16	0.03	-0.02
Mar.	-0.94	-0.74	-0.68	-0.07	0.68	0.89	0.74	0.43	-0.11	-0.50	-0.65	-0.72
April	-0.44	-0.50	-0.56	0.21	0.67	0.91	0.86	0.76	0.35	-0.11	-0.24	-0.26
May	-0.74	-0.64	-0.23	0.57	0.68	0.86	0.72	0.89	0.42	-0.27	-0.64	-0.90
June	-0.72	-0.64	-0.15	0.64	0.64	0.87	0.52	0.70	0.41	-0.21	-0.66	-0.74
July	-0.94	-1.00	-0.63	0.05	0.11	0.13	0.03	-0.02	-0.18	-0.49	-1.00	-1.02
Aug.	-0.90	-0.89	-0.70	0.22	0.50	0.60	0.59	0.46	0.25	-0.15	-0.57	-0.81
Sept.	-0.61	-0.58	-0.67	-0.21	0.40	0.48	0.47	0.39	0.12	-0.35	-0.48	-0.65
Oct.	-0.80	-0.80	-0.92	-0.76	0.17	0.54	0.53	0.23	-0.10	-0.45	-0.67	-0.81
Nov.	-0.52	-0.54	-0.46	-0.59	0.05	0.28	0.27	0.14	-0.06	-0.25	-0.33	-0.45
Dec.	-0.50	-0.35	-0.32	-0.27	-0.21	0.15	0.28	0.09	-0.07	-0.18	-0.22	-0.35

Table 11-2. Temperature lapse rate (°C/100 m) for the air layer from 1.1 to 15.2 m, giving the diurnal and annual variation in the form of monthly averages for the three years 1945-1948 at Rye, England. (Corresponding humidity gradients are given in Table 15-3)

Hour	2	4	6	8	10	12	14	16	18	20	22	24
Jan.	-6.1	-6.6	-5.6	-5.2	-1.6	-0.1	-0.4	-3.7	-5.7	-5.2	-6.0	-5.5
Feb.	-3.9	-3.2	-2.9	-1.8	0.6	0.6	0.6	-0.9	-3.1	-3.9	-3.9	-4.6
Mar.	-7.6	-7.0	-6.1	-1.0	2.6	4.4	3.9	1.4	-2.1	-5.9	-6.6	-6.7
Apr.	-8.4	-7.4	-3.7	2.2	4.7	5.5	4.7	2.5	-0.2	-5.2	-7.3	-8.3
May	-6.2	-6.9	-0.2	4.1	5.6	6.2	5.6	4.2	1.6	-3.3	-6.0	-6.5
June	-7.2	-6.1	0.8	3.1	5.6	5.6	5.2	3.1	0.4	-3.5	-6.6	-6.8
July	-5.2	-5.6	0.7	3.5	5.4	5.9	5.5	3.4	0.4	-3.2	-5.7	-6.2
Aug.	-5.6	-5.2	-0.3	3.7	5.6	6.3	5.5	3.3	0.1	-4.6	-6.5	-5.8
Sep.	-6.8	-6.3	-4.4	0.8	3.7	4.4	4.1	1.8	-2.1	-6.8	-6.8	-6.8
Oct.	-7.1	-6.9	-7.1	0.1	2.7	3.4	3.3	0.6	-5.0	-8.1	-8.4	-7.0
Nov.	-3.7	-2.8	-3.2	-1.8	1.3	1.5	0.7	-2.2	-4.4	-5.1	-4.8	-4.4
Dec.	-3.5	-2.5	-3.0	-2.7	-0.8	0.8	0.2	-3.0	-4.6	-4.4	-4.6	-4.4

Tables 11-1 and 11-2 show the variation of the temperature lapse rates through the day and year based on the observations of A. C. Best, et al. [1101]. For the sake of uniformity and comparison, all lapse rates have been expressed in °C/100 m. For the smaller lapse rates at higher levels (Table 11-1), the data are expressed in hundredths of a de-

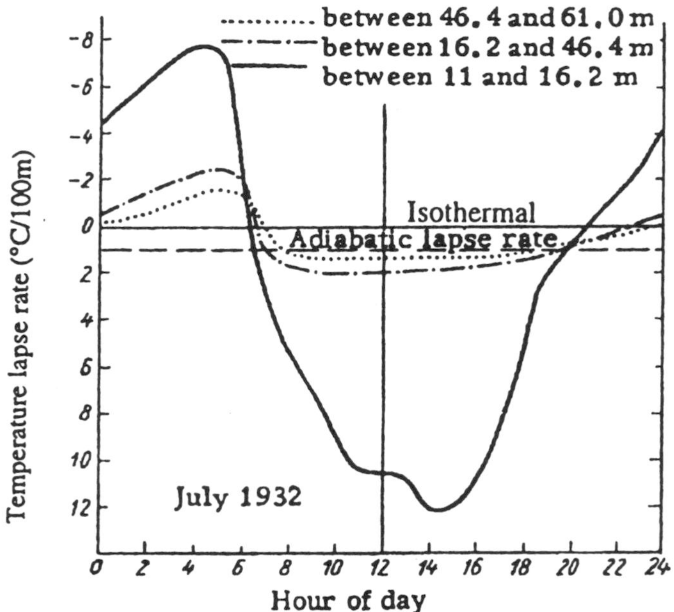

Figure 11-3. Diurnal variation of temperature lapse rate. (From W. D. Flower's [1105] observations in Egypt)

gree, corresponding to the accuracy of measurement. For the larger gradients at lower levels (Table 11-2) it is sufficient to express the data in tenths. Undoubtedly, even with 3 yr averaged values, chance effects of weather can still be noticed. Nevertheless, the increased magnitude of the gradients as the ground is approached, and the change of sign from positive to negative, controlled by the time of day and season, can still be clearly seen.

The graphs in Figure 11-4, from the measurements of N. K. Johnson [1107], show the effect of cloudiness within the lowest 17 m layer by contrasting clear and cloudy days of a summer and winter month. In June, the mean temperature in both situations is approximately the same, but the daily temperature fluctuation, the temperature gradient, and the displacement of the maximum temperature with height are substantially greater on clear days than on cloudy days. In December, the mean temperature is higher and the temperature gradient smaller on cloudy days.

The increase of temperature fluctuation, as the ground is approached, is shown in Figure 11-5 for a typical summer and winter month, from the results obtained by N. K. Johnson and G. S. P. Heywood [1108] in Leafield. Table 11-3 gives the displacement of the time of maximum temperature at various heights, calculated from the averages over five years. The time of occurrence of the maximum temperature occurs later as the distance from the surface increases.

Finally, the tautochrones in Figure 11-6 show the changeover from a positive to a negative surface net radiation balance for the first 100 m layer above the ground, using the measurements made in Rye, England. From average values of 19 clear summer days obtained from three years of data, tautochrones have been drawn with full lines to show the transition from the negative net radiation balance at 04:00 through the isothermal

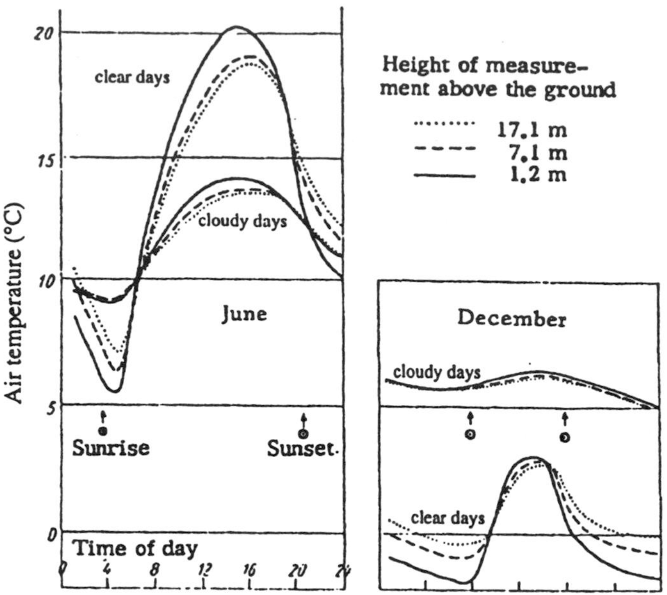

Figure 11-4. Influence of clouds on temperature variations near the ground during summer and winter. (After N. K. Johnson [1107])

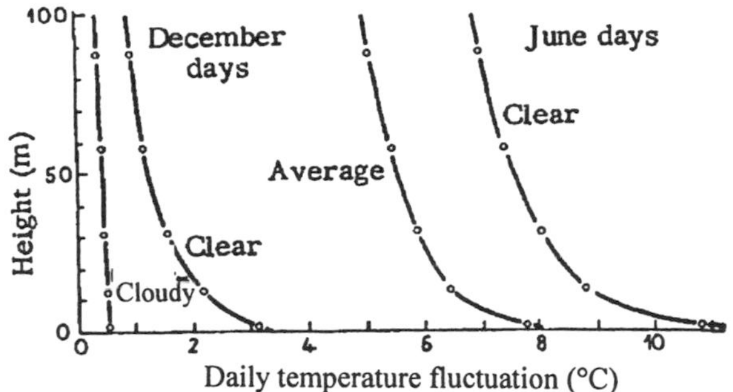

Figure 11-5. Daily temperature fluctuation as a function of height above ground, season and cloud cover. (After N. K. Johnson and G. S. P. Heywood [1108])

stage (06:00) and the rapid early morning establishment of surface heating (08:00), to the strong positive net radiation balance at 14:00. The broken tautochrones show the opposite transition, which takes place during the afternoon and night, during which surface radiative heat loss determines the temperature profiles.

Table 11-3. Time (hours:minutes) of maximum temperature at various heights.

Height above ground (m)	1.2	12.4	30.5	57.4	87.7
December (mean value)	14:05	14:26	14:34	14:42	14:50
June (mean value)	14:55	15:35	15:30	16:06	16:20
June (clear days)	15:45	16:35	17:00	17:14	17:24

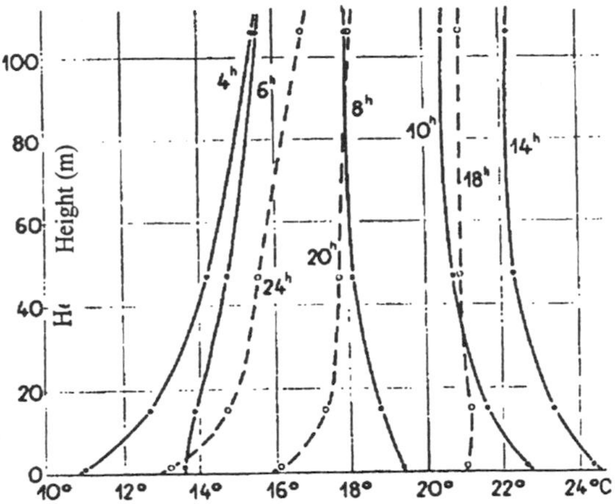

Figure 11-6. Tautochrones of daily temperature sequence in the lowest 100 m of the atmosphere on clear summer days. (After A. C. Best, et al. [1101])

12 The Unstable Sublayer and the Inversion Sublayer

K. Brocks [1102] has analyzed observations made in the lowest 100 m of the atmosphere. The method of representation is a double logarithmic coordinate system as shown in Figure 12-1. As before, T (°C) is the air temperature, z (m) is the height above the ground, and γ (°C/100 m) is the lapse rate. The abscissa is log γ, and the ordinate log z. The hourly mean values of the lapse rate for every month were expressed in this coordinate system for measurements made in Central Europe and in Egypt. If the observed gradients had been assigned to the mean height of the layers of air in which they were measured, the values obtained near the ground would be too high on account of the rapid increase of gradient in that region. It was therefore necessary to introduce a suitable reduction. The hourly values were averages for all weather situations. The results are indicative of the temperature structure in the lower atmosphere associated with a strong positive net radiation balance. They provide us with an insight into the normal or average situation prevailing in the air layer near the ground. In individual cases, of course, there may be substantial deviations from this.

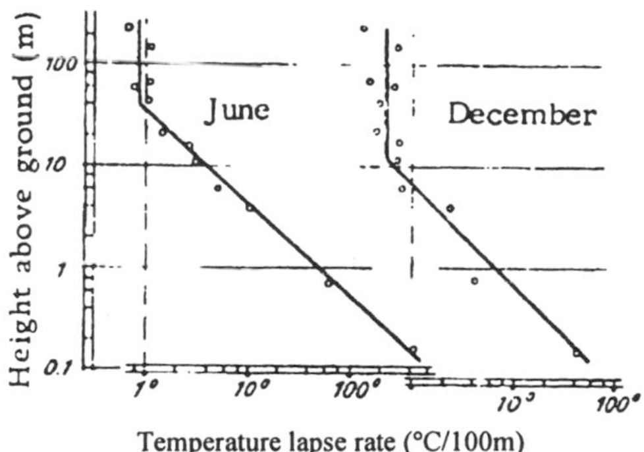

Figure 12-1. Temperature lapse rate as a function of height, at midday. (After K. Brocks [1102])

The example shown in Figure 12-1 is at noon for summer and winter months. These two observations, so different in time and place, indicate that the layer of air can be divided into two distinct parts. Above the ground in summer (June) and in winter (December), there is a layer in which the rate of change of the temperature gradient is log-linear and may be represented by a straight line (on the double logarithmic plot). In this layer temperature decreases rapidly with increased height. K. Brocks [1102] called this the unstable sublayer. In summer, it is higher in elevation than in winter, and the slope of the line is somewhat greater in summer. Above this, there is a second layer in which the temperature gradient is constant with height; its value is less than the dry adiabatic lapse rate (shown by a vertical dashed line). K. Brocks [1102] called this layer the adiabatic intermediate layer. The adiabatic intermediate layer is a few hundred meters in height.

The unstable sublayer begins forming when the sun's elevation reaches about 10°. Then it increases in thickness, as do the temperature gradients observed in it. During this period, a substantial amount of energy is transferred from the ground to build up this sublayer. When the sun reaches an elevation of about 30°, the temperature gradient continues to increase, but the height of the unstable sublayer no longer increases because, by then, the convective motion has become so vigorous that further supplies of heat from the ground are transferred mainly into layers higher up.

The height of the unstable sublayer is a minimum of 4 m at midday in December, increasing to a minimum of 30 or 40 m in June. The yearly average is 21 m. The temperature gradient at a height of 1 m changes seasonally in a similar way from 6°C/100 m in December to 45°C/100 m in June with a yearly average of 27°C/100 m. It is a linear function of the sun's elevation.

The straight line in the unstable sublayer of Figure 12-1 is represented by the equation: $\log \gamma = b \log z + \log a$, or

$$\frac{dT}{dz} = az^{b} \qquad (^{\circ}\mathrm{C\ cm^{-1}})$$

in which a and b are constants. Since the temperature gradient decreases with height, b is negative. The equation governing the distribution of temperature with height is,

$$T = a\int z^{b}\,dz + C \qquad (^{\circ}\mathrm{C})$$

In order to determine the constant of integration, it is assumed that at height z_1 the temperature has the value T_1. Then the solution of the equation becomes, for $b = -1$,

$$T = T_1 + a\ln\frac{z}{z_1} \qquad (^{\circ}\mathrm{C})$$

which gives a logarithmic distribution of temperature with height. When b is not equal to -1,

$$T = T_1 + \frac{a}{1+b}\left(z^{1+b} - z_1^{1+b}\right) \qquad (^{\circ}\mathrm{C})$$

The dot-dashed curve in Figure 12-2 shows the frequency distribution of the index $(1 + b)$ for the time when the unstable sublayer is present and the net radiation balance is positive. The curves are derived from observations made by K. Brocks [1102]. They show a wellmarked maximum for the value $1 + b = 0$. This, however, corresponds to a logarithmic distribution of temperature with height ($b = -1$).

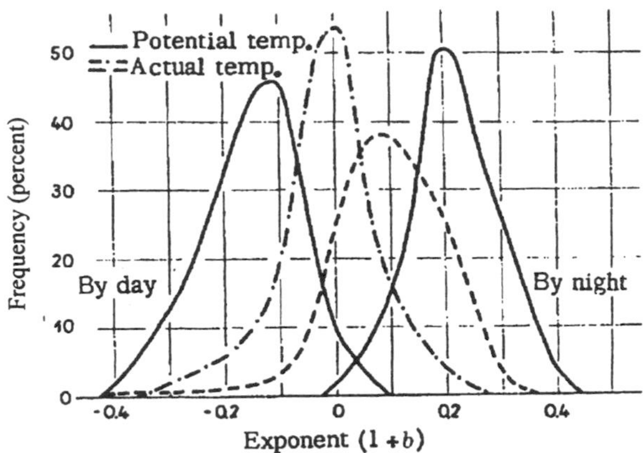

Figure 12-2. Frequency distribution of the observed exponents in the temperature-height function, for observed and potential temperatures. (After K. Brocks [1104])

At night, an inversion builds up from below because of surface radiative cooling; it increases in height until sunrise, often reaching 100 m in height and, in exceptional cases, extending beyond 1000 m. Within this inversion lies an inversion sublayer in which temperature gradients develop similar to those shown in Figure 12-1 but with a negative value of γ. With increased height the temperature gradient gradually decreases until, at the upper limit of the inversion, it becomes isothermal.

In contrast to the unstable sublayer during the day, the inversion sublayer does not exhibit a marked yearly variation in its vertical extent. Temperature gradients observed at a height of 1 m are much less at night than during the day (Table 12-1 evaluated from K. Brocks [1102]). Although weaker, the inversion sublayer may even occur on overcast nights.

The air temperature as directly observed, usually termed the actual air temperature, is unsuitable for use when questions of eddy diffusion are considered. In a layer of air 100 m thick (Section 7), the change of temperature resulting from the change of pressure with height is no longer negligible. Instead of the actual temperature, one must use the potential temperature, that is, the temperature that the air would have if reduced adiabatically to a pressure of 1000 mb.

K. Brocks [1104] has also plotted the variation of potential temperature as a function of height; it is shown in Figure 12-2 by the solid lines. The exponent $(1 + b)$ is -1/7 during the day and 1/5 at night.

Our discussion of temperature conditions in the neighborhood of the ground may, therefore, be summed up in the following way. Between the warm or cold surface of the ground and the free atmosphere, there is a layer of air. During the day, the exponents are negative; the unstable sublayer is built and instability increases in proportion to the incoming radiation. The difference in temperature between the ground and the air is, at first, mainly retained within this layer; then, when the sun's elevation reaches 30°, heat is transported to higher levels by convective mixing.

Table 12-1. Mean values for the inversion sublayer.

Period	Vertical thickness (m)		γ (°C/100 m)		Exponent 1+*b*
	Of the inversion	Of the inversion sublayer	At 1 m height	At upper boundary of inversion sublayer	
Yearly average	104	19	-19.0	-1.6	0.10
Seasons:					
spring	100	19	-21.4	-1.7	0.10
summer	90	18	-16.9	-1.9	0.20
autumn	130	15	-16.6	-1.8	0.10
winter	100	25	-17.1	-1.2	0.04
Clear nights:					
December	>100	20	-31.3	-3.8	0.18
June	>100	21	-34.5	-2.9	0.14

During the night, the exponents in the inversion sublayer temperature-height function are positive. Differences in height and strength occurring over the year are small due to the small seasonal differences in the intensity of the outgoing radiation.

A transition from one sublayer to the other occurs twice daily. Figure 12-3 shows the change from day to night (left) at sunset, and the change from night to day (right) at sunrise. The graphs are based on observations collected by K. Brocks [1103] for the periods from 2 hours before to 12 hours after sunset, and from 6 hours before to 2 hours after sunrise.

The curves at the top (a) show how the inversion becomes deeper during the night. On the average, a positive surface net radiation balance changes to a negative net radiation balance between 1-2 hours before sunset. While the vertical extent of the inversion (a) increases steadily during the night, the inversion sublayer (c) builds up very quickly, at first, and then increases more slowly or becomes relatively stable. The temperature gradient of the inversion sublayer (d) reaches its maximum strength 1-2 hours after sunset and then decreases somewhat through the night (because it is averaged over a greater depth). Through the night the depth of the inversion sublayer (c) increases only slowly from around 12 to 20 m, while the inversion itself (a) increases from 20 to almost 200 m. The exponent of the height function (b), a measure of the strength of the inversion, also remains nearly constant during the night. The destruction of the inversion sublayer (c) takes place quickly following sunrise. On the average, a positive surface net radiation regime has been reestablished about 1-2 hours after sunrise.

In the following discussion, a twofold task still remains: first, to evaluate the unstable sublayer and the inversion sublayer in the region of the ground with respect to temperature and gradient values; and second, to examine anomalies that occur in proximity of the surface of the earth.

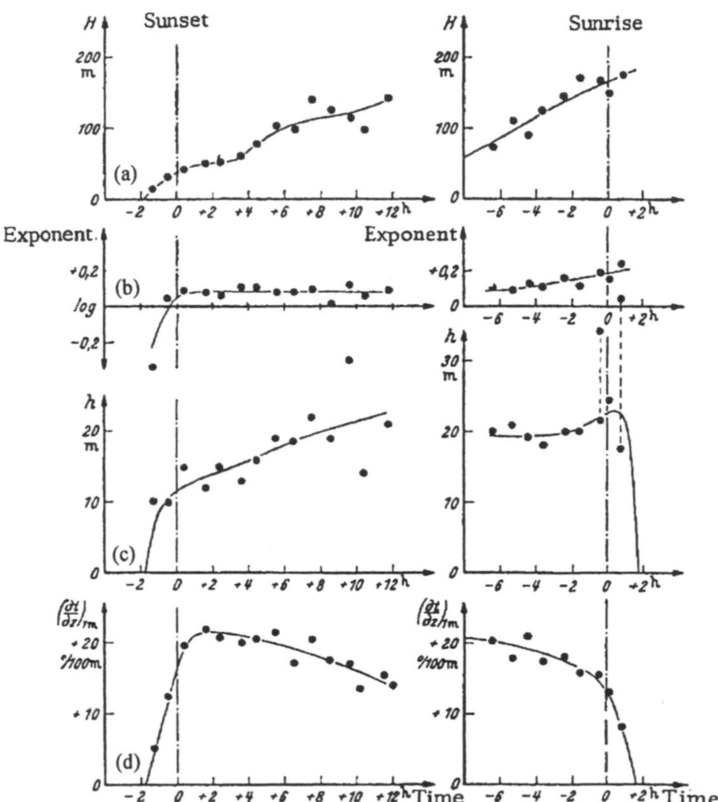

Figure 12-3. Formation and destruction of the inversion sublayer in the evening and in the morning. (After K. Brocks [1103])

13 Daytime Temperature of the Air Layer near the Ground

Whereas a good series of measurements is available for the first 100 m of the atmosphere, the situation is different for the 2 m layer of air close to the ground. The increasing gradient of all characteristics near the ground surface demands instruments that are as small as possible. The shielding of these instruments against radiation from above and below presents considerable difficulty. It is difficult to generalize the results because of the chance influences of surface conditions at the measurement site such as soil type, soil condition, and vegetative cover, which further complicate the mechanical and technical difficulties.

A. C. Best [1302] made a set of temperature records at Porton, England, for the layer of air close to the ground, from August, 1931 to July, 1933. Measurements were made at heights of 2.5, 30, and 120 cm above grass, which was kept short. Tables 13-1 and 13-2 are based on these results.

A considerable number of shorter term observations are available for the air layer near the ground which cover several characteristics, and provide general insight on the surface energy balance and humidity variation near the ground. One comprehensive set of observa-

Table 13-1. Temperature lapse rate (°C/100 m) in the layer of air from 30 to 120 cm over short grass at Porton, England; 2 yr averages from 1 August 1931 to 31 July 1933.

Hour	2	4	6	8	10	12	14	16	18	20	22	24
Jan.	-25	-33	-21	-17	2	7	6	-22	-34	-31	-33	-26
Feb.	-25	-23	-21	-10	15	30	20	-7	-31	-32	-31	-30
March	-46	-41	-39	10	45	60	49	17	-35	-58	-52	-46
April	-34	-32	-15	27	57	62	50	29	-7	-38	-39	-38
May	-30	-27	7	42	56	64	56	34	5	-27	-34	-36
June	-29	-25	9	44	66	77	60	43	10	-25	-36	-34
July	-23	-18	2	26	38	48	46	28	6	-21	-26	-25
Aug.	-20	-16	0	27	49	59	54	36	-1	-33	-34	-31
Sept.	-29	-26	-15	14	36	36	28	12	-15	-33	-28	-34
Oct.	-35	-33	-28	4	25	32	25	-1	-48	-46	-44	-39
Nov.	-21	-19	-21	-13	14	15	7	-22	-30	-26	-26	-23
Dec.	-18	-18	-20	-19	2	–	1	-23	-30	-27	-31	-25

Table 13-2. Temperature lapse rate (°C/100 m) in the layer of air from 2.5 to 30 cm over short grass at Porton, England; 2 yr averages from 1 August 1931 to 31 July 1933.

Hour	2	4	6	8	10	12	14	16	18	20	22	24
Jan.	-94	-89	-69	-58	46	103	36	-105	-143	-121	-115	-109
Feb.	-119	-117	-101	-61	123	220	125	-77	-208	-191	-173	-155
March	-187	-167	-139	99	315	397	325	46	-195	-258	-228	-197
April	-135	-113	-48	260	426	442	359	151	-46	-151	-159	-171
May	-75	-54	115	357	460	480	408	212	12	-101	-101	-89
June	-97	-85	214	513	622	682	519	327	42	-133	-165	-127
July	-54	-30	71	361	456	502	420	244	28	-105	-109	-83
Aug.	-63	-42	46	252	432	492	428	202	-42	-127	-111	-101
Sept.	-61	-73	-32	165	321	315	199	89	-93	-123	-93	-105
Oct.	-141	-107	-97	65	242	262	197	-67	-222	-187	-173	-111
Nov.	-97	-103	-115	-77	97	121	36	-127	-147	-121	-139	-135
Dec.	-99	-105	-107	-111	10	65	0	-147	-143	-133	-155	-161

tions was carried out in August and September, 1953 by the joint effort of sixteen universities and state institutions in O'Neill, Nebraska on a uniform area of prairie grass (H. H. Lettau and B. Davidson [104]).

To illustrate the temperature field in the neighborhood of the ground, we shall use the comprehensive series of observations made by the Climatological Laboratory in Seabrook, New Jersey (39° 34' N, 75° 13' E), by C. W. Thornthwaite and his assistants [1314]. From this series of hourly measurements of temperature at 10, 20, 40, 80, 160, 320, and 640 cm above the ground, five spring days were selected from 1951 (17 and 18 March, 8-10 May) for which there were a minimum number of gaps in the observation record.

The daily sequence of temperature is shown in Figures 13-1 and 13-2. The curves run at almost the same distance from one another because the measurement heights were selected logarithmically (Section 12). In Figure 13-1 the minimum occurs before sunrise at 10 cm and about an hour later at higher levels (640 cm); an isothermal state sets in almost simultaneously at all levels 1 to 2 hours after sunrise. At a height of 10 cm, the maximum temperature is reached immediately after the midday radiation maximum, but is delayed until 14:30 at 640 cm. The transition from a positive to a negative net radiation balance occurs about 1-2 hours before sunset (Section 12) and the temperature gradient becomes isothermal again. Since almost cloudless spring days are involved, the temperature at midnight is noticeably higher than at midnight 24 hours earlier. The near constant temperatures throughout the day at a height of 1000 m and depth of 20 m are in sharp contrast to the rapidly changing conditions in the near surface air and soil layers. The isothermal conditions near sunrise and sunset are clearly shown.

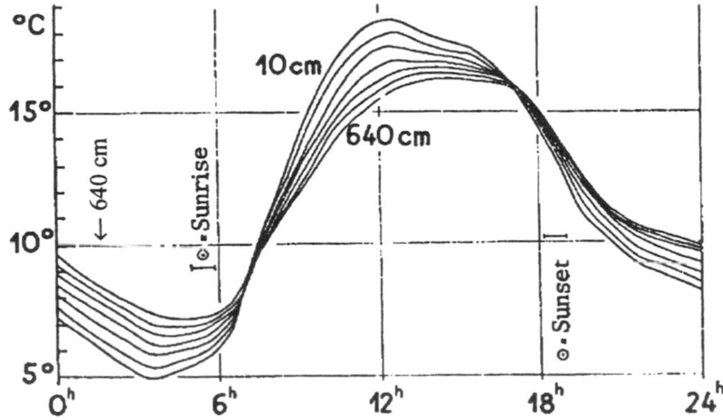

Figure 13-1. Temperatures at 10, 20, 40, 80, 160, 320, and 640 cm in the air layer near the ground on clear spring days at Seabrook, New Jersey. (After C. W. Thornthwaite [1314])

The tautochrones in Figure 13-2 show the relation between temperatures in the lowest 150 cm of the atmosphere and measurements made at depths of 2.5, 5, 10, 20, and 40 cm in the ground. Between 10 cm in the air and -2.5 cm in the soil, the tautochrones can only be estimated. The tautochrone for 03:00 represents a nocturnal negative net radiation balance. The beginning of diurnal heating is clearly seen at 07:00 by the flow of heat into the ground down to 5 cm. There is also a weak indication of heating in the air as indicated by the temperature decrease with height. At 09:00, the preceding night's cool temperature has penetrated to 25 cm. At 12:00 the maximum temperature is associated with the maximum solar radiation. The lighter, broken tautochrones for 19:00 and 23:00 show a similar return from a positive to a negative net radiation balance for the surface.

Comparison of Figure 13-2 and Figure 11-6 shows considerable similarity. On a small scale, in the layer close to the ground, there is a substantial increase in the temperature gradient toward the surface. From Tables 11-1 and 11-2 it can be seen that this is the case throughout the year. This is further illustrated in Table 13-2 which show temperature gradi-

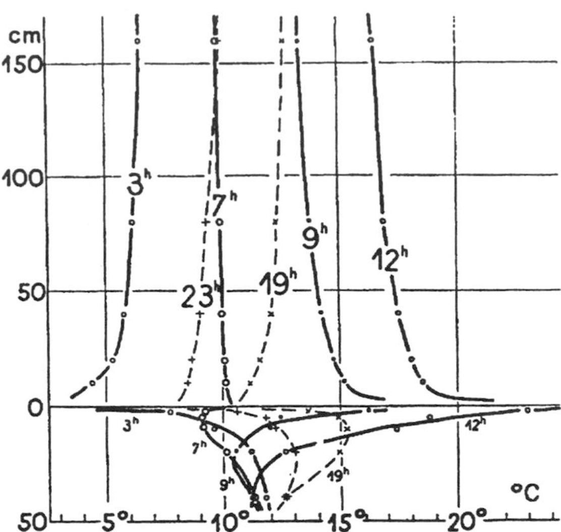

Figure 13-2. Tautochrones above and below the ground surface at Seabrook, New Jersey. (After C. W. Thornthwaite [1314])

ents obtained by A. C. Best [1302] for the layer from 2.5 to 30 cm over short grass. From these mean monthly values, gradients in summer can be observed to be several hundred times the adiabatic rate in the lowest 0.3 m of the atmosphere. When $\gamma = 3.4$°C/100 m, the air density is constant with height. Thus, with gradients of this magnitude, air density increases with height in the lowest meter (about 2 percent).

Figure 13-3 shows the temperature distribution in both the air and soil for a summer day at O'Neill, Nebraska. W. A. Baum [1301] adopted an idea from D. Brunt and showed that even with gradients of this strength, it was still possible to have stable conditions. Lord J. W. S. Rayleigh proved, in 1916, that thin layers of fluid might still remain stable, in spite of increasing density with height, if certain conditions were fulfilled. In addition to temperature, conductivity, variation of density, and thickness of the layer, these equations also contain a kinematic viscosity. By applying these results to the layer of air close to the ground, and by making certain probable numerical assumptions, W. A. Baum [1301] theoretically determined that gradients of 170°C/100 m in the lowest 10 m of the atmosphere or of 2100°C/100 m in the lowest 2 m may still represent stable conditions.

However, the distribution of temperature close to the ground cannot be entirely explained from the thermodynamic point of view alone. G. Falckenberg and his pupils [1404-1407] pointed out the significance of longwave radiation, particularly its importance in the energy budget at night. Much of the longwave radiation (particularly outside the water vapor windows) emitted from the surface is absorbed close to it. The exact height and amount depend not only on the water vapor content and wavelength but also on air pressure and temperature. For example, in an atmosphere with a water vapor pressure of 13.3 mb, 50 percent of the radiation in the band from 6.25 to 6.75 μ is absorbed within a distance of 1 to 2 m. When temperature differences are large, this means that within the air layer near the ground, there must be a radiation exchange that cannot be neglected. A mutual influence of Earth's surface and the layer of air close to it act on each other to give the process which G. Falckenberg

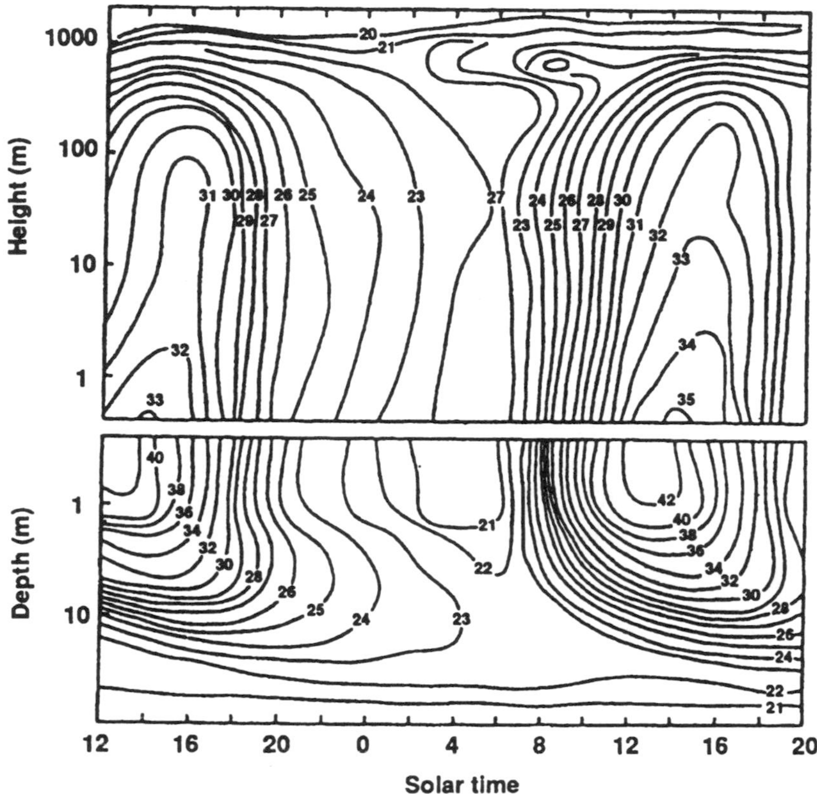

Figure 13-3. Atmosphere (upper) and soil (lower) temperatures observed at O'Neill Nebraska 24 August 1953. (From R. D. Graetz and I. Cowan [1307] after H. H. Lettau and B. Davidson [104])

[1404] has called wavelength transformation. The reason for this mutual influence is that H_2O and CO_2 in the atmosphere emit radiation in bands, while the solid ground emits radiation in a continuous spectrum.

The laminar boundary layer at the ground (Section 7) is a special case because, although energy is being radiated upward, little radiant energy is being received from below. Thus, this surface layer experiences very strong radiative cooling. Using the Thornthwaite measurements mentioned earlier, and selecting from them three fair weather days in the spring of 1952, F. Möller [1311] calculated the temperature change that would take place in thc afternoon period from 13:00 to 14:00 through longwave radiation (Table 13-3).

Temperatures, shown in parentheses in the table, were extrapolated from measurements within 25 mm of the ground surface. At first, the results looked surprising; the rate of cooling through radiation of the air in contact with the heated surface of the ground amounts to 22.8°C hr^{-1}. But this cooling is limited to a laminar boundary layer only 1.2 mm in height. In the layer of air close to the ground above this layer, the air is strongly heated through longwave radiation. In this case, the amount of heating in the first meter is between 8.4 and 16.5°C hr^{-1}. Using a different method involving a radiosonde ascent in August, O. Czepa [1303] calculated the gain of energy through radiation in the lowest meter of the atmosphere to be 7.1 J hr^{-1} which corresponds to a change of temperature of 60°C hr^{-1}. This was based

Table 13-3. Afternoon cooling at various heights by longwave radiation. (After F. Möller [1311] based on the C. W. Thornthwaite [1314] measurements)

Height above the ground (cm)	0	0.06	0.27	0.89	2.65	10	20	80	640
Observed (extrapolated) air temperature (°C)	(25.5)	(24.5)	(23.5)	(22.5)	(21.5)	20.3	19.8	18.9	18.1
Radiative heating (+) or cooling (-) (°C hr^{-1})	-22.8	-5.1	+11.9	+16.5	+15.7	+10.6	+8.4	+8.6	+1.3

on an assumed surface temperature of 45°C, or 20°C higher than F. Möller [1311] had extrapolated from C. W. Thornthwaite's [1314] measurements. There can be no doubt, however, that the amount of heating due to radiative exchange is very substantial when a positive net radiation regime is in effect.

The layer of air close to the ground is, thus, characterized as the lowermost edge of the atmosphere, in which eddy diffusion increases rapidly with height (almost in a linear fashion according to theory). The flow of energy from the warm surface of Earth penetrates through this layer of rapidly changing eddy diffusion into the atmosphere. It does so with its strength more or less undiminished provided the temperature gradient changes rapidly with height, that is, if d^2T/dz^2 is large.

W. A. Baum [1301] has shown that the temperature gradients near the ground, in the middle of the day, could not be explained by the suggestion that they were always being regenerated by the enormous supply of energy, in spite of continual overturning of the layers of air. No such overturning has been shown by observation. This may be recognized by a study of temperatures recorded above the ground in desert areas such as those given in Figure 9-2. In spite of the great unsteadiness of temperature (Figure 9-2), parcels of air from a height of 1 mm, which have temperatures from 6° to 9°C, do not reach a height of 1 cm since the temperature at this level fluctuates from 0° to 4°C. This is explained by the coupling of the Austausch coefficient and the temperature gradient. As soon as the Austausch coefficient decreases for any reason, the temperature gradient immediately increases and therefore maintains a constant flow of energy. Horizontal temperature contrasts within very small distances are also probably important. Compensatory processes such as this are always in action, both in space and in time. The flow of energy oscillates to and fro by eddy diffusion. The facts we observe with our instruments are nearly all average values of the sum of individual events.

The larger scale overturning of superheated layers of air close to the ground in desert areas is an exceptional occurrence, and is often accompanied by dust devils and sand devils. The upward swirling of the air becomes visible by the dust, sand, leaves, twigs, and paper that it drags with it. Wandering off, usually slowly, it sucks overheated layers into its orbit and so maintains itself. In arid zones of the world, dust devils are a regular feature during hot hours in the afternoon. They have been investigated rather closely by W. D. Flower [1305] in Egypt, and by R. L. Ives [1308] in the North American prairie. B. D. Kyriazopoulos [1310] found this striking micrometeorological phenomenon portrayed even on the Corinthian pillars of St. Sophia in Thessalonika, dating back to the fifth century BC, where acanthus leaves are shown being picked up by a whirlwind.

Figure 13-4, based on Egyptian measurements from 1926 to 1932, shows the frequency distribution of dust devils during the day. The diagram also shows temperature gradients, measured for 1932 on days when dust devils were present. The dust devil is initiated by some chance happening, perhaps at a pile of crushed rock at the side of a highway, by an automobile, by a gust of wind at the edge of a forest, or by the confluence of two trails of dust at a street intersection.

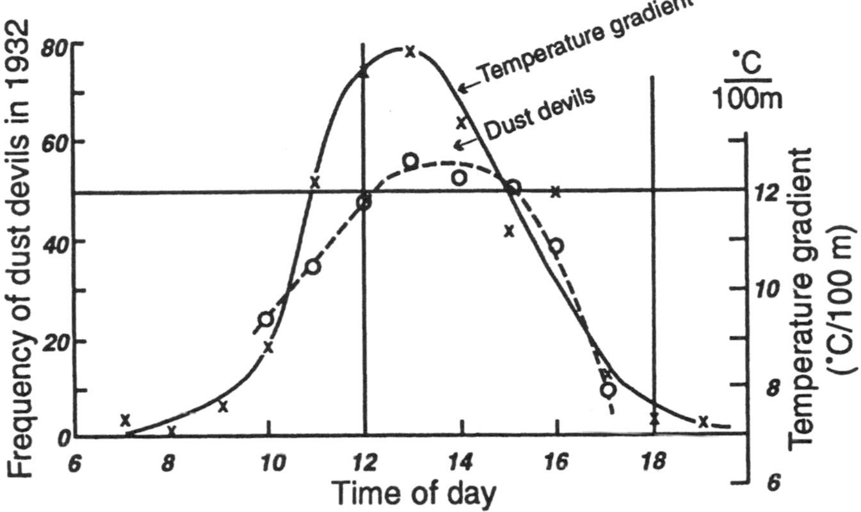

Figure 13-4. Incidence of dust devils and vertical temperature gradients in Egypt. (After W. D. Flower [1305])

Contrary to popular belief, dust devils rotate in either direction. For example, W. D. Flower [1305] reported that the direction of rotation of 175 dust devils were in a clockwise direction, with 200 rotating in the opposite direction. This follows from the fact that the Coriolis parameter has units of radians per second and, given the short life cycle of dust devils, is unable to gain sufficient strength to influence rotational direction. R. L. Ives [1308] was able to observe that a dust devil, which had almost come to the end of its existencc, might revive itself in contact with an obstacle and rotate in the opposite direction. This is what happened to a Potsdam dust devil, after collision with an apple tree, as observed by L. Klauser [1309] in 1950. For a further discussion about this and other misconceptions in physical geography, the reader is referred to B. D. Nelson, et al. [1312] and R. H. Aron, et al. [1300].

In arid zones, dust devils can reach a height of 1000 m and last for several minutes. In Utah, a dust devil 800 m high was observed continuously for 7 hours along a 60 km path. Inside the dust devil, there is a lowering of pressure and a rising current of air. One dust devil passed right over the weather station in Phoenix, Arizona, and in the report on it H. L. De Mastus [1304] states that the barograph trace fell by 1.73 mb, and the disturbance lasted for 30 sec. Shortly before this, another dust devil passed by at a distance of 23 m, producing a pressure drop of 1.6 mb in 6 sec followed by a rise of 1.07 mb. The true decrease of pressure in the interior of these dust devils is certainly much greater. This was established by R. L. Ives [1308] who chased one in a Jeep, holding an instrument inside it on the end of a long stick. Upward currents of 10 to 15 m sec^{-1} have been estimated by seeing rats picked up, and later measuring their actual rate of fall.

From time to time, dust devils can cause damage. H. Schlichtling [1313] writes of one observed on 19 May 1934 in Lübeck. "It was interesting to see how three people were caught up in the whirlwind. A woman managed to get out of it quickly, but two men were in it for some time. Their clothes were flapping about vigorously, and they had to hold on tightly to their hats. They had great difficulty in keeping [on] their feet."

The optical effects associated with superheated layers of air close to the ground will be discussed in Section 18.

14 Nighttime Temperature of the Air Layer near the Ground

The temperature distribution that develops above the ground that is experiencing radiative cooling has already been shown in Figure 11-6 for the 100 cm layer by 4 hr tautochrones, and in Figure 13-2 for the air layer near the ground by 3 hr tautochrones. S. Siegel [1419] made a more detailed study of the changing nocturnal temperature pattern above the ground at the Meteorological Institute of the University of Hamburg by placing 23 thermoelements between a height of 0 and 4 m. Figure 14-1 shows a typical isotherm pattern for a clear summer night depicted by 1°C intervals. The decrease of wind during the night is shown at the top of the diagram. The night inversion, which builds up slowly, reaches its maximum height just before sunrise. After sunrise, it is destroyed quickly (Figure 12-3).

On any individual night, the temperature may behave much more irregularly, of course, because of variations in wind speed. Figure 14-2 shows a typical example: this is a record of the isotherms on the night of 9-10 July 1935. An increase of wind shortly before 02:00, shown at the top of the diagram, is followed by a weakening of the inversion, which forms again as soon as the wind subsides. The wind speed gives an indication of the magnitude of eddy diffusion. This brings heat to the ground surface, which is cooling because of a net longwave radiation loss.

It was pointed out as early as 1932 by L. A. Ramdas and S. Athmanathan [1416] that in India, especially in the area of the dark cotton-growing soil, the lowest temperatures at night frequently were observed not at the ground surface but a few centimeters, and sometimes even as much as 1 m, above it. This was confirmed in India on many occasions. As an example, Figure 14-3 gives the record of measurements made by K. R. Ramanathan and L. A. Ramdas [1415] in Poona on a January night in 1933. The arrowheads pointing upward show the temperature of the ground surface; the circle at zero height gives the temperature of the air immediately above ground. It is much lower than the ground temperature and the temperature continues to decrease with height, in this particular case, to between 10 and 30 cm, depending on the time the observation was made. T. R. Oke [1414] observed the raised minimum over grass, snow, and bare smooth and rough soil surfaces. He discusses a number of possible causes for this phenomenon, but noted that a small amount of cloud cover caused this minimum to disappear. When the clouds had passed the minimum was often re-established in a few minutes. This, he suggested, is "a clear indication that the phenomenon is strongly affected by radiation."

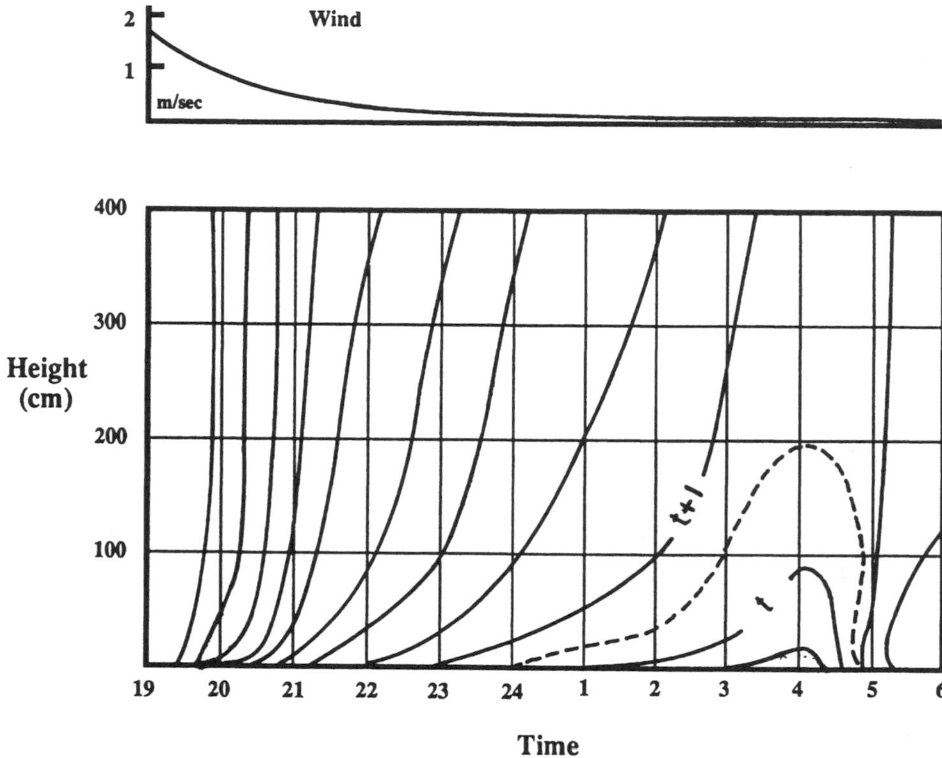

Figure 14-1. Typical temperature stratification and wind speed on a clear summer night. (After S. Siegel [1419])

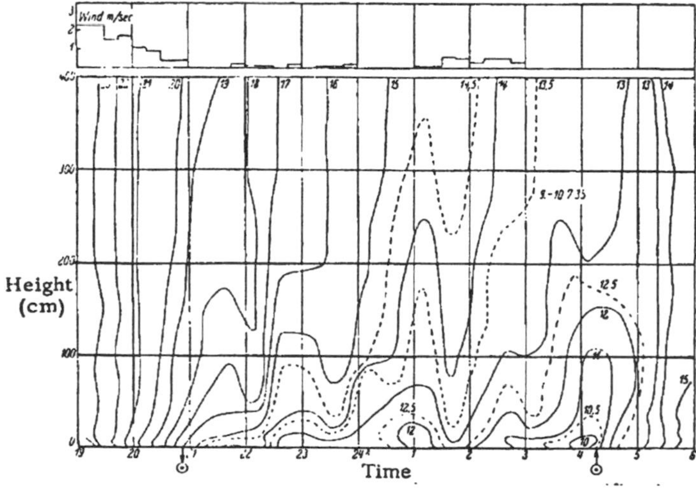

Figure 14-2. Relationship between air movement and temperature stratification on a night in July. (After S. Siegel [1419])

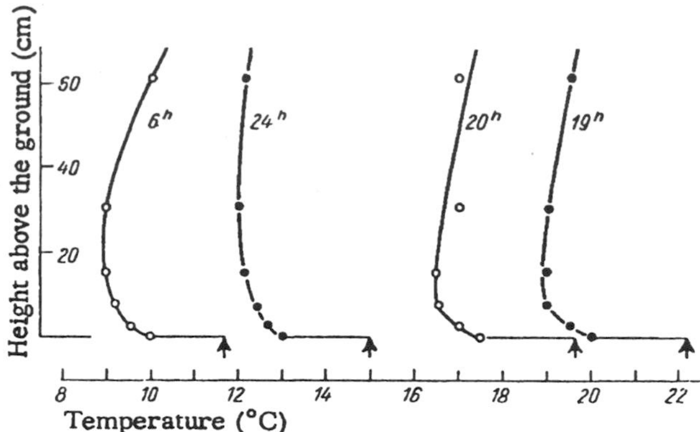

Figure 14-3. The night minimum temperature above ground surface in India. (After K. R. Ramanathan and L. A. Ramdas [1415])

Figure 14-4 shows the temperature in the early morning hours of 29 December 1954 at 0.1, 1, 5, 10, 100, and 1000 cm above the ground. The trace at the bottom shows the wind speed recorded at 20 cm, and gives an indication of the magnitude of eddy diffusion. At 01:30 and 02:30, when wind and eddy diffusion were at their greatest, the lowest temperatures occurred at the lowest measurement position, and the inversion between 1 mm and 10 m was 4 to 5°C. However, as soon as the wind speed slowed to below 0.5 m sec^{-1}, an elevated minimum again becomes established at a height of 1 to 5 cm both around 02:00 and after 03:00.

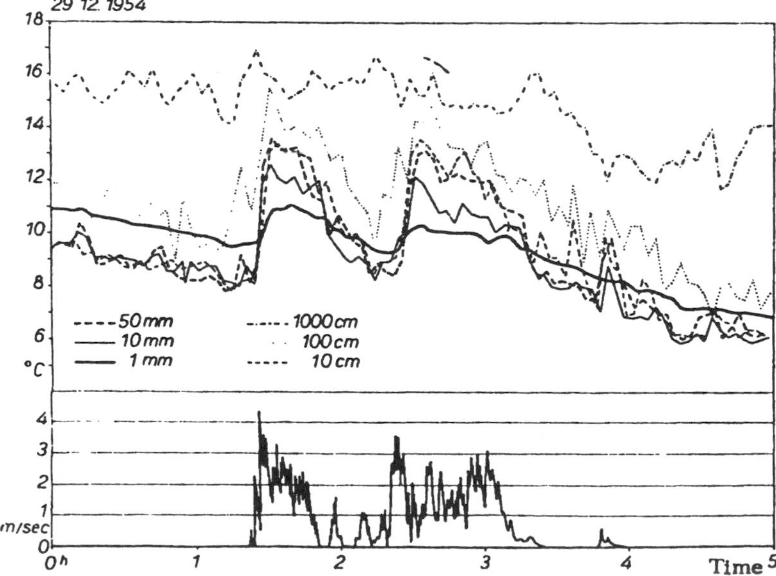

Figure 14-4. Temperature sequence in the morning of 29 December 1954 over bare ground in Poona, showing the change in temperature distribution due to radiation and eddy diffusion. (After K. Raschke [1417])

G. Falckenberg [1404] has suggested that at night the radiative exchange in the air near the ground is not negligible. There is, at the time of strong outgoing radiation at night, the apparently paradoxical state of radiative heating of a millimeter thick boundary layer. Above this, the layers close to the ground are cooled by a net loss of longwave radiation. F. Möller [1311] has evaluated this for C. W. Thornthwaite's [1314] records, from Seabrook, to give the following average values of radiative cooling for the period from 03:00 to 06:00.

Height above the ground (cm)	O	10	40	160	640
Radiative cooling (°C hr^{-1})	+5.4	-2.0	-1.3	-1.2	-1.4

It was calculated theoretically by R. G. Fleagle [1408], during his studies in 1953 on the formation of fog, that above a cold and dark surface longwave radiative exchanges would give rise to considerable heating close to the surface, and that the greatest cooling would take place at a height of about 1 m. In 1956, he made optical measurements of light refraction to measure temperature gradients near the ground. There was radiative heating at night immediately at the radiating surface (water in this case) of 10°C hr^{-1}, a maximum radiative cooling of 6°C hr^{-1} at a height of 10 cm, and from 30 cm up to 1.5 m a practically constant rate of cooling of 3°C hr^{-1}.

The nocturnal distribution of temperature is, therefore, a function of both longwave radiation exchange and eddy diffusion. If mixing is particularly low, which is usually the case at night close to the ground surface, the influence of radiation predominates, and the minimum is found at about 10 cm above the ground. With light air movement, convective mixing results in the sinking of the coldest air at 10 cm, and the minimum descends to the surface or remains only as a secondary minimum. R. Narasimha and A. S. V. Murthy [1413], A. S. V. Murthy et al. [1411] and R. Narasimha [1412] suggest that the elevated minimum is also affected by surface emissivity, roughness and soil thermal conductivity. With a higher surface emissivity more energy flows into the air weakening the minimum. Increased surface roughness weakens the elevated minimum both by promoting increased eddy diffusivity and because of its higher emissivity. Decreasing soil thermal conductivity also weakens the elevated minimum by increasing the surface rate of cooling (Table 19-2).

In view of these facts, it is of interest to obtain quantitative values of the Austausch coefficient *A* at night as a function of height above the ground. The calculation of *A* from observations is beset with uncertainty, and to a great extent depends upon the theoretical assumptions made. A set of values based on various methods of measurement and calculation are summarized in Figure 14-5.

Line M gives the values of *A* from midnight to 04:00, calculated by F. Möller [1311] from the observations made by C. W. Thornthwaite at Seabrook in the spring of 1952. The values obtained by H. Hoinkes and N. Untersteiner [2509] (line H) in August 1950 over the Vernagtferner were averaged for the hours from 18:00 to 07:00 when there was a considerable amount of wind (1 to 3 m sec^{-1} at a height of 28 cm). H. Sverdrup's measurements [1420] on the Isachsens Plateau in Spitsbergen over a period of seven summer weeks in 1934 made it possible to compute *A* from the distribution of temperature and wind. The group average for the lowest wind speed (0.44 m sec^{-1}), corresponding most closely to night con-

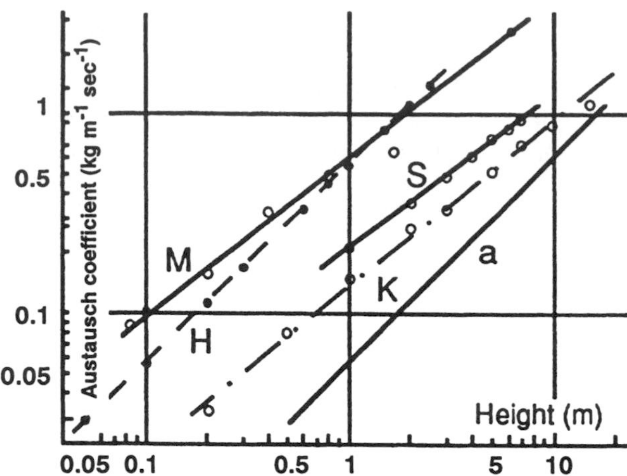

Figure 14-5. The value of the Austausch coefficient A at night, as a function of height.

ditions, is shown by line S. Values of A made by H. Kraus [2510] on a cloudless night on 23-24 September 1954 from 18:30 to 01:00 at the Munich-Riem airport are shown by line K. The average wind speed at a height of 2 m was 0.74 m sec^{-1} and the inversion between 0.2 and 15.0 m amounted to 2.7°C.

As might be expected, there is appreciable scatter in the values obtained, with values varying by approximately a factor of 10. This is primarily due to the influence of wind speed. It can be seen from the graphs that higher wind speeds lead to higher Austausch values. In all cases, however, the computed values lie along a straight line in each set. If these lines had the slope of line a, it would mean that there was a linear increase of the Austausch coefficient with height, such as adiabatic layering of the atmosphere would require in theory. With stable layering at night, the exponent of z is less, and lies between 0.75 and 0.99.

The Austausch coefficient will not continue to increase indefinitely at the rate shown in Figure 14-5. At higher levels these lines must begin to level off. If we extrapolate backward toward the ground, the values at 1 mm above the surface would be 0.05 to 0.24 kg m^{-1} sec^{-1}, which is about 5 to 10 times the amount of molecular diffusion ($2 \cdot 10^{-5}$ kg m^{-1} sec^{-1}). This is quite likely and proves that there is a gradual change from strong eddy diffusion at a height of 1 to 2 m to molecular transport of energy in the vicinity of the surface.

With the normal nocturnal temperature distribution, the ground is colder than the temperature indicated by the thermometers in the instrument shelter (Figure 5-3). This is important in the spring and autumn periods of frost hazard. For this reason, weather forecasters giving frost warnings have the responsibility of advising farmers, nurserymen, market gardeners, fruit growers, builders engaged in concrete work, and other commercial interests that they should not just consider the shelter minimum temperature but should also take into account the minimum temperature near the surface. This is called for short, if not quite correctly, the "ground minimum" or, since it is usually on a grass plot, the "grass minimum." G. Schwalbe [1418] investigated the difference between shelter and ground minima. The frequency distribution of this temperature difference, averaged over the years 1937 to 1944 by F. Witterstein [1422] for the Agricultural and Meteorological Experimental and Advisory Center in

Geisenheim on the Rhine, is shown in Figure 14-6. In extreme cases, the temperature at 5 cm was 6.5°C colder than in the shelter. On cloudy nights, with advection of cold air, the reverse can be the case. Figure 14-6 is for nights when the wind speed was less than 4.4 m sec^{-1}. With stronger winds the temperature difference between 5 cm and shelter height would be much less. Although the temperature differences would still be greater under clear sky conditions than with cloudy skies, the frequency distributions would tend to merge with increased wind speeds. This difference was at a maximum in the spring months of April and May, which is the most critical time for frost danger. A second, less pronounced maximum appears in November. F. Hader [1409] also found a sharp maximum in spring. The monthly average differences for March, April, and May were 3.6, 4.1, and 3.6°C, respectively. J. van Eimern and E. Kaps [1403] recorded a distribution similar to that of Figure 14-6. A. Bottsma [1400] found that with low wind speeds (<1.8 m sec^{-1}), the amount of cloud cover alone was the best estimator of the temperature difference between the grass and shelter heights. When wind speeds were between 2.2 and 5.8 m sec^{-1}, however, wind made a significant contribution in explaining this difference, reducing the residual error of cloud cover alone by 11 percent. L. Dimitz [1402] has shown that for the Austrian meteorological network, frost sets in at 5 cm on the average 10 days earlier and finishes 14 days later than at shelter height.

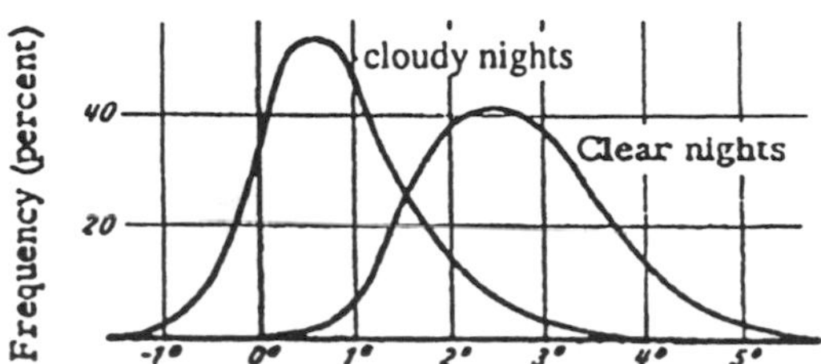

Figure 14-6. Frequency distribution of the difference in temperature (°C) between the shelter and the ground minima on clear and cloudy nights at Geisenheim. (After F. Witterstein [1422])

These low night temperatures, taken along with the extremely high midday temperatures, mentioned in Section 13, produce a microclimate of extreme temperature fluctuation at the ground. F. de Quervain and M. Gschwind [1401] show this facet of the microclimate close to the ground in the Figure 14-7, which is a photograph of the sandstone balostrade of the city hall of Winterthru, Switzerland. It shows how the weathering of the stone, due to temperature fluctuations, becomes more pronounced near the ground. The effect of large temperature fluctuations is enhanced by the influence of water. Splashing water and falling snow make the stone nearer the ground more subject to frequent changes from wet to dry. In winter, the splitting effect of frost also needs to be considered, since water on freezing at 0°C expands by about 9 percent and a greater frequency of melting and refreezing may occur at the ground or immediately above a snow surface.

Taking the values from C. W. Thornthwaite's [1314] measurements for the warmest and coldest hourly observations, the following increase in daily fluctuation is found as the ground is approached:

Height (cm)	10	20	40	80	160	320	640
Temperature fluctuation (°C)	14.4	12.7	11.7	10.8	10.1	9.5	9.1

An important consequence of this temperature fluctuation is the frequency of frost changes, or the passage of the temperature through 0°C, regardless of the direction in which it takes place, whether from + to - or vice versa. A "frost change day" is a day on which one or more movements through 0°C occur. The number of frost changes occurring on a frost change day is called the density of frost change. On days with a frost change this number is at least 1 and in the mid-latitudes is usually 1.5 to 2. At high altitudes in the tropics, where general temperatures do not change appreciably with the time of year, the temperature is almost always above 0°C by day and below 0°C by night, giving a frost change density near 2.

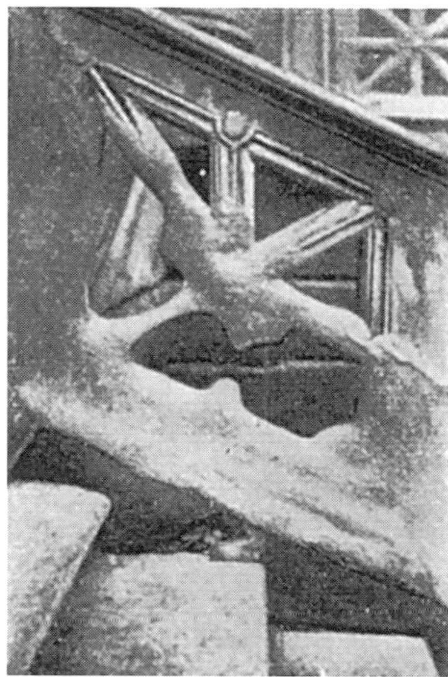

Figure 14-7. Weathering of Berne sandstone on the city hall in Winterthur. (From F. de Quervain and M. Gschwind [1401])

The importance of this number as one of the significant elements of microclimate was recognized by C. Troll [1421]. He expressed numerically the great difference between high latitudes, where frost changes are restricted to spring and autumn, and high altitudes in the tropics which have been observed to experience up to 337 frost changes per year (El Misti in southern Peru). As expected, the density of frost change increases with proximity to the ground, as shown by L. Dimitz [1402] for six Austrian stations. It can reach very high values close to the surface because of the high fluctuation of temperature there. The frequency of frost change is greatest at the surface and decreases rapidly with depth in the ground. E. Heyer [1410] gives the following values for Potsdam observations from 1895 to 1917:

Depth in the ground (cm)	0	2	5	10	50	100
Number of frost changes per year	119	78	47	24	3.5	0.3
Average density of frost change	1.8	1.8	1.7	1.5	1.1	1.0

In the instrument shelter (at 1.9 m) the yearly total was 131; on the observation tower (34 m) it was 95; at both positions the frost-change density was 1.8.

15 Distribution of Water Vapor above the Ground

The ground surface is as important in the water budget of the atmosphere as it is in the energy budget. Evaporation takes place either from the ground surface or from its vegetation cover. This stream of water vapor is directed upward. Water returns to earth in quite a different fashion; it is precipitated in liquid or solid form. Only at night can there be a downward transport of water in the form of vapor, that is, when there is a "fall" of dew or frost. This situation will be termed a "humidity inversion" by analogy with the nocturnal temperature inversion.

The annual precipitation in Germany amounts to 800 mm, almost half of which is returned to the atmosphere in the upward stream of water vapor through surface evaporation and transpiration. The amount of dew per year is on the order of 30 to 40 mm. It follows that the downward stream of vapor through surface deposition is only about one-tenth of the upward stream.

Just as eddy diffusion plays a preponderant role in the transport of energy (compared with molecular diffusion) beyond the laminar boundary layer, the influence of mass exchange overshadows that of diffusion, as far as the transport of water vapor is concerned. In the eddy diffusion equation (Section 7), the quantity of a characteristic per unit mass was discussed. In the problem now at hand, this becomes q, the mass of water contained in 1 kg of moist air. In the following discussion, vapor pressure (e) is measured in millibars.

In nature, the quantity of greatest significance is often relative humidity. The distribution of vapor pressure will be discussed first and relative humidity afterward.

As with temperature, the discussion will begin on a large scale, considering the first 100 m of the atmosphere. The 3 yr series of observations made in England by A. C. Best, et al. [1101] included humidity measurements. In Figure 15-1, the abscissa gives the variation of the gradient of water vapor content in 10^{-3} kg m^{-3}/100 m. Negative numbers signify increases of water vapor with height (humidity inversion), while positive figures indicate a situation in which water vapor decreases with height.

The flow of moisture is directed upward in 60 percent of the cases, and this applies at all levels; while on 40 percent of the occasions, there is less water vapor below than above. At first, this appears to contradict the fact that water vapor is transported upward most of the time. However, the total movement of water vapor in the vertical direction z is proportional to the product $A\ dq/dz$ or approximately proportional to $A\ de/dz$. The negative values of de/dz in Figure 15-1 occur during the night, when A is very small, while the positive values are associated with values of the Austausch coefficient that are ten times greater during the day.

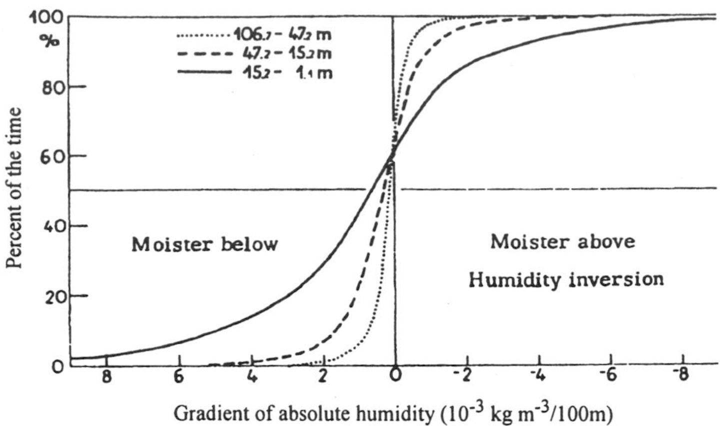

Figure 15-1. Frequency distribution of the absolute humidity gradient in three layers of air below 107 m, from measurements made over three years in England. (After A. C. Best, et al. [1101])

Figure 15-1 also shows, as did Figure 11-2, that the water vapor gradient increases near the ground surface in a way similar to that of the temperature gradient. The maximum gradients observed over the three years were:

Height in air layer (m):	1.1-15.2	15.2-47.2	47.2-106.7
Decrease in water vapor (10^{-3} kg m^{-3}/100 m):	40.8	13.1	7.5
Humidity inversion (10^{-3} kg m^{-3}/100 m):	-22.2	-11.3	-5.5

In the layer nearest the ground, the water vapor gradient is closely related to the time of day. Humidity decreases begin about 08:00 to 10:00 and lasts through the day and early evening, and humidity increases (inversions) begin about midnight and last through the rest of the night. As distance from the ground increases, other factors become involved, causing variation in the timing of these phenomena.

Figure 15-2 shows vapor pressure tautochrones for 19 clear summer days (calculated from the mean temperature and absolute humidity) for the same hours as were used for the temperature tautochrones in Figure 11-6. Before sunrise, there is a flow of water vapor from a height of about 40 m toward the ground to form dew. Vigorous evaporation is initiated after sunrise by surface heating, as may be seen from the increase in vapor pressure in the layer nearest the ground until 06:00. Since eddy diffusion is still restricted, this supply of water vapor is trapped near the ground, and the daily maximum of surface vapor pressure is reached about 08:00 with strong vapor pressure gradients. As the day progresses, the tautochrones (dashed line) then become progressively displaced toward lower vapor pressure because of increasing eddy diffusion. This transport out of the layer close to the ground continues until a minimum value is reached at about 14:00. At 18:00, vapor pressure still decreases with height. By 20:00, the decreasing eddy diffusion and the resulting increased vapor content of the surface air layer have begun to reestablish the water vapor inversion near the ground, and as time goes on, it gradually increases in height in a similar way to the temperature inversion.

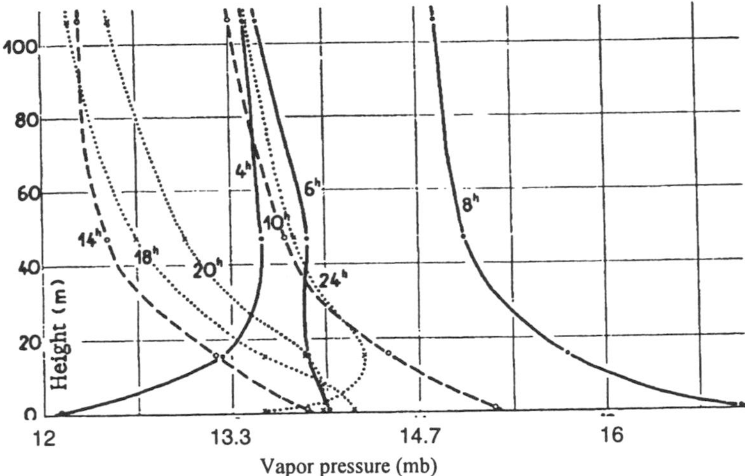

Figure 15-2. Tautochrones of water vapor stratification in the lowest 100 m on clear summer days. (After A. C. Best, et al. [1101])

These results were demonstrated by E. Frankenberger [2506] at Quickborn in Holstein for the 70 m layer of air (Table 15-1) on clear July days in 1954. Here, too, the flow of water vapor at night is directed toward the ground surface (covered with dew) and is remarkably large in the hours before sunrise. The transition from day to night conditions can be recognized by the similar values for the period 20:00 to 22:00.

Table 15-1. Vapor pressure (mb) at Quickborn on clear July days. (After E. Frankenberger [2506])

Height (m)	Time of day					
	22-2	2-6	6-8	8-14	14-20	20-22
70	11.6	12.0	11.9	11.5	12.3	12.3
28	11.6	11.6	12.0	11.7	12.4	12.4
13	11.5	11.3	12.1	12.1	12.7	12.4
2	10.9	10.4	12.4	13.1	13.3	12.3

The daily pattern is shown in Figure 15-3 for heights of 2, 13, and 70 m. The double wave of vapor pressure is easily recognized and appears at all levels. The diurnal amplitude of fluctuation increases as the ground is approached. In all these layers, the evening maximum is higher than the morning value due to the evaporation of water during the day. The morning peak of the bimodal water vapor distribution is the result of relatively large amounts of evaporation and limited eddy diffusion. As the sun's elevation becomes higher in the sky (past approximately 30°), the unstable sublayer depth and strength no longer increase because by then convective mixing has become too vigorous. This vigorous convective motion results in the entrainment of dryer air from aloft, resulting in the midday dip (Figure 15-3). Since convective motion is reduced as the ground is approached, the midday dip becomes less pronounced (Figure 15-3) and very close to the ground is replaced by a single maximum at 5 cm (Figure 15-4). In the late afternoon (Figure 15-3), a second maximum occurs due to the relatively large amounts of evaporation and the reduction of convective mixing. An increase in the fluctuation of water vapor content occurs as the ground is approached, which is similar to the increase in the fluctuation of diurnal temperature.

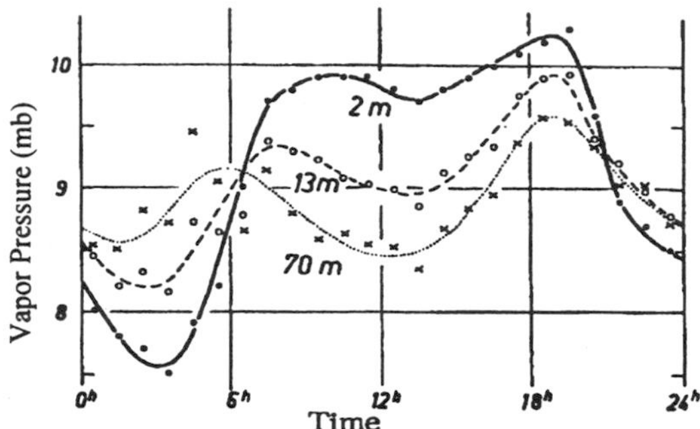

Figure 15-3. Diurnal variation of vapor pressure at Quickborn on clear July days. (After E. Frankenberger [2506])

The mean distribution of vapor pressure (mb) for a dry climate is shown by observations by L. A. Ramdas [1502] for 1933 to 1937 in Poona (18.5°N) as shown in Table 15-2.

Table 15-2. Vapor pressure at sunrise and midday in a dry climate. (After L. A. Ramdas [1502])

Height (cm)	0.8	2.5	7.5	15	30	61	91	122	305
At sunrise	10.0	9.9	9.9	10.0	10.1	10.4	10.7	10.9	11.7
At midday	13.3	12.8	12.5	12.0	11.9	11.6	11.5	11.3	11.1

The night inversion is well marked, and the daily fluctuation of 3.3 mb at 0.8 cm is five times its value of 0.6 mb at 305 cm. Observations made at Pretoria, South Africa (26°S), by E. Vowinckel [1507], gave the average results for cloudless days, from 21 to 27 June 1950, shown in Table 15-3. In this dry climate, the increase in vapor pressure at night is more noticeable than the normal decrease of vapor pressure during the day.

Table 15-3. Twelve-hour averages of vapor pressure (mb). (After E. Vowinckel [1507])

Height (cm)	08:00-19:00	20:00-07:00	Diurnal fluctuation
130	7.08	7.17	1.87
5	7.15	6.15	3.07

The situation is quite different in damper climates. The lower half of Figure 15-4 shows the average daily variation of vapor pressure for three dry August days in 1934 in Finland (about 61°N). M. Franssila [2108] made readings at three different heights above ground level. The night increase in vapor pressure with increasing height is only weakly marked, in contrast to the decrease in vapor pressure with increasing height during the day. The midday dip in vapor pressure, due to convection, observed in Figure 15-3 is somewhat noticeable in Figure 15-4 at 100 cm, but decreases in strength as the ground is approached because of the reduction in mixing. Relative humidity is calculated by dividing the vapor pressure by the saturation vapor pressure. The saturation vapor pressure is a function of the air temperature. While the vapor pressure increases during the day and decreases at night, the saturation vapor pressure typically shows an even greater increase and decrease respectively, resulting in a higher relative humidity at night and a lower relative humidity during the day (Figure 15-4). Over bare soil, the relative humidity is usually higher near the surface and decreases with height. This difference is greater by day than at night.

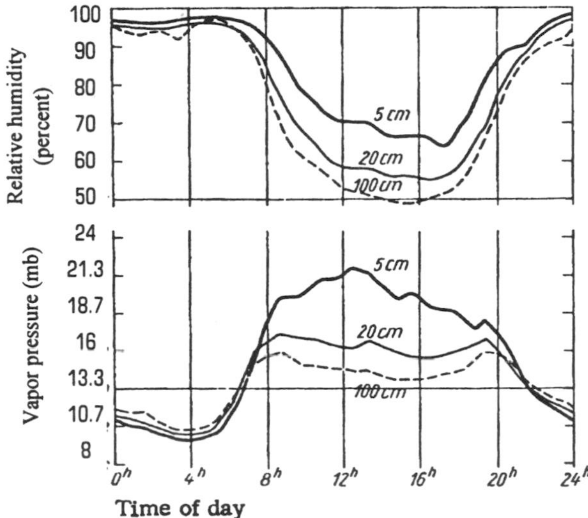

Figure 15-4. Diurnal pattern of humidity in the lowest meter of the atmosphere. (From observations by M. Franssila [2108] in Finland)

There seems, however, to be cases in which a water vapor inversion is no longer established near the ground at night. C. W. Thornthwaite [1314] in Seabrook (36.6°N) observed this even during the clear spring days, for which the temperature conditions were mentioned earlier (Section 13). Unfortunately, the observations between 23:00 and 05:00 are missing, but from 05:00 to 23:00 the vapor pressure decreases continuously from 5 cm upward. During the night, however, the decrease is noticeably lower than by day, as may be seen from the following mean values (mb):

Height (cm)	2.5	5	10	20	40	80-640
14:00-17:00	8.5	8.4	8.0	7.9	7.7	7.6
20:00-23:00	9.5	9.5	9.3	9.3	9.3	9.2

The vapor pressure graphs published for Seabrook also show that the water vapor inversion is an exceptional feature in Nebraska. The measurements made in O'Neill, Nebraska (42.5°N) also show, almost without exception, a decrease of vapor pressure with height.

This is surprising because dew also forms at night in these places. The conclusion must be that, in certain instances, water present in the atmosphere is hardly used at all in the formation of dew, as enough water comes out of the damp ground for this purpose. This is supported, for example, by the fact that in Germany more dew is observed in autumn, when the ground is warmer, than in spring when the ground is colder, under similar conditions. P. Lehmann and H. Schanderl [1500] proved as long ago as 1942 that evaporation and the deposition of dew can take place simultaneously.

The deposition of dew takes place when the surface temperature is lower than that of the saturation temperature of the overlying air or underlying soil. Since at night the ground surface cools more quickly than the air above or the ground below, water vapor may flow toward the surface in either or both directions. Evaporation depends on the difference between the vapor pressure of the evaporating surface and the momentary vapor pressure of the air, and is proportional to this difference. Dew may be deposited on the tips of plants while the ground, or lower parts of the plant, are still giving off water vapor. Even with the ground free of vegetation, poorly conducting soil can be receiving a covering of dew while better-conducting soil is still evaporating. This may be observed directly when dew is deposited on the grass while driveways are free of dew. P. Lehmann and H. Schanderl [1500] were able to register a marked fall of dew while at the same time and in the same place a lysimeter still showed a loss of weight. If a surface is covered with dew in the evening, it does not follow that the evaporation process has necessarily ended. Water vapor may be flowing from the warmer soil toward a cooler surface, resulting in the formation of dew, and, at the same time, some of this dew can be evaporating into the air. Thus, the amount of dew at the surface depends, in part, on the upward flow of water vapor in the soil and, in part, on the flow of water vapor in the air.

J. L. Monteith [1501] provided an analysis of the sources of water in dew formation over an extensive short grass surface in southern England. By using a soil balance sunk in the ground to continuously record dewfall on the natural soil or plant surfaces, and filter paper to measure moisture on the grass, he was able to distinguish between two types of dew formation processes. In one case, the soil balance continued to record a loss of weight even though moisture began to appear on the grass surfaces. This resulted in what he termed distillation, or the upward flux of water vapor from the soil. Distillation rates between 0.01-0.02 kg m^{-2} hr^{-1} were recorded on very calm nights, with clear skies, rapid surface radiative cooling, and wind speeds of less than 0.5 m sec^{-1} at 2 m. On these nights molecular processes dominated the exchange of heat and mass. In the second case, the soil balance recorded a weight gain approximately equal to the rate of condensation on the grass leaves. This process, termed dewfall, was characterized by the downward flux of water vapor to the surface. Dewfall rates ranged from as low as zero with wind speeds as low as 0.5 m sec^{-1}, but increased to as high as 0.03-0.04 kg m^{-2} hr^{-1} at greater wind speeds. On these nights turbulent processes dominated the exchange of heat and mass.

Since the formation of dew depends on outgoing radiation at the particular site and is affected by the humidity of the air, local air currents, horizontal shielding, cloudiness, and temperature gradient, it is not a continuous process, even if the quantity of dew present usually increases with the passage of time during the night. There is a short term changing rhythm of evaporation and deposition of dew, depending on the type of plant cover and structure of the ground. Plants are not the only recipients of dew. Hygroscopic particles present in the atmosphere can also extract considerable quantities of water. Several authors (C. R. V.

Raman, et al. [1503] and S. E. Tuller and R. Chilton [1505]) have shown that in some areas and situations dew plays an important role in the regional and crop water balance.

Near the ground the nocturnal decrease of temperature with increased height on calm nights may partially explain how A. R. Subramaniam and A. V. R. Kesava Rao [1504] observed an increase in dew deposition with increased height. However, further research may be necessary to explain why dew deposition reached a maximum at 50 cm at one location and at 100 cm or higher at two other locations.

16 The Wind Field and the Influence of Wind near the Ground

The horizontal transport of an air mass is a consequence of large scale differences in air pressure, that is, of the horizontal pressure gradient. The ground surface acts as a brake on the movement of air. The effect of surface friction on the strength of surface winds is confined to the lowest layer of the atmosphere known as the friction layer. In general, the friction layer is 1000 to 1500 m thick. Above this layer the gradient winds are unaffected by surface friction.

Wind speed usually increases with height. The Coriolis effect causes the wind to veer to the right in the Northern Hemisphere (and to the left in the Southern Hemisphere). Since the Coriolis effect is directly related to the wind speed, winds tend to veer to the right with increased height. This variation of wind speed and direction with height (Eckman spiral) depends, to a marked degree, on the nature of the underlying surface. Since rougher surfaces cause increased friction and reduced wind speeds, the variation of wind direction with height is at a maximum over irregular terrain and a minimum over flat land and water surfaces.

The increase of wind strength with height is a factor of some considerable importance for the layer near the ground. This may be observed directly by looking at the rime in Figure 16-1. Rime is formed when supercooled water droplets, carried by the wind in fog or cloud, come in contact with a solid surface and freeze immediately. If an even distribution of such droplets in the lowest meter is assumed, then the amount of rime deposited on the fence post will be greater as the number of drops transported in a given time period increases, that is, the greater the wind speed. The length of the banner of rime is, therefore, a measure of the strength of the wind, blowing from the left.

Figure 16-2 shows the diurnal variation of wind speed at various heights above the ground, for observations made over a year (mean values from September 1953 to August 1954) by E. Frankenberger [2506] in Quickborn, Holstein. At lower levels there is a daily sequence with a well marked maximum in the hours around noon and a minimum at night. The curves lose this distinct form with increasing height until at 70 m the trace is approximately uniform. The maximum wind strength at the ground near midday is due to solar input to the surface, convective mixing, the maximum values of the Austausch coefficients and the resulting coupling and exchange of momentum from higher elevations. This corresponds to a minimum of wind speed at heights over 100 m because at this time there is more momentum being carried downward (Figure 16-3). There is a maximum of wind speed aloft at night due to the reduced surface coupling and momentum exchange, which may be seen in Figure 16-2 for a height of 70 m and in Figure 16-3. The reduced

Figure 16-1. The banner of rime shows the increase of wind with height. (Photograph from Mount Washington)

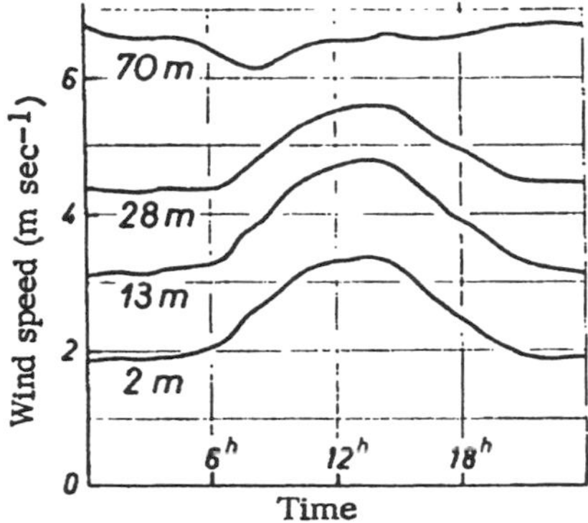

Figure 16-2. Diurnal variation of wind speed in Quickborn, at four different heights above the ground. (After E. Frankenberger [2506])

momentum exchange is also the reason for the nocturnal surface reduction in wind speed. The height of this transition layer varies. It is relatively high when either free or forced convection is great, that is, in summer, in cloudless weather, in low pressure troughs, or in gales; on the other hand, it is low in winter, with overcast skies, in high pressure areas, and with light winds. The height of this intermediate layer varies between 50 and 100 m, but these limits may be exceeded in both directions. With greater coupling and momentum exchange during the day, the variation of the surface wind direction with height from the gradient winds tends to be greater at night (Figure 16-4).

Low level jets have been observed in numerous locations (C. Walters [1610]). C. Walters [1610] and Y.-L. Chen et al. [1602] divide these into two types (synoptic and boundary layer) based upon the factors that most strongly control their origin. Boundary layer wind maximums typically occur around 700 m at night and to a lesser degree on cloudy days. These boundary layer wind maximums are a result of low momentum exchanges. They are typically

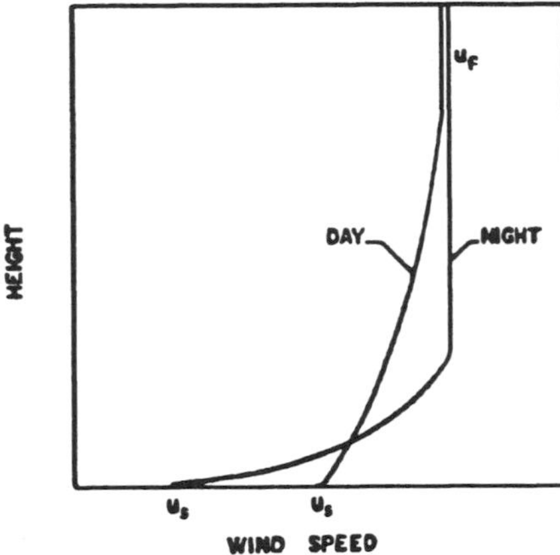

Figure 16-3. Typical variation of wind speed with height in the friction layer. u_s represents the surface wind and u_f the gradient wind. (After E. W. Hewson [910])

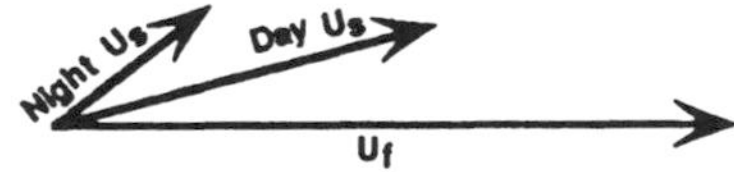

Figure 16-4. The surface wind u_s in relation to the gradient wind u_f by day and by night. (After E. W. Hewson [910])

attenuated with daytime heating, convection, momentum exchanges and a recoupling of the surface and near surface boundary layers. For more information about low level jets the reader is referred to A. K. Blackadar [1600], W. D. Bonner [1601], H. Wexler [1611] and particularly C. Walters [1610].

H. Henning [1106] measured the number of calm periods expressed as a percentage of a year's observations from April 1953 to March 1954 at the Lindenberg Observatory, Berlin (Table 16-1). As expected, there was an increased frequency of calm periods at night and close to the ground.

Table 16-1. Number of calms as percentage of a year's observations, at Lindenberg Observatory.

Height of measurement (m)	Hour of day 2	4	6	8	10	12	14	16	18	20	22	24
10	4.1	5.0	2.8	2.2	0.8	0.6	0.6	1.4	2.8	6.6	6.1	5.7
5	9.1	9.7	8.0	3.3	1.7	1.1	1.7	3.0	4.4	10.2	10.0	9.4
1	15.7	16.3	10.2	4.4	2.5	1.9	1.9	3.0	8.8	17.9	16.8	17.6

As early as 1918, G. Hellmann [1604] made a series of measurements close to the ground over a period of several months at heights of 5, 25, 50, 100, and 200 cm on the Nuthe-Wiesen near Potsdam. Figure 16-5 shows the number of hours of observation when the wind was calm, expressed as a percentage.

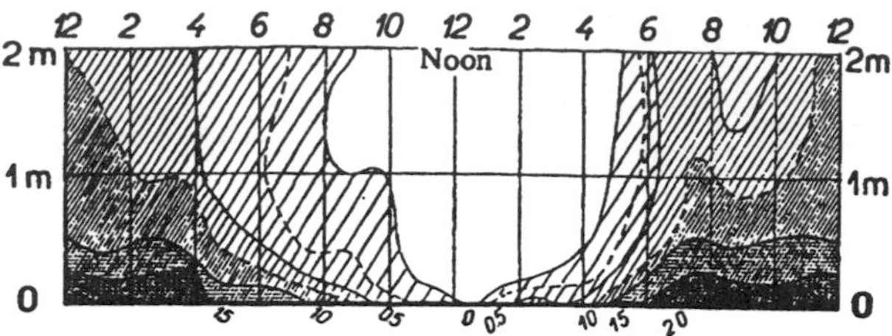

Figure 16-5. Percent of hours of calm in the lowest 2 m. (After G. Hellmann [1604])

The features of the vertical wind profile and knowledge of the laws that govern the increase of wind speed with vertical height are of great interest, both for the theory of air streaming and turbulent mass exchange and for the energy and water budgets of the atmosphere.

C. W. Thornthwaite [1314] at Seabrook observed average values from five undisturbed spring days for four hourly intervals (Figure 16-6); the numbers on the lines are the hours for which these mean values apply. On the left (A), the wind profiles are shown with arithmetic scales. It can be seen that the wind speed slows quickly as the surface is approached which implies a downward transfer of horizontal momentum. Also note that the strong wind gradients near the ground are similar to the gradients of temperature and vapor pressure. The simplest way in which the relation between wind speed u (m sec^{-1}) and height z (m) can be expressed is by a power law:

$$u = u_1 z^a \qquad (\text{m sec}^{-1}),$$

in which u_1 is the wind speed at a height of 1 m. With the aid of logarithms, the equation becomes

$$\log u - \log u_1 = a \log z$$

from which it follows that, in a log-log coordinate system, the relation between wind speed and height will become linear. This can be seen in the center diagram (B). Although 5 hr averages were used, the observations approximately fulfill these conditions.

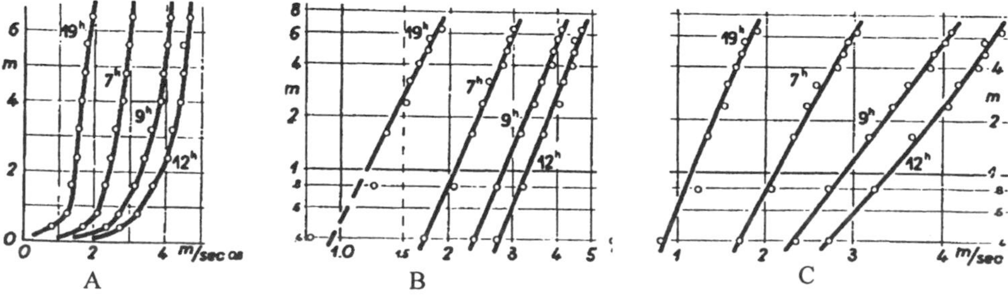

Figure 16-6. Wind profiles at Seabrook, in three different methods of representation. (Corresponding temperature profiles are shown in Figure 13-2) (After C. W. Thornthwaite [1314])

The differing slopes of these four straight lines make it clear that the exponent *a* in the foregoing law is not a constant; however, since *a is* equal to the tangent of the angle the line makes with the log *z* axis, it must become smaller as wind speed increases. Since we have seen that there is a diurnal variation of wind speed, it follows that the value of *a* must change with the time of day. In addition to this, *a is* a function of height, and increases with proximity to the ground. O. G. Sutton [1609] used the observations made by G. S. P. Heywood [1605] for the lowest 100 m to compute the daily variation of this exponent subdivided for summer and winter readings. From April to September, *a* varied between 0.07 at noon and 0.17 at night, and from October to March, between 0.08 and 0.13. H. Henning [1106] found seasonal averages deduced from the measurements made in Lindenberg to be 0.32, 0.39, 0.53, and 0.28 during the day in spring, summer, autumn, and winter, respectively; while by night, the values were 0.38, 0.49, 0.59, and 0.28. These values were for the lowest 10 m, however the autumn values were for the first 5 m only and for that reason are somewhat high. The actual value of *a* also depends on the atmospheric stability, which changes throughout the day and explains much of the scatter about the lines in Figure 16-6(B). The roughness of the ground surface varies on a seasonal basis, which also influences the value of *a* and explains the seasonal dependence of the values.

A quick comparison of wind readings taken at different heights in the air layer near the ground can be made by using the power law. If a basic value for *a* of 0.25 is used, then wind speed increases as a fourth root function of height.

A more realistic view of the connection between wind speed and height, however, can be obtained from laboratory experiments on wind speed profiles. As a preliminary simplification, the assumption is first made that there is no effect due to the layered structure of the temperature distribution, that is, that there is no instability which might favor vertical transfer of momentum, nor is there a stable layering which might suppress it. For conditions of neutral stability, we have L. Prandtl's [1608] logarithmic law:

$$u = \frac{u_*}{k} \ln\left(\frac{z}{z_0}\right) \qquad (\text{m sec}^{-1})$$

In this equation u_* (m sec^{-1}) is the shear velocity (or friction velocity), which is indicative of the amount of turbulence, and its value is independent of height for a given wind profile; k is the von Kármán constant, which has been independently evaluated as 0.4 (a dimensionless number) for the layer of air close to the ground, and z_0 is the roughness length, which has the dimensions of length and provides a quantitative measure of the aerodynamic roughness of the ground. Changing to common logarithms and collecting all the constants together, we arrive at the law of wind change with height:

$$u = c \log\left(\frac{z}{z_0}\right) \qquad (\text{m sec}^{-1})$$

This equation is a straight line on a semilog graph with coordinates u and log z. C. W. Thornthwaite's [1314] observations are plotted in this manner on Figure 16-6(C). From Figure 13-2, the temperature distribution between 07:00 and 09:00, after the night inversion has been broken down and before diurnal heating has set in, corresponds most closely to the conditions of neutral stability that were presupposed in formulating the foregoing law.

Many researchers have tried to extend L. Prandtl's [1608] law to other temperature structures. They have done this by improving the technique of measurement, by eliminating the influences of the environment, and by increasing the number of observations. The dependence of wind speed upon temperature profiles may be readily seen in Figure 16-7. W. D. Flower [1105] made these observations in the winter of 1931-32 in the desert near Ismailia on the Suez Canal. The abscissa of Figure 16-7 is the wind speed at 62.6 m at the top of the observation mast. The ordinate is the difference in wind speed between 15.2 and 62.6 m. Generally, as the wind increases at 62.6 m, it will also increase at 15.2 m, but as shown in Figure 16-7, not as much. Thus, as wind speed increases at 62.6 m, the difference in wind speed between these two heights increases. The amount of the increase is strongly influenced by the temperature structure. The four curves shown in Figure 16-7 are for four different temperature gradients observed at the same heights. The value of the gradient, in °C/100 m, is given alongside the curve. The four curves tend to converge toward the origin since a calm aloft is accompanied by still air below. When there is an inversion, the cooler air will remain near the ground, and an increase in the gradient wind primarily increases the difference between the winds above and below the inversion. Thus, the less the lapse rate, the less the momentum exchange, and greater the potential difference of wind speed with increased height.

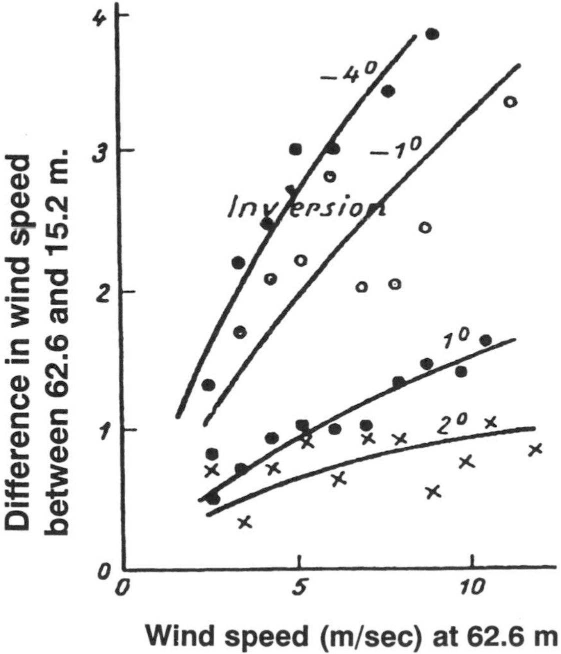

Figure 16-7. Influence of temperature stratification on the increase of wind with height. (After measurements by W. D. Flower [1105] in Egypt)

If the simple form of L. Prandtl's [1608] equation is differentiated we get:

$$\frac{du}{dz} = cz^{-1}$$

E. L. Deacon [1603], who carried out a series of measurements at Porton, England, determined a simple form of the law:

$$\frac{du}{dz} = cz^{-\beta}$$

The factor β depends on the temperature structure of the lower atmosphere. If the atmosphere is neutrally stable, β = 1, and the equation reduces to the L. Prandtl [1608] equation. When the atmosphere is stable and layered, the increase of wind strength with height is greater than with neutral stability, and β becomes less than 1. In these circumstances, the $u = \log z$ line develops a concave curvature toward the u-axis. This can be seen on Figure 16-6(C) for 19:00. When the atmosphere has an unstable structure, β is greater than 1 and the curve is convex toward the u-axis. The 12:00 data on Figure 16-6(C) show this configuration. The range of values for β in the lowest 10 m of the atmosphere is $0.7 < \beta < 1.2$.

The effect of temperature structure on wind speed profiles is further illustrated in Figure 16-8, which shows results of observations made over long grass in the summer of 1941 (E. L. Deacon [1603]). To allow for the effect of vegetation cover on the wind speed profile, 25 cm is subtracted from all heights before they are used in Figure 16-8 (see Section 30). Four examples, using a relative scale for u, are shown. On the left are stable temperature structures, characterized by a negative lapse rate, concave curvature of the wind profile toward the u axis, and low wind speeds. On the right are unstable temperature structures, characterized by a positive lapse rate, a wind profile convex toward the u axis, and higher wind speeds. The temperature gradient (°C/100 m) between 1.2 and 17.1 m are written beside each profile. The greater degree of curvature of the wind profile with a stronger inversion is illustrated in the two profiles at the left, as is the more rapid rate of momentum exchange in the lower 10 m under stable atmospheric conditions. This dependence of the wind speed profile on the temperature structure is greater in higher layers than it is in the lowest 2 m, where it is almost nonexistent. The

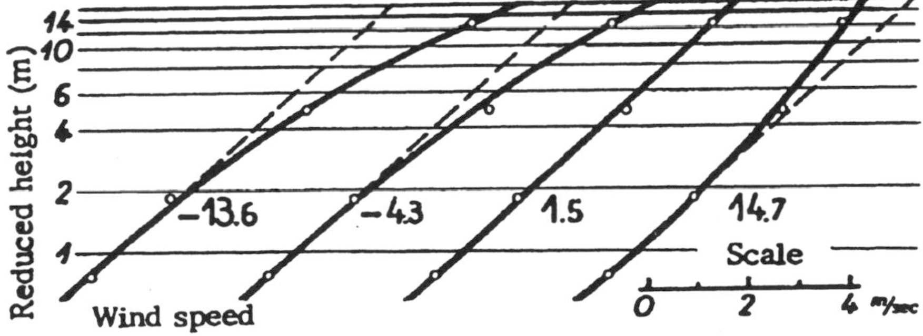

Figure 16-8. Dependence of the wind profile on temperature stratification. (After E. L. Deacon [1603])

assisting or restricting influences of stable or unstable temperature structures requires a certain mobility of the air, which is lacking in these layers near the ground. In this layer, the simplified Prandtl equation can be used, even with temperature profiles that are not adiabatic, without excessive errors.

In applied climatology it must always be borne in mind that laws governing wind increases with height only apply to mean values over a long period of time, or, if applied for short terms, it must be with observations from which other influences have been eliminated.

Figure 16-9 shows temperatures recorded on three consecutive nights in Kentfield, California by A. G. McAdie [1606]. The increased wind speed in the middle of the second night resulted in warmer air from aloft being brought to the surface, which mitigated the cold temperatures. This resulted in the second night's minimum temperature being warmer than the other two nights. For the layers near the ground at night, mixing means heating, which in this extreme case amounted to between 10-12°C. For this reason, the farmer and the fruit grower have less fear of night frosts when the wind continues to blow in the evening.

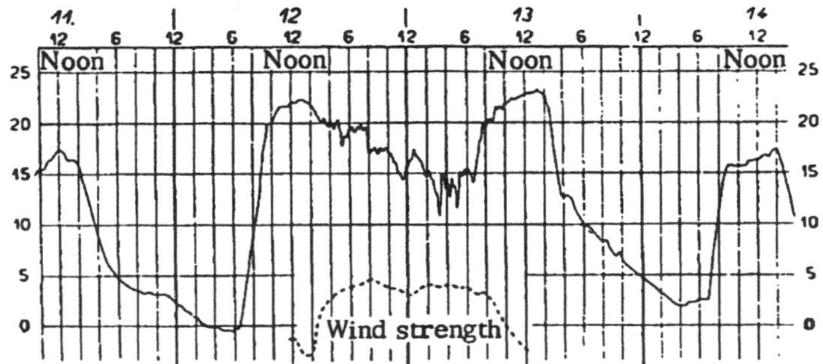

Figure 16-9. Night temperatures at Kentfield, California, on 11-14 December 1911. (After A. G. McAdie [1606])

The differential utilization of incoming and outgoing radiation has a large effect on the microclimatic conditions on days with light winds or calm conditions. Stormy conditions, however, minimize these variations by mixing the air. Those who wish to carry out microclimatological investigations and identify microscale variations should, therefore, try to find weather situations with light winds or still air when these influences are most clearly developed.

As previously discussed two relationships exist between wind strength and the temperature structure for a given wind speed: (l) due to the increased stability the dependence is substantially greater by night than by day, and (2) the dependence increases with proximity to the ground.

N. K. Johnson and G. S. P. Heywood [1108] reported the mean temperature gradients shown in Table 16-2. They were evaluated for two groups of wind speeds from 5 yr records of observation in Leafield. At noon above 10 m, they found only a minor effect of wind speed on temperature gradients; however, in the lowest layer, it was stronger. At night the influence of wind is easily recognized, and is especially strong in the lowest 10 m. It should be pointed out that as the air becomes more thoroughly mixed, the lapse rate approaches neutral stability (1°C/100 m). This may explain the nocturnal increase in the lapse rate between 57.4 and 87.7 m with increasing wind speed (Table 16-2).

Table 16-2. Mean temperature lapse rate (°C/100 m) at Leafield.

Hour of day	Wind speed	Height (m)								
	(m sec^{-1})	1.2		12.4		30.5		57.4		87.7
Noon (12:00)	2.5		8.0		1.6		1.2		1.2	
	7.4		6.4		1.5		1.1		1.1	
Night (02:00)	2.5		-8.0		-1.8		-0.8		0.0	
	7.4		-2.4		-0.4		-0.1		0.3	

For clear days (from 10:00 to 14:00) M. Franssila [1306] found that there was "no noticeable effect" of wind speed on the temperature gradient for the layer from 240 to 5 cm in Finland. However, at midday (11:00-13:00), A. C. Best [1302] found a decrease in the temperature gradient with increasing wind speed (Table 16-3).

Table 16-3. Wind effects on the temperature gradients (°C/100 m) during the day (11:00 - 13:00) in England. (After A. C. Best [1300])

Wind speed (m sec^{-1})	0		1		3		5		7		>7
2.5-30 cm, March		412		501		522		364		327	
2.5-30 cm, June		–		707		599		568		452	

Table 16-4 shows the effect of a nocturnal increase in wind speed on the temperature gradient. Spring nights show the influence of wind more clearly than do summer nights, because of the greater daily fluctuation of temperature earlier in the year.

Table 16-4. Wind effects on the temperature gradients (°C/100 m) at night (23:00-01:00) in England. (After A. C. Best [1302])

Wind speed (m sec^{-1})	0		1		3		5		7		>7
30-120 cm, March		-112		-74		-38		-31		-27	
30-120 cm, June		-98		-48		-38		-10		–	
2.5-30 cm, March		-396		-203		-183		-167		-160	
2.5-30 cm, June		-315		-196		-197		-92		–	

In Figure 16-10, E. Frankenberger's [2506] observations from Quickborn are for clear summer midday periods (dashes) and clear nights (solid lines). The temperature gradient is positive during the day and negative at night. The influence of wind speed on the temperature field is practically negligible on clear days with strong convection. At night, in contrast, the dependence of the temperature gradient on the wind speed is large and increases as the ground is approached. While the initiation of wind always brings about a substantial reduction of the temperature gradient, additional increases in wind speed result in a smaller reduction of the temperature gradient. The ordinate scale on the right of the diagram for this curve is for the 0-2 m nocturnal observation.

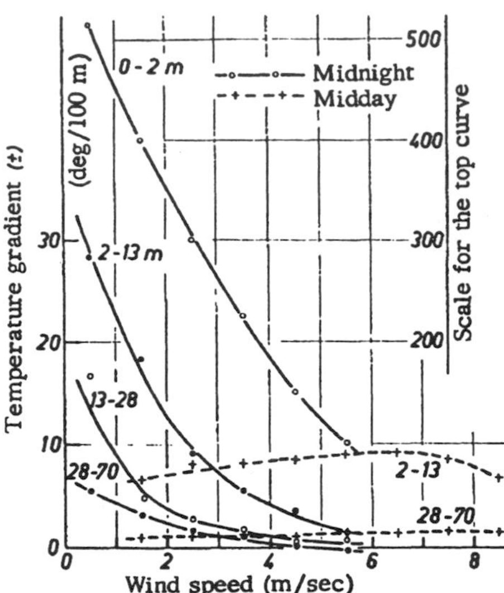

Figure 16-10. Dependence of the temperature gradient on the wind speed at Quickborn on summer days. (After E. Frankenberger [2507])

In summation, during the night, an increase in wind speed will result in increased mixing and a reduction in the strength of the inversion. Close to the ground where mixing of all types is suppressed, increasing wind speed, both during the day and night, results in a decrease of the temperature gradient (or at least a smaller departure from neutral stability). During clear days, however, when convective mixing is already quite strong beyond the surface layer, an increase in wind speed will have a smaller or in some cases no effect on the temperature gradient.

In taking wind measurements, great care must be exercised. Unless the horizontal plane in which a wind vane lies is free in all directions from large obstructions, erroneous values of wind direction are likely to be measured. An obstacle may cause a wind direction at the instrument, which is not representative of the general air flow in the area. For the same reason, a wind vane mounted near a cliff or the side of a hill will not give representative values.

A useful rule-of-thumb is that the instrument should be located at a distance from any obstruction equal to at least ten times the height of the obstruction (W. M. O. [917]). Thus, a wind vane should be at least 150 meters from a grove of trees 15 meters high. The standard height for measuring wind direction is 10 meters. If the instrument is to be mounted on a building, it should be at least 10 meters above the highest part of the building. If the building is large, however, even this height may not be sufficient to ensure representative measurements (E. W. Hewson [910]). E. W. Hewson [910] suggests that:

> "When a vane is to be attached below the top of a steel tower, however, it should be mounted at the end of a horizontal boom at a distance which is at least equal to the width of the tower at that height, and preferably twice that distance."

For additional information on mounting wind vanes on the top of stacks or on meteorological towers, the reader is referred to E. W. Hewson [910] and W. M. O. [917].

17 Distribution of Particulates and Trace Gases

Besides heat, water vapor, and wind, which have already been discussed, there are other atmospheric elements such as solid particles suspended in the atmosphere or sinking slowly through it, and trace gases, which are a constituent part of the air. These particles and gases generally have sources at or near the ground surface, and their distribution is controlled from the ground in a way similar to that of water vapor or heat. The vertical distribution of the particulate content of the air, carbon dioxide, and radioactive gases is determined by the strength of their surface sources, the state of the ground, and the conditions prevailing in the air near the ground.

Particulates and trace gases have surface sources and sinks which influence their distribution in the surface layer. Dust and sand are transported by storms from prairies and deserts; industrial plants pour out particulates and waste gases; carbon dioxide is liberated in combustion; and ozone can be entrained from the upper atmosphere. Besides eddy diffusion, the spread of these materials is also affected by advective dissemination.

Advective processes near the ground are illustrated by the movement of sand and snow in the layers close to the ground, and by the driving of spray in the layer close to the water surface in heavy seas. For all three substances the surface is the primary source and sink. In these cases, a strong wind is able to remove particles of sand, snow, or water from the surface and carry them into the air. The concentration of the particles in the air depends on the wind speed, the size and shape of the particles, and on their specific gravity. The term "sweeping" is used when the horizontal visibility is not affected, while "driving" is used for cases in which the surface particles are lifted high enough to cause a marked reduction in visibility. These processes are of some considerable importance, since, along with the particles, the properties of the underlying surface are also transported into the layer of air near it. For example, a desert sandstorm heats the air when the particles of sand, at temperatures above that of the air, are picked up.

First, we shall consider the sweeping and driving of sand. As wind speed increases, some grains of sand begin to roll along the surface (traction). Through collision, they set other grains in motion. When the wind speed reaches about 5 m sec^{-1}, the grains that have been hit tend to fly into the air. They rise at an angle between 30° and 70° and fall again in a flat curve, meeting the ground at an angle of 2° to 15°. If the wind speed becomes very high, the particles of sand remain suspended in the air.

Within the ground layer, both the size of the grains and the quantity of sand decrease rapidly with height. The following distribution was measured during a sandstorm at Colomb-Bechar on 14 April 1956 by L. Demon et al. [1709]; the wind speed was 11 m sec^{-1} at 1.4 m and the figures give the number of grains of sand in 1 m^3 multiplied by 100. The maximum frequency is displaced toward grains of smaller size as the height above the ground increases; the quantity of sand transported was 89 mg cm^{-3} at 60 cm and 32 at 140 cm:

Diameter of grains (μ)	40	50	60	70	80	90	100	120	140	160
At height of 60 cm	25	24	30	45	57	59	52	25	8	2
At height of 140 cm	50	77	65	53	46	41	33	10	2	1

Investigations made by K. H. Sindowski [1730] on the island of Norderney have shown that grains are stratified by shape. The spherical grains are carried highest. The normal distribution of sand within the air layer near the ground can be obtained from Table 17-1. Measurements were made at the upper edge of the steep beach in Norderney with a wind of 11 m sec^{-1} at a height of 10 cm with conditions called "smoking sand dunes." The mass of sand, rounded off to the nearest gram for each section and unit of time, is expressed as a percentage of the maximum value at the ground surface. The main quantity of sand is carried into the first few centimeters above the ground. Using the simplified version of the equation on page 94 and a value of $a = 0.25$, we obtain the wind profile on the third line (Table 17-1). The amount of kinetic energy transported is equal to one-half the product of the mass of the sand and the square of its speed. Since the mass of sand decreases with height but the wind speed increases, the maximum of kinetic energy was found at around 15 cm. The kinetic energy distribution of the sand grains as a percent of the maximum kinetic energy is shown in the last line. In this case, a height about 15 cm above the ground constitutes the zone of maximum wind erosion. This result is in good agreement with the observations of E. Blissenbach that in the Egyptian desert telegraph poles were subjected to the greatest wear at a height of about 10 cm.

Table 17-1. Effect of wind on sand, Norderney Island.

Height above ground (cm)	0	5	10	15	20	25	30	40	85
Quantity of sand carried (kg m^{-3} hr^{-1})	1170	340	330	310	270	220	180	110	10
As percentage of maximum value	100	30	28	26	23	19	15	9	0
Wind speed (m sec^{-1})	0.0	9.2	11.0	12.2	13.1	13.8	14.5	15.6	18.8
Relative kinetic energy (percent)	0	85	97	100	95	84	65	45	4

The incidence of blowing sand has been shown by W. Haude [515] for the Gobi Desert to be a function of the season of the year, the time of day, and the water content of the sand. Studies from other geographical locations on the meteorological conditions accompanying dust storms include W. G. Nickling's [1724] work in the Slims River Valley, Yukon Territory, Canada, I. Y. Ashwell's [1701] study in central Iceland, N. J. Middleton's [1721] examination of South Asia, A. J. Brazel and W. G. Nickling's [1703] research on the Sonoran-Mojave Desert region in the United States, and E. Jauregui's [1718] work on Mexico City.

In the mid-latitudes soil erosion by wind can be a problem. Fertile top soil is blown away and the exposed seed may grow up defective or not at all. Young shoots suffer mechanical damage and young plants are engulfed in sand. R. von Gehren [1713] showed that "medium sand," with grains 0.1 to 0.5 mm in diameter, were most liable to be blown away. It was more difficult for the wind to loosen smaller grains from the soil, while heavier grains offered a greater resistance because of their weight.

The blowing of snow occurs in a similar way in the winter landscape. A striking example of this phenomenon comes from F. Loewe [2430] who found that in the Antarctic, at least 20,000 tons of snow were swept over each meter of the coast of Adélie Land into the sea each year, which is an amount equal to half the precipitation falling in the 200 km coastal strip. From 1086 measurements of blowing snow, he obtained the following mean vertical distribution of the snow transported:

Height above snow cover (m)	10	20	50	100	200	300
Quantity of snow (10^{-4} kg m^{-3})	12	6.4	2.6	1.3	0.56	0.22

In severe gales (35 m sec^{-1}), 1 m^3 of air at a height of 1.5 m contained 6 to 10 g of snow (the visibility was under 10 m).

The following discussion will be limited to particulates and gases that come from the soil.

The term particulates also covers the dry, rough, microscopic yet visible components of airborne biological material. Diameters of particulates are between 1 and 50 μ, which would give terminal velocities between 0.1 and 200 mm sec^{-1}. These particulates, which are usually very fine pulverized soil particles, can be lifted up into the atmosphere by wind and eddy diffusion and fall back to the ground very slowly.

The decrease in particulate content with height can be measured either directly by a particulate counter or indirectly by optical means. In the latter case, it is not easy to allow for the additional influence of water droplets in mist. H. Goldschmidt [1714] was able to prove, by using a search light, that the atmospheric turbidity near the ground was considerably greater than the calculated value. From direct particulate counts made in aircraft flights over Jena, H. Siedentopf [1729] showed that 10 to 60 percent as many particulates were present in the air near the ground as were encountered at a height of 1 km. The percentage is greater in summer, with its strong convection, than in winter.

Figure 17-1 is from records by E. F. Effenberger [1711] for clear summer days from 28 to 30 July 1939. They show a maximum particulate content at night and a minimum during the day. The daytime minimum is a result of increased mixing with "cleaner" air being brought down from aloft. The nocturnal maximum is the result of reduced mixing and particulates settling from aloft.

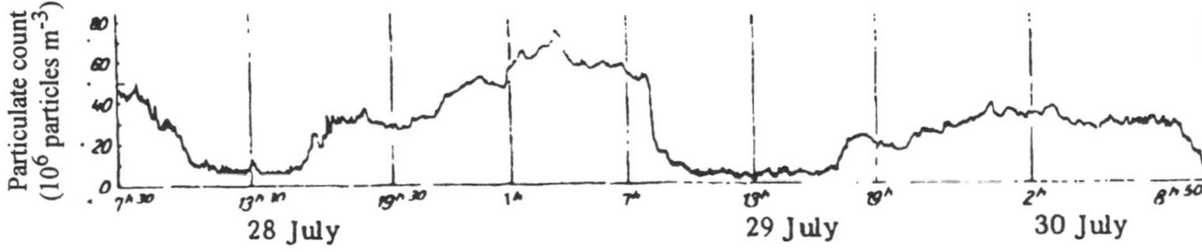

Figure 17-1. The diurnal variation of particulate content of the air over a meadow near the ground on 28-30 July 1939. (From recordings by E. F. Effenberger [1711] on the Collmberg)

Naturally, the particulate concentration in the atmosphere varies greatly from place to place and from time to time. Normal background particulate concentrations are between 10^6 and 10^8 particles per m^3. Lowest concentrations are found over the open sea or over a winter snow cover, and the highest concentrations occur in industrial areas and large cities. As soon as convective mixing sets in with morning insolation, particulates are carried upward, and air aloft with low particulate concentrations descends, resulting in a decline in the concentration near the ground. This is shown particularly well in Figure 17-1 at 07:00 on 29 July. This is consistent with the fact that the particulate content of the air on high mountains is greatest about noon because of the transport of material by valley breezes from the low-lying country to these levels (section 43). Pollen is also a particulate and, for the reasons given above, has the same diurnal fluctuation pattern as other particulates. Most people with pollen allergies (including R. H. Aron and well over 100 people to which he has spoken) notice that their pollen allergies are worse at night. Failure to realize that the pollen concentration is usually higher at this time has led to some unnecessary actions. Returning home at night from work or school, the search begins for what is in the home that is causing the allergies to become more bothersome.

The concentration of carbon dioxide (CO_2) in the surface layer displays a very complex and dynamic temporal pattern. Both the ground surface, through soil respiration, and the atmosphere, through entrainment of CO_2-enriched air, act as CO_2 sources. Furthermore, the major CO_2 sink, photosynthetic uptake of CO_2 by the vegetation canopy, occurs at an uneven rate which varies with atmospheric/soil conditions and occurs over a zone whose depth varies with the age and species of the canopy. For example, E. Ohtaki and T. Oikawa [1725] discuss carbon dioxide and water vapor fluxes over a rice paddy as a function of its stage of development and radiation conditions.

The atmospheric CO_2 concentration displays distinct long-term, annual, and short-term variations, which are superimposed upon the daily cycle. Figure 17-2 illustrates some of these variations. It presents data derived from hourly CO_2 observations at Barrow, Alaska (71°3'N) by J. T. Peterson et al. [1726]. The slightly curved line shows the long-term trend in CO_2 over the period 1973 to 1982, during which time the CO_2 concentration rose steadily from 332.6 to 342.8 ppm. The wavy solid line in Figure 17-2 shows an annual amplitude of nearly 15 ppm, from a monthly maximum of 348.4 ppm in April 1982 to a

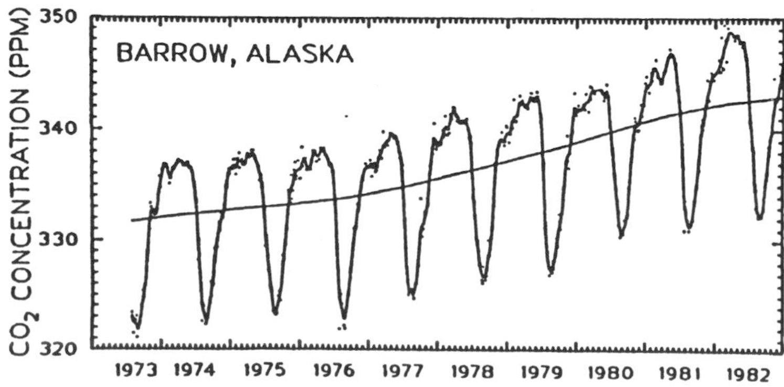

Figure 17-2. Variation in CO_2 concentration (ppm) at Barrow, Alaska based upon 5 day averages from 1973 to 1982. (After J. T. Peterson et al. [1726])

monthly minimum of 333.2 ppm in August/September. This asymmetric pattern results from the changing relative strength of the hemispherical CO_2 sources and sinks. Stronger uptake of CO_2 by the ocean and terrestrial biosphere results in an annual minimum at the end of the growing season (autumn), while a more active biospheric source produces an annual maximum at the end of the dormant or less active growing season (spring). The dots in Figure 17-2 are 5 day averages, and reveal substantial short-term CO_2 variations due to variable atmospheric transport and CO_2 exchange between the atmosphere and ocean/biosphere.

A marked diurnal variation in the CO_2 concentration normally occurs near the ground. Measurements made over the past decade give higher values at night than by day. Using an infrared absorption recorder which is capable of detecting a change in carbon dioxide concentration of 0.0001 percent, B. Huber [1715] was able to chart the daily variation as a function of height above the ground. Figure 17-3 shows the average of ten investigations made between 20 June and 1 August 1952 above a potato field near Munich. The diurnal variation is nearly 100 ppm at 0.5 m; even at a height of 100 m it is still about 35 ppm.

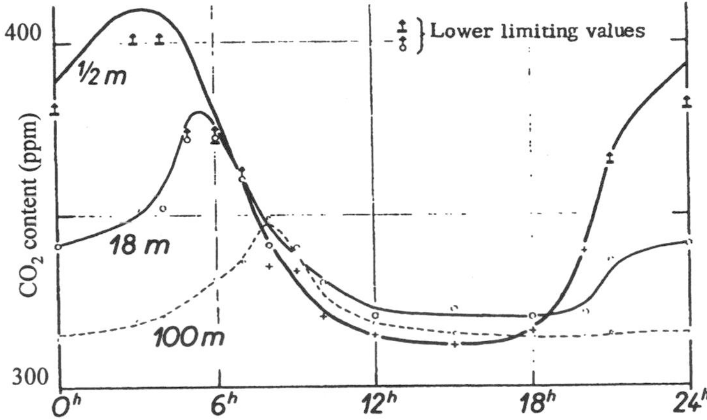

Figure 17-3. The marked diurnal variation of carbon dioxide content in the lowest layer of the air. (From observation by B. Huber [1715] near Munich in summer)

At night, increased atmospheric stability, reduced surface wind speeds (Figures 16-2 and 16-3), and continued CO_2 production by soil and plant respiration results in an increase in the atmospheric CO_2 concentration. This increase will be particularly marked with strong inversions. I. Iizuka [1717] has demonstrated that there is a rapid increase in the evening with the onset of the nocturnal inversion. The carbon dioxide content decreases when the wind increases, as shown by B. Huber [1716] through simultaneous measurements of wind and carbon dioxide content. When heating starts in the morning and eddy diffusion and photosynthesis set in, there is a sharp decrease in the concentration. While the general reduction in the carbon dioxide concentration during the day is due to convection, the increase of carbon dioxide with height is because of photosynthesis. The vertical distribution is similar from late forenoon until evening, due to active mixing throughout the lowest 100 m. The diurnal distribution of carbon dioxide, therefore, follows a pattern similar to that of particulates (Figure 17-1).

The strength of this diurnal variation changes throughout the year. In early spring during the period of the opening of buds, flowering, and rapid leaf growth and development, more CO_2 may be released in respiration than is absorbed in photosynthesis. (H. E. Garrett et al. [1712], M. Schaedle [1728] and P. M. Dougherty [1710]). Figure 17-4 shows the daily variation of summer monthly mean CO_2 concentration measured at 6.5 m above the ground during 1971 in southern Saskatchewan (50°42'N) by D. L. Spittlehouse and E. A. Ripley [1731]. The measurements were made over grassland with a 0.3 m tall canopy. The mean daily maximum and minimum CO_2 concentrations decreased because there is more photosynthesis than respiration. The mean daily maximum and minimum concentrations decrease through the summer growing season. The amplitude of the mean daily cycle decreases from June through September. This reflects the general decrease in the rates of soil and plant respiration and photosynthesis which occur with the end of the growing season. For further information concerning the effects and implications of the changing global carbon cycle on the soil, plants, and the atmosphere, the reader is referred to R. L. Desjardins et al. [4805].

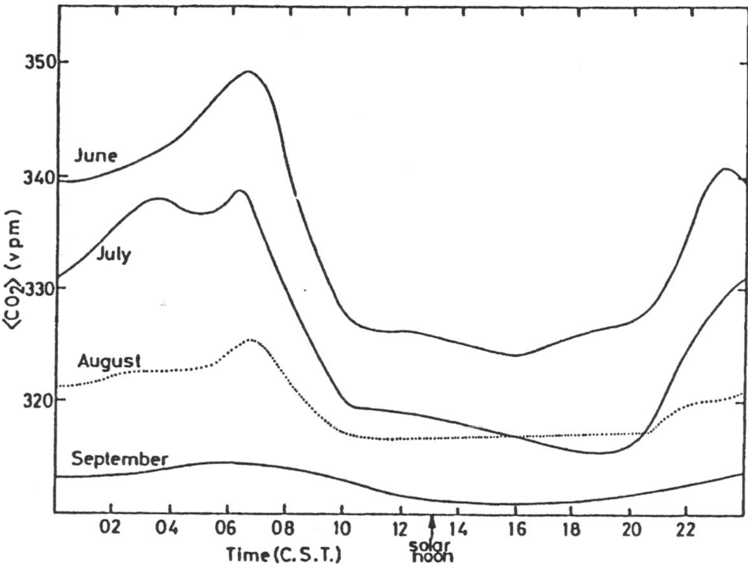

Figure 17-4. Average diurnal variation in CO_2 concentration at 6.5 m over a grassland in southern Saskatchewan in 1971. (After D. L. Spittlehouse and E. A. Ripley [1731])

The vertical distribution of radioactive gases is controlled by the strength of the surface source, the emanation rate of the gases from the soil, the strength of the atmospheric sink, which arises from the radioactive decay rate (half-life) of radium and thorium and their associated daughter products, and the degree of convective and frictional mixing. Other things being equal, a gas with a long half-life will be found at higher concentrations than one with a short half-life. According to R. Mühleisen [1723] the air above land usually contains $20\text{-}400 \cdot 10^{-9}$ p*Ci* l^{-1}, while over the sea, where the surface source is much less, the value is about $1 \cdot 10^{-9}$ p*Ci* l^{-1}. If the Austausch coefficient *A* is taken as 2.3 kg m^{-1} sec^{-1} at 100 m, and the power law for the increase of wind with height applies (Section 16) where $a = 0.14$, then the decrease in content of radioactive gases with height, expressed as a percentage of its value at a height of 1 cm above the ground, according to J. Priebsch [1727], will be as shown in Table 17-2.

Table 17-2. Content (percent of value at 1 cm) of radioactive gases in air.

Height (m)	0.1	1	10	100	1000	13,000
Radon	91	80	66	44	18	3
Thoron	62	19	1	0	–	–
Thorium B	89	74	51	28	4	0

The emanation rate in the layer close to the ground depends on the state of the ground. Wet or frozen ground gives off less than dry ground, and a few centimeters of snow can slow the rate of emanation tremendously. Soil temperature also influences the emanation rate through its influence on soil aeration. The state of the weather, the type of soil (permeability), and especially the nature of the parent material also affect emissions.

The radon concentration in the surface layer is dependent upon four factors: (1) the emanation rate from the soil, which depends upon the condition and temperature of the soil; (2) the atmospheric diffusion rate, which varies with the wind speed, surface roughness length, and atmospheric stability; (3) the advective rate of horizontal wind; and (4) the radioactive decay rate of radon, which has a half-life of 3.825 days.

F. Becker [1704] has measured the amount of radium (radon) emanation in the air layer near the ground. Figure 17-5 shows the diurnal variation for 4-5 April 1943 at heights of 1 m and 13 m above the ground. On the average, as might be expected, the amount is greater at the lower level. The similarity of this graph to that of the daily variation of carbon dioxide content (Figure 17-3) is also a result of reduced mixing at night. There is about a 3 hr lag between the maximum and minimum radon levels at the 1 and 13 m levels.

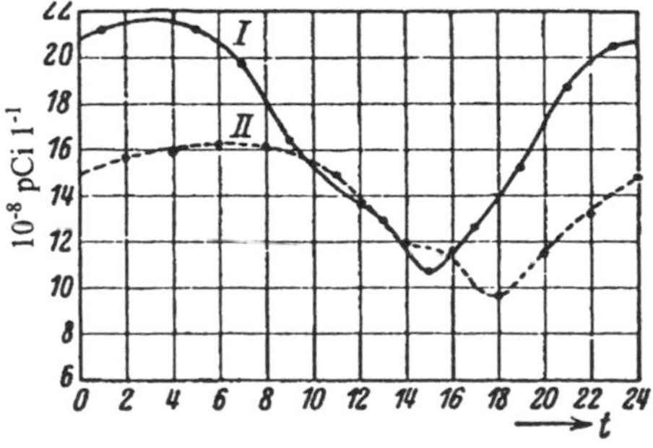

Figure 17-5. Diurnal variation of radon content of the air at 1 m (I) and 13 m (II) above the ground. (After F. Becker [1704])

H. Moses et al. [1722] used a 39.9 m tall instrumented tower at the Argonne National Laboratory in Lemont, Illinois, to provide a detailed investigation of the relationship between meteorological conditions and surface radon concentrations. Their results, shown in Figure 17-6 and summarized in Table 17-3, were obtained for two clear days in May and July which were characterized by strong daytime surface heating and a well-developed nocturnal inversion. A third cloudy day in August had weaker daytime surface heating and a weaker nocturnal inversion.

Table 17-3. Summary of radon concentration measurements (pCi l^{-1}) in air at Argonne National Laboratory in 1958.

	20-21 May			
	0.97 m	5.72 m	23.8 m	39.9 m
Maximum concentration	0.960	0.850	0.650	0.445
Time of maximum	03:00	03:00	02:00	02:00
Minimum concentration	0.053	0.039	0.039	0.039
Time of minimum	10:00	11:00	10:00	10:00
	16-17 July			
	0.032 m	0.97 m	5.72 m	39.9 m
Maximum concentration	3.930	0.928	0.582	0.276
Time of maximum	24:00	24:00	03.00	06:00
Minimum concentration	0.223	0.047	0.038	0.034
Time of minimum	10:00	12:00	10:00	10:00
	27-28 August			
	0.032 m	0.97 m	23.8 m	39.9 m
Maximum concentration	1.070	0.488	0.383	0.355
Time of maximum	23:00	01:00	06:00	06:00
Minimum concentration	-	-	-	-
Time of minimum	-	-	-	-

Because radon emanates from the ground, highest radon concentrations were always found closest to the ground and decreased with height. A direct relationship was found between atmospheric stability and radon concentration. Lowest concentrations at all levels occurred during the day when eddy diffusion was strongest. The time of occurrence of the daytime minimum concentration was nearly uniform for all levels, reflecting intense daytime vertical mixing. The highest concentrations occurred at night due to the reduced mixing associated with the nocturnal inversion. The time of occurrence of the nighttime maximum concentration took place earlier at lower levels, and lagged by as much as 6 hours at 39.9 m, reflecting the deepening of the nocturnal inversion throughout the night. Cloud cover on the night of 27-28 August weakened the nocturnal inversion, resulting in lower nighttime maximum concentrations. In general, an inverse relationship was found between wind speed and radon concentration. Horizontal advection of radon from a stronger source area can be seen in the rapid increase in radon concentration at all levels that took place between 01:00 and 02:00 on 21 May when the wind directions suddenly changed. Observations made by M. H. Wilkening [1734] at 0.8 m

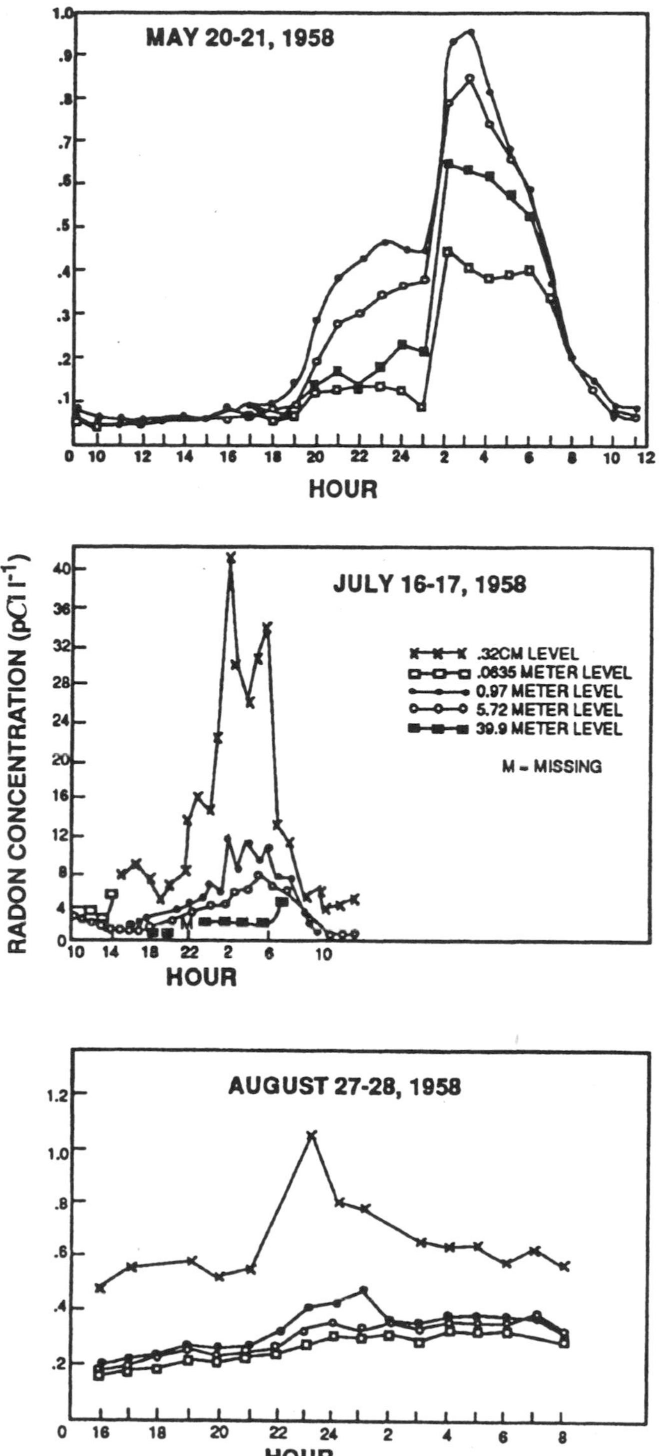

Figure 17-6. Hourly averages of radon concentration at indicated levels on two clear days and one cloudy day during 1958 in Lemont, Illinois. (After H. Moses et al. [1722])

in Socorro, New Mexico, over a 6 yr period, provide information on the average diurnal variation in radon concentration for an annual period.

The reverse distribution is usually found with background levels of surface ozone (O_3). Ozone is naturally formed in the atmosphere and found at its highest concentrations between 15 and 35 km (S. Manabe and R. T. Wetherald [1720]). From these heights, it slowly disperses toward Earth's surface. Near Earth's surface, the average background concentration is typically around 20 ppb. Since the primary source for the natural background surface ozone is diffusion from the stratosphere, ozone concentrations often increase with elevation.

F. Teichert [1732] made comparative measurements for various heights above the ground. During the summer of 1954, 222 separate observations were made at a height of 80 m on a tower and at table height at the foot of the tower. The mean value on the 80 m tower was higher throughout than the value below. For example, in August, with vigorous mixing, it was 37 at the tower and 32 γ m^{-3} (micrograms) on the table below; in September, the figures were 23 and 22 γ m^{-3}. Figure 17-7 is for 7 July 1954, and illustrates a typical daily fluctuation during the warm season. In spite of the unavoidable scatter of the readings made on a single day, it can be seen how low values of ozone occurred at the end of the night, and how, when convection set in, an increase occurred at the top of the tower before it was observed on the ground. The surface maximum occurs in the late afternoon when vertical differences are smoothed out by vigorous mixing. Background ozone concentration decreases sharply following sunset with the reduction of eddy diffusion. The increase occurring before midnight is associated with an increase in wind speed (it started to rain at 01:00).

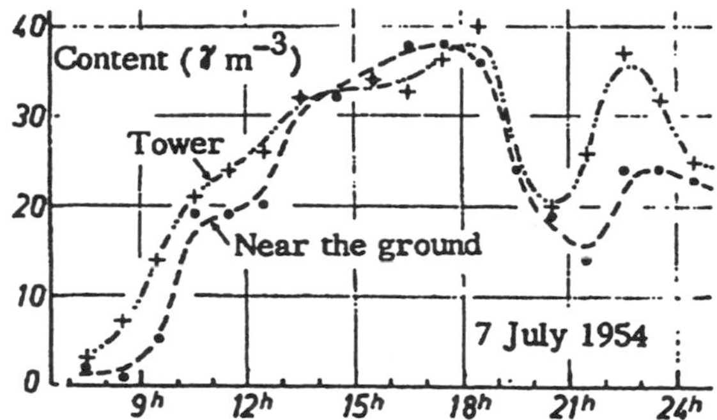

Figure 17-7. Diurnal variation of ozone content at two different heights above the ground. (After F. Teichert [1732])

While most instances of high surface oxidant concentrations are a result of photochemical reactions, a few cases of unusually high concentrations have been reported that are thought to have resulted from stratosphere intrusions associated with unique meteorological conditions. For example, R. G. Lamb [1719] reported that during the predawn hours of 19 November 1972, the air pollution monitoring station at Santa Rosa, California recorded five consecutive hours of high oxidant concentrations. The highest of the hourly averages was 230 ppb. From a detailed analysis of the meteorological conditions surrounding this incident, R. G. Lamb [1719] showed that the ozone responsible for this anomalous concentration originated in the stratosphere and was not a result of anthropogenic sources. Other authors, including R. D. Davis [1706], W. Attmannspacher and R. Hartmannsgruber [1702], and R. Chatfield and H. Harrison [1707], have also reported high surface concentrations of ozone that were at least partially thought to have been of stratospheric origin. It has long been known that ozone diffuses from the stratosphere and is the primary factor responsible for tropospheric background concentrations. R. G. Lamb [1719] and others cited above have shown that under special conditions high concentrations of ozone can be brought near the surface. Y. S. Chung and T. Damm [1708] provide a detailed description of the meteorological conditions which accompanied a peak ozone content of 228 ppb at 20:00 on 27 December 1980 in Regina (50°N), Saskatchewan. A stratospheric intrusion occurred as a result of convergence to the west of an upper-level trough, which produced subsidence behind a cold front. The resulting downdrafts provided a break in the tropopause, which was then intensified by sinking along the forward edge of the trailing anticyclone. S. Wakamatsu et al. [1733] profiled a similar stratospheric intrusion which led to an elevated ground level ozone content in excess of 100 ppb during 10-19 May 1986 in the northern Kyushu district of Japan. Because these stratospheric intrusions result from meteorological conditions in the upper troposphere, rather than surface layer effects, they can produce peak surface ozone concentrations at any time of the day, as indicated by the time of peak occurrence at Regina (20:00).

What is still unknown, and would be very fruitful areas for further research, include: 1) The meteorological or other factors that can affect the rate of ozone transport from the stratosphere, and 2) On a given day with given meteorological or other conditions, how much of the observed surface ozone concentrations are of stratospheric origin. For more information dealing with surface ozone formation, the reader is referred to R. W. Boubel et al. [903]. For meteorological and other factors that influence ozone concentrations in a large city, the reader is referred to R. H. Aron and I-M Aron [902].

18 Optical Phenomena and Acoustics near the Ground

The density of the air layer near the ground is not uniform because of large contrasts of temperature, water vapor content (Tables 11-2, 13-1, and 13-2), and small amounts of eddy diffusion. This vertical density gradient gives rise to optical unsteadiness.

If, on a warm summer day, one looks along a hot country road, over a sandy surface, or along the embankment of a railroad, the lower parts of distant objects appear to flicker, and the edges of objects seem to waver. Usually, optical phenomena like this are observed from eye level. The increase in the temperature gradient, decrease in eddy diffusion and resulting unsteadiness is made evident when one bends down and observes the increased flickering of the layer of air close to the ground. A similar effect occurs when we look at a star. The air above us is in constant motion, which brings air of varying density (temperature) across our line of sight. This causes slight variations in refraction and results in scintillation (twinkling) of the starlight. This is particularly noticeable for stars near the horizon.

Mirages are a result of refraction as light passes through air of varying density. When the lapse rate is $\gamma = 3.42$°C/100 m, the density of air is constant with height, and a light ray will travel in a straight line. During the day close to the ground, however, the lapse rate is often considerably greater than this, thus density frequently increases with height.

When the lapse rate is greater than 3.42°C/100 m, the curvature of the light rays will be in a direction opposite to Earth's curvature. This leads to a mirage in which objects appear below the actual position of the object (inferior mirage). The number of times this happens can be deduced from Tables 13-1 and 13-2. If the lapse rate is only moderately greater than 3.42°C/100 m, sinking may occur. In sinking, the object appears to be below the surface but in an upright position (Figure 18-1). When the lapse rate is substantially greater than 3.42°/100 m, as in the desert during the day, objects may appear twice, once in the actual location and again as an inverted image below the actual object (Figure 18-2). In Figure 18-2, as ray B comes directly toward the observer, it takes a straight path because it travels through air with a constant density (the lapse rate would be 3.42°C/100 m) and has a constant refractive index. Ray B', that left the same point N on the pole, travels downward and is refracted upward, and can also be seen by the observer. Thus, the same position of the pole would be seen twice. First, it would be seen when the eyes are directed at point N, and second when the eyes are directed downward toward point N'. Ray A' leaving from point M will be refracted by a greater amount as it passes through a layer of very hot air near the ground, and appears to come from point M'. The effect will yield both an upright and an inverted image with some vertical distortion. Only one ray enters the eye from point P. Rays originally leaving toward lower angles are refracted sharply upward and never reach the observer. This height marks the vanishing point below which no portion of the object can be seen because none of the rays reach the observer. As objects become more distant, the vanishing point becomes higher and very distant objects may not be seen at all (from S. Williamson and H. Cummins [1826]). Following ray A' beyond the telephone pole, one would see the sky. Therefore, one would see the sky both by looking up and down to point M'. Thus, the blue mirage on the desert floor or on an asphalt road is actually an inferior mirage. The blue of this mirage is usually, however, somewhat different in color than the blue of a lake.

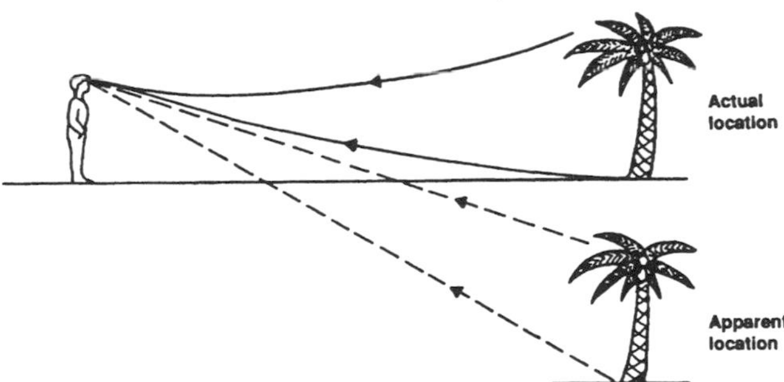

Figure 18-1. Sinking occurs when the lapse rate is moderately greater than 3.42°C/100 m.

We usually associate inferior mirages with sunny days and high temperatures as in the desert or on the asphalt road. However, the presence of this type of refraction does not depend on the actual temperature but only on change of density. J. Fényi [1805] observed this type of mirage with temperatures below 0°C and K. Brocks [1803] showed it could occur, although not as strong, on cloudy days. Figure 18-3 shows the diurnal variation in the coefficient of refraction for a mid-latitude climate, for a horizontal line of sight. The values were obtained by K. Brocks [1803] for heights up to 300 cm, for

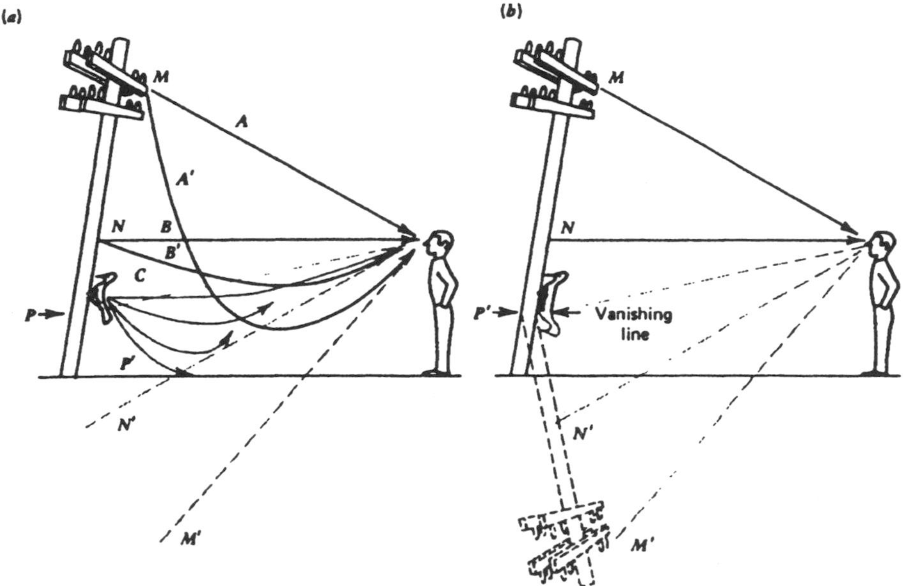

Figure 18-2. An inferior mirage occurs when the lapse rate is substantially in excess of 3.42°C/100 m. (After S. Williamson and H. Cummins [1826] Reprinted by permission of John Wiley & Sons, Inc. Copyright © 1983)

clear and overcast days in June. Anomalous negative curvature of the observed light path reaches 50 times the curvature of Earth shortly before noon on clear days at a height of 10 cm. The change to a curvature of 0 does not take place until about 12 m above the ground. Even with overcast skies, the curvature is negative during most of the day but becomes zero at a height of about 10 m during midday. Because of reduced surface heating on overcast days, however, the degree of curvature is naturally less. Positive coefficients occur when the lapse rate is less than 3.42°C/100 m. They typically begin to occur from an hour or two before sunset to an hour or two after sunrise and are strongest on clear nights with calm conditions. Positive coefficients are more common at low sun angles (late autumn and early winter), and negative coefficients are strongest at high sun angles. Tables 11-1, 11-2, 13-1 and 13-2 suggest the variation of positive and negative coefficients throughout the day and year at various heights.

When the lapse rate is less than 3.42°C/100 m, a superior mirage may occur. In this case, the curvature of the light rays will be in the same direction as Earth's curvature. The refractive index will be greatest near the surface and decrease with height (Figure 18-3). These mirages are called superior because the image is seen above its true position. Superior mirages are most dramatic with strong inversions such as in polar regions during summer, over cool ocean currents, ice covered lakes in spring, or when warm air is advected over a snow covered surface.

The classic view of a superior mirage is of a ship looming or suspended above the horizon (Figure 18-4). In this case, the light rays coming from M and N are bent so as to appear to come from M' and N'. However, even when there is no inversion, densities frequently decline with increased height (lapse rate less than 3.42°C/100 m). Since human per-

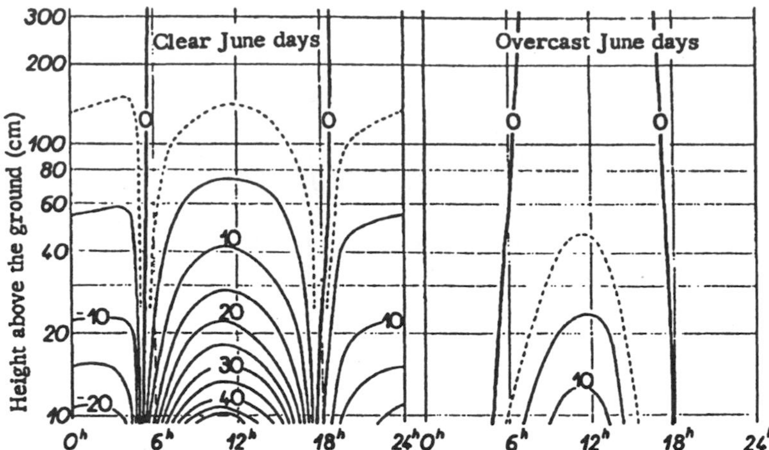

Figure 18-3. Refraction of light rays in the layer near the ground on clear and cloudy days in June. Positive values indicate a lapse rate greater than 3.42°C/100 m and a curvature of light in a direction opposite to the earth's curvature. The values are the coefficient of refraction which is the ratio of the radius of curvature of the light to the radius of curvature of the earth. (After K. Brocks [1803])

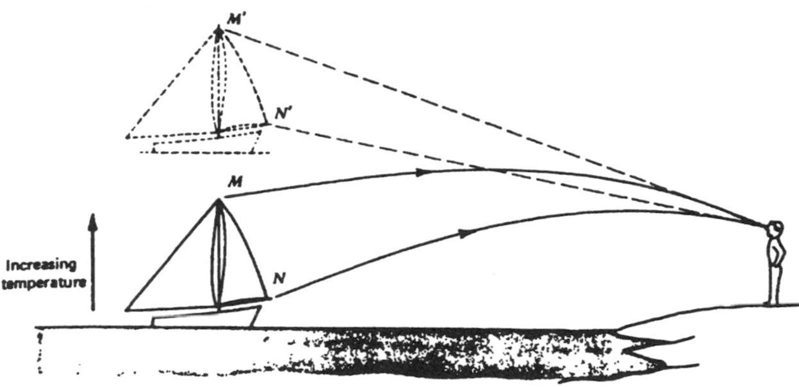

Figure 18-4. A superior mirage. (After S. Williamson and H. Cummins [1826] Reprinted by permission of John Wiley & Sons, Inc. Copyright © 1983)

ception is based on the assumption that a light ray travels in a straight path, objects may frequently appear to be slightly higher than they actually are. This illusion would be frequent for a couple of hours around sunrise and sunset, on windy days, or on days with clouds and medium to low angles of incident solar radiation. Sometimes a superior mirage will result in a stretching of the image. Figure 18-5 is two photographs taken by W. H. Lehm and I. Schroeder [1813] at the edge of Lake Winnipeg, Manitoba on May 28, 1980 at 15:15. The surface of the lake was still mostly covered by ice, and a warm breeze was blowing from the land toward the lake. This resulted in a strong inversion and the resulting superior mirage. W. H. Lehm and I. Schroeder [1813] quite convincingly further suggest that mermen and mermaids are not, as many suggest, simple sailors' (with too many days at sea) view of a dugong or manatee, but a superior mirage. Monsters at sea or in Loch Ness may have a similar

a.

b.

Figure 18-5. a. Lake Winnipeg, May 28, 1980. The boulder in the center foreground was identified as the source of the image in Figure 18-5b. Figure 18-5a photo was taken at 15:14. The boulder is about 68 cm wide and 35 cm high. b. Photo was taken at 15:17. The boulder is distorted into a merman shape. The image subtends 5.4' vertically and 2.2' horizontally. (After W. H. Lehm and I. Schroeder [1813]. Reprinted with permission from *Nature*, Vol. 289, pages 362-366 copyright © 1981 Macmillan Magazines Ltd.)

origin. Fata Morgana or Castles in the Sky are also superior mirages. Fata Morgana can transform a fairly uniform terrain to one with vertical cliffs and floating islands.

Another situation in which an object's size is magnified is called towering. In this case, the layer near the ground with a lapse rate greater than 3.42°C/100 m is relatively shallow and the lapse rate increases as the ground is approached. In this case, the bottom of the object will be displaced downward more than the top, resulting in a downward stretching of the image (inferior mirage). Above eye level the lapse rate is less than 3.42°C/100 m (superior mirage), resulting in the top of the object being stretched upward. Towering can be quite spectacular. Minor objects can appear to have attained great size.

Varying densities in the atmosphere can lead to numerous additional variations of the basic types of mirages. For those who wish to pursue this topic further see Section 18, Bibliographic entries.

The temperature, lapse rate, wind speed and direction, and relative humidity affect acoustics in the atmosphere. This discussion will focus both upon factors that affect speed and the distance sound can be heard.

Due to turbulence sound intensity fluctuates near the ground. These fluctuations increase with increased frequency. For example, over a distance of 40 m at 0.25 kHz the intensity was almost uniform, but at 4 kHz it fluctuated as much as eight fold within a few seconds (H. Neuberger [1818] and L. P. Delasso and V. O. Knudsen [1804]). H. Neuberger [1818] referred to this fluctuation as "acoustical scintillation."

Air temperature, relative humidity, lapse rate, and wind direction affect the speed of sound. The equation $V_S = 20.06\sqrt{T}$ expresses the variation of the speed of sound with temperature where V_S is the speed of sound in m sec^{-1} and T is the temperature in Kelvin. Table 18-1 shows the speed of sound at various temperatures. The change in the speed of sound associated with changes in relative humidity is shown in Table 18-2. For more information on the relation between relative humidity and the speed of sound, refer to A. D. Pierce [1820]. Table 18-3 shows the change in the speed of sound as a result of temperature and humidity changes combined.

Table 18-1. Speed of sound in dry air at various temperatures. (After D. A. Bohn [1802])

Temperature (°C)	Speed (m sec^{-1})
0	331.45
10	337.46
20	343.37
30	349.18
40	354.89

Table 18-2. Percentage increase in speed of sound (compared to 0°C) due to variation in relative humidity. (After D. A. Bohn [1802])

Temperature (°C)	Relative humidity (%) 10	20	30	40	50	60	70	80	90	100
5	0.014	0.028	0.042	0.056	0.070	0.083	0.097	0.111	0.125	0.139
10	0.020	0.039	0.059	0.078	0.098	0.118	0.137	0.157	0.176	0.196
15	0.027	0.054	0.082	0.109	0.136	0.163	0.191	0.218	0.245	0.273
20	0.037	0.075	0.112	0.149	0.187	0.224	0.262	0.299	0.337	0.375
30	0.068	0.135	0.203	0.272	0.340	0.408	0.477	0.546	0.615	0.684
40	0.118	0.236	0.355	0.474	0.594	0.714	0.835	0.957	1.08	1.20

Table 18-3. Total percentage increase in speed of sound (compared to 0°C) due to temperature and humidity combined. (After D. A. Bohn [1802])

Temperature (°C)	Relative humidity (%)					
	0	30	40	50	80	100
5	0.91	0.952	0.966	0.980	1.02	1.05
10	1.81	1.87	1.89	1.91	1.97	2.01
15	2.71	2.79	2.82	2.85	2.93	2.98
20	3.60	3.71	3.75	3.79	3.90	3.98
30	5.35	5.55	5.62	5.69	5.90	6.03
40	7.07	7.43	7.54	7.66	8.03	8.27

Next we focus on the consequences of the change in the speed of sound caused by variations in temperature near the ground. The speed of sound increases with increasing temperature (Table 18-1). When the air is isothermal, the speed of sound is constant with height and sound travels in a straight line. When the temperature decreases with height, the speed of sound also decreases. Sound waves will be refracted upward and will not carry (heard at ground level) as far. If, however, temperatures increase with height, the speed of sound will also increase with height, and sound waves will be refracted downward, and can be heard at greater distances. The stronger the inversion the further the sound waves will carry. The sound of a plane or rocket taking off in the morning may decrease quickly as it passes through the surface inversion (R. G. Fleagle and J. A. Businger [2006] and H. Neuberger [1818]).

Wind can either retard or enhance the speed sound travels. If sound is traveling in the same direction as the wind it will progress faster (downwind). Since wind speed increases with height, sound waves that move in the direction of the wind are refracted downward. This refraction generally increases with increasing wind speed (K. B. Rasmussen [1819]). If sound is moving against the wind (upwind) its speed will decrease with height. Sound waves moving against the wind are refracted upward; this upward bending increases with both wind speed and frequency. If the wind and distance are sufficient no sound will reach the observer (K. B. Rasmussen [1819]). Thus, coastal foghorns are audible at great distances out to sea with offshore winds, but inaudible at short distances with onshore winds (R. G. Fleagle and J. A. Businger [2006]). Near the ground, large changes in both wind speed and temperature can result in large changes in the speed and refraction of sound. H. Klug [1808] measured the speed of sound near the ground. He confirmed that the speed of sound depends on the lapse rate and wind direction (neglecting the minor variations due to relative humidity). He found that with an inversion or when sound waves are being carried downwind they are refracted downward. This results in less attenuation of sound near the ground and the carrying of sound waves a further distance.

Around the turn of the 20th century it was noted that sound from explosions from volcanoes or man-made sources propagated anomalously. Near the source the sound is heard. This is surrounded by a zone of silence followed by a zone where the sound could be heard again (zone of anomalous audibility). There can be several of these oscillating zones (H. Neuberger [1818]). D. Thomson [1823] suggested that this anomaly was because "Normally the ambient sound speed gradients are not constant. The consequence of inconstancy is that there are zones of shadow and enhancement produced. Thus, asymmetric "rings" of relatively enhanced impact are formed. These rings may be incomplete when one takes the refractive effects of the vector wind into account ..."; with very loud explosions the zone of silence may be eliminated by the sound wave moving along the ground.

Loud sounds from explosions or volcanoes can propagate via the stratosphere (this was first proposed by F. J. W. Whipple [1825]. Thus, at great distances the sound may be heard twice, once from the direct wave traveling along the ground and again from propagation via the stratosphere. An example of this was the volcano Krakatau, which erupted on the 27 August 1883. The sound from this eruption was heard great distances (as far as Rodriguez in the Mascarene Islands 4775 km away (O. Meisser [1816])) and sent out a pressure wave that had an amplitude of 67 millibars near the volcano and in Europe the amplitude was still 1.7 millibars (H. Neuberger [1818]).

Divergence of sound causes a reduction in sound intensity due to the spreading of the wave through the air. The intensity of a sound wave is inversely proportional to the square of the distance from the source. In addition to divergence and the dispersion of sound due to wind and temperature gradients, several properties of air combine to attenuate sound waves. One results from molecular absorption and dispersion by polyatomic gases involving an exchange of energy between colliding molecules. Others are due to viscosity and heat conduction (L. L. Beranek [1801]).

Absorption has an inverse effect on sound propagation. Table 18-4 gives the total absorption of sound by the atmosphere at different frequencies and relative humidities. Table 18-5 shows the increase in sound absorption due to changes in relative humidity as a function of frequency.

Table 18-4. Total sound absorption in dB/km versus relative humidity as a function of frequency at 20°C. (After D. A. Bohn [1802])

Frequency	Relative humidity (%)										
(kHz)	0	10	20	30	40	50	60	70	80	90	100
2	4.14	38.2	17.4	10.9	8.34	7.14	6.55	6.28	6.19	6.21	6.29
4	8.84	102	62.3	38.9	28.0	22.2	18.7	16.6	15.2	14.2	13.6
6.3	14.9	154	135	90.6	65.6	51.3	42.5	36.7	32.7	29.8	27.7
10	26.3	202	261	205	155	123	102	87.3	77.0	69.3	63.5
12.5	35.8	224	338	294	232	187	156	134	118	106	96.6
16	52.2	250	428	423	355	294	248	214	189	170	155
20	75.4	281	511	564	508	435	374	326	289	261	238

Snow is the most absorptive naturally occurring ground cover (D. G. Albert [1800]). Absorption of sound varies with sound frequency and snow thickness (Table 18-6). The coefficients in Table 18-6 represent the fraction of the sound incident to the surface which is absorbed. However, since sound is measured logarithmically, at 0.5 kHz with 10 cm of snow and 90 percent absorption, for example, a sound wave with an intensity of 60 dB would be

Table 18-5. Increase in sound absorption in dB/km, due to relative humidity as a function of frequency compared with absorption at a temperature of 20°C and a relative humidity of zero percent. (After D. A. Bohn [1802])

Frequency	Relative humidity (%)										
(kHz)	0	10	20	30	40	50	60	70	80	90	100
2	45.7	34.1	13.3	6.76	4.20	3.00	2.41	2.14	2.05	2.07	2.15
4	65.6	93.2	53.5	30.1	19.2	13.4	9.86	7.76	6.36	5.36	4.76
6.3	71.2	139	120	75.7	50.7	36.4	27.6	21.8	17.8	14.9	12.8
10	74.7	176	235	179	129	96.7	75.7	61.0	50.7	43.0	37.2
12.5	77.2	188	302	258	196	151	120	98.2	82.2	70.2	60.8
16	79.8	198	376	371	303	242	196	162	137	118	103
20	84.6	206	436	489	433	360	299	251	214	186	163

Table 18-6. Absorption coefficients of snow as a function of frequency and snow thickness. (After G. W. C. Kaye and E. J. Evans [1809])

Frequency (kHz)		0.125	0.250	0.500	1.0	2.0	4.0
Snow Cover	2.5 cm	0.15	0.40	0.65	0.75	0.80	0.85
	10 cm	0.45	0.75	0.90	0.95	0.95	0.95

reduced to 50 dB. D. G. Albert [1800] found peak pulse amplitudes at 0.5 kHz decayed much faster over snow than grass, recording an order of magnitude difference at 100 m.

Two other atmospheric acoustic phenomenon of interest are thunder and eolian sounds. Thunder is a result of the rapid heating of the air by lightning, which produces a compression wave. Thunder often has a rolling or rumbling sound. This is because each point along the path of a lightning stroke is a source of sound. Since different parts of the lightning bolt are different distances from the observer, and travel through air with different characteristics, such as wind speed and direction, the arrival times and intensity of the sound waves vary accordingly. In addition, a lightning stroke appears to flicker because each bolt is actually made up of from 2-27 individual strokes. Each individual stroke produces its own thunder. When wind moves past objects such as telephone wires, ears or tree limbs, eddies form in the wake of the obstacle. The frequency of this sound is directly proportional to the velocity of the wind and inversely proportional to the size of the eddies. Thicker wires or branches produce larger eddies.

For further information on environmental effects on sound propagation D. A. Bohn [1802], K. B. Rasmussen [1819], U. Kurze and L. L. Beranek [1812], H. Klug [1810], A. D. Pierce [1820], and H. Neuberger [1818] are recommended.

CHAPTER III

INFLUENCE OF THE UNDERLYING SURFACE ON THE ADJACENT AIR LAYER

19 Soil Type, Soil Mixtures, and Soil Tillage

The extreme diversity of microclimates can be attributed to the varied nature of the surfaces underlying the air layer near the ground. The surface may be solid ground with or without vegetation, a snow cover, a water surface, or ice in the form of frozen lakes or glaciers. In Sections 19 to 21, the discussion will be limited to solid ground without vegetation. The laws governing conduction of heat in the soil have been dealt with in Section 6, and the average temperature variation in the soil in Section 10.

In spite of the great contrasts in the distribution of heat, water vapor, wind, particulates, and so forth, pointed out in Chapter II, it is much simpler to achieve a balance in air than in the ground. The ground consists of particles of different soil types, differing in density, grain size, thermal conductivity, and water content, which also varies with time. Embedded stones, tree roots, dead organic matter, earthworms, soil organisms, and water passages all combine to transform the soil into a veritable mosaic. Many samples must be taken before it is possible to determine the type of soil or its representative moisture content over an area of any size. The same is true for soil temperature. In contrast to methods used to measure air temperature, simultaneous readings must be taken at several neighboring points to ensure adequate sampling of the area. Where plants are growing, "the whole volume of the ground," as G. Winter [1917] writes, "becomes split up into smaller areas each centered about a root. Every plant residue that penetrates into the soil creates about itself, by the inhibiting and activating substances it contains and other specific components, a special set of conditions for the existence of soil microflora, an effect that becomes further complicated by the interaction of colonizing bacteria."

These minuscule variations in soil properties will be ignored in the discussion that follows. Only typical and homogeneous soil types will be dealt with because the influence of the ground on the air overlying it is quite complicated. All factors involved in the energy balance of Earth's surface (Section 3) play a part in shaping the distribution of energy in the soil and in the air layer near the ground.

In the energy balance equation $Q^* + Q_G + Q_H + Q_E = 0$ (Section 3), in which $Q^* = S + D + L{\downarrow} - L{\uparrow} - K{\uparrow}$ (Section 4); only direct-beam solar radiation S and the longwave counterradiation $L{\downarrow}$, are independent of the nature of Earth's surface. The nature and state of the soil influences all other factors. The albedo of the ground surface directly affects the reflected solar radiation $K{\uparrow}$ and indirectly affects diffuse solar radiation D, because atmospheric backscattering depends in part on the surface albedo. Because of the great importance of soil surface characteristics, they will be dealt with in greater detail in Section 20.

The quantities Q_G and Q_E depend on the properties of the ground as well as on its temperature and surface roughness length. The thermal conductivity λ and the thermal capacity $(\rho c)_m$ will be dealt with in the present section in so far as they are determined by the nature and state of the ground. The water content of the soil, which also affects conduction Q_G, evaporation Q_E, and the surface albedo, will be treated separately in Section 21.

The quantities $L\uparrow$ and Q_H of the energy balance depend primarily on the soil's surface temperature. For a given level of longwave counterradiation $L\downarrow$, the nocturnal longwave radiation loss is less when the soil surface temperature T is lower. For example, F. Sauberer [1914] found that the radiation balance Q^* for summer nights with three-tenths cloud cover had the following values: hard roadway, 95 W m^{-2}; sandy soil, 72; bare ground 68; meadow, 50. The first two (possibly drier) surfaces give off more of their large energy storage by longwave emission because of their higher surface temperature than the poorly conducting, cooler (and probably also wetter) surfaces. Since the surface of earth actually emits as a gray body rather than as a black body (Section 4), surface longwave emission shows a small dependence upon microscale variations in the surface emissivity ε, a surface radiative property that also exhibits small variations with surface water content.

The factor Q_H shows a similar response to the soil's surface temperature. The hotter the surface becomes in the middle of the day, the greater the temperature contrast of the ground surface and air, and therefore the stronger the conduction and convective mixing, and the greater the quantity of energy transferred from the ground surface to the adjacent air. Similarly, by night, the colder parts of the ground surface will have more energy transferred to them from the air through conduction and convective exchanges than the warmer parts. Like evaporation Q_E, convection Q_H shows a minor secondary dependence upon the surface roughness length z_0.

This survey shows how difficult it is to isolate the pure influence of the soil itself upon the surface microclimate, and how carefully all these quantities mentioned must be assessed in trying to determine the causes for temperature measurements.

As already indicated in Section 6, every soil consists of three elements: the soil substance, water, and air. The water content, which varies with rainfall, snowmelt and irrigation, and the air enclosed in the soil, varies with the soil structure and the state of cultivation. The influence that the type of soil exerts on the energy balance of the ground, and hence on that of the adjacent air layer, will now be considered.

To get an idea of the influence of the type of soil, apart from all other factors, a comparison may be made between soils with high and low thermal conductivity, by assuming that both absorb equal quantities of energy at the surface during the day, and give off equal quantities of radiation by night (equal values of Q^*). According to Section 6, the daily and the annual energy variations penetrate deeper into soils which are good conductors (Table 6-3) than into soils which are poor conductors. Correspondingly, at the surface, the daily maxima will be lower for a soil with a high thermal conductivity, and the daily minima higher. Thus, the surface of the soil with high thermal conductivity will have more uniform temperatures; by analogy with macroclimatologic processes, one might say it has a more marine climate. Plants in a soil with high conductivity are not stimulated to premature growth by too much heat during the day, and the higher night temperatures experienced for the same reasons reduce the danger of late frosts. A further consequence is that soils with high conductivity give off less energy during the day to the adjacent air, and, therefore, the Q_H term in the energy balance is smaller during the day and larger at night than a soil with a low thermal conductivity. For good conducting soils, more of the radiant energy received is retained within the soil; the average temperatures are higher and diurnal temperature fluctuations penetrate deeper into such soils than into poorly conducting soils.

These statements are supported by the comparison of granite and sandy heath illustrated in Figure 19-1. This is a classical experiment carried out in 1893 in Finland by T. Homén [1905]. Simultaneous temperature measurements were made in three different kinds of soil. For an average of three clear August days, the tautochrones of the warmest and the coldest hours are shown, and the curves are extended beyond the surface to indicate the temperature extremes measured at the level of the top of the grass cover.

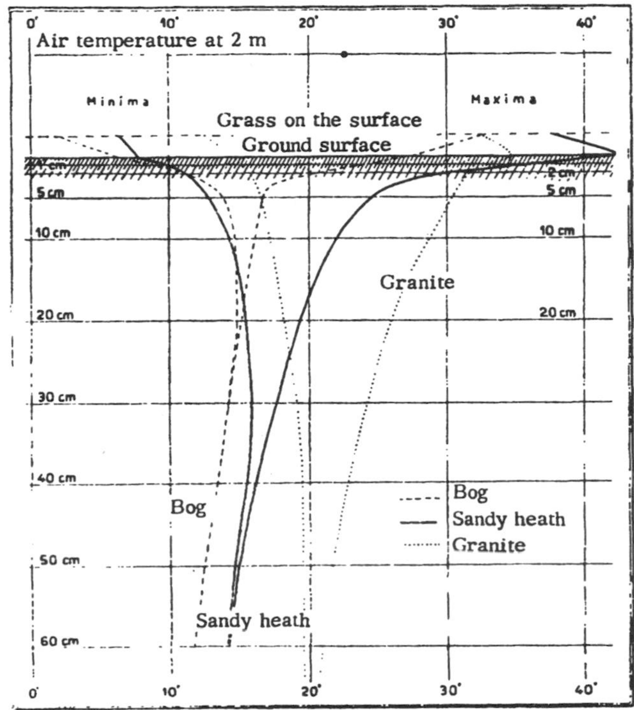

Figure 19-1. Average range of temperature fluctuation in three different soils on three clear August days in Finland. (After T. Homén [1905])

Granite, which is an excellent conductor, shows measurable diurnal variation at a depth of 60 cm. It is warmer, on the average, than the other soils, but in the surface layer the maximum is nearly 10°C lower, and the minimum 5°C higher, than in the poorly conducting sandy soil. The bog soil conforms only partially to the rules mentioned. It could be recognized as the poorest conductor of these three soils by its generally lower temperature, shallow penetration of diurnal temperature fluctuations, which hardly get beyond 20 cm, and by the steep increase of diurnal fluctuations in the top few centimeters. However, bog soils, which are rich in moisture, have substantial energy losses by evaporation, which accounts for its surface being 15°C cooler than the sandy soil.

There have been many attempts to elucidate theoretically what effect soil type has on the temperature decrease at night. One theoretical approach, by H. Philipps [1912], has been amplified by H. Reuter [1913]. This theory takes account of the interaction of exchange, air-temperature stratification, and the nature of the soil on the modification of ground surface temperature at night. As before, A (kg m^{-1} sec^{-1}) is the Austausch coefficient and γ (°C/100 m) the lapse rate, both of which had to be considered as constant with height in order to carry out the calculations; ρ (kg m^{-3}) is the air density, and c_p (J kg K^{-1}) is the specific heat at constant pressure; λ (W m^{-1} K^{-1}) is the thermal conductivity and $(\rho c)_m$ (J m^{-3} K^{-1}) is the thermal capacity. If Q^* (W m^{-2}) is the radiation balance of the ground surface, which is negative at night, the change of temperature dT in time t, according to H. Reuter, would be given by the characteristic equation:

$$dT = \frac{2}{\sqrt{\pi}} \frac{(Q^* + \gamma c_p A)}{\sqrt{\lambda(\rho c)_m} + c_p\sqrt{A\rho}} \sqrt{t} \qquad (°C)$$

At night, Q^* is negative, and, therefore, dT is normally also negative. It may happen, however, that with large values of A and $-\gamma$ (strong inversion), the expression in parenthesis in the numerator may become zero and, therefore, $dT = 0$. In this situation, there is a flow of energy from the air above the ground down to the surface, brought about by intense mixing, which is equal to the energy loss from the surface by radiation. The decrease in surface temperature is proportional to the square root of time. Thus, after 4 hours of night conditions, dT is twice and 9 hours later, three times as high as at the end of the first hour. To determine the influence of the remaining factors, the drop in temperature during a 10 hr night for four types of soil, four temperature gradients, and two different exchange coefficients were calculated. The results are given in Table 19-1.

Table 19-1. Calculated nocturnal drop of temperature (°C) at the ground surface during 10 night hours. (After H. Reuter [1913])

Austausch coefficient A			1				10			
Temperature lapse rate °C/100 m			+1	0	-1	-3	+1	0	-1	-3
Type of ground	Thermal conductivity λ	Thermal capacity $(\rho c)_m$								
Rock	46.1	2.18	5.1	5.1	5.0	4.9	4.7	4.2	3.7	2.7
Wet sand	16.8	1.68	8.9	8.8	8.7	8.5	7.1	6.4	5.6	4.1
Dry sand	1.7	1.17	22.5	22.3	22.0	21.5	12.3	11.0	9.7	7.0
Bog soil	0.6	0.38	35.2	34.8	34.4	33.5	15.5	13.8	12.2	8.8

Thus, besides atmospheric moisture (Figure 5-3) and clouds (Section 5), four factors affect surface nocturnal cooling. Of these, the two most important are the soil's thermal capacity $(\rho c)_m$ and thermal conductivity λ. In soils with high thermal capacity $(\rho c)_m$, temperature increases slowly during the day, and since their surface temperature remains lower, less energy is lost by $L\uparrow$. At night, these soils may give off substantial amounts of energy as they cool. Likewise, soils with higher thermal conductivity (λ) can absorb larger amounts of energy during the day, much of which returns to the atmosphere at night. Thus, the higher a soil's thermal capacity and conductivity, the more energy that can flow from within the soil toward the surface, and despite a larger loss of energy to the atmosphere, its surface nocturnal temperature drop will be less.

The remaining two factors that can affect the surface nocturnal temperature drop are the amount of atmospheric circulation as measured by its Austausch coefficient and the lapse rate. Increased mixing at night usually results in warmer air from aloft being brought to the surface, thereby retarding surface cooling. The smaller the lapse rate (increased stability) the greater will be the atmospheric counterradiation (Figure 5-3) and the more energy that may be brought to the surface by increased mixing. Thus, both increased mixing and a smaller lapse rate will retard surface nocturnal cooling (Table 19-1).

The water content of the soil will, by altering its thermal conductivity and capacity (Section 20), also affect the nocturnal temperature drop. This can be seen in Figure 31-4 in which the wet soil remains warmer than dry soil because of its higher thermal conductivity and capacity. The grass surface experienced the largest nocturnal temperature drop due to its low thermal conductivity.

From these discussions on the different properties of soil, it can be concluded that adding constituents in cases where conditions are disadvantageous can bring improvements. Everyone who owns a garden has already made use of this process when they have mixed peat into a heavy soil or added sand to a rather compact loam.

The process of adding constituents is used for improving bog soils perhaps more than any other type of soil. This is not a question of putting a layer of sand on top (Section 20), but of mixing it in. In Japan, R. Yakuwa [1918] mixed peat and loam in five different proportions, placed the mixtures in 16 m^2 boxes, and measured the temperature variation to a depth of 20 cm during two summer months. In 1951, in the Donaumoos, H. Kern [1906] compared an old cultivated bog with a neighboring soil, which had a 50 percent mixture of sand. Y. Pessi [1910] made a detailed comparison of four mixtures containing different proportions of sand and peat soil, each with an area of 100 m^2, over the years 1952-1954 in central Finland (Pelsonsuo, 64°N). Table 19-2 has been extracted from these sources. Adding sand to the soil resulted in more heat being absorbed, increasing the thermal diffusivity, and reducing evaporation. This resulted in an increase in the mean and minimum temperatures. The difference was greatest in spring and early summer.

W. Baden and R. Eggelsmann [1901] made observations in the air layer over an uncultivated bog and a meadow reclaimed from this bog soil by 40 years of cultivation in Königsmoor near Lüneburg. Table 19-3 gives monthly means from readings taken daily at 14:00 at a height of 5 cm above the ground and of the night minimum between 8 May and 17 August 1951. As might be expected from the previous results, temperatures above the bog were more extreme, being warmer by day and colder by night. The relative and absolute humidity of the air was also lower over the bog. From continuous measurements of the water content, it appeared that the uncultivated bog soil consisted of 80-90 percent water. It, however experienced less evaporation than from the grass covered meadow, where the water content varied much more with changing weather and, in the hot days of summer, dropped as much as 50 percent by volume. The air above the meadow was, therefore, more humid and had a higher water vapor content. The evaporation was greater over the meadow because of the larger evaporating surface area of the grass as opposed to the uncultivated bog. As shown in Table 19-2, in reclaiming the meadow its thermal diffusivity increased, resulting in cooler temperatures during the day but increasing them at night.

Table 19-2. Effect of adding sand to bog soil.

Quantity measured	Depth (cm)	Quantity of sand added to bog soil (m^3 ha^{-1})			
		0	200	400	800
Thermal conductivity λ (W m^{-1} K^{-1})	0-10	3.4	4.5	5.3	7.3
Thermal diffusivity *a* (10^{-7} m^2 sec^{-1})	0-10	1.17	1.45	1.78	2.58
Heat absorbed by ground in 24 hour (MJ m^2) (9 June 1953)		3.31	3.52	3.60	3.98
Mean temperature (°C) June 1953	5	15.7	16.0	16.0	17.0
	20	7.1	10.7	11.1	12.9
July	5	16.0	16.6	16.5	17.3
	20	12.5	13.8	14.2	15.3
Temperature minima (°C) within plant cover (oats), mean of					
14 frost nights in May 1951		-3.6	-3.2	-2.9	-2.7
10 frost nights in June 1951		-3.9	-2.7	-2.6	-2.0
Mean snow cover (cm) in 1954					
January		17	24	21	18
February		26	32	30	19
March		27	32	28	28
17 April		20	20	15	10

Cultivation also changes the nature of the soil since it increases the proportion of air in it, and, therefore, reduces its ability to conduct heat. Gardeners and farmers, therefore, should avoid loosening the soil in spring, when there is danger of frost, in order not to increase the diurnal temperature fluctuations at the surface, which would lower the night minima. K. Bender [1902] writes, "I will always remember how, after a night of frost in a potato field, the plants in the part that was weeded the day before suffered the most frost damage without exception, while the part that, fortunately in this case, had not been worked showed no signs of damage." In a forestry report, R. Geiger noticed the advice that weeds in the ground around young oaks should not be cleared in autumn, since this would increase the risk of damage to the shoots of the sensitive young plants. In farming, when there is a danger of

Table 19-3. Monthly air temperature (°C) and humidity 5 cm above an uncultivated bog (B) and a reclaimed meadow (M).

	Air temperature						Air humidity			
Month	Maxima		Minima		Diurnal Variation		Relative (percent)		Vapor pressure (mb)	
	B	M	B	M	B	M	B	M	B	M
May	23.6	21.4	2.9	3.4	13.3	12.4	65	79	11.19	13.73
June	22.4	20.8	5.6	6.2	14.0	13.5	63	77	17.07	20.27
July	23.5	20.5	8.0	8.2	15.8	14.4	71	84	19.73	20.40

spring frost, grain should not be harrowed, nor should the soil be piled up around potato plants.

In a given soil with large grains, such as sand, there will be a few large pores or voids, while a clay soil, which has small grains, will have many small pores. In contrast to what a person might think, the proportion of air in the soil decreases with increasing grain size. D. A. de Vries [1916] showed that the dependence of thermal conductivity on grain size can be arrived at theoretically, by assuming that soil particles are ellipsoidal in shape. Using his results, R. H. A. van Duin [1903] calculated to what extent the thickness of the loosened layer of the surface soil would affect the diurnal temperature fluctuation. The fluctuation rises sharply at first and then reaches a constant value when the depth of loosening reaches 10 cm. Bearing in mind the danger of night frosts, including alterations in the energy flow, we can estimate that loosening of the upper soil layer to a depth of 2 cm, in a midlatitude climate, would reduce the night minimum by 2°C for moist soil and by 3°C for dry soil. During the course of a year, tilling of the soil in autumn and winter seems to lower the mean temperature of the plowed layer, but to raise it in spring and summer.

W. Schmidt [1915] found as long ago as 1924 that, at the end of an August night, the surface temperature of a freshly plowed field was 2°C colder, and at a depth of 10 cm, 1°C colder than in an untilled field. During the day, the surface temperature of the loosened soil at 15:00 was 5.5°C warmer, because of its reduced thermal conductivity. It follows that increasing the density of the surface layer will help in guarding against frost. This is why soils are rolled (a heavy roller is moved over the soil to compact it). The effect of several types of rollers has been investigated in Sweden by A. Olsson [1909], by taking temperature readings 5 cm above and 10 cm below the surface of three different soils. The comparative measurements of Y. Pessi [1911] in Finland showed that in addition to increasing soil temperature by 1.5°C, there was also an increase in the yield of oats.

The energy budget of the soil, finally, can be improved by heating it artificially. A. Morgen [1907] showed how difficult it is to distribute heat evenly in soil, if an economical number of heating ducts is to be used. The increase in yield that resulted from raising the temperature at root level was of the order of 10 percent for a 1°C increase in temperature.

20 Ground Color, Surface Temperature, Ground Cover (Mulching) and Greenhouses

Natural surface albedos have already been listed in Table 4-2. Soil moisture affects both the amount of evaporation and the surface albedo (Figure 4-2). This, together with the color of the surface, affects the amount of solar energy absorbed and thus the surface temperature.

F. Sauberer [2032] obtained the following figures for the net radiation balance on a clear September day at 13:30. The highest values of 427 W m^{-2} were found over water surfaces above the Lower Lunz Lake which reflect only slightly; while a tributary with colder water, which consequently emitted less longwave radiation, had a value as high as 450. Additional values were: 367 over a potato field, 338 over Alpine sorrel, 300 over a mossy meadow, 281 over a gravel path, and 257 over a hard road. Light-colored soil free of vegetation had the lowest value.

Figure 20-1 from research by L. A. Ramdas and R. K. Dravid [2026] in India shows the temperatures for 40 days at the experimental surfaces. The figure on the right illustrates temperatures for an unaltered control surface. Five days after observations were started (time A), the dark cotton growing soil of the test patch was covered with a thin layer of white powdered lime. The immediate cooling effect can be observed in the sharp rise of the isotherms. Before wind and water were able to weaken the effect of the white coloration, the difference in temperature between the surfaces reached 15°C, and the effect was felt at the greatest depths for which readings were taken. After the powder was removed (time B), a week elapsed before the experimental surfaces reached the same temperatures as the control surface.

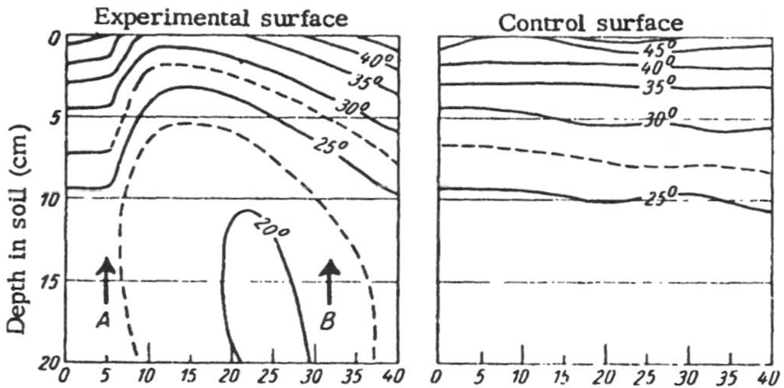

Figure 20-1. Changes in soil temperature caused by whitening the surface. (After L. A. Ramdas and K. Dravid [2026])

The effects of pasture fires on soil temperatures are less than might be anticipated. B. E. Norton and J. W. McGarity [1908] found that only in the top 15 mm of the soil were temperature changes in excess of 10°C. They found no clear relationship between the amount of material burnt and the temperature increase. The amount of moisture (which strongly affects the thermal conductivity of the soil) was positively correlated with the extent of the temperature increase. For more information dealing with the effect of grass fires on soil temperature, soil properties, and microbiological dynamics, the work of P. de V. Booysen and N. M. Tainton [1904] is recommended.

In Israel, high soil temperatures during summer often result in inadequate germination of a number of vegetable crops. G. Stanhill [2036], in an effort to reduce these temperatures, applied a surface dressing of magnesium carbonate. He found that this resulted in a reduction of the mean maximum temperature of 7-10°C.

T. R. Oke and F. G. Hannell [2025] conducted a similar experiment over short grass. Three plots were used: one whitened with talc powder, the second blackened with carbon black, and the third remained untreated as a control. During a summer day, the maximum temperature at a depth of 10 mm in the blackened plot was more than 6°C warmer and in the white plot 8°C cooler than the control (Figure 20-2). Although the night temperature differences were less, the white plot remained cooler.

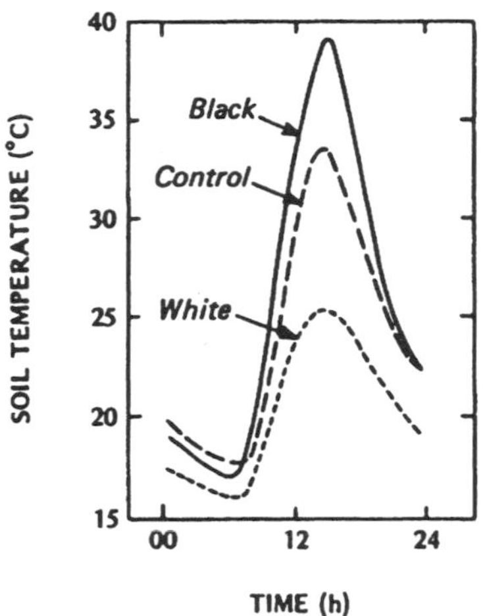

Figure 20-2. Effect of soil cover treatments on temperature. (After T. R. Oke and F. G. Hannell [2025] from T. R. Oke [3332])

P. G. Aderikhin [2001] made comparative measurements of air temperature 50 cm above natural and artificially darkened and lightened surfaces were made from 1948 to 1951. On windless days, the air temperature was 3 to 5°C higher over the darkened soil than over the light-colored surface and 2 to 3°C higher than over the natural surface. The difference increased as the water content of the soil decreased.

Temperature differences of up to 10°C were measured 1 cm below the surfaces of dry soil that had been made white and black by dusting, in Mendoza (32°53'S), by W. W. Ehrenberg [2005]. He examined the theoretical and technical possibility of making practical use of soil coloration on a grand scale. This investigation combined a color test with a humidity test. A white and black field were moistened to the same extent. Since at first the increased evaporation from the black field was in opposition to the influence of color, the temperature difference was only 5°C. After a few weeks without rain, the black field dried out, but the white field was still giving off water, and the difference rose to 14°C. Blackening the soil with coal dust as a possible method to minimize frost-related problems will be discussed in Section 53.

In models of irrigation water tanks of dimensions 40 x 40 x 10 cm, R. Yakuwa and F. Yamabuki [2049] measured water temperature as a function of the color of the paint. While the night minimum temperature was independent of the color, the maximum temperature in the white painted tank was 31.8°C, in the dark red 35.2°C, and in the black 36.3°C. In one tank, which was protected against evaporation by a thin floating black zinc plate, the temperature reached 44.9°C.

It is of interest to note the findings of S. Sato [2030] and S. Sato and Y. Funahashi [2031] that when a black powder is scattered over the water surface in rice fields, the amount of heat in the ground and in the air close to it is reduced in comparison with a field with a clear water surface. Here also, the radiation at night is independent of color. The turnover of energy during the day may be adduced from the example of 8 August 1955, from 12:00 to 13:00. With a global solar radiation $K\downarrow$ of 942 W m^{-2}, more energy went into the darkened water, 94.2 compared with 66.3 (W m^{-2}). But a greater amount of evaporation Q_E was caused by the darkening, 548 against 486 W m^{-2}. The energy loss due to conduction and longwave radiation was also greater, being 229 for the darkened water against 200 W m^{-2}. The soil of the rice field under the water absorbed only 71 W m^{-2} below the black surface, in comparison with 160.5 below the clear water surface.

The influence of paint on the temperature of wood has been investigated by C. Dorno [2003]. For this purpose, he set up four cylindrical pieces of wood, 3 cm in height and 2.5 cm in diameter, on a balcony with a southern exposure in Davos. He found that the incoming solar radiation of 698 W m^{-2} produced the following increase in temperature of the wood over that of the surroundings: blue-white paint, 10.8°C; rose paint (white zinc and Dammar varnish), 11.0°C; yellow-ochre paint, 14.4°C; red oil paint, 15.7°C; and finally for a soot-coated piece of wood, 16.9°C. K. Schropp [2035] found that, under similar conditions, black paper rose to 45°C, black enamel to 55°C, white surfaces to 15° to 20°C, while a polished aluminum foil reached only 15°C. At night, all the surfaces had a temperature 2 to 4°C lower than the temperature of the air at shelter height.

The importance of the color of vertical surfaces has also been recognized and evaluated. H. Schanderl and N. Weger [2033] investigated the effect of coloring trellised walls in Geisenheim over a period of two years. Some of these trellises were left with their natural light-brown color, and other parts were painted white or black. The growth and yield of tomatoes planted in front of a wall 3 m high, orientated in a SW direction, were observed. At a distance of 10 cm, differences in air temperature could no longer be detected, but radiation conditions had been altered. On a sunny June day, the amount of reflected shortwave radiation in front of the white wall was 56 percent higher than in front of the black wall. This increased the yield of tomatoes. The longwave radiation of the black wall, which was warmer, allowed the plants to grow more rapidly, but their yield remained lower. The higher yield in front of the white wall justified the expenditure on paint.

N. Weger [2045] repeated these tests in 1943 with peach trees and in 1947 and 1948 with vines. With the peach trees, the growth of wood was greater in front of the black wall, 31 percent more than in front of the white and 24 percent more than in front of the natural-colored trellis wall. The number of buds on a 1 m length of the young shoots in two tests was 75 percent and 190 percent more, respectively, in front of the white wall than in front of the black. With vines the best growth of wood was found in front of the white wall and there was (not verified) an increase in the weight of grape juice.

Previous reports on the effect of the surface color of soil showed that an ideal black surface would have the highest temperature. The temperature of natural surfaces is controlled in part by solar elevation, sky conditions, slope of the land, thermal conductivity and capacity of the underlying soil as well as the movement of air over the surface. A. F. Dufton and H. E. Beckett [2004] demonstrated that surface temperatures could exceed that of a black bulb thermometer (a blackened sphere in a vacuum). Roofing paper on top of an insulating base, for example, heated to 65.5°C, while the black bulb thermometer reached 56.1°C. When the temperature was measured in a box, insulated with blackened sides and closed by a glass plate, it was found to be 120°C.

From observations made by B. Huber [2014], it appears that surface temperatures up to 70°C have been recorded under natural conditions in Germany. When taking careful account of all possible errors in measurement, O. Vaartaja [2040] found surface temperatures from 50° to 60°C, with a maximum of 63°C, in south Finland. If the temperature measurements made by M. Andó [2002] and A. Kiss [2016] in Hungary at depths of 2 cm and below in sandy soil under extreme local conditions are extrapolated to the surface, the same order of magnitude is derived.

F. Firbas [2007] has shown that in spring, before the trees break into leaf, the temperature of the forest floor in oak and beech forests may be very high in Central Europe because of extremely poor conduction of the leaf litter on the ground. When the spring sunshine was able to penetrate unhindered between the bare stems and twigs, he was able to measure in the first few days of May temperatures up to 43°C within the litter.

Newly germinated seedlings are at risk from heat damage when midday soil tempertures reach such extreme levels. This is particularly common in early to late spring when lack of complete canopy foliage allows intense radiative heating of the soil surface. Heat damage will most commonly occur within northern hemisphere mid-latitude locations on flat or south-, southeast- or southwest-facing slopes, and on bare soils with a low heat capacity and conductivity, or on dark soils which have a high organic matter content or have recently been burned. Increased seedling mortality can also be found on dry or sandy/gravelly soils (O. T. Helgerson [2012]).

During winter in The Netherlands, P. Stoutjesdijk [2037] also observed that the temperatures above a pine litter with a low thermal conductivity were substantially higher than that of the surrounding air. He found daytime pine litter temperature excesses of as much as 28°C in December and 37°C in February. Over decaying grass tussocks that had dried out, he found temperature excesses of up to 50°C in late February. This is consistent with H. Schmeidl [2034], who also found temperature excesses of up to 50°C with similar sun angles in early October over moss tussocks. P. Stoutjesdijk [2037] pointed out that many insects and small animals take advantage of these localized, unseasonably warm temperatures.

Figure 20-3 is a result of measurements made by A. Vaupel [2042] in Palermo on Monte Pellegrino (38°10'N) 425 m above sea level, above brownish red soil. The temperature was measured in three positions on a fixed horizontal flax thread at a height of 8 cm above the ground. The midday surface temperature of the ground was 33°C warmer than the air temperature at a height of 2 m. The highest temperature at the surface was in excess of 60°C; on the *Opuntia* leaf it reached 50°C, slightly above (2 mm) the plant it was 3-5°C cooler and in the air at 15 and 200 m it only reached 35 and 29°C, respectively. The highest temperatures at all levels were associated with maximum solar radiation. H. F. Neubauer [2021] took 800 readings of ground temperature in Afghanistan, and in the desert to the north of Jalalabad (34°27'N) he recorded the highest temperature of 60.2°C at 14:20 on 4 August 1951, at a depth of 1.5 cm. This corresponds at most to a surface temperature of 70°C. The

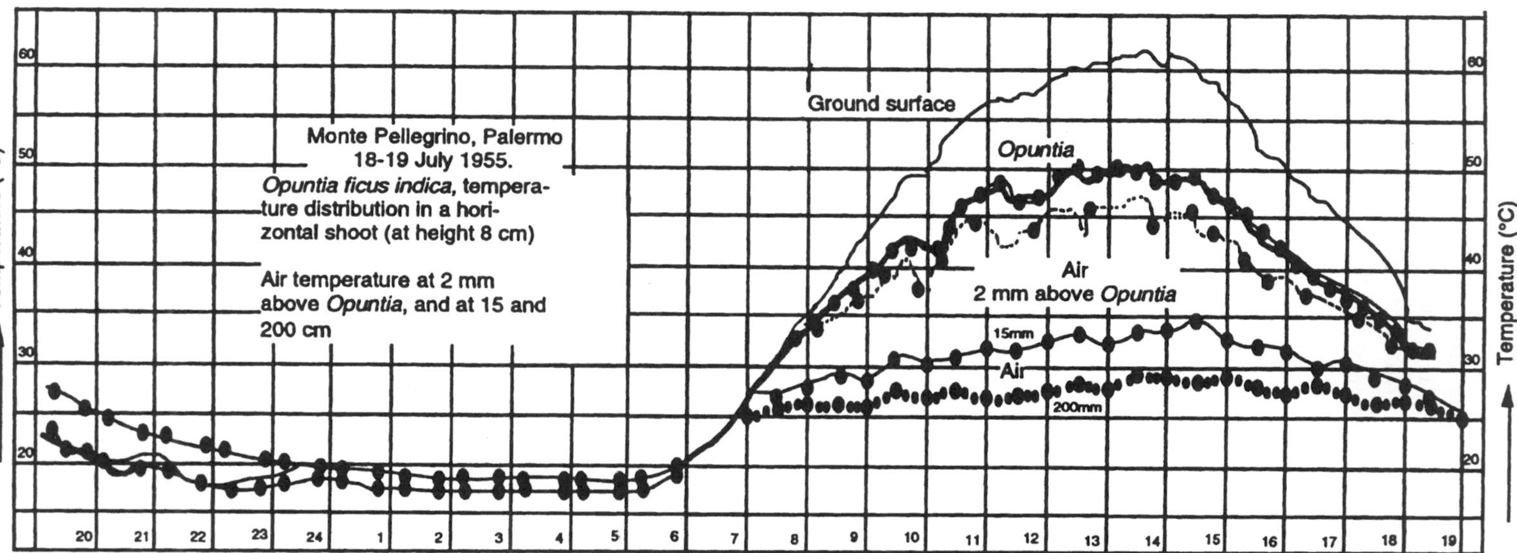

Figure 20-3. Ground and plant temperature in a Mediterranean climate. (From thermo-element records near Palermo, made by A. Vaupel [2042])

highest surface temperature recorded in North America is 93.9°C at Furnace Creek in Death Valley, California (P. Kubecka [2016a]).

Surface temperatures are high even in a polar climate. The German Antarctic Expedition of 1938-1939 under A. Ritscher was surprised to find a group of unfrozen lakes in the ice continent at 70° 41' S. H. Regula [2028] showed that volcanic warming is not necessary to explain this phenomenon. The unfrozen water is due to the warming of the dark red-brown rock by solar radiation (air temperatures were about 0°C).

Unfavorable soil properties can often be improved by relatively simple means. Besides alternation of soil coloration, which has been discussed, numerous other methods can be used. These include superimposing a soil layer with the desired properties over the ground surface, mixing in elements to the soil to improve its quality (mulching), or removing an undesirable layer.

Mulching is the application or creation of any soil cover that creates a barrier to the flow of heat or water vapor. The purpose of mulching may be to conserve soil moisture, prevent excessive cooling of the soil or even enhance soil warming. Mulches have been constructed out of weeds, lawn clippings, hay, wood chips, leaf litter, aluminum foil, gravel, various types of plastic covers, paper, and foam.

The temperature variation in a soil with an artificial covering has been investigated theoretically by D. A. de Vries and C. T. de Wit [2043] on bog soils that had been sanded over to reduce the frost hazard. The experience was that, out of this two-layer soil, there soon developed a soil with three layers, namely, dry sand, moist sand, and bog soil, all of which differed in their thermal conductivity and thermal capacity. Calculations showed that a sand cover 10 to 15 cm thick is sufficient to reduce the very high night frost danger of bog soil to that of pure sand. If the sand dries out, which is usually the case in the first 3 cm with Western European spring weather, then even less than 10 cm of sand will be enough to nullify the negative effects of a bog soil. This top-sanding procedure is, of course, more successful than if the sand is mixed into the soil, which is often done for other reasons (Section 19).

N. Weger [2045] investigated the influence of a 5 cm thick layer of peat or sawdust spread on top of a mineral soil to protect germinating plants. There was a large rise in the temperature of the covering layer in the middle of the day because of its low thermal conductivity. For example, on a fair summer day (24 July 1947) it rose to 44°C at 2 cm below the surface of the sawdust, and to 53°C under the dark peat. At a depth of 2 cm into the soil, the maximum temperature on the same day was only 30°C in the peat-covered soil and 29°C in the sawdust-covered soil, but reached 44°C in the uncovered soil. The delicate parts of the plants emerging from the soil, consequently, were protected from overheating and drying. M. Ludlow and M. Fisher [2017] found that the amount of leaf death of *Macroptilium atropurpureum* was positively correlated with the quantity of leaf litter surrounding the plants. They suggested that the litter impedes the flux of energy into and from the soil, resulting in lower nocturnal surface and leaf temperatures and greater frost damage to leaves over the litter than occurred over bare soil.

K. Keil [2015] gives an impressive example of the removal of an unfavorable top layer. The ground in Yakutsk in northern Siberia is permanently frozen at depth, with only the surface layer thawing out in summer. A thick, poorly conducting layer of moss prevents the radiant energy, which is not used up for evaporation, from reaching the lower levels. During World War II this layer of moss was removed from a number of experimental areas. This led to heating, which became more noticeable from year to year. The yield from plants was so improved that the experiment was extended to wide areas.

The effectiveness of mulching in increasing crop yield has been tried and repeatedly demonstrated. The existing climate of a place and the nature of the local site will decide the success and profitability of the scheme. The selection of a particular mulch material depends on which factor of the climate needs to be maximized. A few examples will illustrate possible solutions.

In Israel, the second planting of potatoes usually takes place in September. It is desirable to advance this date, but this would involve a risk of damage from heat from the high temperatures prevailing in August. During the summer of 1949, J. Neumann [2022] tested the effect of a 5 cm layer of straw over the soil. The air temperature at a height of 10 cm showed an increase in the maxima of 1 to 6°C because of the higher albedo of the straw, which means greater heating of the plants. The minima were 1 to 3°C lower than for the untreated areas of the ground. The straw slowed the nocturnal flow of energy from the soil into the air. The climate of the ground under the straw was greatly moderated as long as the plant foliage only over-shadowed the ground to a slight extent. The average of two series of readings taken between 23 August and 31 October showed that the temperature fluctuation at a depth of 15 cm below the ridges was only 57 percent of that found in the untreated fields (5.1°C against 9°C for a daily fluctuation of 13.9°C in shelter temperature).

A. M. Gurnah and J. Mutea [2009] and R. R. Van Wijk, et al. [2041] found that grass and other organic mulches insulate the soil, generally resulting in lower soil temperatures during warm periods and warmer soil temperatures during cool periods. A. M. Gurnah and J. Mutea [2009] also found that polythene mulch could have a strong effect on soil temperatures. A white polythene mulch left the soil temperature essentially unchanged, while black and transparent polythene mulches resulted in a large increase in soil temperatures.

S. Sato [2029] was faced with a similar task in looking for a method of reducing the high temperatures to which young plants were exposed in the water-covered rice fields in Japan. He put rice straw or grass clippings on the surface of the water, and observed the temperature maxima given in Table 20-1 for 12 August 1953. Since most of the solar radiation is absorbed within the rice straw or grass clippings mat on the surface of the water, the water beneath the mat and the soil surface are noticeably cooler.

Table 20-1. Temperature maxima in a rice field, 12 August 1953.

Measurement site	Open rice field	Grass Mulch	Rice Straw Mulch
In water	41.4	38.5	38.2
Ground surface	42.3	37.5	36.6
5 cm in ground	38.0	34.5	33.0
20 cm in ground	30.3	29.3	29.0

It has become more common in many farm regions to practice either no-till farming, where the crop residue following harvest is left undisturbed on the surface of the soil, or conservation tillage, where the crop residue is only tilled to a shallow depth (<20 cm). Maintaining these crop residues upon or near the soil surface produce a variety of changes to the soil microclimate and result in a number of physical and biological improvements to the soil environment. J. L. Hatfield and J.H. Prueger [2011] report the results of a study of cotton planted in standing wheat stubble in the semi-arid environment near Lubbock, Texas, and corn sown into a flat residue field in a humid environment near Ankeny, Iowa. The standing residue produced a number of significant changes. Total water use was reduced by 5 mm during the first 30 days after planting, soil water evaporation was reduced by half, and total leaf area and water use efficiency were increased. Mean soil temperatures were warmer in the standing wheat residue at all depths, with a corresponding decrease in the diurnal soil temperature range and level of soil temperature extremes. Air temperatures over the standing wheat field were generally higher during the day and night. The reduced wind speed created by the standing stubble also reduced the daytime vapor pressure deficits surrounding the cotton canopy. In general, the standing residue produced a number of positive changes in the growing environment at the semi-arid site. Microclimatic effects on the corn grown in the flat residue field of the humid site were much less significant. A review of the effect of crop residues on the soil surface radiation and energy balances, soils water content, and soil temperature can be found in R. Horton et al. [2013].

Within a mulch winds can still occur. M. D. Novak et al. [2023] reported that during the daytime (high wind) wind speeds within a straw mulch are strongly affected by the winds above the mulch. At night (low wind), however, there is a decoupling of the wind within the mulch from that above. At this time wind speeds, while highly correlated with each other, are not highly correlated to the wind speeds above the mulch. Unstable conditions within the mulch at night lead to free convection.

N. Weger [2046] used glass panes laid on either side of rows of tomato plants, inclined at an angle of 5°, so that the water would run in. Below the glass, in a free space of 3 cm, the air temperature reached 40° to 55°C at midday. Even at night, the enclosed air remained 4 to 6°C warmer than the open air. This procedure had a very useful side effect in that only one weed *(Portulaca oleracea)* was able to survive this "murderous microclimate." On 3 June 1952, the soil temperatures at a depth of 5 cm were as given in Table 20-2. This heating is primarily due to the effect of the glass in reducing mixing (as in a greenhouse).

Table 20-2. Effect of glass covering on soil temperature 5 cm below surface.

Temperature (°C)	Bare soil	Rows of plants between glass sheets	Below glass
Highest	24.4	25.7	38.1
Diurnal mean	19.2	21.1	27.3
Lowest	15.6	18.5	20.3

It was formerly believed that a greenhouse is warmed by the same process as the atmosphere; namely, that the glass allows shortwave radiation to pass through but absorbs outgoing longwave radiation. This was proven to be essentially false in 1909 by R. W. Wood [2048]. He built two small model greenhouses, one constructed of glass, the other of rock salt; the latter was transparent to both short and longwave radiation. When placed in the sun, both models reached about the same internal temperature level. This proved that the greenhouse's higher temperature is not primarily a result of absorption of outgoing longwave radiation by glass (R. G. Fleagle and J. A. Businger, [2006]). The primary reason a greenhouse or a car with closed windows is warmer than the external air is due to the reduction of mixing. When the sun heats the ground, and the ground in turn heats the air, this warmed air may rise many thousands of feet. In a greenhouse or a car, the walls and the roof limit mixing and thus the heat is confined to a relatively small volume. The reduction of mixing by the walls is around four or five times as important as the absorption of longwave radiation by the glass in explaining the daytime temperature excesses found within greenhouses. In Section 19 it was clearly shown that a reduction of mixing resulted in lower nocturnal temperatures near the ground. At night, the glass of a greenhouse allows a substantial amount of longwave energy to pass through and, by emitting energy, its surface cools below the ambient air temperature (the emissivity of most types of uncoated glass is around 0.84). By retarding the nocturnal mixing of warm air, a greenhouse's (or car's) temperature may become lower than that of the ambient air. Figure 29-11 indicates that putting various types of bags around flowers will, by reducing mixing, result in nocturnal temperatures within the bags being cooler than the ambient air temperature. R. M. Whittle and W. J. C. Lawrence [2047] have shown that nocturnal temperatures inside a greenhouse (throughout the year) are equal to or slightly lower than those outside. K. J. Hanson [2010], T. Takakura [2038] and C. Manera et al. [2018] also observed lower nocturnal temperatures inside greenhouses. However, since the ground in a greenhouse absorbs substantially more energy during the day, one might expect that this would keep it somewhat warmer at night. In a later study, T. Takakura [2039] observed that while the main portion of the greenhouse's nocturnal temperatures were lower than that of the temperature outside, near the ground they were warmer. This explains N. Weger's [2046] observations (cited above) that nocturnal temperatures below a glass 3 cm above the ground remained above those in the open. Thus, the reduction of mixing, which was the primary factor responsible for the temperature excesses during the day, also results in the temperatures in an unheated greenhouse being lower than the ambient air temperatures at night, except near the ground where the heat absorbed during the day keeps the air warmer at night.

In an unheated greenhouse the nocturnal temperatures will be dependent in part on the infrared absorbing qualities of the glass (and glass coatings), the amount of energy absorbed by the floor and contents in the greenhouse during the day, and the cloud cover. In a somewhat related study J. D. Martsolf et al. [2020] covered the sides and top of a citrus orchard with a porous cover. As expected wind speed was reduced by at least 80 percent. These covers reduced incoming solar radiation by around 66 percent. Nocturnal leaf temperatures were, on the average, between 1.2-1.7°C cooler than leaves outside of the covered orchard. This was probably due to both a reduction in mixing and daytime heating of the soil. Utilizing heaters inside the covered orchard raised the temperature about 0.6°C. This resulted in the covered orchard temperature being 0.6-1.1°C lower than those that remained uncovered. In an attempt to minimize the problems associated with nocturnal heat loss, greenhouses have been clad with thermal screens. Y. Zhang [2050] found that double polyethylene claddings reduced nocturnal heat losses by 23-24%.

B. Mason [2019] presented an excellent explanation of why temperatures in a greenhouse are cooler at night. He states "At night, the longwave radiation from the plants and soil will be trapped inside the glasshouse only if the albedo of the glass is high – and the albedo of glass for longwave radiation is so close to zero that it may be ignored . . . If the longwave radiation is not reflected, then it is either transmitted or absorbed by the glass. The transmittance ratio of glass for longwave radiation is low. The absorption is high, so the temperature of the glass should rise; as long as the glass surface is warmer than the underlying surfaces (plants and soils) then the resultant radiation flux will be downwards and the fall of temperatures inside the house will be slow. However this ignores the fact that the glass is also radiating to the sky. The glass used is thin with a low thermal capacity. The heat conductivity of the glass is good. Hence – the inner surface of the glass is heated by the longwave radiation – this heat is rapidly conducted to the outer surface where it is re-radiated. The only effect of the interposition of the glass is that the radiation from the plants, etc., instead of passing directly to space, passes to space via the glass as it were. With the glass becoming very cold, the air near to the glass is cooled by conduction and thus the enclosed air in the glasshouse is rapidly cooled. The effectiveness of the glasshouse lies in the fact that it cuts off entirely, or markedly diminishes, the turbulent mixing of the air in the glasshouse with air in the free atmosphere; which [may] serve a very useful purpose during the daytime," but is the opposite of what we want to encourage at night (as far as frost is concerned).

Rainwater was used better by the uncovered beds. In the plots covered by glass the length of the side roots of the tomato plants was 27 percent greater; the yield of ripe fruit was 17 percent more and unripe harvested fruit was 104 percent greater than in the uncovered field. For information dealing with plastic covers over the ground, the reader is referred to P. E. Waggoner, et al. [2044].

J. Seemann [5039] published a comprehensive exposition of greenhouse climate and the numerous possibilities of influencing it. The method of glazing with clear glass (windowpanes), ground on one side, has only a small effect on incoming shortwave radiation. The amount lost was 7 to 8 percent by reflection and 8 to 15 percent by absorption for the whole solar radiation spectrum (0.32-2.8 μ), but only 2 to 3 percent for visible radiation (0.4-0.7 μ). However unavoidable, soiling of the glass must be taken into account in practice, for which A. Niemann [5031] has given figures. It is possible to keep this loss down to 5 percent, but in industrial areas it may exceed 50 percent and even 70 percent in extreme cases.

The microclimate of greenhouses is also influenced by the way they are built and used. Those built in an E-W direction have been shown to be superior to those lying N-S in the winter and in the transitional seasons for a European climate. Steep glass roofs are better in winter than flat ones. The position and number of the parts of the framework that cast shadows are of importance. The gardening glass, with its ability to scatter light, is preferable to window glass, since it reduces the sharp contrast of light and shadow from the roof struts. For more information regarding the thermal properties of a number of substances from which greenhouses can be constructed and the advantages of single versus double pane greenhouses, the reader is referred to K. J. Hanson [2010] and J. Nijskens, et al. [2024].

Practical measures employed to control climate include shading by movable mats or tarps, coir materials, blinds, and so forth, to cut down incoming radiation; ventilation, for which there are very many different methods; artificial heating of the air or soil, or direct heating by radiation; and finally, covering to prevent loss of heat.

Principles of greenhouse microclimates can also be observed within smaller enclosed spaces. E. King [5023] made measurements inside an automobile. After the first set of measurements had been made, its color was changed from black to white. The car was left standing in the palace yard at Tubingen with two people inside it. The excess of temperature, compared with a shelter nearby, and the actual temperature of the air in the interior, were as given in Table 20-3. Although in the second case the weather was warmer and the car was exposed to more radiation, it was cooler inside. The temperature of the sunny side of the automobile was 2.5 to 3.5°C higher than that of the shady side. With the automobile in motion, the excess of interior temperature over that of the air measured simultaneously above the street, is given in Table 20-4. The influence of color remains considerable, with the stationary white car being cooler than the black car, even when it is traveling at 80 km hr^{-1}.

Table 20-3. Temperature conditions in an automobile.

Color	Measurement	Time of day		
		08:00	10:00	12:00
Black	Global radiation (W m^{-2})	90.7	237.3	286.1
(9 May)	Temperature excess (°C)	3.0	10.6	19.4
	Automobile temperature (°C)	15.6	28.0	40.6
White	Global radiation (W m^{-2})	132.6	279.1	314.0
(29 May)	Temperature excess (°C)	1.2	3.0	12.0
	Automobile temperature (°C)	21.0	26.2	37.4

Table 20-4. Excess of temperature (°C) in a moving car over that of the outside air.

Speed (km hr^{-1})		0	20	40	60	80
	Closed	14.0	12.5	12.1	11.9	11.3
Black Sedan	Window open	9.2	7.9	6.8	6.1	5.4
White Sedan	Closed	7.3	6.6	6.0	5.7	5.6
	Window open	4.4	3.9	3.5	2.6	2.5

Fewer measurements have been made in trains and aircraft. The microclimate of trains, truck trailers, and aircraft are important because the goods they carry may be sensitive to weather conditions. A. Cagliolo [5006] measured maximum and minimum temperatures in three types of cars designed for transport of fruit in Argentina; W. L. Porter [5035] in Arizona recorded temperatures up to 67°C under the roofs of covered freight cars when temperatures outside were 46°C.

21 Soil Moisture and Ground Frost

The water content of soil may be expressed as a percentage either by weight or volume. The first of these is defined as the ratio of the weight of water to the weight of dry soil; the second is the ratio of the volume of water present to the volume of soil. It is only possible to convert from one to the other if the volume of air spaces in the soil and the density of the inorganic components of the soil are known. S. Uhlig [2129] presents a guide to the determination and conversion of these quantities. The term "field capacity" is the water content remaining after drainage of the soil water by gravity. This is not an exact term since water will continue to drain slowly from the soil for long periods.

The mosaic structure of soil is also related to its water content. For example, when S. Uhlig [2129] took 25 equally distributed samples from an area 0.4 x 0.4 m of a loamy soil near Bad Kissingen, he found that the water content varied at depths between 10 and 20 cm from 19.9 to 21.9 percent by weight. Significant microscale soil moisture variations can arise due to small-scale heterogeneities in soil profile characteristics. T. Zhang and R. Berndtsson [2134] examined the temporal and spatial variability of soil moisture content along both horizontal and vertical scales at 7 sites near Lund, Sweden over the period 1971-81. They found large vertical soil moisture content differences, with the soil moisture content generally increasing with depth. Temporal variations in soil moisture content decreased with depth while spatial variations increased. They found that the mean soil moisture content time series didn't vary within a distance of 10-20 m, although greater spatial variability was found in summer. A. G. Price and B. O. Bauer [2122] have investigated similar microscale variations near Chalk River, Ontario.

Figure 21-1 shows how the tracks of cartwheels in dry sandy soil near Konin were deeply impressed and remained visible over a period through the growth of *Panicum lineare*. The soil below the cartwheels had become firmer, thus providing a better base for germination of the seeds pressed into it. The channels that formed were able to collect not only more wind-blown seeds but also rain water as well. In analyzing the effect of wheel tracks on the temperature of the soil M. J. Liddle and K. G. Moore [2116] found that the overall effect of track creation was to increase the diurnal surface soil temperature range by as much as 15°C. The increase was greater in summer than winter. They attribute this increase to the removal of vegetation, which results in both a decrease in evapotranspiration, more solar energy being absorbed, and more longwave radiation being lost by the soil surface. Soil in the wheel track becomes compressed there by increasing its thermal conductivity and capacity. In a companion experiment they showed that in isolation this would reduce the daily temperature but in concert only ameliorated the effect of the removal of vegetation.

Figure 21-1. The tracks of a wagon are marked by a dense growth of weeds. (Photograph by R. Tüxen)

R. Geiger reports seeing the spoor of a fox, the paw marks projecting 1 to 2 cm above the sandy surface of the dunes, near Hiddensee, on the island of Rügen in the Baltic, one summer forenoon. The animal, in pressing down on the still damp sand, had consolidated the soil. When the morning sun dried out the sand and the wind began to blow, these more solid and sheltered parts gradually stood out in relief, so that on the windward side of the slope (and only there) a negative of the spoor appeared.

The amount of water a soil can hold is related to its porosity. As soil grain size decreases, its porosity increases. The many small pores of a fine grained soil have a larger total volume than the few pores of a coarse grained soil. The fraction of the soil volume actually containing water is called the water content (ν_W). Since a clay soil has smaller grains than a sandy soil, it will be more porous and have a larger water content.

The amount of water in the soil can be represented by lines of equal soil moisture. Figure 21-2 gives measurements made by K. Unger [2130] in Quedlinburg in 1950. The soil was a loam with a light layer of humus, and grass was growing on it. Beyond a depth of 80 cm there was a gradual change to a gravel base at 120 cm. The water content was measured from core samples every 20 cm of depth, three times a month. The amount of precipitation is shown for 10 day periods. The normal annual soil moisture variation shows up very clearly for this particular year, with a soil moisture maximum at the end of the winter and a minimum in late summer.

Annual soil moisture variations also depend on the type of soil. From 1948 to 1951, E. Unglaube [2131] averaged over the first 50 cm the weekly water content for the loess soil of an orchard and the stony soil of a vineyard near Geisenheim. The annual variation, based on these weekly values, is shown in Figure 21-3. These curves are relatively smoothed since they give the averages of a 4 yr period that contained both a dry year and a wet year. The time of least water content for both of these soils occurs from the end of August to the end of September, with a maximum in February. This is consistent with Figure 21-2. Not only is the water content of the loess always higher than that of the stony soil, but the difference between the two values is greater in winter than in summer. The annual variation is smaller in the stony ground, with its poorer water holding capacity.

From a microclimatological viewpoint our primary interest is the effect the varying water content has on the energy budget of the soil. First, the color of the ground, and therefore its albedo, is altered by soil moisture (Figure 4-2, Table 4-2). The density, thermal capacity and thermal conductivity (and therefore the thermal diffusivity) of the natural soil also change with varying water content (Table 21-2). Finally, the partitioning of surface net radiation into the latent heat, sensible heat, and soil heat fluxes is very sensitive to the surface soil moisture content.

The effect of soil moisture upon these three factors can be illustrated in the investigation by S.E. Tuller [2101]. He examined the dependence of the soil surface microclimate upon the soil moisture content at a southerly exposed coastal beach in southern California. His observations were taken on a clear summer day when the microclimatic gradient was most strongly developed. The microclimatic gradients observed along the beach throughout the day within a completely dry sand zone and a wet sand zone were almost entirely due to variations in surface soil moisture content. A summary of the daily energy budgets for the two sites is given in Table 21-1. The presence of soil moisture reduced the wet sand zone albedo to only 71% of the dry sand value. The lower thermal conductivity of the dry sand site resulted in reduced soil heat flux (Q_G) in spite of a significantly higher surface temper-

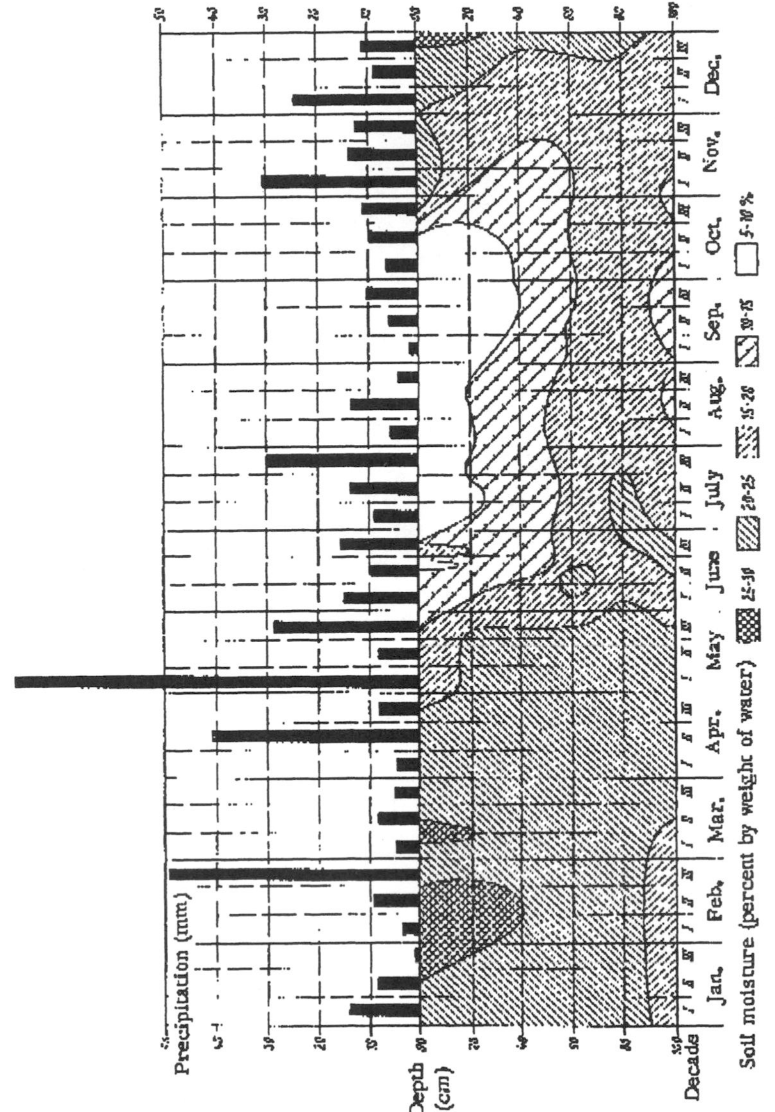

Figure 21-2. Annual soil moisture variation at Quedlinburg in 1950. (After K. Unger [2130])

ature. The absence of evaporative cooling from the dry sand site led to higher sensible heat flux (Q_H) and greater outgoing radiation ($L\uparrow$). The presence of soil moisture, in turn, allowed the wet sand site to remain nearly 15°C cooler than for the dry sand site during midday, despite the higher net radiation totals for the wet site. These results reveal the significant impact local-scale variations in soil moisture content can have upon the local climate. Since there is no evaporation in the dry soil ($Q_E = 0$) the Bowen Ratio ($\beta = Q_H/Q_E$) in Table 21-1 under dry sand is α.

D. A. de Vries [1916] has shown that the thermal conductivity of a moist granular soil can be calculated approximately if the assumption is made that the grains have the shape of

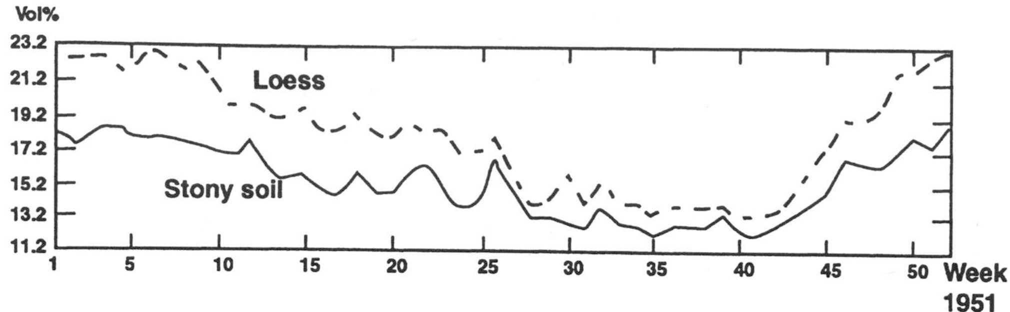

Figure 21-3. Annual moisture variation in two different types of soil at Geisenheim. (After E. Unglaube [2131])

an ellipsoid. For a fine sand comprised of 89 percent quartz and 11 percent feldspar, with grains occupying 57.3 percent of the volume, the relation between the thermal conductivity λ and the water content in percent weight is shown in Figure 21-4. At temperatures around 20°C, λ first increases very quickly, because even a very thin film of water on the grains substantially increases the surface area in contact. As more water is added, the area of contact increases much more slowly. Thus λ increases quickly until around 7 percent water content, and then increases at a much slower rate (Figure 21-4, Table 21-2).

Table 21-1. Daily energy balance totals (MJ m^{-2} day^{-1}) at two sand beach sites in southern California. (After S.E. Tuller [2128]).

	dry sand zone	wet sand zone
$K\downarrow$	26.8	26.8
$K\uparrow$	-7.1	-5.0
K^*	19.7	21.8
$L\downarrow$	28.1	28.1
$L\uparrow$	-37.7	-35.0
L^*	-9.6	-6.9
Q^*	10.1	14.9
Q_E	0	-8.3
Q_G	-0.8	-1.1
Q_H	-9.5	-5.6
β	α	0.68

De Vries [2132] used the relation between the water content v_w (percent volume) of a clay soil and the thermal conductivity λ, to find v_w from λ. The increase of λ with increasing v_w is similar to its increase with water content expressed in percent weight. This increase is shown in Table 21-2. If the values for thermal capacity $(\rho c)_m$ as a function of v_w

Table 21-2. Change of thermal conductivity, thermal capacity and thermal diffusivity with increasing water content in a clay soil.

Water content ν_w (percent volume)	0	10	20	30	40
Thermal conductivity λ ($W\ m^{-1}\ K^{-1}$)	0.25	1.0	1.5	1.68	1.8
Thermal capacity $(\rho c)_m$ ($10^6\ J\ m^{-3}\ K^{-1}$)	1.25	1.67	2.09	2.51	2.93
Thermal diffusivity a ($10^{-6}\ m^2\ sec^{-1}$)	0.2	0.6	0.72	0.67	0.61

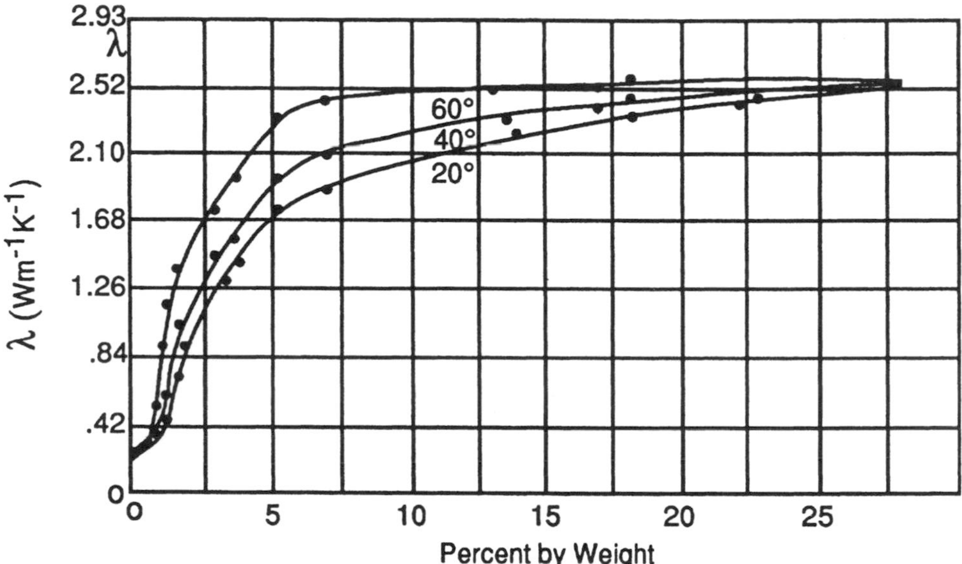

Figure 21-4. Dependence of thermal conductivity λ on water content (percent weight) of a sandy soil, at different temperatures (°C). (After D. A. de Vries [1916])

(given in Table 6-1 and are also approximately valid for this soil) are used, then the thermal diffusivity a of the soil can be determined:

$$a = \frac{\lambda}{(\rho c)_m} \qquad (m^2\ sec^{-1})$$

Since $(\rho c)_m$ increases at a constant rate with increasing water content and λ increases very rapidly in dryer soils but more slowly as the water content increases (Table 21-2), thermal diffusivity a increases rapidly in drier soils, shows a maximum for soils of medium values and decreases as soils become wetter. Thus in dry soils, a temperature wave passes through the soil more slowly because λ is small and in wetter soils more slowly because $(\rho c)_m$ is large. Therefore, a soil temperature profile will respond most rapidly to a change in surface heat energy in soils with intermediate moisture contents. This discussion has assumed a homogeneous soil. A. M. B. Passerat de Silans [2121] discusses methods of measuring thermal diffusivity in nonhomogeneous soils.

A daily variation in λ is to be expected since the upper layer of the soil is dried to some extent during the day (in fair weather) and gains water again at night. M. Franssila [2108] was able to measure a maximum value of λ in the uppermost 2 cm of the soil toward the end of the night, which was 16 percent higher than the minimum in the late afternoon. This diurnal variation can also be recognized from the pulsation of the curve in Figure 6-3 for a 1 cm depth during the precipitation-free period at the beginning and in the second half of July.

Figure 21-5 shows the influence of watering on the temperature distribution in the soil. L. A. Ramdas and R. K. Dravid [2026] observed soil temperatures at two identical surfaces at 14:00 in India. At time *W* (Figure 21-5) one of the surfaces was watered. The steep change in the slope of the isotherms shows that the added water had the effect of a "cold shower." This, however, is not the result of the lower temperature of the water, which soon reaches equilibrium temperature with the soil, but of the energy required for evaporation.

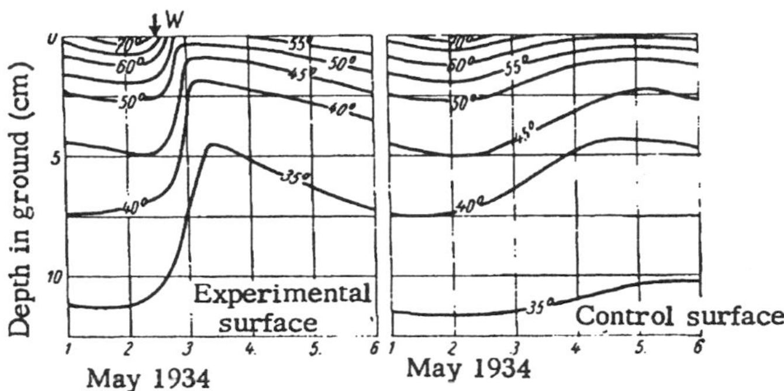

Figure 21-5. Effect of watering on soil temperatures. (After L. A. Ramdas and R. K. Dravid [2026])

The flow of energy into and out of wet soil is typically much greater than in dry soil due to the increase in both the thermal capacity $(\rho c)_m$ and thermal conductivity λ of the soil as water is added. F. A. Brooks and D. G. Rhoades [2103] carried out an experiment in a California pear orchard at 39° 04' N, in which they measured the flow of energy in two experimental areas of 2 hectares each, one watered and the other unwatered. The flow of energy was 2 to 3 times as much in the watered as in the unwatered plot. The highest noon value in the moist soil was 251 W m^{-2} min^{-1}, against 118.6 for the dryer soil. As might be expected, at a depth of 3 mm the noon temperature in the dry soil was 54.4°C in comparison with only 33.9°C in the wet.

From Figure 21-5, it can be seen that the watered soil does begin to warm up after a few days but retains its features for the duration of the experimental period. F. A. Brooks and D. G. Rhoades [2103] found in their experiments that it took 1 to 2 weeks before the two soils had the same temperature again.

In a horizontal soil environment isolines of temperature and soil water content will tend to be horizontally-layered, with mean soil water content increasing with depth and mean soil temperature decreasing with depth. These spatial patterns will be preserved even in the presence of infiltrating water. Significant variations in both soil water content and soil temperature can occur, however, with only minor changes in surface topography. R. Berndtsson et al. [2101] measured soil water content and soil temperatue profiles along a sand dune in the Tengger Desert in northwestern China (37°27'N). The sand dune extended 60 m horizontally from crest to crest, and 15 m from dune crest to dune bottom. They found that the mean soil water content was not uniform with depth, but was higher along the dune crest and in the dune bottoms. The dune slopes were consistently drier due to reduced infiltration resulting from the increased slope and reduced inclination of falling rainfall. Following a rainfall event on 11-12 August 1992 the zone of highest soil water content moved over several days to a depth of 1.0 m. By the time the infiltrated water reached this depth the surface 0.1-0.2 m had already become depleted of water. The rainfall affected soil temperatures to only a depth of 1 m. In general, the rainfall changed the temperature patterns in the soil from a predominantly horizontally-layered pattern before the preciptation to a more vertically-shaped pattern. The soil temperatures also returned to their pre-rainfall condition within about one week.

Not only does the variable water content of the soil determine its temperature, but also the temperature stratification in the soil affects the water distribution in it. Soil water can be transported by gravity, capillary action, soil tension, and differences in the vapor pressure. When the air pores in the soil are not filled with water and the air can circulate between soil particles coated with a film of water, a transport of vapor takes place in the soil. This transport is directed toward the region of lower vapor pressure. Since the saturation pressure for water vapor increases with temperature, and the air in the soil is usually saturated or very nearly so, through its evaporation in the one region and condensation in the other, water in the soil moves from regions of higher toward regions of lower temperature, other things being equal.

Generally, the downward movement of rain or irrigation water and the capillary rise of water masks this method of water transport. There are times, however, when the relation between soil temperature and moisture alone is able to account for the observed data. Figure 21-6 by H. Rettig [2123], shows measurements made at the German Weather Service's experiment station in Neustadt an der Weinstrasse. The upper curve *(a)* shows the variation of the daily mean air temperature in comparison with the long-term temperature; the hatched areas show periods of above average temperatures. Section (*b*) shows warm and cold temperature penetration into the soil in the form of geotherms. Precipitation, shown in (*c*), did not produce any deep percolation but was taken up by the top 25 cm of the soil, except in one instance, that of 13.3 mm on 24 May 1956. The daily evaporation from bare soil is shown by (*d*) together with the saturation deficit of the air. At the bottom of the diagram *(e)*, the downward flux of water vapor (stippled area below the line) and the upward water vapor flux (solid area above the line) are shown. It can be seen that the downward water vapor flux is restricted to warm periods. It follows that the small amounts of precipitation, which fall in cool periods, do not immediately flow downward into the soil. The downward movement takes place in subsequent warm intervals when the temperature and vapor pressure gradients are directed downward.

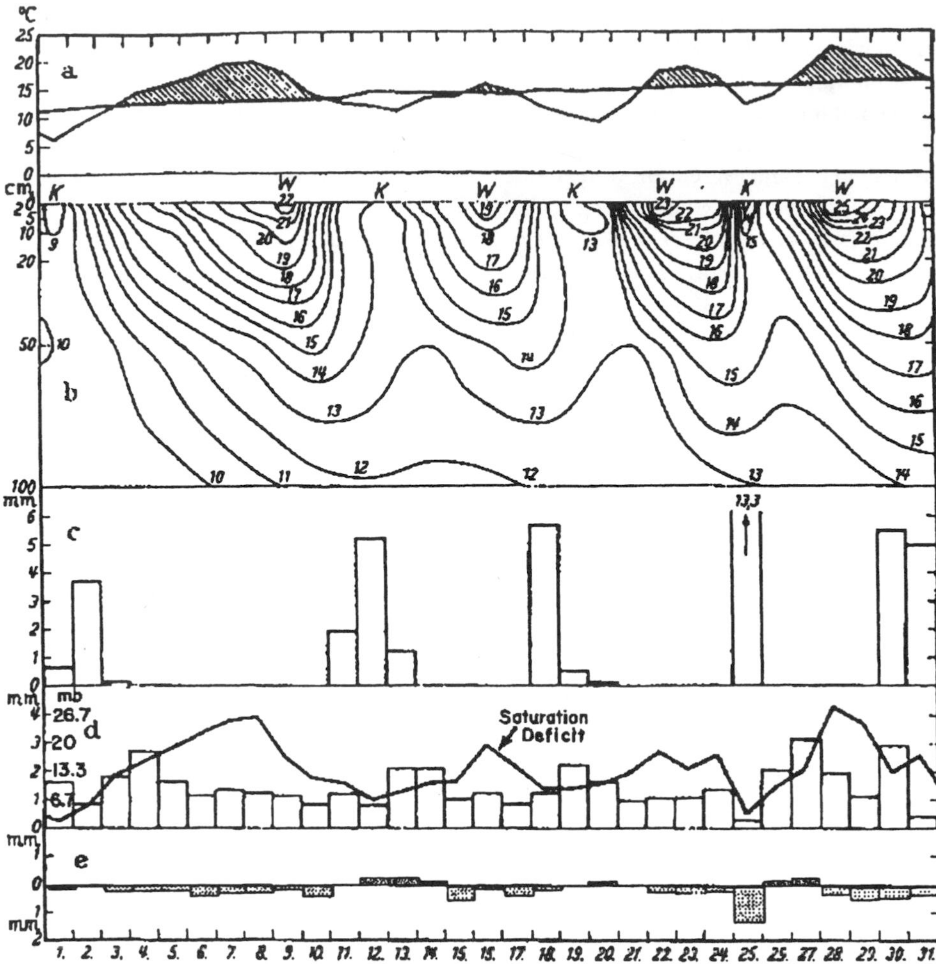

Figure 21-6. Weather and water transport in the soil at Neustadt an der Weinstrasse in the relatively dry month of May 1956. (After H. Rettig [2123])

The movement of water vapor to colder parts of the soil becomes significant in magnitude and of considerable practical importance when the ground freezes. Soil freezing in winter is welcomed by the farmer to the extent to which it helps, by repeated freezing and thawing, to break up the coarse blocky soil in bare areas into a more crumbly structure, which might be called "tilling by frost." On the other hand, farmers fear frost's action on the planted land, since it draws the young plants out of the ground or breaks their roots and dries out the soil. Surface freezing is also a problem in the maintenance of highways. Upon freezing, the movements of the ground may break the surface of a highway. After a surface thaw, a waterfilled roadway forms over the remaining ice with a greatly reduced load-bearing capacity. Once cracks develop in a highway and water gets in, repeated freezing and thawing (and heaving, Section 21) result in the formation of potholes. Frost danger in road construction has been studied in works published by A. Dücker [2106], R. Ruckli [2124], J. Schmid [2126], and L. Schaible [2125].

In addition to vapor pressure differences, moisture in the soil also moves by differences in soil moisture tension. Soil moisture tension is the energy necessary to remove the next unit of water from the soil with any given degree of dryness. W. P. Lowry and P. P. Lowry [2117] have defined soil moisture tension as a combination of:

1. The gravitational force that acts downward and resists movement of soil water upward toward the soil surface;
2. The hydrostatic and pressure forces that are exerted by air pressure;
3. The forces of surface tension that act to "draw water" into narrow passageways between grains by "capillarity";
4. The osmotic forces that tend to move relatively pure water from the outer edges of the film to inner portions of the film where the concentration of dissolved salts are larger; and
5. The adhesive, electrochemical forces that bind water molecules directly to those of mineral soil.

W. P. Lowry and P. P. Lowry [2117] further suggest that the importance of the soil grain size can be appreciated when one considers the force necessary for removing water. In a saturated sandy soil, a large fraction of the water can be removed before the thickness of the residual water layer becomes very small. In clay, by contrast, very little water can be removed (due to the small grain size) before the film thickness is the same as with the afore mentioned sand. Since the resistance offered for further removal of water from a soil is inversely proportional to the film thickness, small grained soils like clay for all their porosity will, in losing only a small fraction of their total moisture, have a high moisture tension and resist further moisture loss (Figure 21-7). Soil moisture, thus, moves in response to gradients of soil moisture tension and vapor pressure rather than by relative humidity and water content. Plant roots extract moisture from the soil by creating a higher moisture tension than found in the neighboring soil. The more salts a soil has, the higher its tension. If a soil contains too much salt, plants may have difficulty extracting water from the soil.

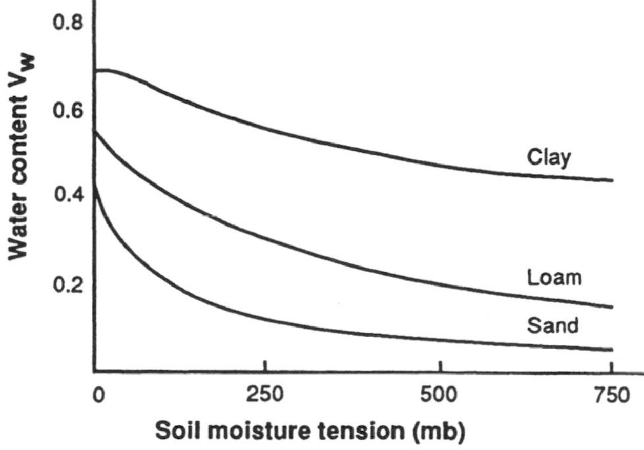

Figure 21-7. Water content vs. soil moisture tension in three typical soil types. (From W. P. Lowry and P. P. Lowry [2117])

G. A. Nakshabandi and H. Kohnke [2120] examined clay, silt loam and fine sand soils and found very different thermal conductivity and thermal diffusivity values when each of the three soils contained similar soil moisture contents (Figure 21-4, Table 21-2). However, the thermal conductivity and thermal diffusivity values of the three soils were very similar when their soil moisture tensions were the same.

When the air is below 0°C but the ground is not yet frozen, a striking phenomenon can occasionally be seen – the formation of mush frost. This is also called stalk ice, hair frost, ice needles, or ice fibers, and is know in Sweden as "Pipkrake." This is formed in loose damp soil, mostly loam or sandy loam, on drainage slopes, at road cuttings, on pine needles in forests or on decaying wood. Ice needles about 1 mm in diameter grow perpendicularly, or in a slightly bent form, out of the ground. They form close together in such great numbers that they look like a comb. In growing, they lift straw, particles of soil, and stones (sometimes as big as a fist) to heights of several centimeters.

Mush frost forms most easily when (warm) rain is suddenly followed by a sharp frost without further precipitation. The needles grow most quickly at the beginning. Occasionally it is possible, as seen in H. Fukuda's [2109] photograph in Figure 21-8, to see growth on successive days. The first 15 cm below the surface of the soil plays the greatest part in producing this effect. The water content of the soil below 30 cm has little effect. Mush frost may also form under a snow cover. Usually the ice needles are a few centimeters in height. However, in rare cases, they have been observed as high as 50 cm.

Figure 21-8. Mush frost, showing diurnal layers. (Photograph by H. Fukuda [2109])

Water in the soil does not freeze at 0°C. The depression of the freezing point, investigated closely by R. Ruckli [2124], is greater with lower water content and with fine grained soils. When the temperature decreases, the capillary water in the soil pores freezes first, then the hygroscopic water films around the particles, and lastly, or sometimes not at all, the strongly adsorbed films in the smallest pores. In freshly poured concrete, for example, the water in the pores freezes at approximately -3° to -4°C.

If, as winter progresses, the ground freezes, a number of important changes in its physical state may occur. The first of these is the change in specific volume from 0.001 for water at 0°C to 0.00109 m^3 kg^{-1} for ice at 0°C. This 9 percent increase in volume causes heaving of the ground. The thermal capacity of ice at 0°C is 1.92 x 10^6 J m^{-3} K^{-1}, which is less than half that of water. The thermal capacity of the soil, therefore, decreases when freezing occurs, to an extent dependent on the amount of water present. The thermal capacity for a frozen and unfrozen sandy soil with density $\rho_s = 2630$ kg m^{-3} and a specific heat $c_s = 0.84 \times 10^3$ J kg^{-1}K^{-1} at various water contents is shown in Table 21-3 from the equation:

$$(\rho c)_{m(ice)} = 0.01\ (v_s \rho_s c_s + 0.505\ v_w) \qquad (\mathrm{J\ m^{-3}\ K^{-1}})$$

For a sandy soil with $\rho_s = 2630$ kg m^{-3} and $c_s = 0.838.10^3$ J kg K^{-1} (Section 6), we have the values given in Table 21-3.

Table 21-3. Thermal capacity of unfrozen and frozen soil.

Water content, v_w (percent volume)	0	10	20	30	40
Ice content, v_e (percent volume)	0	11	22	33	44
$(\rho c)_m$ unfrozen (10^6 J m^{-3} K^{-1})	1.25	1.67	2.09	2.51	2.93
$(\rho c)_m$ frozen (10^6 J m^{-3} K^{-1})	1.25	1.47	1.67	1.88	2.09

From Table 6-2, it can be seen that the thermal conductivity of ice is roughly four times that of water, and the thermal capacity of ice is less than half that of water. Thus, the thermal diffusivity a of ice is approximately ten times that of water. When soil (which is only partially made up of water) freezes, its thermal diffusivity will typically increase from 20 to 50 percent. Freezing of the soil will, therefore, result in an increase in the flow of energy from within the soil toward the surface and a freezing of the soil to a greater depth than might otherwise be expected.

Of great importance in the total exchange of energy is the latent heat released when water freezes (0.334 MJ kg^{-1} at 0°C). It was calculated, for example, by J. Keränen [2112] that during the winter of 1915-16 in Sodankylä (67° 22' N), for every 1 cm^2 that froze to a depth of 1.1 m, 6721 J were released. Of this total, 69 percent (4623 J) came from latent heat, 24 percent (1629 J) from deeper layers in the soil and 7 percent (469 J) from the cooling of the first 1.1 m of the soil.

Because the saturation vapor pressure is lower over ice than water (Table 21-4), another effect of soil freezing is that the vapor pressure gradient in the unfrozen ground below is always directed upward and is independent of the diurnal temperature variation at the surface. Therefore, both the water rising under capillary action and the water vapor moving upward through the soil pores arrive at this frozen layer, freeze to an extent dependent on the rate of the latent heat of fusion, and sublimated heat can be conducted toward the surface. Experience shows that, on the average, the temperature increases linearly with depth in a homogeneous frozen layer.

The enrichment in water of the frozen layer can be seen in Figure 21-9 from observations made by N. Weger [2133] in Geisenheim on the Rhine in January 1953. The water content in the fine sandy loam soil of an orchard was measured continuously for 10 cm layers. The average value for the topmost layer was assigned to the depth 5 cm with which the diagram begins. The bar graph shows the depth of freezing. There was only 2.0 mm of precipitation

Table 21-4. Saturation vapor pressure (mb) over ice and water. (From R. J. List [4015])

Temperature (°C)	0	-5	-10	-15	-20	-25	-30	-35	-40
Sat. vap. pressure over ice	6.11	4.02	2.60	1.65	1.03	0.63	0.38	0.22	0.13
Sat. vap. pressure over water	6.11	4.22	2.86	1.91	1.25	0.81	0.51	0.31	0.19

during the month, which had no significant influence on water movement in the soil (Figure 21-6). The soil had been well moistened by rain in the second half of December. The lines of equal water content in percent by weight show the increase in the frozen layer at the expense of the unfrozen layer below. If the change in the distribution of soil moisture is worked out from 30 December 1952 to 16 January 1953 for a column of area 1 cm^2, a gain in water of 2.14 g is found for the frozen layer and a loss of 2.08 g for the layer below down to a depth of 50 cm. Transport of water in the first half meter of the soil in January was, therefore, ten times the precipitation falling on the surface. When the ground thaws at the end of the month, the lines of equal moisture diverge; some of the water seeps downward (20 percent line), but most evaporates from the soil into the atmosphere. Repeated freezing and thawing thus dries up the ground. This process can be a problem for farmers.

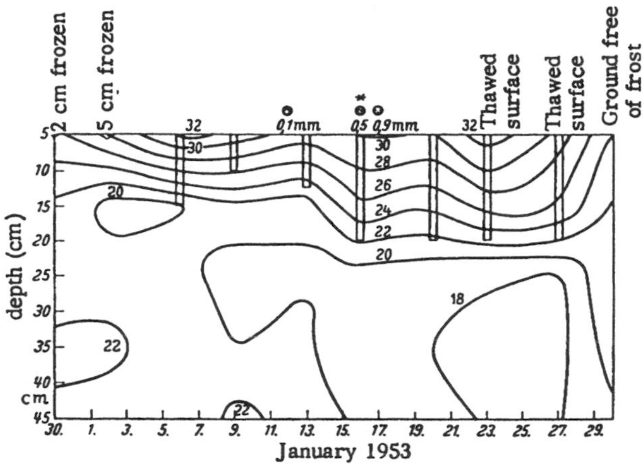

Figure 21-9. Soil water migration in the frozen layer (isolines are water content of the soil in percent by weight). (From observations by N. Weger [2133] in Geisenheim)

Homogeneous freezing occurs in noncohesive soils down to a grain size of 0.05 mm where the grains are surrounded by a crust of ice and become linked together. Sometimes, however, for a number of reasons, the soil gets displaced from its original position by the formation of (mainly horizontal) ice lenses. These lenses consist of clear ice, generally with a vertical fibrous structure, occasionally containing air inclusions. They are found in all sizes from microscopic up to 20 cm (up to 35 cm in rare cases).

The depth of freezing in winter is important for the laying of pipes, such as water mains, that might be affected by freezing. There are great differences from year to year. W. Kreutz [2115] has compared the time and depth of freezing of the same soil during four mild and four severe winters between 1939 and 1949. The most important factor in determining the depth of freezing is the depth and duration of snow cover. Winters with small amounts of snow are the most dangerous with respect to freezing. Snow consists mainly of air and, therefore, has a very low thermal conductivity (Table 6-2). When the ground is covered with snow, energy from the ground surface is not lost as rapidly and the ground remains warmer. With the first fall of snow, the soil surface may actually warm as heat from greater depths is trapped (Figure 24-6). The experimental area of clean gravelly sand used in the long series of observations at Potsdam from 1895 to 1948 was deliberately kept clear of snow. During this period, as shown by G. Hausmann [1008], the temperature at a depth of 2 m never dropped below 0°C. At 1 m, frost was observed 8 times in the 54 years, shown chronologically in Table 21-5. These results show the maximum depth of freezing for ground without snow protection in this area. The type and moisture content of the soil affects the depth of soil freezing. Damp soils freeze more slowly and to a lesser depth than dry soil, because of the latent heat of fusion released, but they also thaw more slowly in the spring.

Table 21-5. Duration of ground frost and minimum temperature at two depths.

Winter	Frost duration (days)		Lowest temperature (°C)	
	1 m	0.5 m	1 m	0.5 m
1894-95	6	32	-0.2	-5.1
1900-01	8	21	-0.3	-6.5
1916-17	2	38	-0.1	-4.7
1921-22	10	33	-0.3	-5.2
1928-29	36	63	-2.7	-9.6
1939-40	41	58	-1.6	-7.4
1941-42	40	64	-1.1	-7.6
1946-47	51	80	-1.7	-8.1

Figure 21-10 shows the duration and depth of ground freezing in four different types of soil at the agrometeorological experiment station at Giessen, for the cold winter of 1939-40 (W. Kreutz [2114]). The lighter shaded areas indicate where the temperature was below 0°C and the darker areas where it was less than -4°C. In the dry basalt gravel, the frost penetrates at an average rate of 2 cm per day and reaches a depth of 67 cm; however, the whole soil thaws by 25 February. The rate of frost penetration into moist humus, due to its low thermal conductivity, is only 0.6 cm per day and reaches only 32 cm, but it lingers until 22 March under a surface, which is already thawed out. The higher the thermal diffusivity of the soil the greater will be the depth of freezing. However, soils with high thermal diffusivity also will thaw quicker in spring. Attempts have been made to calculate the depth of frost penetration theoretically, for purposes of road construction, taking all the various factors into account and including both the case of homogeneous soil freezing and that of ice-lens formation (R. Ruckli [2124]).

The 9 percent expansion of water upon freezing and ice-lens formation both contribute to soil heaving. Hard surfaced roads are subject to heaving, develop frost fissures, buckle,

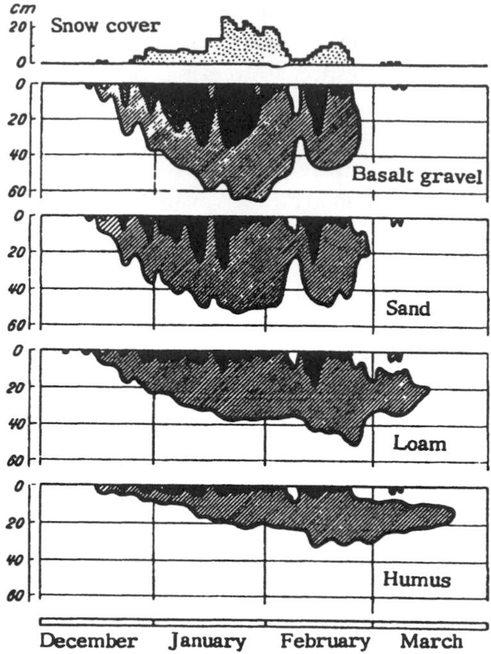

Figure 21-10. Duration and depth of ground frost in four different soils in winter of 1939-40 at Giessen. (After W. Kreutz [2114])

and suffer other forms of frost damage. Heaving can amount to some tens of centimeters. In natural soil, which is the main concern here, the amount of movement is much smaller, as shown by the following frequency table for various amounts of displacement evaluated by R. Fleischmann [2107] from observations made in Hungary from 1931 to 1935:

Heaving (mm/24 hr)	0-5	5-10	10-15	15-20	20-25
Number of cases	57	38	4	3	2

G. Kretschmer [2113] and J. Schmid [2126], found that the movement of arable land did not exceed 24 mm a day, and reached a maximum of 36 mm for the whole winter.

Figure 21-11 shows three typical periods taken from a continuous series of observations made by G. Kretschmer [2113] in loose plowed soil, near Jena, during the winter of 1954-55. The first strong frost heaving occurred when there was a dry frost at the end of December and beginning of January. On 1 January, there was a fall of snow, and the snow cover reached a depth of 12 cm by 3 January. The snow insulated the soil, and frost heaving came to an end. There was freezing weather from 22 to 25 January with a short thaw in the middle of the day. Each time it thawed, the ground settled a little. During a period of thaw from 28 to 31 January, with short night frosts, the surface of the ground was subjected to marked periodic movements. All these movements can damage the young shoots of grain or woodland plants by pulling the roots out of the soil so that they dry up or freeze, or by tearing roots at greater depths.

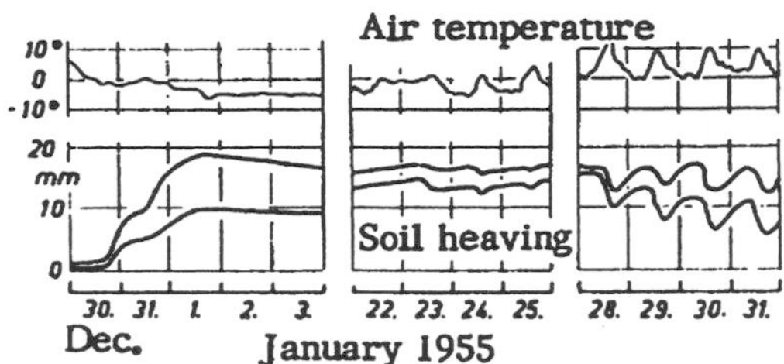

Figure 21-11. Typical frost heaving of the surface of plowed land. (From measurements by G. Kretschmer [2113])

In the subarctic regions of Alaska, Canada and Russia, permanently frozen ground (permafrost) occurs which shows substantial microscale differences in spatial extent and vertical depth due to microclimatic variations. The occurrence of permafrost has become especially critical due to development of energy, mineral, and lumber resources of the far North. R. J. E. Brown [2104 and 2105] and R. D. Thompson [2127] provide reviews of macroclimatic and microclimatic factors which control the distribution of permafrost. In Canada, for example, permafrost covers 50 percent of the land, ranging in thickness from 2 m at the southern boundary of the permafrost zone, to 300 m at its northern boundary. A zone of continuous permafrost roughly parallels the -9°C mean annual temperature isotherm, while the southern boundary of the discontinuous permafrost zone is weakly correlated with the -1.1°C isotherm. In the discontinuous zone, frozen ground is juxtaposed with unfrozen ground in a very complex pattern controlled by the surface microclimate. The upper layer of permafrost normally thaws during the summer to create a saturated active layer whose depth varies from 3 m in the south to a few centimeters in the high Arctic. Favorable sites for permafrost in the discontinuous zone include north-facing slopes, frost hollows, and surfaces covered with sphagnum moss, thick accumulations of peat which possess low values of thermal conductivity and thermal diffusivity and isolated patches along shaded river banks. The depth of the permafrost zone and the active layer depend upon a complex array of microclimatic factors, including surface albedo, vegetation cover, soil moisture content, surface drainage, thermal conductivity, topography, and snow cover depth.

With the increased economic development of these regions human-induced permafrost changes can result, particularly thermokarst subsidence due to surface disturbances such as clearing vegetation, drainage, and pipeline, road, and building construction. In summer, road construction leads to reduced surface albedo, increased solar radiation absorption, reduced evaporation, and increased thermal conductivity, all of which lead to increased conduction of heat into the soil. This produces a deeper active layer, which is accompanied by increased drainage and ground subsidence. In winter, snow removal operations along roads remove the insulating snow cover, resulting in increased heat conduction to the surface, and a permafrost table rise beneath roads.

H. Hayhoe and C. Tarnocai [2110] illustrates the effects of natural vegetation disturbance on soil temperatures in an area of discontinuous permafrost (Table 21-6). The two sites, less than 20 m from one another, were examined during the period from May 1988 through September 1991 at Manner's Creek, Northwest Territories, Canada (61°36' N). The first site was undisturbed and had an 18 cm thick organic soil layer with a vegetation covering of white and black spruce and moss. The second site was in a 25 m wide right-of-way along the Norman Wells pipeline, which had been cleared of all forest cover. The organic soil layer at the disturbed site had been left alone, albeit in a somewhat disturbed condition, and the area had been reseeded with grasses. The disturbed site was warmer than the adjacent undisturbed site throughout the year, and had a deeper active layer. The warmer disturbed site winter temperatures appeared to be due to greater snow accumulation along the open right-of-way, while the higher summer soil temperatures were due to the greater thermal conductivity of the disturbed site surface layer.

Table 21-6. Measured soil temperature (°C) at 50 cm depth at Manner's Creek, NWT for the period 1988-1991. (After H. Hayhoe and C. Tarnocai [2110])

	Mean annual soil temp. °C	Mean summer soil temp. °C	Jan.	Feb.	Mar.
Undisturbed site	0.2	2.1	-1.6	-1.6	-1.3
Disturbed site	1.3	4.8	-0.9	-1.1	-0.9

	Apr.	May	Jun.	Jul.	Aug.	Sep.	Oct.	Nov.	Dec.
Undisturbed site	-0.4	0.0	0.9	2.3	3.1	2.1	0.5	-0.4	-1.1
Disturbed site	-0.4	0.0	2.0	5.4	6.9	4.2	1.1	-0.1	-0.4

Before concluding the discussion of Section 19-21 on soil temperature and moisture, three meteorological processes should be mentioned that show the connection between soil characteristics and the surface energy exchange. These are the melting of freshly fallen snow, the formation of hoarfrost, and the formation of glaze.

Snow is formed in the upper levels of the atmosphere, and is, therefore, independent of microclimate. When wet snow is falling in mountain areas where temperatures are a little above 0°C, the lower limit of the snow cover coincides with an elevation contour. As soon as the snow accumulates, however, microclimatic differences quickly begin to emerge due to the influences of radiation, wind, and the conduction of energy from the soil. These differences are more noticeable with thinner snow cover. Snowmelt begins first at lower elevations.

An example of this is shown in Figure 21-12, from a photograph by H. Mayer [2118], who states that there was a snow cover about 10 cm thick over the soil. The snow and the air temperature were 0°C, while the soil was warmer. The snow stayed longest over the loosely consolidated soil with a low thermal conductivity. Mayer writes: "Continuous snow cover about 10 cm deep and 0°C lay over ground above 0°C. Air temperature was also 0°C. The cloud ceiling was very low and the only light was from weak diffused sky radiation, so that the snow was melted primarily by heat coming from the ground. It melted first on rocks with growth on them, then on grassland and overgrown debris fallen from the slopes. It had all disappeared completely when we saw, in front of the Staubach Valley, a part still completely snow-covered among the green slopes. The Staubach Waterfall, tumbling down 300 m of perpendicular wall left by an old glacier trough, had carved a small erosion valley in the old consolidated and overgrown debris from the cliffs. The debris in the new valley still had a complete snow cover. The rock forming this debris cone, brought out by the water in comparatively recent times, almost certainly had the same high thermal conductivity as the soil. The conductivity of the entire debris cone, however, was determined by the air present in the hollow spaces between the rocks, and was, therefore, much less than that of the solidified ground. The snow began to melt only when the weather cleared and the sun broke through."

Figure 21-12. Snow melts last on a heap of stones with low thermal conductivity. (Photograph by H. Mayer [2118])

J. L. Monteith [2119] compared two photographs, one of an area paved with uneven flagstones, with small plants growing between the stones, and the other of the same view after a snowfall. The snow had melted on the well-conducting stones but remained on the plants, outlining the pattern with white bands.

Equally conclusive observations can be made when there is hoarfrost. Whereas snow falls everywhere indiscriminately and microclimatic differences become visible only when it melts, the formation of frost depends on the characteristics of the place where it forms. The time at which frost melts should also be observed. Piles of wood are still white in the morning long after the ground around them, which is a good conductor, has become dark. Water pipes, which slow the movement of heat, become visible through white frost in an otherwise uniform street. R. Geiger reports that once in Bad Kissingen he saw the complete structure of the smooth tin roof of a large shed outlined in snow white lines a few minutes after sunrise. Where there was only a thin foundation below the tin, the frost melted quickly in the morning sun; but where there were joists and cross-ties underneath, these had taken up some of the sun's heat, so that the frost did not melt until a little later.

Figure 21-13 is an early morning view of a roof after a night with hoarfrost. The supporting timbers are clearly outlined by hoarfrost. The timbers have cooled through the night and their larger mass takes slightly longer to heat the next day.

A few trees were removed and their stumps completely uprooted from an avenue in the Hofgarten in Munich in the autumn. After the holes had been filled in and smoothed over, their position had become quite invisible. The following spring, A. Schmauss [705] noticed

Figure 21-13. Roof after a night with hoarfrost. (Photograph by R. H. Aron)

that, after cold nights, the whole area that had been dug out was white with frost. The soil in the holes was looser than elsewhere and, therefore, had a lower thermal conductivity.

Figure 21-14 is an air photograph taken by the Royal Air Force in Lincolnshire, England, at the suggestion of O. G. S. Crawford. It shows the land around the medieval village of Gainsthorpe during frost. In this photograph, published by J. Herdmenger [2111], the walls of the old village, which had been lost from sight since 1610, show up in frost because of differences in conductivity of the ground below the surface. Archaeological researchers have frequently made use of such processes.

Figure 21-14. Traces of the foundations of an old village are sketched by frost. (Air photograph by R. A. F. From J. Herdmenger [2111])

Glaze is very sensitive to changes in soil characteristics. Glaze forms in two ways, either by the freezing of supercooled rain drops on the (warm) ground, or by the freezing of rain drops (above 0°C) on very cold ground. Microscale variations are quite large, with every street, wall, and type of stone having its own style of glaze. The roughness of the surface, the thickness and the type of stone surfacing, and the slope of the ground, all have their effects.

The investigator of microclimate should always study the local effects of various weather situations carefully. Where does fog occur? Where is dampness first noticeable on the ground? Where do the first cracks appear in dry periods? What is the pattern of dew, hoarfrost, glaze, or rime? Where does moist vegetation dry out first? By this method, insight can be gained into local variations within a study area.

22 The Air Layer above Small Water Surfaces

If the lower boundary of the atmosphere is water, rather than solid ground, the nature of the surface-atmosphere interactions is affected by the changed surface state. In water, there is an exchange of mass that is not present in soil. Solid ground and water react differently toward shortwave radiation. In the ground, in the absence of an adequate water supply, actual evaporation falls below the potential level. In water, the potential rate of evaporation is determined by the temperature of the evaporating surface and the state of the air above it.

Even a slight roughness in water surfaces modifies the wind field over water as compared with that in the air layer near the ground and thus alters the energy exchange. Energy exchange in the ground is controlled almost exclusively by molecular conduction. In water, which is mobile, there is mass exchange. The action of wind on the surface causes a certain amount of mixing in the uppermost layers. To this frictional exchange there is added, as in air, convectional exchange. Cooler parcels of water sink, while warmer parcels rise. Water, therefore, has the properties of extremely conductive soil. Following the rules established for good conductors (Section 19), the daily temperature fluctuation at the surface will be small, amounting in the open sea, for example, to only a few tenths of a degree (Figure 23-1).

In certain circumstances this mass exchange may be very small. Water reaches maximum density around 4°C. Thus, as the surface of a water body cools in autumn, the denser water sinks until the vertical temperature profile becomes uniform, at which point convectional mixing ceases. Spontaneous freezing can set in during calm periods in winter nights, since frictional mixing is also minimal. K. Keil [2205] describes a Swedish invention, which was able to eliminate the difficulties caused to shipping. Long pipes with small holes in them were laid out in the lanes to be kept free. When there was a danger of the surface freezing, compressed air was fed into the pipes. It escaped through the holes causing an artificial mixing of the surface layers with water at 4°C from lower levels. This delayed ice formation for a short time until the ships still at sea could reach their berths. If there is sufficient frost to inhibit mass exchange completely, as over the polar seas, the climate adopts continental traits. The layer of air above water then becomes similar to air over snow or ice.

In making a comparison with solid ground, there is also the different behavior of water to long and shortwave radiation. Water behaves much the same way as ground toward longwave radiation; both are subject to the same conditions of absorption and emission at night. Table 4-2 showed that reflection of incoming solar radiation from water surfaces was less than that of most other natural surfaces. The albedo of water varies with the sun's altitude, state of the water's surface, and sky conditions as shown in Figures 4-2 and 22-1.

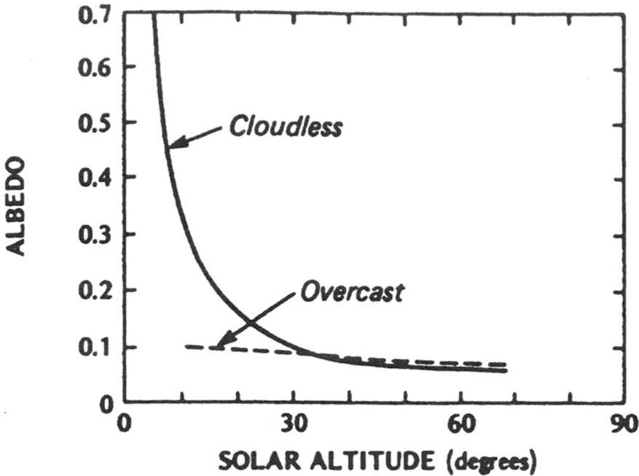

Figure 22-1. Relation between the albedo of water and solar altitude for calm clear and cloudy days. (From T. R. Oke [3332]; after M. Nuñez, et al. [2206])

The specular reflection, which occurs under clear sky conditions at low solar altitudes, is of practical importance at the seashore and on the banks of lakes and rivers. The microclimate of vineyard terraces may be influenced by reflected light from a river below. In February, at midday, O. H. Volk [2217] measured light intensity and found that direct-beam, diffuse, and reflected solar radiation from the Main River were in the ratios 42:11:41; the average of five sample tests made in March showed that ground reflected radiation amounted to 65 percent of light from above. "The best situations for vine terraces," writes O. H. Volk [2217], "are those that enjoy this extra light. To the east and west of the Main Valley, there are within a few kilometers from the river, many slopes that have the same exposure, inclination, geologic structure, and soil characteristics. These slopes are not recognizably different in any respect from the south and west slopes of the Main Valley. The latter, however, either no longer grow vines or produce inferior wines. There is no question of there being microclimatic differences between the Main and Wern Valleys. I was unable to account for the difference in wine production until I became aware of the difference in the amounts of light from above and below." This difference can also be clearly seen in the distribution of wild flowers.

Western shores receive a detectable amount of reflected radiation from the morning sun and the eastern shores from the evening sun. The principal difference in the behavior of water toward radiation, however, is the depth to which radiation penetrates. Every swimmer knows that objects can be recognized to considerable depth and that visible light, therefore, must penetrate there. In Table 22-1, W. Schmidt [2213] gives for three ranges of wavelengths, the intensity of solar radiation at various depths in clear water expressed as a percentage of the surface radiation. In the range from ultraviolet to orange (0.2-0.6 μ), almost three-fourths of the radiation penetrated as far as 10 m, and about 6 percent reached 100 m. From red to infrared (0.6-0.9 μ) penetration is poorer, as longer wavelengths are absorbed in the topmost layer of the water. Penetration of the near infrared wavelengths (0.9-3.0 μ) was poorer still.

Table 22-1. Percentage of incident solar radiation reaching various depths in clear water.

Wavelength μ	Depth					
	1 mm	1 cm	10 cm	1 m	10 m	100 m
0.2-0.6	100.0	100.0	99.7	96.8	72.6	5.9
0.6-0.9	99.8	98.2	84.8	35.8	2.6	0.0
0.9-3.0	65.3	34.7	2.0	0.0	0.0	0.0

Differences in the transmission of different wavelengths in the solar spectrum may be seen from Figure 22-2. The top curve, after W. R. Sawyer and I. R. Collins [taken from 2215], shows the percentage of solar radiation of various wavelengths penetrating a layer of clear water 1 m deep. In the infrared range, penetration decreases sharply beyond 0.85 μ. Toward the longer wavelengths of solar radiation water surfaces may, therefore, be considered as behaving in the same way as solid ground.

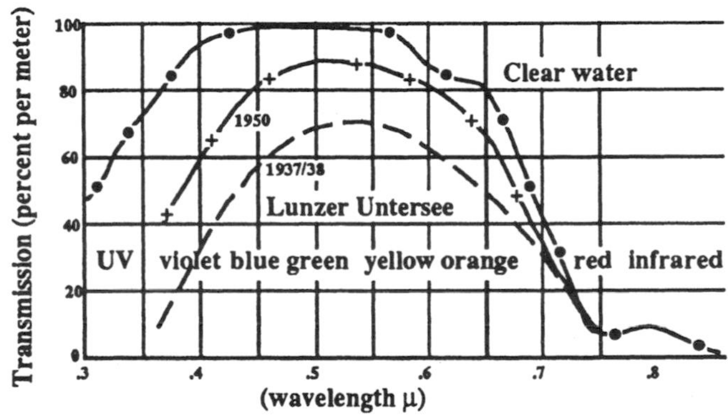

Figure 22-2. Transmission of light in pure water and a lake as a function of spectral range. (After W. R. Sawyer and I. R. Collins [taken from 2215])

These clear water values are not valid for natural bodies of water such as lakes, ponds, or even pools. Dissolved and suspended substances not only color the water, but also affect its transparency. In comparison with clear water the transparencies of natural water bodies are usually lower as a whole, and the optimum penetration of rays is shifted toward longer wavelengths. Figure 22-2 contains two curves from I. Dirmhirn [2202], typical of the Lunzer Untersee in Austria, in two different years. The total quantity of suspended matter has a pronounced effect on relative penetration, while the spectral distribution of the radiation remains essentially unaltered (see also Figure 24-2).

On closer examination, the radiation exchange taking place in water is much more complicated, since, in addition to absorption by pure water, the amount of absorption by dissolved substances and foreign particles in suspension has to be taken into account. The effect of light refraction is such that, viewed from inside the body of water, all light from outside comes from within a cone of semivertical angle 48.6°, the critical angle for total reflection. Besides being absorbed, light undergoes scattering. Scattered radiation comes primarily from the surface of the water, but light is also scattered from below and may be increased considerably by reflection from the bottom, provided this is not too far down. Weakening of radiation intensity with depth, therefore, is not merely a result of absorption; taking all of these factors into account, it is properly referred to as extinction.

Table 22-2 shows the wide range of variation in natural waters of the transmission of light within the visible spectrum. Values already given in Figure 22-2 for pure water are included for comparison on the first line. The results for Austrian waters are from F. Sauberer [2210, 2211] and those in Brandenburg from O. Czepa [2201]. Among the lakes, the Neusiedlersee, southeast of Vienna, is unique because it is only 40 to 80 cm deep, with a

Table 22-2. Light transmission by water (percent per meter). (After F. Sauberer and O. Czepa [2201, 2210, 2211])

Wavelength (μ)	0.375	0.4	0.45	0.5	0.55	0.6	0.65	0.7	0.75
Spectral range		Violet	Blue	Green	Yellow	Orange		Red	
Pure water	84	93	98	98	97	87	81	43	7
Achensee	51	65	80	85	82	73	57	33	8
Lunzer Untersee	18	33	56	68	70	63	50	31	7
Lunzer Obersee	2	9	26	39	46	47	41	27	6
Müggelsee near Berlin	–	–	8	23	34	36	36	28	5
Arm of Danube, clear	3	8	15	21	26	25	21	16	4
Danube, slightly turbid	0	1	5	11	16	20	15	8	2
Kalksee I, near Berlin	–	–	4	10	15	17	15	13	3
Neusiedlersee									
Area with reeds	–	0	2	8	13	17	17	12	–
Open water	–	0	1	3	6	7	6	4	–
Flakensee, near Berlin	–	–	1	3	6	5	5	5	1
Danube, very turbid	–	0.0	0.1	0.1	0.3	0.7	0.8	0.5	0.1
Flat moorland pool	–	0.0	0.1	0.2	0.8	1.8	2.8	4.8	1.3

surface area varying between 250 and 300 km^2. Even with moderate winds a layer of mud about 0.5 m thick on the bottom is stirred up. In the wide shore area, which is covered with reeds that hinder the mixing process, penetration of light is often deeper at these times than in the open water. The small lakes Kalksee and Flakensee in Brandenburg have thick growths of algae, and the Flakensee is made even more turbid by the dirty water of the Löcknitz, which flows into it.

The way in which absorbed radiation is used, in the energy budget of water, is best seen from an example from I. Dirmhirn [2202], for the Lunzer Untersee. Assuming a daily global solar radiation of 25.1 MJ m^{-2} in high summer, and excluding other types of energy loss, the first meter of water would heat up by 4.3°C, the second by 0.6°C, and the third meter by 0.3°C. The temperature maximum will lie below the surface if, in addition to penetrating shortwave radiation, only the longwave radiation of the water surface is considered, that is, if evaporation and mixing below the surface are eliminated from consideration. H. Reuter [2443] was able to demonstrate this.

To gain insight into the state of the air layer close to a water surface, a distinction must first be made between small water surfaces, lakes, and the sea. Besides these stationary bodies of water, which have time to reach some form of thermal equilibrium with their environment, we have to consider flowing water, which may be colder or warmer than the environment into which they are flowing. The following classification was developed by W. Pichler and amplified by W. Höhne [2204]:

I. Stationary waters

1. Small bodies of water (dealt with in Section 22).

(a) Puddles: small, flat, mostly temporary collections of water, the temperatures of which are determined by the ground, so that there is no temperature stratification within the water (maximum depth about 10 cm).

(b) Pools: permanent or temporary collections of water which show heating from the ground but in which there is thermal stratification subject to daily variations (depth 10 to 70 cm).

(c) Ponds: usually permanent bodies of water, in which there may be diurnal variation of stratification but in which there is no development of a summer discontinuity layer (depth up to several meters).

2. Lakes, in which a summer discontinuity layer develops (Section 23). The term discontinuity layer or thermocline is used to describe a zone at some depth in the water through which there is a particularly great change in temperature. It separates the surface layer, the temperature of which is determined by daily weather changes, from the thermally stable deep waters. Since the thermocline position changes considerably as the weather changes, it usually cannot be identified from mean values.

3. The open sea.

II. Flowing water (Section 23)

In winter, puddles always, and pools often, lose their characteristics as bodies of water when they freeze. In the midlatitude climate, only at times of exceptional cold will ponds and large volumes of flowing water freeze. O. Pesta [2207] has shown how the lives of various creatures in high alpine pools develop according to whether they become inactive in winter, or whether activity can be maintained under an ice cover.

Figure 22-3A shows tautochrones for a 5 m deep clear puddle on a sunny day. The whole layer of water is practically isothermal with its temperature controlled by the air above and ground below. Evaporation results in lower surface temperatures throughout the day. Nocturnal cooling is experienced at the water surface and spreads throughout the layer during the night. The water temperature at the ground is slightly higher during the day due to the absorption of solar radiation and at night due to flow of heat from the ground. Figure 22-3B shows a turbid puddle on the same day. While the nocturnal temperature distribution is not substantially different, during the day the maximum temperature is found within the puddle where much of the incoming solar radiation is being absorbed. Normally, temperatures at the center of the puddle will be slightly cooler than at the shore.

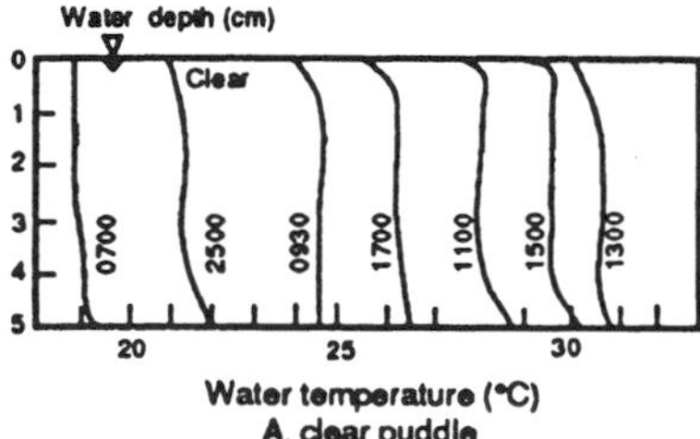

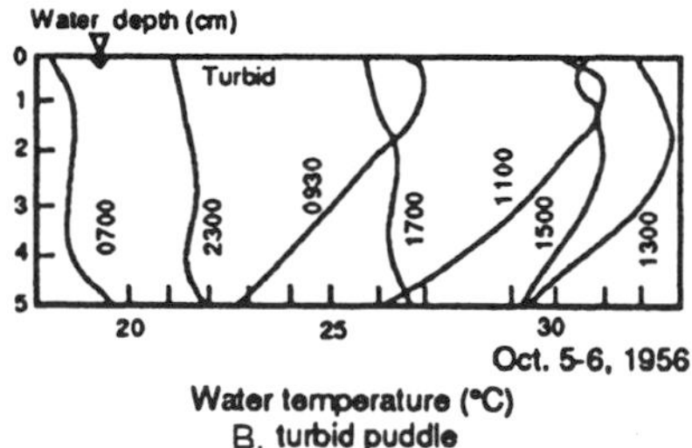

Figure 22-3. Tautochrones for a (A) clear puddle and (B) turbid puddle. (After S. Sato [2209] from Z. Uchijima [2216])

The shape of isotherms is very different in pools. In Figure 22-4 the diurnal temperature variation of the water surface is considerable (8°C on clear days). The daytime heating penetrates slowly into the 30 cm deep mass of water, so that the bottom remains almost 8°C colder than the surface. It is not possible to identify any heating due to absorption of residual radiation at the bottom of the pool in this example. The lower temperatures at the air-water surface of the pool during the day is probably a result of evaporation, while at night these lower temperatures are a result of both evaporation and cooler air temperatures. At night, the warmer temperatures at the ground/water surface are due to energy flowing upward within the soil.

In ponds, inertia of the water mass causes the fluctuation of the surface temperature to be small. The daily temperature fluctuation at the surface decreases with depth in the water, the rate of decrease being inversely proportional to the fourth root of the depth. This can be illustrated by measurements made by J. Herzog [2203] in the Kirchenteich near Leipzig. This is a stretch of water 1.1 km long, averaging 200 m in breadth and 2 m in depth. Figure 22-5 shows temperature measurements made at seven different depths, on a clear summer day with light winds (17 July 1934). Remarks on the type of weather are at the top of the diagram. The isotherms, which are largely horizontal below 1 m, indicate that the daily variation in deeper water is small. Even at the surface, the difference between day and night is only 2°C.

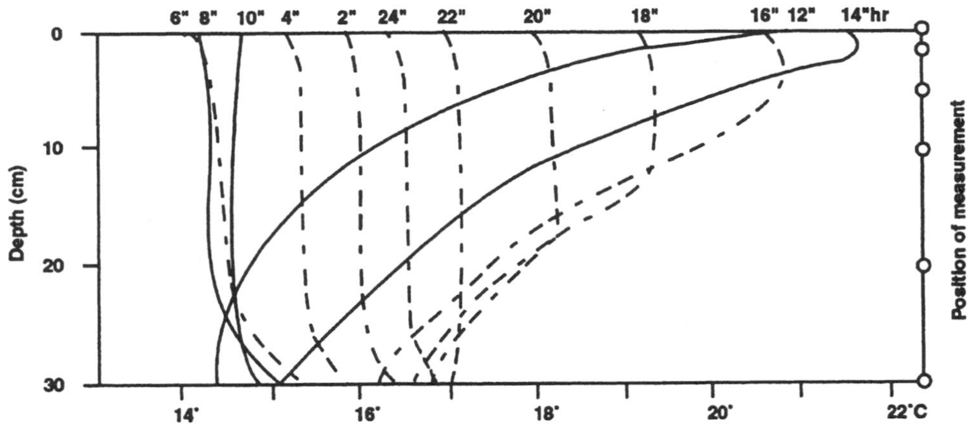

Figure 22-4. Tautochrones in a pool on a clear summer day. (After W. Höhne [2204])

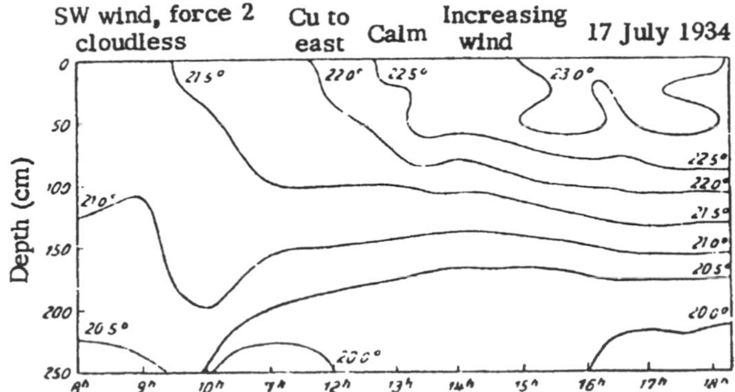

Figure 22-5. Diurnal temperature variation in a 2 m deep pond, the Kirchenteich near Leipzig. (After J. Herzog [2203])

The air temperatures close to the water are determined by water surface temperature in much the same way as the temperature of the ground surface controls those in the layer near the ground. With small water surfaces, advective influences from surrounding areas can be significant. The difference between air temperatures at the center of the pond and the shore areas increases with the size of the body of water. The temperature difference responds slowly to changes in radiation and depends strongly upon wind speed.

Ordinarily, an increase of humidity might be expected over the surface of small areas of water. However, in the middle of the day, the water is cooler than the surrounding land and, therefore, evaporates less. W. Höhne [2204] consequently found that vapor pressure was somewhat lower in the air layer near the water than at the neighboring shore.

At the Zeppelin factory on Lake Constance, engineers took their tools, which were liable to rust, out to the water in the floating dock at night. Even if the warmer lake water experienced more evaporation than the land, the relative humidity increased much more over land because of the sharp fall in temperature, and it was, therefore, relatively drier over the water and less damaging to the tools.

The general impression that during the day, there is often a higher vapor pressure around small bodies of water is correct but for the wrong reason. Evapotranspiration from the vegetation around the lake rather than the lake itself is usually the source of this water vapor. The vegetation evaporates more than the drier parts of the shore or drier land and more than the surface of the water itself because of a larger evaporating surface area and increased atmospheric mixing associated with the vegetation canopy. The only water surfaces that provide rich supplies of water vapor are warm waters (thermal springs or gaps in polar ice) and the spray of waterfalls, as shown by E. Rathschüler [2208] for the great Krummler waterfall in Salzburg area.

Ice formation in small bodies of water does not begin at the banks but on small solid objects at the surface. The surface water, to a depth of 1 cm, becomes supercooled (at most by -0.5 to -0.9°C). When it freezes, the temperature rises spontaneously to 0°C. Melting begins from below. Observations on both of these processes have been published by W. Höhne [2204] for the first 5 cm above and below the water surface.

If a small body of water has a growth of reeds, water plants or algae, they will absorb shortwave radiation that penetrates the water and rise to a higher temperature than the water itself. H. Schanderl [2212] was able to measure excess temperatures of up to 6.3°C in plants compared with the surrounding water in areas of thickly growing algae and pondweed in Lake Geneva. Strong incoming radiation, calm conditions, and transparent water are necessary for such a temperature difference to occur. Excess temperatures are usually only 1 or 2°C because the water normally quickly removes the heat through mixing.

Figure 22-6 shows W. Schmidt's [2214] results for the shore zone of the Lunzer Untersee on a calm, sunny and warm autumn day, 13 November 1926. The positions above and below the surface at which readings were taken are marked on the right side of the diagram. The 10°C line is drawn for reference. The letters *a* through *d* indicate the successive times of day shown below each graph; *a* illustrates positive radiation exchange at 11:12 hr, and *d* a negative radiation exchange at 16:27 hr.

The upper row of graphs gives the measurements made in a shallow bay 20 cm deep. The surface of the ground, which absorbs radiation, shows a temperature maximum (a_1) at midday. There is a second maximum just below the evaporating water surface. With reduction (c_1) and cessation (d_1) of shortwave radiation, water temperatures become uniform. The temperature of water near the shore drops sharply under the advective cooling influence of the land (d_1).

In the lower row of diagrams, the heavy solid line refers to a pool, which dried out a few weeks earlier. The tautochrone a_2 shows the positive net radiation at the surface with cool temperatures from the previous night at 6 cm. The dry pool bottom temperature fluctuations are much greater than the bay. As the cold pulse from the preceding night (a_2) penetrates deeper into the bottom of the dry pond through the day (b_2, c_2, d_2) it becomes attenuated. At 15:03 (c_2) the onset of the surface negative net radiation balance can already be seen.

The dotted lines in Figure 22-6 show the temperature distribution in a dried out area covered with reeds that were mostly dead. Shielding of the surface by the reeds (b_2) makes it slightly cooler than the vegetation-free soil. Night cooling (d_2) is considerable, perhaps because there is still an evaporation influence. The fine solid lines are for an area where reeds are growing in 10 cm of water. This area responds more like dry land than like free-standing water.

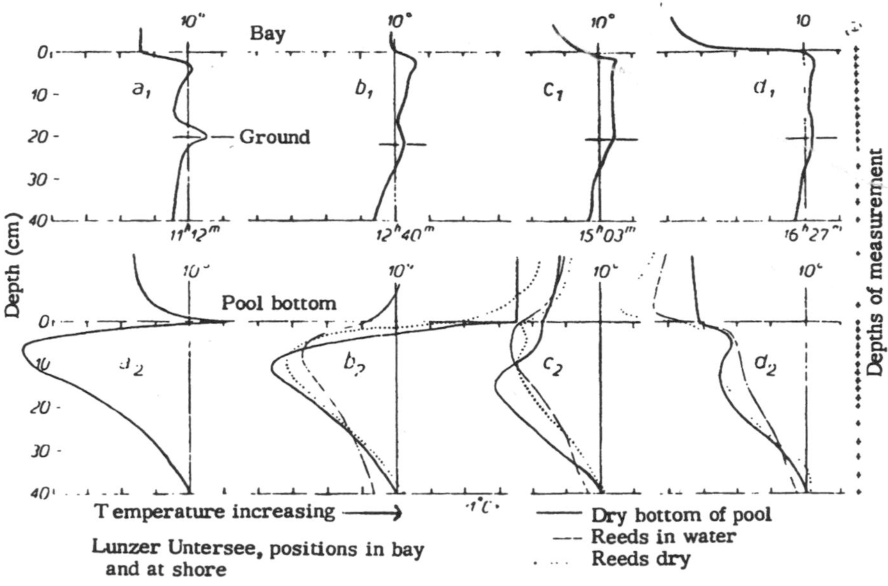

Figure 22-6. Temperature measurements by W. Schmidt [2214] near the shores of the Lunzer Untersee in autumn.

23 The Air Layer near the Water Surface of Lakes, Seas and Rivers

The first noticeable difference found in large water bodies, such as lakes or seas, as compared to the small water bodies discussed in Section 22, is that the daily temperature fluctuation becomes smaller in comparison with the annual variation.

V. Conrad [2303, 2304] found that the daily range of temperatures in the surface waters of Alpine lakes was on the average only 1 or 2°C, even in midsummer. The June average for the Wörther Lake was 2.6°C. In winter, it decreases to a few tenths of a degree. This small diurnal variation has superimposed on it the irregular changes in water temperature caused by currents in the water, upwelling of colder water, wind influence (especially in spring and early summer), unequal vertical mixing caused by changes in the wind field, and cold precipitation. For Lake Constance, W. Peppler [2312] found the day-to-day temperature change in the surface waters averaged 0.18°C for the calmest month (January or February) and 1.22°C for the most unstable month (April or May). V. Conrad [2303] recorded 0.23 and 0.99°C for Lake Gmund, and 0.07 and 1.12°C for Lake Pressegger in Austria. Similar results are given in a collection of data published by O. Eckel [2305] for 14 Austrian lakes.

In the open sea, the daily temperature variation is reduced to a few tenths of a degree. Figure 23-1 shows the daily temperature variation in surface waters. Temperatures were measured by taking samples of water with a sea-bucket made of sail cloth or tin, thus giving only mean values for a surface layer about 1 m deep. The solid line (K) gives the results obtained on the German Meteor. Expedition in the South Atlantic Ocean by E. Kuhlbrodt and J. Reger [2311]; the dotted curve (*W*) is for the North Sea from measurements by E. Wahl [2319]. In spite of the vast geographic differences between the two areas, the temperature fluctuates within about ±0.1°C of the mean value over a daily period for both sites.

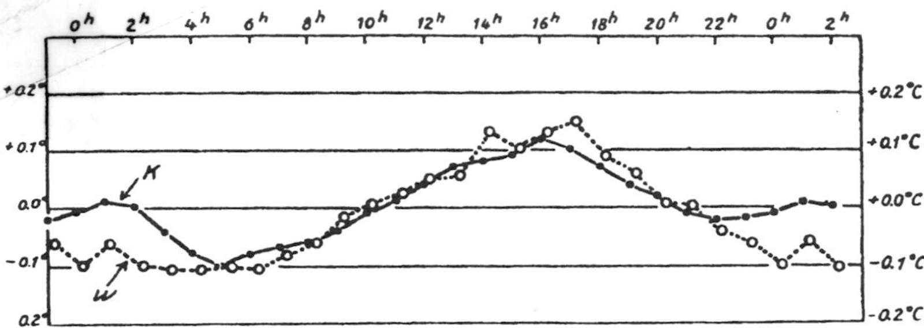

Figure 23-1. Diurnal temperature variation in the surface water of the sea. Solid line is for the South Atlantic; dashed line is for the North Sea. (After E. Wahl [2320])

It should be noted that the daily temperature fluctuation in the air near the surface is greater because of the influence of direct absorption of radiation. Meteorological observations of maximum air temperature, between 12:00 and 13:00, were 0.25 to 0.45°C higher than the minimum; while the maximum for the surface water was only 0.26°C higher and this was not reached until about 18:00. H. U. Roll [2315] showed that the diurnal fluctuation increased with height. He found (in August) that when there was a range of 0.4°C in the surface water temperature, the air temperature increased from 0.6°C at 20 m to 0.9°C at 150 m.

So far as the influence of the underlying surface is concerned, air temperatures near the water, in great lakes and at sea, are determined by the annual temperature variation. Figure 23-2 illustrates water temperature observations (from 1927 to 1950) in the Hallstatt Lake. This lake is 8 km long, 1 to 2 km wide, and 125 m deep; it lies in the Salzkammergut district of Austria. In winter, water temperature is lowest near the surface and increases with depth until it is around 4°C. If ice forms the temperature of the water immediately below the ice will be 0°C, and will increase with depth until it is around 4°C. This temperature corresponds to the maximum density of the water. In spring, the warming of the water takes place from the surface and progresses downward. Within the top 5 m August is the warmest month. From a depth of 5 to 30 m it is September; and at still greater depths, the full effect of summer heat is not felt until October. In November and December as the surface water cools it sinks, resulting in an isothermal pattern. In spring, when the lake is warming up, even the monthly average figures show substantial temperature gradients in the top few meters. Other lakes show similar typical temperature variations.

In these larger water masses, the temperature difference between surface water and the air over it is no longer determined by the regular temperature variation of the underlying surface but rather by the irregular change of air masses. If the water is warmer than the overlying air, the air near the surface behaves in much the same way as the air near the ground at a time of positive net radiation. An unstable sublayer is built up (Section 12) in which gradients are superadiabatic. K. Brocks [2301] investigated this layer at the German

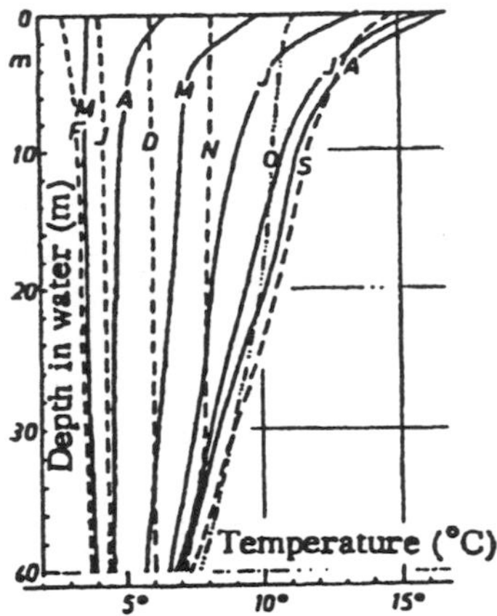

Figure 23-2. Mean monthly tautochrones of water temperature in the Halstatt Lake in Austria. (After O. Eckel [2306])

North Sea coast and found that the thickness h of the superadiabatic layer increased with increasing temperature difference ΔT between the water surface and the air at a height of 5 m in the following way:

ΔT (°C)	0.5	1.0	1.5	2.0	2.5	3.0
h (m)	6	11	15	19	21	23

If, on the other hand, the water is colder than the overlying air, the layer close to the surface behaves in a way similar to the negative net radiation situation above the ground, with an inversion developing above the water surface.

H. Bruch [2302] made temperature measurements in the Baltic Sea. They were made on both sides of the water surface on a clear summer day at the positions and times shown by the small circles in Figure 23-3. Heights above and below the surface are shown on a logarithmic scale to enhance the processes occurring near the surface.

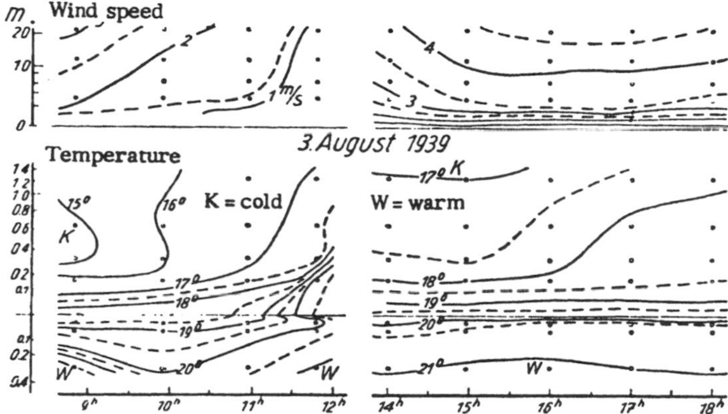

Figure 23-3. Wind and temperature stratification on a summer day above the sea near Greifswald. (From observations by H. Bruch [2302])

The bottom right half of the diagram shows that the warmer water gave off heat to the cooler air in conditions of normal wind stratification (top section, right). The largest gradients were found in the narrow boundary layer in which there is little mixing. A significant decrease in wind can be observed in the upper left half of the diagram after 11:00. This caused a marked rise in the temperature of the surface layer of air, because of reduced mixing, and the formation of a cold film on the surface of the water. This is due to the radiative exchange, and is reinforced by evaporative heat losses. This cold layer is clearly shown for two July days in Figure 23-4. The temperature differences between air and water were small and, unlike Figure 23-3, the water was colder than the air. The isotherms show the cold surface layer of water and the way it influences the air close to it. The effect is very marked in the calm conditions of the left half of the diagram, while with winds of 2 to 4 m sec^{-1}, as on the right half, it still exists but is recognized only when isotherms are drawn at smaller intervals (0.2°C).

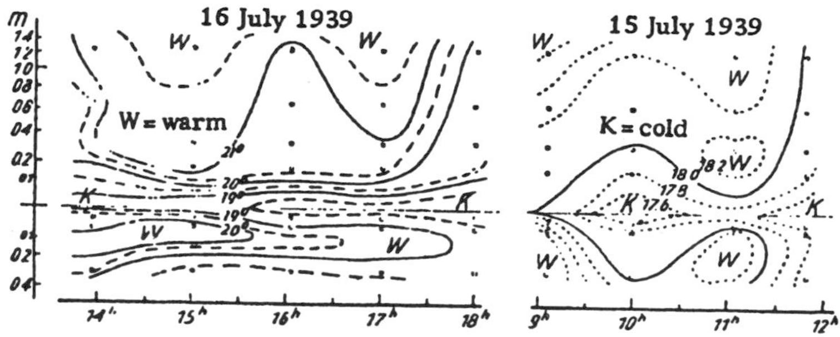

Figure 23-4. The cold layer at the boundary surface of air and water. (From observations by H. Bruch [2302])

H. U. Roll [2317] measured the temperature profile in salt water pools left behind at ebb tide, from 8 cm to -6 cm, under changing conditions in August 1950. In practically all sets of observations, the same surface cooling was found in a layer 2 to 10 mm thick. The degree of cooling may be seen from Table 23-1 for observations on 10 August 1950, between 15:30 and 16:05:

Table 23-1A. Temperatures at the air water boundary.

	Water				Air		
Height (cm)	-6	-2	-0.5	0	0.2	1	10
Temperature (°C)	26.1	26.2	26.1	25.8	26.4	26.5	26.8

H. U. Roll [2316] indirectly demonstrated the existence of this cold water film by another method. Evaluation of 165 temperature profiles taken from three sources in different places, gave, when extrapolated, a water surface temperature 0.5 to 1.5°C lower than that obtained simultaneously by the above methods. Since these figures cover a wide area, the second method probably more closely represents the amount of cooling due to evaporation in the boundary layer between water and air.

The temperature, water vapor, and wind profiles are closely linked over the ocean. For additional information, about air-sea interaction the journal "Ocean-Air Interactions: Techniques, Observations and Analysis, an International Journal" or books by E. B. Kraus and J. A. Bussinger [2310], B. A. Kagan [2309] and G. L. Pickard and W. J. Emery [2313] should be consulted.

Our attention now turns from conditions above stationary water to consideration of the microclimatic features of flowing water in brooks, streams, and rivers.

All streams start off with the temperature of the water or melting snow at the source. As a rule, particularly in summer, in the midlatitudes, this is lower than the temperature the water would have in equilibrium with the environment. In winter, spring water may be at a higher temperature, so that it is cooled when it first emerges, especially if the source is in the mountains. This is especially true of warm springs.

The diurnal temperature variation in river water flowing in a completely level plain has been evaluated for a sunny July day by O. Eckel and H. Reuter [2308] to illustrate the influence of initial temperature and depth of the water. The daily water temperature variation is shown for four initial water temperatures (Tw_A): 10°, 15°, 20°, and 25°C, and for four depths: 30, 60, 100, and 300 cm (Figure 23-5). The daily temperature variation observed downstream will be a function of the distance from the source, speed, the amount and depth of the water, difference between the air and water temperature, and any systematic changes that may occur in riparian shading or adjacent land use along the river. The magnitude of the daily temperature cycle will be smallest in winter, when water temperature variations are caused by daily fluctuations in weather, and greatest in summer when the effects of variable net radiation dominate.

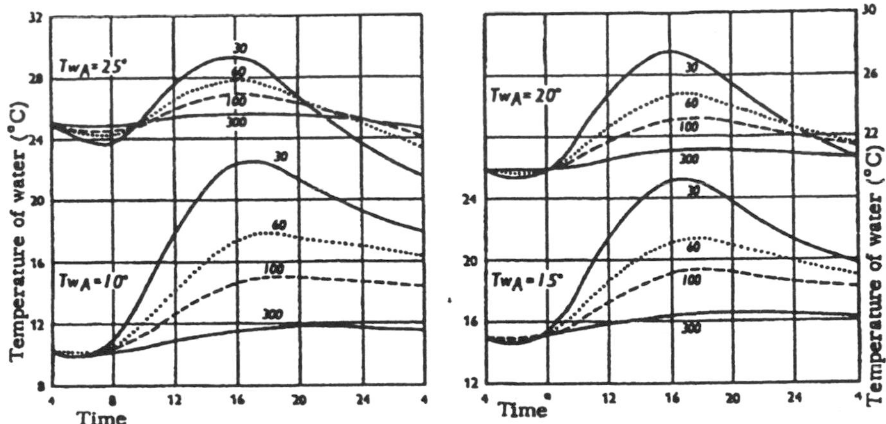

Figure 23-5. Summer temperature variation in rivers as a function of source temperature Tw_A and river depth. (After O. Eckel and H. Reuter [2308])

B. W. Webb and D. E. Walling [2321] examined the long-term thermal behavior of three small tributaries in Devon, UK based upon hourly observations of water temperature over a 14 year period. Average seasonal water temperature regimes were very similar for the three streams, with each exhibiting a peak toward the end of July and a minimum in February. The annual cycle approximated a sinusoidal curve with a steeper spring rise and a more gradual fall decline in water temperatures. This pattern is consistent with the normal variation in the annual soil temperature regime (Section 10). The general absence of spring foliage and presence of fall foliage, combined with seasonal changes in the temperature of groundwater, with its high thermal capacity, further enhance the asymmetric pattern. Significant short-term variations in water temperature are found in all seasons, however, due to synoptic weather disturbances and associated air mass contrasts.

O. Eckel [2307], who made a detailed study of Austrian river temperatures, established that the rivers of the eastern Alps in Austria did not generally reach the equilibrium temperature even after traveling 100 to 400 km. In its early stages, the temperature of water may rise by several degrees on a summer day; but 30 to 80 km from the source, the increase is only about 0.6°C and at 150 to 350 km only 0.15°C for every 10 km of the river bed. The modification of the river temperature along its course is inversely proportional to the distance from the source. This rule, which was derived from observations of the Austrian hydrographic network, is valid for the early morning water temperature reading, which is approximately the daily minimum.

The drag of the flowing water on the neighboring air above it also should be taken into account. This drag effect has been investigated on the Inn River by E. R. Reiter [2314], on the Main near Frankfurt, and on the Rhine near Bonn and Cologne by K. O. Wegner [2323]. Wind speed must be less than 4 m sec^{-1}, otherwise frictional mixing by the gradient wind reaches down to the surface. E. R. Reiter [2314], therefore, carried out his research during calm early mornings, while K. O. Wegner [2323] waited until evening and night.

E. R. Reiter [2314] dropped smoke bombs into the Inn River from a suspension footbridge crossing it, and filmed the track of the smoke. Figure 23-6 shows the smoke trails from 4 of the 63 experiments made at intervals of 1 second. In experiment *a* (top left) the air was stationary at 3.3 m. Below this level, it was entrained by the river with increasing speed as the surface was approached. A few turbulent eddies can be seen disturbing the movement of the smoke, and above the stationary level, there is evidence of a light wind blowing up the valley. Although experiments *b* to *d* followed within a few seconds, the height of the stationary level shifted, moving down to 2.5 m. In all measurements, the water surface had the same degree of roughness. When the upvalley winds became stronger, there was a sudden jump in roughness because of capillary waves created by the wind.

In K. O. Wegner's [2323] observations on the Main and Rhine, the air dragged along with the water showed no temperature decrease with increased height. This layer was from a few tens of centimeters to 150 cm thick. Above this layer, there was a sudden change in wind direction and a transition to a gradient wind, with speed increasing with height as over the ground. K. O. Wegner [2323] suggested this was due to increasing frictional exchange which accompany stronger winds and makes it possible for impulses received from the moving water to remain effective up to higher levels. The substantial differences in wind profile within this layer of air, accompanying the flowing water, may have their origin both in differences in width of the rivers, the type of surrounding country and in the time of day at which the measurements were made.

Because of the generally narrow width of rivers and the frequent presence of riparian vegetation, rivers often have their radiation input reduced by horizontal shielding from trees along their banks. O. Eckel and H. Reuter [2308] established, for a river 30 m wide, the relative amount of radiation received, as a percentage of unobstructed solar radiation (shield-

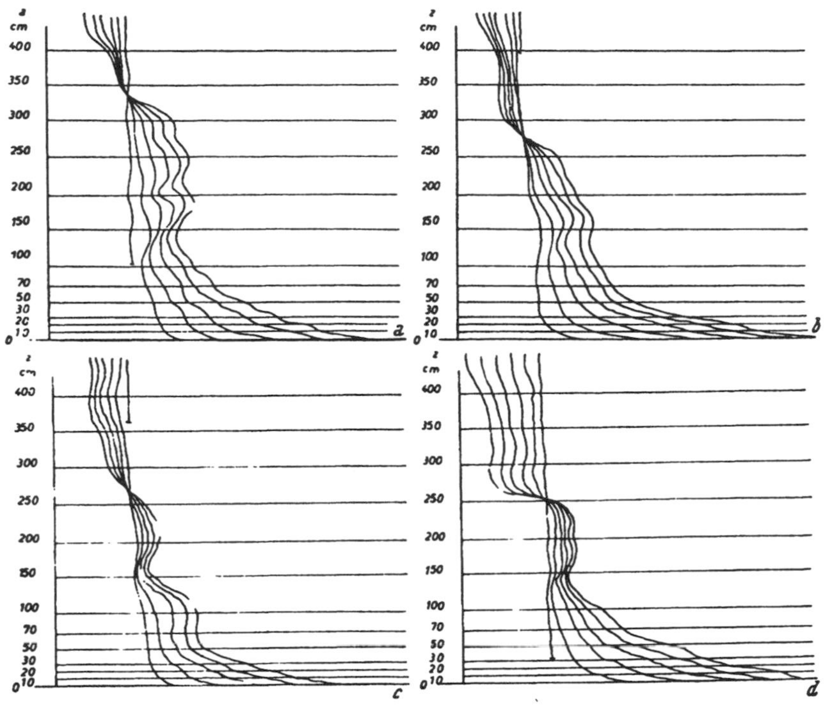

Figure 23-6. The movement of smoke trails at intervals of 1 second in the air close to the surface of a river, showing the drag effect of the flowing water and the influence of a slight wind up the valley. (After E. R. Reiter [2314])

ing angle 0°) for the March and September equinoxes and for midsummer (Table 23-2). The effect of shielding on outgoing radiation by night has already been dealt with in Section 5.

Table 23-2. Relative solar radiation received by rivers as percentage of unobstructed solar radiation.

Greatest angle of shielding		6°	11°	17°	22°	26°
Corresponding to tree heights(m)		3	6	9	12	15
	Equinoxes					
	N-S	86	75	67	59	54
Direction of valley	NW-SE or NE-SW	88	77	68	60	54
	W-E	91	80	71	62	54
	Midsummer					
	N-S	88	78	70	65	57
Direction of valley	NW-SE or NE-SW	91	83	76	70	64
	W-E	93	88	84	79	76

The temperature of stream and river water is determined by the exchanges of energy across the channel surface and bed. Under non-advective conditions characterized by the absence of tributary inflow, groundwater flow, and channel precipitation, the energy budget of a short reach of a stream or river is: $Q_N = Q^* \pm Q_E \pm Q_H \pm Q_{Gb} \pm Q_F$ where Q_N = total net heat exchange, Q^* = net radiation, Q_E = latent heat transfer due to evaporation or condensation, Q_H = sensible heat flux between the atmosphere and water surface, Q_{Gb} = heat transfer across the channel bed, and Q_F = heat transfer due to friction.

B. W. Webb and Y. Zhang [2322] monitored these variables in the Exe Basin, Devon, UK, along short 20-30 m reaches which encompassed a wide range of channel morphologies, sizes, and riparian characteristics. Results of energy term gains and losses from the summer of 1992 are given in Table 23-3. Values are given as a percentage of energy gains (+) and losses (-) due to the variable size of the three stream reaches. Black Ball Stream covers a small drainage area (1.89 km^2), has a W-E valley orientation, and is adjacent to moorland and roughly grazed land. Iron Mill Stream covers a larger drainage area (32.9 km^2), has a W-E valley orientation, and is surrounded by dense deciduous woodland. River Culm 2 drains 180.5 km^2, has an E-W valley orientation, and is adjacent to land grazed by cattle. These results provide insight to the nature of stream energy exchanges, and the variation that can occur due to channel size, channel morphology, and the nature of the local channel environment.

Table 23-3. Percentage of average daily gains (+) and losses (-) of non-advective heat energy in three rivers in the Exe Basin, Devon, UK during the summer of 1992. (From B.W. Webb and Y. Zhang [2322])

Energy term	Gain (+) or loss (-)	Black Ball Stream	Iron Mill Stream	River Culm 2
Net radiation, Q^*	+	83.0	70.1	86.9
	-	35.7	66.2	53.1
Evaporation, Q_E	+	0.7	2.9	0.3
	-	23.4	21.5	19.4
Sensible heat, Q_H	+	11.7	16.4	9.8
	-	5.8	6.3	3.4
Bed conduction, Q_{Gb}	+	0.5	0.6	1.3
	-	35.1	6.0	24.1
Friction, Q_F	+	4.1	10.0	1.7

Net radiation (Q^*) comprises the most important form of both energy gain and loss. Considerable variation in net radiation gain can occur, however, because of differences in channel shading by riparian vegetation, channel beds, and valley topography. Seasonal variations can also be significant. Wide and unobstructed channels receive their greatest net radiation input in summer, while narrow channel reaches heavily shaded by deciduous vegetation can actually receive their greatest net radiation input in fall or winter. Large variations in net radiation input also occur due to changes in daily cloud cover. Net radiation also normally contributes the largest percentage of energy loss, particularly for reaches with a high degree of shading.

Evaporation (Q_E) generally provided the second greatest loss of energy, approaching energy loss totals by net radiation in some cases. Open and exposed streams experience the greatest loss of energy by evaporation, while heavily shaded reaches, with reduced exposure to high winds and higher humidity, have the lowest Q_E percentages. Evaporative losses are enhanced in shallow channel and decrease as channel depth increases. Heat gain by condensation is only a small source of energy.

Q_H is the second most important source of non-advective energy input after net radiation. Sensible heat flux normally provides a source of energy during spring and summer when the air temperature is greater than the water temperature, and an energy loss during fall and winter when water temperatures often exceed air temperatures. Sensible heat becomes a more important source of energy as a channel reach becomes more shaded from the sun by vegetative cover.

Conduction of heat into the channel bed comprised a significant source of energy loss, rivaling the total loss by evaporation and sensible heat flux on occasion. Energy loss by Q_{Gb} was highest in the summer and for shallow unobstructed channels when daytime net radiation can penetrate the water and be absorbed along the channel bed. Energy input by friction of flowing water, Q_F, was a significant term along channel reaches with a large channel gradient. Seasonal variations in Q_F also occurred due to fluctuations in river discharge.

Advection of heat can also occur in stream reaches due to heat transfer associated with direct precipitation onto the channel surface and groundwater inflow. Heat input by channel precipitation was found by B. W. Webb and Y. Zhang [2322] to be very small. Heat transfer associated with groundwater flow into or out of a stream reach, however, was much more significant. Whether heat transfer by groundwater flow represents an energy input or loss will depend upon whether the water is seeping into or out of the river reach, and whether the groundwater is warmer or cooler than the river water.

For an introduction to current work on surface energy exchanges and stream temperature variations in flowing streams, the reader is referred to B. A. Sinokrot and H. G. Stefan [2318].

24 Snow and Ice Properties and Interaction with the Environment

Snow cover as an underlying surface is as varied as the ground. Snowflakes can be deposited so loosely that the density of the snow cover is only 10 kg m^{-3}. In general, new snow has a density of 50 to 200, and old snow 300 to 500 kg m^{-3}. The density and cohesion of new snow depends largely upon its temperature. Dry snow occurs at lower temperatures and is characterized by lower density and poorer cohesion, while wet snow occurs at higher temperatures and has a higher density and viscosity. Rain falling into snow cover increases its water content and therefore its density. The density of permanent snow cover on high mountains increases to 800 kg m^{-3} as it changes to firn snow and then firn ice; glacier ice can have a density in excess of 900 kg m^{-3}.

The many different types of snow cover and avalanches are discussed in W. Paulcke [2440a] and D. McClung and P. Schaerer [2431]. They describe the dozen different types of snow that an observant walker or skier should be able to distinguish, as well as the many changes snow undergoes when it is transported, redeposited by wind, settles on aging, melts and refreezes. W. Paulcke [2440a] was the first to construct a snow profile from actual measurements depicting the state of the snow cover at a given moment as a result of the alternation of new snowfalls and the aging processes.

Figure 24-1 shows the accumulation, melting, and structure of the snow cover on the Hohenpeissenberg in the Alps from measurements made by J. Grunow [2420] in the winter of 1951-52. The observation was on a westward trending ridge of the 977 m high mountain. Daily readings by the Weather Service nearby were slightly different and are included in Figure 24-1 for comparison with average snow density (above) and snow depth (below). The upper half of Figure 24-1 shows the new snow (bars) between November and March, and the increase in average snow density for the whole snow cover (trace), which is related to the aging of the snow. The depth of the snow cover, its inner structure, and the quantity of meltwater are given in the lower part of the diagram. After a new snowfall, the older snow settles under the weight above it. Density should, therefore, increase with depth. Experience shows, however, that the maximum density is often not at the bottom but close to

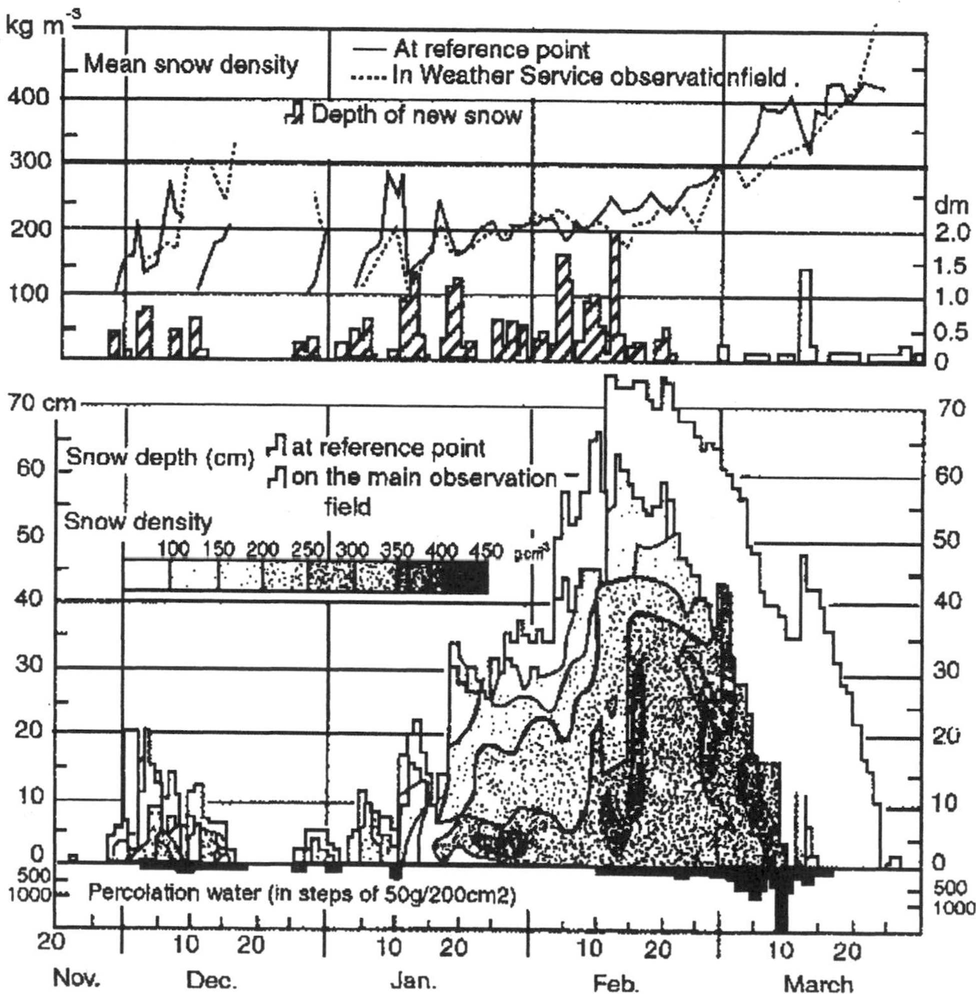

Figure 24-1. Height and structure of snow cover on the Hohenpeissenberg in the winter of 1951-52. (After J. Grunow [2420])

the top. There are three reasons for this. First, when the surface snow melts under the influence of solar radiation (or advected warm air), the water stays near the snow surface (and often refreezes) rather than migrating to the bottom of the layer of snow. Second, the decrease of temperature in winter from the warmer, deeper layers of snow to the colder snow surface (Figure 24-3) produces, in addition to the capillary rise of water, an upward transport of water vapor by diffusion, similar to that described for frozen ground in Section 21. This contributes to both the surface crust and the frequently observed relatively hollow spaces beneath the crust from which the snow has sublimated and the water vapor diffused toward lower vapor pressure at the surface. Rain freezing at the surface is a third factor which may contribute to the surface crust and higher surface snow density.

Temperature gradients within a glacier or snow pack result in vapor pressure gradients and a resulting flow by sublimation and condensation from areas of higher (warmer) to areas of lower (cooler) vapor pressure. In autumn, when the surface is cooling rapidly and the underling layers are still relatively warm this flow can be large and results in what W. S. B. Patterson [2440] called depth hoar, which is a hoar frost with the snow pack. The direction of the flow of water vapor within the snow pack is reversed in spring. In addition, water vapor flows from smaller particles to larger ones because the vapor pressure is higher over convex than over flat surfaces and higher over flat than concave surfaces. This process is sometimes called "destructive metamorphism" because the original crystal forms are destroyed, resulting in a simplified form (N. J. Doesken and A. Judson [2415]).

During the interval between dry snow deposition and the melting of wet snow, a snowpack undergoes continuous changes in a number of important physical properties. This transformation process, called snow metamorphism, is completed before snowmelt runoff occurs, and involves changes in the density, albedo, strength, temperature, water and impurity content, insulating quality, and the shape and size of snow grains. The process of snow metamorphism is driven by the surface compaction of wind, surface melting, infiltration, and subsurface refreezing of snow, as well as recrystallization of snow grains. In general, the snow near the surface is heated either by absorbing solar radiation and/or by contact with warm air. As it melts it sinks into the colder snow below and refreezes, releasing latent heat of fusion, and raising the temperature and heat content of the snow pack. Through this process of energy redistribution from the surface of the snow cover to the snow pack interior the snow cover gradually warms and becomes more dense. When the snow pack reaches an isothermal state of 0°C, ablation by snow melt production begins. Prior to this ripened condition most ablation is by densification. The snow cover shown in Figure 24-1 reached peak snow depth on 10 February but did not reach maximum snow density until about 1 March. The ablation of the snow cover between 10 February and 1 March was due to densification. Ablation after 1 March was due to snow melt production. After a snow pack reaches a ripened condition nearly all the positive net radiation is converted into latent heat of fusion, resulting in relatively rapid snow pack removal.

W. S. B. Patterson [2440] suggests that in the initial stages of snow melting the grains typically pack together. Melting increases the rate at which grains become rounded, because the snow melts earliest at its extremities. The average grain size increases because smaller grains melt first and join the larger ones. Joining of snow is particularly rapid in the surface layers influenced by a daily freeze-thaw cycle. For additional information dealing with snow packing, the transformation of snow to ice or about glaciers, W. S. B. Patterson [2440] is recommended. S. C. Colbeck [2412] provides a review of the process of metamorphism, and E. J. Langham [2427] summarizes the different physical properties of dry and wet snow.

The property that is most relevant to the nature of the boundary surface of snow and air is the high albedo of snow for shortwave radiation. Table 4-2 shows that new snow has an albedo of 75 to 98 percent. H. Hoinkes [2422] gives a value of 40 to 60 percent for the albedo of firn snow. This decreases still further as firn gradually changes to ice. Glacier ice has an albedo of 20 to 40 percent, depending on how dirty it is. F. Sauberer [2445] found a value of 7 to 8 percent for clear ice from the Lunzer Untersee. The reflected light seen by an observer consists not only of radiation reflected from the surface, but also of spectral light filtering through the surface from below (Section 22). I. Dirmhirn [2414] has shown this makes up a significant, sometimes even the preponderant, part of the reflected solar radiation measured by instruments. Numerical values which she obtained for ice, firn, and old snow are:

Albedo (percent):	10	20	30	40	60	80
Fraction (percent) of light from below:	31	48	80	92	88	64

As already shown by F. Sauberer [2445], reflected solar radiation within the visible spectrum is only slightly dependent on wavelength, and then only to the extent that the middle range of the visible spectrum is reflected most strongly. This has been confirmed for Antarctic snow fields by G. H. Liljequist [2428] at Maudheim station, where the reflectivity on overcast days varied between 92 percent (red) and 97 percent (green, yellow). I. Dirmhirn [2414] found the following reflectivities for Alpine glaciers in 1950:

Wavelength (μ):	0.4	0.5	0.6	0.7	0.8
For clean ice (percent):	44	54	56	48	32
For dirty ice (percent):	24	53	36	31	19

When snow first falls, it has a high albedo. As time progresses, the albedo decreases. D. G. Baker, et al. [2407] analyzed the relationship between temperature, heat sums, solar altitude, the number of days since the last snowfall, and the decrease in the albedo of snow. They found that the number of days since the last snowfall resulted in the simplest and strongest relationship explaining the decrease of the snow albedo. This was substantially better than the (expected) relationship between this decrease and heat sums (which by their nature include time as a factor).

The depth of snow required to mask the underlying surface varies with the extinction coefficient of the snow, the albedo, the underlying surface, and height of the vegetation. D. G. Baker et al. [2408] examined the relationship between mean daily albedo and snow depth over three surfaces – bare soil, sod, and alfalfa – during 19 winters (1969-87) at St. Paul, Minnesota, and found that the albedo was about 70% when bare soil was covered by 5.0 cm, sod by 7.5 cm and alfalfa by 15 cm of snow. Additional amounts of snow cover above these levels resulted in only a negligible albedo increase.

The high albedo of snow has a particularly strong influence upon air temperature, resulting in what is often referred to as the temperature-albedo feedback. Although K. F. Dewey

[2413] demonstrated that this influence can be detected in daily temperatures, its influence is most noticeable when climate data are aggregated over longer periods of time.

D. G. Baker, et al. [2406] systematically examined the temperature and radiation differences observed between days with and without snow cover at St. Paul, Minnesota [45°N, 296 m). Their measurements included 23 years (1963-85) of snow depth, and maximum, minimum, and average daily temperatures, and 11 years (1975-85) of global solar radiation, reflected solar radiation, counterradiation, and outgoing longwave radiation observations. Measurements were made from 16 December through 15 March, and were organized into three categories for analysis: days with no snow cover (10.4% of all days), days with <10 cm of snow cover (62.4% of all days), and days with ≥10 cm of snow cover (17.2% of all days). Mean daily temperature and radiation totals for the entire study period are given in Table 24-1. Only minor differences in incoming shortwave and longwave radiation were noted among the snow cover groups. In comparison to days with no snow cover, however, days with ≥10 cm snow cover had an albedo which was higher by 59.3%. This produced significant differences in the remaining variables. Mean reflected solar radiation for the ≥10 cm snow cover group was 3.8 times greater than the no snow cover category, while mean absorbed solar radiation totals were only 25% as large. The smaller outgoing longwave radiation totals observed for the ≥10 cm snow cover case was primarily due to lower surface temperatures, which, (assuming an emissivity of 0.98 for snow and 0.92 for bare soil), were found to be 15°C lower than for bare soil. Differences in net radiation observed among the sites were primarily due to the effect of albedo in controlling the absorption of solar radiation, although moderate differences in net longwave radiation were also found. The impact

Table 24-1. Mean daily air temperture (°C), radiation totals (MJ m^{-2} day^{-1}), and albedo (percent) for days with no snow cover, <10 cm snow cover, and ≥10 cm snow cover for St. Paul, MN. See text for explanation of period of measurement. (After D. G. Baker, et al. [2406])

Variable	No snow cover	<10 cm snow cover	≥10cm snow cover
Maximum air temperature	2.8	-3.7	-5.6
Minimum air temperature	-6.7	-13.0	-15.1
Average air temperature	-1.9	-8.3	-10.4
Surface temperature	4.0	-6.0	-11.0
Global solar radiation ($S + D$)	9.11	7.49	8.63
Reflected solar radiation ($K\uparrow$)	-1.78	-4.34	-6.77
Net solar radiation (K^*)	7.32	3.15	1.86
Albedo	20.2	56.3	79.5
Counterradiation ($L\downarrow$)	22.07	21.49	19.58
Outgoing radiation ($L\uparrow$)	-26.60	-24.35	-22.66
Net longwave radiation (L^*)	-4.53	-2.86	-3.08
Net radiation ($K^* + L^*$)	2.79	0.29	1.22

of snow albedo upon the surface microclimate can be observed in the mean air temperatures in Table 24-1, which show that the mean maximum and minimum air temperatures for the 10 cm snow cover group were both 8.4°C lower than the no snow cover group. This was due to both the higher albedo and the reduction of the flow of heat from the soil with increased snow cover.

K. G. Bauer and J. A. Dutton [2410] observed a relationship between temperatures and snow cover at Madison, Wisconsin. In the spring, as long as snow remains on the ground, the air heats slowly, since only about 30-40 percent of the incoming radiation is absorbed. Once the snow cover is removed, however, approximately 85 percent of the incoming solar radiation is absorbed, and air temperatures increase much more rapidly.

A. Ångström [2404] first showed that global solar radiation increased over snow-covered terrain in comparison to similar terrain in a snow-free condition. This results from an enhancement of diffuse solar radiation (D) due to backscattering of reflected solar radiation ($K\uparrow$) toward the surface. This produces a series of multiple reflections of diffuse solar radiation between the surface and the atmosphere which is proportional to $(K\downarrow)\ (1 - R_T B_S)^{-1}$, where $K\downarrow$ = global solar radiation ($S + D$) over a horizontal surface, R_T = regional albedo (fraction), and B_S = atmospheric backscatterance (fraction). The regional albedo is not identical to the surface albedo, but is representative of the surrounding terrain. The atmospheric backscatterance is a function of atmospheric conditions, and varies with the cloud type and amount, and the aerosol concentration. The effect is greatest with an extensive new snow cover and partly cloudy conditions, and decreases in magnitude with aging of the snow, reduction of snow cover, and clear skies. W. T. Kierkus and W. G. Colborne [2424] have shown, using data from eight Canadian stations, that this multiple reflection process can lead to a 30-40 percent increase in mean daily diffuse radiation. J. E. Hay [4008] presents historical monthly mean values of R_T and B_S for five stations in Canada, while F. Möller [2433] summarizes earlier empirical work on the process. The winter observations in Figure 40-4(a) above the maximum curve of $K\downarrow$ are a result of backscattering. At night, in cities with a low cloud cover after a fresh fall of snow, repeated backscattering of city light from clouds and reflection from the snow can be particularly intense, resulting in unusually bright nocturnal conditions. P. Todhunter has observed conditions in Grand Forks, North Dakota that were bright enough to enable a person to read outdoors by this enhanced illumination.

Shortwave radiation may also penetrate into snow but not as easily as into water (Section 22). While in the latter case penetration was measured by the percentage penetrating each meter, it is more useful here to use the extinction coefficient ν (cm^{-1}), which is defined by the relation

$$I_z = I_0 e^{-\nu z} \qquad (\text{W m}^{-2})$$

in which I_0 is the radiation received at the surface and I_z is the radiation penetrating to the depth z cm. The relation between the transmission D and coefficient ν is given by $D = 100e^{-100\nu}$; for example:

Extinction coefficient ν (cm^{-1}):	0.005	0.01	0.02	0.03	0.04	0.05
Transmission D (percent m^{-1}):	60.7	36.8	13.5	5.0	1.0	0.7

Extinction coefficients of about 0.07 and 0.23 cm^{-1} are commonly recorded in snow although in extreme cases, values as high as 1.5 cm^{-1} can occur. It varies within wide limits; however, according to F. Löhle [2429], this is not related in any ascertainable way to the amount of water in the snow. Figure 24-2 shows the percentage of the radiation penetrating the snow surface to various depths, for values of ν within the range just mentioned.

Figure 24-2 has been extended to include ice, which will be discussed later, and also lake and pure (distilled) water for comparison with Sections 22 and 23. The value of ν for distilled water is about 0.003 cm^{-1} in the green-yellow range. The curve of $\nu = 0.005$ corresponds roughly to the water of the Untersee in 1937-38 for the same spectral range (Figure 22-2).

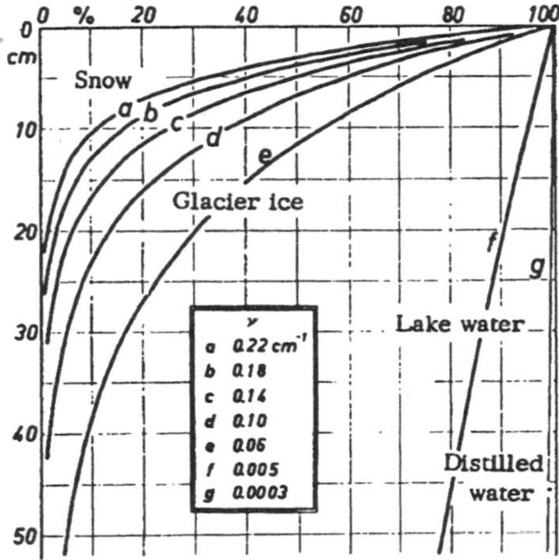

Figure 24-2. Penetration of light into snow, ice, and water.

F. Sauberer [2445] demonstrated that there were considerable differences in the penetration of light in the range 0.38 to 0.76 μ into snow. However, no systematic relation with the wavelength could be found. If, however, the amount of water contained in the snow increases, then the extinction coefficient increases, with increasing wavelength. G. H. Liljequist [2428] found, at the Antarctic station of Maudheim that with snow of density 400 kg m^{-3} extinction coefficients were:

Wavelength (μ):	0.42	0.52	0.59	0.65
Color of light:	blue	green	orange	red
Extinction coefficient, ν (cm^{-1})	0.066	0.083	0.114	0.172

Shortwave radiation can penetrate ice more easily than snow. F. Sauberer [2445] found an average extinction coefficient of 0.03 cm^{-1} for snow plates 4 cm thick with a negligible dependence on wavelength. Measurements by W. Ambach [2403] on the Hintereis Glacier in the Ötztal Alps gave a mean value of $\nu = 0.057$ cm^{-1} for 228 individual measurements. This value corresponds approximately to the *e*-line in Figure 24-2. Therefore, at a depth of 20 cm, we still find 30 percent and at 40 cm, almost 10 percent of the shortwave radiation penetrating the ice surface. Dependence of the extinction coefficient on wavelength is much the same for ice as for water, as shown by F. Sauberer [2446]. Ice from the Lunzer Untersee had the following extinction coefficients:

Wavelength (μ):	.313	.35	.4	.45	.5	.55	.6	.65	.7	.75	.8
Color of light:			violet	blue	green	yellow		orange	red		
Extinction coefficient, ν (10^{-3} cm^{-1})	1.00	0.5	0.4	0.5	0.8	1.3	2.0	3.4	6.0	10.6	17.7

There is a sudden jump in the value of the extinction coefficient in the near-infrared part of the spectrum, as shown above. Thick ice has a bluish hue because of the low extinction coefficient at these wavelengths.

Snow and ice are practically black bodies for longwave radiation. There is no surface in nature that approximates so closely the ideal hollow-box radiator as the porous surface of a snow cover. The longwave reflectivity of snow (1-ε) is only around 1.4 percent. Hence the Falckenberg paradox that a snow surface can be made to reflect better (for this spectral range) only by spreading soot on it. K. Ya Kondratyev [212] mentions that the transition from snow as a good reflector in the visible spectrum, to snow as a near blackbody in the infrared spectrum is very gradual. The reflectivity of snow for shortwave radiation decreases steadily down to 2.6 μ where it amounts to only a few percent.

The thermal conductivity of snow is a function of snow density ρ. This relation has been expressed by several writers in a number of different equations. H. Abels [2401] proposed a simple and useful relation:

$$\lambda = c\rho^2$$

The constant c has the value 28.4 J kg K^{-1} according to H. Abels [2401], while F. Loewe [2430] obtained a value of 27.6 from measurements made in Adélie Land, while measurements by J. Bracht [2102] gave a value of 20.5 for snow of density 190 to 510 kg m^{-3}. Thermal conductivity is given in Table 24-2 for three different values of c and various snow densities. The larger values probably are more correct than the lower ones. As density increases, λ gradually approaches the value for ice, given in Table 6-2 as 20.9-29.3 W m^{-1} K^{-1}.

Table 24-2. Thermal conductivity λ (W m^{-1} K^{-1}) of snow for various values of specific heat *(c)* (J kg K^{-1}) and snow density.

	Snow density (kg m^{-3})							
c	100	200	300	400	500	600	700	800
21	0.02	0.08	0.19	0.34	0.52	0.75	1.03	1.34
25	0.025	0.10	0.23	0.40	0.63	0.90	1.23	1.61
29	0.03	0.12	0.26	0.47	0.73	1.06	1.44	1.88

H. Reuter [2442] demonstrated that in less dense snow there is a distinct mixing process of a convective nature at work in the air spaces, which increases thermal conductivity in the upper layers to 7 or 8 times the value of molecular conduction. If computations are made according to the rules laid down in Section 6 for the ground, which do not take this type of exchange into account, incorrect values are obtained for the decrease of diurnal temperature fluctuation with depth and to a smaller extent for the phase lag. This results in an appreciable uncertainty in the values of λ given in Table 24-2. The thermal conductivity of ice increases as the temperature decreases (Table 24-3).

Table 24-3. Variation in the thermal conductivity and specific heat of ice as a function of temperature.

Temperature (°C)	0°C	-20°C	-40°C	-60°C
Thermal conductivity λ ($W\ m^{-1}\ K^{-1}$)	2.2	2.4	2.7	2.9
Specific heat c ($10^3\ J\ kg\ K^{-1}$)	2.09	1.97	1.80	1.68

The low thermal conductivity of snow gives rise to large vertical temperature gradients within the snow covering. Figure 24-3 shows temperature variations in snow varying in depth between 32 and 65 cm in the winter of 1937-38, measured by O. Eckel and C. Thams [2416]. It should be noted that the isotherms are drawn at a larger contour interval near the air-snow surface than near the ground-snow surface. When the air-snow surface temperatures are -33°C, the temperature of the ground surface seldom falls much below 0°C. This shows the high degree of protection afforded by the insulating effects of snow cover. Plants covered by snow are not only protected from extreme cold, but also from the drying out of the soil by repeated freezing and thawing (Section 21). Portions of the plant which protrude through the snow are subject, however, to both abrasion by blowing snow and ice, and by the very low temperatures which occur at or slightly above the snow surface.

Figure 24-4 shows the daily temperature variation in snow for 16 January 1932 from observations made by E. Niederdorfer [2514]. The reading for 09:45 shows a temperature difference of 9°C between the surface and a depth of 20 cm and is indicative of nocturnal radiation loss. Warming during the forenoon brings the snow surface up to the melting point. It should be noted that the maximum temperature lies 1 to 2 cm below the snow surface. In summer in Antarctica, T. W. Schlatter [2447] found that the highest temperatures and initial snowmelt occurred around 10 cm below the surface. H. Reuter [2443] evaluated the temperature variation that would occur in a penetrable medium, assuming initial isothermal conditions, that only shortwave radiation was able to penetrate the snowpack, that a constant amount of longwave radiation was emitted from the surface only, and that no other influences were at work. The results are shown for one computed example in Figure 24-5. The temperature maximum slightly below the surface coincides with the occurrence of maximum net radiation at that depth, and is where the maximum solar radiation is being absorbed. By absorbing penetrating shortwave radiation, objects buried in snow and ice can become warmed and melt a cavity. This can result in a very irregular micro-relief as melt holes break through to the surface.

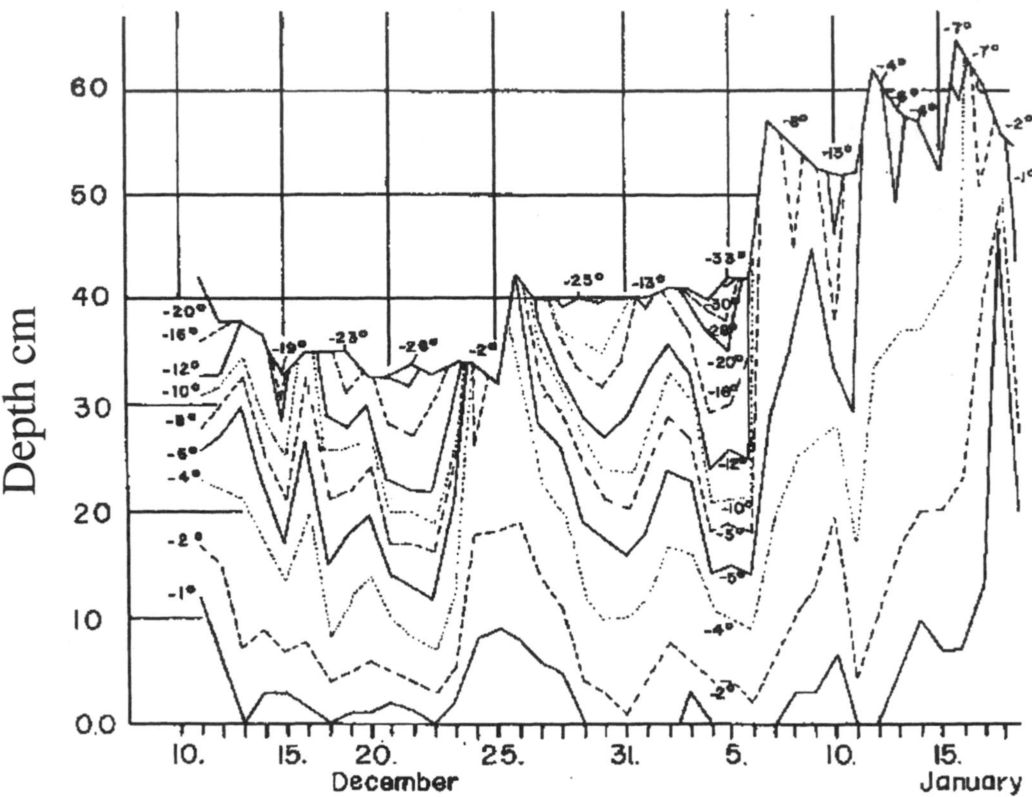

Figure 24-3. Temperatures in winter snow cover at Davos. (After O. Eckel and C. Thams [2416])

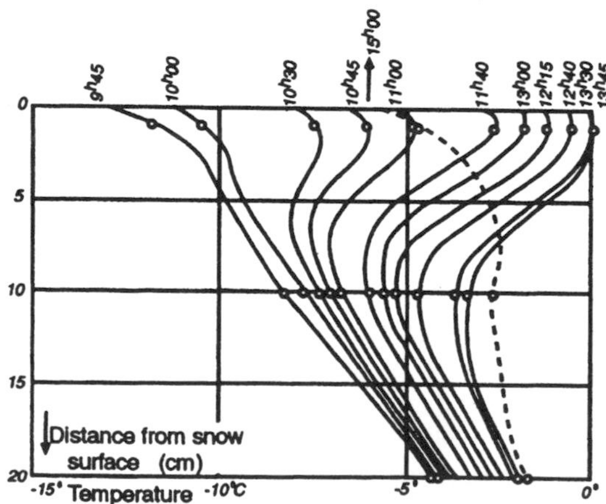

Figure 24-4. Tautochrones in snow cover on a sunny winter day. (After E. Niederdorfer [2514])

Three different underlying surfaces are compared in Figure 24-5; moist sand: R = 10 percent, $(\rho c)_m = 1.42\ 10^6$ J m^{-3} K^{-1}, λ = 13.4 W m^{-1} K^{-1}; old snow: R = 60 percent, ρ = 400 kg m^{-3}, c = 2115 J kg^{-1} K^{-1}, λ = 4.6 W m^{-1} K^{-1}; and ice: R = 0, ρ = 917 kg m^{-3}, c = 2115 J kg^{-1}K^{-1}, and λ = 21.4 W m^{-1}K^{-1}. Isothermal conditions were initially assumed for all three cases, followed by 3 hours of incoming shortwave radiation of 551.3 W m^{-2}, and outgoing longwave radiation of 69.8 W m^{-2}.

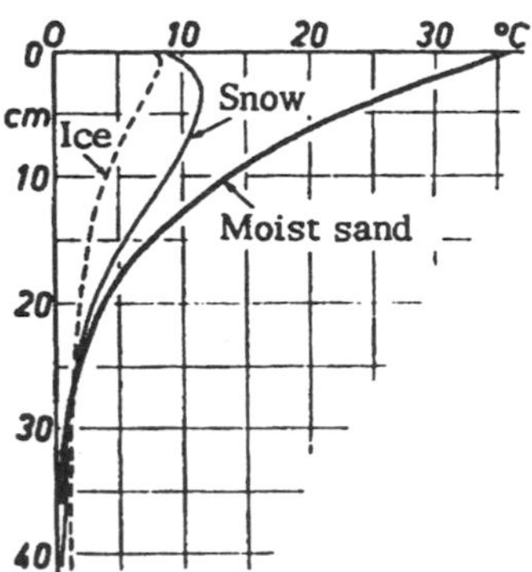

Figure 24-5. Theoretical temperature distribution in ice, snow, and ground. (After H. Reuter [2443])

Since it is assumed that ice has the same selective absorption of radiation as water (Section 22), the dashed curve (Figure 24-5) will also give the temperature distribution in water in which no form of mass exchange is taking place and which is determined only by radiation. It would appear that even if evaporation is not taken into account, the temperature maximum will lie just under the surface of water as well. The sandy soil shows the positive net radiation pattern, somewhat similar to Figure 10-2. The theoretical maximum for snow, at a depth of a few centimeters under the surface, is in good agreement with Figure 24-4 and with similar measurements made by J. Keränen [2423]. This also provides an explanation for F. Loewe's [2430] observation that the annual average temperature in Antarctic snow at Adélie Land was 0.6°C higher at a depth of 5 cm than on the surface. It follows that the position of the temperature maximum in the stationary state does not depend on the thermal conductivity or the thermal diffusivity of the snow, but on the intensity of the incoming and outgoing radiation and on the extinction coefficient ν. Table 24-4 gives the theoretical depth in centimeters of the maximum temperature below the snow surface, for the four values of the extinction coefficient used in Figure 24-2. The theoretical depth of the maximum temperature is inversely proportional to the extinction coefficient. The maximum temperature difference with respect to the surface depends on the thermal conductivity, and increases very quickly as the density of the snow decreases.

Y. Takahashi, et al. [2450] have measured temperature profiles in a snow cover at night. They found that the night minimum, like the day maximum, did not occur at the surface, but was 7 mm below it. This could be explained theoretically by assuming that the porous structure of the snow allowed outgoing radiation to be emitted down to a certain depth under the surface. It was only when the amount of mixing was small, and, therefore, the transport of energy Q_H from the layer of air close to the snow was also small, that the minimum was found at the surface.

Table 24-4. Theoretical depth (cm) of maximum temperature below snow surface.

Radiation ($W m^{-2}$)		Extinction coefficient ν (cm^{-1})			
Net shortware	Net longwave	0.10	0.14	0.18	0.22
279	105	4.7	3.4	2.6	2.1
279	70	2.9	2.1	1.6	1.3
558	105	2.1	1.5	1.2	0.9
558	70	1.3	1.0	0.7	0.6

Measurements made by W. Ambach [2402] in the ice of Vernagt Glacier in the Ötztal Alps showed that the latent heat of fusion of ice (0.334 MJ kg^{-1}) played an important role in determining the behavior of night temperatures. When the glacier ice, which has been warmed up to 0°C during the day, cools at night, the latent heat released on freezing retards the drop in temperature. Calculation of the thermal balance on three nights in July 1952, gave the following (Table 24-5) average values for the period from 18:00 to 06:00:

Table 24-5. Nocturnal energy exchange (MJ m^{-2}) in a glacier. (After W. Ambach [2402])

energy loss:	
through net longwave radiation	1.93
through evaporation and sublimation	0.04
energy received:	
from the air	0.71
from latent heat of fusion of ice	0.67
amount lost on cooling	0.59

Let us now turn our attention to temperatures in the layer of air close to snow or ice. Figure 24-6 shows qualitatively the temperature pattern for three different types of weather situations observed at the Munich airport.

The upper record is for 9 January 1935 when a fresh snowfall during the night covered the thermometer lying on the ground. Its temperature increased, while the air temperatures above the snow continued to decrease until morning. The middle record, for 20 January 1935, in clear frosty winter weather, shows the temperature variation under the 9 cm snow cover (this shows only minor short-term fluctuations); also shown is the air temperature at three different heights above the snow. In the middle record, there is a slight temperature decrease with height during the day and a strong inversion beginning in the early evening and continuing through the night (not shown). The traces in the lowest part of the diagram, for 12 February 1935, are for melting snow that is still 6 cm deep. Temperature at the snow-ground surface is constant at 0°C, but the layer of air close to it exceeds 10°C because of the effects of solar radiation and warm air advection. Between 14:00 and 15:00 the positive

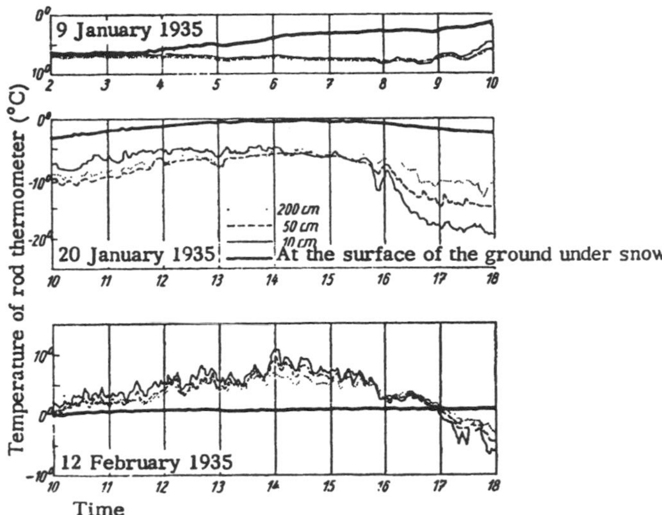

Figure 24-6. Temperature measurements at the snow-ground surface and in the air above the snow for the Munich area; fresh snow (top), winter frost (middle), and thaw (bottom).

net radiation balance is well marked, changing at sunset with the onset of frost to a negative net radiation regime.

The low thermal conductivity of snow (Table 24-2) is one of the reasons for the sudden cold that often follows a snowfall (particularly in fall or early winter). The snow reduces the flow of heat from the soil, and while this results in the soil remaining warmer, it contributes to the cooling of the air. The high emissivity of snow also contributes to the cold following a snowfall.

In comparing the rate of cooling over a grass and a fresh snow surface, L. C. Nkemdirim [2438] found that under clear skies the actual rate of cooling was twice as large over fresh snow as over grass. This result was attributed primarily to the near blackbody emissivity of a fresh snow cover in contrast to the graybody emissivity of grass. He suggested that a secondary factor in explaining the differential rates of cooling was greater roughness of the grass as compared to the snow cover. The greater roughness resulted in more mechanical turbulence, more heat being brought to the surface (Table 19-1), and thus contributed to the slower rate of cooling over the grass.

It is difficult to make accurate measurements of true air temperature close to snow during the day on account of the strong reflected solar radiation from the surface. Even small polished thermoelements require some form of radiation shielding, as shown in experiments by G. Band [2409]. Things are simpler at night. A. Nyberg [2439] has made measurements of the night inversion in the lowest 25 mm above snow cover in Sweden. These observations give the relation between wind speed and the increase of temperature with height (Table 24-6) and show the decrease in the lapse rate and the rise in nocturnal temperature associated with increasing wind speed.

The inversion found at night over the air layer near snow can also occur during the day in high latitude climates in winter and in midlatitude climates in summer over glaciers. This feature has been studied by H. Hoinkes and N. Untersteiner [2509] on the Vernagt Glacier

Table 24-6. Temperature (°C) as a function of wind speed and height above snow surface at night.

Wind speed ($m\ sec^{-1}$)	Number of observations	Height above snow surface (mm)						
		1	5	10	15	20	25	1400[a]
Absolute calm	37	-17.6	-17.0	-16.4	-16.1	-15.9	-15.7	-12.1
0.3-0.6	30	-11.5	-10.7	-10.1	-9.8	-9.4	-9.2	-6.7
0.9-1.2	21	-9.3	-8.7	-8.4	-8.2	-8.1	-8.0	-6.4
1.8	20	-4.1	-3.7	-3.5	-3.4	-3.3	-3.3	-2.7

[a]Extrapolated.

in the Ötztal Alps and by H. Hoinkes [2421a] on the Hornkees in the Zillertal Alps and on the Hintereis Glacier [2508] in the Ötztal Alps. Table 24-7 gives a few values extracted from the observations made on the Hornkees in September 1951 at an elevation of 2262 m. The increase of vapor pressure with height follows (as temperature) a logarithmic pattern.

Table 24-7. Air temperature and vapor pressure above glacier ice.

3-9 September 1951		08:00	12:00	15:00	18:00	Absolute Maxi-mum	Absolute Mini-mum	Average 09:00-16:00
Air temperature (°C)	130 cm	6.3	7.6	8.1	7.1	12.0	–	7.41
	10 cm	4.2	5.9	6.2	4.6	–	1.2	5.70
Vapor pressure (mb)	130 cm	6.8	7.2	7.5	7.6	–	–	7.29
	10 cm	6.5	6.8	7.1	7.2	–	–	6.92

In the plant world, a snow cover means both protection and danger. The temperatures given in Figure 24-3, for the interior of the snow, show that plants within the cover, or to the extent that they are partly below the snow surface, are protected against severe winter cold and are completely shielded from the wind. As long ago as 1902 W. Bührer [2411] advocated the rule, based on measurement of the minimum temperature at the snow surface and at the ground surface below it, that even 1 cm of snow would give some protection. Above 20 cm, additional snow cover affords little further protection.

The amount of carbon dioxide emanating from the ground and exhaled by plants at night can reach high though not dangerous levels under the snow, as shown by F. Pichler [2441]. He measured the carbon dioxide content in a rye crop under snow from 35 to 135 cm deep, lasting from 17 to 92 days, and only once obtained a value of 0.21 percent by volume. He ascribed this low value to the permeability of the snow cover, to low plant respiration, and to absorption of carbon dioxide by the snow.

The situation is much more dangerous for the plant parts projecting above the snow. They are exposed during the day to both direct-beam and diffuse solar radiation as well as the reflected solar radiation from the snow surface. They are strongly heated, a fact that is immediately evident from the melt holes, which develop around projecting twigs or trunks in sunny weather. Figure 24-7 is a photograph taken by A. Baumgartner in March 1954 on the southeast slopes of the Gross Flakenstein in the Bavarian Forest, which illustrates this feature. At night, outgoing radiation from the snow surface can produce very low temperatures. In addition, when winds are strong, physical damage to plant parts exposed above the snow can result from the hard crystals of drifting snow (Section 17).

Figure 24-7. The melted areas around each trunk are the result of radiative melting. (Photograph by A. Baumgartner)

Figure 24-8 is a reproduction of a photograph taken by P. Michaelis [2432] of a pine tree at the tree line in the Allgäu Alps. The broken line shows the level of winter snow cover. The configuration of the tree is controlled by the double surface. Above the winter snow, lower branches are missing from the right (north) side as a result of the abrasive action of the wind-driven snow. This part of the trunk is also lacking in lichens, which are prevalent elsewhere; often the bark is deeply scored. On the left (south) side of the picture, however, the branches are withered but still present as dead wood and are overgrown with lichens. The very large temperature fluctuations over the strongly reflecting snow surface are responsible for the dead branches.

A. Niemann [2437] published excellent color photographs of similar damage on a small scale. One is of a cherry laurel of which the leaves above the snow have a reddish-brown dead appearance, while those below the snow are a fresh green color. For an excellent summary of the causes, types and shapes of tree deformation caused by the wind and weather, the reader is referred to M. M. Yoshino [2712], J. Hennessey [2421], S. L. Backhouse and R. L. Pegg [2405], and R. C. Musselman [2436].

In midlatitude climates sublimation is the principal cause of the decrease of snow cover when the temperature is lower than 0°C. Sublimation is possible only when the vapor pressure of the air is less than that of the snow surface; if the reverse is the case, frost will be deposited on the snow. Since the latent heat of sublimation is high (2.835 MJ kg^{-1} at 0°C), only about 0.2 to 1.0 mm (equivalent depth of water) is normally sublimated even on sunny days. Even on days with extremely favorable conditions, the amount seldom exceeds 6 mm.

H. G. Müller [2435] has demonstrated that the humidity of the air plays an important role in the melting process. The saturation vapor pressure over melting snow (0°C) is 6.11 mb. With decreasing temperatures, the saturation vapor pressure over ice decreases quickly as shown in Table 21-4. If the vapor pressure of the air is lower than this, sublimation will take

Figure 24-8. Damage to a fir tree on a mountain above the level of winter snow (broken line) through drying-out (left) and drifting snow (right). (After P. Michaelis [2432])

place. If, however, it is higher than this value, water vapor will condense from the air onto the snow surface, and by releasing 2.501 MJ for every kilogram condensed (0°C), will be able to melt about seven and one-half times the quantity of water condensed. The moister the air, the more quickly the snow melts. If radiation is discounted, the energy supplied from the ground is omitted, and if the snow is not melting, then the snow's surface will behave as a wet-bulb thermometer and be cooled (by sublimation) below the air's temperature. In the following table, the energy used in sublimation is equal to the flow of energy by conduction from the air to the snow's surface resulting in no melting snow.

Relative humidity (percent):	100	80	60	40	20	0
Air temperature (°C)	0.0	1.2	2.5	4.2	6.3	9.4

From this, it can be seen that snow melts more rapidly with increasing air temperature and higher relative humidity under otherwise similar conditions. With the unusually high temperatures that are found in association with föhn winds in the foothills of the Alps, the snow may disappear very quickly. J. Grunow [2420] reports that on 13 March 1951, the snow cover on the Hohenpeissenberg decreased 24 cm in depth and lost 60 mm of meltwater.

Substantial micro-scale differences in radiation, temperature, and humidity patterns are common in most snow-covered environments. An example has already been given in the photograph in Figure 21-12, which shows that snow melts at different rates. On that occasion melting was caused by differences in the thermal conductivity of the ground. T. Fukutomi [2418] analyzed 144 observations made in January and February 1950 (while searching for warm springs), and established that there was a good correlation between the ground temperature and the thickness of the snow cover (temperatures ranged from 1° to 10°C and snow cover from 70 to 0 cm). J. L. Monteith [2434] observed on 7 January 1955 that snow melted more quickly over the short grass on a cricket pitch than on the longer grass in the vicinity because the thermal conductivity was better, but it melted later on 14 February, in the same year, because this time melting was caused by warm air to which the longer grass responded more readily. The effect of the soil temperature and its thermal properties on melting snow is greater in fall when the soil temperatures are higher but persist throughout the winter. Table 19-2, for example, shows the relationship between different soil thermal conductivities and diffusivities on snow depths throughout the winter and for a snow fall in spring.

In comparing winter soil conditions with corn stubble either standing, lying prostate (no tillage) or removed, B. S. Sharrat et al. [2448] found that standing stubble trapped more snow. This resulted in warmer soil temperatures (as much as 2°C), shallower frost penetration (as much as 0.5m less) and earlier soil thaw (up to 20 days). Winter soil temperatures, and depth and duration of soil freezing were the same in soils where the stubble was removed or lying prostrate.

It is, therefore, possible to indicate the microclimatic peculiarities of a restricted area by charting the state of its snow cover. H. Slanar [2449] was probably the first to plot maps of the melting process. K. Kreeb [2426] plotted the area on 22 February 1952 and showed that the mapped state of snow cover was in good agreement with phenologic charts for the same area. G. Waldmann [4511] used snow maps to delineate the boundaries of similar plant communities in the Bavarian Forest. Figure 45-6 is an example of such a map. The melting of snow cover in the Alps at the end of winter follows a similar pattern, as shown by H. Friedel [4615]. H. Friedel [4615] writes, "We have only to remember the figure of the 'Falconer', which appears about the middle of May on the slopes of the north chain, below the Rumerspitze, often seen from Innsbruck, and the shape called the 'white scythe' which appears on the Monte Veneto toward the end of June, and which is a sign to the farmers in the Innster district that it is time to begin mowing hay." The surprising thing is that this melt pattern is so strongly influenced by microclimate that photographs or charts of the same piece of ground present the same appearance year after year as the snow melts, even if the date on the calendar may be different.

As the snow cover melts, ice plates are often formed. The process by which they form is illustrated in Figure 24-9. Freshly fallen snow is piled up in a mound over the tufts of grass (1). During the following day, there is a certain amount of melting on the "southern slope," and at night this water often freezes again, forming ice (2). The next day, solar radiation penetrates almost without hindrance through the thin ice plate and continues the melting process at the heated grass (3). The ice plates remain in position above the disappearing snow, often like piles of broken glass on top of the grass (4). Relatively high temperatures can often be produced in the hollow spaces under the ice plates because, supported on the blades of grass, they are insulated from the warmer ground.

A second situation in which ice plates form is over a depression in the snow such as caused by a footprint. With calm conditions, the minimum temperature is typically found from 5 to 10 cm above the surface (Section 12). As water vapor condenses at these colder temperatures, a thin plate of ice "grows" from the edges covering the depression.

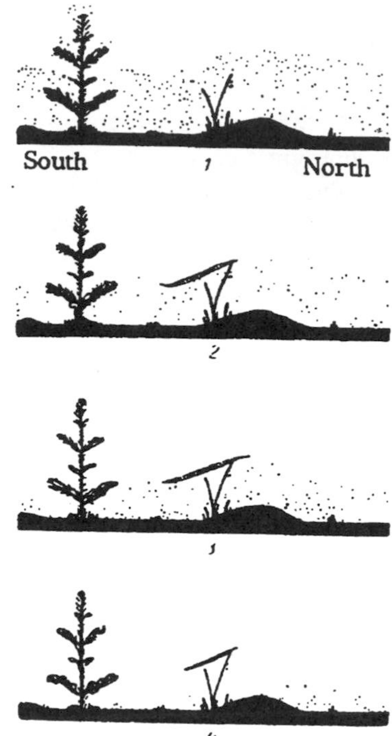

Figure 24-9. The formation of ice plates when snow melts.

Once large open spaces form, snowmelt proceeds rapidly because of intense heating of the exposed ground. Figure 24-10 records temperature measurements made by M. Köhn [2425] on 29 May 1937 in the Black Forest. Conditions of winter cold still prevail under the patch of snow, which is a few meters in diameter, while 2 m away from its edge soil temperatures reach 15°C.

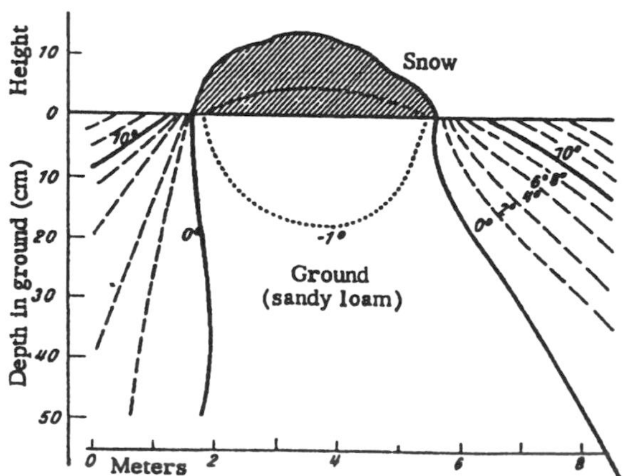

Figure 24-10. Ground temperature (°C) around a melting mass of snow.

In situations like this, "snow smoke" may be observed. F. Rossmann [2444] stated, if air temperature and humidity are high and the wind is light, a very fine misty veil becomes visible above the snow. When there is a breath of wind, this misty veil forms on the windward side of the snow patch and evaporates again not far beyond the limit of the snow on the lee side. The average of a number of windward readings gave 18.1°C and 82 percent relative humidity for the air, while in the lee it was 15.2°C and 89 percent. Cooling of the warm flow of air near the ground by the snow surface causes this phenomenon.

A more unusual occurrence is the formation of snow rollers or cylinders. When there is a strong wind, which must reach at least 20 m sec^{-1} in gusts, snow rollers or cylinders may be formed in a continuous snow cover on smooth fields and meadows. Figure 24-11 was photographed by D. White [2451]. The term roller is used when there is a hole through the axis of the cylinder, so that one can see through it, and it is called a cylinder when completely filled. The width of the rollers varies between 15 and 80 cm and the diameter between 8 and 50 cm; sometimes they form in hundreds and then look like molehills when seen from a distance. The tracks that they leave behind during the course of formation may be up to 1 cm deep and are usually very clearly marked; their length is 10 to 20 m; the highest value recorded being 34 m.

Figure 24-11. Snow rollers. (After D. White [2451])

W. Friedrich [2417] demonstrated that temperature conditions played an important part in their formation in addition to the preponderant influence of wind. Powder snow does not have the power of cohesion necessary for the formation of snow rollers, while heavy soggy snow cannot be whirled up and have its form altered by wind action. The unusual conditions necessary for their formation are only transitional stages, which are present only for brief periods. A warm wind will make the loose snow, whisked up into the air, moist and gluey enough to be able to increase in size with further rolling. Topography also has some effect, since small irregularities disturb the wind field and lead to lee eddies that initiate the process. It seems likely that they all start as rollers and are changed by pressure and subsidence into cylinders. More details, references, descriptions, and photographs can be found in W. Gressel [2419].

Snow cover will be mentioned again later when the microclimate of high mountains is described in Section 46.

CHAPTER IV

FURTHER ANALYSIS OF THE ENERGY BALANCE

25 Basis and Methods of Evaluation

Earth's radiation equilibrium with the surrounding universe was reviewed in Section 2. The processes which determine the energy budget at the base of the atmosphere, that is, at the ground, were mentioned in Section 3, and the laws governing these processes were discussed in Sections 4 to 8. Knowledge of these matters is a prerequisite for understanding the state of the air near the ground, the changes taking place in it, and in the underlying surface. Discussion of the energy balance, which has been described qualitatively, can now be subjected to quantitative examination. This chapter will focus upon the energy balance.

As in Chapter I, the problem will be considered first in a broad sense. Figure 25-1 presents the earth's average annual energy budget. This diagram was first developed by J. M. Mitchell, Jr. [2512] and has been modified and updated with data from R. M. Rotty. The left side of Figure 25-1 illustrates the shortwave energy balance. Assuming that 100 units of energy (percent) arrive from the sun, about 3 units of the total incoming solar radiation are absorbed by ozone in the stratosphere, and about 17 units are absorbed in the troposphere primarily by water vapor, carbon dioxide, and particulates. On the average, 39 units of the incoming solar energy are intercepted by clouds, of which about half (20 units) are reflected and scattered back to space, 4 units are absorbed by the clouds and 15 units are scattered by clouds to the surface. On the average, about 49 units of total solar radiation reach the surface (24 units as direct-beam and 25 units as diffuse). A small amount of this radiation (3 units) is reflected, particularly in areas covered by snow or at low sun angles. About 30 units of the total incoming solar radiation are not absorbed by the earth-atmosphere system and are reflected or scattered back to space; this is referred to as the planetary albedo. Reflection from clouds accounts for nearly two-thirds of the planetary albedo. For comparison, the planetary albedos of Jupiter and Venus are 41 and 49 percent, respectively.

The longwave radiative exchanges are illustrated on the right side of Figure 25-1. Earth's surface emits about 114 units of longwave radiation of which 108 are absorbed in the atmosphere by water vapor, trace gases, clouds and particulates, and 6 units are lost from the earth-atmosphere system through the atmospheric window (Section 5). On the average, more energy is conducted-convected (We use the term conduction-convection because at the boundary layer of the air near the surface energy is gained or lost by conduction, yet most of the energy flow to and from the atmosphere is a result of convection (as described in Section 7)) from the ground to the air (primarily during the day) than from the air to the ground (primarily during the night). The surface of Earth thus acts as a net source of sensible heat (6 units). In addition, more (24 units) energy flows from the surface into the air as latent heat

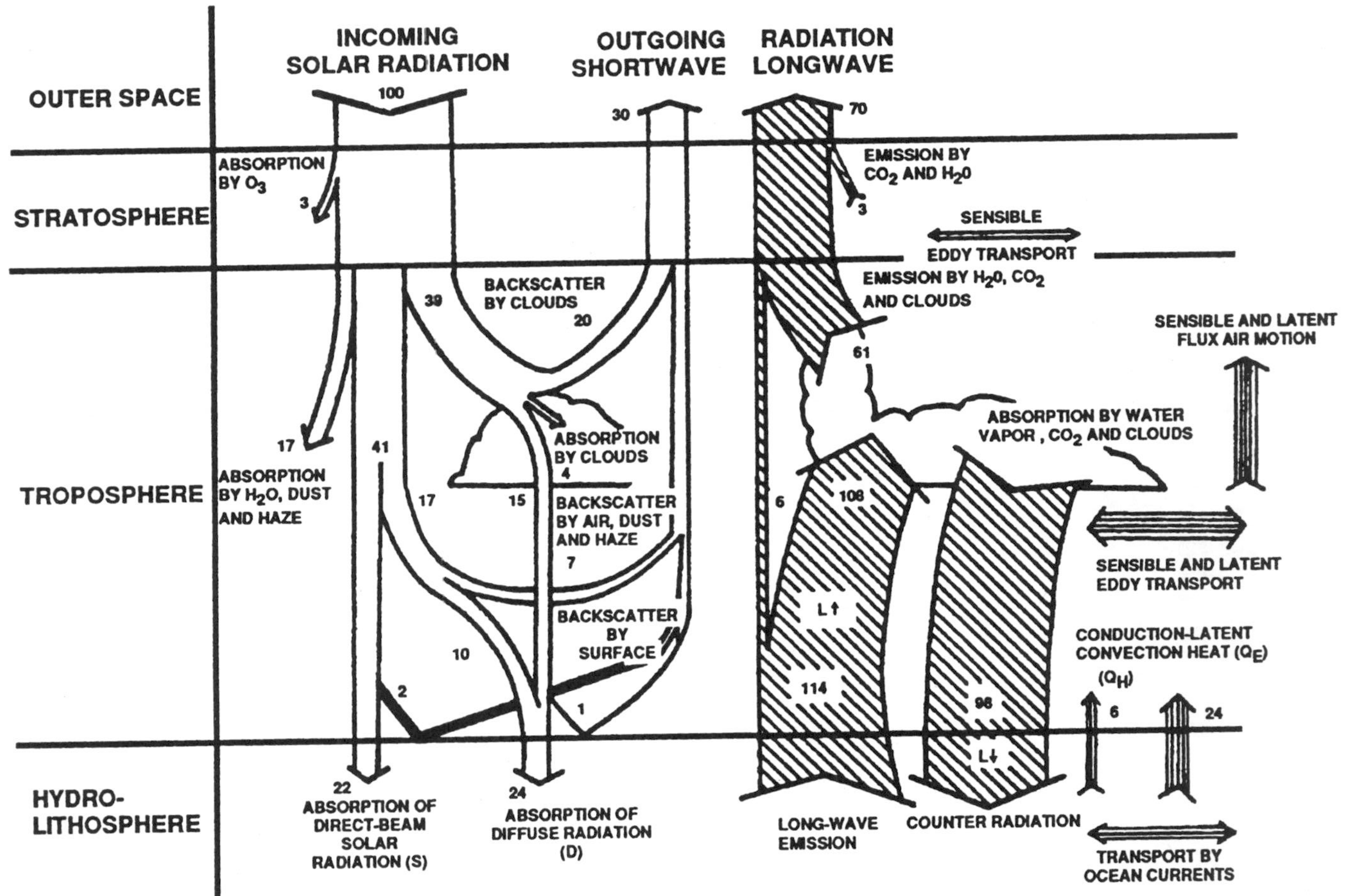

Figure 25-1. Average annual energy balance of earth and its atmosphere. (Modified from J. M. Mitchell, Jr. [2512] as adapted by R. M. Rotty, Institute for Energy Analysis)

than condenses out as frost or dew. About 70 units of longwave radiation is emitted to space by the earth-atmosphere system.

Several features of the average annual energy balance of Earth shown in Figure 25-1 merit specific mention because of their relevance to the surface microclimate. First, scattering of solar radiation by clouds, gas molecules and particulates is so prevalent that over 50 percent of the average global solar radiation at the surface of Earth is diffuse radiation. Second, clouds play the dominant role in explaining geographical and seasonal variations in surface shortwave and longwave radiation regimes. Clouds account for 67 percent of the planetary albedo, 17 percent of atmospheric absorption of solar radiation, and 60 percent of the diffuse solar radiation reaching the surface. Third, although the surface emits 114 units of longwave radiation, the input of longwave radiation to the surface by counterradiation is also large, resulting in a surface net longwave radiation loss of only 16 units. Conduction-convection and evaporation ($Q_H + Q_E$) total 30 units, accounting for 65 percent of the 46 units of net surface energy loss. Finally, 24 units of latent heat are lost from the surface, primarily through evaporation, which represents 52 percent of the 46 units of net surface energy loss. This shows the paramount importance of evaporation in the surface energy exchange.

Let us now consider the ground surface energy balance (Section 3) in more detail. The energy balance equation for the ground-air interface is:

$$Q^* + Q_G \text{ (or } Q_W) + Q_H + Q_E + Q_A + Q_R = 0$$

Since the present concern is with solid ground, the factor Q_W can be neglected. It is always very difficult to evaluate the gain or loss of energy caused by transport of warmer or colder air from surrounding areas. Thus, Q_A will be considered to be zero for the research cited. This assumption requires that the surrounding environment have the same characteristics as the actual point of measurement, such as equal roughness, same type of soil and vegetation cover, etc. The first quantitative measurements of the energy balance, such as those of H. U. Sverdrup [2516] or E. Niederdorfer [2514], were carried out over level snow surfaces which satisfy these requirements. In the "Great Plains Turbulent Field Program," O'Neill, Nebraska was selected because it was relatively homogeneous. According to H. Lettau and B. Davidson [104], the ground surface was homogeneous 1300 m downwind, and the observation site was surrounded by a circle 16 km in radius that was uninterrupted except by a few trees along the river bank, and these were 8 km distant. As a rule, however, the assumption that $Q_A = 0$ is not fulfilled, and for this reason Section 27 is devoted to these advective effects.

The quantity Q_R, representing the gain or loss of energy due to precipitation, has little direct influence. Precipitation, which usually falls from higher levels, mostly has a cooling effect (Figure 6-1). Its indirect effects are great, however, because it changes the moisture status of the ground, which, in turn, affects all the other elements in the energy budget. If precipitation occurs in a solid form, the resulting snow or ice cover completely transforms all energy relations at the ground-air interface. An attempt, therefore, is made to have $Q_R = 0$ as well.

After eliminating the extraneous factors, the energy balance is given by the equation

$$Q^* + Q_G + Q_H + Q_E = 0$$

Two of these factors, namely Q^* and Q_G, can be measured directly. In Section 4, it was shown that the radiation balance Q^* consisted of several shortwave and longwave radiation components. Figure 4-1 showed the proportion of these quantities for two selected times. Four-year monthly averages from 1 March 1954 to 28 February 1958 are shown in Figure 25-2. Analyses of the figures for two individual years have been published by R. Fleischer [2504a, 2505].

The top curve in Figure 25-2 gives the longwave radiation emitted by the ground. In addition to the small amount of reflected longwave radiation (Section 4), this consists principally of radiation emitted by the ground ($L\uparrow$) (Stefan-Boltzmann law) and, therefore, increases and decreases as the temperature of the ground rises and falls. This radiation would cause a tremendous loss of energy, in comparison with incoming solar radiation, even in summer, if it were not compensated to a considerable extent by counterradiation $L\downarrow$. The vertically hatched area between the two curves represents the (negative) longwave radiation balance ($L\downarrow$ - $L\uparrow$); the loss of energy is greater in spring and summer than in winter. On individual days, it is a maximum with clear skies and at a minimum with low clouds or fog.

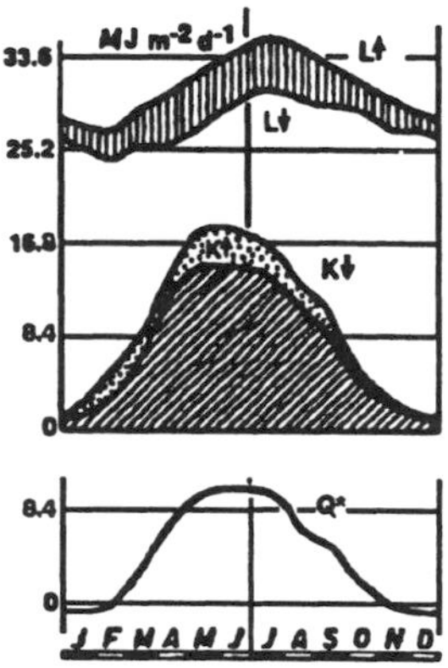

Figure 25-2. Individual components of the radiation balance throughout the year, from measurements made at the Hamburg Meteorological Observatory. (After R. Fleischer [2504a, 2505])

In the middle section of Figure 25-2, the annual variation of global solar radiation $K\downarrow$ (i.e. $S + D$) is shown, from which the quantity reflected by Earth's surface $K\uparrow$ (dotted area) is subtracted. The area with oblique hatching indicates the value of the shortwave radiation balance, varying with the noon solar altitude of the sun during the year. The albedo of the Hamburg observation field averages 16 percent, but varies during the year under the influence of changing weather. In winter, when snow is present, it may rise to 85 percent. The influence of snow cover in February may be detected even in the average values for the 4 yr period.

The difference between the net solar radiation $K\downarrow$ - $K\uparrow$ (oblique hatching), and the net longwave radiation, vertically hatched area ($L\downarrow$ - $L\uparrow$) at the top of Figure 25-2, gives the net radiation balance Q^*, which is shown separately at the bottom of Figure 25-2. Only in winter is the net radiation term negative. This curve shows the values of the net radiation balance obtained by direct measurement.

In evaluating the energy exchange in the soil (Q_G), if the ground is covered with vegetation, it is important to use a higher level called the "active surface" or outer effective surface

(Section 30) as a reference surface in discussions on energy balance, and not the ground surface. For example, when A. Baumgartner [2502] investigated the energy balance of a pine thicket near Munich, he selected the average height of the crowns of young pines at 5 m as the reference surface. The factor Q_G then acquires a slightly altered meaning and comprises the flow of energy entering or leaving the ground, the vegetation on it, and the air layer below 5 m.

Since the two quantities Q^* and Q_G can be measured directly, the value of the two remaining quantities can be determined, since $Q_H + Q_E = -(Q^* + Q_G)$. From Section 7:

$$Q_H = c_p A_H \frac{d}{dz} \qquad (\mathrm{W\ m^{-2}})$$

in which c_p (1009.8 J kg^{-1} K^{-1}) is the specific heat of the air at constant pressure, Θ the potential temperature, z (m) the height above the ground, and A_H (kg m^{-1} sec^{-1}) the appropriate Austausch coefficient for the transport of sensible heat. In a similar way, the energy required for evaporation Q_E is given as:

$$Q_E = r_v A_E \frac{dq}{dz} \qquad (\mathrm{W\ m^{-2}}),$$

in which, r_v is the latent heat of vaporization of water, and varies with temperature (being 2.50 MJ kg^{-1} at 0°C and 2.45 at 20°C); q is the specific humidity or the mass of water vapor per gram of moist air (Section 15), and A_E is the Austausch coefficient for transport of water vapor.

In making his first calculations of the energy balance, H. U. Sverdrup [2516] made use of the assumption first postulated by W. Schmidt that the same mixing process would transport all characteristic properties in the same way, so that $A_H = A_E$. Using this similarity assumption, the ratio

$$\frac{Q_H}{Q_E} = \frac{c_p d\ /dz}{r_v dq/dz}$$

can be calculated from the temperature gradient and water vapor gradient of an air layer of equal thickness near the ground. This ratio (Q_H/Q_E) is known as the Bowen ratio (β), and is frequently used as a measure of energy partitioning at the earth's surface. From this, it follows that

$$Q_E = -\frac{Q^* + Q_G}{1 + \beta} \text{ and } Q_H \quad -\frac{\beta\ (Q^* + Q_G)}{1 + \beta}$$

The Bowen ratio facilitates the calculation of the energy balance because the ratio can be determined from temperature and humidity data at two elevations, and is not dependent upon corrections for atmospheric stability. If Q_H or Q_E is known, the Austausch coefficient can be calculated. Examples of measurements of Q^*, Q_G, Q_E and Q_H for different climates are presented in Section 26.

D. S. Munro [2513] provides an overview of the strengths and limitations of the Bowen ratio approach to the estimation of the turbulent fluxes of heat and water vapor within the boundary layer near the surface of earth, as well as a discussion of two other approaches, the aerodynamic and eddy correlation methods. His bibliography also lists the major references covering theoretical developments in the field over the past few decades. The works of S. P. Arya [100] and R. B. Stull [108], mentioned earlier, should also be consulted for a more formal treatment of the subject.

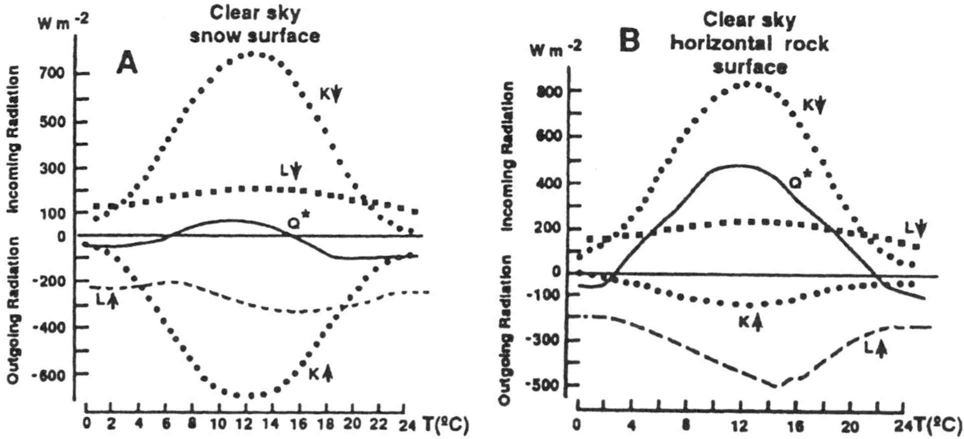

Figure 25-3. Radiation budget over a snow (A) and a rock surface (B) in Queen Maud Land. (After Y. Gjessing and D.O. Øvstedal [2507])

Y. Gjessing and D.O. Øvstedal [2507] measured radiation over both a rock and a snow covered surface (Figure 25-3) during the summer in Queen Maud Land, Antarctica. Over the snow (albedo 80-95 percent) most of the incoming global radiation is reflected. There is a net loss of longwave radiation ($L\uparrow > L\downarrow$) for the entire 24 hour period. Net radiation (Q^*) is negative except during the middle of the day. Over the rock surface the albedo is only 17 percent, resulting in surface temperatures occasionally exceeding 30°C. Because the air temperatures are still relatively cool $|L\uparrow - L\downarrow|$ is much larger over the rock than the snow surface. Net radiation (Q^*) is much more positive over the rock surfaces. Bare rock surfaces on nunataks in particular can have warm temperatures and support a variety of algae, lichens and a moss (Figure 25-4). As expected, there is a time lag in the occurrence of the maximum temperature on the south and north slopes of the nunatak.

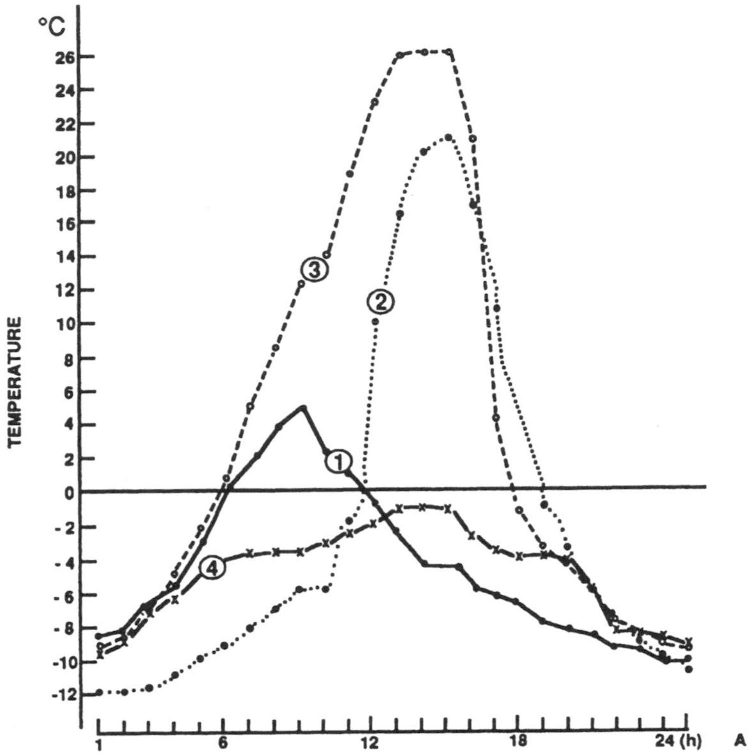

Figure 25-4. Temperatures on a south slope 1, north slope 2, top of a nunatak 3, and air temperatures 4 in Queen Maud Land. (After Y. Gjessing and D.O. Øvstedal [2507])

26 Results of Previous Energy Balance Measurements

We will begin with the diurnal variation of the energy balance terms for a representative humid mid-latitude climate. E. Frankenberger [2506] measured and evaluated temperature, water vapor, and wind from 1 September 1953 to 31 August 1954, up to a height of 70 m above a flat meadow near Quickborn, Holstein. Figure 26-1 gives the results for clear days with light winds for three seasons. The figures for the small energy exchanges at night and in winter were not sufficiently reliable, and thus are not included. Since only clear days were considered, the net radiation balance Q^* is a smooth, balanced curve, nearly symmetric about the 12:00 line. During the day energy flows into the ground (Q_G), the air (Q_H), and is used for evaporation (Q_E). Elements shown below the line in Figures 26-1 through 26-3 transport energy away from the surface. In the humid mid-latitude climates, particularly in summer, most of the net radiation is used in evaporation.

Figure 26-2 by A. Baumgartner [2502] is for a dry period in midsummer for a spruce thicket 5 to 6 m in height near Munich (24 hours). The quantity Q_G comprises not only the energy exchange in the forest soil, but also in its protective covering of trees. The value of Q_G is comparatively small. A chance variation in the weather during the recording period from 29 June to 7 July 1952 causes the curves in Figure 26-2 to appear somewhat more irregular than those in Figure 26-1. Initially, after the sun rises, only a small amount of solar radiation goes into heating the air; most is used to evaporate dew. Once the dew has been evaporated, an increasing amount of energy from solar radiation goes into heating the air (Q_H). After the air temperature increases and relative humidity decreases, evaporation again begins to increase. The ground heats slowly through the morning. Maximum rates of Q_G tend to occur in the pre-noon period. At about 15:00, Q_G becomes positive as the ground surface reaches its maximum temperature. The ground then begins to cool. This can also be observed in Figures 26-1 through 26-3.

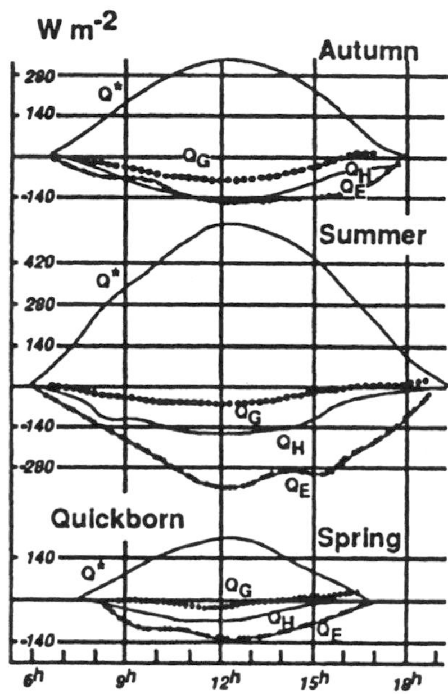

Figure 26-1. Daily energy exchange on clear days with light winds at Quickborn, Holstein. (After E. Frankenberger [2506]).

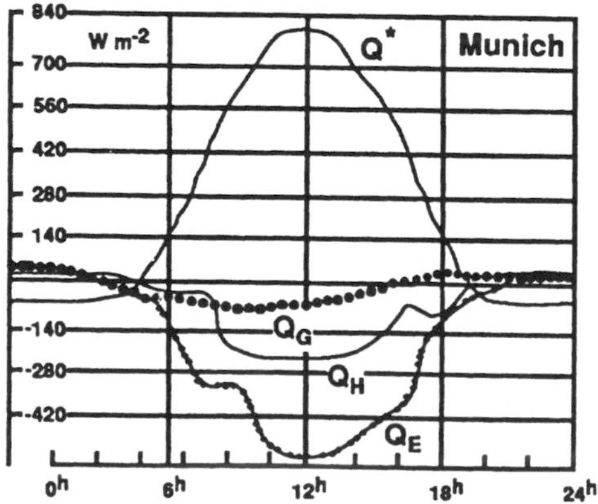

Figure 26-2. Daily energy exchange during a dry period in midsummer in a stand of young firs near Munich. (After A. Baumgartner [2502])

Figure 26-3 by M. Franssila [2108] presents the mean of three August days in 1934, for a meadow near Tauriala, Finland. The ground, which had stored a surprising amount of energy before noon, begins releasing energy after 16:00. The nocturnal exchange of energy appears large, perhaps due to the longer daylight duration in high latitudes or because of the inaccuracy of the radiation measuring instruments available at that time.

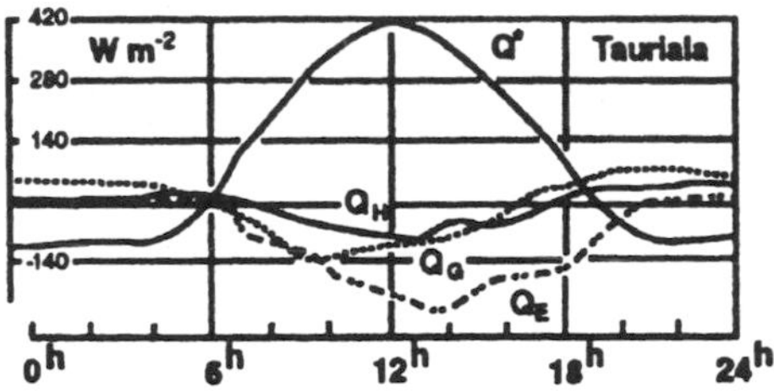

Figure 26-3. Daily energy exchange on clear summer days in Finland. (After M. Franssila [2108])

The situation is different in areas of low precipitation (Figure 26-4 and 26-5). The top of Figure 26-4 shows the energy budget averaged for 4 August and 2 September 1953 by H. Lettau and B. Davidson [104]. The bottom of Figure 26-4 shows the results from the Gobi Desert for 11-20 May 1931, from W. Haude's [4407] observations in Ikengung during the Sven Hedin's China Expedition (evaluated by F. Albrecht [1001]). The curves appear smooth because only 2 hr averages were published. Evaporation is minor in the North American prairies, and in the Central Asian desert steppe of the Gobi (about the same latitude) it is even lower. In contrast to the nocturnal condensation observed in Figures 26-1 through 26-3, in O'Neill, there is a small amount of nighttime evaporation.

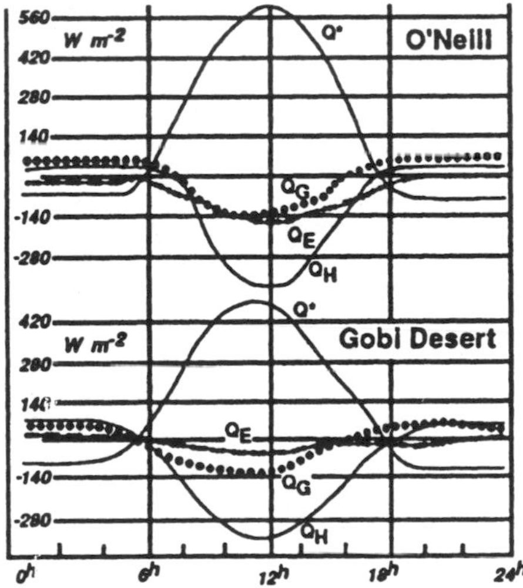

Figure 26-4. Daily energy exchanges in the semi-arid O'Neill, Nebraska (above), and the arid Gobi Desert (below). (After F. Albrecht [1001] and W. Haude)

The upper half of Figure 26-5 gives the values of the various factors by day, while the lower half shows the night values. Site details are given in Table 26-1. Since the quantities

Table 26-1. Details of the data used to illustrate the energy balance in Figure 26-5.

Country	Observation site				Observer and analyst	Data used in Fig. 123		Year of publication
	Latitude	Longitude	Height (m)	Situation		Period of recording	Weather conditions	
China	41.9°N	107.8°E	1500	Gobi Desert near Ikengung	F. Albrecht, from W. Haude's [4407] observations [1001]	11 to 20 May 1931	Fair weather	1941
USA	42.5°N	98.5°W	603	Prairie grassland near O'Neill, Nebraska	H. Lettau, from the observations of several teams [104]	9, 13, 19, 25 August; 1, 8 September 1953	Fair weather	1957
South Germany	47.9°N	11.7°E	645	Fir thicket 5-1/2 m high, 30 km SE of Munich	A. Baumgartner [2502]	29 June to 7 July 1952	Summer drought	1956
England	51.5°N	0.3°W	5	Grass plot at Kew Observatory	N. E. Rider and G. D. Robinson [2515]	20 to 24 June 1949, 11:00 to 15:00	Fair weather	1951
North Germany	53.7°N	9.9°E	12	Flat meadow, 1 m above ground water at Quickborn, Holstein	E. Frankenberger [2506]	1 September to 31 November 1953, and 1 March to 31 August 1954	Clear days with light winds only	1955
Austria	46.5°N	14.6°E	560	Snow field near Eisenkappel, Carinthia	E. Niederdorfer [2514]	14 to 15 January 1932	Clear and calm nights	1933
Finland	61.2°N	24.4°E	Low	Meadow on a plain near Tauriala	M. Franssila [2108]	Three August days in 1934 [2511]	Fair weather	1936
Greenland	70.9°N	40.8°W	3000	Ice cap at "Eismitte" station	D. H. Miller, from observations by the A. Wegener and R. E. Victor expeditions [2511]		Clear summer day (14 daylight hours) and winter night	1956
Austria	46.9°N	10.8°E	2973	Glacier ice on the Vernagt Glacier in the Ötztal Alps	H. Hoinkes and N. Untersteiner [2509]	21 to 31 August 1950	Period of fair weather	1952
Austria	47.0°N	11.8°E	2262	Glacier ice on the Hornkees Glacier in the Zillertal Alps	H. Hoinkes [2508]	3 to 9 September 1951	Period of fair weather in midsummer	1953

are per minute, averaged over the whole day or night, the values are small. Except for the alpine glaciers, on the extreme right, the locations have been arranged in order of increasing latitude. Haude's [4407] measurements in Ikengung were in May, those at O'Neill in August and September. Munich recorded the highest net radiation value because the observations were made near the sun's highest elevation and during a midsummer drought. The measurements by N. E. Rider and G. D. Robinson [2515] are averages of seven individual sets of readings taken between 20 and 24 June 1949 about noon (11-15). The Quickborn figures are arranged according to the season as in Figure 26-1. Since these contained no night values, the nocturnal energy balance measurements of E. Niederdorfer [2514] over a snow surface were inserted. Greenland values are the work of the French and German Expeditions to the Eismitte station (70°54'N, 40°42'W, 3030 m) taken from D. H. Miller [2511].

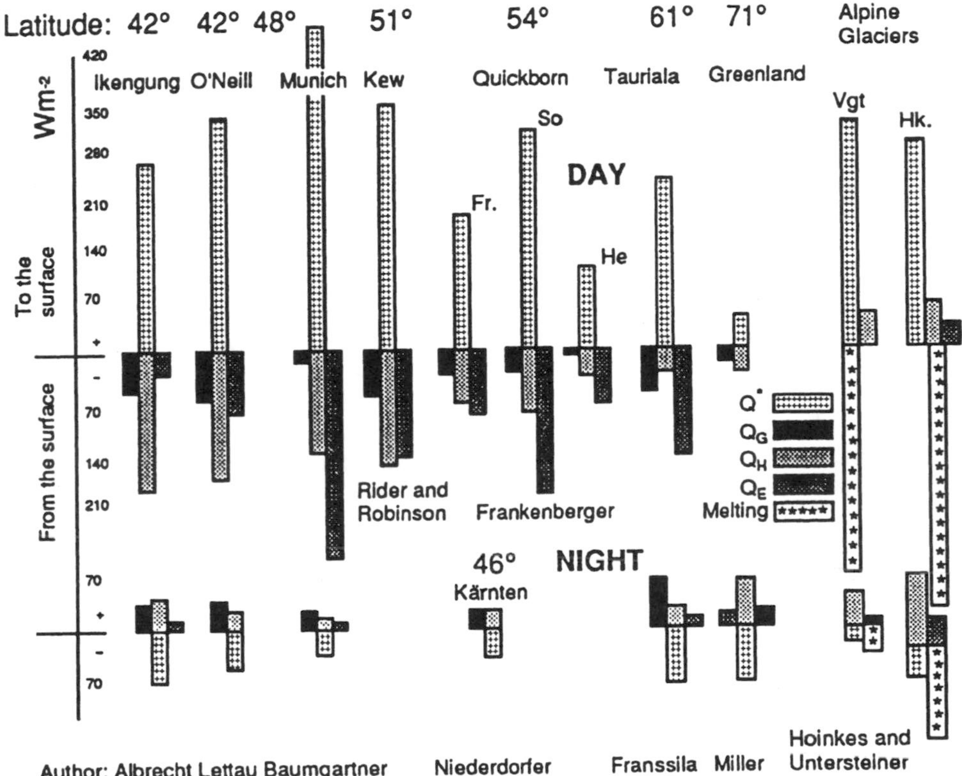

Figure 26-5. Summary of all microclimatologic energy balance measurements cited in Table 26-1, separated into night (below) and day (above). (From F. Albrecht [2501])

Taken as a whole, Figure 26-5 provides a good picture of the energy exchange at the earth's surface. In humid temperate climates, most of the positive net radiation during the day is used in evaporation. Evaporation in lower latitude dry climates and higher latitude humid climates is limited by lack of available surface moisture and net radiation, respectively. At night the radiation balance (Q^*) is negative. This is compensated for by a flow of energy to the surface from the air and soil. Evaporation and/or condensation are minimal at night.

The energy balance of alpine glaciers in summer was determined by H. Hoinkes and N. Untersteiner [2509] for the Vernagt Glacier in the Ötztal, and by H. Hoinkes [2508] for the Hornkees Glacier in Zillertal. It was tested by making simultaneous measurements of the quantity of water produced by melting. It follows from the great temperature contrast between the glacier and its environment in midsummer that Q_H remains positive both day and night. Even Q_E is nearly always positive, which means that energy flows to the glacier, both by cooling the air and through condensation (Table 24-6). The main energy source for melting these glaciers is radiation. N. Untersteiner [2517] found similar results for the Chogo-Lungma Glacier in the northwest Karakorum at latitude 37°N at an elevation between 4000 to 4300 m.

The factors Q^*, Q_H, Q_G (or Q_W), and Q_E for the 12 months of the year, subdivided into two half-years by a vertical line, are shown in Figure 26-6 for six land stations, and in Figure 26-7 for five ocean stations. In both diagrams, latitude increases from left to right. Thus, net radiation Q^* changes from being positive throughout the year at lower latitudes to negative in winter at higher latitudes. Land and ocean do not differ substantially in this respect because cloudiness primarily has a zonal distribution, and the influence of temperature and surface properties is not substantial. However, the ways in which the sun's energy is utilized are quite different.

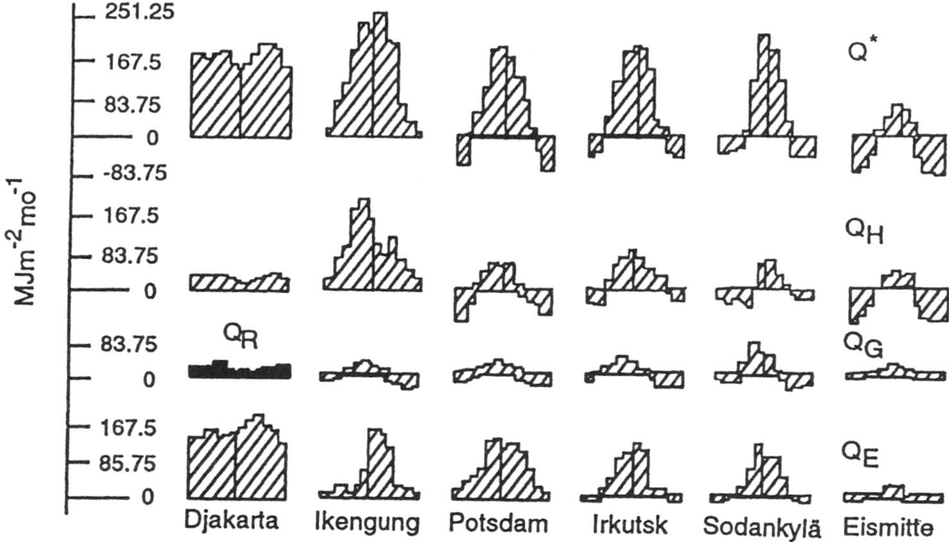

Figure 26-6. Monthly variation (January-December) of the energy balance at six land stations at a range of latitudes. (After F. Albrecht [2501])

At land stations (Figure 26-6), the seasonal energy exchange in the ground (Q_G) is comparatively small. At Djakarta in Java, there is practically no difference from month to month, and it can be neglected. The quantity of energy used to warm the cold precipitation water Q_R is substituted for Q_G and is always cold in the tropics. In humid mid-latitude climates, with their changing air masses, Q_R may be neglected for a yearly period since precipitation may be either warm or cold. The greater the contrast between the seasons, the greater the seasonal energy flowing into or out of the ground. The maximum in Figure 26-6 is for Sodankylä in Finland (67°N), which is in good agreement with the particularly large value of Q_G for Tauriala in Figure 26-5.

Evaporation Q_E is high in the tropics and low in the polar regions. The actual amount of evaporation, however, also depends on the quantity of available water; therefore, Q_E is lower at Ikengung (42°N), apart from the short summer monsoon period, than it is at Potsdam (52°N). In the far north (Eismitte), during the polar night condensation may even be greater than evaporation (as seen from Figure 26-5).

The quantity Q_H is the amount of sensible heat gained or lost by the surface to the atmosphere. The figures for Ikengung, where limited amounts of energy is used in evaporation, show that the arid parts of the earth act as intense sources of sensible heat for the atmosphere. In higher latitudes, large quantities of energy are withdrawn from the air in winter and can be replaced by advection from warmer latitudes (Figure 26-8). The energy exchange at a given location, and the large-scale horizontal exchanges of energy, are closely linked with one another.

The magnitude of the energy exchange Q_W with the changing seasons at the ocean stations in Figure 26-7 is at a maximum in the Straits of Florida (25°N) and near the Scilly Isles (50°N), decreases toward the tropics, which are always warm, and decreases toward the poles. A comparison between Q_G and Q_W, in Figures 26-6 and 26-7, shows the difference between continental and maritime climates. At sea, the actual evaporation is always equal to the potential evaporation and, therefore, decreases in a fairly regular manner toward higher latitudes. It can be neglected over polar oceans because of the small amounts involved. In its place, the quantity of energy used to melt ice in the summer months has been inserted in the diagram.

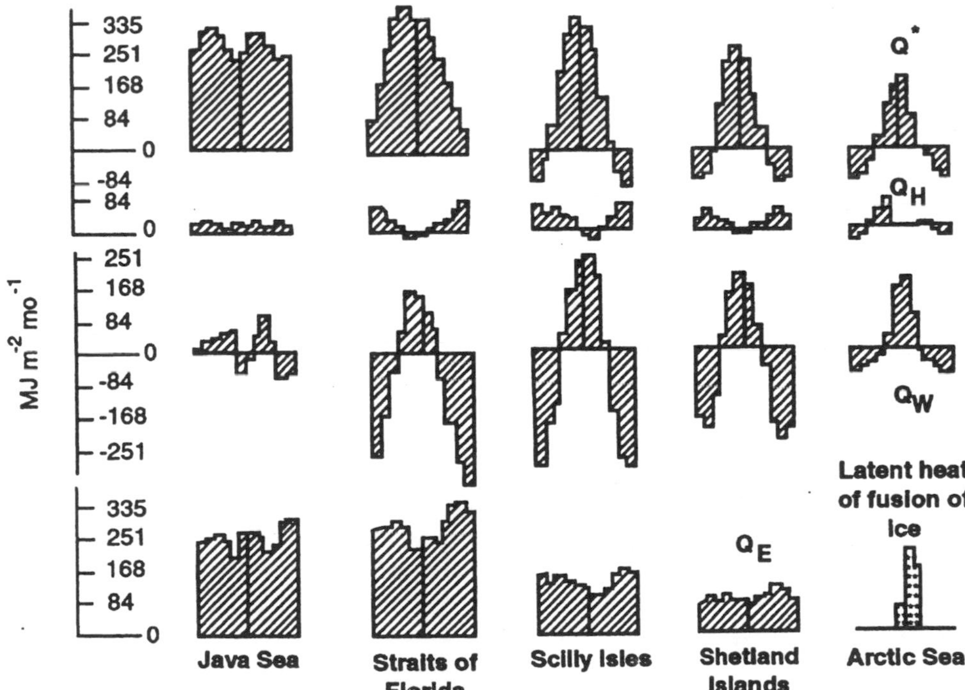

Figure 26-7. Monthly variation (January-December) of the energy balance at five ocean stations at various latitudes. (After F. Albrecht [2501])

The average annual energy balance for the earth-atmosphere system has already been presented in Figure 25-1. Figure 26-8 shows the latitudinal variation in the average annual energy balance, for combined land and ocean surfaces, based on the results of M. I. Budyko [2503]. The scale for latitude is contracted toward the poles in order to compensate for the different area between latitudinal zones. Results from a later calculation by M. I. Budyko [2504], given in Table 26-2, show essentially the same features, and include separate totals for the land and ocean surfaces. Q_M represents energy transfer by ocean currents.

Table 26-2. Latitudinal annual means of energy balance components for the surface of Earth, in 10^2 MJ m^{-2} yr^{-1}. (After M. I. Budyko [2504])

Latitude	Land			Ocean				Earth as a whole			
(degrees)	Q^*	Q_E	Q_H	Q^*	Q_E	Q_H	Q_M	Q^*	Q_E	Q_H	Q_M
North											
70-60	9.2	-6.7	-2.5	9.6	-13.0	-9.2	+12.6	9.2	-8.4	-4.6	+3.8
60-50	13.4	-9.6	-3.8	18.0	-19.7	-7.9	+9.6	15.5	-13.8	-5.5	+3.8
50-40	18.9	-10.5	-8.4	26.8	-28.1	-6.7	+8.0	22.6	-18.9	-7.5	+3.8
40-30	24.3	-9.6	-14.7	37.7	-40.2	-5.9	+8.4	31.8	-27.2	-9.6	+5.0
30-20	26.8	-8.0	-18.8	46.5	-45.7	-2.9	+2.1	39.4	-31.4	-8.8	+0.8
20-10	31.0	-13.4	-17.6	50.6	-49.0	-2.9	+1.3	45.7	-39.8	-6.7	+0.8
10-0	33.1	-23.9	-9.2	51.9	-43.6	-2.9	-5.4	47.7	-38.9	-4.2	-4.6
South											
0-10	33.1	-25.6	-7.5	53.2	-41.5	-2.5	-9.2	48.6	-37.7	-3.8	-7.1
10-20	31.4	-18.8	-12.6	51.1	-47.3	-3.8	0	46.9	-41.0	-5.9	0
20-30	29.7	-11.7	-18.0	45.6	-44.4	-4.6	+3.4	41.9	-36.9	-7.5	+2.5
30-40	26.0	-12.2	-13.8	38.5	-34.3	-4.6	+0.4	36.9	-31.8	-5.9	+0.8
40-50	18.4	-9.2	-9.2	30.2	-21.4	-2.5	-6.3	29.7	-20.9	-2.9	-5.9
50-60	14.7	-9.2	-5.5	19.3	-14.7	-3.8	-0.8	19.3	-14.7	-3.8	-0.8
Earth as a whole	20.9	-11.3	-9.6	38.1	-34.3	-3.7	0	33.1	-27.6	-5.4	0

The radiation balance Q^* for the surface of Earth is 33.1 • 10^2 MJ m^{-2} yr^{-1}, and is matched by an equally large radiation loss by Earth's atmosphere. The surface radiation balance is positive at all latitudes except near the poles. Within 25 degrees of the equator, where the noon sun is always near its zenith, the radiation balance is approximately 42 • 10^2 MJ m^{-2} yr^{-1}. From 25 degrees poleward, the annual total of Q^* decreases in a steady manner. Where a positive radiation balance exists, the mean annual surface temperature will be higher than that of the air, resulting in a transfer of energy from the surface to the air by conduction-convection Q_H. In the two arid zones, at approximately 30° latitude, most of the net radiation provides energy for heating the air layer near the ground, as indicated in Figure 26-4. Q_H is greater over land than over the ocean, except poleward of 50°N where poleward moving warm ocean currents lose large amounts of sensible heat to the atmosphere. The energy consumed in the evaporation process (Q_E) exhibits a strong dependence on latitude.

The factor Q_G (or Q_W) does not appear in Figure 26-8 or Table 26-2 since over a yearly period the energy gain approximately balances the energy loss, as may be deduced from Figures 26-6 and 26-7.

To properly evaluate the global surface energy balance, the advective transport of energy must be considered. Advection of sensible heat by ocean currents Q_M equalizes the latitudinal imbalance of net radiation shown in Figure 26-8 by transporting energy from equatorial regions to mid-latitude and polar regions.

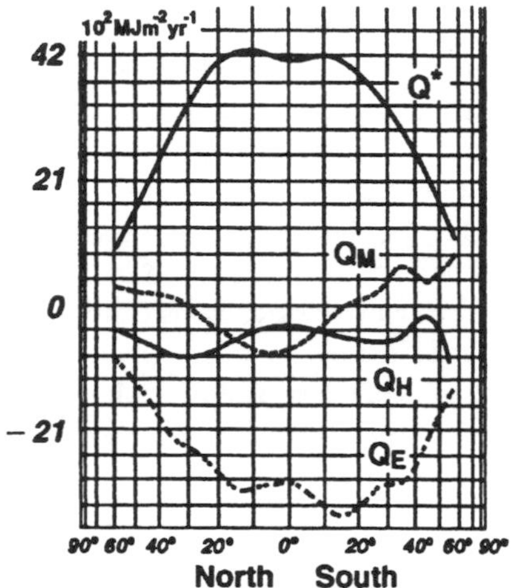

Figure 26-8. Latitudinal variation in the annual energy balance for the surface of Earth. (After M. I. Budyko [2503])

The surface of Earth, therefore, experiences a two-fold pattern of energy transport. First, energy must be transferred from the surface to the atmosphere to compensate for the 33.1 x 10^{-2} MJ m^{-2} yr^{-1} surplus in the former and deficit in the latter (Figures 26-1 through 26-7). Second, energy is transported along a latitudinal transect from lower to higher latitudes. There are three means of accomplishing this: (1) sensible heat transport by the atmospheric circulation, (2) sensible heat transport by the oceanic circulation, and (3) latent heat transport associated with the hydrological cycle. According to A. Henderson-Sellers and P. J. Robinson [3922], the ratio of these three processes at a global scale is 60:25:15.

27 Advective Influences and Transitional Climates

The energy balance of the ground surface was analyzed in Sections 25 and 26 with respect to the vertical flow of energy toward or away from the surface boundary. Besides vertical gradients, horizontal gradients may also have an effect. Advective processes do not concern the ground surface, but apply to the layer of air close to it. Since the ground surface is a horizontal reference surface, it cannot be affected by horizontal (advective) energy transfer. However, we may postulate a volume of air bounded by the ground surface, a surface parallel to it, and sides formed of vertical surfaces and then consider the energy budget of this volume.

The equations thus derived enable us to distinguish advection proper from the vertical flow of energy by frictional and convective exchange (Section 8). Just how much energy finds its way by advection into the volume of air being considered depends on the magnitude of the horizontal energy flow and on the height (volume) above the surface being considered.

There are many studies from which we can assess the significance of advective processes, and the nature and magnitude of their effects. These are most easily appreciated where two different kinds of surfaces abut on each other (land, water, and snow), or where the ground is put to different uses (road, meadow, plowed land, and forest). Because ground cover is normally heterogeneous with respect to the surface properties which govern the microclimate, we must usually contend with some kind of advective influence among microclimates.

All such boundaries develop a transitional or boundary climate. A coastal climate is one example of such a transitional climate, which often extends its influence inland for many miles. As the coastal air travels further inland, it becomes less distinct. R. A. Craig [2702] has investigated the reciprocal process by which air up to a height of 300 m passing from land to sea undergoes transformation in Massachusetts Bay.

Consider now the dimensions of microclimate. The small amount of mixing near the surface ensures that those characteristics of the air that depend on the underlying surface (Chapter III) are, at first, retained *in situ*. The extraordinary contrasts that may arise within a very small space will be illustrated for the boundary between land and water.

It was previously mentioned (Section 22) that water and land surfaces exert a mutual influence on each other. H. Berg [2701] and A. Mäde [2707] have investigated the boundary zone on either side of the line of discontinuity. Figure 27-1 shows the temperature and water vapor profiles over a flat sandy part of the bank of the Rhine, from 09:00 to 10:30 on 2 July 1952 when winds were light. In the lowest 10 cm, there is in juxtaposition a cool moist layer of air over the water, and only 8 m away a layer 13°C warmer over the sand. A little higher, between 10 and 40 cm, temperatures are practically identical, but at this height over the water the vapor pressure is slightly less than it is over the warmer bank. Further measurements and observations led H. Berg [2701] to conjecture that during the day cool moist parcels of

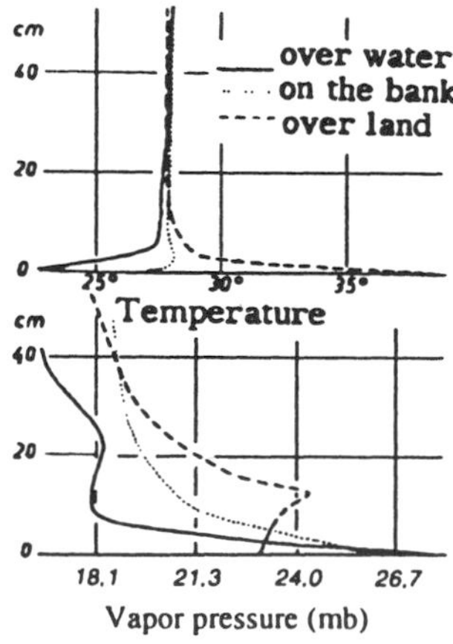

Figure 27-1. Temperature (°C) and vapor pressure (mb) profiles on the banks of the Rhine late in the morning in July. (After H. Berg [2701])

air from over the water flow over the hotter air near the land surface, but during the night cold air descends toward the water. A zone of weak exchange is set up even in the absence of a fully developed land and sea breeze circulation. In observing temperatures associated with sea smoke, W. Vieser [1506] was able to show that the cold air flowing out from the coast at night was heated and enriched with moisture only in the lowest 10 cm.

A. Mäde [2707] conducted observations on Lake Süssen near Eisleben. This lake lies in a shallow saucer of land, between 2 and 3 km in area, and its vegetated shores are only about 5 cm above water level. Recordings were made up to a height of 1 m and for a distance of about 100 m horizontally on both sides of the water's edge. Figure 27-2 shows the different distributions of isotherms at 3 hr intervals for two series of observations, from 16 to 18 August 1955 (left), and 4 to 7 September 1955 (right). Relatively warm areas are hatched with lines sloping up to the right, and relatively cold areas with lines perpendicular to these,

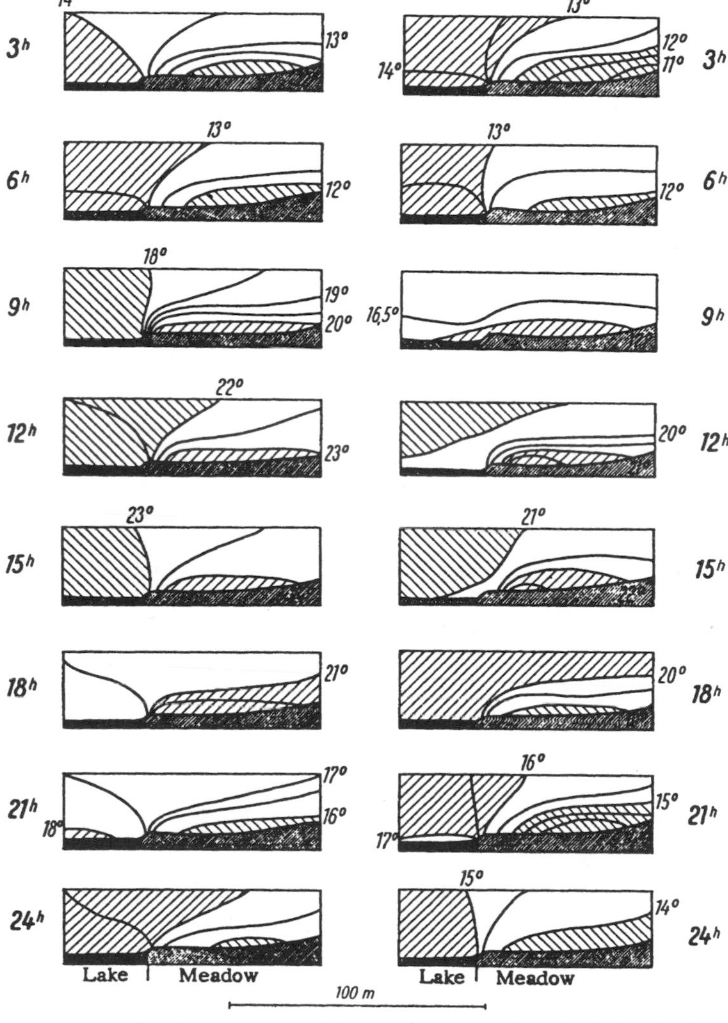

Figure 27-2. Average temperature fields near the banks of Lake Süssen near Eisleben. (After A. Mäde [2707])

sloping up to the left. Although there is no fringe of reeds here that might hinder the equalization of temperature, in the lowest half-meter, the gradient is very steep at the boundary between land and water. S. S. Visher [2713] reported that in the Great Lakes in January, the average temperature was 2.8°C and the lowest temperatures were from 5.6-8.3°C warmer on the lake shore than further inland. The Great Lakes also reduced exceptionally high temperatures in summer by about 1.7-2.8°C. The frost-free period near the lake was also reported to be between 30-40 days longer. L. J. Verber [2712] reported the decreasing influence of Lake Erie as the distance from the lake increased.

The deciduous fruit belt on leeward coasts of the Great Lakes in Michigan and Ontario are in large part a result of the lakes' moderating effects. The lakes have two effects, the first in spring and the second in fall. In spring, the lakes heat slowly keeping areas on the leeward side of the lakes relatively cool. The deciduous fruit buds remain frost resistant (cold hardy) in these areas through winter and into spring as long as temperatures remain cool (Section 53). Warm temperatures stimulate renewed growth and increasing frost sensitivity. Thus, the lakes, by keeping temperatures cool, minimize the danger of spring frosts. In fall, the lakes cool slowly and, as mentioned above, increase the frost-free growing season. This increase is primarily the result of a retarding of first frost in fall. Thus, the lakes assist in protecting both trees and fruit from early frosts in fall.

A. Nyberg and L. Raab [2708] investigated the eddy diffusion process in air flowing from sea to land along the west coast of Öland Island, Sweden. Thermistors were fixed on four masts at the heights shown by small circles in Figure 27-3. Temperature, wind speed, and direction were recorded on two clear days. Isotherms in the first 20 m of the atmosphere are shown by solid lines and the calculated Austausch coefficients A (kg m^{-1} sec^{-1}) in broken lines.

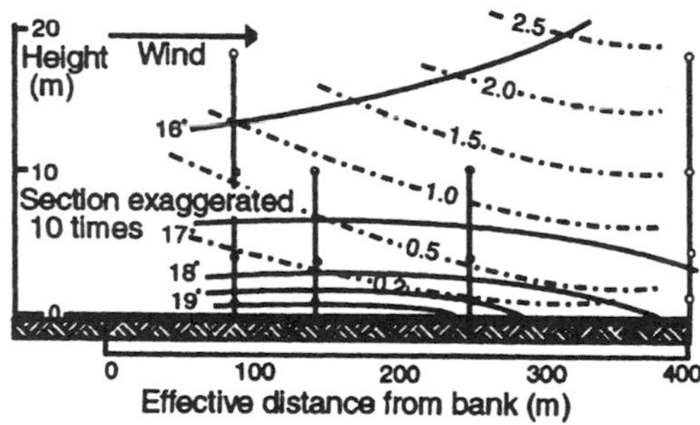

Figure 27-3. Temperature and eddy diffusion in an air mass passing from a lake over heated land (After A. Nyberg and L. Raab [2708]). Solid lines are temperature °C, dot dashed line are the Austausch coefficients.

The eddy diffusion component due to free convection initiated by vigorous ground heating becomes more significant as the air mass travels inland. The ground is heated everywhere roughly to the same extent throughout the area shown (the ground surface was heated to more than 35°C). Near the shore, where the Austausch coefficient is small, the energy transferred to the air is restricted to a shallow layer. The air over the water is stable (cooled from below). As it moves inland over the warm sand it becomes more unstable, there is more convection and mixing, and the Austausch coefficient increases. This explains the observed results which are at first surprising, namely that the temperature at a height of 1 m inland from the coast first increases, then decreases. As air moves on shore there is a sharp increase in roughness. The increased friction results in a rapid reduction in the wind speed.

S. E. Tuller [2711] analyzed summer onshore flows in Victoria, British Columbia. The waterfront, under the direct influence of the onshore flow, had lower air temperatures, vapor pressures, and higher wind speeds than inland locations. He pointed out that the lower waterfront vapor pressure was a result of limited evaporation from the cool waters (11-11.5°C) around Victoria and from high evaporation from the warm lawns (21°C) with greater surface roughness and evaporating area. These lower vapor pressure observations are contrary to onshore flow observations where the offshore water temperatures are higher (S. Zhong and E. S. Takle [2715] and C. G. Helmis et al. [2703] or where the inland surfaces were drier (B. Krawezyk [2705]). (For an excellent discussion of the land and sea breeze, the reader is referred to *Climate in a Small Area* by M. M. Yoshino [2714] or *Sea Breeze and Local Winds* by J. E. Simpson [2710]).

The strong contrast between snow cover and bare patches of soil has already been covered in Figure 24-10. This feature is often best observed in mountainous regions where radiation is strong, and spring fields of flowers may be seen blooming next to deep snow.

Over land there are many kinds of transitional climates, owing to differences in soil type, surface albedo, and the type and height of vegetation growing on it. Figure 27-4 shows measurements made at four different heights above the ground on a clear, calm night after 22:00 when a slight dew began to be deposited. These readings were taken by W. Knochenhauer [2704] over the concrete landing strip (left) and the grass beside it (right) at Hanover airport. The air close to the building is warm and dry, while over the grass it is cool and moist. Over the grass there is a layer of cool moist air, and over the concrete the air is warm and dry. The influence of the latter is stronger (perhaps a wind effect), and an equilibrium is established over the grass only after a distance of 30 to 40 m from the boundary of the two surfaces. At this distance, the lines of equal temperature and humidity are horizontal. There is a region of maximum temperature at a height of about 1 m over the grass, which is also a zone of lower relative humidity due, at least in part, to the higher saturation vapor pressure. It appears as if the warm dry air that has built up over the concrete is flowing out over the grass at this height and sliding over the cold, moist surface air in the process.

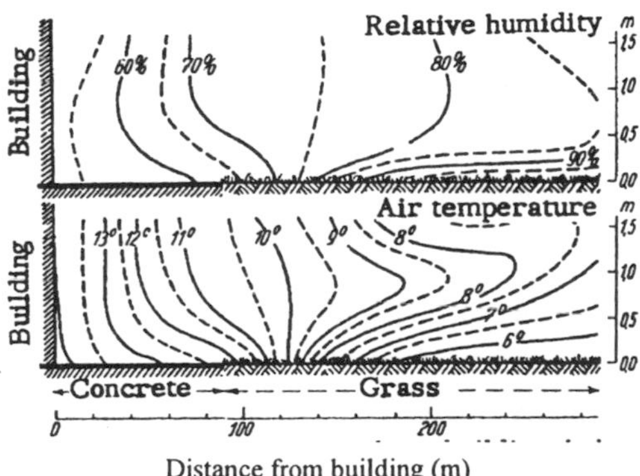

Figure 27-4. The transitional climate of a landing strip and meadow on an airfield. (After W. Knochenhauer [2704])

H. Runge [2709] found that in thick fog there is usually a shallow layer of air over a highway that is clear, or only slightly turbid, rising to a height of about 35 cm. Visibility, therefore, can be improved substantially when driving at night in fog, if an extra headlamp is fitted down low on an automobile in such a way that the beam is directed downward at an angle such that upward scattering of the light is avoided. Here the microclimatic effect of the dry road surface extends upward to about 35 cm and is in contrast with the microclimate of the surrounding cultivated land.

Very often the contrast of microclimate can be seen directly. J. Landeck and S. Uhlig [2706] made the following observation on 7 March 1950. "A field of young grain in which the ground was partly shaded by the small plants and covered with heavy dew, and a piece of fallow land that had been plowed but not yet harrowed, lay under the same strong morning sunshine, the latter warming up more quickly. About 13:00 a cold NW wind set in. The vertical temperature decrease, and, therefore, also the amount of eddy diffusion, became greater over the unsown land than over the field with crops. Now there formed over the open unplanted soil because of the vigorous transport of water vapor, a shallow ground fog like sea smoke, which was swirled up into streamers and veils by the turbulent motion of the air, rising from the ground, swaying, swirling, and dancing to a fast tempo, providing a fascinating spectacle for almost an hour." Further details, particularly the question of why this fog never crossed the boundary of the cultivated field in spite of the wind blowing in that direction, can be found in the original work.

The term transitional climate or boundary climate may appropriately be restricted to places near the boundary line between the various underlying surfaces. Advective elements, originating some distance away, may also, however, have their effects. This is particularly true in areas where topographic inequalities give rise to local winds, which may transport the characteristics of the locality for some considerable distance. Every change in type and height of vegetation produces similar transitions in climate. This subject will be discussed further in Chapter V. Sections 37 and 38 deal with the kind of transitional climate typical at the edge of a forest.

28 Remarks on Evaporation

It was shown in Sections 4 and 26 that after radiation, evaporation was the most important factor in the energy balance. Evaporation is also one of the most important factors in the water balance since, globally, approximately two-thirds of the precipitation falling on the land surface of earth is returned to the atmosphere by evaporation. Since a fixed amount of energy is required to evaporate each unit mass of water, the surface water and energy balances are closely linked, and one can be used to verify estimates of each transfer process against the other.

The term evaporation includes two processes. The customary interpretation of the term evaporation refers to the vaporization of water from an environment in which water is directly in contact with the atmosphere, such as a moist sandy soil, a wet concrete roadway, or a wetted leaf surface. This process is primarily controlled by physical laws. Transpiration refers to the vaporization of water from the moist inner tissues of plants which then pass through plant stomata and into the atmospheric environment. When a living plant transpires water, however, the processes of plant physiology also play an important role, in addition to the purely physical laws. When water is in short supply, plants can decrease the amount of transpiration by closing their stomata. At the same time, the plant must give off water in order to live, since moisture and minerals are carried from the soil in the stream of sap, and carbon must be assimilated from the outside air. For every gram of carbon assimilated, several hundred times the quantity of water must be absorbed through the plant's root system, transported to the leaf surfaces, and given off to the air through the same stomata by which carbon enters the leaf tissue. A surface covered by natural vegetation loses water by soil evaporation, from transpiration by the living plants, and by evaporation of water on wetted plant surfaces. Since most natural land surfaces include both evaporation and transpiration, the term evapotranspiration is normally used when both processes are taking place.

As with the energy balance, this study will start with a short survey of the global water budget. Water supply estimates for earth as summarized by D. H. Speidel and A. F. Agnew [2817] are given in Table 28-1. Values by other researchers (F. Van der Leeden, et al. [2818]) of the same features may vary slightly due to uncertainties in the estimating procedure. The 13,000 km^3 of atmospheric water, which comprises only 0.001 percent of total water, is almost entirely contained within the troposphere. During the course of a year, approximately 496,000 km^3 of water evaporates from the land and ocean surfaces of earth and return in the various forms of precipitation. If this amount is compared with the water content of the atmosphere, it is seen that the atmospheric reservoir of water turns over every 9 to 10 days.

Table 28-1. Estimates of global water reservoirs. (After D. H. Speidel and A. F. Agnew [2817])

Reservoir		Total (km^3)	Percent of Total (%)
Ocean		1,350,000,000	97.40
Atmosphere		13,000	0.001
Land			
	Ice caps and glaciers	27,500,000	01.984
	Ground water	8,200,000	0.592
	Lakes	205,000	0.015
	Rivers	1,700	0.0001
	Soil moisture	70,000	0.005
	Biota	1,100	0.00008

Figure 28-1 provides estimates of the reservoirs and vertical and horizontal fluxes within the global hydrologic cycle. These values were developed by the National Research Council [2812] and are based upon a total reservoir volume of 1.46×10^9 km^3 of water. The values given are mean totals expressed in thousands of cubic kilometers per year. When making comparisons between land and ocean surfaces, it should be remembered that oceans comprise 70.8 percent of the surface of Earth.

The general water balance for any period of time t can be expressed in the form:

$$r = E + f + b \qquad (\text{cm } t^{-1})$$

where r is precipitation, E is evaporation, f is runoff, and b is the rate of soil moisture change in the upper layer of soil. The quantity b may vary substantially over short periods of time. For an annual period, however, b will be small, and will approach 0 for periods significantly longer than a year. For the oceans b will, of course, always be zero. The value of f is considered negative for land surfaces, and positive for ocean surfaces, and is equal to 36,000 km^3 yr^{-1}. The difference between E and r must also equal the net advection of atmospheric moisture, and is positive for land surfaces, and negative for ocean surfaces.

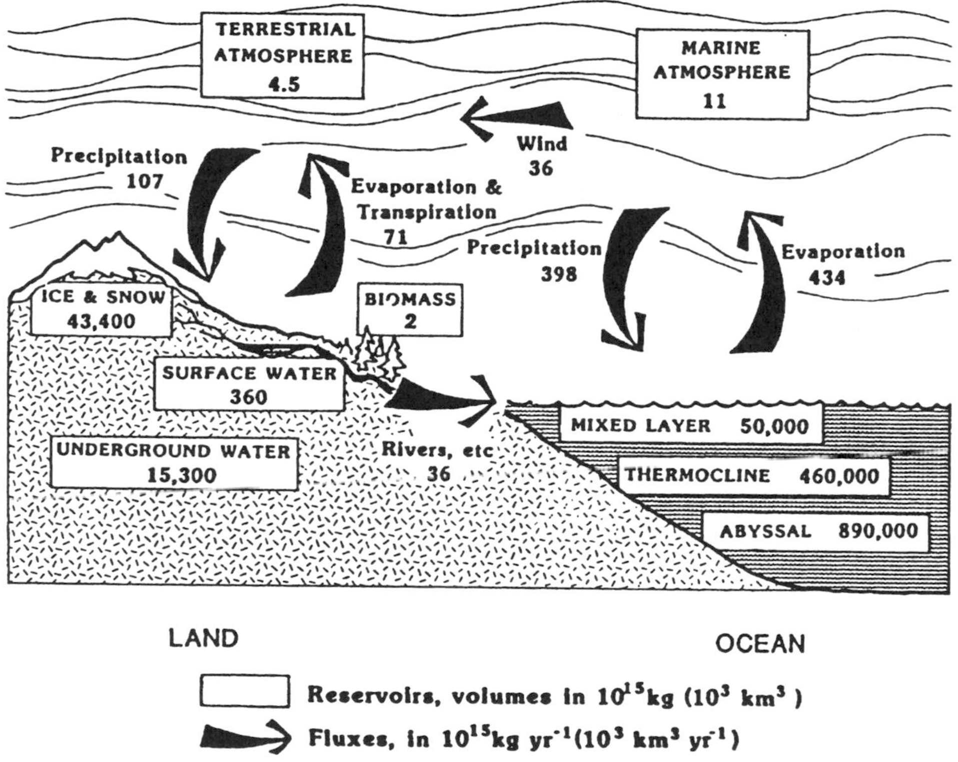

Figure 28-1. The global water cycle, showing estimates of contents of major reservoirs and rates of transfer between them. Reprinted with permission from *Global Change in the Geosphere-Biosphere: Initial Priorities for an IGBP*, National Academy Press, Washington, D.C., 1986 [2812].

Table 28-2 presents a summary of the global water balance for the continents and oceans based upon the work of O. A. Drozdov and cited by M. I. Budyko [2504].

The preceding information, which shows the importance of the part played by evaporation in the global water budget, forms the foundation for the study of evaporation as a factor in microclimate.

Table 28-2. Water balance of the continents and oceans. (After M. I. Budyko [2504])

Continents and oceans	Precipitation ($cm\ yr^{-1}$)	Evaporation ($cm\ yr^{-1}$)	Runoff ($cm\ yr^{-1}$)
Europe	79.0	50.7	28.3
Asia	74.0	41.6	32.4
North America	75.6	41.7	33.9
South America	160.0	91.5	68.5
Africa	74.0	58.7	15.3
Australia and Oceania	79.1	51.1	28.0
All the land	80.0	48.5	31.5
World Ocean	127.0	140.0	13.0
The earth as a whole	113.0	113.0	0

Both water and energy are needed for evaporation. If there is a freely available supply of water, one can speak of potential evaporation. If its supply is restricted there is an actual or effective evaporation, which is less than or, at most, equal to the potential evaporation. Where there are free water surfaces, these two quantities are generally equal.

The dependence of evaporation on temperature has already been indicated in Section 3. With snow and ice, the term sublimation is used in place of evaporation. The latent heat of sublimation is typically between 2.835-2.839 $MJ\ kg^{-1}$ (Table 3-1).

To illustrate some of the fundamental issues involved in determining the rate of evapotranspiration from different types of surfaces we will present three basic approaches. These will be presented in order of increasing surface complexity and physical rigor. The three approaches are the mass transfer method for open water surfaces, the combination method for open water surfaces, and the combination method for vegetated surfaces.

Mass transfer equations estimate evaporation (E, $mm\ hr^{-1}$) as a product of the saturation deficit E_0-e and an empirically determined constant:

$$E = c(E_0 - e)$$

where E_0 (mb) is the saturation vapor pressure of the air, e (mb) is the vapor pressure of the air, and c is an empirically determined constant. Because mass transfer approaches provide limited insight to the physical basis of evaporation they are of limited usefulness beyond the specific areas and conditions for which the empirical constant has been determined.

If the temperature of the evaporating surface is known, the saturation vapor pressure for this surface is also known and will be given the symbol $\acute{E}$. Proceeding by using the Dalton evaporation formula:

$$E_0 = c(\acute{E} - e) \qquad (mm\ hr^{-1})$$

The quantity $\acute{E}$-e is often called the saturation deficit, although as a rule it is substantially different from E_0-e. E_0-e is the saturation deficit of the air; while $\acute{E}$-e is the saturation deficit of the air and the evaporating surface. When a surface is warmer than the air, $\acute{E}$-e will be positive and if there is water available, evaporation will occur even if air is saturated (E-e = 0). As moisture is added to the air, and the surface air is mixed and cooled with the air aloft, steam ("smoke") may be observed. If a surface is colder than the air, $\acute{E}$-e may become negative and condensation (dew) will occur even if the air is not saturated (E_0-$e > 0$). In the above equation, the constant c is determined empirically from measurements of the potential evaporation. It has the following approximate values, depending on the wind speed u:

u (m sec^{-1}):	0.1	0.5	1	2	5	10
c (mm hr^{-1} mb^{-1}):	0.0053	0.0120	0.0173	0.0240	0.0375	0.0533

This indicates the large effect increasing wind speed has upon evaporation. The Dalton formula helps us to assess the importance of the temperature difference between the ground and the air. Table 28-3 shows the rate of evaporation in millimeters per hour for four different air temperatures as a function of the relative humidity and the temperature differences between the air and the evaporating surface. T represents the air temperature and the addends ($\pm$) the temperature difference of the air and the evaporating surface. Table 28-3 indicates how the potential evaporation increases as air temperatures increase, relative humidities decrease, and the evaporating surface becomes warmer than the air layer above it.

Table 28-3. Rate of evaporation (mm hr^{-1}) from a surface at different temperatures, as a function of air temperature and relative humidity.

Air temperature, T(°C)	Temperature (°C) of evaporating surface							
	T-3				T+0			
	100%	80%	60%	40%	100%	80%	60%	40%
0	con-	0.00	0.03	0.06	0.00	0.03	0.06	0.09
10	den-	0.01	0.06	0.12	0.00	0.06	0.12	0.18
20	sation	0.02	0.13	0.24	0.00	0.11	0.22	0.34
30		0.04	0.24	0.45	0.00	0.20	0.41	0.61
	T+3				T+6			
0	0.04	0.06	0.09	0.12	0.08	0.11	0.14	0.17
10	0.07	0.12	0.18	0.24	0.14	0.20	0.26	0.32
20	0.11	0.23	0.34	0.45	0.25	0.36	0.47	0.58
30	0.19	0.39	0.60	0.80	0.41	0.61	0.82	1.02

Initial insight into the physical processes governing evaporation can be gained from G. Hofmann's [2806, 2807] research. He developed a combination equation for open water evaporation that subdivided evaporation Q_E (W m^{-2}) into a radiation fraction Q_{Es} and a ventilation fraction Q_{Ev}, based upon the sources of energy supply which affect evaporation. We then have

$$Q_E = Q_{Es} + Q_{Ev} = -r_w\omega_s\left(Q^* + Q_G\right) - \frac{r_w\omega_v}{E}a_L\left(E - e\right) \qquad \text{(W m}^{-2}\text{)}$$

where Q_E (W m^{-2}) is the energy used in evaporation which corresponds approximately to a rate of evaporation in millimeters per hour; a_L is the energy transfer coefficient which expresses the amount of energy flowing from the surface to the air per unit area and unit time for a 1°C temperature difference between them; ω_s and ω_v are coefficients that are functions of the air temperature T_A only, and have, according to G. Hofmann [2806, 2807], the values given in Table 28-4. The advantage of the G. Hofmann [2806, 2807] method is that the temperature of the evaporating surface does not enter into the discussion. Although the temperature of an evaporating surface has a great effect on the rate of evaporation, it is very difficult to measure on a routine basis.

Table 28-4. Coefficients in equation for the energy used in evaporation.

Coefficient	Air temperature (°C)									
	-20	-10	0	0	5	10	15	20	25	30
	Evaporation from ice			Evaporation from water						
$r_w\omega_s$	0.24	0.31	0.49	0.43	0.51	0.58	0.65	0.70	0.75	0.80
$\frac{r_w\omega_v}{E}$ (°C / mb)	1.62	1.32	0.975	0.96	0.83	0.71	0.59	0.49	0.40	0.33

Let us assume that the air is saturated with moisture, so that the saturation deficit E-e = 0. Then the fraction Q_{Ev} = 0, and evaporation is determined by $Q^* + Q_G$ only. When the radiation balance is positive, evaporation will take place even if the air is in a saturated state. This will lead to the formation of mist or fog in the layers near the ground. As shown in Section 26, in general, Q^* is far in excess of Q_G during the day, so that Q_{Es} is called the radiation fraction. At times, however, Q_G can play an important part in evaporation. When the streets are "smoking" after a cold thundershower, this evaporation is caused by the energy stored in the material of the roadway (Q_G) and the increased radiation from the sun breaking through again (Q^*). When very warm water is in the presence of cold air, the factor Q_G may also have the major role, as for example, with hot springs which "steam," open seas in the Arctic ice "sea smoke," or when warm ocean currents flow into cold air layers.

The importance of the surface temperature is illustrated in comparing evaporation from a shallow pond and a deep lake (Figure 28-2). The shallow pond temperature is closely associated with the air temperature and therefore evaporation increases in summer and decreases in winter. In large lakes, however, the water temperature lags behind the air temperature and thus evaporation will be highest in autumn, when the water is relatively warm and the air cool (when $\acute{E}$-e is at a maximum), and lowest or negative in spring (when $\acute{E}$-e is at a minimum or even negative). Evaporation (or condensation) will tend to be at a maximum when the air and water surface temperature differences are greatest.

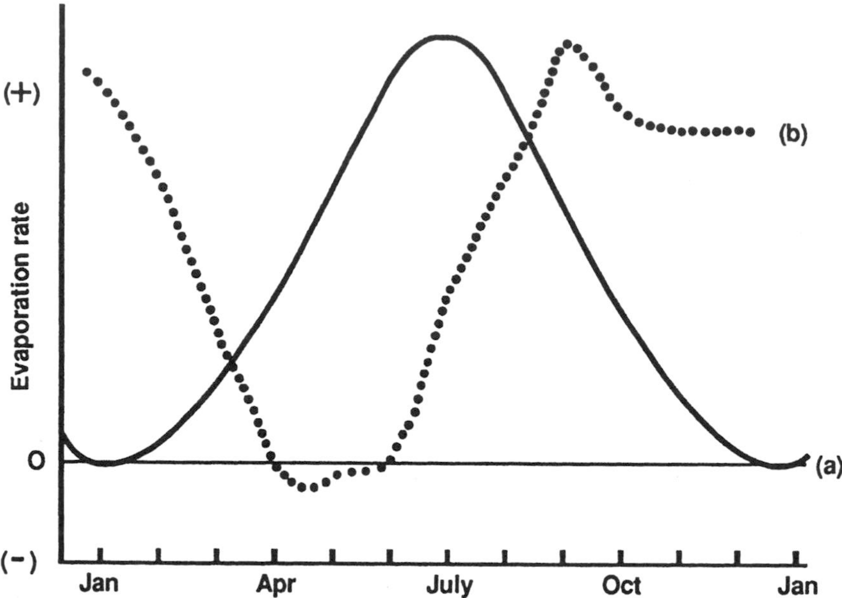

Figure 28-2. Seasonal trends of evaporation from: a) shallow pond; b) a deep lake (based upon Lake Ontario). (After D. H. Miller [2926])

Estimation of water loss from a vegetated surface cover is complicated by several unique factors not encountered with open water surfaces. These include the evaporation of intercepted water from wetted plant surfaces, plant physiological control over the transpiration process, the presence of multiple canopy layers in forests, conditions of limiting soil moisture content, and the presence of partial or incomplete canopy layers. Each of these factors will be briefly reviewed.

Some of the precipitation falling onto a vegetated cover or through a plant canopy is intercepted by the vegetation and adheres to the plant surfaces due to the surface tension and adsorptive properties of water. Evaporation of this intercepted water proceeds at a very rapid rate and water stored on the vegetated surfaces is normally consumed before significant transpiration loss begins.

Measurements made from a 45 m instrumented meteorological tower over a 25-month period above a 35 m tall continuous undisturbed tropical rain forest canopy in Manaus, Brazil (2°57'S) illustrate several of these processes (W.J. Shuttleworth, [2816]). Figure 28-3 gives the monthly totals (mm month^{-1}) of precipitation input, total evaporation, and interception loss. Precipitation input for this equatorial climate station shows a strong seasonal dependence with a maximum in March and a minimum in August. Interception loss from the wet canopy was determined as the difference between above canopy measured precipitation minus below canopy measured throughfall (direct precipitation, leaf drip and stemflow). Interception loss from the wet canopy surfaces represented 9% of gross precipitation, with percentages ranging from 8% of gross precipitation during dry months to 32% during wet months. Absolute interception losses were slightly higher during the wet season and somewhat lower during the dry season. Interception loss as a percentage of total

Q_E ranged from less than 10% in the driest months to more than 50% during the wettest months.

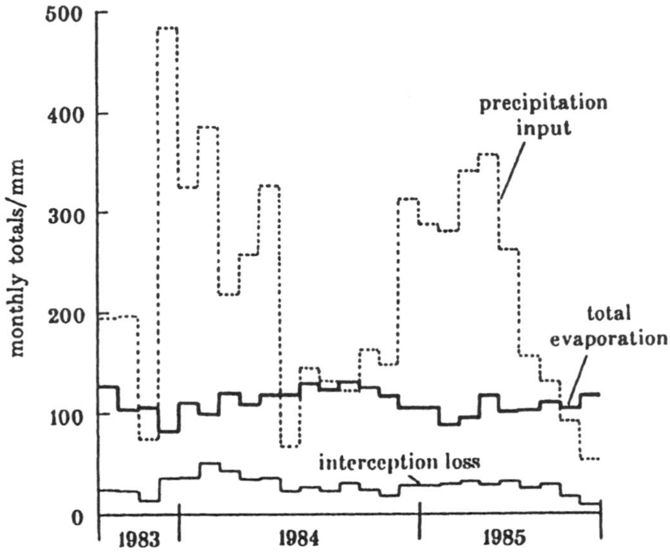

Figure 28-3. Monthly precipitation input, total evaporation, and interception loss from an Amazonian tropical rainforest over a 25 month period. (After W. J. Shuttleworth [2816])

Interception loss from a wet canopy is commonly modeled using a Rutter type model described in references 3627-3629. Total interception loss depends on numerous variables. Canopy characteristics, including leaf area index, leaf shape, and canopy density, define the canopy storage capacity and fraction of precipitation falling directly through the canopy. The amount of canopy interception is directly related to precipitation amount, while the fraction of precipitation intercepted is inversely related to precipitation intensity. Interception loss is enhanced by high levels of net radiation and strong winds, and by low air vapor density and aerodynamic resistance (R_A). This latter variable is related to the aerodynamic properties (roughness length and zero-plane displacement) of the vegetation surface.

Transpiration from the plant cover increases as the intercepted water is consumed. Plants exhibit substantial control over the rate of transpiration through regulation of the size of the opening of individual plant stomata. Plant stomata are small openings on the epidermis of a leaf through which carbon dioxide diffuses in and water vapor diffuses out from the soft leaf mesophyll. Plants regulate the size of these openings by way of guard cells on the leaf surface and thus regulate the diffusion of gases between the ambient and leaf environments. This direct physiological control of leaf transpiration is a dynamic process that responds to microclimatic conditions. Measurements of the resistance to the diffusion of water vapor from individual plant stomata can be measured in the field with a leaf porometer. The restriction of the guard cells on the diffusion of water vapor through the leaf stomata is usually expressed in terms of the stomatal resistance (R_{ST}, sec cm^{-1}) or its reciprocal, the stomatal conductance (G_{ST}, cm sec^{-1}).

Plant stomata maintain relatively uniform stomatal conductance values within a preferred range of selected environmental conditions, but have clearly defined threshold values above and/or below which the stomatal conductance values show an abrupt decrease. R. Avissar et al. [2801] conducted laboratory experiments with *Nicotiana Tabaccum* var. "samsun" leaves over different plant developmental stages that illustrate typical responses of relative stomatal conductance to selected environmental variables. A relative stomatal conductance value of 1.0 implies no stomatal closure, while a value of 0.0 implies complete stomatal closure. The results shown in Figure 28-4 indicate the response of this species to selected environmental conditions and suggest that numerous variables can limit transpiration, although the specific threshold values would differ for other species. *Nicotiana Tabaccum* leaves exhibit no stomatal closure until global solar radiation totals fall below 100 W m^{-2}. Relative stomatal conductance is 1.0 between 20-32°C, but drops off at leaf temperatures below and above this range, becoming 0.0 at 0°C and 40°C. The dropoff is more rapid at the upper end of the leaf temperature range. It is important to note that stomatal conductance responds directly to leaf temperature and only indirectly to air temperature through the effect of that variable upon the leaf energy budget. Relative stomatal conductance falls below 1.0 above a vapor pressure difference of 2000 Pa, but shows no drop in response to atmospheric CO_2 concentration between a range of 10-1000 ppm. High negative soil water potential is also thought to control relative stomatal conductance through its control on the internal leaf water content but was not directly measured in this study. At a relative stomatal conductance of 0.0, transpiration only occurs through the leaf cuticle. For this species cuticular transpiration losses were only 5% of the maximum water loss through plant stomata. These results apply to a single species under laboratory conditions in which only one variable is systematically varied. In a natural field setting different species will show separate response patterns, and stomatal conductance for an individual plant species will vary in response to the combined effects of multiple environmental forcings.

Estimation of evapotranspiration from a uniform well-watered vegetated surface, such as a grass turf, must take account of this physiological control if an accurate determination of Q_E is to be made. The Penman-Monteith equation has become the most popular method of estimating Q_E from various land surfaces. It is a combination equation that includes both a radiation term and an aerodynamic term in the numerator as well as a resistance term for plant physiological control of transpiration. It is given as

$$Q_E = (\Delta(Q^* + Q_G) + c_p(e_s\text{-}e_a)/R_A)/(\Delta + c_p[1 + R_S/R_A]/\lambda)$$

where Δ = the slope of the saturation specific humidity curve with temperature, Q^* = net radiation, Q_G = substrate heat storage, c_p = specific heat of air at constant pressure, (e_s-e_a) = specific humidity deficit, R_A = aerodynamic resistance to the transport of water vapor from the evaporating surface to the reference height z, R_S = surface resistance to the diffusion of water vapor from within the surface as a whole, and λ = latent heat of vaporazation of water. For simple uniform surfaces, such as a grass turf, pasture, or small grain field, the surface resistance R_S (sometimes called the bulk stomatal resistance) can be taken as the average of a sample of individual stomatal resistances R_{ST}.

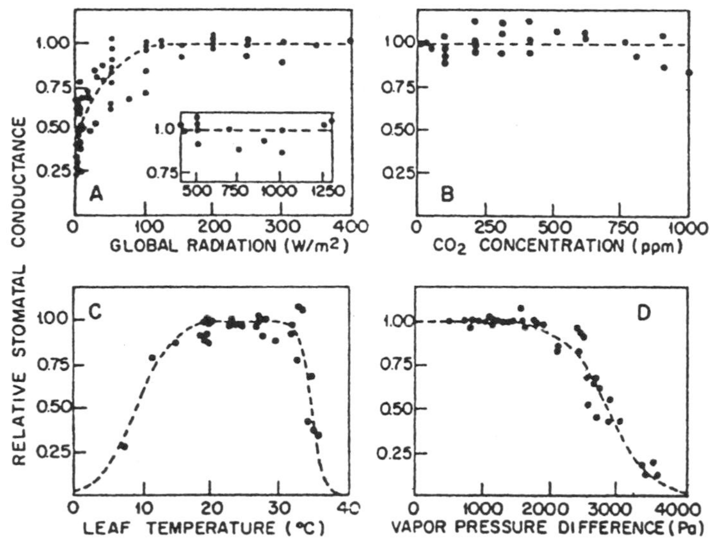

Figure 28-4. Effects of global radiation (A), CO_2 concentration (B), leaf temperature (C), and leaf-air vapor pressure difference (D) upon the relative stomatal conductance of *Nicotania Tabaccum* as measured in the laboratory (•) and as fitted by an empirical model (dashed line). (After R. Avissar et al. [2801]) Reprinted from Agricultural and Forest Meteorology with permission from Elsevier Science.

K. Beven [2803] conducted an analysis of the Penman-Monteith evapotranspiration equation to determine its sensitivity to meteorological input data and model parameters. He examined two vegetation types (grassland/pasture and Scots/Corsican pine forest) using meteorological data from several temperate maritime climate stations in the United Kingdom. The aerodynamic resistance term R_A is expected to vary over time due to the dependence of the roughness length and zero-plane displacement upon wind speed. He used a constant value of 46 sec m^{-1} for the grass surfaces and 4 sec m^{-1} for the forested surfaces, however, and found little change in the results. Surface resistance R_S, however, does show a strong diurnal variation due to the influence of environmental variables upon the behavior of individual plant stomates (Figure 28-4). A seasonal variation in R_S can also occur due to the increase in the leaf area index over the growing season. A life cycle variation in R_S associated with leaf senescence may also occur. Under well-watered conditions surface resistance normally decreases from a maximum at sunrise to a minimum at mid-day, and back toward a maximum again at sunset. K. Beven [2803] used R_S values between 50-200 sec m^{-1} for the grass surface, cover and between 100-400 sec m^{-1} for the pine forest cover. He found that for the grass surfaces, Q_E was most sensitive to estimation of Q^* because the radiation term dominated the aerodynamic term in the equation. For the pine forest sites Q_E was dominated by the aerodynamic term and showed greatest sensitivity to the surface resistance R_S parameter. Because the two model parameters differ significantly between varying vegetation types, successful estimation of Q_E using the Penman-Monteith equation requires accurate measurement of these two variables.

When a canopy is wetted following a precipitation event evaporation of intercepted water will proceed at a faster rate than does transpiration from the vegetation because the canopy resistance R_{ST} for the wet canopy is zero. Water stored on plant surfaces within the canopy is thus quickly consumed. When the canopy water storage reaches zero, transpiration from the vegetation will supply the water lost through Q_E, with only minor contributions from soil evaporation. Under such dry canopy conditions Q_E for a well-watered surface is strongly controlled by variations in R_S.

Complications arise when dealing with a tall forest canopy because of the microclimatological variations that occur with increasing depth in the canopy. A.J. Dolman et al. [2804] report that micrometeorological conditions within the upper two-thirds of the tropical rain forest canopy at Manaus, Brazil were well-mixed, and had air temperature, humidity, and humidity deficit conditions that were generally close to the ambient conditions observed at the meteorological tower above the canopy. The lower one-third of the canopy, however, was essentially decoupled from the upper canopy and featured lower air temperatures, higher humidity, and a smaller humidity deficit. Wind speed and solar radiation both decreased with depth into the canopy. Because stomatal resistance R_{ST} responds to variations in these environmental conditions, stomatal conductance G_{ST} will vary significantly with depth in the canopy. A. J. Dolman et al. [2804] found that typical G_{ST} values in the upper canopy were three times greater than in the lower canopy, and exhibited a strong diurnal variation but were relatively constant in the lower canopy. B. Saugier and N. Katerji [2815] report that G_{ST} normally decreases with depth into the canopy but that variability of G_{ST} increases with canopy depth due to the reduced solar radiation totals but greater variability of sunlight present as the forest floor is approached.

Application of the Penman-Monteith equation to a forest canopy requires accurate specification of a canopy resistance R_C term (or canopy conductance G_C) that integrates the behavior of the individual stomatal resistances. This is a challenging task because R_{ST} varies between species, within an individual leaf, among different leaves in the same canopy layer, between different layers of the same canopy, and in response to varying environmental conditions throughout the canopy. This estimation is normally accomplished by either taking an adequate horizontal and vertical sample of stomatal resistance from many leaves within the canopy to determine the integrated canopy resistance, or by solving for the canopy resistance from the Penman-Monteith equation based upon micrometeorological measurements made above the canopy top. The determination of canopy resistance for various forest and plant canopies from either micrometeorological measurements or a scaling up from individual leaf stomatal resistances is an active area of current micrometeorological research.

Frequent rains, a large potential wet canopy surface area, abundant net radiation, a low aerodynamic resistance due to the rough forest canopy, and high canopy conductances arising from the moist soil conditions combine to produce extremely high levels of Q_E in tropical rain forest environments. W. J. Shuttleworth [2816] found that average Q_E was nearly 90% of Q^* over his 25-month study period, with Q_E/Q^* ratios ranging from 75-80% on clear days, to greater than 100% on rainy days due to the supply of sensible heat from the air $-Q_H$.

B. Saugier and N. Katerji [2815] state that G_C values for a species may vary by an order of magnitude during conditions of water stress or plant senescence. Because severe water stress in unlikely in tropical rain forest environments, W.J. Shuttleworth [2816] found that canopy resistance R_C was most strongly influenced by variations in the incident solar radiation and the humidity deficit. He did, however, find a small seasonal dependence of average canopy conductance upon the soil water tension within the top 1 m of soil. Forests are able to extract soil moisture from a greater soil volume because of their deeper and denser root systems and therefore generally exhibit less dependence of G_C upon soil moisture tension than do more shallow rooted plant systems. Conditions of limiting soil moisture will lead to a reduction in G_C due to the link between soil water potential and leaf water potential. In general, the ratio of actual evapotranspiration to potential evapotranspiration will be 1.0 until a critical level of soil water depletion has been reached. Below this critical level the ratio will decrease linearly with soil water content until a value of 0.0 is reached at the soil wilting point. This simple relationship, however, varies with the plant rooting depth and density, plant type, evaporative demand, and soil texture.

A final condition to consider is the presence of incomplete canopies where, for example, a broken forest canopy layer is intermixed with a grass/herb layer or bare soil. In such cases the Penman-Monteith equation can be used to estimate evapotranspiration for each cover type. These separate estimates can then weighted by the fractional coverage of each type to arrive at a first approximation of the total Q_E for the incomplete canopy. Such a simple treatment, however, ignores enhancement of Q_E due to the advection of sensible heat produced on open ground.

Evaporation will vary with the type of soil or vegetative cover. Through intuition, however, one may not always be able to anticipate the effects of a change in the environment. For example, one might normally assume that reeds growing in a lake might result in increased transpiration and thus increase the water loss, particularly in dry climates. Quite to the contrary, however, E. T. Linacre, et al. [2808] found that in dry climates, it is likely that the growth of reeds in a lake or other water body will reduce rather than increase water loss. This appears to be due to a combination of the sheltering effect of the water surface from dry winds by the reeds, the shading effect created by the reeds, their relatively high albedo in comparison to open water, and the internal resistance of the plant stomates in the reeds to transpiration losses. Thus, E. T. Linacre, et al. [2808] among others (W. S. Eisenlohr [2805], D. A. Rijks [2814]) found that evaporation is less over a swamp than a lake.

It is not easy to determine the value of potential evapotranspiration, and the difficulties multiply when an attempt is made to measure the actual evapotranspiration. Measurements made by lysimeters, in so far as they are of interest to microclimatology, will be given in Section 30. Modern developments in micrometeorological theory have led to significant advances in our understanding of the physics of evaporation, while similar new insights in botany have produced a new appreciation for the role of plant physiological control over transpiration. The benchmark work by W. Brutsaert [2802] should be consulted for additional information dealing with evaporation.

Methods have been developed for reducing the amount of evaporation from open water surfaces by artificial means. These methods, which have the potential to slow evaporation, depend on covering the surface of the water with a thin layer of a substance that will hinder, to a great extent, the passage of water into the air. The substance requires a high viscosity so that it can resist the action of wind and waves; yet it must not interfere with water biology and be economically practical. Films of oil reduce evaporation but are too easily broken up.

According to W. W. Mansfield [2809], this monomolecular layer offers such resistance to the escape of water molecules from the surface into the air that evaporation is reduced by 75 percent. However, the reduction of E brings about changes in the energy budget of the water surface and its temperature must rise substantially. It follows that $\acute{E}$-e and E will increase, so that the influence of the protective layer will be offset to some extent. In experiments with flat basins of water with a surface temperature of 30°C, a relative humidity of 23 percent, and a wind of 3 m sec^{-1}, the expected evaporation was reduced by about 75 percent in the absence of sunshine. In the presence of midday sunshine, the reduction was about 30 percent. With a wind of only 0.6 m sec^{-1}, the reduction was reduced to only a few percent. More favorable results were obtained with deeper water where the extra heat could spread over a great mass of water. According to J. L. Monteith [2811], 37 percent of water was saved over a 14 week period in Australian reservoirs, at a cost of less than 1 cent per hectoliter of water saved.

For a monomolecular layer to be maintained over a large surface, which is in constant motion as a result of wind and waves, there must be a small excess of the substance. This is also required because of continual losses at the shore.

CHAPTER V

THE EFFECT OF LOW PLANT COVER ON THE SURFACE AIR LAYER

Vegetation occupies the space between Earth's surface and the atmosphere. Plant cover not only takes up space, but forms a transition zone, because the individual plant parts, such as leaves, needles, twigs, and branches, behave like solid ground, absorbing and emitting radiation, evaporating and transpiring water, and playing their part in the exchange of energy with the surrounding air. However, the air is still able to circulate within the plant cover more or less freely. Thus, vegetation forms a new component part of the air layer near the ground.

The total mass of plant components taking part in radiative, conductive, and moisture exchanges with the environment has a remarkably small thermal capacity. The thick, 5 m high stand of pine trees for which A. Baumgartner worked out the energy budget (Figure 26-2) was shown by K. Mauerer [3324] to contain a plant mass equivalent to a layer of wood only 19 mm thick.

Plant parts differ from solid ground in that shortwave radiation can pass through them. Below a closed forest canopy there is a subdued greenish light. Plant organs are not part of a dead physical system. The process of living implies having an active role in relations with the environment, as in the orientation of leaves and flowers toward the direction of incident light, or the closing of leaf stomata to reduce transpiration.

Plants, by being anchored to the place where they grow, are dependent on the climatic conditions of the locality. These may be advantageous or injurious to the life of the plant. On the other hand, the effect of plants on the microclimate of the locality increases as the plants grow.

This chapter deals with the influence of vegetation cover on the climate near the ground. In humid areas, vegetation-free level ground is rare. Thus, the consideration of plant cover on the surface microclimate provides a more representative analysis of the climate near the ground.

Before discussing the mutual influence of plants and microclimate, we must first consider the energy budget of plants and how plant temperatures become adjusted to the temperature of the ground and the air.

29 Energy Balance and Temperature of Plant Components

When radiation falls on a leaf or a needle, part of it is reflected from the surface. The reflection coefficient or albedo R expresses the reflected radiation as a percentage of the incident radiation. Another part passes through the leaf or needle, the quantity reemerging being expressed as a percentage D known as the transmission coefficient. The remaining radiation, represented by the absorption coefficient A, is absorbed and therefore converted into heat which raises the leaf temperature. In all circumstances $R + D + A = 100$ percent.

Table 4-2 shows that for vegetation, R lies between 5 and 30 percent for shortwave radiation. Sometimes the albedo may rise to 60 percent for lighter surfaces of variegated leaves. The value of R is smaller in the ultraviolet range; K. Büttner and E. Sutter [203] found it to be 2 percent for heather on sand dunes.

If photographs of the countryside are taken on film that is sensitive to near-infrared radiation, trees appear very light, almost white, which was demonstrated by E. V. Angerer [2901] as early as 1930. In the near-infrared range leaves and needles reflect more strongly, with a reflectivity of 35 to 50 percent, in contrast with solid ground. Figure 29-1 shows the results of F. Sauberer's [2933] measurements of the reflectivity R_λ of a concrete surface (dashes) and a meadow (solid line). With the concrete surface, the reflectivity shows very little dependence on wavelength, but the meadow has a weak maximum at 0.5 μ (green light), then increased reflectivity in the range from 0.75 to 1.0 μ. Most leaves appear green when illuminated by visible light due to the weak maximum of reflectivity around 0.5 μ. For near-infrared radiation K. Egle [2905] found that at 2.4 μ the reflectivity varied between 5 and 16 percent for the green leaves of five different types of plant.

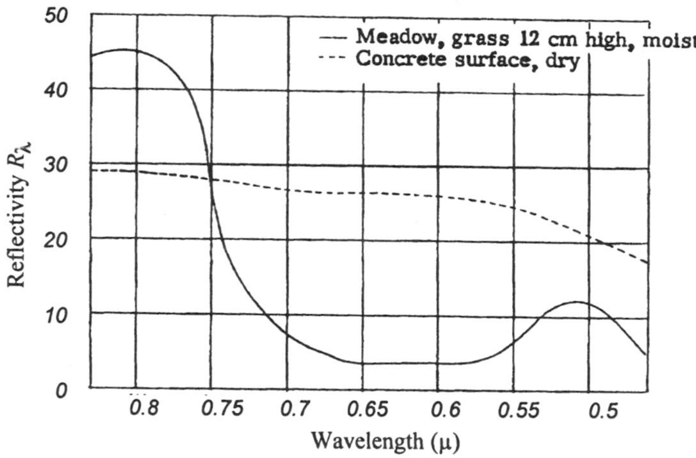

Figure 29-1. Change in reflectivity R_λ (percent) with decreasing wavelength for meadow (solid line) and concrete surfaces (dashed line). (After F. Sauberer [2933])

Plant canopies usually reflect solar radiation in a diffuse manner, although specular reflection may occur at times, as is evident from the bright appearance of the leaves of some Mediterranean evergreens. The radiation that is transmitted through the leaves is, however, always diffuse. The reflection (R), transmission (D), and absorption (A) coefficients vary with the type of plant. Table 29-1 shows these coefficients for a number of plant leaves. The variation of these coefficients with wavelength for a typical hardwood leaf is shown in Figure 29-2. At low sun angles the reflectivity of leaves is greater. For more detailed information about reflectivity and absorbtivity from leaves at high and low sun angles and how these vary through the growing season, D. H. Miller [2926] is recommended. Transmissivity D_λ is also affected by the movement of chloroplasts and is dependent on wavelength. Figure 29-3 illustrates the measurements made by F. Sauberer [2933, 2934] with a young red beech leaf (full line), a primrose (dashed line), and a hellebore (dot-dash). There is a very striking increase in transmissivity from about 10 percent to four or six times that value as wavelengths increase beyond around 0.7 μ. The marked maximum is maintained until about 1.5 μ after which it declines slowly. We would describe the light within forests as preponderantly near-infrared if our eyes were as sensitive to it as they are to the weak maximum in the yellow-green range (0.55 to 0.58 μ). Between 1.0 and 2.4 μ, K. Egle [2905] obtained D_λ values of 25 to 47 percent.

Table 29-1. Typical shortwave reflection *R*, absorption *A*, and transmission *D* coefficients for plant leaves. (After D. H. Miller [2926])

	R	*A*	*D*
American Beech (*F. Grandifolia*)	0.24	0.52	0.24
Ash (*F. Pennsylvanica*)	0.31	0.51	0.18
Banana (*M. Paradisiaca*)	0.26	0.55	0.19
Bird of Paradise (*Streliztsia sp.*)	0.23	0.62	0.15
Black Cherry (*P. Serutina*)	0.25	0.51	0.24
Black Oak (*Q. Velutina*)	0.24	0.50	0.26
Cotton (*G. Hirsutum*)	0.22	0.52	0.26
Cottonwood (*P. Deltoides*)	0.24	0.50	0.26
Liverwort (*Reboulia sp.*)	0.16	0.81	0.03
Peach (*P. Persica*)	0.25	0.59	0.16
Pepper (*C. Annuum*)	0.21	0.53	0.16
Silver Maple (*A. Saccharinum*)	0.23	0.48	0.29
Sunflower (*H. Annus*)	0.22	0.52	0.26
Tulip Tree (*L. Tulipifera*)	0.24	0.52	0.24
White Oak (*Q. Alba*)	0.22	0.44	0.34

According to Sauberer [2933], it makes a difference whether the upper or the lower sides of the leaves are exposed to radiation. A leaf of white poplar, for example, gave $D = 22$ percent for the upper surface and $D = 15$ percent for the lower. This difference changes with the wavelength. There are also large differences between leaves from the same plant. K. Raschke [2929] found a mean value of $D = 10$ percent for the leaves of *Alocasia indica,* with values for individual leaves varying from 4 percent to 29 percent.

Transmissivity produces a spectrally-filtered green light in forests, which is spoken of as green shade. According to A. Seybold [2935], this is different from the blue shadow found on the north side of walls. Since blue light is scattered to a greater degree than other colors, the diffuse energy on the north side of walls has a higher percentage of blue light. On the north side of vegetation, however, P. Stoutjesdijk [3632] suggested the term blue infared

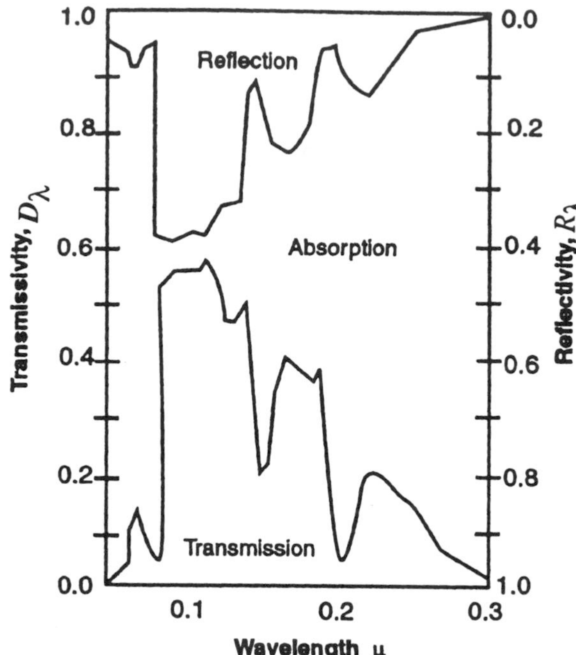

Figure 29-2. Transmissivity, reflectivity, and absorptivity of a typical hardwood leaf. (After D. M. Gates [2910])

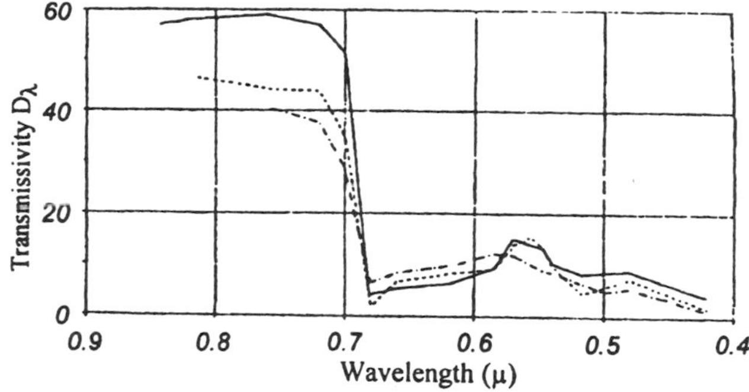

Figure 29-3. Change in transmissivity D_λ (percent) with wavelength for various types of leaves. (After F. Sauberer [2933])

shade would be more appropriate because incoming energy is concentrated in both these wavelengths. For additional information on the spectral modification of light in plant canopies refer to C. Varlet-Grancher et al. [2940].

During the daytime chlorophyll reflects energy in both the red and blue spectral regions. There is a strong inverse relationship between this reflection and photosynthetic activity. This reflection is not visible because it is only a small part of the light within the canopy. For more information dealing with the natural emission of chlorophyll the reader is referred to G. Guyot [2911] and F. B. Salisbury and C. W. Ross [2932].

In the shortwave radiation budget, absorbed radiation is retained inside the leaf and is used to either heat the leaf, or in some other way, such as transpiration or photosynthesis.

The leaf behaves almost as a blackbody toward longwave radiation, absorbing from below and above and in turn emitting longwave radiation appropriate to its own temperature ($L\uparrow$). A quantitative evaluation of the radiation exchange of leaves, illustrated in Table 29-1, is taken from the research of K. Raschke [2929]. The radiation exchange was determined, using a horizontally placed *Alocasia indica* leaf, with R = 21 percent and D = 10 percent, under climatic conditions representative of India.

Table 29-1. Radiation balance of leaves.

	By Day	By Night
Temperature of ground surface	65°C	35°C
Temperature of leaf	46°C	34°C
Shortwave radiation budget ($W\ m^{-2}$)		
From direct-beam solar radiation (S)	+558	--
From diffuse solar radiation (D)	+105	--
From ground reflected radiation ($K\uparrow$)	+105	--
Net shortwave balance (K^*)	+768	
Longwave radiation budget ($W\ m^{-2}$)		
Incoming radiation from the atmosphere ($L\downarrow$)	+440	+377
Incoming radiation from below ($L\downarrow$)	+684	+468
Radiation emitted by the leaf ($L\uparrow$)	-1124	-956
Net longwave balance (L^*)	0.00	-111
Net radiation balance (Q^*)	+768	-111

During the day the leaf gains radiant energy by absorption of direct-beam solar radiation (S), diffuse radiation (D) and counterradiation ($L\downarrow$). The last two energy terms, however, are absorbed both by the top and bottom of the leaf. Energy is lost by longwave radiation emitted ($L\uparrow$) by both sides of the leaf. The excess in radiation gain during the day is lost by conduction-convection to the air, evapotranspiration (latent heat) and to a small extent by a gain in heat storage within the leaf. At night the leaf has a net radiation loss (Table 29-1) which is made up primarily by conduction-convection from the air around the leaf and to a very small extent by a loss in heat storage within the leaf.

A plant can influence the radiation balance of its own leaves. This is not merely a question of the structure of the plant, by which it avoids excessive heating by corrugations on the leaves, by a field of thorns in cacti, or by pubescence. For more information dealing with plant structure and temperature regulation, the reader is refered to the work of B. Huber [2912]. Plant leaves can also reduce the amount of radiation falling on them by changing their leaf orientation. This is done by movement of the peculiar joints of the leaf so as to reduce the area exposed to the sun. O. W. Kessler and H. Schanderl [2916] have published photographs of white melilot with its leaves at various angles. I. N. Forseth and A. H. Teramura [2907] discuss how Kudzu changes its leaf orientation to avoid excessive temperatures. In dry Mediterranean areas and hot prairies, there is "a peculiar change in the physiognomy of the landscape at certain hours of the day" in which such reorientation can be recognized.

J. Ehleringer [2906] studied adaptations to heat stress of two varieties of *Encelia*. *Encelia Californica* grows in the coastal areas of southern California and *Encelia Farinosa* grows in the much hotter Mojave and Sonoran Deserts. The two species differ primarily in leaf pubescence. *Encelia Farinosa* has smaller, more reflective heavily pubescent (hairy) leaves; *E. Californica* leaves are glabrous. While the main temperature controlling factor was thought to be the difference in the leaf albedo, leaf angles also assisted in controlling temperatures. Midday spring air temperatures at Pt. Magu, California averaged around 20°C while those in Tuscon, Arizona averaged around 30°C. As a result of these leaf adaptations, however, the differences in leaf temperature between the sites were not as extreme. *E. Californica* leaf temperatures (17.3-25°C) were warmer, while those of *E. Farinosa* (21.6-29.6°C) were cooler than ambient air temperatures. In summer, *E. Californica* becomes dormant (leafless) during the dry season, while *E. Farinosa* maintains some leaves whose temperatures were always cooler (5°C or more) than ambient air temperatures.

The factor Q_G corresponds in leaves to the storage of energy within the leaf and to the energy supplied through the stalk of the leaf. However, since the thermal capacity of herbaceous plants is small, these can be neglected. For example, in the fairly fleshy leaves of *Alocasia* just mentioned, it is only 0.0837×10^6 J m^{-3} K^{-1}. Leaves, therefore, adjust themselves within a few seconds to changes in their surroundings. The amount by which the temperature of a leaf rises above that of the air when there is a positive radiation balance, or the amount of cooling with a negative radiation balance, depends primarily on the other three factors in the energy balance, namely Q^*, Q_H and Q_E.

The exchange of energy with the surrounding air (Q_H) is determined, as we have already seen in Section 28, by the energy transfer coefficient a_L. Thus, $Q_H = -a_L (T_l - T_A)$, where T_l is the temperature of the leaf surface, and T_A is the temperature of its environment. A boundary layer a few millimeters thick is formed at the surface of the leaf, in which T_l gradually changes to T_A. When the air is completely still, a_L has a value of about 6.98 W m^{-2} K^{-1}. It increases roughly proportional to the square root of the wind speed, and depends on the size, shape, and position on the leaf at which measurements are made. The peak values, which are reached with small leaves and high winds on exposed parts of the leaves, amount to about 209.4 W m^{-2} K^{-1}.

In assessing the factor Q_E, it must be remembered that the leaf can take part in both evaporation and transpiration. All plant canopies have a canopy storage parameter *S* that represents the maximum amount of water stored on the canopy surfaces when they are fully wet. Any water on the canopy in excess of *S* will freely drain and not be available for evaporation. Evaporation of intercepted water from a wet canopy normally proceeds at the potential evaporation rate as determined by, for example, the Penman-Monteith equation. The actual amount of water stored on the plant canopy *C* will range from zero, for a completely dry canopy, to *S* for a completely wet canopy. When *C/S* equals 1.0, the canopy conductance is set to zero and all latent heat transfer occurs by evaporation from the canopy surface. In such cases, evaporation is limited by the availability of energy, the vapor density deficit, and the aerodynamic resistance term. When *C/S* equals 0.0, all latent heat transfer occurs by transpiration and the canopy conductance becomes a principle control on water loss. For values of *C/S* between 0.0 and 1.0, the rate of evaporation of intercepted water is set equal to a proportion (*C/S*) of the wet canopy evaporation rate, and transpiration is determined as a proportion

(1- (*C/S*)) of the dry canopy transpiration rate. W. J. Shuttleworth [2816] discusses the implementation of this approach for a tropical rainforest environment. The temperature of a plant is, therefore, determined by the ambient conditions surrounding the plant (net radiation, air temperature, humidity), the physical dimensions of the leaf surfaces (through the aerodynamic roughness and aerodynamic resistances), and the degree of physiological control exerted by the plant on transpiration (through the canopy resistance).

K. Raschke [2928, 2929] shows the diurnal variations of the elements in the energy budget of a leaf in Figure 29-4. Energy input (above) and output (below) are placed in juxtaposition, and are of equal magnitude on either side, since the thermal capacity of a leaf is negligible. The measurements are for a horizontally placed leaf of *Alocasia indica* on clear spring days in Poona, India (18°31'N). The curves are drawn through half-hour average values, so that short-term variations are eliminated. Since stomata are generally located on the lower surface of the leaf, the difference in transpiration loss from the upper and lower surfaces is fairly large.

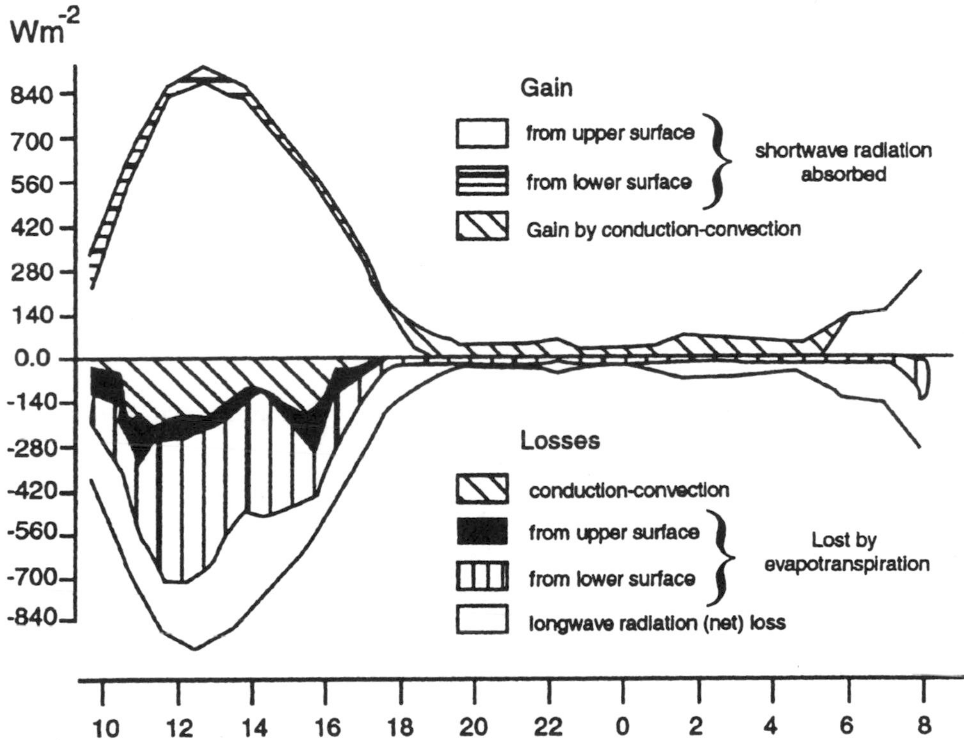

Figure 29-4. Diurnal variation of the energy budget of an *Alocasia indica* leaf. (After K. Raschke [2928, 2929])

As with the energy exchange between the ground and air (Section 25), the flow of energy to and from the air (Q_H) is referred to as conduction-convection. While all the heat flowing to and from the leaf in its laminar boundary layer is by conduction, beyond this thin layer, convection is the primary mode of heat transfer. Even under the calmest conditions, convection is far more effective in the total transport of energy to and from the leaf than conduction. During the day energy is gained by the absorption of shortwave radiation on the upper and lower surfaces of the leaf. At night it is gained by conduction-convection. During the day the leaf looses energy by conduction-convection, evapotranspiration and longwave radiation. At night, energy is lost primarily by emission of longwave radiation and secondarily by evapotranspiration.

Since Q_H, in Figure 29-4, appears as an energy loss during the day and an energy gain at night, it follows that during the day the temperature of the leaf must be higher, and at night lower, than the air temperature. Figure 29-5 shows the temperature difference of a plant and the surrounding air and gives the excess (+) or deficit (-) of leaf temperature as a function of a_L (shown on a logarithmic scale) and of the net radiation balance Q^* (from K. Raschke [2929]). The daytime figures (positive Q^*) above the zero line are based on an air temperature of 22°C, 61 percent relative humidity, vapor pressure of 16.0 mb (summer day). The nocturnal values are based on a temperature of 13.2°C, relative humidity of 95 percent, vapor pressure of 15.2 mb, and a water coverage factor of 0.01.

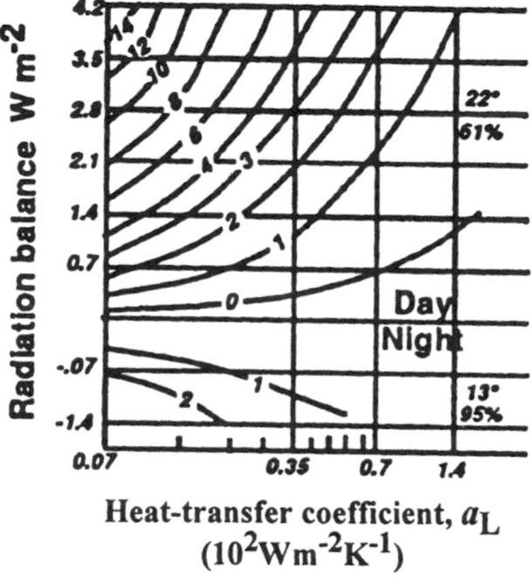

Figure 29-5. Temperature excess (°C) by day and deficit by night of a leaf, as a function of net radiation balance and a ventilation factor a_L. (After K. Raschke [2929])

During the day, if there is a brisk wind (large a_L) and a small amount of incoming solar radiation (cloudy skies), transpiration may reduce the leaf temperature a few tenths of a degree below the air temperature. On the other hand, with strong solar radiation and calm conditions, the temperature excess may be more than 10°C, although the rise in temperature will cause increased transpiration which may produce a noticeable cooling.

Copiapoa haseltoniana (acatus) hairy apex were 20°C and its surface temperature 13°C above the ambient air temperature. One of the main functions of the hairy surface is to provide an insulating layer protecting the plant from excessive warm or cool temperatures (G. Krulik [2918]). Since the radiation exchange is relatively small during the night, the fall of leaf temperature below that of the air seldom exceeds 2°C. Nocturnal leaf temperatures are controlled primarily by the loss of thermal radiation and by the gain of sensible heat from the surrounding air. In their study of radiative frosts, B. Ithier, et al. [5330] observed nocturnal temperatures of the plants from 0.5-2.0°C cooler than the ambient air temperatures with clear skies, and from 0.1-0.2°C cooler under foggy conditions. D. N. Jordan and W. K. Smith [2914] in subalpine conditions found that as a result of decreased counterradiation (due to the decreased atmospheric density and low moisture content) under relatively calm conditions nocturnal leaf temperatures were in excess of 6°C below ambient air temperatures. R.

Leuning and K. W. Cremer [2922] found nocturnal leaf temperatures from 1-3°C lower than the air temperatures. Horizontally held leaves were from 0.5-1.5°C cooler than leaves oriented vertically. They also observed that small leaves generally had a higher temperature at night than larger leaves (Figure 29-6). This follows from the fact that leaf size is inversely related to the energy transfer coefficient a_L. In comparison with large leaves, increased conduction-convection from small leaves results in greater cooling during the day and warming at night. Thus, not only are small leaves warmer at night, they are also cooler during the day. Consistent with this finding D. N. Jordan and W. K. Smith [2915] observed, in a subalpine environment in Wyoming during the summer at a height of 8 cm, 41 nights of frost on broadleaf plants but only 25 nights of frost on needleleaf seedlings. Small leaf size promotes heat loss by conduction-convection and, consequently, leads to reduced evaporative losses. This water saving adaptation can be noticed in the small leaves of desert herbaceous plants. R. Leuning [2923] found that nocturnally leaves with condensation on them (dew or frost) were 1-2°C warmer than leaves without condensation under otherwise similar conditions.

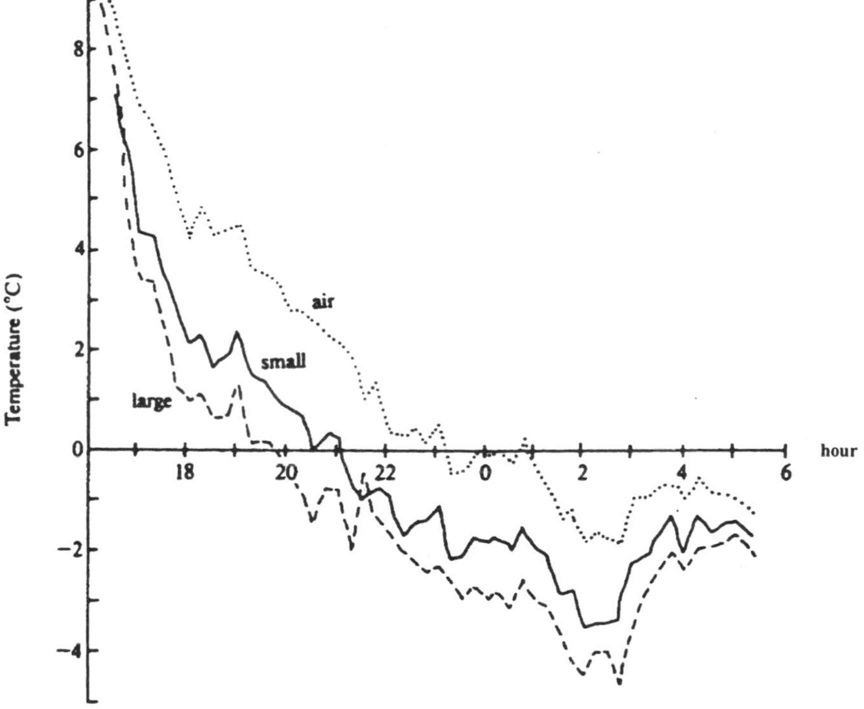

Figure 29-6. Time series of temperatures of air, small leaves (*E. viminalis*) and large leaves (*E. pauciflora*) at 100 mm above grass on 20-21 July 1984. (From R. Leuning and K. W. Cremer [2922])

According to B. Huber [2912], the temperature of a leaf, needle, or bud will lie between that of the surrounding air and the ground surface. The larger and fleshier the leaf or bud, the more closely the plant part will approach the temperature of the ground. K. Kunii [2919] found that the twigs of cherry blossoms fluctuated by 2–22°C during the day in Japan, whereas the variation in the air was only 1-13°C. D. S. Brown [2902] found the maximum temperatures of buds in winter were up to 8.5°C warmer than air temperatures. J. A. Clark and G. Wigley [2904] measured the temperature variations of the surface of a leaf at varying wind speeds. At a given wind speed, the temperature difference between the flower or leaf and the ambient air may be affected by any of the many factors that control the intensity of the solar radiation to which it is exposed. These include shade, fog, haze, cloud cover, incidence angle of the sun, position of the leaf, flower, or bud, etc. For example, S. Lu, et al. [2924] showed that on calm, clear nights, ovaries on flowers facing upward were on the average 0.33°C cooler than those facing down. This difference was much less on cloudy nights and not observed on windy nights.

K. Takasu [2937] published a series of recordings, an example of which is shown in Figure 29-7. This presents the record of observations made on the 2720 m summit of Mount Shirouma during the forenoon of 16 July 1943. The high elevation, and the presence of light fog patches around the summit at the time the readings were taken, account for the large variations and the high peak values of radiation intensity, shown in the lowest curve. The temperature of the lower surface of a *Lagotis glauca* leaf, a perennial plant with leaves near the ground, was measured at a height of 5 cm above the ground. Close analysis of the two curves shows that the leaf temperature responds more rapidly to a sudden increase than to a decrease in solar radiation. The air immediately below the leaf shows a greater temperature instability than does the leaf itself because of its negligible thermal capacity, but follows fluctuations of leaf temperature very closely. Air temperature at a height of 1 m, in all its short-term unsteady movement, reflects to a large extent the variations in radiation. As was expected, the temperature of the ground surface is highest and shows the greatest lag in responding to changes in the radiation regime. The lower mass and greater ventilation of the leaf are responsible for its greater temperature fluctuations as compared to the ground surface. The leaf temperature is lower than that of the ground but higher than that of the air surrounding the leaf or at one meter above the ground.

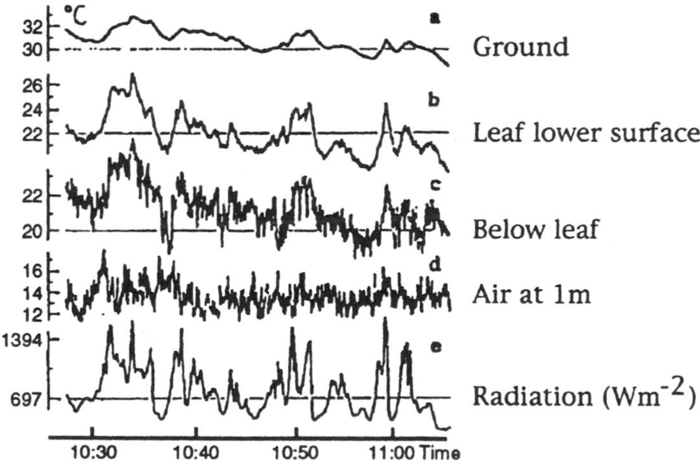

Figure 29-7. Fluctuations of temperature and radiation for a *Lagotis glauca* leaf. (After K. Takasu [2937])

Besides the intensity of solar radiation, one of the main factors affecting the difference in air and leaf temperatures is wind speed. If for a given radiation balance the temperature difference between a bud and the ambient air is known for any given wind speed, it can be calculated for any other wind speed using the following equations from D. M. Gates [2908]:

$$h_C = 6.17 \cdot 10^{-3} \frac{u^{1/3}}{D^{2/3}}$$

where u is the wind speed, D is the bud diameter, and h_C is the rate of energy transfer for conditions with wind and

$$\frac{dQ}{dt} = h_C A \Delta T$$

where A is the area of the surface in contact, dQ is the calories transferred, dt is the unit of time and $\Delta T = T_S - T_A$ where T_S is the bud surface temperature and T_A is the ambient air temperature. Substituting

$$\frac{dQ}{dt} = 6.17 \cdot 10^{-3} \frac{u^{1/3}}{D^{2/3}} A \Delta T$$

or

$$u^{-1/3} \Delta T = \frac{dQ}{dt} \frac{1}{A} \frac{1}{6.17} \cdot 10^3 D^{2/3}$$

Under conditions when the wind speed is equal to one unit (one mile or kilometer per hour or m sec^{-1}) ΔT = Constant (K), therefore,

$$\frac{dQ}{dt} \frac{1}{A} \frac{1}{6.17} \cdot 10^3 D^{2/3} = Constant(K)$$

thus,

$$\Delta T = \frac{K}{u^{1/3}}$$

If the wind speed changes (as long as the potential ΔT under radiative conditions with a wind speed of one unit (mile or kilometer per hour or m sec^{-1}) does not), a new ΔT can be estimated. Figure 29-8 illustrates estimates of the differences in the temperature of the ambient air and a bud at various wind speeds. These figures assume a given ambient air-bud temperature differential under the conditions of a wind speed of one unit (mile) per hour. Thus, when wind speeds are low, a slight increase will result in a large decrease in the ambient air-bud temperature differential. However, increasing wind speeds will have a decreasing effect in reducing this temperature differential. Further, regardless of the wind speed, a bud's temperature will remain substantially different from that of the air so long as the radiative conditions responsible for creating the air-bud temperature differential have not changed. This conclusion is equally correct for any other part of the plant.

Lichens, which lie directly on the ground or on stones and can dry out, may reach temperatures of 70°C. This was investigated experimentally by O. L. Lange [2920, 2921]. On the

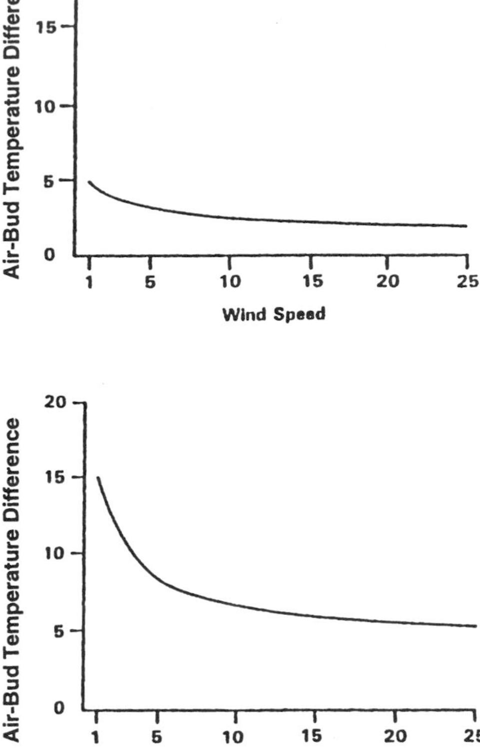

Figure 29-8. Effect of wind on air-bud temperature differential with initial conditions of 5° (top) and 15°C (bottom) at one mile per hour.

Kaiserstuhl (350 m elevation) at Freiburg (near Emmandingen) in the Black Forest, he obtained readings in *Cladonia furcata,* for three branches at different heights, of 42°C at 4 cm from the ground, 46°C at 1 cm, and 66°C on the thallus covering the ground. The color of these lichens, which may vary considerably, has a large effect on their temperature. In greenhouses, when the air becomes saturated with water vapor and evaporation Q_E is consequently low, plant temperatures may rise to lethally high levels.

The temperatures shown in Figure 29-9 were not measured in a leaf, but in the bark of a twig of green alder by G. and P. Michaelis [2925]. There was still snow in the tiny Walser Valley in the Allgäu, at a height of 1670 m on 16 March 1933. The origin of the ordinate indicates the position at which the twig emerged from the snow. With air temperatures between -2° and 4°C, the temperature of the bark reached 30°C; however, as a result of the cooling effect of the melting snow, this maximum was not at the snow surface but some 15 cm above it. At night the minimum was always found at the snow surface.

A. Büdel [2903] investigated the temperature variation in the blooms of spring flowers which are of significance to bee keepers since the secretion of nectar depends on the temperature of the flower. Figure 29-10 is for a sunny day; the dotted line is the temperature recorded by a thermoelement inserted between the petals of the dandelion flower (*Taraxacum*

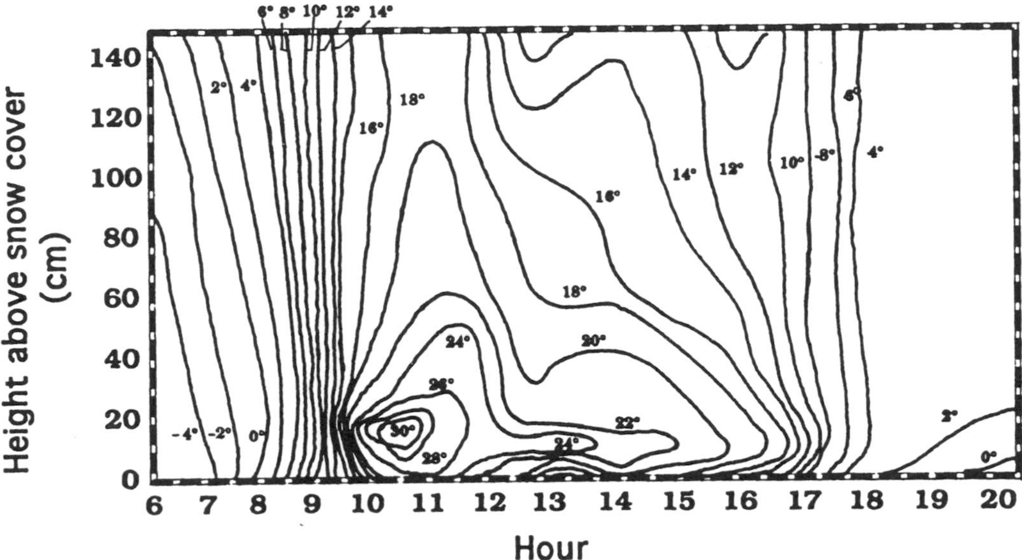

Figure 29-9. Temperature variations (°C) in a twig of green alder, projecting through snow on a March day. (After G. and P. Michaelis [2925])

officinale); the crosses give the air temperature in the immediate neighborhood of the flower; and the dashed line records the air temperature at a height of 1 m. The rapid response of this comparatively large flower is striking. The large fluctuations are due to the oscillating shielding provided by neighboring trees. With midday temperatures of 17°C, the flower temperature rose to 30°C.

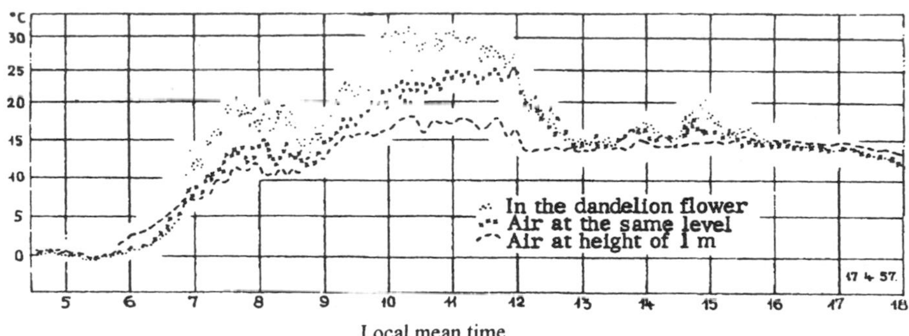

Figure 29-10. Temperature variations in a flower of *Taraxacum officinale* on a spring day (17 April 1957). (After A. Büdel [2903])

Plant temperatures are of particular interest when there is danger of freezing. Living winter buds of *Anemone pulsatilla* and *Helleborus niger* were found by K. Hummel [2913] to be up to 10°C warmer than the surrounding air, when the air temperature dropped to -12°C. Dead buds did not exhibit any excess temperature. R. Knutson [2917] found internal temperatures of eastern skunk cabbage (*Sympolocarpus foetidus L.*) to be 15-35°C above the ambient air temperature during February and March. The air temperatures varied from -15 to 15°C. "The tissues of the inflorescence (spadix) are not frost-resistant and escape freezing by maintaining a high respiratory rate" (R. Knutson [2917]). The spadix begins to cool immediately if cut from the plant. The root of this cabbage stores large quantities of starch which is the source of the energy needed to maintain elevated temperatures. K. A. Nagy, et al. [2927] reported that the inflorescence of *Philodendron sellourn* spadix temperatures remained between 38.6° to 45.8°C even when air temperatures were near freezing. The philodendron inflorescence maintain relatively high temperatures by regulating their rate of oxidative metabolism. "The heat is produced primarily by small, sterile male flowers that are capable of consuming oxygen at rates approaching those of flying hummingbirds and sphinx moths."

Another type of temperature change occurs upon freezing. H. Ullrich and A. Mäde [2939], S. Suzuki [2936], and K. Takasu [2937] have published measurements of the temperature variation inside leaves, which show clearly the sudden increase in temperature when latent heat is released at the onset of freezing. S. Suzuki [2936] and T. Takasu [2937] subdivided the increase into two parts, with an interval of 2 to 4 minutes between. The first increase results from freezing of the dew on the leaf, and the second from freezing of the sap in the cells.

When flowers are covered in breeding experiments with protective pollen proof-bags, they are subjected to quite a different type of microclimate. Figure 29-11 from N. Weger [2941] shows the average daily temperature variation for five sunny days in May and June 1937 inside four different types of bags. Compared with the ambient air temperature (heavy continuous curve), the interior air temperatures reach an excess of 15°C. N. Weger, et al. [2942] observed that, due to the higher temperatures, enclosed buds on a pear tree blossomed 2 to 4 days earlier. At night the temperature falls 1 to 2°C below air temperature. The temperature excesses during the day and deficits at night are due to the reduction of mixing with the environmental air.

Inside plant coverings E. Rohmeder and G. Eisenhut [2931] also found excess temperatures, at midday, of up to 18°C, and deficits of even more than 5°C at night. Inside the coverings, there was a marked thermal stratification. In the middle of the day the side receiving radiation was 6°C warmer than the opposite side. At night the side radiating to the sky was 2°C colder, which explains the uneven nature of the damage often found inside the protective coverings. Inside the bags during the day the air is saturated with water vapor with a relative humidity about 35 percent higher than outside.

The size of the dusting bag is of very little influence. The material of which it is made, on the other hand, has a marked influence, as may be seen from Figure 29-11. The intensity of light in the interior varies between 95 percent (polyethylene) and 54 percent (silk) of the light outside. Polyethylene bags are cheap and the flowers can be easily seen inside them, but they prevent any diffusion of gases; despite considerable condensation water collecting inside (up to 800 g in 14 days), there is still substantial nocturnal cooling. Woven bags are more resistant to wear and tear, the mesh of the weave can be adjusted for pollen size, and gases are able to diffuse through the material; variations of the temperatures are thus moderated. Imitation parchment bags are cheap, but get easily damaged, and show extremes of microclimate in the absence of any gaseous diffusion.

There are means whereby the often lethal microclimate within these bags can be moderated. Adding perforations to cellophane bags help somewhat (Figure 29-11). When E. Rohmeder and G. Eisenhut [2931] attached the protective bag rigidly to a wire basket, 10 to 15

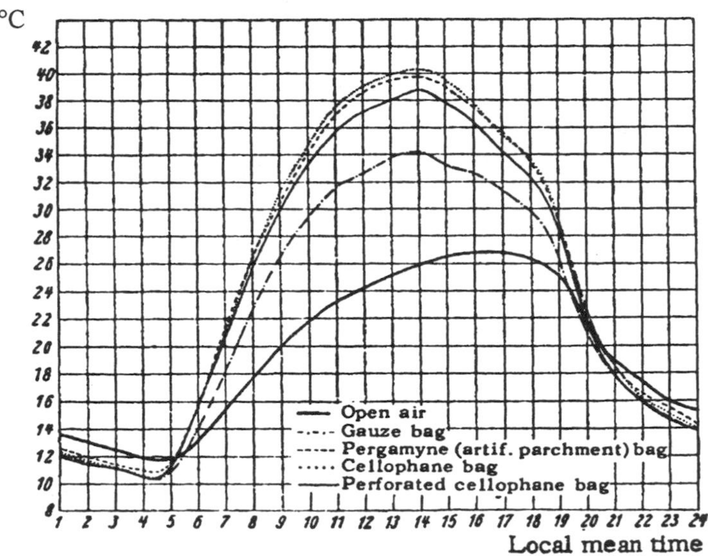

Figure 29-11. Air temperature variation in flowers enclosed in different types of bags. (After N. Weger [2941])

cm larger in diameter, the shade of the basket reduced excess temperatures by half and raised the nocturnal temperatures somewhat. W. Tranquillini [2938] found, in experiments with glass bowls in high Alpine radiation conditions, that the temperature excess inside was 25°C (maximum temperature of the leaves were 52°C). These conditions could be alleviated by using a heatfilter or by artificial ventilation with a current of air of at least 0.5 cm sec^{-1}.

The danger of frost in the bags can be delayed for half an hour by wetting them, since the latent heat released when the bag freezes delays freezing inside the bag. E. Rohmeder and G. Eisenhut [2931] were able to save some valuable breeding material by introducing tiny electric heaters (weight 30 g) into the bags. The fins of these heaters had a maximum temperature of 40°C and were able to protect flowers, without any difficulty, with ambient temperatures as low as -8°C.

30 Radiation, Eddy Diffusion, and Evaporation in a Low Plant Cover

The total surface area of vegetation growing on a meadow is 20 to 40 times the area of the ground on which it grows. The ratio of the total leaf area contained within a one m^2 column extending from the soil surface to the top of the vegetation canopy is called the leaf area index (LAI). It has units of m^2/m^2, and is used to quantify the density of the plant canopy. This magnification of surface area by vegetation has no influence on the amount of incident solar radiation. Both types of surfaces also emit, per square meter, the same amount of radiation by day and night, provided their surface temperatures are equal. Differences in surface emissivity, shortwave albedo, and surface temperature bring about a difference in the radiation balance between bare and plant-covered soil.

There is, however, a basic difference in the vertical distribution of the radiation gained or lost. The amount of solar radiation (Figure 30-1) incident upon a horizontal surface above bare ground and the top of a plant canopy are naturally equal. At half the height of the growth on a meadow of lush grass and *Dactylis glomerata*, A. Ångström [3001] found the radiation value was somewhat reduced, because the tips of the grass blades had absorbed a small part of the incident radiation. The amount of absorption increased farther into the canopy as the grass became denser and only one-fifth of the original radiation reached the ground.

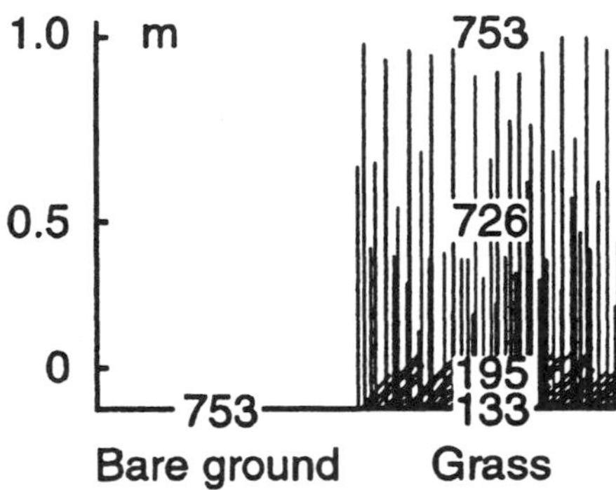

Figure 30-1. Attenuation of solar radiation (W m^{-2}) within a meadow canopy. (After A. Ångström [3001])

A similar distribution is found with outgoing nocturnal radiation. Since the grass cover has a similar temperature to the soil surface, it emits about the same amount of radiation as the soil. Over the grass-covered soil, however, the amount of longwave radiation which escapes from the ground surface is low. As the height within the grass cover increases, the sky view factor also increases, leading to increased net loss of longwave radiation. With reduced shielding from the night sky, outgoing radiation increases until, at the upper surface of the canopy, it reaches a value equivalent to that of bare ground (at the same temperature). This results in both the highest temperatures during the day and the coldest temperatures at night being within the mass of the vegetation (outer effective surface). The relatively still air within the mass of the vegetation acts as an insulating layer, moderating the energy exchange at the soil surface. This is illustrated in Figure 30-2, which shows the soil heat flux Q_G before and after the harvest of a barley crop (*Hordeum vulgare*). The removal of the barley crop produced larger diurnal surface temperature extremes, which resulted in greater conduction of energy into the soil during the day and from the soil at night.

The individual parts of plants may receive significantly different levels of solar and longwave radiation, depending upon the density of the vegetation, the structure of the canopy, and the resulting sky view factor (amount of shielding) of the individual plant parts. As a consequence, substantial differences in temperature within a plant canopy are quite common. In the relative calm that normally prevails inside a vegetation cover, these temperature differences produce a lively exchange of longwave radiation, in addition to some turbulent mixing of sensible and latent heat.

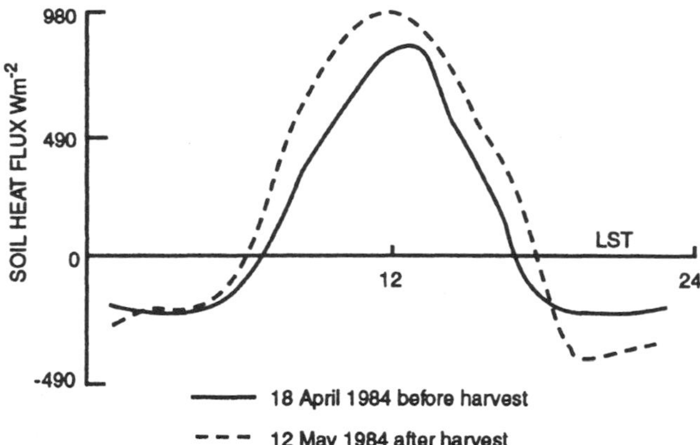

Figure 30-2. Soil heat flux (W m^{-2}) for a barley field in Syria for 18 April and 12 May 1984. (After S. A. Oliver, et al. [3413])

The penetration of global solar radiation into a uniform plant canopy is often modeled using a Beer's Law approximation:

$$K{\downarrow}_z = K{\downarrow}e^{-\nu \mathrm{LAI}_z}$$

where $K{\downarrow}_z$ = global solar radiation at depth z into the canopy from the top of the canopy, $K{\downarrow}$ = global solar radiation above the plant canopy, ν = an extinction coefficient that varies with plant species, canopy architecture, plant age, wind movement of leaves and branches, and leaf shape, form, inclination, and orientation, and LAI_z = cumulative leaf area index at depth z into the canopy from the top of the canopy. The value of ν is between 0.3 to 0.5 for plant canopies with a predominantly vertical leaf structure, and between 0.7 to 1.0 for canopies with a predominantly horizontal leaf structure (N. J. Rosenberg, et al. [5264]). Figure 30-3, from F. Sauberer [2933], shows the diurnal variation of visible light at the ground underneath various types of plants in comparison with bare ground. It is easy to distinguish the different degrees of shielding of the sky provided by summer barley, only 12 to 15 cm high, winter rye, nearly 1 m high, and clover with its dense umbrella of leaves. The quantity of light reaching the ground is important for the growth or the suppression of weeds. K. Unger [3012] measured the degree of illumination inside a stand of peas, from 23 to 25 June 1950, for 0.65 μ (the wavelength of maximum photosynthetic assimilation) and found it to vary between 38 and 84 percent of the outside value. The ability of different kinds of plants to provide a screen against radiation is an important property for weed control as anyone who has even grown a vegetable garden can attest.

The extent to which illumination can be reduced is illustrated by F. Sauberer's [2933] measurements in a 3 m high thicket of young elms with dense undergrowth, and overgrown with clematis (Table 30-1). When the low plants became fully developed by the beginning of July, only 0.01 percent penetrated to within 1 cm of the ground and only 2 percent came through the first 100 cm of the growth. Ground illumination increased when the undergrowth died, but it was only when the leaves fell (15 November) that the situation changed significantly. In winter, 6 to 7 percent of the illumination above the elm thicket reached a height of 1 cm above the ground.

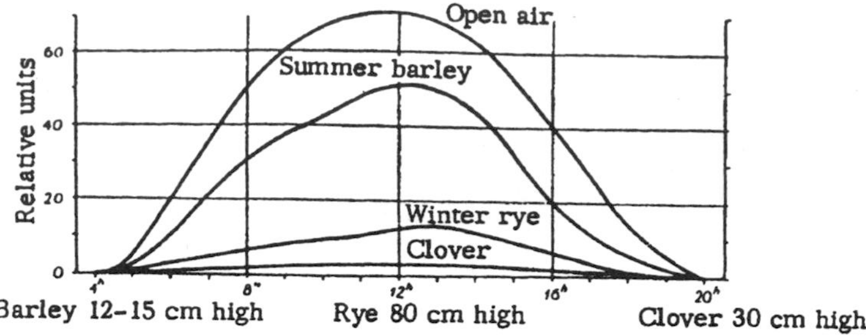

Figure 30-3. Diurnal variation of light at the ground below different types of vegetation in May. (After F. Sauberer [2933])

Table 30-1. Percentage of outside illumination within an elm thicket.

Date	Height above ground (cm)			
	1	10	25	100
5 July 1936	0.01	0.06	0.13	2.1
19 July 1936	0.03	--	2.17	2.2
15 November 1936	0.50	22	30	59

With vegetation of such thickness, the ground surface no longer serves as the boundary surface with the atmosphere. A. Woeikof discusses how an outer effective surface is developed, corresponding approximately to that of the upper surface of the vegetation, which in many respects assumes the role of the ground surface.

The increased vertical distribution of radiation received and emitted by vegetation results in smaller surface temperature extremes than when the ground is bare. The surface temperatures beneath a vegetation cover are lower by day and higher by night.

Besides radiation conditions, eddy diffusion also influences the microclimate in growing plants. When strong winds blow over the tops of plants that are easily swayed, as in a lush meadow or a field of grain, a wave motion develops at the boundary surface. Less pliable or more irregular plant surfaces, on the other hand, facilitate the development of large turbulence elements. Figure 30-4 illustrates the wind structure over a field of wheat stubble (above) and over a beet field with a growth of leaves reaching 40 to 50 cm (below). The area was flat and the wind had sufficient fetch to have adjusted itself to the roughness of the surface.

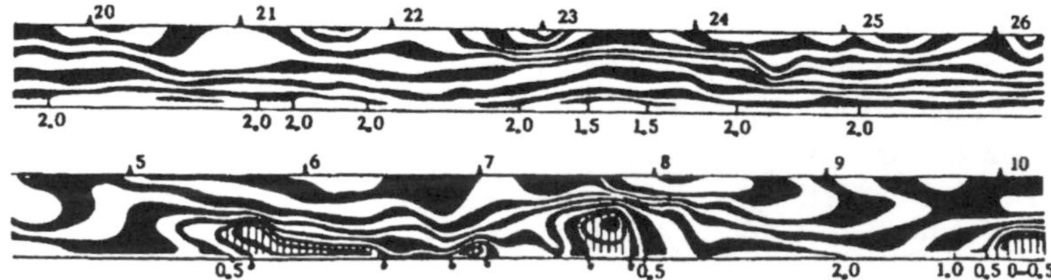

Figure 30-4. The different wind patterns over a wheat stubble field (above) and a beet field (below). (After W. Schmidt [706])

Figure 30-4 is for a vertical section 1.5 m high above both the stubble and the beet fields. There is a time scale in seconds at the top of each profile, so that rapid turbulent variations can be captured. Lines of equal wind speed are 25 cm sec^{-1} and the spaces between are alternately black and white. If the wind blows in a direction opposite to that of the main wind direction, it is indicated by vertical hatching. While the wind increases fairly regularly with height over the stubble, major disturbances appear over the taller beet field. Three times within 6 sec there is a reversal of wind direction immediately above the leaves, with intervals of relative calm in between.

For the increase of wind speed over vegetation-free ground for the case of an adiabatic temperature distribution, one can use the simple relationship $u = c \log (z/z_0)$ (Section 16). Figure 30-5 presents wind profiles measured by W. Paeschke [1607, 3010] in the neighborhood of Gottingen, during six different periods of the day, arranged in order of increasing wind speed. Since both axes are on a linear scale, the profile above a height of about 0.5 m resembles that of the bare ground in Figure 16-6A.

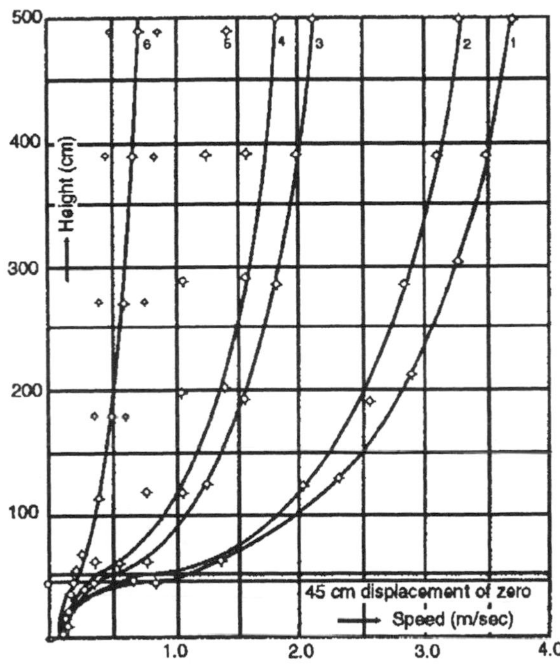

Figure 30-5. Wind profiles in and above a beet field. (After W. Paeschke [1607, 3010])

The equation for wind increase with height, under an adiabatic temperature lapse rate and with vegetation cover, becomes

$$u = c \log \frac{z-d}{z_o}$$

In Section 16, the constant c was discussed and z_0 (cm) was called the roughness parameter (or roughness length). A reduced height (z-d) is inserted in the equation where the parameter d is the zero-plane displacement (cm). The quantity d is not the average height of the vegetation surface, but represents the apparent level at which momentum has been absorbed by the individual elements of the plant canopy. It is evaluated from wind profile data and vegetative characteristics, and has been found to vary with plant type, plant growth stage, canopy height, spacing of canopy elements, and wind speed. For beet fields shown in Figure 30-5 the value of $d = 45$ cm was used. With the recordings made by E. L. Deacon [1603], on which Figure 16-8 was based, a value $d = 25$ cm was used, although the vegetation in question was grass with flower stalks 60 to 70 cm high. The high degree of pliability of grass under the pressure of wind is responsible for the lower value of d than was observed for the beet field.

Figure 30-6 gives roughness length values summarized by E. L. Deacon [1603]. The abscissa, on a logarithmic scale, shows z_0. The ordinate is the wind speed in meters per second that must exist at a height of 1 m for the flow to become turbulent. Since natural surfaces are extremely varied, the diagram can only indicate the approximate mean values for the quantity z_0. Starting with snow and water surfaces (top left), which offer little frictional resistance, the scale progresses to forest cover (bottom right), the most aerodynamically active natural surface cover. Individual measurements may differ considerably. For grass, E. L. Deacon [1603] found the following relation:

Height of the grass (cm):	1	2	3	4	5
Roughness parameter (cm):	0.1	0.3	0.7	1.6	2.7

As the ordinate values show, flow over vegetation-covered surfaces is nearly always turbulent.

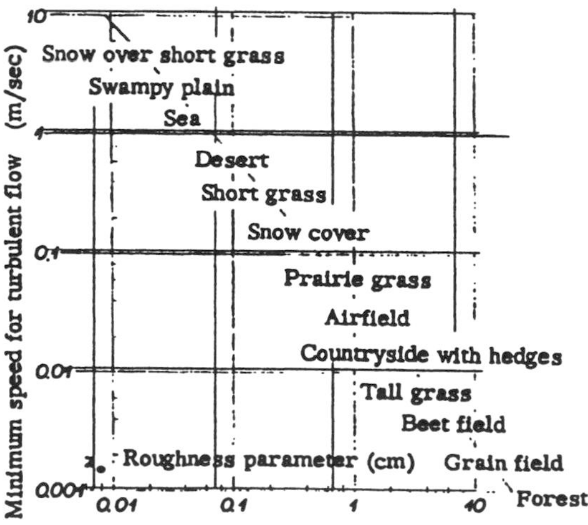

Figure 30-6. The roughness parameter (cm) of various natural surfaces. (After E. L. Deacon [1603])

Relative humidity is high among the transpiring plant leaves, turbulent fluctuations and eddy diffusion are low, and a quiet, moist climate prevails. R. Kanitscheider [3006] measured the air temperature at intervals of 2 seconds in a stand of dwarf pines 2 to 3 m high at an elevation of 1600 m on a southern slope near Innsbruck. Figure 30-7 is a record of the mean difference of two consecutive readings (in tenths of a °C) as a function of time of the day and height above ground (the darker the shading, the greater the fluctuations). Instability is greatest when the sun is at its highest. Here also the outer effective surface can be clearly seen to be at height of about 2.5 m.

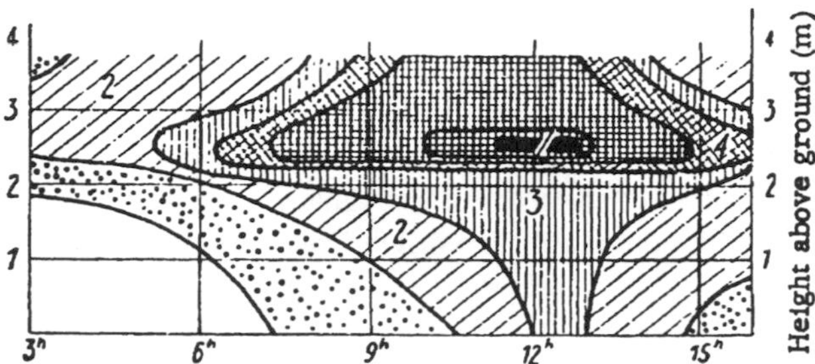

Figure 30-7. Temperature fluctuations in a stand of dwarf firs near Innsbruck for 28 July 1931, a clear day. (After R. Kanitscheider [3006])

A third distinction between the microclimate of bare and vegetation-covered ground arises from differences in surface moisture availability. Photosynthesis requires that plants absorb a large volume of water through their root systems and return it again to the atmosphere by transpiration. The transpiration ratio expresses the number of liters of water a plant consumes in order to produce 1 kg of dry matter. According to J. N. Köstler [3007], this quantity depends on the type of plant; values for trees range between 170 (pine) and 400 (beech). H. Walter [3013] found average values for agricultural crops to be between 400 and 600, with extremes of 300 for millet and 900 for flax. In humid regions plants have been shown to exhibit a linear relationship between yield and the total amount of water consumed, with the slope of the line depending upon the plant species and the presence and timing of water stress during the life cycle of the plant. The term water use efficiency (kg t^{-1}) is also used. It represents the number of kg of dry matter production by photosynthesis per ton of water consumed in evapotransporation. In the best subsistence farming systems water use efficiencies range between 0.1 - 0.2 kg t^{-1}. For the best modern agricultural systems, with their improved cultivars, improved crop management practices, and increased inputs of fertilizers, water and chemicals, water use efficiencies of 0.7 - 1.2 kg t^{-1} have been realized. On modern experimental farms, where nonproductive soil evaporation can be reduced even further, values between 1.0 - 1.8 kg t^{-1} are achievable (N. J. Rosenberg, et al. [5264]).

Plant cover evaporates more than bare ground (Table 19-3). This conclusion can be reached by merely considering the surface areas involved. The surface area of the canopy can be estimated by making a leaf count and measuring the area of a sample of leaves. The ratio of the leaf surface area m^2 divided by the area of ground m^2 results in the leaf area index m^2/m^2. A sparse forest has a leaf area index of 4 to 12 times that of the ground on which it is growing. P. Filzer [3003] gives a leaf area index value of 20-40 for grassland, and shows on the basis of experiments, that the quantity of water transpired increased proportionally as the surface area increased. This is true as long as eddy diffusion is able to remove the water vapor from the moist boundary layer of the leaves and the plant is able to resupply the moisture lost through transpiration. A. R. G. Lang [3008] discusses various techniques for measuring leaf area but recommends the use of the surface area indices method.

The increased evaporation from vegetation-covered ground can be determined experimentally through lysimeter measurements. The results given here are the work of J. Bartels [3002] and W. Friedrich [3004] at the Forestry College in Eberswalde. Three similar boxes of capacity 1.5 m^3 with surface area 1 m^2 were placed in the ground, level with the surface and 3 m apart. Their weight could be measured from a cellar with an accuracy of within 100 g, which is equivalent to 0.1 mm of precipitation. Percolation water could be collected and measured, or the boxes could be sealed from below to give a ground water level of any desired height. Table 30-2 gives annual averages for evaporation and percolation at a depth of 1.5 m taken from comprehensive evaluations by K. Göhre [3005]. To facilitate comparison, evaporation and percolation are both expressed in percent of the annual precipitation for the same period. The two figures add up to only approximately 100, since soil moisture storage changes in the soil enters as a third, although insignificant, factor. Since the figures are averages for a number of years, their soil moisture storage changes are never more than ± 5 percent of the annual precipitation. The table includes two observation series for ground that has been artificially kept free of vegetation by the removal of all weeds.

Table 30-2. Annual average evapotranspiration and percolation at a depth of 1.5 m.

Vegetation cover	Period	Average yearly totals				
		Precip-itation (mm)	Evapotrans-piration (mm)	Perco-lation (mm)	Evapotrans-piration (%)	Perco-lation (%)
Bare ground	1929-1932	674	178	484	26	72
	1950-1953	533	270	266	51	50
Short grass	1929-1937	615	356	259	58	42
Pines (3 yr old in 1932)	1932-1937	576	450	149	78	26
Oaks (3 yr old in 1949)	1950-1953	533	454	117	85	20

In the period from 1929-1932, which was a moist period, a quarter of the precipitation was lost from the bare ground as biologically unproductive evaporation. In the drier period from 1950-1953, evaporation was nearly half of the precipitation. Variations with time are not surprising, since the amount of bare soil evaporation is very responsive to weather changes. Vegetation, however, by means of its root system, is able to overcome the water conduction barrier created by the dry upper layer of the soil. The three plant covers shown, arranged in order of increasing water requirement, take up an ever-increasing proportion of precipitated water. If the growing period from May to September is considered, instead of the whole year, the portion of precipitation used by the vegetation is even greater. For example, the young oaks consumed 145 percent of the precipitation during this period, which was possible only by drawing on the water stored in the soil during the cool non-growing season.

From 1933 to 1937 the ground water level in a lysimeter on which grass was growing was kept artificially at 40 to 50 cm below the surface. With an annual precipitation of 706 mm, evaporation amounted to 121 percent. This is approximately the value of the potential evaporation and shows not only the requirements of vegetation but also the value of watering.

E. Malek [3009] investigated the relationship between night-time (sunset to sunrise) and daytime (sunrise to sunset) evapotranspiration for an irrigated alfalfa field at Logan, Utah (41°45'N, 1460 m). His results, based upon a complete growing cycle (4 August to 7 September 1991), showed that average night-time evapotranspiration was only 1.7 percent of the 24 hr evapotranspiration total. Even on windy nights characterized by more vigorous vertical mixing, the night-time totals were at most 14 percent of the 24 hr total. Maximum daytime evapotranspiration over the period was 7.85 mm, while maximum night-time evapotranspiration was 1.05 mm. Dew deposition was observed on only 37 percent of the nights, and even then for only a few hours per occurrence.

J. Bartels [3002] computed daily evaporation values for well defined meteorological conditions in the months of May to August from 1930 to 1932 (Table 30-3). The bare sandy soil restricted evaporation considerably in clear and dry periods because of the inability of the soil to conduct water upward to replace the moisture lost by surface evaporation. The grass area could do this only to a limited extent during dry periods, and showed the results of water deprivation by yellowing and dying. On clear days with ample soil water the plants' root systems were able to extract water from considerable depth, resulting in greater evapotranspiration than that of the open water.

Table 30-3. Surface evapotranspiration (mm day^{-1}) under various conditions.

Meteorologic condition	From bare ground	From short grass	From a water surface
On a day after rain	2.38	2.80	2.24
On clear days	0.47	2.15	3.61
On drought days	0.26	1.14	3.80

Figure 30-8 shows the relationship between the leaf area index and the amount of dry matter produced (growth). As might be expected this relationship is curvilinear because there is only a finite amount of energy available for photosynthesis. Due to partial shading of lower leaves each unit of additional surface area will absorb a decreasing amount of remaining sunlight. In row crops, such as raspberries, apples, peaches and sour cherry, fruit quality, and set and flower initiation decline with increased shading. J. W. Palmer [3011] discusses efforts to minimize this problem through the use reflectors, thinning and planting row structure.

In conclusion, it must be pointed out that temperature and moisture distributions within the soil change considerably when the soil has a vegetation cover. This will be discussed in more detail in the following sections.

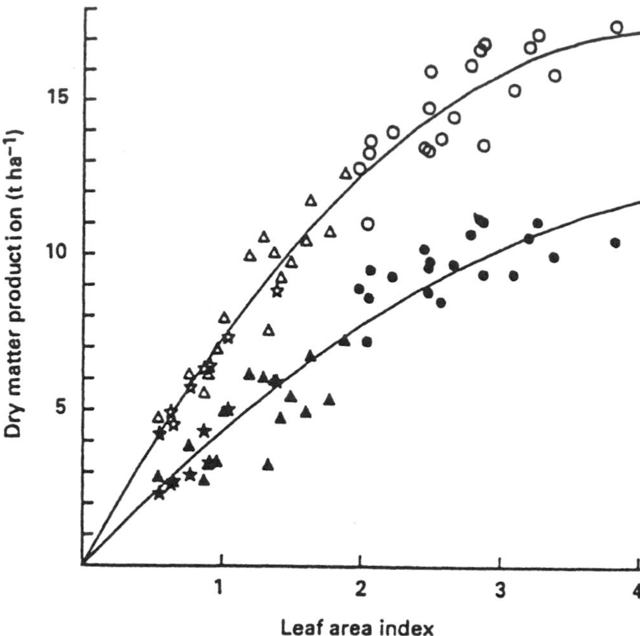

Figure 30-8. The relationship between maximum seasonal LAI and total (open symbols) and fruit (closed symbols) dry matter production from 'Crispin'/M27 bed system trees, 1982-1984; (O); 'Golden Delicious'/M9 spindlebush orchards at East Malling, 1974-1975, (Δ); and at Wageningen, 1969, (Π). (From J. W. Palmer [3011])

31 The Microclimate of Meadows and Grain Fields

Variations in the surface microclimate become subdued as the surface vegetation cover develops and leaf area index, canopy depth, percent ground cover, and surface shading increases. As the vegetation cover increases in density and becomes more continuous the daily surface temperature variation, and the maximum surface temperature decrease, while the surface soil moisture content increases. The outer effective surface which is the zone of greatest temperatures fluctuation also moves upward from the ground.

This section examines plant communities with a predominantly vertical (erectophile) structure and leaf distribution, such as those found in meadows and grain fields. Individual plants are tall relative to their horizontal extent. Their vertical leaf orientation allows for significant transmission of sunlight into the plant canopy.

In July 1953 W. R. Müller-Stoll and H. Freitag [3109] measured temperatures in the ground and in six different types of meadow communities in the Spree Forest, which were near enough to each other in the undulating countryside for comparison. The measurements for 23 July 1953, a sunny day, were made in three plant communities. The first was grass loosely arranged in thickets about 15 cm high on a sandy soil above a low water table. The grasses consisted of *Nardus stricta* and *Festuca rubra* with rosette plants in between (25 percent of the ground remained bare). The second was a moist pipegrass meadow which was growing on humic sand with the water table at 55 cm. It was a thick growth 20 cm high with individual grasses of *Molinia* towering to 40 cm. The third was a wet sedge only 35 cm above the water table on partially decomposed peat with a dense growth 80 cm high (Table 31-1). These measurements illustrate how the microclimate becomes milder as the soil covering increases.

Table 31-1. Temperatures and daily temperature fluctuation (°C) in three types of vegetation (asterisks indicate maximum values).

Type of growth (arranged in order of increasing ground shielding)	Maximum temperature		Daily temperature fluctuation			
	Ground surface	Vegetation surface	-10 cm	0 cm	5 cm	100 cm
Grass thickets	43	35	*4	*26	10	*14
Pipegrass	27	31	1.5	9	*13	12
Sedge	24	31	0.5	7	10	*14

The increasing amount of shade reduces the daytime maximum temperature at the ground surface. The zone of greatest temperature fluctuation (the 06:00 observation was used as the minimum temperature) moved from the ground up to the outer effective surface (asterisks). The change in the ground from dry to moist is closely associated with the change in the type of plant on the meadow and exerts the same kind of influence on temperature.

Figure 31-1 illustrates the typical vertical distribution of meteorological elements within grass 50 cm tall, which F. L. Waterhouse [3116] recorded in Scotland on a sunny June day between 15:00 and 16:00 following precipitation. The maximum temperature (*T*) is at 30 cm which is the outer effective surface. In the grass layer nearest the ground it is quite cool, and insects that are sensitive to radiation find shelter there. Near the ground the relative humidity is close to saturation, vapor pressure (*e*) is at a maximum, and the air is calm. The curve for the saturation deficit E_0-e follows that of the temperature. The saturation vapor pressure (E_0) is dependent upon the temperature. The variations in vapor pressure (*e*), while modifying the E_0-e curve, are too small to dominate it in this meadow. If the temperature variation through the vegetation was small, as on a cloudy day, the variation in *e* would have a stronger effect on the saturation deficit (E_0-e) curve. The differences between this protective zone close to the surface and the air above the grass, therefore, shows considerable variations with time.

Temperature recordings were made in winter by J. T. Norman, et al. [3110] in uncut grass, and grass that had been cut short, near the Thames Valley. The two meadows were only about 5 m apart. The grass that had been cut a number of times in the previous growing period was 2 to 3 cm tall; the other grass was 30 to 45 cm tall. The temperature at 2.5 cm above the ground was recorded continuously in both fields during the winters of 1953-54 and 1954-55. Table 31-2 gives the temperature differences recorded for four-week periods of the first winter 1953-54. The differences were positive when the thermometer in the long grass was warmer than the one in the short grass, and negative when colder.

The most striking result is the degree of protection afforded against cold by the tall grass. The number of nights with temperatures below 0°C (nights with frost) is only half that in the short grass. During winter afternoons, it is only a little cooler under the tall grass. Toward

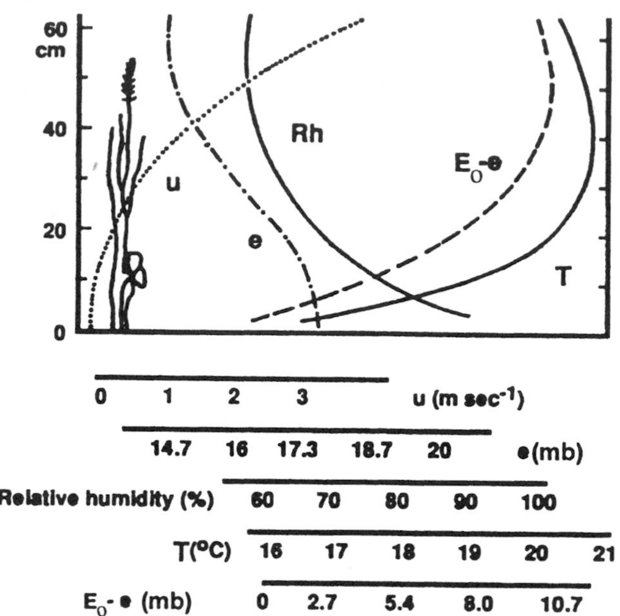

Figure 31-1. Temperature, humidity, and wind in a meadow on a summer afternoon. (After F. L. Waterhouse [3116])

Table 31-2. Temperature difference between tall and short grass.

Winter of 1953-54	Temperature difference (°C) (Tall–Short grass)			Number of nights with frost	
	Average maximum	Daily average	Average minimum	Tall grass	Short grass
4 November-1 December	-0.9	1.0	2.2	0	3
2 December-29 December	-0.1	1.1	2.1	0	6
30 December-26 January	-0.3	1.1	2.1	11	20
27 January-23 February	-2.7	0.8	2.4	18	20
24 February-23 March	-4.6	-0.3	1.6	7	13
Mean (or sum)	-1.8	0.7	2.1	36	62

spring, with the quick growth of grass in the mild English climate, the difference increases and by March the daily average temperature in the tall grass is lower than that in the short grass. These results show how the microclimate becomes milder as the grass grows taller.

M. C. Ball et al. [3101] analyzed the maximum and minimum temperatures in winter and spring at 5 cm above the ground, in the center of bare patches of soil surrounded by a low grass cover. The bare patches ranged in diameter from 0 to 120 cm. While they found no difference in the maximum temperatures there was a steady increase in minimum temperatures with the increasing size of bare patches. The minimum temperature was on the average 2°C warmer in the largest bare area. Seedings surrounded by grass experienced lower minimum temperatures with more frequent and severe frosts. In this situation grass acts as insulation not allowing the heat in the soil to escape.

The temperature variation in a field of winter rye near Munich was measured by R. Geiger [101]. The upper part of Figure 31-2 contains the midday temperature profile. As the rye grows at first the zone of maximum heating was substantially below the upper surface of the grain, since both solar radiation and wind are able to penetrate into the slender vertical structure. As the grain grows and becomes thicker the zone of maximum heating moves upward. However, as the grain ripens, dries, and becomes lighter in color, the maximum falls again toward the ground and is at the surface after reaping. At night (lower part of Figure 31-2) the cold air forming at the surface of the grain sinks slowly, so that just before reaping the lowest temperatures are found at a height of 1 m above the ground.

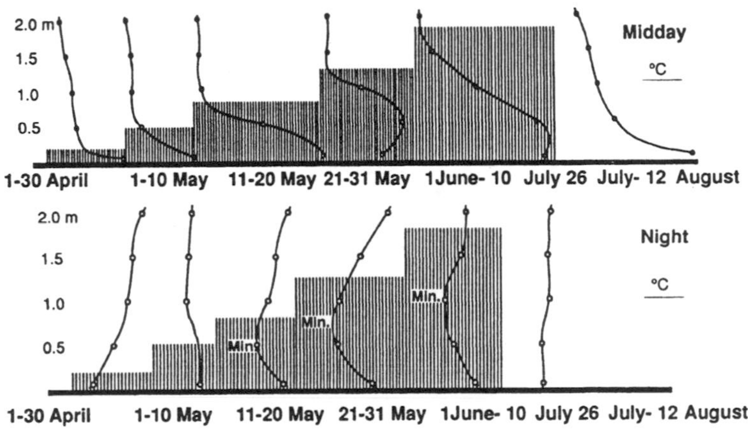

Figure 31-2. Temperature profiles in a field of winter rye near Munich. (After R. Geiger [101])

The variations in temperature in a field of rye, potatoes, Jerusalem artichoke, and over a lawn are shown in Figure 31-3. This figure shows the deviations of temperature from a height of 2 m above grass during sunny weather in August 1936. The height of each stand is marked B.H. The density of the vegetation and its height affect the distribution of temperature. The zone of daily maximum heating extends much higher over the crops than the lawn. With potatoes and lawn, the maximum is at the surface. Rye and Jerusalem artichokes are dense enough to result in the maximum temperature being within the mass of the vegetation (Figure 31-3), and during the day, the ground surface is distinctly colder than the outer effective surface. At night, the minimum temperature above the lawn is at the ground. With the crops, the minimum (while affected by their density) clearly occurs within the mass of the vegetation. This method of making comparisons with a reference crop (lawn) at a base station is to a large extent independent of the actual temperature and changes in the weather.

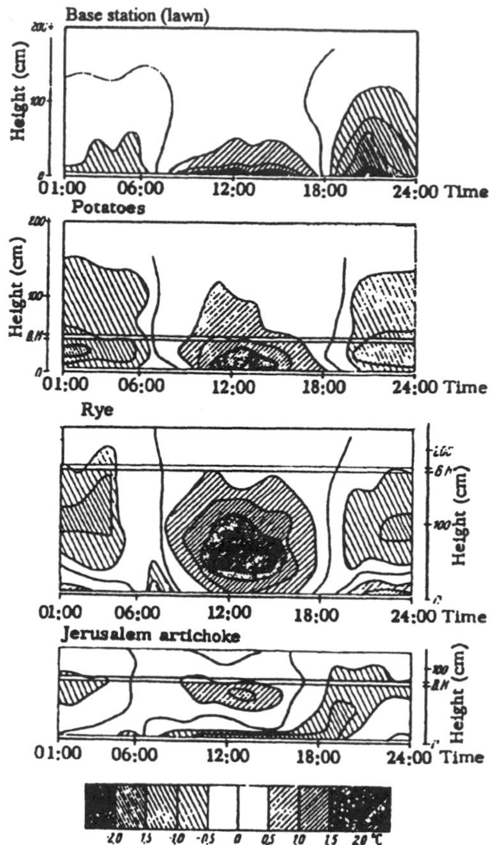

Figure 31-3. Temperature deviations of various types of crops from a base station or reference crop (lawn) microclimate. (After A. Mäde [3107])

R. Leuning and K. W. Cremer [2922] observed nocturnal temperatures above grass, dry soil, and wet soil (Figure 31-4). Temperatures were generally cooler above the grass, with the grass providing an insulating layer between the cool air and warmer soil. This can sometimes also be observed after early snow falls in autumn. The snow on streets or bare soil disappears quickly, while snow on the top of the grass (insulated from the soil's heat) lasts much longer. R. Leuning and K. W. Cremer [2922] also observed higher nocturnal temperatures over wet soil (Figure 31-4). The wet soil has a higher thermal capacity and conductivity and provides a greater nocturnal heat flow to the surface.

On an August day in 1953 E. Tamm and H. Funke [3115] measured the temperature in a field of corn 2.1 m tall and adjoining bare land, at 16 and 9 different heights, respectively. During the period of positive net radiation, the corn at 40 to 60 cm above the ground was 0.5 to 2.5°C cooler, and above this up to 2.1 m about 0.5 to 1.5°C warmer than the bare land. The outer effective surface was not very well marked, between 80 and 180 cm, and changed its appearance with different amounts of radiation and weather conditions. During periods of negative net radiation, the corn field was 0.5 to 1.0°C warmer to a height of 80 cm and

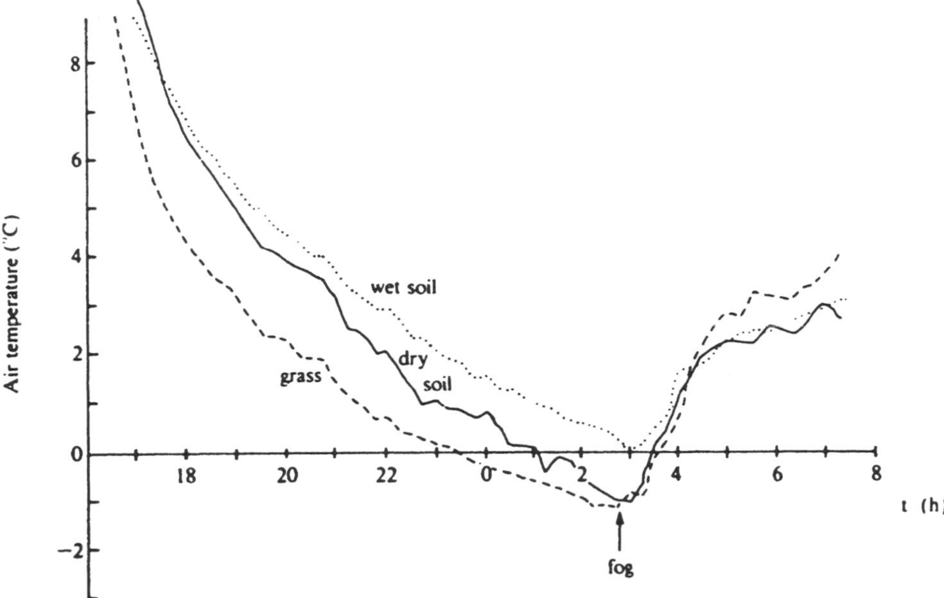

Figure 31-4. Time series of nocturnal air temperatures measured at 5 mm above grass (dashed line), dry soil (solid line), and wet soil (dotted line) on 15-16 August 1984. (From R. Leuning and K. W. Cremer [2922])

above this 0.5°C cooler; the outer effective surface was better marked than by day, although differences remained small.

The density of growth is also an important factor controlling humidity. P Filzer [3003] experimented with planting corn of different density. He specified density by the area in square centimeters of leaf surface per cubic centimeter of air space. The average relative humidity he found for a series of measurements were: in the densest corn (1.81 cm^2 cm^{-3}), 73 percent; in a medium dense crop (0.82), 64 percent; in a thin crop (0.38), 51 percent; the relative humidity in the open was 40 percent.

In fields that are watered artificially, the distribution of moisture and temperature are substantially different. L. A. Ramdas, et al. [3111] investigated this in irrigated sugarcane fields near Poona (18°N). Their measurements were extended in 1951 by K. M. Gadre [3102]. Figure 31-5 compares the temperature profiles during the dry period of the Indian northeast monsoon, over bare ground, a field of millet 150 to 180 cm high, and an artificially watered sugarcane field 2.5 m high. Around noon, the sugarcane uses the incoming radiation mainly for evaporation; the air temperature over the sugarcane is, therefore, 14°C lower than over the bare ground, and 8°C lower than under the shade of the millet. At night, the sugarcane field experiences warmer temperatures up to 1.2 m than over bare ground, due to daytime storage of heat and shielding of the surface. In the loose growing millet field the nocturnal temperatures are warmer than the bare ground up to 0.5 m and cooler above this height.

The corresponding humidity distribution is given in Figure 31-6 (vapor pressure) and Figure 31-7 (relative humidity). The highest humidity, both absolute and relative, is found

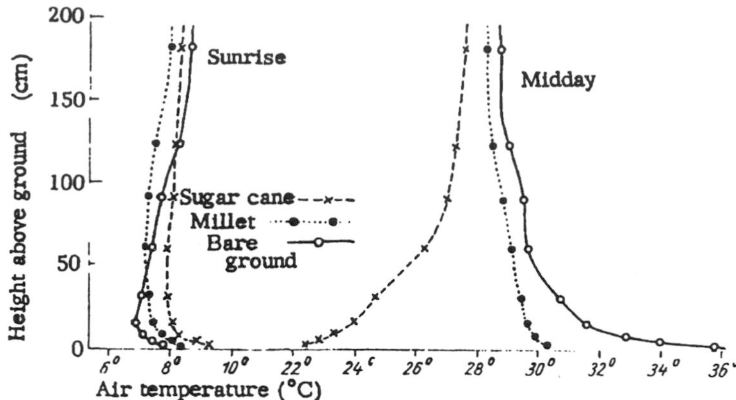

Figure 31-5. Stratification of temperature over bare ground, a millet field, and an irrigated sugarcane field near Poona. (After L. A. Ramdas, et al. [3111])

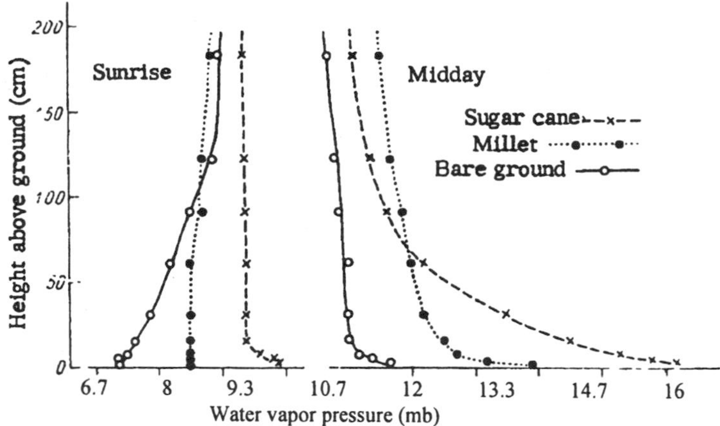

Figure 31-6. Stratification of vapor pressure at the same places as in Figure 31-5. (After L. A. Ramdas, et al. [3111])

over the watered sugarcane field, followed by the millet, and then the bare ground. The vapor pressure gradient in the sugarcane field is very high around noon.

K. M. Gadre [3102] planted thirty-three rows of sugarcane 1.2 m apart running east to west. Once a week, at noon, psychrometer readings were taken at six different heights in the middle of the field. The same readings were taken simultaneously in a neighboring unwatered field, free of growth. Noticeable differences began to appear only 3 to 4 months after planting, when the sugarcane had reached a height of 70 to 90 cm. In November, when the sugarcane was nearly 5 m high, air temperature, vapor pressure, and relative humidity were measured (Table 31-3). The boundary effect (Section 27) extended to row 4 (maximum values are indicated with an asterisk (*)).

K. L. Khera and B. S. Sandhu [3104] differentially irrigated a sugarcane field to investigate what effect crop stress would have on canopy temperature. They found that the mean

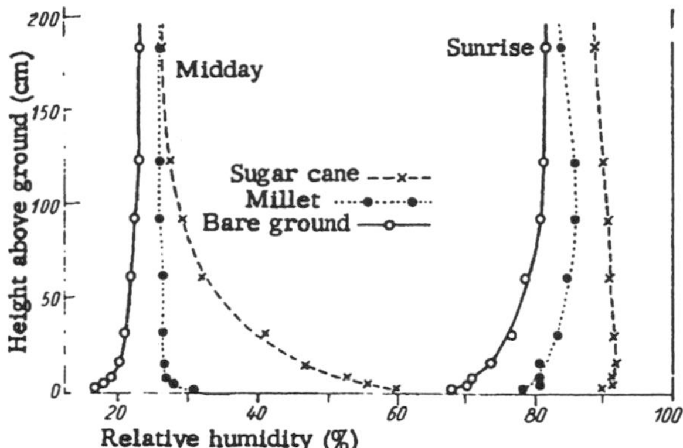

Figure 31-7. Stratification of relative humidity at the same places as in Figure 31-5. (After L. A. Ramdas, et al. [3111])

Table 31-3. Temperature T_A (°C), vapor pressure *e* (mb), and relative humidity *RH* (percent) over bare ground and a sugarcane field.

Height (m)	Bare ground			Sugarcane field			Difference		
	T_A	*e*	*RH*	T_A	*e*	*RH*	T_A	*e*	*RH*
6	30.2	18.7	43*	29.1	20.0	49	-1.1	1.3	6
5	30.3	18.7	43*	29.6	22.0	53	-0.7	3.3	10
4	30.5	18.7	42	30.2*	24.5	57	-0.3	5.9*	15
3	30.7	18.7	42	29.7	22.9	55	-1.0	4.3	13
2	31.0	18.7	41	29.0	22.3	55	-2.0	3.6	14
1	31.6	18.7	40	28.0	23.9	63	-3.6	5.2	23
0	38.0*	24.6*	37	26.1	28.0*	83*	-11.9*	3.3	46*

canopy temperature of the stressed crop was always higher than the unstressed crop. During the hotter part of clear days canopy temperatures were 2-7°C less in the unstressed crop. B. R. Gardner, et al. [3313] conducted a somewhat similar study on the effect of water stress on temperature of a differentially watered corn crop. They found that at any level within the stressed canopy, plants were warmer than at the same height in the non-stressed canopy. The midday temperatures of sunlit leaves of non-stressed and moderately stressed plants were generally from 1-2°C below, and that of stressed plants, as much as 4.6°C higher than, that of the ambient air. Similar results were found by B. S. Sandhu and M. L. Morton [3113] in oats and by D. C. Reicosky, et al. [3112] in soybeans. As water stress increases, individual plant stomata begin to reduce the size of their openings, which causes the rate of plant transpiration to decrease. Thus, a larger proportion of the net radiation is stored in the plant or used to heat the air by convection (Q_H). In all probability, this causes the higher air temperatures observed in the fields with stressed plants.

A vegetation cover exerts a strong influence on the temperature and water content of the soil. In England, J. L. Monteith [3108] made direct measurements of energy flow just under the ground surface and determined the quantity of energy absorbed during times of positive radiation balance and emitted during times of negative radiation balance. Measurements were made for a meadow, a wheat field, and a potato field; the quantities of energy on a clear day in July 1956 are shown in Table 31-4. During the daytime, Q_G was between 13 percent (wheat) and 16 percent (potatoes) of the radiation balance Q^*. The average of all daytime observations in June and July gave a higher mean value: nine-tenths of all values of Q_G were between 18 and 23 percent of Q^*. At night, when Q_G was positive and Q^* was negative, the corresponding figures were 39 to 46 percent; there was an extraordinary scatter of individual values with changing weather (30 to 140 percent!). The three fields showed little difference in this respect. In the meadow the ratio $Q_G:Q^*$ decreased as the wind speed increased, because the stronger wind removed more energy through turbulence thus balancing radiation losses. This relation did not exist with the wheat and potatoes. The decisive factor here was that soil moisture was large after a rainfall, and thus more incoming radiation may have been stored by the soil due to increased thermal capacity and conductivity.

Table 31-4. Energy exchange (MJ m^{-2}) in fields with different crops.

Period of time	Radiation balance, Q^*	Energy exchanged in the soil, Q_G		
		Meadow (2 cm)	Wheat (100 cm)	Potatoes (75 cm)
Day	14.2	2.2	1.9	2.3
Night	-1.93	-0.84	-0.75	-0.88
24 hours	12.3	1.3	11.3	1.4

J. R. Wright et al. [3117] found large variations in global solar radiation and temperatures in a 12.3 m transect. Temperatures varied by as much as 35°C between open interspaces and under a sagebrush (*Artemisia tridentada ssp.wyomingensis*) canopy (Figure 31-8) in response to shading and surface soil moisture variations.

The water content of the soil as well as its temperature is influenced by the type of plant cover. The dryness of soil in the neighborhood of the tap roots of trees is well known. Comparative measurements of soil moisture were made in the botanical gardens at Cologne between April and July 1951, by R. Knapp, et al. [3106] under *Lolium perenne* and in black fallow. Records published for four depths in both areas show a considerably greater fluctuation of moisture under the lawn than under the bare plowed surface.

The Australian botanist R. L. Specht [3114] discussed an extreme example of the influence of plants on soil moisture in 1956. Figure 31-9 illustrates the distribution of soil moisture (percent) on a summer day in 1955 under desert vegetation, from measurements made at six depths 30 cm apart horizontally. The previous day, 24 mm of rain had fallen after a long dry spell. This dry spell can still be recognized at the middle depths shown with large dots. The plants' different structure is shown in the sketch. The plants absorbed and evaporated some of the rain water and also directed some rainwater into the soil directly beneath the plants by stemflow. The day after the rain, the highest moisture was not found under the spaces between the plants where it fell unhindered, but in their root zone. P. E. Gruver et al. [3103] found that in the African Savannah rain water is concentrated by flowing down the

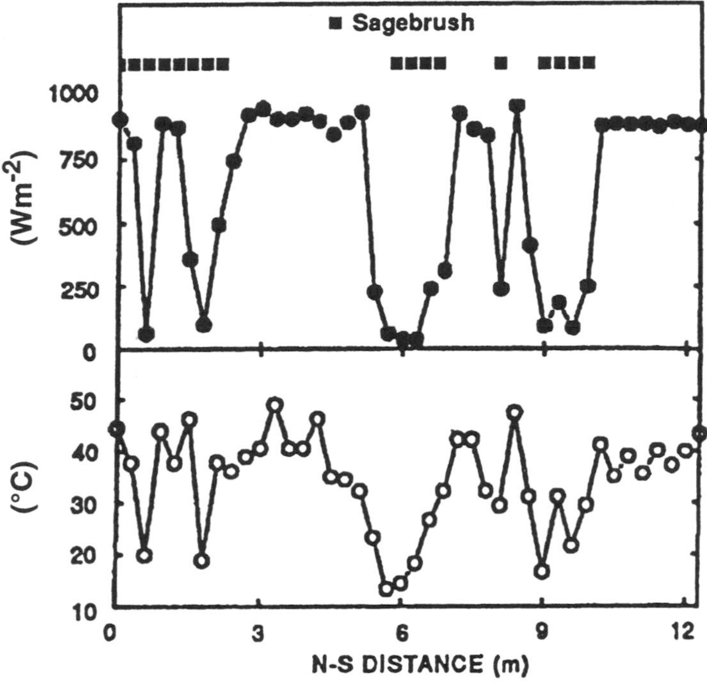

Figure 31-8. Global solar radiation at the soil surface (W m^{-2}) and temperature (°C) at the ground surface measured at solar noon, March 4, 1989, along a 12.3 m transect at the Quonset site on the Reynolds Creek Experimental Watershed, Reynolds, ID. (After J. R. Wright et al. [3117])

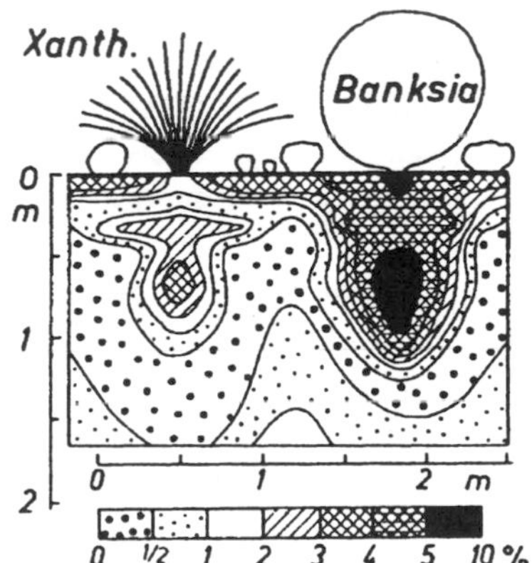

Figure 31-9. Soil moisture below Steppe vegetation in Australia after a rainfall. (After R. L. Specht [3114])

stems of isolated trees (B. *aegyptiaca*). This water infiltrates around the base of the trees resulting in deeper wetting. The amount of this flow and water accumulation is positively correlated with plant height. They suggest that this funneling of water toward the trees may play a part in maintaining drier conditions and grassland vegetation patterns around the trees.

The energy balance characteristics of grass surfaces under contrasting water content conditions are illustrated in the measurements of the surface energy balance of tall-grass prairie at Manhattan, Kansas (39° 03' N) from May-October 1987 by J. Kim and S. B. Verma [3105]. The totals in Table 31-5 represent midday averages from 1230 to 1430 hours and the energy balance terms are given as a fraction of net radiation Q^* to express the results in a less site-dependent manner. The volumetric water content of the surface layer (0.0-0.3 m) was 0.35 $m^3\ m^{-3}$ in mid-May but progressively decreased to around 0.15 $m^3\ m^{-3}$ by mid-October. Values were above 0.25 $m^3\ m^{-3}$ throughout most of the growing season and were generally adequate for plant growth, with the exception of a dry spell in mid- to late-July when moisture stress conditions were present and the surface volumetric water content fell to 0.15 $m^3\ m^{-3}$. The green leaf area index LAI peaked at 3.2 during the peak growth stage late-June and gradually fell to 1.0 at the senescent stage in October.

Table 31-5. Mean midday (1230-1430 hours) energy budget terms and meteorological conditions for tallgrass prairie at Manhattan, Kansas, 1987. (After J. Kim and S. B. Verma [3105]).

	T_A (°C)	D (kPa)	u (m sec^{-1})	Q_E/Q^*	Q_H/Q^*	Q_G/Q^*	Green LAI	G_A (mm sec^{-1})	G_C (mm sec^{-1})
11 July	31	1.8	8	0.67	0.21	0.09	2.8	48	13.1
30 July	37	4.3	6	0.35	0.48	0.11	2.6	36	1.5

The energy balance totals for 11 July in Table 31-5 are representative of a grass surface experiencing favorable soil moisture conditions. The average midday net radiation total of 601 W m^{-2} was primarily used to support high levels of evapotranspiration, with much lower levels of convection and conduction. Bowen ratio values β were actually negative in the early morning due to dew deposition, but quickly rose to 0.31 soon after sunrise and remained at that value throughout the day. The high levels of Q_E cooled the plant surface canopy during the day, resulting in a reduced vapor pressure deficit D and lower midday air temperatures T_A. Values of G_C ($G_C = 1/R_C$, where R_C is the canopy resistance) were determined from the Penman-Monteith equation. When grass surfaces are well-watered, canopy conductance values closely follow the availability of net radiation throughout the day. On 11 July G_C values rose from 1.0 mm sec^{-1} at sunrise to 13.1 at midday before falling again as sunset approached. If water is readily available, Q_E is primarily controlled by meteorological conditions such as net radiation Q^*, wind speed u, vapor pressure deficit D, and air temperature T_A.

Under conditions of water stress canopy conductance is instead controlled by plant physiological parameters. On 30 July the midday net radiation of 587 W m^{-2} was primarily used to heat the air and ground, with daytime Bowen ratios ranging from 0.17 at sunrise to 1.37 at midday. Plant canopy temperatures increased due to the reduced Q_E, the vapor pressure deficit increased significantly in response to the reduced evaporative cooling, and air temperatures rose in response to the increased convection Q_H. On 30 July the maximum canopy conductance of 3.4 mm sec^{-1} occurred shortly after sunrise and gradually fell throughout the day to a low of 1.5 near sunset. Under conditions of limiting water, plant leaves begin to experience loss of leaf turgor more quickly and will initiate stomatal closure in order to maintain adequate internal water conditions. Loss of leaf turgor is reflected in the reduced LAI values for 30 July in Table 31-5. Thus, the daily pattern of G_C is essentially decoupled from its dependence on Q^*. For this reason, maximum rates of photosynthesis often occur in the early morning for water stressed plants or for arid environments in general. Aerodynamic conductance G_A values varied directly in response to wind speed for all wind speeds greater than 1.5 m sec^{-1}.

32 The Microclimate of Gardens and Vineyards

We now turn our attention to plant communities in which the growth form and leaf distribution present a more horizontal structure (planophile). These include ornamental plants, agricultural crops planted in more widely spaced rows that require a longer time to achieve full ground cover, and vineyards and orchards that never achieve full ground cover. In more open canopies the ground surface plays a more important role in the surface energy balance and microclimate.

The first example to consider is a flower bed planted with antirrhinum in a Munich garden (R. Geiger [101]). Figure 32-1 shows the temperature profile. In July, when the plants are small and form an open type of cover, the midday temperature profile (upper diagram) is still similar to that over bare ground. In August, however, when the plants are fully grown, the dense leaf structure raises the height of the zone of maximum temperatures much more markedly than happens in grain, with its vertical type of leaf orientation and canopy structure (Figure 31-2). The outer effective surface lies just under the upper surface of the stand of plants. At night, when the upper surfaces of the plants radiate, the cooler air can sink to the ground more easily than in grain crops; therefore the minimum often lies at the ground surface in flower beds.

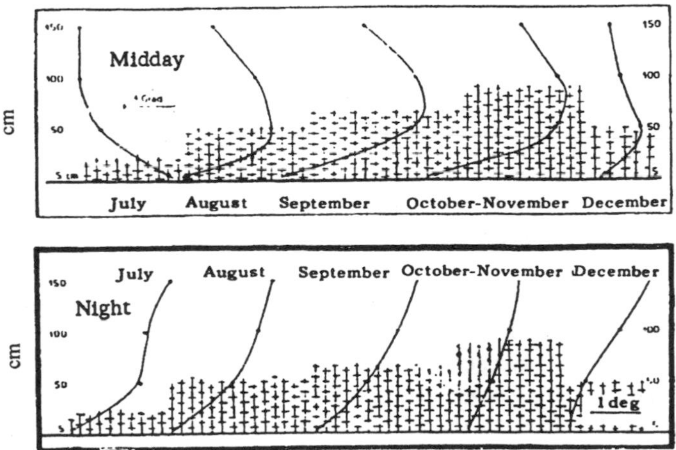

Figure 32-1. Temperature profiles above a flower bed near Munich. (After R. Geiger [101])

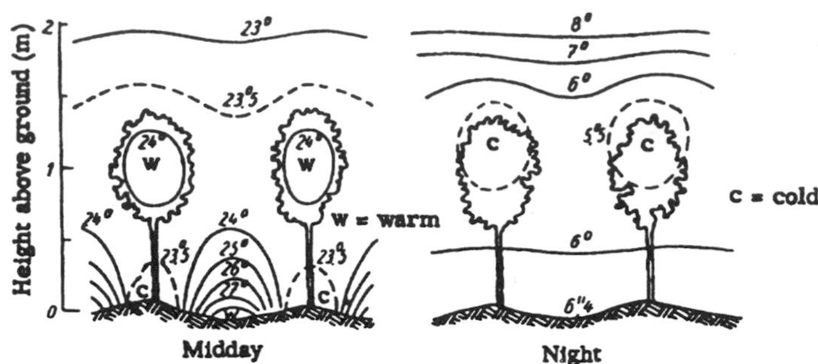

Figure 32-2. Temperatures at midday (left) and at night (right) in a vineyard on 17 September 1933. (After K. Sonntag [3209])

L. Broadbent [3202] was able to show, in investigations on a potato crop at Rothamsted, England, that the midday maximum temperature was at a height of 30 cm in a densely planted crop 60 cm high, and at 10 cm when planted less densely.

J. M. Hirst, et al. [3204] measured temperature and humidity relations at night at six different heights between 10 and 320 cm over a potato field. In a normal night, with deposition of dew, the gradient of vapor pressure was directed both from above and below toward the upper surface of the crop.

From a macroclimatological point of view, German vineyards lie near the northern limit of vine cultivation. As a result, they must be planted in areas where the microclimate is especially sunny, warm, and free of frost. An introduction to microclimatic conditions on a vineyard terrace has been given by O. Linck [3206].

The microclimate of a "wine mountain," as it is called in German, is made up of many individual factors. To begin with, the sunny slopes of hills are selected for the cultivation of grapes. The climate of a terrace vineyard is therefore that of a slope, which is dealt with in Section 42. In many places the vineyards are divided by stone walls. These walls run down into the valley and form a shelter against the wind, hence providing warm spaces. They do not, however, impede the downward flow of colder air into the valley. These walls become strongly heated by the sun during the day and by emitting longwave radiation act as heat sources for the vines.

The type of country in which the hill is situated is of equal importance. If its foot lies near a river or a lake shore, additional energy is obtained through specular reflection, which was described in Section 4. If the hill is topped by a cold plateau, there will be an increased risk of night frosts through cold air draining downhill at night (Section 42). As a protection against this, thick hedges or woods are often planted around the boundary of the plateau to minimize cold air drainage.

But the vines themselves, and the manner in which they are trained, also play a decisive role in creating the microclimate of the wine mountain. K. Sonntag [3209] recognized that a basic distinction had to be made between the climate of the rows of wine grapes and that of the open lanes between them. Figure 32-2 shows the temperature distribution for midday (left) and at night (right) on and after a sunny September day in 1933. The sun is able to penetrate to the ground in the lane, which runs N-S, producing high surface temperatures and a large temperature gradient close to it. In the rows of grape vines high temperatures occurred below the outer effective surface where the foliage gives protection from the wind, while between the rows of grape vines the highest temperatures occurred at the ground surface. At night, the lowest temperatures are found at the level of the radiating leaf surfaces (not at the ground which is partially shielded from the cold sky), thus dew formed to a greater extent within the plants. "Even outside the vineyard," K. Sonntag [3209] wrote, "an iron pole was dry from the ground up to the level of the stems, but was covered with water droplets above the level of the leaves."

The more open structure of these types of plant canopies result in evapotranspiration totals and energy balance patterns that are a complex function of the plant canopy and adjacent soil surface. This is evident in the extreme case of a vineyard investigated by J.L. Heilman et al. [3203] and whose daytime (sunrise to sunset) energy balance totals are summarized in Table 32-1. Energy balance totals for a vineyard with a row azimuth 160° from north were taken from 31 May through 7 June 1992 at Lamesa, Texas (33°30'N). The compact hedgerows were spaced 3 m apart. Individual plants were 1.6 m in height, 0.4 m in width, planted 1.7 m apart from one another, and had little vegetation below 1.25 m. The vineyard canopy was quite open, as indicated by the leaf area index values that increased from only 0.7 to 1.1 $m^2\ m^{-2}$ over the 10-day study period.

Separate energy balance totals and energy terms are given in Table 32-1 for the entire vineyard (a), the soil surface (b), and the plant canopy. The structure of the vineyard, with its tall plants and widely spaced rows, produced significant variation in the exposure of the soil and canopy to solar radiation throughout the day. Canopy absorption of $K\downarrow$ peaked during the early morning and late afternoon hours when the sun's rays were oriented more directly to the canopy, while penetration of $K\downarrow$ to the soil surface was dominant during the middle portion of the day. The greater penetration of solar radiation to the soil surface, combined with the low LAI totals, resulted in soil evaporation being a large contributor to vineyard evapotranspiration. The open canopy also allowed for greater production of Q_H at the soil surface that was locally advected within the rows to provide an additional energy source for canopy transpiration. Soil surface temperatures exceeded canopy temperatures during midday by as much as 17°C. Consequently, vineyard Bowen ratios were generally high, compared to well-watered grass surfaces (Section 31), and were even negative for the canopy which acted as a sink for sensible heat produced at the soil surface. As a result, canopy Q_E exceeded canopy Q^* on all days.

Less open canopies, such as those for soybean and potato fields, would exhibit less significant differences from the energy balance patterns of a complete grass surface, but would exhibit some similarities to the vineyard, particularly during the early growth stage before complete ground cover is established.

The crown area of the vines was subject to radiative heating and evaporation loss by day. In the higher vines radiative heating was greater because of its greater leaf area. The temperature was therefore higher in the taller vines than in the shorter vines. However, since the grapes in the shorter vines are closer to the ground, this balances out somewhat. Grapes of both taller and shorter vines experience similar temperatures in the warm hours at midday. This explains why the fear of the vintner, that the yield of grapes will be reduced if they are trained too far from the warm ground, is not justified. Vapor pressure is, however, often higher over the vines that are trained higher because of the greater mass of transpiring leaves.

Table 32-1. Daytime (sunrise to sunset) energy balance totals and energy terms for a vineyard at Lamesa, Texas. (After J.L. Heilman et al., [3203])

Day	Q^* MJ m^{-2}	Q_G MJ m^{-2}	Q_H MJ m^{-2}	Q_E MJ m^{-2}	Q_G/Q^*	Q_H/Q^*	Q_E/Q^*	β Bowen ratio
(a) Vineyard								
31 May	13.3	-3.2	-3.3	-6.8	-0.24	-0.25	-0.51	1.00
4 June	15.4	-4.5	-2.6	-8.3	-0.29	-0.17	-0.54	0.31
5 June	17.1	-5.0	-4.3	-7.8	-0.29	-0.25	-0.46	0.55
6 June	12.5	-1.8	-3.5	-7.2	-0.14	-0.28	-0.58	0.49
7 June	16.6	-3.2	-3.3	-10.1	-0.19	-0.20	-0.61	0.33
(b) Soil Surface								
31 May	11.9	-3.2	-4.1	-4.6	-0.27	-0.34	-0.39	0.89
4 June	12.9	-4.5	-3.6	-4.8	-0.35	-0.28	-0.37	0.75
5 June	14.2	-5.0	-4.9	-4.3	-0.35	-0.35	-0.30	1.14
6 June	10.9	-1.8	-5.9	-3.2	-0.17	-0.54	-0.29	1.84
7 June	13.8	-3.2	-4.1	-6.5	-0.23	-0.30	-0.47	0.63
(c) Canopy								
31 May	1.4	-	0.8	-2.2	-	0.57	-1.57	-0.36
4 June	2.5	-	1.0	-3.5	-	0.40	-1.40	-0.29
5 June	2.9	-	0.6	-3.5	-	0.20	-1.20	-0.17
6 June	1.6	-	2.4	-4.0	-	1.50	-2.50	-0.60
7 June	2.8	-	0.8	-3.6	-	0.29	-1.29	-0.22

This result agrees with the observations of R. Weise [3211], which show that the higher form of training (Frankish stem training) does not show a loss of heat during the day, in comparison with the lower form (Frankish head training). The higher form, however, allows the cold night air to flow away more easily in the comparatively foliage-free space (Figure 32-2), and therefore runs less risk of frost damage than the lower form. This has been proved by R. Weise [3212].

There is substantial research dealing with the impact of vine training, spacing, and leaf removal on grape production, quality, and disease. These topics are beyond the scope of this book. For additional information please refer to A. G. Reynolds et al. [3207], and R. E. Smart [3208]. For additional climatic information on the factors that determine the suitability of different varieties of grapes to particular locations, the reader is referred to R. H. Aron [3201] and the classic book *General Viticulture* by A. J. Winkler et al. [3213]. W. M. Kliewer and J. A. Woldert [3205] found that in dense vineyards leaf removal decreased fruit shading, relative humidity and fruit rot and increased air movement and evaporation. This is an excellent work discussing vineyard microclimate, row spacing, trellising, pruning, shoot positioning etc.

CHAPTER VI

FOREST CLIMATOLOGY

When tree seedlings are planted, the young plants are particularly sensitive to hazardous weather. Late frosts, winter cold, spring dryness, summer droughts, and persistent winds are the causes of most of the damage to the young plants.

When the forest has grown to maturity, new problems arise. Possible weather damage due to high winds, rime, breakage under the weight of snow and ice, destruction by lightning, and so forth will not be covered. This book will confine itself to the problems arising from forest climate and microclimate. Chapter VI will deal with the climate of the forest. In Sections 33-36, the climate of a mature forest will be studied. The microclimate of its border where most of the young growth takes place will be covered in Section 37, and the microclimate of clearings in Section 38. Finally Section 39 deals with the age old question of the influence of forests on the regional climate.

33 Radiation in a Forest

A mature stand of trees with a closed crown is an extension of the types of vegetation cover discussed in Chapter V. It differs in that the trunk area and enclosed air space is larger and, therefore, has a greater heat storage capacity. The climate of the trunk area is transitional between that of the open air above the dense crown and the climate of the forest floor. In a mature stand, the outer effective surface is situated at the top of the crown. In this narrow vertical zone, radiation is absorbed and emitted and the wind is allowed some degree of penetration. The top of the tree canopy is an irregular surface, characterized by a pronounced clumping pattern. This produces an irregular surface at the top of the forest canopy in which the separation distance between the clumped tree crown tops and the relative depth of the tree crown gaps varies. This irregular crown top pattern greatly affects the reflection, transmission, and absorption characteristics of the canopy and in particular increases the efficiency of solar radiation absorption.

The microclimate of the crown is characterized by a great degree of instability. The contrast between it and the equilibrium in the trunk area is very noticeable. The moderating effect of the trunk area is most obvious on entering a forest from the blistering summer heat, a howling gale, or biting winter cold of the open country. The characteristic properties of the microclimate of the crown, trunk area, and forest floor will be covered, beginning with the primary factor, radiation.

The bulk transmission, reflection, and absorption properties of a forest canopy depends upon canopy leaf area index, leaf spectral properties, solar altitude (Figure 33-1), tree height (Figure 33-2), and the size, shape, and orientation of the canopy leaves. In an investigation of the albedo of a mixed deciduous forest in Ontario, Canada, J. H. McCaughey [3325] found that the forest albedo in summer varies from 12-15 percent. W. J. Shuttleworth [3338] reports average albedos of 12 percent for full-leaf tropical evergreen and temperate evergreen forests, and slightly higher values of 14 percent for temperate deciduous forests. The enhanced absorption provided by a forest canopy in comparison to grassland and agricultural canopies is clearly apparent in Figures 33-1 and 33-2.

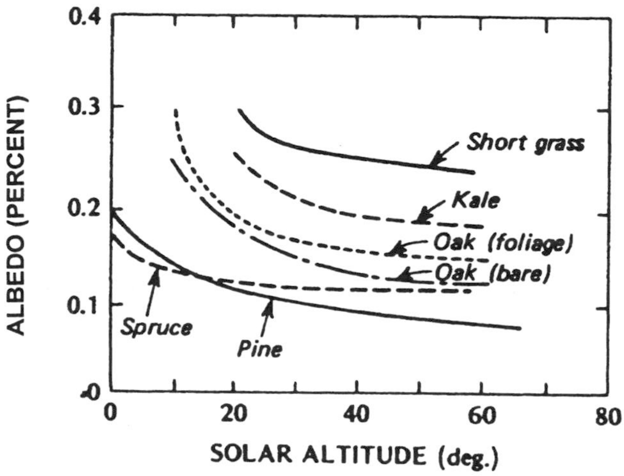

Figure 33-1. Relationship between the albedo and solar altitude of different types of vegetation. (After T. R. Oke [3332]; grass and kale from J. L. Monteith and G. Szeicz [3329]; oak forest from J. V. L. Rauner [3333]; spruce forest from P. G. Jarvis, et al. [3320]; and scotch pine forest from J. B. Stewart [3343]).

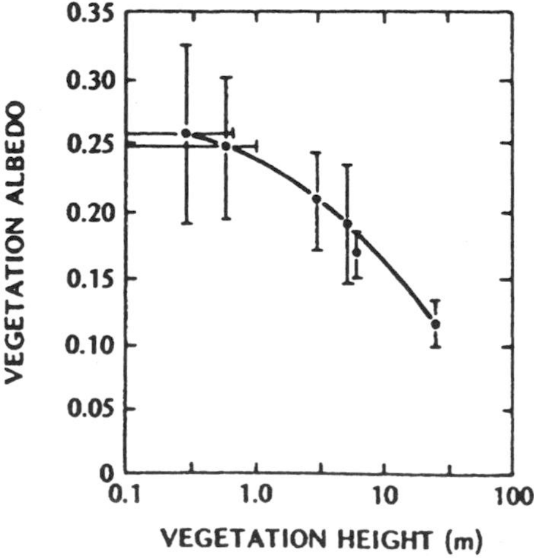

Figure 33-2. Relationship between the mean albedo and vegetation height. (From T. R. Oke [3332], after G. Stanhill [3342])

In winter, the albedo of a temperate deciduous forest falls to a minimum of 10 percent in the absence of snow. In general, the albedo of a bare canopy is less than a canopy in full-leaf (Figure 33-1). The albedo of a temperate deciduous forest increases to around 50 percent immediately after a snowfall, but decreases quickly and stabilizes around 20 percent while snow is on the ground. The deciduous forest albedo changes quickly in spring with leaf development and in autumn with leaf fall.

Forest albedo shows a weak dependence on solar altitude (Figure 33-1), except at very low solar altitudes where the effect is more pronounced. The less marked dependence of albedo on solar altitude in comparison to grassland and agricultural canopies arises from the clumped pattern of the tree crown. J. H. McCaughey [3325] found that forest albedo was also affected by cloudiness and canopy wetness.

Although the irregular topography of the forest crown top has a dominant influence on the initial scattering of solar radiation at the outer effective surface, the subsequent transmission of solar radiation through the forest canopy is controlled by the canopy leaf area index and canopy architecture.

Shortwave radiation decreases within a forest canopy in a manner similar to a meadow (Figure 30-1). A. Baumgartner [3703] measured relative light intensity within various types of vegetation as a function of the relative height of the plant cover. He found that two different types of vegetation could be distinguished (Figure 33-3). In one type, the light intensity curve is convex upward, and in the other, it is linear or concave. These correspond to the planophile and erectophile leaf distribution patterns, respectively. The canopy of leaves or needles causes a sharp decrease in light intensity at the top. This may be seen in Figure 33-3 with the 2-3 m tall young pines and beeches and is also recognizable in the moss cover, which is only 4 cm thick. As with most plants, different varieties of grasses differ in their location for the absorption of photosynthetically active radiation. L. M. Fliervoet and M. J. A. Werger [3311], for example, found that in *Cirsio-Molinietum*, phytomass, leaf area and ab-

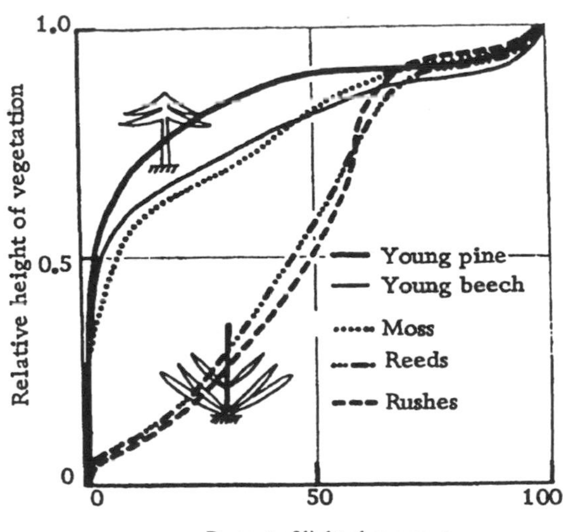

Figure 33-3. Two different light attenuation patterns within a forest canopy. (After A. Baumgartner [3703])

sorption of solar radiation is concentrated close to the ground much as the rushes and reeds in Figure 33-3. On the other hand, phytomass, leaf area and absorption of solar radiation for *Seneciomi-brometum*, is located higher on the plant and its absorption curve more closely resembles that of mosses and trees (Figure 33-3). There are of course many intermediate stages between these two leaf distribution types.

Figure 33-4 gives the light curve, after E. Trapp [3345], for a 31 m high stand of 120 to 150 yr old red beeches, intermingled with a few pines, situated at 1000 m above sea level on a 20° slope with a SE exposure, near Lunz in Austria. The full line gives the distribution of light above the forest floor on sunny days. When the weather is cloudy and a greater percent of solar radiation is diffuse, the absolute light intensity is naturally lower in the forest, but its rate of decrease is slower (dashed line). Y. Harada [3318] similarly showed that there was a relatively greater proportion of light in a stand on foggy days than on days without fog. The transmission of full-spectrum solar radiation through a clumped forest canopy can be adequately described by a Beer's Law approximation (Section 30).

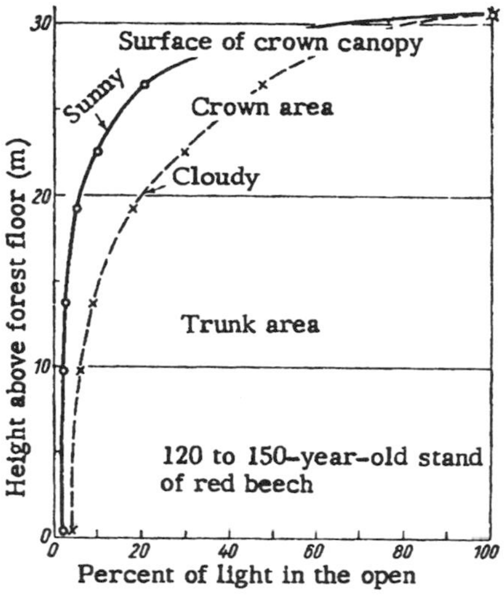

Figure 33-4. Decrease of light in a stand of red beech with dense foliage. (After E. Trapp [3345])

The decrease of shortwave radiation in a stand depends, to a large extent, on the type of tree, the character of the stand, its age, and its productivity. Many examples are available from R. Geiger and H. Amann [3407], F. Lauscher and W. Schwabl [3321], F. Sauberer and E. Trapp [3336], E. Trapp [3345], W. Nägeli [3331], G. Scheer [3337], G. Sirén [3340], D. H. Miller [3326, 3327], and A. Baumgartner [2502, 3301]. Some values for light intensity in different kinds of tree stands are found in Table 33-1. W. J. Shuttleworth [3339] reports bulk extinction coefficients v of approximately 0.47 for both coniferous and deciduous forest stands. These values are low in comparison to coefficients for grassland and agricultural canopies (Section 30) and are illustrative of the more rapid extinction of solar radiation through forest canopies.

Because of the great importance of light reaching the ground for the plants growing there, several attempts have been made to deduce a relation between the age of the stand or its den-

Table 33-1. Light intensity (percent of that outside) in stands of trees.

Type of tree (old stand)	Without foliage	With foliage
Deciduous trees		
Red beech	26-66	2-40
Oak	43-69	3-35
Ash	39-80	8-60
Birch	--	20-30
Evergreen trees		
Silver fir	--	2-20
Spruce	--	4-40
Pine	--	22-40

sity and the light intensity at the forest floor. G. Mitscherlich [3328] measured the intensity of light at the floor of many fir stands in relation to the age of the stand and its yield class. Figure 33-5 shows his results for 87 different stands. As the young stand grows less light reaches the forest floor. When the stand is approximately 17 years old only about 10 percent of the external light reaches the forest floor. As the forest ages, more energy begins to pass through to the forest floor. The better yielding classes have fewer, but stronger trunks and, therefore, let more light penetrate than do the poorer yielding types. No undergrowth was found on the forest floor when less than 16 percent of the outside light was able to penetrate; from 16 to 18 percent, the first undemanding mosses appear; with 22 to 26 percent berry plants are seen; and only with 30 percent are the first naturally seeded firs observed.

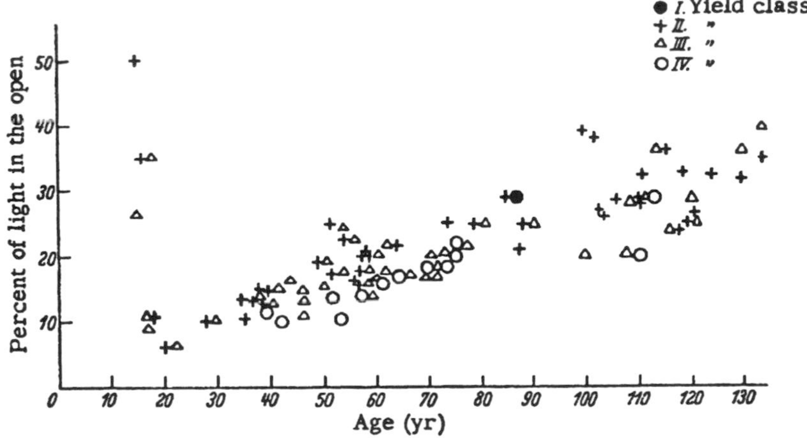

Figure 33-5. Relation between the age of a stand of fir and the light penetrating to the forest floor. (After G. Mitscherlich [3328])

The nine different series of measurements from German and American sources, illustrated with different symbols in Figure 33-6, show a strong relation between a stand's density and the amount of light reaching the forest floor. Figure 33-6 also includes very sparse tree densities found in a parkland.

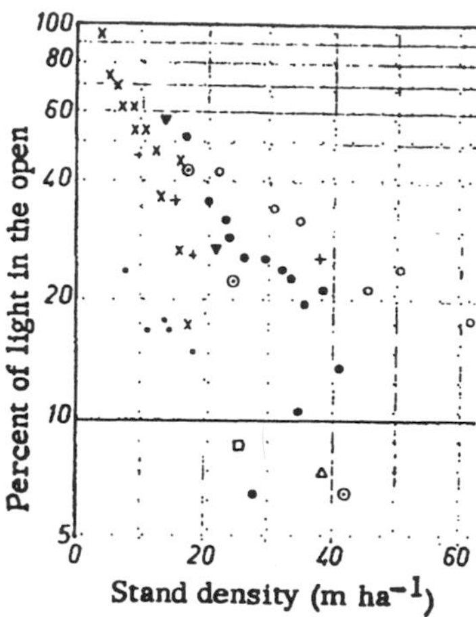

Figure 33-6. Light penetration to the forest floor as a function of stand density. (After D. H. Miller [3326, 3327])

In tropical forests, the amount of light penetrating to the ground is extremely small. The darkness, moist heat, and the odor of decay are common characteristics of the tropical rain forest. M. Gusinde and F. Lauscher [3317] found, in a 30 m high primeval tropical forest in the Congo, only 1 percent of the outside light at 2 m above the ground. H. Slanar [3341] recorded only 0.5 percent, H. Eidmann [3308] observed 0.4 percent and P. M. S. Ashton [3803] observed 1 percent in a Sri Lankan tropical rain forest. Figures at various heights in Panama, measured by W. C. Allee, are found in a book by P. W. Richards [3334] on the tropical rain forest. According to this source, the quantity of light as a percentage of the outside light was:

Position	Upper crown area	Tops of small trees	Trunk area	Forest floor
Height (m)	25-23	18-12	9-6	0
Light (percent)	25	6	5	1

For the light at the ground below the rain forest, P. W. Richards [3334] gives a figure of 0.5 to 1 percent, exceeding 1 percent for only short periods, but never more than 5 percent. W. J. Shuttleworth et al. [3339] found ground level global solar radiation was only 1.2 percent of above canopy solar radiation for a tropical Amazonian forest. In general, ground level solar radiation in tropical forests is about 3 percent, and about 5 percent in temperate forests. The percentages are less variable for full-canopy tropical forests than for temperate forests. W. J. Shuttleworth et al. [3339], for example, report values of 14 percent for some temperate coniferous forests, and values as low as 1 percent for a Corsican pine forest. These percentages increase and become more variable as the forest canopy structure becomes more open. The penetration of solar radiation to the forest floor beneath a full canopy is very low because of the high leaf area index and low extinction coefficient of the forest canopy.

Measurements made by G. C. Evans [3405], using spectral filters, show that in the tropical forest, 8 percent of the solar radiation reaching the forest floor was in the violet-blue range from 0.32 to 0.50 μ, 22 percent in the green between 0.47 and 0.59 μ, and 45 percent in the red beyond 0.60 μ. The preponderance of red wavelengths in the internal radiation in forests is also encountered in central European forests. K. Egle [3307] determined the illumination penetrating to the forest floor, expressed as a percentage of that incident on the foliage for a number of wavelengths (Table 33-2). As the amount of foliage increases, the decrease in the intensity of the penetrating light takes place more slowly in the longer wavelengths. As a result, the spectral distribution of solar radiation changes with depth into the forest canopy so that the light in the trunk area contains a decreasing proportion of blue wave-lengths.

Table 33-2. Light intensity (percent of that at the top of the foliage) reaching forest floor.

Date	Wavelength (μ) and color					
	0.7 Red	0.65 Orange	0.57 Yellow	0.52 Green	0.45 Blue	0.36 Violet
12 March (buds still closed)	61	54	51	48	46	44
15 April	59	39	35	33	32	30
10 May	19	6	7	6	6	5
4 June	14	4	5	4	3	3

The difficulty of measuring the quantity and spectral distribution of light in forests is largely due to its special structure under these conditions. At the forest floor light patches alternate with deep shadows and the mosaic pattern of darkness and light changes with the movement of the sun. M. I. Sacharow [3335] proved that the interior illumination of Russian forests increased as the wind speed increased. G. Sirén [3340], on the other hand, was unable to prove this in the birch forests of Finland. W. Nägeli [3331] made a distinction between basic illumination, resulting from diffuse radiation in the forest, and the contribution due to direct-beam radiation, which is more variable and of short duration. K. Brocks [3304] investigated the differential illumination at surfaces inclined at different angles inside an oak and fir forest near Eberswalde. In each position, he made 84 individual measurements of the illumination at a hemispherical surface 1 m above the forest floor. In comparison with the value outside the forest, measured simultaneously, the ratio of the lightest to the darkest area on the hemisphere was 10:1; in a stand of oaks without leaves, it was 180:1; and after the trees were in leaf, 17:1. Much of the solar energy received by the understory is contributed by "sunflecks" (the momentary shining of light through the upper canopy). Researchers have found that sunflects contribute 24 to 70% of the global solar radiation received by the understory (P. M. S. Ashton [3803], G. C. Evans [3310], T. C. Whitmore and Y. K. Wong [3346],

P. J. Grubb and T. C. Whitmore [3316], R. L. Chazdon and N. Fetcher [3305]). As the wind speed increased from 1 to 7 m sec^{-1} H. Tong and L. E. Hipps [3344] found that sunflecks and photosynthetically active radiation (PAR) increased in a linear fashion. They conclude that sunfleck density can be evaluated from the wind speed. V. L. Barradas [3302] found that short sunflecks (less than 100 seconds) depend primarily upon solar zenith angle, plant height and leaf and stem area index, whereas long sunflecks (more than 100 seconds) depend primarily on the zenith angle. These sunflecks contribute to the large variation in the extinction coefficients as reported by W. J. Shuttleworth [3338] for both coniferous (±0.11) and deciduous (±0.13) forest stands.

Figure 33-7 illustrates a series of measurements made by H. Ellenberg [3309] on two 100 m^2 test areas in two different types of forests. The left two parts of Figure 33-7 show the light intensity measured at the tops of the shoots of the flora growing on the floor of the forest. Black areas represent tree trunks. Lines of equal light intensity are drawn for every 0.5 percent, values for some of these are shown along the margins. The top left of Figure 33-7 is for a forest of oak and hornbeam, while below it on the left is a pure beech forest. The intensity of light in the beech forest is 4 to 5 percent of that in open country, and its distribution is quite even. While in the oak-hornbeam, it is noticeably darker, and great differences of illumination occur because of the steplike build-up of the stand. The diagrams on the right show the flora on the surface of the test areas. Their distribution is very closely correlated with the distribution of light shown on the left side of Figure 33-7. The original work can be consulted to find the meaning of the symbols used to designate the flora.

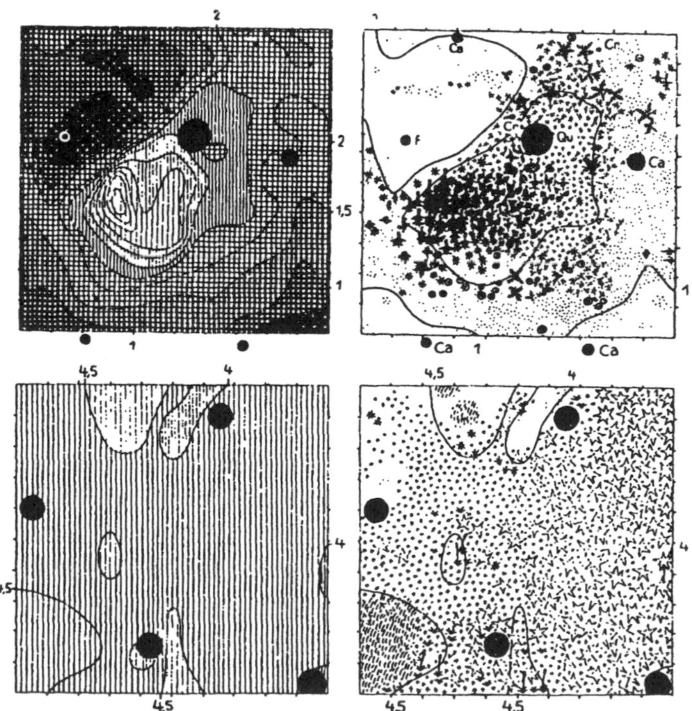

Figure 33-7. Light intensity (left) and ground flora (right) in a forest of oak and hornbeam (above) and a beech forest (below). (After H. Ellenberg [3309])

In arid and semi-arid environments, however, shading by taller plants can create conditions in which a dense understory of plants can grow, creating what M. J. Moro et al. [3330] called "islands of fertility." In Spain they found cooler surface temperatures and less radiation stress as the center of the vegetation island was approached.

Figure 33-8, from W. J. March and J. H. Skeen [3323], shows the annual pattern of global solar radiation in a small clearing (15 m by 25 m) and under a forest canopy in a woodland in Atlanta, Georgia. The primary peak in spring occurs during the bud break. The intensity of global solar radiation is increasing rapidly, but with the renewed growth of leaves in spring, decreasing amounts of solar radiation penetrate the canopy. In autumn, a secondary peak occurs because of leaf fall despite the declining solar radiation. Figure 33-9, from B. A. Hutchison and D. R. Matt [3319], shows the annual variations in global solar radiation above and within a forest in eastern Tennessee.

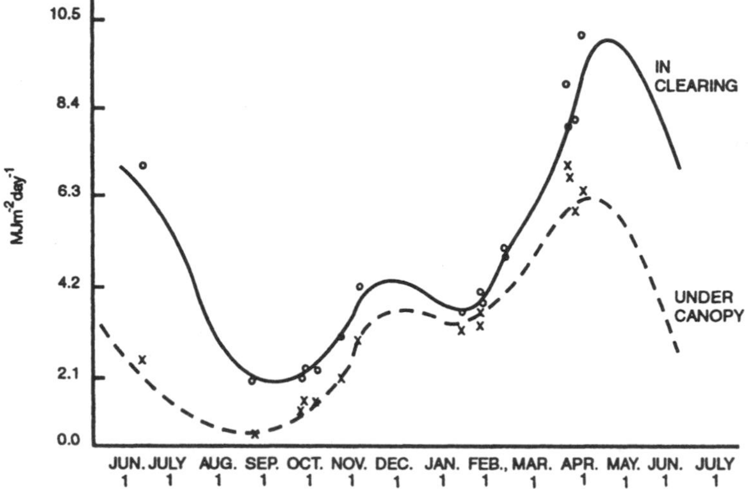

Figure 33-8. Annual variations of global solar radiation in a clearing (solid line) and under the canopy (dashed line). (After W. J. March and J. H. Skeen [3323])

Weakening of light intensity in forests means that twilight ends later in the morning and starts earlier in the evening. J. Deinhofer and F. Lauscher [3306] found that with a cloudless sky, the end of civil twilight (no longer possible to read a newspaper) was 16 min earlier in a deciduous forest, 20 min earlier in a coniferous forest, and 28 min earlier in an old tall forest than in the open. With overcast skies, the difference reached 45 min, and 54 minutes in one case when it was raining. G. Scheer [3337] estimated the seasonal variation of the intensity of global illumination in various tree shaded areas of the Darmstadt Botanical Gardens by observing the time when birds started to sing in the early morning.

The exchange of longwave radiation is greatest at the outer effective surface in the crown of the trees. This is made visible by the heavy deposition of dew in the upper part of the crown area after a night of strong outgoing radiation (Section 36). In the trunk area, temperatures are relatively uniform, so that the net internal exchange of longwave radiation is usually insignificant. If the trees are thinly planted, however, significant net longwave radiation exchanges may also take place from the trunk area and the ground below with the air above the trees.

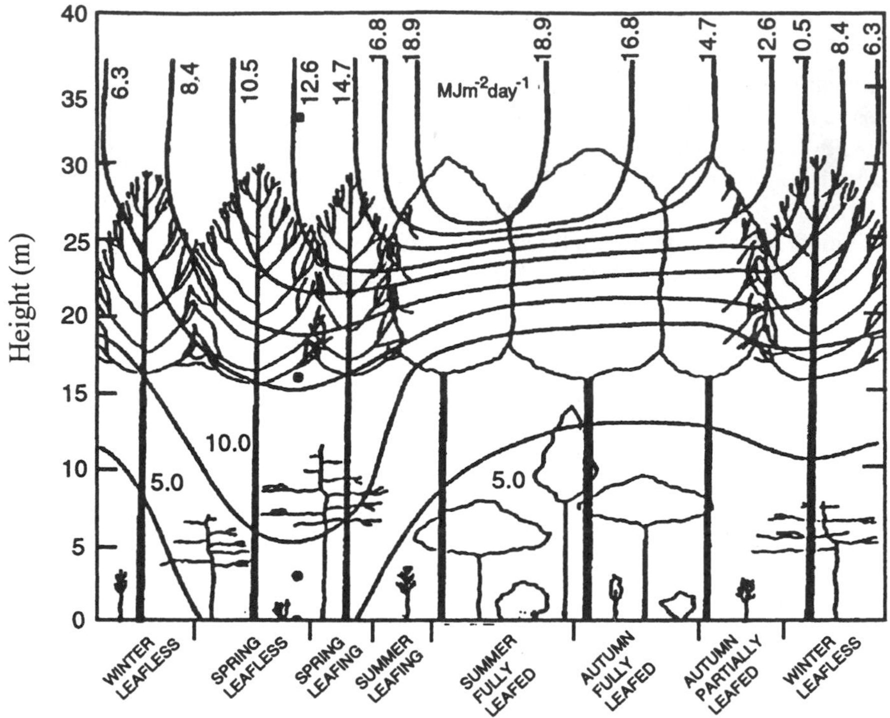

Figure 33-9. Annual variation in the average global solar radiation received above and within a forest. (From B. A. Hutchison and D. R. Matt [3319])

If there is a snow cover in the forest, so that the forest floor temperature does not exceed 0°C, there will be a net exchange of longwave radiation from the tree trunks to the snow cover, provided that the tree trunks are warmer than the snow cover. This process, which contributes to snow cover melting, was investigated by D. H. Miller [3326], for the thinly wooded area of the Sierra Nevada in California.

Short and longwave radiation together determine the net radiation balance. A. Baumgartner [2502, 3301] made measurements of net radiation in a young growth of firs 5 to 6 m high, 30 km southeast of Munich. These measurements will be referred to frequently in the subsequent discussion, and it is, therefore, appropriate that some general information about the measurement site be given. A full description is to be found in K. Mauerer [3324].

The area was level and had a 20 yr old natural growth of young firs. On the ground was a 2 cm deep layer of needles and brushwood. The trunk area was filled with dead underbrush to a height of about 2.5 m. The crown area, which began there, extended upward to about 4 m, above which there were numerous tops, and it reached to a height of 6 m. The quantity of wood was 189 m^3 ha^{-1}, and there were 349 trees per 100 m^2 of surface in the crown area.

Figure 33-10 shows how the instruments were set up. Cup anemometers (W) were placed up to nearly three times the height of the stand; thus, a good picture of the wind field above the forest was obtained. Thermographs and hygrographs (T, H) were set up at five different levels, up to 10 m above the ground. Net radiation balance meters Q^* were fixed in three positions at the end of long horizontal poles so as to be completely exposed. Soil thermometers, Leicke dew plates, and Piche evaporimeters were also used.

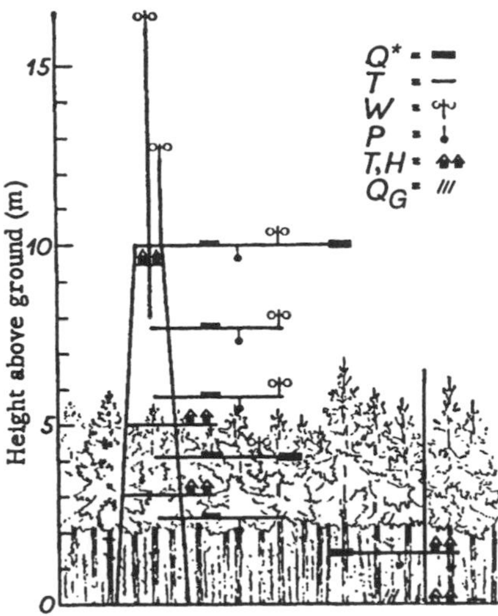

Figure 33-10. Arrangement of instruments for studying stand climate in young fir near Munich. (After A. Baumgartner [2502, 3301])

Figure 33-11 shows the typical variation of the net radiation balance for a clear day in the late summer of 1951. Above the stand (heavy line), the uniform, negative balance at night changes about an hour after sunrise and rises steeply during the morning. The weather was clear until 11:00, at which time clouds appeared on the southern horizon up to 13:00, and isolated cumulus clouds appeared after 16:00. Both sets of clouds resulted in unsteadiness in the net radiation balance and changes in evaporation. One and a half hours before sunset,

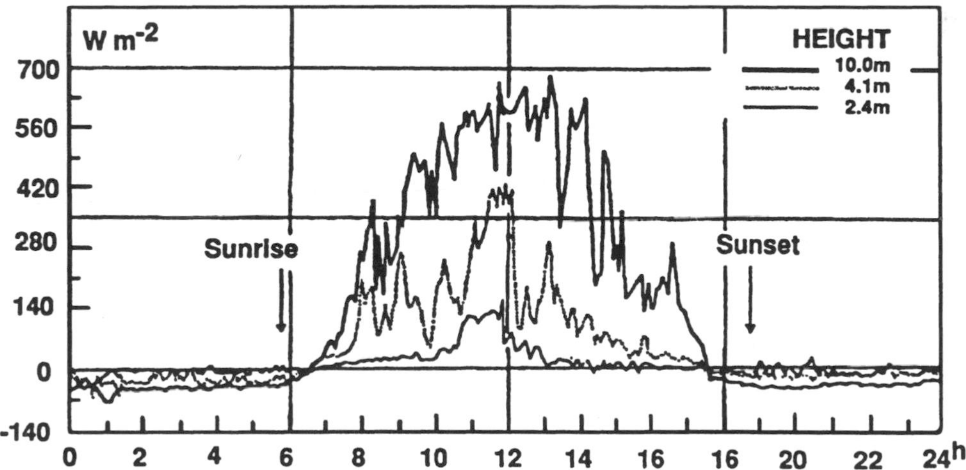

Figure 33-11. Diurnal variation of net radiation in young fir on 11 September 1951. (After A. Baumgartner [2502, 3301])

the net radiation balance became negative again. Below the crown area at 4.1 m (dotted line), net radiation varied with random shadow effects from the treetops and branches. Only when the sun was high did significant amounts of net radiation penetrate to 2.4 m within the forest.

The daily variation of net radiation for the young fir plantation during a dry period in mid-summer, from 29 June to 7 July 1952 is shown in the form of isopleths in Figure 33-12. In the early morning radiation is absorbed in the treetops and as the day progresses spreads to the crown area. The height of the sun can be noticed in the shallow dip in the crowded isopleths about noon when the sun reaches its maximum altitude. The heated tops and crown areas radiate longwave radiation upward into the air and downward to the cooler ground. This net exchange of radiant energy continues until the radiation emitted from the treetop zone has lowered the temperatures, resulting in a reversal of the net radiant exchange. The forest floor gains energy during the day when the crown is warmer. However, when the crowns are cooler, the net longwave radiation exchange is from the ground toward the crowns. Some of this longwave radiation escapes through gaps toward the night sky.

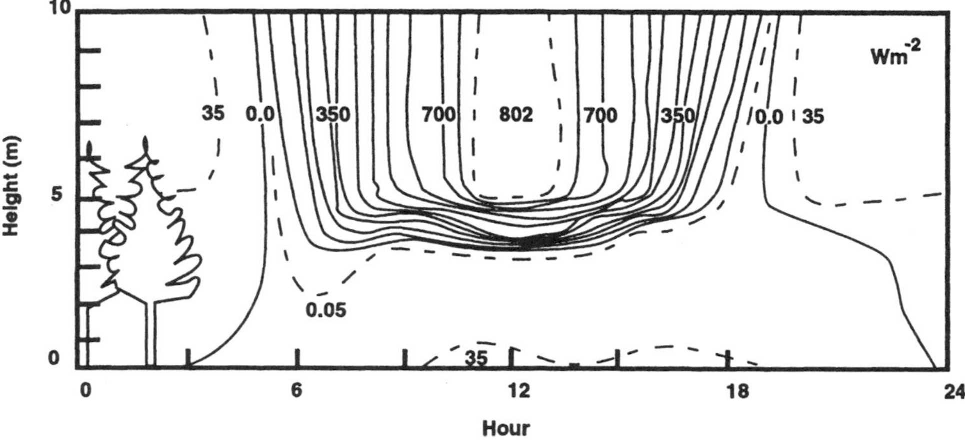

Figure 33-12. Isopleths of net radiation for a young fir plantation near Munich.

Averages for the hourly values of the net radiation balance for the five different heights for 2, 4, 5, and 7 July 1952 are given in Table 33-3 (A. Baumgartner [3303]). The term "night" refers to the period of negative radiation balance, and "day" that of a positive radiation balance.

In the radiating treetop area, the period of negative net radiation balance is longest, while close to the ground, it lasts only for 3 hours before sunrise. At 5 m the maximum temperature is reached in the early afternoon. After this more radiant energy is emitted than received. In the afternoon when the radiation exchange at 5 m is negative, it usually is still warmer than the ground (0.2 m) and thus will be emitting more energy toward the ground than it is receiving from it. It is not possible for the positive net radiation balance at 0.2 m to be greater than at 3.3 m, as shown in the above figures and indicated by the 35 W m^{-2} isoline at the forest floor in Figure 33-12. The accidental positioning of the instruments might have resulted in more sunflecks being received at 0.2 m than in the trunk area where the higher instrument was located.

Table 33-3. Net radiation at different heights in a stand of young firs.

Height of measurement (m)	Duration (hr)		Average balance (W m^{-2})		Daily balance	
	Day	Night	Day	Night	MJ m^{-2}	%
10.0, above the forest	15.0	9.0	258.2	-37.7	25.3	100
5.0, tree-top area	13.6	10.4	258.2	-32.1	23.1	91
4.1, crown area	15.0	9.0	114.4	-9.8	10.9	43
3.3, trunk area	17.0	7.0	9.1	-7.7	1.2	5
0.2, on forest floor	21.0	3.0	14.0	-3.5	1.8	7

In comparing the outer effective surface at the top of a pine shelterwood with the ground surface at a nearby clearing H. B. Granberg et al. [3314] found on calm nights initially a slightly greater radiative loss of energy from the clearing (Figure 33-13). For most of the night, however, radiative energy loss from the tops of the trees was greater than the clearing. They suggested that the increased air movement and resulting increased convective heat fluxes, provided a greater source of heat to the tree tops than to the surface of the clearing.

On windy nights (Figure 33-14) "the measurements were initially less negative in the shelterwood (a) than in the clear-cut (b). However through the night the difference between the two sites gradually declined to essentially zero by sunrise." The radiative losses from the outer effective surfaces (treetops and clearing surface) were greater on the windy night due to the larger convective heat fluxes.

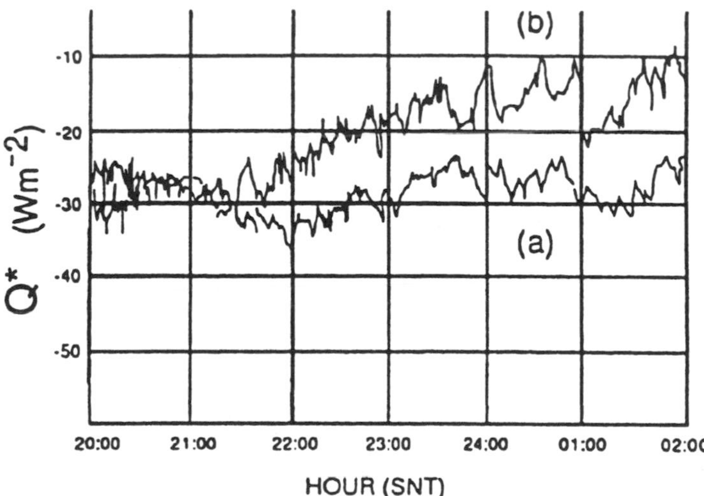

Figure 33-13. Average net radiation from (a) the shelterwood; (b) the clearcut on a clear and calm night (3-4 July, 1987).

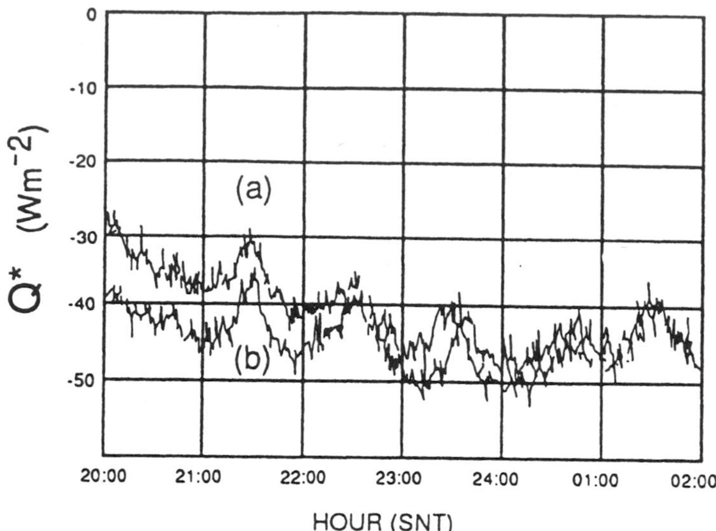

Figure 33-14. Average net radiation from (a) the shelterwood; (b) the clearcut on a clear and windy night (2-3 July, 1987).

34 Metabolism, Energy Storage, and Wind in a Forest

The energy balance at the outer effective surface in the forest crown area can be written as:

$$Q^* = Q_H + Q_E + Q_S$$

where Q_S = the net energy storage within the canopy and crown area between the free atmosphere through the canopy and includes the forest floor. This energy term is made up of five separate components:

$$Q_S = Q_G + Q_B + Q_A + Q_V + Q_P$$

Where Q_G = soil heat storage, Q_B = biomass heat storage (the energy stored in the trunk, branches, twigs, and leaves or needles of the canopy), Q_A = sensible heat storage in the canopy air volume, Q_V = latent heat storage in the canopy air volume, and Q_P = the photosynthetic energy storage. All of the terms are positive when energy is placed into storage, and negative when energy is released from storage.

When coal burns, it releases the energy accumulated by the forest in earlier geologic history, obtained from sunlight and stored in the assimilation process (photosynthesis). In its green chloroplasts the plant produces glucose from the carbon dioxide extracted from the air and water taken from the sap flow. The energy required to complete this chemical process is provided by photosynthetically active radiation (PAR), or that portion of the global solar radiation $K\downarrow$ that can be effectively utilized in photosynthesis. On average, PAR is about 49% of $K\downarrow$, and it takes about 16,750 J of PAR to produce 1 g of glucose. We may, therefore, calculate the amount of energy used from the amount of material produced. D. D. Baldocchi et al., [3402] examined canopy photosynthesis Q_P for an oak-hickory deciduous forest in Oak Ridge, Tennessee (35°57'N) from 24 July to 10 August 1984. The site had received ample recent precipitation, so that canopy photosynthesis Q_P was primarily controlled by PAR. Canopy photosynthesis was determined by summing the CO_2 flux measured above the canopy F_C and the CO_2 efflux measured from the forest floor. Similarly, incoming and outgoing PAR was measured above and below the forest canopy. For this well-watered deciduous forest Q_P exhibited a curvilinear increase with increasing levels of PAR. Unlike many agricultural crops, the deciduous forest Q_P did not display a light-saturation effect in which Q_P levels off at high levels of PAR, suggesting that for a well-watered deciduous forest Q_P is primarily controlled by PAR. Maximum rates of Q_P were between 0.80-1.00 mg m^{-2} sec^{-1}, lower values but of the same order of magnitude as obtained from agricultural crops by E. Inoue [3409, 3410] and E. Inoue et al. [3411]. The lower Q_P rates for forests are explained by their greater stomatal resistance to the diffusion of CO_2, and their weaker coupling of Q_P to PAR. The canopy photosynthetic efficiency, or ratio of the energy fixed by Q_P to the absorbed PAR, was between 0.04-0.08 (or between 0.02-0.04 of $K\downarrow$). The CO_2 exchange above the deciduous forest canopy F_C provided between 60-80% of the CO_2 consumed in canopy photosynthesis, with the remainder provided by the CO_2 efflux from the forest floor (D. D. Baldocchi et al., [3402]). These values are lower than comparable figures for agricultural crops (80-90%), and are due to the lower F_C flux over the forest canopy compared to an agricultural canopy, and the greater CO_2 efflux from the forest floor compared to an agricultural surface. According to A. Baumgartner [2502], the energy used by a forest for photosynthesis is about 3.5 W m^{-2}, or roughly 1 percent of the total incoming solar radiation. D. H. Miller [3326] reached a similar conclusion from examination of the growth of timber in the Sierra Nevada forests in California. For this reason it is normally ignored in energy balance studies.

The amount of energy assimilated during the day varies significantly, and is controlled by the available PAR, air temperature, vapor pressure deficit, and the stomatal resistance. Low light levels such as occur with low sun angle, cloudiness, or shading can significantly reduce Q_P, as can low or high air temperatures. Plants differ both as to the amount of light and to the upper, optimum and lower range of temperatures required for a particular rate of photosynthesis. Figure 34-1 illustrates the relative effect of both of these factors on the rate of photosynthesis for spruce. When water is not limiting, Q_P will be most closely determined by PAR and air temperature, while under limiting water conditions vapor pressure deficit and stomatal resistance become dominant. For example, maximum Q_P does not coincide with the noon maximum of $K\downarrow$ since the stomata tend to close around noon. The daily variation of Q_P shows one maximum in the late morning and a second during the late afternoon. M. Marek et al. [3418] discuss variations in photosynthetic rates during the day and by different canopy layers in an oak-hornbeam forest.

Countering the energy consumed in photosynthesis, there is a gain of energy through cell respiration. When plants breathe (respiration), sugar is broken down into carbon dioxide and water. In contrast to photosynthesis, cell respiration takes place both day and night and leads to a nocturnal loss of weight by the leaves. The gain of energy is of the same order as the energy consumed in photosynthesis, about 3.5 W m^{-2}. This, however, may amount to 10 percent of the total radiation exchange during the night. Energy released in respiration can lead to a temperature increase of a few tenths of a degree, and many believe they have observed this. Since the respiration rate increases exponentially with increasing temperature,

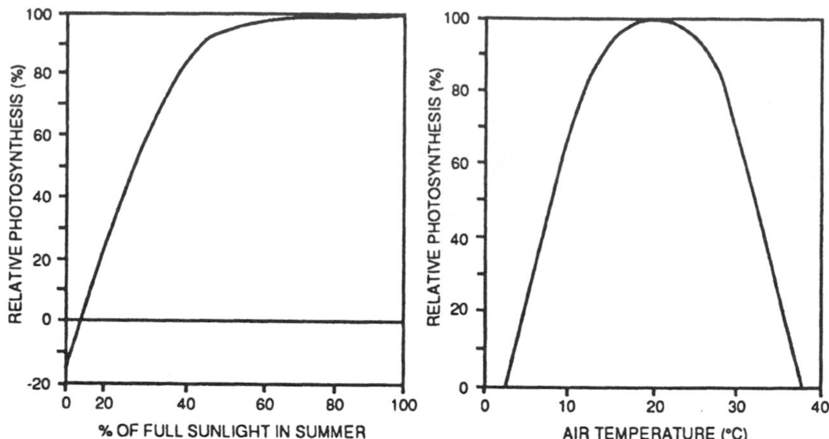

Figure 34-1. The relative effect of light and air temperature on the rate of net photosynthesis for a spruce forest. (From D. L. Spittlehouse and R. J. Stathers [3423])

it is also subject to a diurnal variation with a maximum in the early afternoon at the time of the temperature maximum. The metabolism of the forest is, therefore, made up of the interaction of photosynthesis and respiration. For the young fir plantation near Munich, A. Baumgartner [2502] calculated the values shown in Table 34-1.

Table 34-1. Photosynthesis, respiration and metabolism within a fir plantation ($W\ m^{-2}$). (After A. Baumgartner [2502])

Hour of day	3-4	6-7	9-10	12-13	15-16	18-19	21-22
Photosynthesis	0	-7.0	-9.8	-8.4	-9.1	-7.0	0
Respiration	3.5	4.2	4.9	5.6	5.6	4.9	4.9
Metabolism	3.5	-2.8	-4.9	-2.8	-3.5	-2.1	4.9

In measuring the quantities involved in the energy balance, the small amount of energy involved in metabolism may be neglected. Plant metabolism will, therefore, be left out of all further discussions.

Canopy transpiration T and canopy photosynthesis Q_P are closely related to one another since all plants must lose water in order to gain CO_2. D. D. Baldocchi et al. [3402] found a strong linear relationship between Q_P and T for an oak-hickory deciduous forest. Water use efficiency (WUE) is defined as the ratio of canopy photosynthesis Q_P to canopy transpiration T. D. D. Baldocchi et al. [3402] obtained WUE values of between 6-12 mg CO_2 $(g\ H_2O)^{-1}$ for their deciduous forest, and found that WUE values were nearly independent of net radiation Q^* but exhibited a strong negative curvilinear relationship to vapor pressure deficit.

The importance of heat storage within the forest canopy on the forest microclimate may be readily recognized from a consideration of Figure 34-2 (also for the young fir plantation near Munich). The diagram shows the diurnal variation of the energy exchange for the fair weather period from 29 June to 7 July 1952, evaluated by A. Baumgartner [2502]. The dotted curve marked Q_A shows the energy exchange of the air mass within the forest. The maximum is ± 7 $W\ m^{-2}$, which is not surprising in view of the small thermal capacity of air. The quantity Q_A can, therefore, often be neglected even in an old stand of considerable height. In contrast, the amount of energy Q_B exchanged in the wood and the needles (biomass) is large after sunrise when the biomass is being heated. The biomass reaches its maximum temperature in the mid-afternoon and then begins to emit more energy than is being received. The biomass loses energy at the fastest rate around sunset and reaches its minimum temperature shortly after sunrise. The air (Q_A) within the forest follows a similar sequence. The energy exchanged in the biomass can be greater than the quantity of energy exchanged at the forest floor through conduction Q_G. While the biomass and air mass respond quickly to changes in radiation, the forest floor follows only indirectly the daily variation of the stand that shelters it. Hence, the daily variation of Q_G shows a phase difference of 4 to 6 hr in comparison with Q_B and Q_A. In the early morning the ground is still cooling while biomass is heating at a fast rate and in the late afternoon energy is still flowing into the ground long after the biomass has begun to cool.

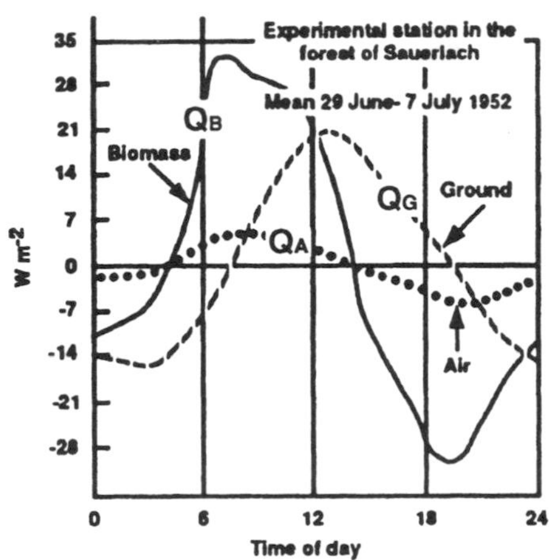

Figure 34-2. Diurnal variation of energy exchanged within a fir plantation and the ground below it. (After A. Baumgartner [2502])

If the energy exchange in the forest as a whole is investigated, Q_A, and Q_B become insignificant; thus, it is more appropriate to consider the energy exchanges in the stand as if it were a part of Q_G. The value of Q_G, after being corrected by Q_A and Q_B, can then be seen in its proper proportions in Figure 26-2, which is for the same forest during the same period. The insignificance of Q_G compared with Q^*, Q_E and Q_H is evident. The average daily energy balance for this period of clear weather is $Q^* = 24.54$, $Q_E = -16.16$, $Q_H = -8.25$, while Q_G is only 0.13 MJ m^{-2} day^{-1}. However, in contrast to the other factors, the factor Q_G involves incoming and outgoing quantities of energy that are practically equal. A better idea of the proportions is, therefore, obtained by adding the flow of energy in both directions, regardless of sign, to obtain: $Q^* = 26.97$, $Q_E = 16.58$, $Q_H = 8.75$, and $Q_G = 2.81$ MJ m^{-2} day^{-1}. From this it may be seen that Q_G accounts for 5 percent of the total exchange of 55.11 MJ m^{-2} day^{-1}.

J. H. McCaughey and W. L. Saxton [3417] investigated the energy balance storage for a mixed forest of 60-70 percent canopy cover near Chalk River (45°58'N), Ontario during May-August 1985. Results were summarized for four sets of atmospheric and canopy conditions: clear sky and dry canopy, partly cloudy and dry canopy, overcast sky and dry canopy, and wet canopy; as well as for four energy storage components: soil heat storage, sensible heat storage in canopy air, latent heat storage in canopy air, and biomass heat storage. For a dry canopy, daily energy storage rarely exceeded 7 percent of net radiation; negative daily totals, however, were observed for wet canopies and during cloudy skies or precipitation. Energy storage increased rapidly following sunrise to as high as 100 W m^{-2} by 09:00, and remained high until noon after which a sharp decline began which reached negative values by 14:00 to 16:00. Nighttime totals were between -35 to -50 W m^{-2}. Cloud cover reduced the daily range of the energy storage term. Soil heat flux (Q_G) was the largest of the four components on an hourly basis and, because of the greater solar radiation penetration due to the incomplete canopy cover, experienced two separate peaks at 09:00 and 12:00. Sensible heat storage in the canopy air followed the most regular pattern, becoming positive at sunrise, peaking at 07:00, and then declining until negative values were reached at 16:00. Biomass heat storage displayed a similar pattern, except that its peak value was reached at 12:00. In contrast, latent heat storage in the canopy air was highly variable, fluctuating rapidly between positive and negative values under all atmospheric conditions. J. H. McCaughey and W. L. Saxton [3417] found that soil heat flux was the largest heat storage term on a daily basis when the soil was moist and had a higher thermal conductivity. When soil moisture became more limiting and thermal conductivity decreased, biomass heat storage Q_B approached soil heat storage in magnitude on a daily basis. Although heat storage normally accounts for only 2-3% of Q^* on a daily basis, it has a strongly nonsymmetrical hourly pattern about noon. Thus Q_S can approach 40-50% of Q^* during the night, when Q_S is negative, and during the transitional times of sunrise and sunset, when Q_S is positive and negative, respectively. Overcast sky and wet canopy conditions reduce the magnitude of Q_S, delay the timing of the individual component peaks throughout the day, and increase Q_S as a percentage of Q^* on a daily basis.

A. R. Aston [3401] conducted a similar investigation of the energy storage term for a young eucalyptus forest, and provided specific information on biomass-air temperature variations and the separate contributions of the litter, trunk, branch, twig, and leaf components of the biomass heat storage term. Because of the reduced penetration of solar radiation to the forest floor due to the completeness of the eucalypt forest canopy the soil heat flux was more symmetrical about noon. An investigation of a tropical forest environment is given in C. J. Moore and G. Fisch [3419]. They found that daily Q_S was between 3-5% of Q^* on dry days, and nearly 6% of Q^* on rainy days. Also, Q_S, Q_A, and Q_V were more nearly equal in magnitude because the increased height of the tropical forest resulted in greater biomass, air volume, and humidity content for the canopy.

The partitioning of net radiation into sensible and latent heat for a well-watered forest canopy differs significantly from agricultural and grassland surfaces. For the latter surfaces (Table 31-5, Section 31) the aerodynamic resistance makes a significant contribution to the total resistance to the diffusion of water vapor, most of the net radiation is partitioned into the latent heat flux, Bowen ratios are generally very low, and canopy resistance is closely coupled to net radiation.

The energy balance measurements obtained by S. B. Verma et al. [3426] during July-August 1984 for a well-watered, fully-leafed oak-hickory deciduous forest in Oak Ridge, Tennessee (35°57'N) illustrate some of these differences. Figure 34-3 gives the diurnal patterns of the aerodynamic and canopy resistances during six days when the canopy was dry, full-leafed, and well-supplied with water. The greater canopy height ($z = 22$ m), larger surface-roughness length and zero-plane displacement, and more extensive leaf area (LAI = 4.9) combine to produce relatively small aerodynamic resistance values R_A that peak in the early morning and late afternoon when surface winds are light, and reach minimum values near noon when wind speed is maximum. Canopy resistance R_C, therefore, exerts primary control over the diffusion of water vapor from the forest canopy. Canopy resistance R_C is lowest in the early morning when leaf surfaces may still be moist from nighttime dew, and values steadily increase until the end of the day. Canopy resistance to water vapor diffusion shows much less sensitivity to net radiation, and much greater sensitivity to the vapor pressure deficit *VPD*. As canopy temperatures increase through the day the vapor pressure deficit *VPD* rises, which increases the canopy resistance and provides a negative feedback upon water loss by Q_E. Consequently, the Bowen ratio β is at a minimum in the early morning and increases during the day, often peaking in the mid- to late afternoon. The peak Bowen ratios of between 0.25-0.65 obtained by Verma et al. [3426] for the deciduous forest were as much as twice as large as corresponding values from a prairie landscape (J. Kim and S. B. Verma [3105]). Latent heat flux Q_E was found to consume between 25-90% of net radiation Q^* for the deciduous forest, while values between 80-120% may be representative of well-watered agricultural surfaces.

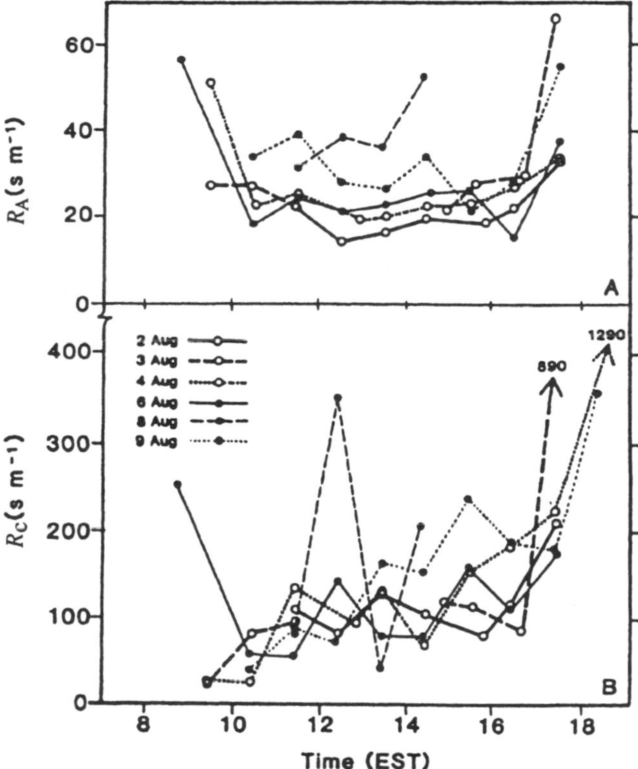

Figure 34-3. Diurnal patterns of aerodynamic resistance (top) and canopy resistance (bottom) to water vapor transfer for a fully-watered oak-hickory deciduous forest in Oak Ridge, Tennessee, USA during six dry days in August 1984. (From S. B. Verma et al., [3426])

Coniferous forests have been found to have even lower aerodynamic resistances, and comparable or higher canopy resistances than in deciduous forests. These effects result from their more rigid structure, somewhat larger stomatal resistances, and year-round presence of needles. Consequently, sensible heat flux is enhanced and latent heat flux diminished in comparison to deciduous forests. A. Lindroth [3414] observed mean daytime Bowen ratios of between 1.0-2.0 for a thin pine forest in Jädraås, Sweden (60°49'N) over the growing season, while C. S. Tan and T. A. Black [3424] found that Q_E consumed at most 58% of Q^* during fine sunny days in a Douglas-fir forest on Vancouver Island in British Columbia, Canada.

The mean diurnal variation of energy balance components for the thin pine forest observed by A. Lindroth [3414] is shown in Figure 34-4 for six dry days during August-September 1978. Latent heat flux Q_E exceeds sensible heat flux Q_H throughout the day. Q_H is concentrated more heavily during the morning hours, and Q_E is more concentrated during the afternoon. Soil heat flux Q_G follows Q^* during the day.

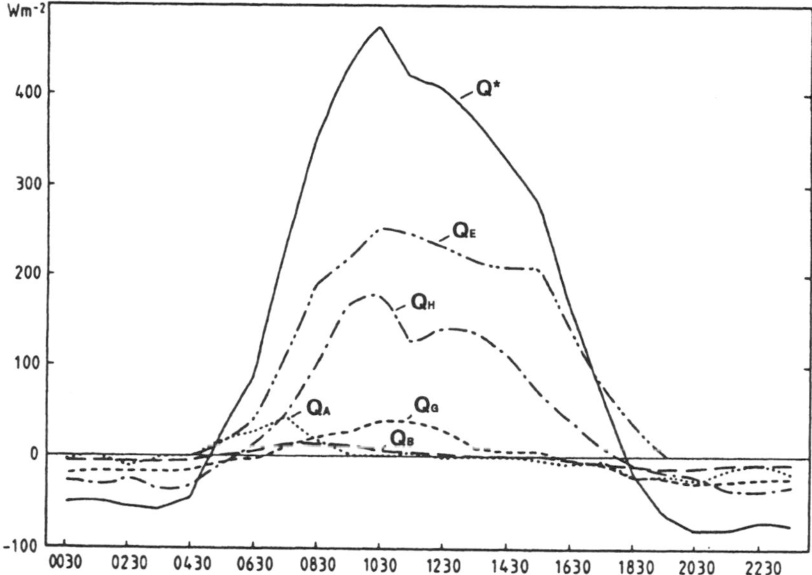

Figure 34-4. Mean diurnal variation of energy budget components for a thin, fully-watered coniferous forest canopy in Jädraås, Sweden during six dry days in August-September 1978. (From A. Lindroth [3414])

The energy budget of a forest canopy is determined by physiological controls, atmospheric environment, soil moisture conditions, and the presence of a wet or dry canopy. A. Lindroth [3413] found that latent heat flux Q_E from a dry pine forest canopy was mainly controlled by canopy resistance R_C and the vapor pressure deficit *VPD*. At low levels of global solar radiation $K\downarrow$, increasing light intensity was found to reduce canopy resistance. Canopy resistance was found to respond much more strongly to changes in vapor pressure deficit. R_C typically is a minimum in the early morning hours due to the cool air temperatures and the onset of $K\downarrow$. Canopy resistance steadily increases, even under fully-watered conditions, as canopy and air temperatures increase during throughout the day. They found canopy resistance to be about twice as sensitive to *VPD* as to $K\downarrow$. Similar patterns were obtained by C. S. Tan and T. A. Black [3424].

The deep and extensive root systems of forests enable their energy exchange patterns to be much less sensitive to short-term variations in soil water content. Some response to longer-term soil moisture changes, however, does occur. C. S. Tan and T. A. Black [3424] noted that as the soil water potential varied between –0.6 and –6.5 bars in the top 45 cm beneath a coniferous forest that the ratio of Q_E/Q^* ranged between 0.58 and 0.27. Canopy resistance also increased with soil drying, and was found to be more sensitive to increased vapor pressure deficit at lower soil water potentials.

If the canopy is wet (nearly) all of the energy is used to evaporate water from the wet canopy and very little/no water goes to transpire water. In terms of the Penman-Monteith model this means that canopy resistance is high (canopy conductance is low), little/no transpiration occurs, and much evaporation occurs. As the canopy wetness decreases with time, canopy resistance decreases (canopy conductance increases), and transpiration becomes a larger fraction of the ET total. Thus, when the forest canopy is wet canopy resistance becomes very large, canopy transpiration approaches zero, and canopy latent heat flux consists nearly entirely of evaporation of intercepted water from wet leaf surfaces. Because of the extremely small aerodynamic resistance R_A of forest canopies, Q_E from a wet canopy will be much higher than Q_E from a dry canopy for the same level of available energy (Q^* - Q_S – Q_G). Additional energy for Q_E from a wet canopy can also be provided by the input of sensible heat $-Q_H$.

The daytime energy budget for the thin pine forest at Jädraås, Sweden is summarized in Figure 34-5 (A. Lindroth [3414]). Because the forest canopy was open, considerable net radiation reached the forest floor. Canopy partitioning of Q^* was dominated by Q_E, while the sensible and latent heat fluxes were more evenly divided at the forest floor. Most of the 9% of net radiation used in heat storage Q_S was in the soil heat flux due to the presence of an open forest canopy.

Considerable seasonal variation in energy exchanges are also observed within forest canopies. These seasonal variations in energy partitioning result from increasing LAI during the growing season, seasonal changes in global solar radiation, vapor pressure deficit, and increasing stomatal efficiency. A. Lindroth [3414] found that diurnal Bowen ratios varied from between –2 to 2 in May to between –1 to 1 in September. He cautioned that these large systematic variations in the Bowen ratio over the growing season make determination of typical growing season energy patterns very problematic.

Large interannual variations in energy budget patterns can also be observed in forests. L. Jaeger and A. Kessler [3412] examined the energy and water balance patterns from a pine forest plantation near Hartheim in the upper Rhine Valley in Germany (47°56'N) for the period 1974-1988. They found significant variations in the forest energy budget over the 15-year study period arising from the effects of variable weather, fluctuations in moisture availability, natural forest growth, and plantation thinning. They conclude that it is very difficult to distinguish the effects of climatic variability, natural forest changes and man-made forest changes upon forest energy balance patterns.

In order to understand the temperature and water vapor distribution in the forest, it is essential to know the wind profile as well as the radiation budget. The amount of eddy diffusion taking place depends primarily on the wind strength and determines whether the radiative energy remains in the forest or is transported elsewhere.

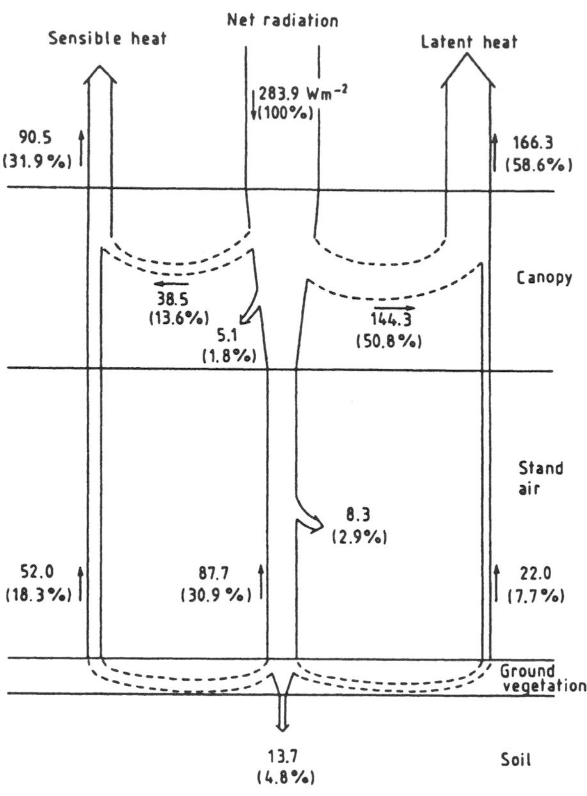

Figure 34-5. Summary of daytime energy budget for a thin, fully-watered coniferous forest canopy in Jädraås, Sweden during six dry days in August-September 1978. Energy terms are shown as both energy totals (Wm^{-2}) and as a percentage of net radiation (parentheses). (From A. Lindroth [3414])

The zero-plane displacement *d* (Section 30) is naturally much greater for a forest than for a grain or a plowed field. The resistance to air flow due to the forest foliage is directly related to its density. In forests air movement is less restricted in a trunk area free of branches, particularly when the wind can blow in through the open borders of the stand. The wind profile at three different wind speeds in a sparse fir stand, in the Bavarian forest, is illustrated in Figure 34-6, using measurements made by R. Geiger [3406]. When winds are strong, there is a secondary wind speed maximum through the trunk area. H. Ungeheuer [3425] was able to show, for a high beech forest on a slope, not only that the katabatic wind at night (Section 43) was able to pass through the forest, but that its passage was even favored by its being protected from external disturbances by the forest crown.

The forest foliage also exerts a unique influence upon the wind field. R. Geiger and H. Amann [3407] measured the wind profiles shown in Figure 34-7 in a 24 m high, 115 yr old oak stand, near Schweinfurt, which had 40 to 50 yr old beech saplings growing underneath. Figure 34-7 shows the wind profile before and after the trees were in leaf. Naturally, the wind is able to penetrate more into the forest before the leaves open. This is of practical im-

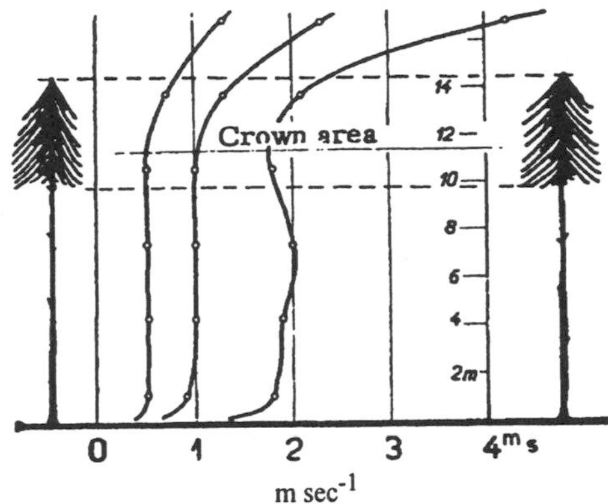

Figure 34-6. Wind profiles in a stand of pine for three ranges of wind speed.

portance in attempting to destroy larvae by dusting with insecticide from an aircraft. Before the trees are in leaf, not only are surfaces lacking on which the poison dust might settle, but also the stronger winds carry it downwind more quickly, and greater turbulence scatters the powder more widely than when the foliage is present.

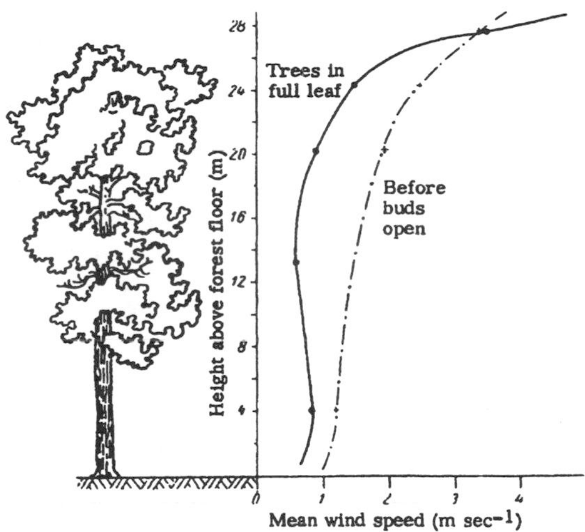

Figure 34-7. Influence of foliage on the wind profile in an oak forest. (After R. Geiger and H. Amann [3407])

The contrast in conditions before and after breaking into leaf shows up most clearly in the frequency of hours with calm ($u < 0.7$ m sec^{-1}); in 206 hr of observation before leaves were present, not a single hour of calm was recorded among the treetops, whereas after leaves were present, calm conditions were recorded in 10 percent of the 494 hours of observation. At a height of 4 m above the forest floor, the number of hours of calm rose from 67 to 98 percent.

The extent to which wind can penetrate a forest depends on several factors. Thickets and growths of saplings slow the air. In old stands, the type of forest and the degree to which the crown is closed are of importance. Figure 34-8, after R. Geiger [3406], shows comparative wind profiles in two stands which were only 86 m apart. The stand called "Lehmlache I" was a 65 yr old pine forest with a uniform, loosely knit crown. In "Lehmlache II" the same kind of stand had an association of fir of different age groups below, so that crowns were found at all levels. This reduced the wind speed, as shown in Figure 34-8, with the immediate result that in the trunk area it was cooler by day and warmer at night. The crown area in the Lehmlache II stand experienced more extreme temperatures than the thinner, better ventilated Lehmlache I stand. Since the wind speed is slowed to a greater extent within and through the canopy of Lehmlache II, the air above the canopy will flow at a faster rate. This same effect can be observed in Figure 34-7 in comparing the wind speed above the deciduous trees before and after the buds have opened. The more the vegetation slows, blocks or funnels the air movement, the faster the air will flow above or through adequate size gaps within the vegetation (Venturi effect). This is considered further in section 52.

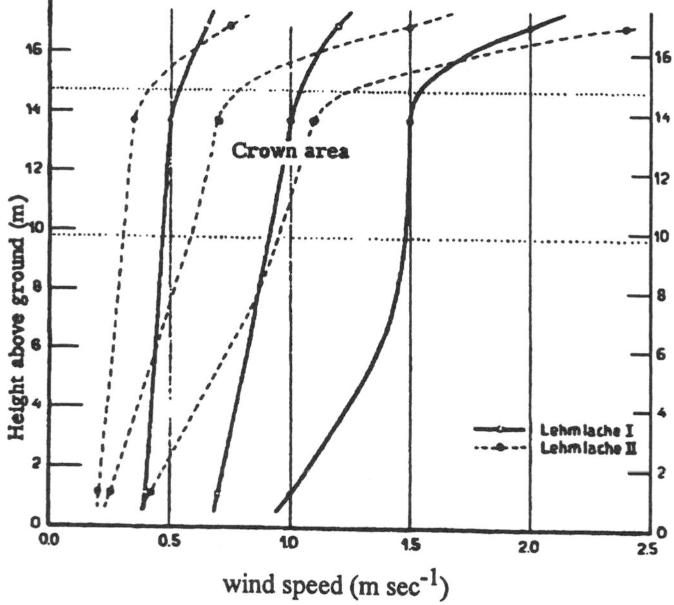

Figure 34-8. Comparison of wind profiles in two stands of pine, one without (I) and the other with (II) an underbrush of fir. (After R. Geiger [3406])

An example of the daily wind variation is given in Figure 34-9 for the plantation of young fir near Munich. In contrast to Figure 33-12 and Figures 35-3 through 35-6, which follow, the height scale is extended well beyond the top of the forest (5 m). The isopleths give wind speeds (cm sec^{-1}) on the clear day selected for this experiment. Due to friction wind speed decreases as the trees are approached. The diurnal variation of wind with its maximum in the early afternoon (Figure 16-2) can be recognized above the stand and is the result of increased convection and momentum exchange during the day (Figures 16-2 and 16-3). A second maximum about midnight is the result of a local downslope wind, which usually blows over the upland plains in Bavaria in areas close to the Alps in summer.

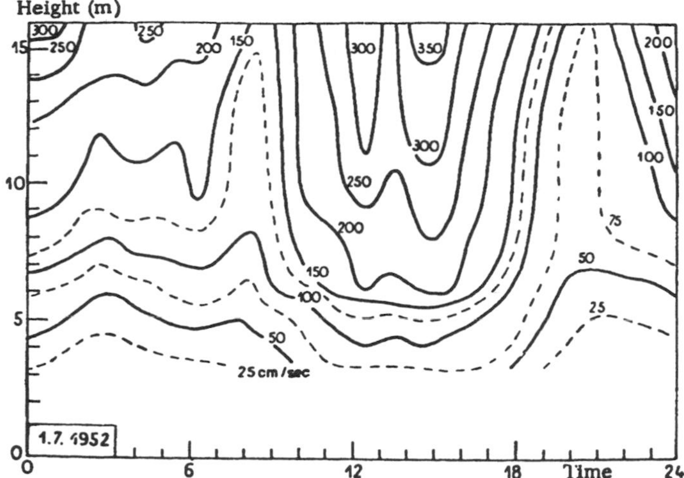

Figure 34-9. Isopleths of wind speed (cm sec^{-1}) in a fir plantation near Munich. (After A. Baumgartner [2502])

A. S. Devito and D. R. Miller [3404] used smoke tracers to determine the nocturnal flow of wind through a cornfield and a leafless oak canopy. They observed that when the winds were light, density air drainage persisted within the canopy even when the air aloft was in the opposite direction (Figure 34-10). When winds above the canopy were equal to or greater than 5 m sec^{-1}, density flow only persisted near the ground. During the leafless phase, ground level drainage flow developed only in the forest. When the oak trees were in leaf, drainage flow developed only above the canopy.

In a maize canopy, A. F. G. Jacobs et al. [3415] found that at night above the canopy the air becomes thermally stable due to longwave radiative cooling at the top of the canopy. Within the canopy, however, the air becomes unstable due to cooling at the top of the canopy and heating from the soil. They found that a state of free convection developed in the lower region of the vegetation where the foliage was less dense. This also undoubtedly also occurs in some dense forests.

A. Local wind direction opposite drainage flow--leafless forest

B. Local wind direction opposite drainage flow--corn

Figure 34-10. Generalized pattern of wind profiles. (After A. S. Devito and D. R. Miller [3404])

35 Air Temperature and Humidity in a Forest

In investigations by R. Geiger and H. Amann [3407] in the Schweinfurt Forest of 115 yr old oaks 24 m high with 40 to 50 yr old saplings below, the diurnal variation on a sunny day, 18 August 1930, is shown in Figure 35-1. They provide us with a vivid picture of a summer day in the forest. Readings were started before sunrise. The temperature was at its lowest at 23 m in the radiating oak crowns. Measurements were also made at five heights at 30 second intervals. At sunrise, the first thermoelement to receive energy was the one above the level of the stand. A marked boundary surface developed within the stand, it is still nighttime and the air is cool and moist. Above it, the incoming radiation is heating the top of the trees and reducing the relative humidity. "The division is clearly perceptible to all who climb the observation ladder on a clear morning; thousands upon thousands of winged insects fill the space above the boundary surface, and this living cloud is sharply cut off at its lower edge" (R. Geiger [3406]). Around 08:00, temperatures in the crown area increase slowly at first because much of the initial energy is used to evaporate the dew. As the day progresses and the dew has evaporated, the crown area heats quickly and the air becomes unstable in contrast to the uniform and slowly increasing temperatures of the trunk area.

About noon, a state of equilibrium is reached; temperature curves then run horizontally and are distinguished from each other by their degree of instability. As is expected with strong solar radiation, temperature decreases from the maximum in the crown area upward into the free atmosphere and downward into the trunk area. The temperature decrease in the evening, in contrast to the increase in the morning, leads to an increasing stabilization in the air layers above the crowns, whereas morning heating produced increasing instability.

Figures 35-2 and 35-3 present results from a sparse and a dense stand, which exhibit basic differences in their temperature variation. Temperature isotherms in Figure 35-2 are based on 24 vertical temperature profiles published by K. Göhre and R. Lützke [3408], measured in a mixed pine beech stand near Eberswalde, on 14 and 15 September 1953 (clear days). The stand consisted of 82 to 100 yr old pines about 20 m tall, with an inter- and undergrowth of 10 to 60 yr old beeches with an average height of 13 m. The measurements were made at eight levels, extending a little beyond the crown of the pines.

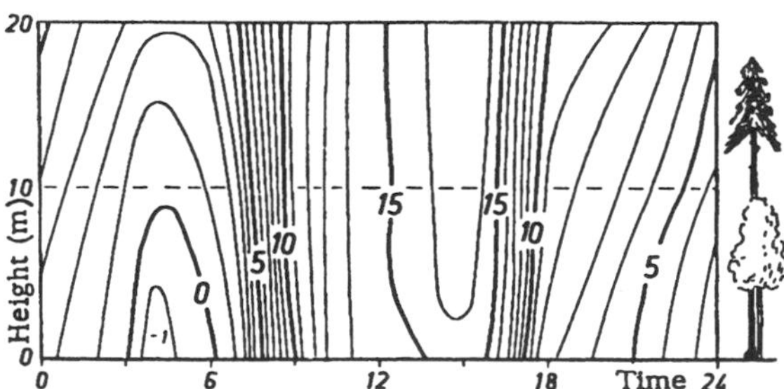

Figure 35-2. Temperature variation during the day in a sparse pine forest near Eberswalde. (After K. Göhre and R. Lützke [3408])

Figure 35-3 is for a dense stand from A. Baumgartner [2502] for the fir plantation near Munich. The readings are averages of a 10 day period in summer (28 June to 7 July 1952), for six levels, extending to twice the height of the trees. In comparing the two diagrams, it

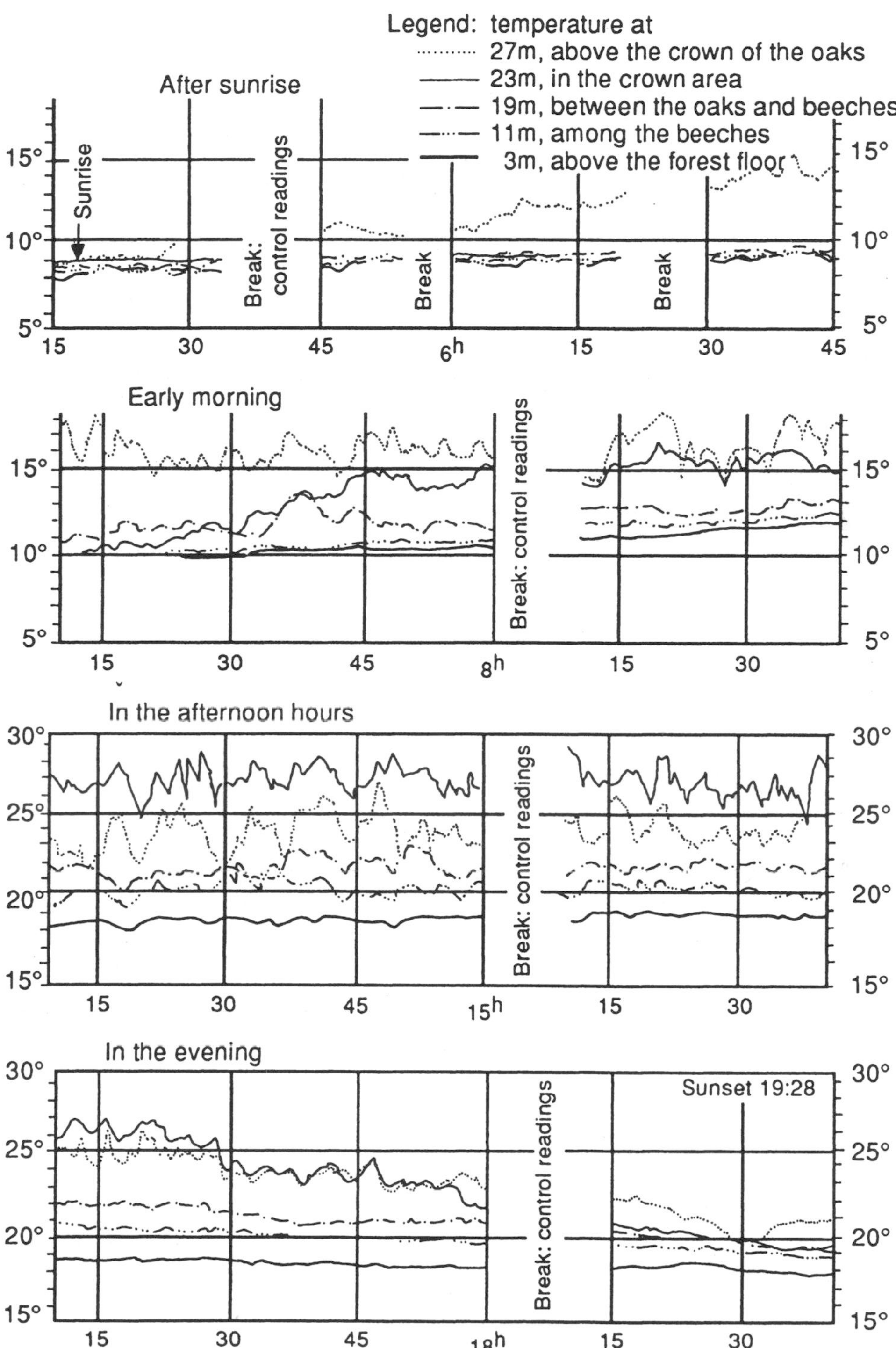

Figure 35-1. Temperature variation at five heights in an old oak stand on a clear summer day: morning, afternoon and evening. (After R. Geiger and H. Amann [3407])

should be remembered that the temperature in Figure 35-3 extends further above the top of the trees.

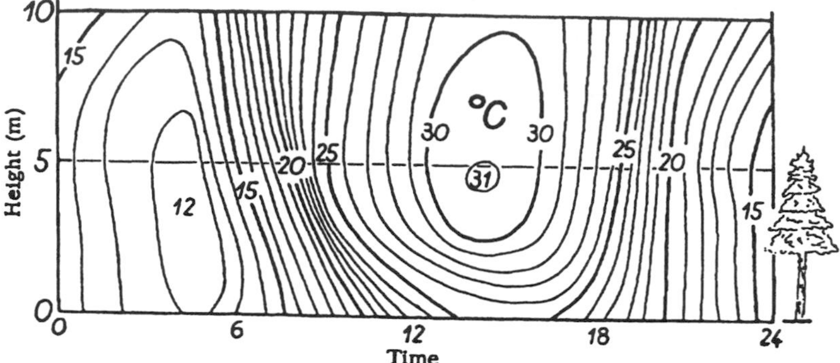

Figure 35-3. Temperature variation during the day in a dense young fir plantation near Munich. (After A. Baumgartner [2502])

The isotherms are crowded at the time of maximum heating a few hours after sunrise and during the cooling period before sunset, the former being more intense than the latter. In the less dense stand (Figure 35-2), the whole air mass of the forest is affected, resulting in vertical isotherms; while in the denser stand (Figure 35-3), heat penetrates only slowly from the crown down through to the forest floor, so that the isotherms are initially tilted, becoming more horizontal with time.

There is a substantial difference in the temperature distributions between the thin (Figure 35-2) and dense (Figure 35-3) stand. In the thin stand, the maximum temperature during the day is at the ground surface and is about 1°C higher than at the top of the trees. At night, an inversion exists with the temperature at the treetops about 5°C warmer than the ground. With the dense stand (Figure 35-3), the treetops become the outer effective surface. The maximum temperature in the crown area is about 6°C warmer at noon but only about 1°C cooler at night than the ground surface. The small nocturnal difference is due, perhaps in part, to the cold air in the crown area flowing downward or because the ground loses energy by long-wave radiation exchange to the colder tree tops. At noon, because the sun's rays are able to penetrate farther into the less dense forest, and more vigorous mixing brings the heated air aloft down more quickly, a somewhat isothermal condition is established. In the denser forest, neither sunshine nor wind can penetrate to a significant extent. Thus during the day, the air in the trunk area remains cool and moist as compared with the warmth of the crown area.

M. L. Löfvenius [3322] found that not only stand density but also stand height affects nocturnal temperatures. He found that a high shelterwood raises the minimum temperature more than a low shelterwood with the same density and view factor.

During periods of strong positive net radiation (in summer), the average day temperature is usually higher in the crown area than within the forest or even at the ground. The daily mean temperatures for the Eberswalde pine forest from 14 to 15 September 1953 were:

Height (m):	0	5	10	15	20
Temperature (°C):	7.1	7.4	8.1	8.4	9.0
Relative humidity (%):	84.0	80.2	79.6	78.4	77.8

The lower relative humidity with increased elevation within the forest is a consequence of higher temperatures and increased mixing with drier air aloft. The high mean temperature in the crown area of the pine forest results from its warmth at night (Figure 35-2). In the dense fir plantation, it is mainly due to the high midday temperatures and is, therefore, more strongly marked in spite of the stand's lack of height.

In summer, the air at the ground is on the average 4°C cooler, and the daily fluctuation between day and night is 5.4°C less at the ground than it is aloft. It may be expected that in winter, when the radiation balance is negative, the opposite will be found, with the air at ground level, on the average, warmer than among the tree tops. This may be seen by a study of the figures for the dry period from 29 June to 7 July, in Table 35-1.

Table 35-1. Temperature and relative humidity in a dense fir plantation. (plus signs (+) represent highest values and asterisks (*), the lowest)

Height of measurement (m)	Temperature °C		Average vapor pressure (mb)	Relative humidity (%)		
	Daily average	Daily fluctuation		Daily average	Daily fluctuation	Average for overcast days
10.0, above the forest	22.3+	16.4	15.9	63*	58	76*
5.0, in tree-top area	21.6	19.4+	14.9*	63*	62+	80
5.0, in crown area	21.1	19.0	16.3	70	62+	84
2.5, in trunk area	20.8	18.4	16.6	69	60	86
1.5, in area of dead branches	19.6	16.5	15.3	71	60	87
0.2, at forest floor	18.3*	14.0*	16.7+	79+	45*	90

E. Schimitschek [3421] has shown the practical importance of this energy distribution in his observation of the development of bark beetles *(Ips typographus)* in a pine forest near Lunz in Austria. He found that the bark beetles only begin to swarm at 20°C. Higher bark temperatures, and more frequent and longer periods of high temperature in the crown area, result in a quicker development of this forest pest in the crown than at the ground. The primary attack on a closed fir stand starts in the crown area and spreads slowly downward.

The radiation distribution in different forest stands also has an effect upon the soil temperature beneath the stand. Figure 35-4 shows the variation of soil temperature with three types of vegetative cover at Berner's Heath in Norfolk in the United Kingdom (52°23'N). The denser the vegetation, the smaller the soil temperature fluctuations and the later the maximum and minimum soil temperatures tend to occur. These measurements illustrate how the microclimate becomes milder as the vegetative covering increases.

The forest transpires and, therefore, acts as a source of water vapor. Table 35-1 also gives the vertical profile of vapor pressure in the Munich fir plantation. The maximum in vapor pressure at the ground is due both to reduced mixing and evaporation from the surface. In the area of dead branches at 1.5 above the ground, vapor pressure is somewhat lower.

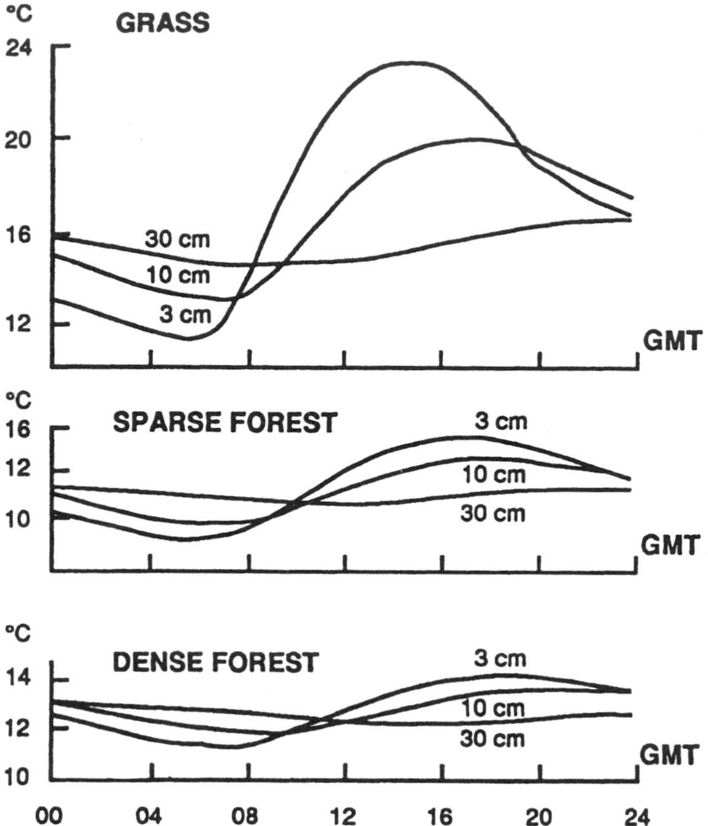

Figure 35-4. Soil temperatures at three depths under three types of vegetation at Berner's Heath on 11 August 1981. (After S. A. Oliver, et al. [3420])

A smaller second maximum in the crown area may also occur due to water vapor being given off by the transpiring leaves. This second vapor pressure maximum is lower than the ground maximum because drier air from above is mixed in.

This process can be seen from the isopleths of vapor pressure in Figure 35-5. The diagram is for the same dry period as in Figures 33-12, 34-9, and 35-3. The night minimum is caused by lower temperatures and dew deposition. The greatest amount of dew and frost are, therefore, found just beneath the crown of the canopy associated with the lowest temperatures. As the sun rises in the morning, the dew and/or frost at the treetops begin to evaporate. Convection is limited and photosynthesis is active, resulting in a maximum vapor pressure around 09:00. As the sun continues to rise and the dew evaporates, further inputs of energy raise the temperatures in the crowns and vigorous convection is initiated. This convection results in drier air being brought down from aloft. By noon, the maximum vapor pressure in the crowns is much less pronounced. In the late afternoon, a minimum of vapor pressure is found in the region of the treetops, primarily due to convective mixing, increased canopy resistance and diminished transpiration. Another minimum is located in the dead branch area, while the forest floor and the crown remain somewhat moister. The main reason for the deep penetration of dry air in the afternoon is the extreme dryness of the weather. K. Göhre and R. Lützke [3408] observed this within the forest even before the trees were in leaf. As the sun begins to set, convective mixing is reduced and evapotranspiration often increases again resulting in a second maximum at 19:00.

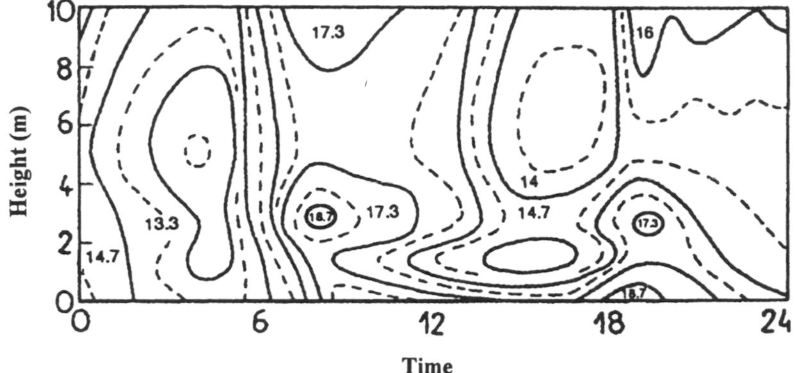

Figure 35-5. Isopleths of vapor pressure (mb) in a young fir plantation near Munich. (After A. Baumgartner [2502])

The distribution of relative humidity is much simpler. With the kind of temperatures found here, a difference of 1.33 mb in vapor pressure is equivalent to a change of about 6 percent in relative humidity. The relative humidity distribution is, therefore, determined mainly by the temperature distribution. The configuration of the isopleths in Figure 35-6 bears a broad similarity to that of Figure 35-3. The relative humidity is high during the day in the cooler trunk area and lower in the treetops. The daily average relative humidity, therefore, decreases with height.

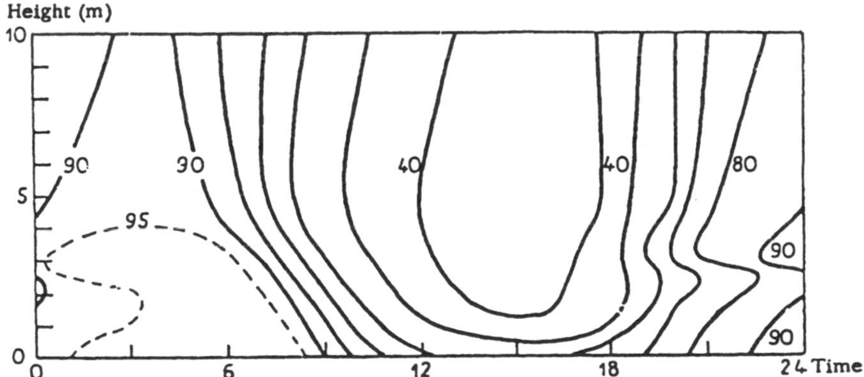

Figure 35-6. Isopleths of relative humidity (percent) in a young fir plantation near Munich. (After A. Baumgartner [2502])

The average relative humidity for a dry period in midsummer, like those for vapor pressure, gives two maxima: at the forest floor and in the crown area. This is a result of averaging for the whole 24 hour period because at night the crown area is often more humid, while the air near the ground is always relatively moist. The protective climate of the trunk area is characterized by the small daily variation in relative humidity.

In a multi-layered oak-hornbeam forest in Czechoslovakia shown in Figure 35-7, E. I. Kratochíova et al. [3416] measured the temperature (Figure 35-8), relative humidity (Figure 35-9), and CO_2 concentrations (Figures 35-10 and 35-11). In agreement with Figure 35-3 temperatures at night were warmest near the ground (Figure 35-8 at one meter) and coldest within the canopy (17 meters). During the day temperatures are highest within the canopy (17 and 22 meters) and coolest at the surface. In partial agreement with Figure 35-6 relative humidity (Figure 35-9) at night is highest within the canopy and lowest at one meter. During the day this pattern is reversed. The carbon dioxide pattern is a result of convective fluxes from the atmosphere, and photosynthetic sinks and respiratory sources within the forest. The forest functioned as a sink for CO_2 from the late morning until early afternoon when the sun's elevation rises above 30°. During this time respiratory sources in the soil produce CO_2. At night, respiration within the forest increases the CO_2 levels. The lowest concentrations of CO_2 were found in the daytime (09:00 – 16:00) and the highest at night (02:00 – 05:00) (Figures 35-10 and 35-11). During the day CO_2 levels were typically much higher at one meter (Figure 35-10) than at higher levels. This is probably due, in part, to CO_2 coming from the soil and the photosynthesis occurring within the canopy above.

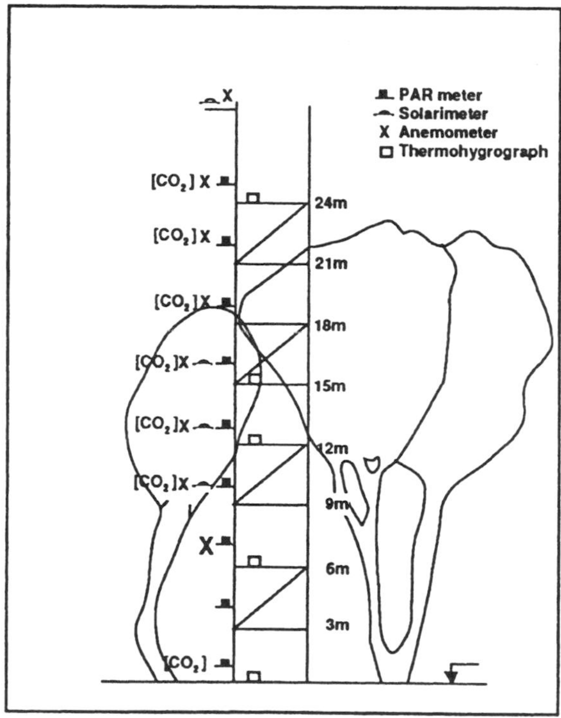

Figure 35-7. Schematic diagram of the positioning of instruments for microclimatic measurements in and above the oak-hornbeam forest (PAR is photosynthetically active radiation). (After E. T. Kratochíova [3416])

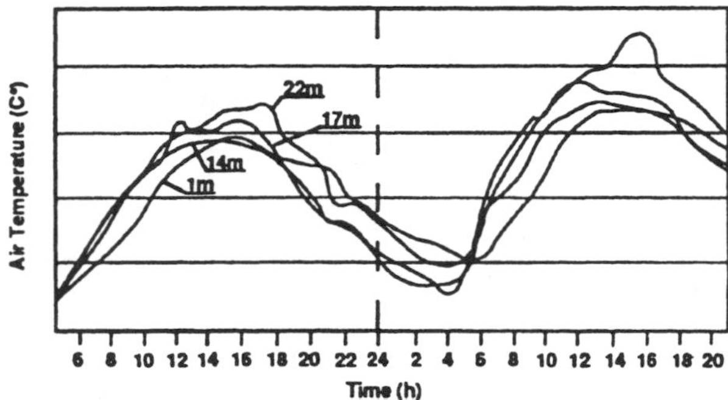

Figure 35-8. Temperature at 1, 14, 17 and 22 m in a oak-hornbeam forest. (After E. T. Kratochíová [3416])

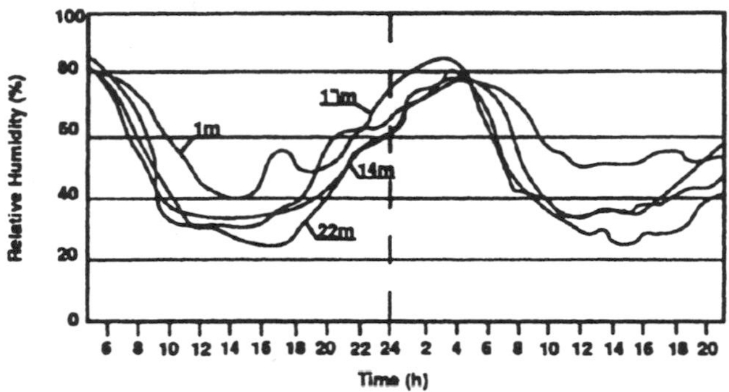

Figure 35-9. Relative humidity at 1, 14, 17, and 22 m in a oak-hornbeam forest. (After E. T. Kratochíová [3416])

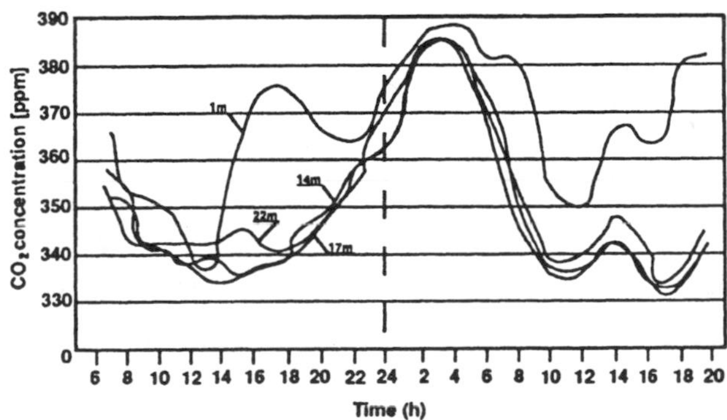

Figure 35-10. Carbon dioxide concentration at 1, 14, 17 and 22 m in an oak-hornbeam forest. (After E. T. Kratochíová [3416])

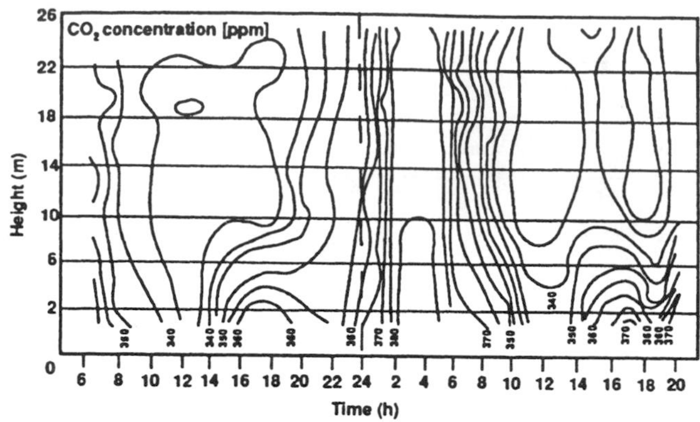

Figure 35-11. Carbon dioxide concentration in an oak-hornbeam forest. (From E. T. Kratochíová [3416])

36 Dew, Rain, and Snow in a Forest

A tree standing alone will shelter the ground below by increasing the shielding of the sky (reducing the sky view factor) and thereby decreasing the outgoing radiation at night. Below a crown covered with dew, there will often be a patch of ground free of dew. A. Mäde [3617] placed a number of dew recording instruments under a beech tree. The lower rim of the canopy of foliage was 5.2 m from the tree trunk, and the instruments were situated in a radial line outward from the tree. Measurements from May to July 1951, when the tree was in full foliage, gave the following quantities of dew, times of deposition, and duration of fall:

Table 36-1. Foliage effect on dew deposition.

Distance from periphery of tree (m)	-3.2	0	2	4	6	8
Quantity of dew on all nights with dew (mm)	0.05	1.1	1.6	2.0	2.0	2.1
Duration of dew deposition (hr)	39	330	370	388	382	368
Duration of dew (hr)	47	420	479	500	499	473

Foliage affects the quantity and duration of dew deposition in part by the amount of shielding from the cold sky at night, which controls the net loss of longwave radiation and the nocturnal temperature drop. Shading during the morning, by retarding surface heating, increases the quantity of dew deposited, the duration of deposition, and how long the dew lasts. A. Mäde [3617] thus observed (Table 36-1) that the quantity of dew deposited increased quickly from the trees' periphery out to about two meters and beyond this somewhat more slowly. Both the duration of dew deposition and how long it lasted increased to 4 m and then decreased slightly. The increase is due to a reduction of nocturnal shielding; the decrease is perhaps due in part both to increased air circulation and a reduction of daytime shading.

Under the circle of foliage, the very small amount that condensed out (0.05 mm) did so because the tree was not fully leafed out in the beginning of May.

P. Stoutjesdijk [3632] defined open shade as the area behind vegetation (or a wall) which receives diffuse radiation but is shaded from direct solar radiation. He suggests that the most obvious effect of open shade is the persistence of dew. He found daytime temperatures in areas of open shade frequently 6-8°C below ambient air temperatures. We usually expect daytime surface temperatures to be above air temperatures. In open shade, however, not only is direct solar radiation zero, but the ground is exchanging radiation with a cold sky (which P. Stoutjesdijk [3631] found to be about 25°C below the ambient air temperature). In open shade he found that the surface radiative exchange could be zero or even weakly negative.

Turning from isolated trees to closed stands, A. Baumgartner [2502] measured dew profiles in a young fir plantation and a nearby afforestation area. Figure 36-1 gives the measurements for all nights at ten levels above the ground. In the forest, maximum dew deposition occurs in the crown a little below the zone of greatest negative net radiation balance, while it occurs close to the ground in the open. When dewfall was slight, only the upper surfaces of needles and twigs were wet, while the entire surface was wet when dew was heavier. It seems likely that some of the water vapor being precipitated as dew in the crown area originates from evaporation at lower levels in the stand (Section 15).

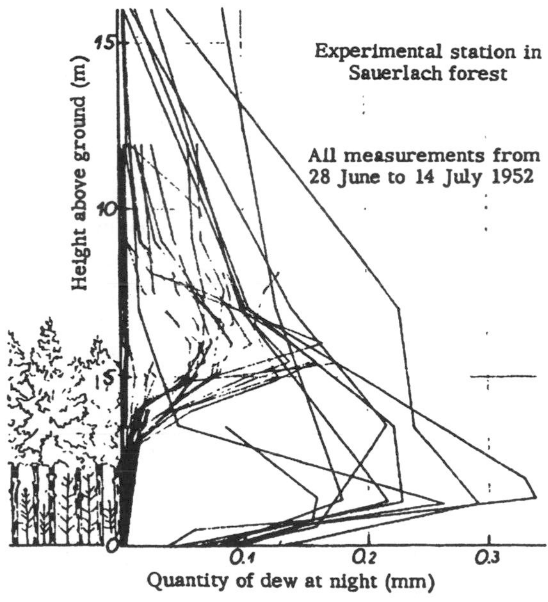

Figure 36-1. Dew profiles measured simultaneously in a young fir plantation and in open land. (After A. Baumgartner [2502])

For the observations in Figure 36-1, the average amount of dew per night was:

Height above ground (m):	0.5	1.0	1.5	2.5	4.0	5.5	8.0	10.0	12.0	16.0
Young fir (mg cm^{-2}):	0.5	0.5	0.5	1	2	12	6	4	3	1

Dew found within a stand may become significant on some nights when there has been a heavy deposit since large drops of dew that form within the crown may drip and release a veritable shower of dew. Additional discussion on the formation of dew along shelterbelts is found in Section 52.

To study the distribution of rain in the forest, it is best to start again with an isolated tree. H. F. Linskens [3614] measured the distribution of rainfall under an apple tree for a whole growing season. The tree selected was a bushy 10 yr old apple tree, standing in loose association with others in an orchard, so as to minimize the effects of the wind. When the tree had no foliage, rainfall was spread evenly under it, except for drops below the forks of branches. As the canopy of leaves developed, they held back the rain. This was observed at first as a general reduction of the rainwater arriving at the ground. It was only after the leaves had unfolded completely, and their rigidity had decreased, that their ability to deflect precipitation increased. Precipitation maxima then occurred below the outer rim of the foliage, which acted as a gutter. The periphery of the tree sometimes received as much as 160 percent of the precipitation in the open which serves as an indication of the ability of trees to concentrate throughfall into selected drip points beneath the crown or canopy. When the leaves began to fall, the distribution of precipitation became quite irregular below the tree, reaching uniformity again when the tree was bare.

H. F. Linskens [3613] also studied the varied behavior of several different types of trees. Figure 36-2 gives a few examples. The top row has sketches of the five types: (*A*) red beech, (*B*) weeping apple, (*C*) pyramid oak, (*D*) maple, and (*E*) Lebanon cedar, to show the structure of the crown. The second row gives a bird's eye view (note the different scales used). The third row shows the distribution of winter rain and the bottom row that of summer rain, both expressed as percentages of the unaffected precipitation in the open.

The conifer (*E*) shows little difference between winter and summer. The thick crown allows only 60 to 90 percent of the rain to pass through, while the drip zone at the periphery receives 10 to 20 percent more than open land. Diagrams for the beech (*A*) show the fairly uniform distribution mentioned above for the winter. The weeping type (*B*) shows where water pours down from the tips of individual twigs. The umbrella type (*D*) even shows a distinct gutter effect when without foliage, but only a few irregularly situated dripping areas when in leaf. The first two types exhibit a marked channeling of water into the drip area below the outer rim of foliage. The pyramid oak (*C*) does not fall into either of these classifications, but in winter acts like a funnel, channeling all the water down the trunk, around which the quantity of water reaches ten times the amount of precipitation. Even when bearing summer foliage, the funnel effect still gives 110 percent in the trunk area, with another high value in the gutter area. K. Haworth and G. R. McPherson [3610] found larger trees af-

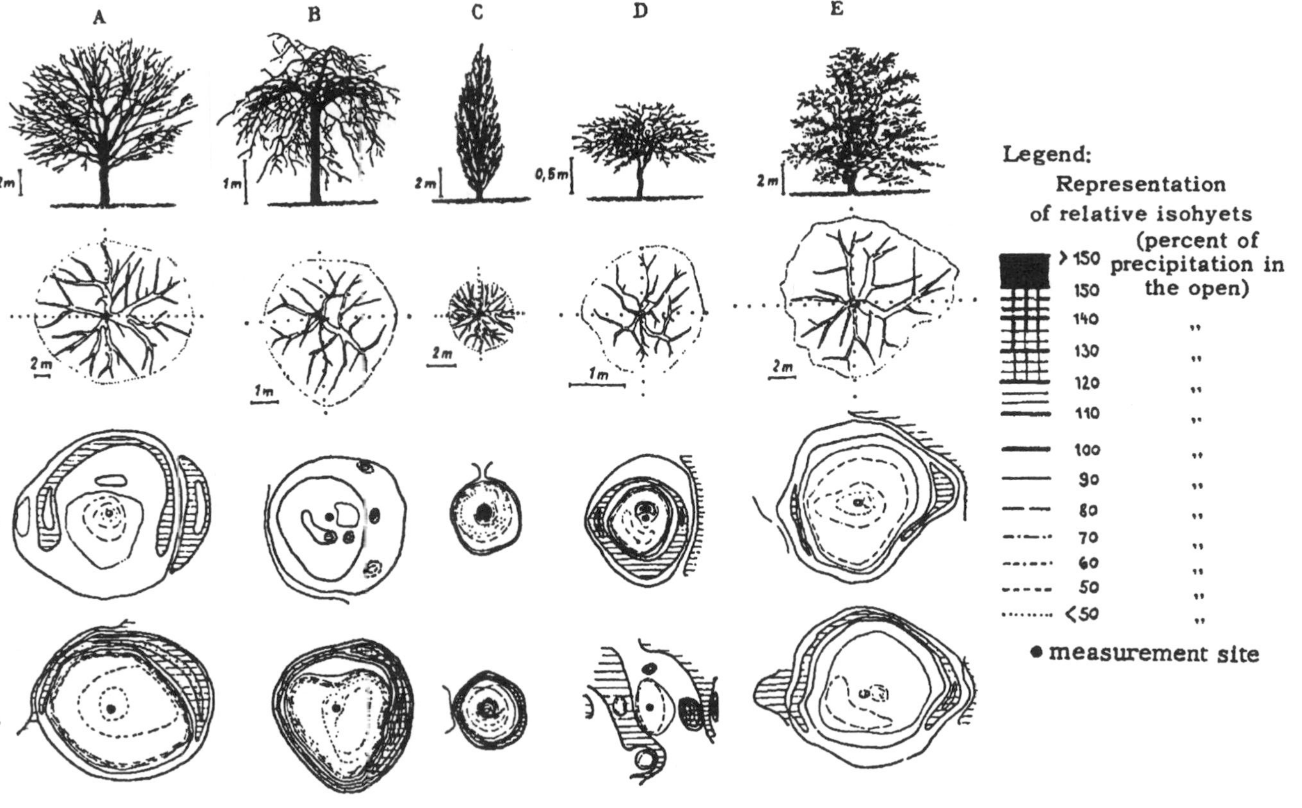

Figure 36-2. Distribution of precipitation under five different tree types in winter (row 3) and in summer (row 4). (After H.F. Linskens [3613])

fect throughfall distribution during both small and large precipitation events while small trees affected throughfall distribution primarily during smaller events.

Spatial variations in throughfall and stemflow beneath individual trees can alter the microclimate, moisture status, and nutrient conditions at ground level and influence understory plant and animal distributions. It has been suggested that mineral-enriched stemflow can affect moisture gradients, mineral gradients, soil properties, and soil accumulation radiating out from the tree bole. K. Haworth and G. R. McPherson [3610] observed that mean soil temperatures beneath individual trees in a semi-arid savanna in Arizona, USA were warmer during the cool season and cooler during the warm season than were adjacent grassland areas. They observed no significant changes in soil moisture or nutrient status, however, in their study.

From isolated trees, the discussion will now turn to a closed stand of trees. In a forest, less rain reaches the ground than in open country; and, therefore, its contribution toward the ground water supply is often less. When the crown canopy is closed and dense, the first flurry of water drops tends to stay on the leaves, needles, and twigs, especially when the rain starts with a gentle fall. J. Delfs, et al. [3603] were able to show, by making simultaneous measurements both inside and outside a stand of old firs, that this initial stage may last for several hours with light rain. The term "canopy storage capacity" is used to describe the amount of rain that can be taken up by the crown before water penetrates to the forest floor. The canopy storage capacity typically amounts to 1 to 3 mm.

As the upper surface of the forest canopy, which is exposed to free winds, is wetted by rain, evaporation begins immediately. A temporary letup in the rain often leads to noticeable evaporation losses and a reduction in canopy storage. Following the cessation of rain, the canopy storage is quickly depleted by drainage and evaporation until the canopy becomes dry. Losses from the wetting and subsequent evaporation of the canopy are called interception (see section 28 for a more complete discussion of interception). Maximum evaporation rates of intercepted water from a wetted canopy may be several times greater than maximum transpiration rates from the same dry canopy because the former is unaffected by physiological resistance to water vapor transfer as measured by canopy resistance. The large leaf surface area and strong degree of aerodynamic mixing above the canopy also contributes to the exceptionally high evaporation rates observed from wetted canopies. The aerodynamic resistance, which acts as the primary control over evaporation is typically an order of magnitude smaller than the canopy resistance for forest canopies.

If it continues to rain, water starts to drip through the crown. If there are gaps in the crown, and the rain is heavy, this effect can be observed from the start. The rain which drips through, called throughfall, consists of raindrops that find a clear path through the treetops, and, to a greater extent, large drops that drip from the tips of needles, leaves, or twigs. The size distribution of these leaf drops is normally different from that of the rain in the open. Figure 36-3 demonstrates this with values from Bedgebury in Kent, England, measured by J. D. Ovington [3620] in growths of various types of saplings. Three 23 yr old stands are compared with open country F (the average of three measuring points). The stands are *Q (Quercus rubra)*, *A (Abies grandis)*, and *L (Larix eurolepsis)*.

The ordinate in Figure 36-3 is the weight of raindrops; the diameters of the drops are also shown to make the curves easier to visualize. These summation curves, which all begin at 0 percent for drops of diameter 0, show that in every case more than half the weight of rain comes from tiny droplets with diameters less than 1 mm. In the open-air *(F)*, drops of more than 2 mm diameter are rare; in this investigation, no raindrops were found in the open measuring more than 2.5 mm. The situation is different within the stand. It is not surprising that the forms of the three curves (*Q, A, L*) are not smooth. In the stand of northern red oak (*Q*), some drop diameters are in excess of 6 mm. Drops of this size in the open cannot last too long, because they acquire such a high fall velocity that they break up into smaller drops. However, drops falling from the canopy of leaves have neither space nor time to reach break-up speed. Figure 36-3 shows that drops of over 5 mm diameter account for about 5 percent of the total weight of throughfall. Since the weight of a drop increases in proportion to the cube of the radius, their number must also be very small. Because drops escape more readily from needle leaves (*A, L*) they are not as large as those from broad leaves.

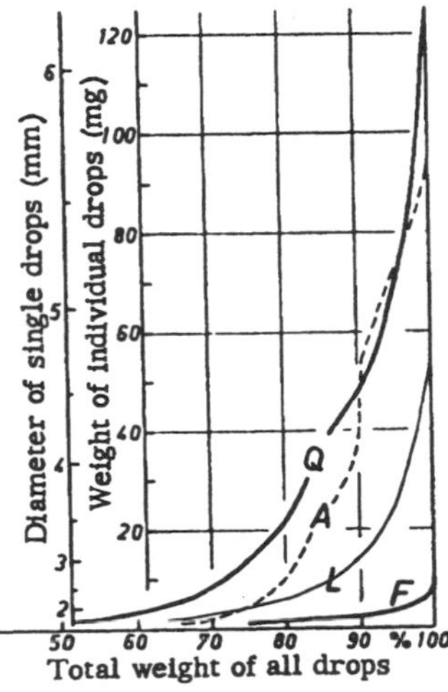

Figure 36-3. Drop size distribution of throughfall in oak *(Q)*, silver fir *(A)*, and larch *(L)* forests compared with the open *(F)*. (After J. D. Ovington [3620])

M. Vis [3633] found that the raindrop size distribution in an open field in Colombia followed a unimodal distribution with a maximum number of raindrops in the 1.5-2.0 mm diameter range, and with 50 percent of the rainfall volume occurring from raindrops less than 2.0 mm in diameter. Throughfall beneath four tropical forest ecosystems, however, all exhibited a bimodal drop size distribution, with 50 percent of the rainfall volume falling from raindrops larger than 4.0 mm, and only 15-20 percent of the throughfall volume occurring from drops less than 2.0 mm. The bimodal throughfall drop size distribution was attributable to the combined influences of the canopy splitting raindrops into smaller sizes and combining drops into larger units through leaf drip.

Vegetation canopies also change the erosional power of rain, which can have important consequences for soils susceptible to erosion. Canopies modify the total storm kinetic energy reaching the surface by reducing the total amount of rainfall via interception losses, changing the drop size distribution of rainfall, and modifying the fall velocity of raindrops. The extent of this change varies with percent of ground cover, forest density, and the shape, size, and orientation of leaves. M. Vis [3633] reported that even accounting for interception losses, the total storm kinetic energy below tropical forest canopies was 4-30 percent greater than in the open. Since terminal velocities for the largest drops can be reached in 8 m, the presence of multiple canopy layers can also be important. J. Brandt [3601] found that the total storm kinetic energy beneath a single canopy layer tropical rainforest in Brazil was 57 percent greater than in the open. Beneath an adjacent multiple canopy layer tropical rainforest, the values were between 9 and 90 percent greater, depending upon the bottom canopy layer mean height. Land cleared of undergrowth and litter beneath a high canopy, therefore, is exposed to increased soil erosion potential.

The spatial distribution of throughfall is always very irregular, particularly in areas of heterogeneous canopy conditions and incomplete ground cover. The highest spatial variability of throughflow is found in tropical forests due to their great species diversity and more varied canopy architecture. Less variability is found in more northern forests with more uniform canopy structure and less varied species composition, or in middle-latitude forest plantations. The accumulation of rain in the drip area of the tree, noticeable in the case of isolated trees, is still conspicuous in a closed stand. E. Höppe [3612] was the first to make a systematic study of rainfall distribution within a forest. He found that near the trunk area, the rain was 55 percent, and at the borders of the crowns 76 percent, of that falling in the open. This uneven spread of throughfall makes it very difficult to measure throughfall distribution in forests and accounts for the wide scatter of values reported for interception as a percent of gross rainfall. After the rain has stopped, dripping continues inside the stand, for up to 2 hours. As an extreme example, C. R. Lloyd and A. de O. Marques F° [3615] found interception totals that ranged between 0 and 410 percent of gross precipitation beneath Amazonian rainforest near Manaus, Brazil (2°57'S). Due to preferential drainage patterns, throughfall totals actually exceeded gross precipitation in 29 percent of the 505 samples, although the average throughfall amount was 91 percent of gross rainfall. For this reason, adequate spatial sampling is critical, and average interception percentages should be reported with standard errors.

Readings of a single rain gauge will be subject to chance. Random influences may be reduced if a position is selected, following the method of J. Grunow [3609], that has been shown by previous investigations of dripping to have average characteristics. In order to obtain a good sample, E. Höppe [3612] placed 20 rain gauges along two lines perpendicular to each other; C. L. Godske and H. S. Paulsen [3608] selected 9 typical points. Another method is that adopted by J. Delfs [3602], who introduced the use of troughs 20 cm long and 5 cm wide, equivalent to 50 normal rain gauges. C. R. Lloyd and A. de O. Marques F° [3615] recommended taking random samples along a 100 by 4 m grid beneath a heterogeneous canopy.

Besides interception in the crown area and throughfall, there is still the stemflow down the trunks to consider. Rain intercepted by twigs and branches is channeled toward the trunk where it flows to the forest floor. Like dripping rain, this begins only after nearly complete wetting of the crown canopy. In coniferous forests, the amount of stemflow is typically less than in deciduous forests. Measurements, however, are also subject to the same high degree of temporal and spatial variability observed for throughfall. Since stemflow normally accounts for only 1-2 percent of gross precipitation, the effects of this variability on the water balance are less important.

The way in which the forest behaves toward falling rain makes it clear that the rain intensity, duration, and variation with time are of vital importance to interception. J. D. Ovington [3620] summed up the results of his observations by saying that interception was "between 6 and 93 percent." He stated that practically any percentage of interception was possible, depending on the character of the rain and the stand. For a single stand, interception will be great in proportion to rainfall when the rain is light, of short duration, or frequently interrupted, and will be smaller for heavy showers or persistent rains. This is shown by a comparison of two readings, both from the same stand of firs in the Harz Mountains. For a rain event of 50 hours duration yielding 70.5 mm, 75 percent passed through the crown area, while for a thundershower yielding 74.6 mm and lasting 3.5 hr, 98 percent got through. In a dry canopy H. Moličová and P. Hubert [3618] found that throughfall density was irregular but became more uniform as the canopy became wet.

A distinction must be made between the types of trees and ages of the stand in order to appreciate the results obtained up to date. In old fir stands, shown in the lower half of Figure 36-4, the relation between interception and throughfall is shown as a function of the amount of precipitation. The heavy curve *M* is the result of measurements made by E. Höppe in a 60 yr old fir stand in Austria. The small crosses, which practically coincide with this line, are the results of the combined work of J. Delfs, et al. [3603], which were conducted in the Harz area between 1948 and 1953. These figures were measured in an 80 yr old, 22 m high fir stand. The dotted curve *H* is the result of readings made on the Hohenpeissenberg by J. Grunow [3609] in an old stand on the steep southern slope, near the peak which is almost 1000 m high; curve *C* is for a NNE slope of 36 percent at 1020 m above sea level.

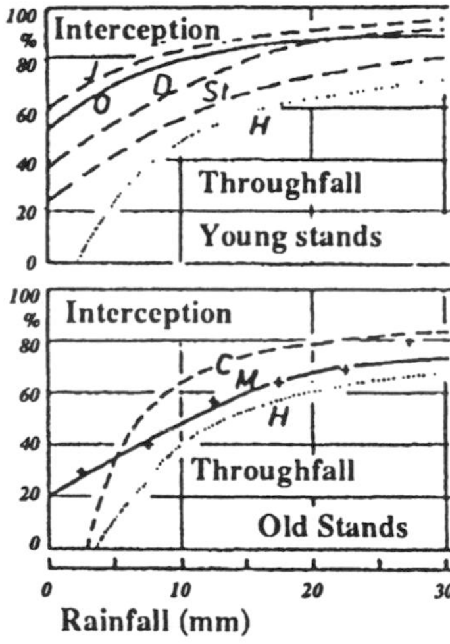

Figure 36-4. Interception and throughfall in fir stands as a function of precipitation amount.

In younger stands more rain falls through to the forest floor, as shown in the upper half of Figure 36-4. The curves are for measurements made in the Upper Harz Mountains in 60 yr old saplings 15 m high (*St*), a 30 yr old 6 m thicket (*D*), and a 15 yr old, not yet closed-up fir plantation (*J*). The average interception for the 4 yr period gave values of *St* = 29 percent, *D* = 24 percent, *J* = 12 percent, in comparison with 37 percent for the old stand. The stemflow was between 0.5 and 3 percent, compared with 0.8 percent in the old stand.

The line *O* is the result of C. L. Godske and H. S. Paulsen's research [3608] in a 35 to 50 yr old plantation of fir saplings at Os near Bergen. The value for interception is somewhat lower, and is due to the thinner growth of Norwegian forests (60° N) compared with German stands of the same age (52°N). The dotted Hohenpeissenberg curve (*H*) is for a 40 to 60 yr old stand of saplings on the north slope of the mountain.

Because of the strong dependence of interception on the temporal distribution of rain, meteorological conditions, the species of trees, and the stand age, it is extremely difficult to estimate an average value for the interception of a forest. Generally, however, interception increases with increased biomass. In the Harz investigation (J. Delfs [3602]), the absolute value of rain (mm) retained in the crown of the fir was determined for a number of individual cases. This increased with the amount of rainfall and the age of the stand as shown in Table 36-2. Thus, with increased rainfall total interception increases while the percentage of rain intercepted by the biomass decreases.

Table 36-2. Interception (mm) as a function of rainfall and vegetation (* missing).

Amount of rain (mm):	0	5	10	15	20	25
Thicket, 30 yr old (mm):	3	6	7	*	*	9
Saplings, 60 yr old (mm):	3	6	8	8	9	12
Old timber, 80 yr old (mm):	4	7	10	10	11	18

Current research on the interception of water by forest canopies has focused upon the development of physical models of the water balance dynamics of wetted canopies. These models compute a running balance of the rainfall, throughfall, evaporation, canopy water storage, and stemflow of a wetted canopy over time, based upon relevant physiological parameters. These include albedo, zero-plane displacement, roughness length, canopy storage capacity, trunk storage capacity, percent of rain passing directly through canopy openings, percent of rain diverted to the trunk, and the drainage characteristics of the canopy. An introduction to early works in this field can be obtained from A. J. Rutter, et al. [3627, 3629] and A. J. Rutter and A. J. Morton [3628].

Table 36-3 provides a summary of interception variables determined using the Rutter model from selected temperate and tropical forests as compiled by W. J. Shuttleworth [3338]. The canopy storage capacity S is the depth of water held in store by the forest canopy, and the free throughfall fraction p is the fraction of rain falling straight through the canopy to the forest floor. Considerable variation is observed both within and between forest types. The average value of S for all temperate coniferous forests was 1.4 ± 0.5 mm. The more effective structure of needle leaves in storing moisture, as compared to broad leaves, is clearly evident. Values of S for deciduous forests in a leafless condition were less than half of the fully-leafed values. Tropical forest S values are more variable and less certain due to the more limited number of studies. The wide range in the openness of temperate coniferous forest canopies results a wide range of values for the free throughfall fraction p. Leafless deciduous temperate forests allow considerable rain to penetrate to the forest floor, while undisturbed tropical forests allow little or no rain to directly reach the ground surface.

In contrast to the needles of conifers, deciduous leaves collect water. Drops from leaf drip can be very large, as we have already seen from the drop-distribution curve *Q* for oak in Figure 36-3. Their leaves are structured so as to channel the water collected via twigs and branches, toward the trunk of the tree. The stemflow is, therefore, of greater importance in deciduous than in coniferous forests.

Table 36-3. Canopy storage capacity and free throughfall fraction for selected temperate and tropical forests. (After W. J. Shuttleworth [3338])

Forest type	Canopy storage capacity S (mm)	Free throughfall fraction p
Temperate coniferous		
Corsican pine	1.05	0.25
Scots pine	1.02	0.13
Sitka spruce	1.73	0.05
Douglas-fir	1.20	0.09
Temperate deciduous		
Oak (summer)	0.80	0.30
Oak (winter)	0.30	0.80
Tropical		
mixed species	1.10	0.00
mixed species	0.74	0.08

The increase of stemflow in deciduous forests can be deduced from comparative measurements made from 1952 to 1958 by F. E. Eidmann [3605] in the forest of Hilchenbach (Westphalia). The fir trees were 70 yr old and 25 to 28 m tall, while the beeches were 95 yr old and 25 to 30 m in height. Both stands were growing adjacent to each other on a southern slope inclined from 25° to 28°, at 600 m. Table 36-4 gives the distribution of precipitation falling on the stands during the 6 years as a percentage of the annual average precipitation (1216 mm). The rain not intercepted by the crown of the fir trees drops through. In the beech stand, in contrast, half the water caught by the leaves in summer is channeled to the trunk. In the deciduous forest in winter, more water flows down the trunk than is retained in the crown. Therefore, interception is less, and the total water reaching the ground is more than in the evergreen forest. The proportion of water dropping through is about the same in both types, and is greater in winter than in summer for both.

Table 36-4. Distribution of precipitation (percent of annual average) falling on two stands of trees.

Type of tree	Season	Interception	Stemflow	Throughfall
Fir	Summer	32.4	0.7	66.9
	Winter	26.0	0.7	73.3
Beech	Summer	16.4	16.6	67.0
	Winter	10.4	16.6	73.0

Interception is proportionally greater when the amount of precipitation is small because of the quantity required to wet the crown, and it decreases in percentage as the amount of rain increases. It follows that stemflow begins only when wetting is nearly complete, and it will increase with the amount and duration of rainfall. E. Höppe's [3612] measurements at Mariabrunn in Austria (Table 36-5) in 1894 for a stand that contained 80 percent beech and 20 percent silver maple and in 1895 for a stand of pure beech, show increased stemflow with increased precipitation. F. E. Eidmann's [3605] mean values are in complete agreement with these results. K. Haworth and G. R. McPherson [3610] found that stemflow increased exponentially with increasing size of the precipitation event and varying from 1-16 percent of gross precipitation.

Table 36-5. Stemflow as a percent of total precipitation.

Amount of rain (mm)	0-4.9	5-9.9	10-14.9	15-20
1894 (percent)	8.4	14.5	15.7	20.2
1895 (percent)	10.1	18.4	18.0	22.5

J. D. Ovington [3620] surveyed 13 different areas of young growth, each 1000 m^2 in area, with trees 22 to 23 yr old. Rainfall was collected in ten rain gages and the stemflow in three containers attached to the trunks. Figure 36-5 shows the interception for seven different types of trees as a function of rainfall depth. Table 36-6 gives details of the trees shown by initials in Figure 36-5. The figures for throughfall are the lowest and highest yearly averages for 1949-1951. The amount of stemflow is insignificant in all cases. The figures in the column headed "Drops" gives the percent weight of precipitation falling in larger drops than were observed simultaneously in the open. These figures serve to amplify Figure 36-3, and at the same time, show the great differences between different types of trees. Xerophytic Mulga (*Acacia aneura* F. Muell.) branches point upward and frequently have a high stemflow. In arid central Australia it was found to be equal to 40 percent of the total rainfall (R. O. Slatyer [3631]), but in more humid Queensland stemflow was only 18 of total rainfall (A. J. Pressland [3621]). A. J. Pressland [3622] found that stemflow was instrumental in storing water at depth in the soil, being particularly noticeable with medium-size (75 mm) rainfall events. Large rainfall events (~160 mm) tended to mask the effect. R. O. Slatyer [3631] suggested that because stemflow penetrates deeper in the soil and the shading of the tree canopy, less water is lost through evaporation and more is available for transpiration.

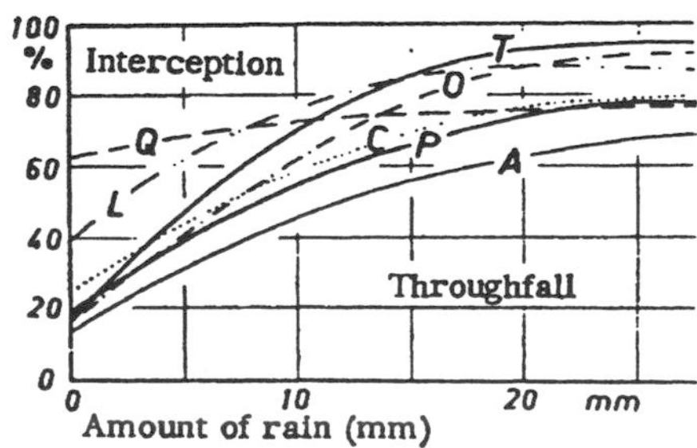

Figure 36-5. Interception in various types of young stands in England. (After J. D. Ovington [3620])

Table 36-6. Interception of various types of trees shown in Figure 36-5.

Description of saplings			Quantities measured			
Type of tree (all 22-23 yr old)	Height of stand (m)	Height of crown canopy (m)	Through-fall (%)	Stem-flow (%)	Drops (%)	Snowfall (mm)
Deciduous:						
Q Quercus rubra	7.3	2.4	68-71	0.3	68	13
Coniferous:						
L Larix eurolepsis	14.6	3.0	70-90	0.1	45	7
T Thuja plicata	7.6	2.1	63-65	0.1	45	7
O Picea omorica	10.1	4.0	59-61	0.2	66	0
C Chamaecyparis lawsonia	8.8	1.8	56-57	0.1	48	--
P Pinus nigra	8.5	2.1	52-53	0.2	61	1
A Abies grandis	14.3	4.9	49	0.1	64	--

In the subtropical forests of Brazil (19° to 23°S, 41° to 45°W, 600 to 900 m), F. Freise [3607] has observed rainfall over many years. Of the total precipitation, 20 percent evaporates in the crown space, 28 percent reaches the ground down the trunk, and 34 percent drops through, making a total of 82 percent. The remaining 18 percent is absorbed by the bark and hollow stems, or is lost by further evaporation. Observations by I. R. Calder, et al. [3604] from a secondary lowland tropical rainforest in West Java (6°35'S, 80 m) revealed that interception losses consumed 21 percent of the gross precipitation. Evaporation of intercepted water from wetted canopies accounted for 40 percent of the total evapotranspiration, resulting in evaporation losses that were two-thirds as large as transpiration losses. Interception losses among the four tropical rainforest ecosystems examined by M. Vis [3633] varied from 24.6 percent in a lowland forest (0-1000 m) to 11.4 percent at a high altitude forest (3750-4700 m). Altitude was considered the primary controlling factor, although structural and compositional differences among the four ecosystems accounted for some of the variation. Increasing altitude led to lower air temperatures and, therefore, reduced evaporation rates, while cloud moisture interception (fog drip) at higher elevations resulted in increased throughfall below the forest canopy.

After rain lasting at least an hour ceases (not earlier and not after snow), forest smoke can be observed on mountain slopes over close stands and also in the plain. This peculiar effect has been described and explained by F. Rossmann [3625]. Small clouds or ragged wisps of fog cling about the treetops just as the rain stops and remain for nearly an hour. In the rain soaked crown area proper, there is no fog, even though the air mass in it is close to saturation. If the warmer air in the crown area is then mixed with the cooler air aloft, the small degree of cooling that results is sufficient to produce a slight condensation, lasting only a short time and not forming a coherent cloud. As soon as the air near saturation in the crown area has been used up in the mixing process, the phenomenon ceases.

Snowfall in a forest often lies below the crown canopy. Protected to a large extent from evaporation, it provides a special form of winter water storage. In the spring, it melts more slowly than snow in the open, thus retaining a significant quantity of water. Being more lightly borne by the wind than rain, snow responds more to variations in the configuration and density of the forest crown.

The type of snowfall strongly affects the amount reaching the forest floor. Large flakes of viscous wet snow cling easily to the crown. In winter, measurements in the Harz Mountains by J. Delfs, et al. [3603], show that 80 percent of a fresh snowfall, 12 cm deep, with a 13.1 mm water content, was retained in the crown of a stand of old firs and saplings. A fir thicket trapped an entire fall of 10 cm of fresh snow in its crown. Catastrophes due to breakage of branches under the snow load usually result from wet snow, since when temperatures are low, powder snow penetrates easily to the forest floor.

In spite of all these facts, measurements show that, on the average, interception of winter precipitation is less than that of summer rain. The cause of this almost certainly lies in the reduction of evaporation at low temperatures, and the tendency for snow to accumulate and then crash down in clumps from twigs and branches. From 120 years of observations at Prussian forest stations with instruments in paired exposures, J. Schubert [3630] obtained a ratio of forest to open land snow depth of 100:90; while for rain the ratio was 100:73. H. Hesselman [3611] found almost the same snow depth in pine stands as in a felled area in Sweden.

P. B. Rowe and T. M. Hendrix [3626] subdivided the figures for the 6 yr series of measurements in California in a 70 yr old pine stand according to whether rain or snow fell. They found that 84 percent of the rain and 87 percent of the snow fell through the crown. In the Upper Harz, it was also found that a higher percentage of precipitation reached the forest floor in winter than in summer. The mean value for a 4 yr period in an old fir stand was 67 percent in winter against 60 percent in summer; in the fir saplings it was 73 percent against 69 percent. Since these figures are for evergreen forests, they would be even greater for deciduous forests in winter. Table 36-6 shows that the average snow depth during an English winter with little snow was a greater depth in the oak than in the coniferous stands. From estimates of the snow load on trees during snow breakage in Upper Silesia, W. Rosenfeld [3624] deduced that in stands of silver fir and spruce, from 25 percent to 55 percent of the snow fell through to the ground, while in a beech stand, it was 60 percent to 90 percent. G. R. Eitingen [3606] observed that in a Russian birch forest, 91 percent reached the ground in summer and 100 percent in winter, with an annual precipitation of 545 mm.

With snow, just as with rain, the ground near the periphery of the tree is more deeply covered with snow sliding from the crown. Large piles of snow develop around individual tree trunks. G. Priehäusser [3623] described their development and influence on stands in the Bavarian Forests. The effect of water filtered out of driving mists by the forest will be dealt with in Section 37.

37 Microclimate at the Stand Edges

In planting seedlings, use is often made of the special microclimatic conditions at the edge of old standing timber. Depending on the orientation of the forest edge, the ground and the air between the old trees may be sunny or shady, windy or still, warm or cold, moist or dry. These qualities may be used to promote growth in natural regeneration or artificial planting. Knowledge of edge climate is, therefore, of great importance in forestry.

According to R. Geiger [3709], the edge climate develops from two sources. First, it is a transitional climate. Influences of the nearby forest affect the outer edge (the strip of open land around the stand). For example, cool forest air from the trunk area may drift over the open land on a summer day. In the inner edge, influences from the outside are very pronounced for trees close to the open area. Second, the edge of a forest forms a step in the topography which influences the variations in solar radiation, wind, and rain at the surface. This second influence is usually more pronounced than the first in modifying microclimate.

The distribution of sunlight and shade is the first factor affecting the edge climate. The greater the proportion of diffuse solar radiation to global solar radiation, as in cloudy weather or winter, the smaller the difference in the solar radiation regime of the forest edges.

Figure 37-1 gives the number of hours of sunshine received by different edges of a stand during the year. The curves, based on a long series of measurements in Karlsruhe from 1895 to 1934 by J. V. Kienle [4014], show the influence of weather. At the southern edge, the summer maximum is split into two by the "European monsoon" in the month of June. In general, however, as far as radiation is concerned (but not the other elements), the symmetry of east and west edges and of spring and autumn is maintained. During winter, edges with a northern exposure are completely devoid of direct sunlight.

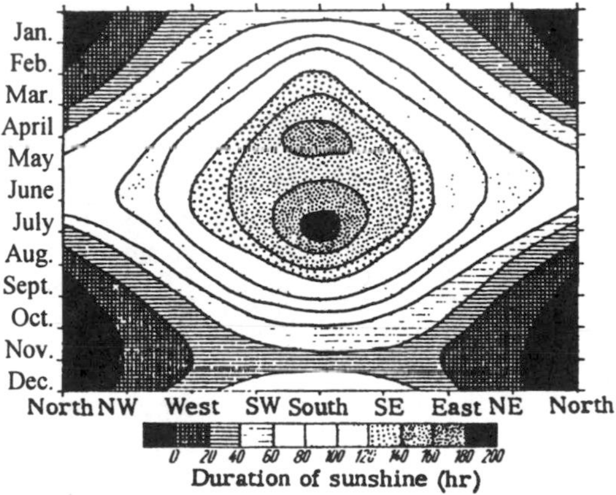

Figure 37-1. Monthly duration of direct-beam solar radiation at stand edges facing in all directions throughout the year. (After J. V. Kienle [4014])

The intensity of the incoming solar radiation is of greater significance than the duration of sunlight. Figure 37-2 gives daily totals of direct-beam solar radiation throughout the year for clear and normal sky conditions. The top half shows the daily totals of direct-beam solar radiation on clear days for the eight cardinal points that the forest edge may face. The figures are based on the calculations of W. Kaempfert and A. Morgen [4013] for clear days at 50°N with average atmospheric turbidity. Diffuse radiation is not included here; its importance will be discussed in Section 40.

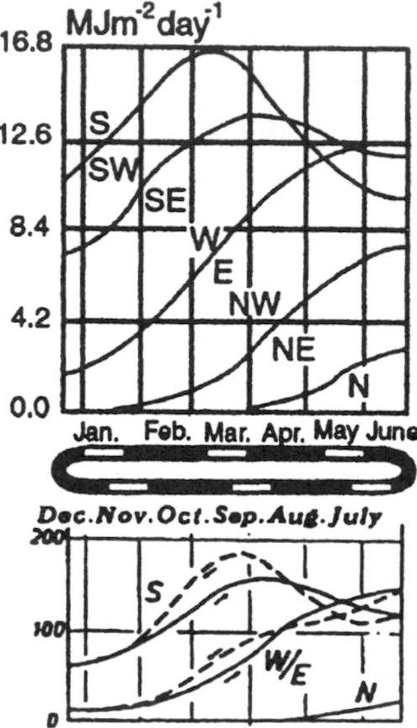

Figure 37-2. Daily totals of direct-beam solar radiation received at stand edges facing in various directions, on a sunny day (above) and on a normal day (below) throughout the year. In the lower diagram, the solid line is for spring; dashed line is for fall.

The southern edges receive the greatest amounts of energy about the time of the equinoxes. In winter, the shorter day reduces the amount of energy received, while in summer, when the sun reaches its maximum noon altitude, the angle of incidence of the sun's rays is very steep. This reduces the quantity of energy received. In summer, the forest edges that face southeast and southwest are more favorably placed. While at the time of the summer solstice, the edges facing directly east and west receive the most radiation.

Cloudiness causes a substantial reduction in the average amount of solar energy received at the forest edge. J. Schubert [3731] has computed average values on the basis of the Potsdam radiation records for the period 1907 to 1923, for S, E, W, and N edges. The daily variation in cloud cover results in the east facing edges receiving more sunlight due to less convection and fewer clouds in the morning. This is in contrast with the western edges, which receive less radiation because of greater afternoon cloudiness. This difference is small and is not indicated in the top part of Figure 37-2 which is for clear sky conditions only. As a result of seasonal cloudiness, the symmetry of the year about the sun's maximum altitude is unbalanced. The lower part of Figure 37-2 shows this difference by a solid curve for spring and a broken line for fall.

F. Lauscher and W. Schwabl [3716] measured the transition in radiation from the open into the forest (edge of the stand). In Figure 37-3, brightness is expressed as a percentage of illumination in the open. The firs (Figure 37-3 top) form a very dense covering, and it is, therefore, always very dark in the inner border. The contrast is greater in full sunlight (curves 2 and 3) than in overcast conditions (1). The curves for a deciduous ash stand (Figure 37-3 bottom) are: winter time bare of foliage (1) and (3), and summertime in leaf (2) and (4). Curve (1), for cloudy skies, shows a fairly uniform distribution of light, while curve (3) shows the effect of the shadows cast by branches. In the summer, curves (2) and (4) show reduced illumination for clear and cloudy skies, respectively, due to the effect of foliage. The difference in illumination characteristics between the outside and inside of the ash stand is greater in summer than in winter, but not as great as the contrast produced by firs.

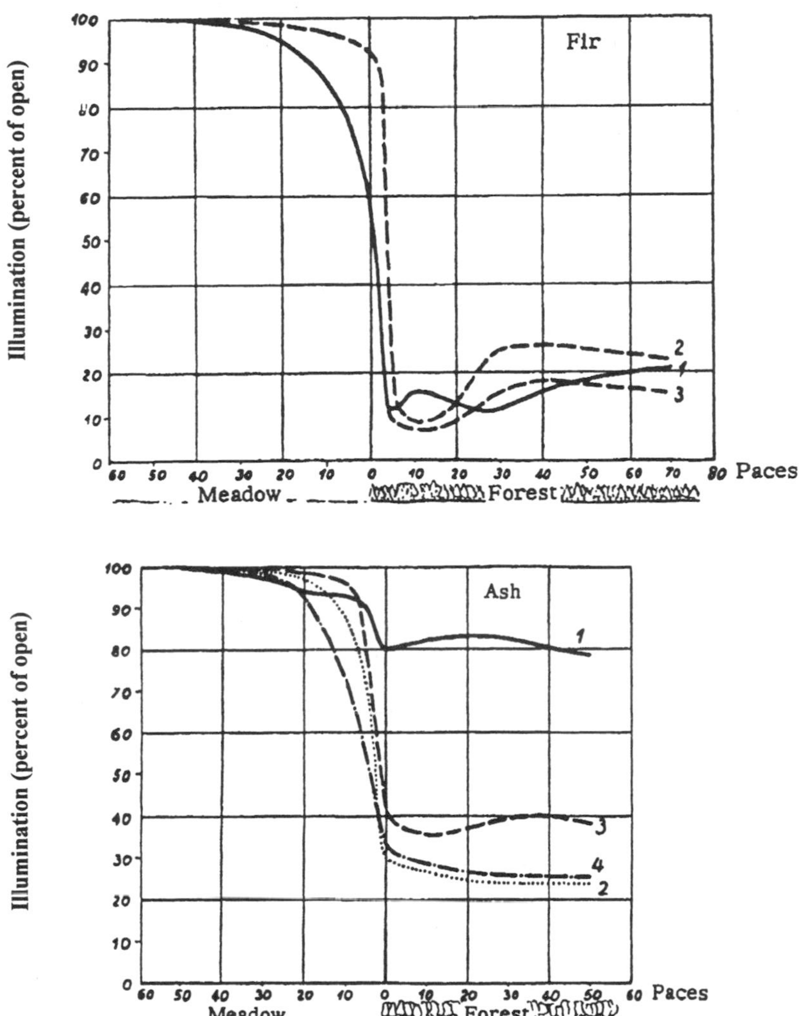

Figure 37-3. Transition of illumination at the edge of the forest. (After measurements of illumination on a horizontal surface by F. Lauscher and W. Schwabl [3716])

What one edge of the forest gains in sunshine, the other edges lose by being overshadowed. The growth of young trees at the outer border depends in part on the length of time they spend in the shadow of the nearby stand. Figure 37-4, after R. Geiger [3709], illustrates the width of shadow at 48°N (Munich) in June(a) and December (b). The abscissa is the direction in which the stand is facing, and the ordinate the time of day. The horizontal dashed lines at the upper and lower edge of each figure give the times of sunrise and sunset. The isopleths are for equal widths of shadow, expressed as a fraction of the height of the stand. While excessive shade often retards growth, it can be beneficial to plants during periods of excessive heat or drought. The heavy zero lines join the times of day (ordinates) and the stand edges (abscissas) for which the sun's rays are tangential. The broad, clear band in the center, running from bottom left to top right, indicates that the edge is receiving direct-beam solar radiation. There are also clear areas in the top left and bottom right corners of Figure 37-4a. In midsummer (Figure 37-4a), the sun rises so far in the NE that direct-beam radiation is received in the early morning on the stand edges facing as far around as NNW. In the same way, stands that face NNE are touched again by the setting sun in the early evening. To determine the time of day at which a 15 m broad clearing will be completely in the shadow of a 20 m high stand facing WSW, we find from Figure 37-4a for a shadow width of 15:20 = 0.75 that on 21 June the clearing is in shadow from sunrise to 09:00. From this time onward,

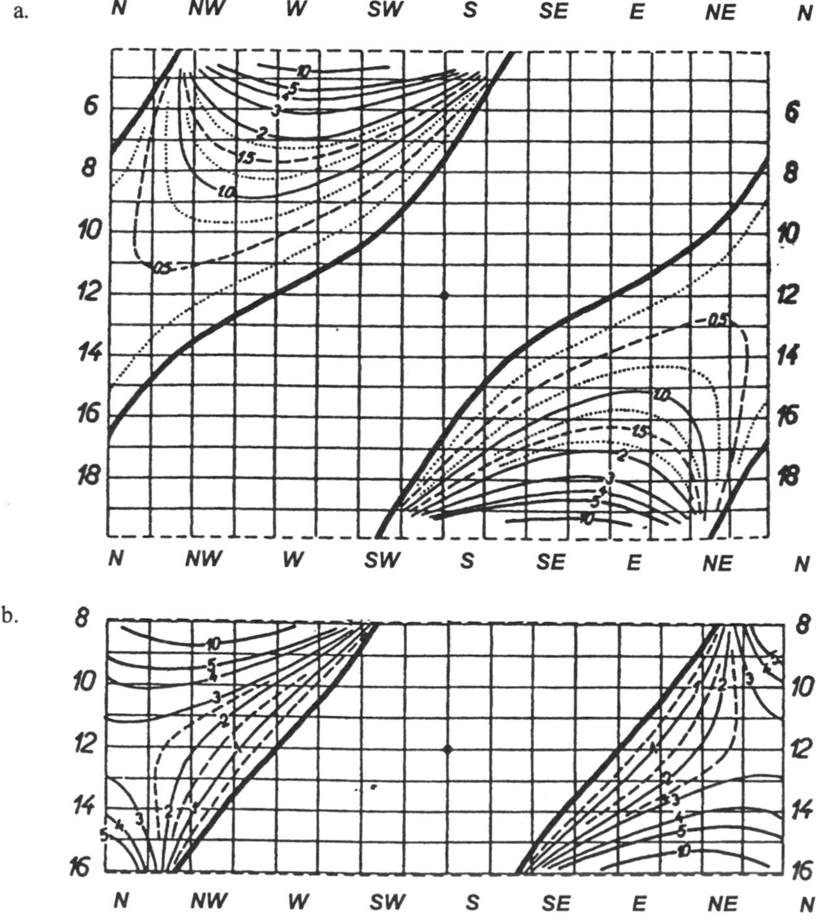

Figure 37-4. Lines of equal shadow width at stand edges during (a) summer solstice; (b) winter solstice at 48°N. (After R. Geiger [3709])

the shadow narrows until shortly after 11:00 when this side of the clearing receives direct-beam radiation. This side continues to receive direct-beam solar radiation until sunset.

J. Schubert [3731] showed that at the equinoxes for a north-facing edge, the width of the shadow is independent of the time of day. The long slanting shadows at sunrise and sunset extend the same distance from the edge of the stand as the shorter noon shadows.

At night, the negative radiation balance of the outer edge is reduced in comparison to the open area since part of the night sky is shielded by the nearby stand. This reduced sky view factor and increased forest view factor leads to a decrease in the net longwave radiation loss, since longwave radiation emitted by the forest edge is much greater than that received from the sky (because the forest edge is at the same or higher temperature than the ground in the outer border). In the following table:

Distance D	0	$0.2h$	$0.4h$	$0.6h$	$0.8h$	h	$2h$	$3h$
A (Lauscher) forest	50	60	70	78	84	88	95	98
B (Bolz) tree	50	79	83	87	89	91	96	98

line A gives the net longwave radiation loss as a percentage of the net longwave radiation loss from the open country, using (instead of the angle γ) the distance D from the edge of the forest as a multiple of the height h of the stand. Since the part of the sky near the zenith makes the least contribution toward counterradiation (Section 5), the protective influence of the forest decreases rapidly with distance from it. H. M. Bolz [3807] has measured the change in the radiation balance with distance from a single tree 10 m high. His figures are shown in line B of the table. Their difference from line A is primarily because a single tree has a smaller shielding influence than the continuous forest edge.

The sheltering effect of the edge of a stand may be observed directly after nights with frost when there is a strip free of frost along the edge of the forest. It can also be seen as the fresh green color of young spruce plants compared with the brown and limp shoots of those further away from the edge of the stand. The protective influence often affects the nature of growth in forest clearings exposed to frost damage; outward from the edge of the stand, the influence of open land gradually becomes more important. Often a group of young plants may be seen growing under a single tree in the middle of growths destroyed by frost.

On summer days in 1943 and 1944, W. Lüdi and H. Zoller [3719], while studying the influence of forests on airfields, observed the temperature variations reported in Table 37-1. The hours selected were those nearest the maximum and minimum daily values; they are considerably later in the soil because of the time lag in penetration. The southern edge is substantially warmer, both by day and by night, than in the forest or the open country. At a depth of 10 cm in the soil, the difference between the stations at the edge and 35 m from the forest was 6°C during the day, and it was even greater at the surface of the ground. The higher temperatures at the edge of the forest (Table 37-1) are due to a reduction of mixing, greater absorption of solar radiation along the edge (than in the forest), partial shielding of outgoing longwave radiation and a reduction of evaporation (Section 53). O. W. Archibold, et al. [3702] in comparing an Aspen Grove with a nearby prairie in Saskatchewan found that the prairie soil at a depth of 10 cm was 6-8°C warmer in summer, but 6° cooler than the grove by midwinter. The annual range of soil temperature was 38°C in the prairie but only 24° in the grove. Winter snow accumulation and summer relative humidity were also greater in the grove.

Table 37-1. Air and soil temperatures (°C) at the southern edge of a forest.

Observation site	20 m in forest	At forest edge	35 m out of forest	100 m out of forest
Air temperature at 10 cm				
at 14:00	18.4	22.2	20.0	18.8
at 05:00	9.0	8.0	6.8	6.0
Difference	9.4	14.2	13.2	12.8
Temperature 10 cm in soil				
at 18:00	11.2	17.8	17.2	17.0
at 08:00	10.6	13.6	13.0	12.8
Difference	0.6	4.2	4.2	4.2

In discussing wind influences on the stand edges, H. Pfeiffer [3724] distinguished between passive and active influences of the forest on the wind. The passive forest influence arises from the fact that the forest edge acts as an obstacle to the flow of air. Even a single tree influences its surrounding wind field. This was measured by M. Woelfle [3735] with a high old oak tree.

On the windward edge of a stand, the flow of air was slowed substantially. A wedge of stagnant air or a lee eddy develops with a horizontal extent of about 1.5 times the height of the stand. Behind the canopy of a dense fir stand, M. Woelfle [3735] recorded wind speeds of only 20 to 30 percent of what they were in the open. As wind speed increased further, this percentage decreased. With an open stand, the wind can penetrate in gusts but is slowed by the tree trunks. This reduction occurs more quickly if undergrowth is present. Streamlines are cramped together in the border zone above the forest. When winds are strong, there is considerable turbulence above the rough forest surface, which can cause treetops to sway. On the lee side of the forest, there is an area sheltered from wind (Section 52). When the wind is blowing at an angle to the stand edge, it acts as a steering line, and wind increases at the outer edge (Figure 52-8). This strong cross wind presents a danger in times of high winds to unwary automobile drivers on roads leading out of the forest at right angles.

R. K. Didham and J. H. Lawton [3706] suggest that fire encroachment into a forest edge eliminates soil seeds, enhances plant mortality and promotes an open edge. In comparing forest edges in the Amazonian Forest they found that there was a marked difference between open and closed edges. Edge penetration was as much as two to five times greater for edges opened up by fire than closed edges with a natural dense growth. The magnitude of these differences suggest that edge structure is one of the main factors of edge penetration. They also found that forest fragments had consistently lower canopy height, higher foliage density, temperature, evaporation drying rate, lower leaf litter moisture and depth than a continuous forest at all distances from the forest edge.

During the day, the air near open ground is heated while the air under the crown of the forest remains cool. When this happens, the cool air may flow out from the trunk area as a daytime forest breeze. L. Herr [3712] and K. Dörffel [3708] believe that this effect can be observed in the cooling and moistening of the air in the outer edge. It is similar in origin to the sea breeze, which blows during the day from the cool sea over the warm land. As early as 1920, A. Schmauss [3730] called it a "sea breeze without a sea."

D. R. Miller [3721], in his summer study of a parking lot at the edge of a forest, found that the air temperatures over the parking lot were warmer by day and cooler at night than the air under the forest canopy (at a height of 2 m). The horizontal temperature gradients across the edge below the canopy top suggested that sensible heat was transported into the stand during the day and out of it into the parking lot at night. The air above the forest canopy heated during the day and cooled at night similar to, although slower than, the air above the parking lot.

K. I. Scott et al. [3732] compared August temperatures of unshaded and tree shaded vehicles in a parking lot (windows closed). For the period from 1200 – 1700 they found unshaded vehicles cabin temperatures averaged 25°C higher than shaded vehicles. Unshaded interior fuel tank temperatures averaged 3°C warmer increasing fuel tank evaporative loss by about two percent.

The temperatures at the boundary (edge) were generally between those of the parking lot and forest and reflected horizontal advection of heat. When the wind was from the parking lot into the forest edge, temperatures at the edge were very similar to the parking lot. When the flow was out of the forest, the edge temperatures were similar to those beneath the forest canopy. The air temperature at the 2 m height was higher at the edge than over the parking lot on calm mornings when the forest edge received direct-beam solar radiation. During relatively calm days, D. R. Miller [3721] also found higher temperatures at the edge of the forest (Figure 37-5). J. Chen, et al. [3705], however, in their examination of the contrasting growing season microclimates of recent clear-cut, edge, and interior old-growth Douglas-fir forest in Washington (46°N), found that with respect to temperature and moisture the most variable microclimate was at the edge of the forest, while for wind and solar radiation it represented a transition between the forest and clearing.

D. R. Miller [3721] found the vapor pressure in the forest to be consistently higher than the parking lot. The steepest horizontal gradients were at the canopy level, which reflects that the upper canopy was the source of water vapor. Figure 37-6 shows the vapor pressure and air temperature profiles. Despite the wind blowing from the parking lot, vapor pressure increases as the forest edge is approached.

A. Young and N. Mitchell [3737] observed that upon entering a forest there is a transition zone. While there was some variability based upon the direction an edge was facing, they concluded that generally the edge zone of a forest could be divided into three parts: (1) an outer edge zone (10 m deep) where photosynthetically active radiation (PAR), air temperature, and vapor pressure deficits (*VPD*) all decrease; (2) an inner edge zone which extends to approximately 50 m into the forest; here air temperature and *VPD* continue to decrease, but PAR has stabilized; (3) an interior zone in which PAR, air temperature and *VPD* remain at relatively constant levels. The *VPD* decreased due to both lower temperatures and higher vapor pressure.

In contrast to the daytime forest breeze, a field breeze at night is observed much less frequently. The strong braking effect of the trees retards its development. In hilly wooded country, a night forest wind blows into the open country, representing the downflow of cooler air forming over the radiating crown area. H. G. Koch [3715] has demonstrated its existence

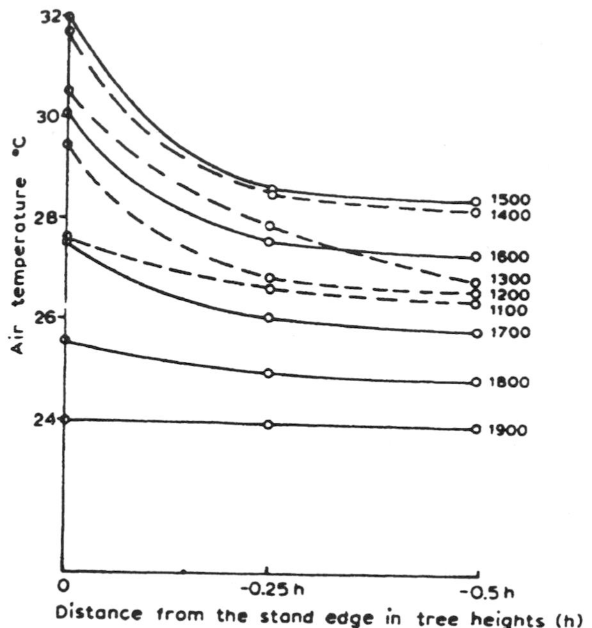

Figure 37-5. Hourly temperature gradients through the day (14 August 1975) with a slight breeze into a forest stand. (After D. R. Miller [3721])

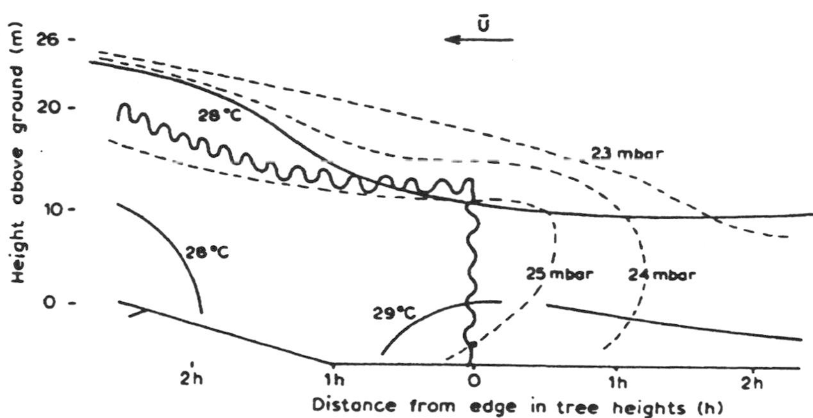

Figure 37-6. Air temperature and vapor pressure across the parking lot-forest edge with the wind blowing into the forest at 12:00 on 24 August 1974. (After D. R. Miller [3721])

by using balloons. M. Woelfle [3736] has pointed out that this is a normal night cooling phenomenon, as described in Section 34 and Section 42, and that it is incorrect to apply the term forest wind to it.

The passive influence of the forest on the wind field at its edge is much more effective than its active influence. The wind field is instrumental in affecting the distribution of particulates (dust) and precipitation at the edge of the forest. Along a country road at the edge of a forest on a hot, dusty summer day, the filtering effect of the trees at the border can be appreciated by observing the particulates over everything. M. Rötschke [3726] found that when winds are perpendicular to the forest edge, there is an increase in particulate content in the inner border zone in addition to the maximum at the edge itself. For example, on 29 January 1935, with a wind of 2-3 m sec^{-1}, the particulate content in thousands of particles per liter in the open in front of (+) and behind (-) the forest edge was:

Distance (m)	-100	-50	-25	+25	+50	+100
Particulate content	10.1	10.2	10.3	14.0	11.8	11.5

Since in this particular case there was a thin snow cover with a temperature of -2°C, the filtering effect could be observed without interference from secondary sources of surface particulates. Further in the forest the air became freer of particulates (Section 39). If the wind blows at an angle to the forest edge, there is a marked increase in particulates associated with the increase in wind speed at the outer border zone.

Rain and to a greater extent snow, because of its lightness, also enter the border area of the stand, reaching the ground in quantities that are determined to a large extent by the wind field. Snow accumulation in the zone of calm at the lee edge of the stand is a familiar sight (Section 53). Rain recordings were made by A. Lammert and reported by O. Ziegler [3738]. In the dry summer of 1947 (May to August), the rain on the east side of a stand of poplars which were oriented N-S (in the shelter of the wind) was as follows:

Distance (m):	4	14	24	outside
Rainfall (mm):	8	33	76	105-110

In this case, there was both a wind shadow and a rain shadow.

Numerous investigations have been made of precipitation from fog, which arises from the filtering of droplets out of blowing fog by the trees. This is sometimes called "horizontal precipitation." R. Marloth [3720] made the first investigation in 1906 on Table Mountain in Capetown. He set up a rain gage with a bundle of brushwood on top of it next to a normal gage, and collected ten times as much rain in the brushwood as in the normal gage.

F. Linke [3718] has shown, by a series of systematic measurements in Germany that precipitation from fog is principally a feature of the edge of the forest. This phenomenon is, however, not confined only to the forest edge. The east coast of the Japanese island of Hokkaido at 43°N has lots of dense sea fogs from the Pacific, similar in character to the Newfoundland fogs. From time immemorial, there has been a belt of forests along the coast, which is said to have a protective influence against the fog. In coniferous forests in 1951 and in deciduous forests in 1952, T. Hori [3713] tested the protective influence and determined which tree types and sizes are best for minimizing fog related problems. The total amount of fog precipitation will depend upon the frequency and duration of fog, the moisture content of the air, and the degree of vertical mixing above the forest canopy. This last factor is related to the wind speed and is inversely related to the aerodynamic resistance R_A of the forest canopy. Turbulence above the treetops resulted in some of the fog droplets falling into the forest and some outside it. The rate of fall caused more drops to fall inside than outside, and the difference between the two shows the amount of water extracted from the fog by the trees.

H. Ooura [3723] measured 0.5 mm of fog precipitation per hour when the wind speed was 4 m sec^{-1} and the water content of the foggy air was 800 mg m^{-3}. That was six to ten times as much as that deposited on grassland under otherwise identical conditions. "Anyone who has observed," writes J. Grunow [3710], "the way a mountain forest pours water down to the forest floor as after heavy rain, when the clouds are lying on it, who has seen how branches and the crowns of trees sway and bend under the heavy load of rime, and how whole areas of the forest can be broken asunder by the weight, or who has seen rime falling to the ground and lying like a snow cover, and even a sledge run being made where the ground was formerly clear of snow, will no longer doubt the great yield of fog precipitation." By comparing rainfall on the Hohenpeissenberg (989 m) with and without fog, J. Grunow [3710] calculated the supplement from fog to be 20 percent of the annual precipitation in this stand. J. Acevedo, and D. L. Morgen, [3701] report fog precipitation of up to 140 mm yr^{-1} for California redwood and Douglas fir forests in a temperate wet climate, while J. Cavelier and C. Goldstein [3704] cite totals of 800 mm yr^{-1} or more for tropical cloud forests in Columbia and Venezuela.

The importance of fog as a source of water for plant growth has been recognized for a considerable time (D. Kerfoot [3714] and R. S. Schemenauer and P. Cereceda [3727]). At the stand edge, fog precipitation makes up a substantially greater proportion. The fog collector of Ooura yielded 20 times as much water on the windward side of the forest as a similar device on the lee side. On the Hohenpeissenberg, J. Grunow [3710] estimated fog precipitation to provide an average supplement of 57 percent to liquid precipitation. F. Linke [3718] found from 1915 to 1919, in fir forests at a height of 800 m, increases in the rain gage readings within the stand, compared with the one in the open (Table 37-2). J. F. Nagel [3722], on Table Mountain near Capetown, from 1 March 1954 to 28 February 1955 observed 3294 mm in a fog precipitation gage of the J. Grunow type [3710] against 1940 mm in a standard rain gage. Particulars will be given in Section 45 on the percentage increase of precipitation due to fog on mountain slopes as a function of height above sea level.

Table 37-2. Precipitation amounts in fir forests as percentage of amount in the open.

Rain gage	Month: Least foggy (June)	Month: Most foggy (Nov.)	Summer	Winter	Year
Immediately on forest edge	104	301	131	184	157
A little farther into the stand	87	259	90	159	123
Average number of fog days	11	24	14	22	18

R. S. Schemenauer and P. Cereceda [3728] used a double layer polypropylene mesh to collect water from fog. This system has been operational since 1992 and provides about 11,000 liters of water a day to Chungungo, Chile in the Atacama Desert. They identified a number of coastal arid regions where there is good potential for fog collection and suggest "one of the most exciting aspects of this resource is that in many regions the supply of water will be limited only by the number of collectors one chooses to install." They also list countries with arid regions where fog collection by artificial collectors or vegetation have been documented (R. S. Schemenauer and P. S. Cereceda [3729]).

When temperatures are below 0°C, the supercooled fog droplets freeze on contact with needles and branches to build dangerous accumulations of rime. The shape and quantity of these ice growths have been studied and described by J. Rink [3725] on the Schneekoppe, and on the Hohenpeissenberg by J. Grunow [3711]. Measurements on the Feldberg (1493 m) in the Black Forest by K. Waibel [3734] indicate the weight of the ice that may accumulate. The trees at the edges of a forest may be crushed under the load of ice. Occasionally the damage becomes catastrophic when heavy snow falls on the already overloaded branches, or when high winds snap the top-heavy trees and toss them about. The amount of ice on high tension power cables was found to depend mainly on wind speed, rather than on air temperature and any accompanying snow precipitation. The maximum hourly growth per meter of cable was 230 g, the highest daily total during two winters was 3.2 kg m^{-1}, and the greatest growth over an extended period was 32.3 kg m^{-1}. M. Diem [3707] has written on the stress to which overhead power cables are subjected. In their investigation of cooling ponds in Illinois, J. Vogel and F. Huff [3733] found that during the colder part of the year steam fogs form over the ponds which move beyond the ponds and coat structures and vegetation with substantial amounts of ice (riming). They found that pond induced riming was associated with the following atmospheric conditions: 1) air temperatures of 7°C or less; 2) a saturation deficit of 0.5 g kg^{-1} or less; 3) a water-air temperature difference of at least 19°C; and 4) winds of at least 1 km hr^{-1}. Over the 19 month observation period they found that of the 185 days with fog, 137 were days when the ponds initiated the fogs. On the remaining 48 days the ponds enhanced existing foggy conditions. On seventy-five percent of the days when steam fogs were initiated they were associated with cold air masses. Frontal activity and low pressure systems were more frequent on days when steam fog enhanced natural fogs.

38 The Climate of Forest Clearings

The interests of forestry are not restricted to the microclimatology of the border areas which form the most favorable sites for regeneration, but are concerned with all areas where new growth occurs and with all conditions that might promote the improvement of older growth. To rejuvenate the forest, the forester may make cuttings or fellings in addition to using forest edge areas. Making circular or elliptical clearings provides the light necessary for the next generation of trees. New growth soon appears as a result of natural reproduction or direct seeding or planting. The young trees enjoy to a large measure the protection afforded by the neighboring stand and thus benefit from the forest climate. They have wind protection, relatively even temperature, and relatively high humidity, all of which promote growth. Extension of the cuttings to give more space to the young trees brings with it a risk of night frost, because of the still air in the open space and the increasing loss of outgoing radiation as the area widens.

The size of a clearing can be defined as the ratio of the diameter D of the clearing (assumed circular) to the mean height H of the surrounding stand. The ratio $D:H$ is called the index of size of the clearing. B. Danckelmann [3809] found, for example, as long ago as 1894 that clearings of index 1.25 in the Brandenburg Forest provided good frost protection, that there was moderate damage with index 1.50, and from index 2.00 onward damage was considerable. These $D{:}H$ ratios correspond to sky view factors of 0.28, 0.36 and 0.50, respectively, for a location in the center of the clearing.

R. Geiger [3810] undertook a systematic study in a mixed stand of pine and beech, averaging 26 m in height, near Eberswalde, by cutting seven circular clearings, which differed only in diameter. Table 38-1 contains details of the dimensions of the clearings and of the measurements made in the centers. The screening angle h is the angle from the horizontal up to the tops of the trees in the surrounding stand, measured from the ground in the center of the clearing. According to F. Lauscher [525] net longwave radiation loss L^* in the middle of the circular clearing, as a percentage of the net longwave radiation loss from open land, can be calculated from the equation:

$$L^* = 100\,(1 - \sin^{r+2} h)$$

in which r is a function of the observed vapor pressure e (mb) that is given with reasonable accuracy by $r = 0.11 + 0.045e$. Values for L^* in Table 38-1 have been computed by this method and show that the net loss of longwave radiation from the largest clearing differed by only 13 percent from the open land.

Table 38-1. Measurements on forest clearings on 8 July 1940. (After R. Geiger [3810])

Measurement	Diameter D (m)						
	0	12	22	24	38	47	87
Size index $D{:}H$	0	0.46	0.85	0.93	1.47	1.82	3.36
Sky view factor Ψ_{sky}	0.00	0.05	0.15	0.10	0.35	0.45	0.74
Screening angle h	90°	72°	59°	58°	48°	40°	26°
Net longwave radiation loss L^* (percent of open land)	0	11	31	33	52	66	87
Midday temperature excess of clearing over stand (°C)	0	0.7	2.0	2.0	5.2	5.4	4.1

The midday temperatures in Table 38-1 show how much warmer the temperature was at 10 cm above the central point of the clearing than in the surrounding stand. Two factors, radiation and wind, affect the midday temperature in a clearing. As the size of the clearing becomes larger, more sunlight comes in and temperatures increase. However, as the size of clearings increase, they become less protected from the wind and mixing increases. As clearing size increases, the increase in solar radiation dominates and temperatures increase until the clearing reaches a $D{:}H$ ratio of around 1.8 (Table 38-1). As the clearing size increases beyond this ratio the increase in mixing dominates over the increase in solar radiation, and the midday temperature excess of a clearing over a forest decreases.

As with hills (Sections 41 and 44), in a forest clearing the maximum temperature usually is somewhat higher on the west than the east-facing side. L. S. Barden [3804] found maximum temperatures averaged around 2°C warmer on the west side. He found that plant growth at the edge of the gap under the overhanging branches was as much as 39 percent higher on the west than the east-facing side, and suggested this was a result of the higher temperatures. This difference in growth decreased rapidly as the distance from the edge increased. L. S. Barden [3804] suggested that these temperature differences may induce a symmetry in species composition.

Nocturnal temperature minima decreased as the diameter of the clearing and longwave radiation loss increased. Figure 38-1 shows the averages of cold nights in the spring and summer of 1940 and for the coldest night with late frosts. Wind, which would keep the minimum higher (Table 19-1), is of less importance since the nights with the lowest temperatures are usually associated with light winds. The nocturnal sheltering effect of a forest can clearly be seen from D. N. Jordan and W. K. Smith's [2915] observations of occurrence of frost in an alpine meadow during the summer. In the understory frost conditions occurred only about half as frequently as at the edge of the clearing. The clearing edge experienced frost conditions only 10-40% as often as the center of the clearing. Two other factors besides nocturnal radiation losses affect the temperature minimum in clearings. The first is a warming influence resulting from the mixing of warmer air from the trunk area. This influence decreases as the size of the clearing and distance from the trees increases. The second is a cooling influence resulting from the flow of cool air down from the crowns of the trees. This influence is not large for small clearings. It increases with clearing size and the potential for more airflow, and decreases again for large clearings as the distance from the trees becomes greater. With respect to nocturnal temperatures M. O. Löfvenius [3616] found that the near-ground temperature differences between a shelterwood and a nearby clear-cut area appear around sunset and remain throughout the rest of the night. M. Nuñez and D. M. J. S. Bowman [3815] measured nocturnal surface temperatures in a forest and nearby open areas (Figure 38-2). Ground temperatures for the bare clear cut and rock outcrop areas were 4°C cooler than for the forested areas. Figure 38-3 shows the degree and frequency to which the forest was warmer than the open areas both on clear and cloudy nights. As expected, temperature differences were greatest on clear nights. Figure 38-1 is the calculated temperature excess of the forest compared to the clearing with respect to stand density and cloud cover. These calculated differences were similar to observations.

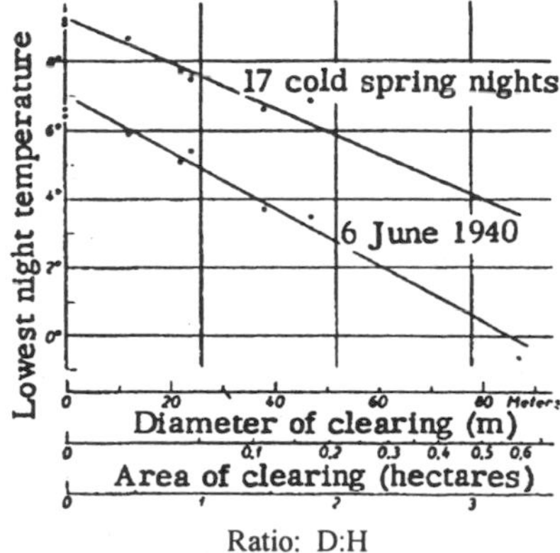

Figure 38-1. The increase in frost danger as clearing size increases.

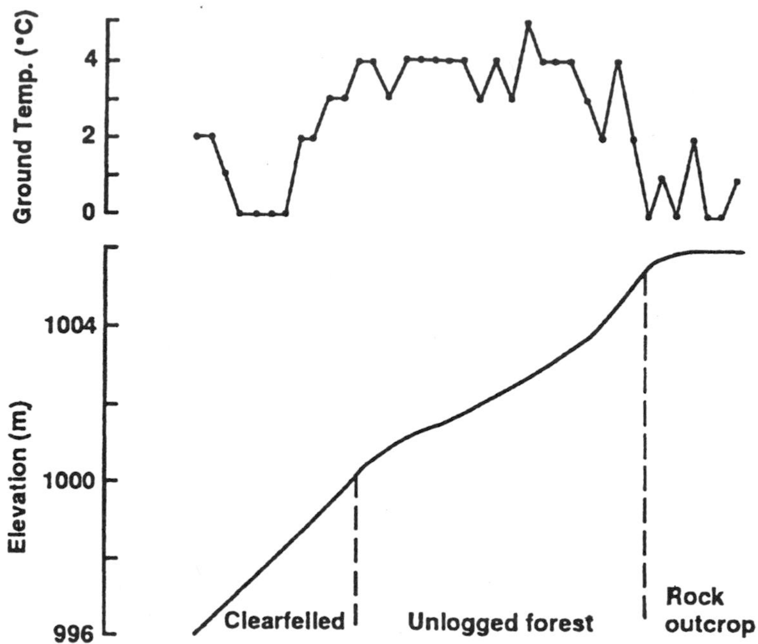

Figure 38-2. Nocturnal surface temperature pattern of a forest and clear areas. (From M. Nuñez and D. M. J. S. Bowman [3815])

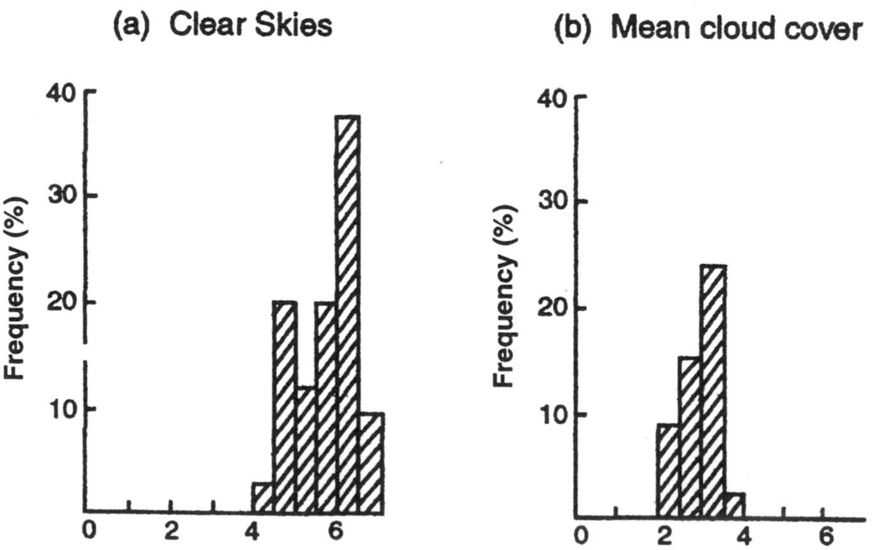

Figure 38-3. Frequency distribution of nocturnal temperature excess (°C) of a forest and nearby open areas for clear (a) and cloudy (b) conditions (mean cloud cover is five-eighths). (From M. Nuñez and D. M. J. S. Bowman [3815])

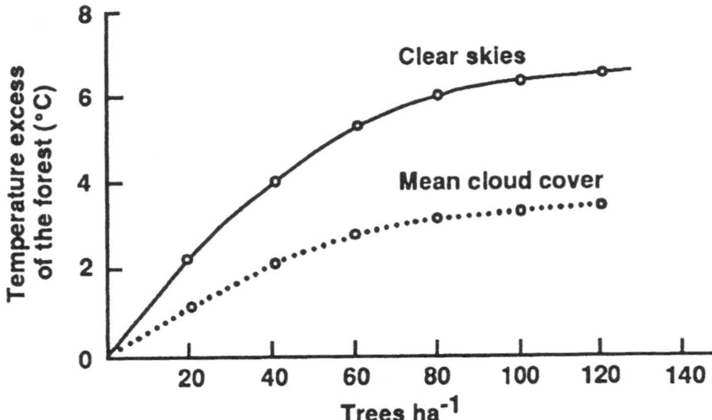

Figure 38-4. Temperature excess of a forest and nearby open areas based on tree density and cloud cover (mean cloud cover is five-eighths). (From M. Nuñez and D. M. J. S. Bowman [3815])

In Nigeria R. Lal and D. J. Cummings [3814] found air temperatures in a clearing in December at 10, 50 and 100 m were from 5-8° C warmer during the day and from 1-2°C cooler at night than those within the nearby forest. Maximum soil temperatures were measured at a depth of 1, 5 and 10 m and were as much as 25, 12 and 7°C, respectively, warmer during the day than within the forest. Relative humidity, while similar at night, was on the order of 15-20% lower in the clearing during the day. The time of occurrence of the minimum relative humidity, during the day, was 2-3 hours later within the clearing.

The microclimate within the clearing is by no means uniform; the most varied local differences are often found next to each other. Temperatures of the ground, air, and bark of trees in sunny and shaded areas were studied from random samples by H. Aichele [3801] in the Black Forest. B. Slavík, et al. [3817] cut a clearing of practically the same diameter as the height of the stand in a mixed forest of 50% oak and 30% beech, with birch, larch, pine, and fir 35 km southwest of Prague. The shape of the clearing is shown by the heavy dot-dash lines in Figure 38-5. During the first 3 years after cutting, numerous environmental factors, and the development of vegetation and young trees, were measured. Figure 38-5 gives a selection of the observations of the local climate.

The lines of equal duration of direct beam radiation are shown in the top left chart for 20 May 1953. The southwest corner receives only the early morning sun and the southeast corner only the evening sun. The north side of the clearing, being a "southern edge," receives the greatest number of direct beam sunshine hours. Daily temperature maxima (not shown) correspond very closely to the distribution of solar radiation.

The bottom right diagram shows isopleths of equal dew deposition (without numerical values in the German summary). The maximum amount of dew is at the center of the clearing. Shielding of outgoing longwave radiation decreases as the distance from the trees increases. Therefore, nocturnal temperatures will be lower and dew deposition greater further from the edge of a clearing.

Precipitation amounts in the two growing periods of 1953 and 1954 are shown on the top right. The prevailing west wind causes the rain to fall into the clearing at an angle, leading to

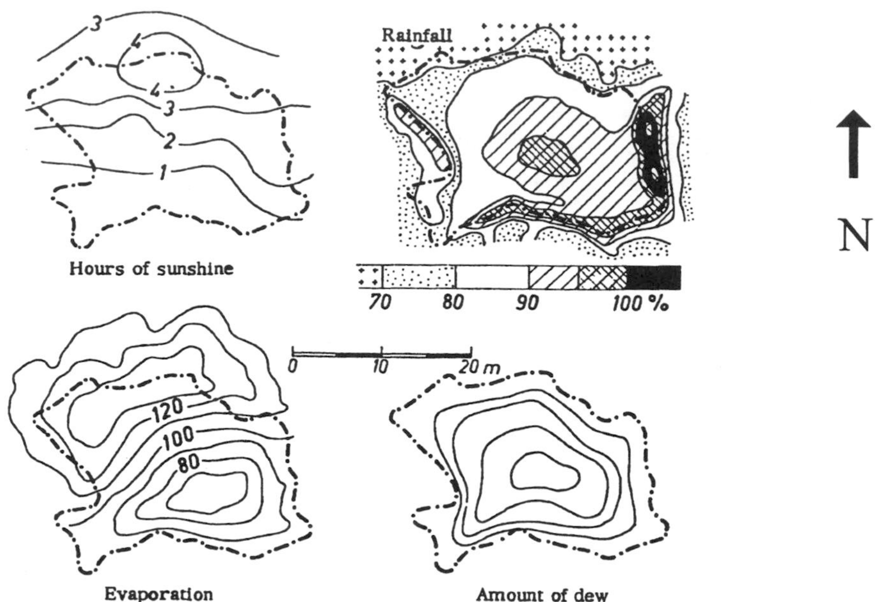

Figure 38-5. Local conditions in a clearing in a mixed deciduous forest. (After B. Slavík, et al. [3817])

an accumulation in the east, where leaf drip from the periphery of the trees further augments it so that it is greater than in the open country (over 100 percent). The effect of the drip line at the periphery of the trees may be seen at other edges of the clearing, while in the calmer area near the center, rainfall is 95 percent of that in the open.

The final chart in Figure 38-5, at the bottom left, shows lines of equal evaporation. The measurements were made at a height of 20 cm above the ground, and the isolines give the mean values, expressed as percentages of the average evaporation in the stand. Three factors affect the variation of evaporation in a clearing. First, the lower nocturnal temperatures at the center of the clearing due to decreased shielding by the trees results in lower evaporation rates or in higher rates of dew deposition. Second, on the southern, southwestern, and southeastern edges daytime shielding by the trees results in lower temperatures and evaporation. This is partially offset at night when this shielding results in higher temperatures and increased evaporation or at least reduced dew deposition. Finally, evaporation is greatest on the northern edges in areas receiving greater amounts of sunlight and experiencing higher temperatures. The contrasting and asymmetrical patterns associated with these moisture input and output processes create significant microscale soil moisture patterns within the clearing that can have important consequences for new growth.

Snowmelt and sublimation patterns within forest clearings also exhibit similar responses. G. J. Berry and R. L. Rothwell [3805] observed snow ablation in a clearing in spring. Snowmelt accounted for 70 to 97 percent of the total ablation (Table 38-2). As the clearing size increased snowmelt increased and sublimation decreased (both in percentage and in absolute amounts). Snowmelt on southern exposures (north side of the clearing) averaged 7 percent higher than northern exposures.

Table 38-2. Snowmelt and sublimation as a percent of total ablation on south and north exposures in circular openings with a forest size index (*D:H*) of from 0 to 5. (From G. J. Berry and R. L. Rothwell [3805])

Opening size (*D:H*)	Percentage of total ablation: Snowmelt South	Snowmelt North	Sublimation South	Sublimation North
0	70	70	30	30
1	87	89	13	11
3	96	84	4	16
5	97	86	3	14

Figure 38-6 shows the N-S cross section of the changes that took place in soil moisture for the conditions shown in Figure 38-5. In June, the soil moisture is similar in the clearing and forest. However, as summer progresses through autumn, the demands made by the adjoining forest's water requirements dried out the soil. Soil moisture did not decrease in the part of the clearing which was beyond the lateral extent of the forest root systems. This observation is consistent with most of the literature (P. E. Black [3806] and D. R. Satterlund [3816]). P. M. S. Ashton [3803], however, in studying tropical forests and clearings in Sri Lanka where there is abundant rainfall, found that the surface soil moisture was almost always greater beneath the canopy than in the center of clearings. He attributed this to the shading by the forest, which more than compensated for the transpiration drawn up by the trees and suggests that the difference may be a result of the depth of the measurements. He measured the moisture at the surface, which in the clearing may dry out as a result of exposure to direct solar radiation. Other studies typically measure the moisture at a greater depth. J. L. C. Camargo and V. Kapos [3808] suggest that in a closed forest the canopy "protects" the lower part of the profile whereas at the edge greater penetration of solar radiation and desiccating breezes result in greater evapotranspiration and depletion of soil moisture. They further suggest that with time growth by secondary species and branch and leaf production by the original trees "seal" the profile along the edge. All these additional layers, however, increase total evapotranspiration at the edge and may further deplete soil moisture. They thus suggest that soil moisture is higher in the undisturbed primary forest than at the edge.

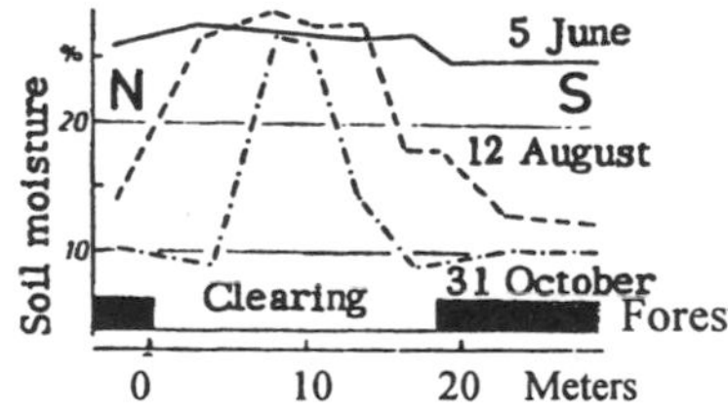

Figure 38-6. Soil moisture during 1953 in the clearing shown in Figure 38-5.

B. S. Ghuman and R. Lal [3811] carried out a series of observations comparing the microclimatic variables in a clearing and a forest in Nigeria. They found that throughfall within the forest was about 12% less in 1984 and 32% less in 1985 than the rain received in a cleared area. The relative humidity was higher in the forest than the cleared area particularly during the rainy season. Open pan evaporation was 4-6 times more in the clearing and wind speeds were about 18 times more than under the forest. On clear days, the maximum temperature at a depth of 1 cm was about 10°C higher in the clearing. This difference dropped to around 3°C on cloudy days. At 50 cm in the soil, they found no appreciable diurnal variations under either the forest or the clearing. However, the 50 cm temperature was always about 3°C lower in the forest. On clear days in the forest, the maximum temperature of the air was about 5°C lower and occurred about 1.5 hours later. They also found that shortwave radiation was 25-30 times more in the clearing.

The procedure of making cuttings has an effect similar to that of thinning on the climate of a stand. Near Vindeln in north Sweden, A. Ångström [3802] observed the progressive gain in heat on the forest floor as the extent of cutting increased. Ground temperatures were 2 to 3°C higher in stands where this had been extensively carried out, and in the spring, the ground thawed 2 to 4 weeks earlier.

In a study of a clearing, C. V. Wrede [3819] found that the direction of the wind in the clearing was often opposite to that in the open country and in the screening stand. This is perhaps due to the creation of a large-scale eddy, which is enclosed on all sides by the old stand (Figure 38-7). H. Pfeiffer [3724] has shown how such eddies develop in the narrow central spaces surrounded by houses in the city (Figure 52-3).

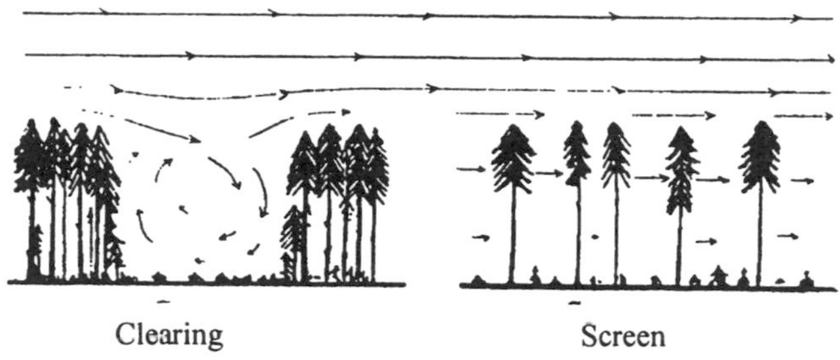

Figure 38-7. Air flow in a regeneration area and under a screen of old trees. (After C. V. Wrede [3819])

K. Göhre [3812] found that cold night air could enter the birch screen from the open surroundings in measurements made in an outer screen of birches that had developed from wind borne seeds in an area left bare by a forest fire. The birches were 2 to 3 m high, and in their shelter, pine seedlings 20 to 40 cm high were growing. Figure 38-6 shows the average temperature minima 10 cm above the ground at 15 observation points (*M*) in the period from 27 June to 10 July 1952. The inner border of the birch screen is the warmest position. Gaps in the outer screen, as well as bare open ground outside, are areas of frost danger. Cold pools develop in these gaps through radiative energy losses. As the trees are approached (M 1-6) the nocturnal temperature increases.

In comparing a clear-cut with a nearby forest M. O. Löfvenius [3616] found that in the pine forest the surrounding trees cut off direct solar radiation earlier than in the clear-cut area. Because of this, the net radiation also becomes negative earlier in the forest (Figure 38-8a). The temperature prior to sunset is warmer in the clear-cut at both 1.5 and 0.1 m above the ground but becomes colder at 0.1 m after sunset. At 1.5 m the temperature is slightly

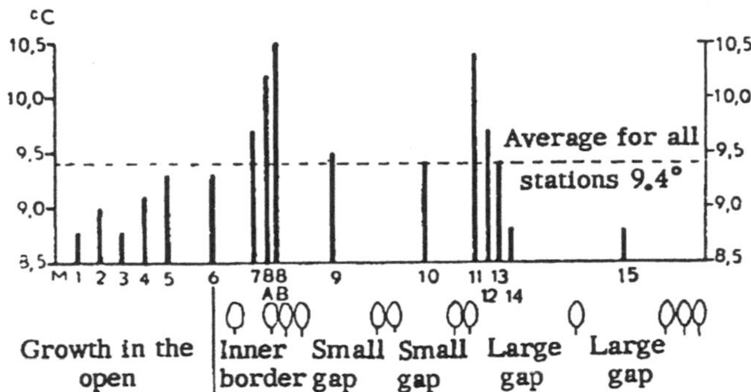

Figure 38-6. Minimum temperatures at 10 cm in an outer forest of birch near Eberswalde. (After K. Göhre [3812])

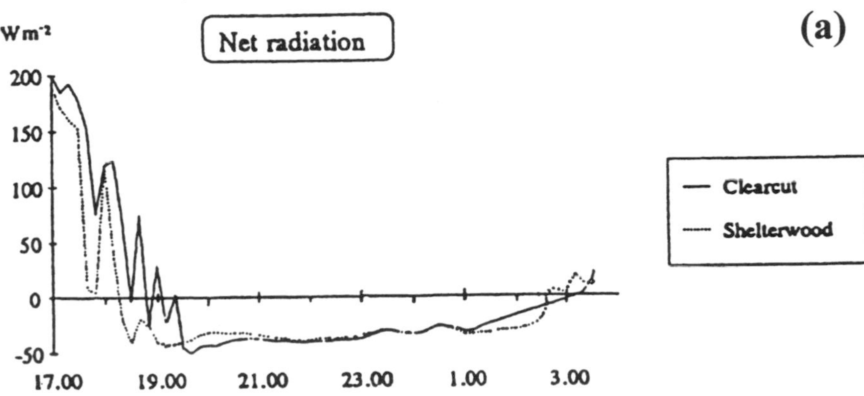

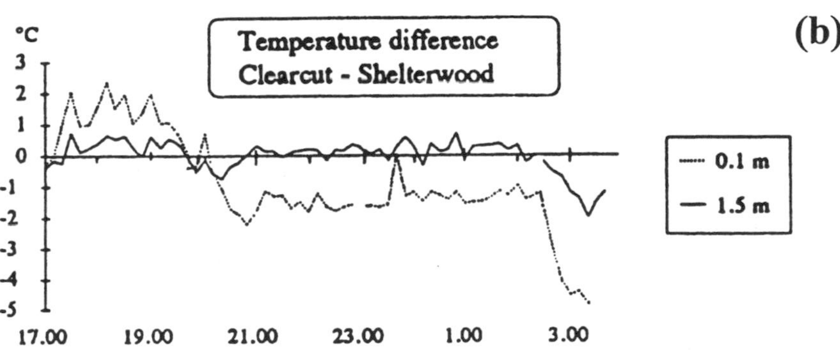

Figure 38-8. Comparative measurements from the center of a shelterwood and a clear-cut. (After M. O. Löfvenius [3616])

cooler in the clear-cut after sunset but later during the night is about the same as that of the shelterwood (Figure 38-8b).

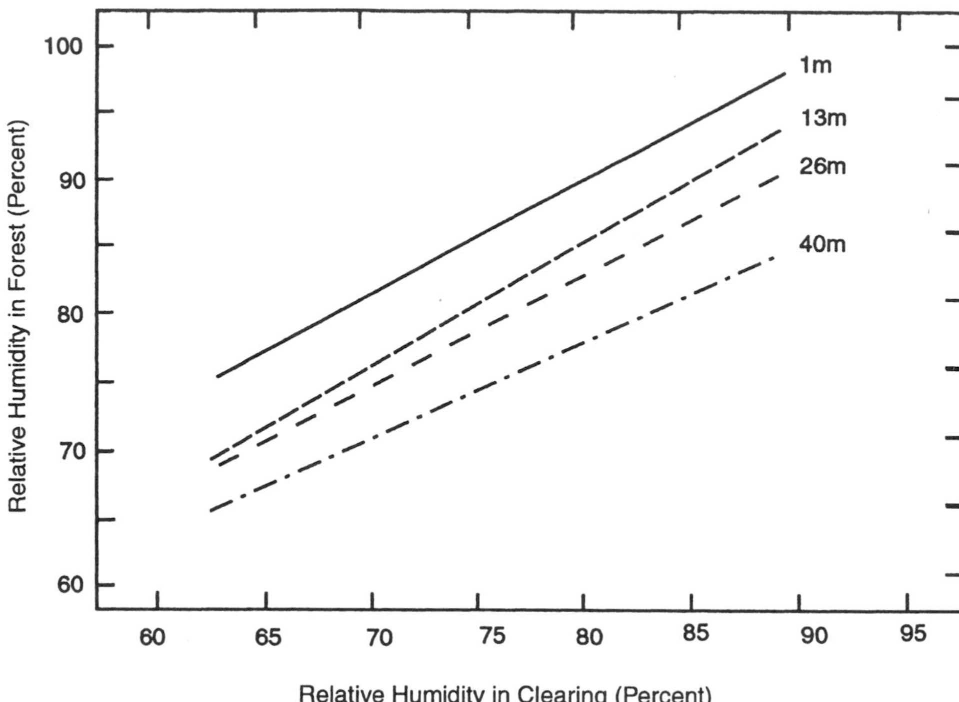

Figure 38-9. Relationship of midday relative humidity at 1, 13, 26, and 40 m in a forest and clearing. (After D. M. Windsor [3818])

D. M. Windsor [3818] found that midday relative humidity in the forest and clearing were closely correlated at all heights (Figure 38-9). In both, relative humidity decreased with height. Throughout the year, however, he found relative humidity about ten percent higher in the forest than the clearing.

There is tremendous variability in the local climate of forests. The work by H. G. Koch [3813] in Figure 38-10 shows the temperature variation measured by an automobile transect on a single day of fair weather in a forest near Leipzig. The 7 km stretch is illustrated in the section at the top of the diagram. Shortly before sunset and about 2 hours before sunrise, the isotherms are closely spaced and primarily horizontal. This indicates that the temperatures change quickly and in a relatively uniform manner along the entire route at these times. The decrease in temperature at sunset and its increase in the morning are such large-scale meteorological features that all differences in the stand are insignificant by comparison. When the energy balance approaches equilibrium, however, the microclimatological effect of local differences become important. This is seen more clearly at night than during the day by the closed isotherms, which are more sharply marked at night. Temperature exceeds 25°C at noon in three places. These are, as seen from the sketch of trees at the top of Figure 38-10, the clearings, the young growth, and the open land. These areas are also coldest at night, falling below 11°C in several places. The isotherms also show the delay in nocturnal cooling in the taller trees.

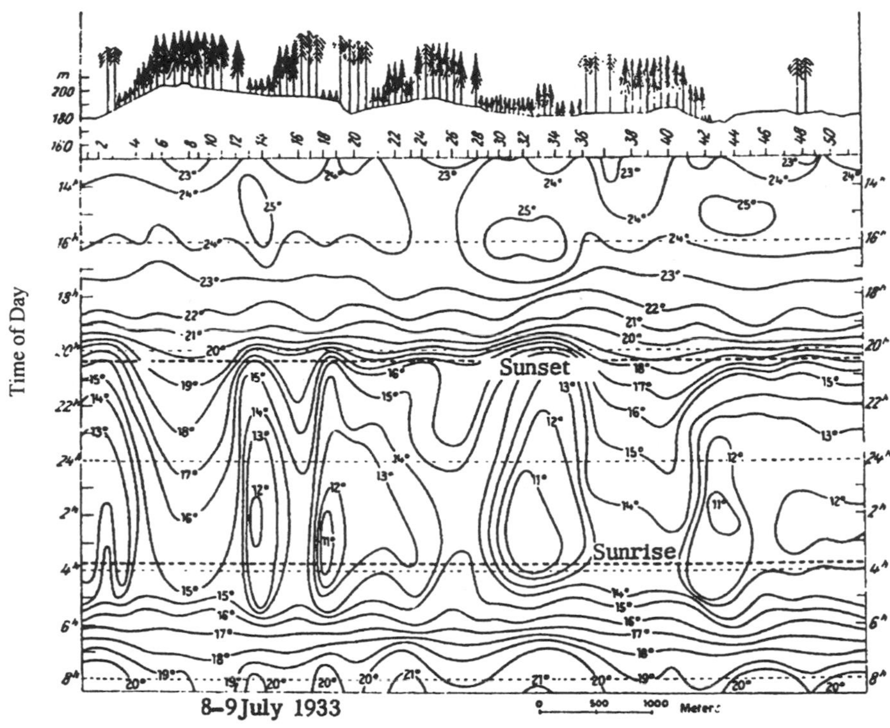

Figure 38-10. Diurnal temperature variation within a closed forest area. (After H. G. Koch [3813])

39 Climatic Influences of the Forest

Sections 33 to 38 dealt with the interaction of vegetation and climate. The question now is whether the sum total of all these vegetation influences, which have such a strong influence upon the microclimate, may not also have a regional scale influence upon the climate.

Deforestation will have a different impact upon the environment, depending on the general climate of the region concerned, the properties of its soil, and the type of vegetation. In steppes and in continental climates, the consequences are different than in a humid, temperate climate. Mountainous areas are more sensitive to the loss of forests than flat areas.

On slopes, forests provide protection against soil erosion by water. In Germany, according to J. H. Schultze [3955], the critical angle of slope at which soil erosion occurs is between 1° and 7° on fields, on roads between 5° and 10°, and in forests between 20° and 30°. In mountain areas the forest helps control flash floods and offers protection against avalanches. Forests provide protection against wind erosion of loose soils. Changes caused in the wind field (discussed in Section 37 for forest edges), and planting of artificial windbreaks (Section 52) are other ways in which trees exert influences.

J. M. Bosch and J. D. Hewlett [3909] provided a comprehensive review of the literature on the effects of deforestation and afforestation on water yield. They examined 75 experimental catchment studies, which included a calibration period and a control basin to separate climatic effects from the effects of vegetated cover changes. Figure 39-1 shows the maximum increases in water yield during the first five years following the reduction of forest cover. They found that reduction (expansion) of forest cover led to increases (decreases) of water yield in all but one experiment. Coniferous and eucalyptus forests yielded approximately 40 mm of water for each 10% forest cover reduction; deciduous forests yielded around 25 mm per 10% forest cover reduction; brush and grass cover resulted in only a 10 mm increase in water yield per 10% ground cover removal. The distinct differences in the effect of forest cover change upon water yield is explained largely by the different physiological characteristics of each forest type, such as leaf area index, albedo, stomatal resistance, aerodynamic resistance, roughness length and zero-plane displacement, which produce varying amounts of interception loss and transpiration. Although trends in water yield changes were found with percent change in vegetation cover, the amount of variance in water yield change explained by forest cover change was low (42% for conifers, 26% for deciduous hardwoods, and 12% for scrub), indicating the complex nature of the relationship. Inconclusive results were also found for small percent changes in forest cover, with changes of less than 20% not producing a measurable stream flow response. The establishment of forest cover on sparsely vegetated land was found to decrease water yield. In dry areas, such as the western United States, it is important that as much precipitation as possible find its way into the ground water supply. There, forests are thinned or replanted with trees which are more economical in their water requirements so that the protective and regulating properties are maintained, but as little water as possible is consumed.

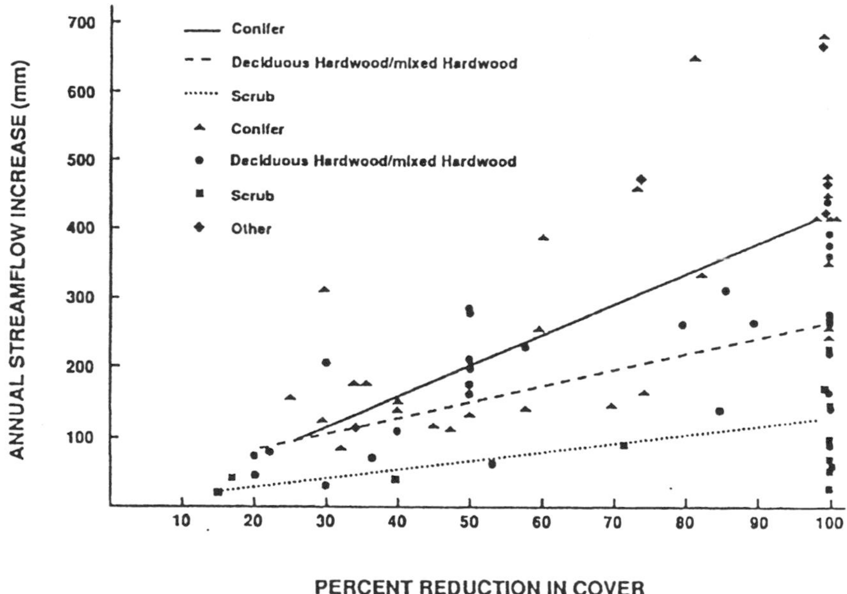

Figure 39-1. Water yield increases following changes in vegetation cover. (After J. M. Bosch and J. D. Hewlett [3909])

Forests are also important to the water balance because they have a regulating influence on stream flow. Forests significantly retard the timing of runoff. The regulating influence of the forest depends primarily on its ability to reduce surface runoff and increase infiltration.

A normal forest floor has a high infiltration capacity. Experiments with artificial sprinkling amounting to 100 mm, made in Switzerland by H. Burger [3913], showed that compact soil under a willow tree took more than 3 hr to fully absorb the water, the soil under another willow took almost 2 hr, a plantation with gaps took only 20 min, while a stand of fir, spruce, and beech needed only 2 min to absorb it. Investigations into mosses and lichens in forests, made by K. Mägdefrau and A. Wutz [3935], showed that they were able to absorb between 3 and 10 mm of rain when the air was dry, depending on the type of moss. Laboratory experiments with 1 dm^2 of various mosses from Bavarian forests gave a quantity of 2.3 to 7.5 mm of water absorption for four fir stands, 8.6 mm in a pine stand, and as much as 14.7 mm for moss from a mixed stand of pine and fir.

The forest floor releases this water slowly. The mosses mentioned above took 16 days. Therefore, depending on the water holding capacity and permeability of the soil, new water is able to penetrate into lower depths, even during dry periods. In the forest region, ground water recharge is more uniform. In winter, a reserve of water is built up in the forest in the form of snow.

These processes result in summer floods being more moderate in forest areas. In the two Harz valleys, for example, rain on 7 July 1950 amounted to 16.4 mm in 37 min. This produced runoff of 200 l sec^{-1} km^{-2} of catchment in the cleared valley, whereas it was only 75 l sec^{-1} km^{-2} in the valley with forests. In winter, however, the water discharge in the two valleys may be interchanged. When the ground water is recharged, floods in spring may be higher in the forest if, for example, the melt waters are carried off along with the rain when the deforested area has long since been snow free. Suppression of surface runoff in the forest improves water quality. Water from the bare valley in the Harz in 1950 contained 56.0 tons of suspended matter, and brought down 2.0 m^3 of pebbles in the stream per square kilometer of catchment area, against 18.6 tons and 0.05 m^3 for the forested valley. During normal flow, both streams contained 5 to 10 mg l^{-1} of suspended matter. This does not change much during flood in the forest valley but rises to 550 mg l^{-1} in the cleared valley.

E. Salati, et al. [3951] in their study of the deforestation taking place in the Amazon Basin suggest that total runoff, and in particular peak flow, may increase, while low river flows may be greatly reduced. J. G. Daniel and A. Kulasingham [3918] reported that Anon [3902], in a study of two small valleys in Malaysia, found that conversion of natural forests to rubber or oil palm tree cultivation doubled peak flows and halved low flows of streams in the study area. The increased peak flow associated with deforestation is usually accompanied by increased soil erosion. Additional information on the effects of forest cover removal and regrowth upon water yield, peak discharge, low flows, and water quality can be obtained from A. R. Hibbert [3931], L. W. Swift, Jr. and W. T. Swank [3957], G. J. Burch, et al. [3911], A. J. Peck and D. R. Williamson [3945], and R. W. Bell, et al. [3906].

Turning now to the climatic benefits arising from forests, we may begin with the often repeated question whether forests increase the amount of precipitation in an area. Assertions of this nature are based on the established fact that the presence of a forest has a favorable influence on the water balance. This was first attributed to increased precipitation. Moist air in the forest, the smoking phenomenon described in Section 36, the observed zone of high humidity in the air surrounding the forest, as described by H. Mrose [3940], and many other observations may all have contributed to provide apparent support for such propositions. The Stalin plan for the transformation of nature was probably the single largest purposeful attempt to change the climate. "Promulgated in October 1948 and somewhat expanded in

1950, the idea was to improve radically the climate and, thereby, the agricultural possibilities of a large portion of the country's arid steppe, semi-desert and desert areas lying within the lower Volga region, Precaspian lowland, Turkmenia, Southern Ukraine and the northern Crimea" (P. P. Micklin [3938]). This was to be accomplished through extensive planting of shelterbelts, creation of ponds and small reservoirs, and irrigation projects. Between 1948-1951 alone, shelterbelts were planted in an area of 13,500 km^2. Following Joseph Stalin's demise, the program was largely abandoned. Most of the trees did not survive the arid conditions in which they were planted (P. P. Micklin [3938]).

Since precipitation formation is a process that takes place primarily in the upper atmosphere, the type of surface below will only have a minor effect on regional precipitation. A study of the African coast of the Mediterranean provides an impressive example of how little effect even a massive supply of water vapor from below may have. Although enormous quantities of water are transferred from the warm sea into the atmosphere by evaporation, the coasts remain arid deserts because the general circulation of the atmosphere in these areas is unfavorable for precipitation.

Early attempts to establish some factual basis for comparing precipitation amounts in forested and deforested areas were inconclusive because of the large number of other active influences present, such as height above sea level, proximity to the coast, type of soil, and relative position to atmospheric pressure centers. Measurements such as those made by J. Schubert [3954] on the Letzlinger moors, which showed an increase in annual precipitation of 5 to 6 percent, could not be easily interpreted because of extreme difficulty in assessing the influence of wind on precipitation. H. Burckhardt [3912] at the meteorological station on the peak of the Erbeskopf (816 m) in the Hunsrück in 1949 found that when the 20 ha beech forest of varied age groups was cut, the depth of precipitation decreased by 15 to 38 percent. An attempt was also made to prove that there was a change in precipitation amounts at the time the island of Mauritius underwent deforestation from 1850 to 1880. H. F. Blanford [3907] made an interesting investigation when a large area of southern central India was reforested as a result of a new forest law in 1875. By comparing precipitation figures before and after, he concluded that he had detected an increase in rainfall. A. Kaminsky [3932], however, later showed that this was due to a climatic variation, which had affected great areas of the country, but had not affected the places H. F. Blanford selected as his control observation stations, from which he intended to verify the accuracy of his before and after reforestation measurements. This example shows how difficult it is to arrive at useful conclusions from experiments covering such a wide range of spatial scales.

Traditional microclimatological methods are inadequate for establishing a relationship between microscale forest and land cover changes, and climate change at the regional and global scales. In the past two decades there has been growing interest in determining whether microclimatic impacts associated with large-scale land cover changes, including deforestation, desertification, agricultural expansion, irrigation and urbanization, may be producing climatic changes at the meso- and macro-scales.

J. Charney [3915], for example, argued that albedo changes resulting from desertification in the Sahel Region of Africa could lead to decreased regional precipitation. R. T. Pinker et al. [3946] observed a mean albedo decrease of 0.03 for a large forest clearing as compared to a nearby forested site for a tropical dry evergreen forest in Thailand (14°31'N) under all-sky conditions. The albedo decrease for the cleared site was observed during all hours of the day, and was most pronounced in mid-summer.

Reviews by Y. Mintz [3939] and P. R. Roundtree [3948] indicate that changes in albedo associated with afforestation or deforestation may affect the distribution of rainfall. Y. C. Sud, et al. [3956] also showed that afforestation or deforestation will affect the roughness of the land, and this could have an affect on the distribution of rainfall. E. Salati, et al. [3951] suggest that deforestation in the Amazon Basin is of such a scale that it may have a significant impact on the local climate. They suggest that since evapotranspiration would decrease, temperatures might be expected to increase and relative humidities to decrease slightly. Precipitation would also be expected to decline, but this decline would be on the order of 10% or less.

These accelerating human-induced global land surface changes have intensified recent interest in basic research concerning the reciprocal relationship between microscale land surface processes and regional and global-scale weather and climate (P. S. Eagleson [3919]). The atmosphere and the land surface are now seen as an interactive coupled system wherein the effects of large-scale land use changes may propagate to distant regions via atmospheric dynamics.

Three developments have been key in our understanding of the interaction between the hydrological cycle and the general circulation of the atmosphere. First, is the increased quantitative understanding of how precipitation and energy is partitioned at the surface of earth. Beginning with the work of A. J. Rutter [3949] and continuing through the more recent work by P. J. Sellers et al. [3952] and D. Entakhabi and P. S. Eagleson [3921], scientists now have quantitative models of how spatial variations in albedo, surface temperature, soil wetness, and vegetative parameters influence patterns of surface and atmospheric heating and therefore affect the dynamic behavior of the atmosphere. Forests, in particular, are known to transfer large quantities of water back to the atmosphere through transpiration, evaporation of intercepted water, and soil evaporation.

A second key factor has been the development of general circulation models (GCMs), which allow scientists to investigate how land surface fluxes from a local area affect the regional climate. P. S. Eagleson [3919] reported on the use of a GCM to trace the movement of evaporated water from model grid cells of interest through the atmosphere. This technique allowed researchers to identify the spatial extent of moisture advection and to determine the degree of recycling of local evaporation via precipitation. He found that 37% of the local evapotranspiration from a grid cell in the Amazon Basin was redeposited as precipitation in the same grid cell. Similar figures for grid cells in Southeast Asia and the Sudd Region of The Sudan were 52% and 19%, respectively. E. Salati and P. Vose [3950] claim that 50% of precipitation falling within the Amazon Basin is returned to the atmosphere as evapotranspiration, and that 48% of this evapotranspired moisture is recycled within the Basin.

A third factor has been the development of a comprehensive global archive of land surface and atmospheric conditions which has enabled climatologists to quantify land surface characteristics, test and develop models of surface-atmospheric interactions. These global data sets can also be used to establish the strength of the feedback between atmospheric conditions and land surface processes. K. L. Brubaker, et al. [3910] used a global aerological data set to estimate the fraction of total precipitation due to local (evaporative) origin. Monthly estimates of this precipitation recycling ratio are given in Table 39-1 for four regions of the world: European Russia, the Mississippi River Valley of North America, the Amazon Basin of South America, and West Africa. The results in Table 39-1 show significant intra-annual and regional variation, and provide additional evidence that forests, through local recycling of precipitation, influence climate at the meso- and macro-scales.

Table 39-1. Estimates of the ratio of local precipitation to total precipitation for selected regions. (After K. L. Brubaker et al. [3910])

Month	Eurasia	North America	South America	West Africa
January	0.07	0.18	0.27	0.14
February	0.00	0.19	0.23	0.41
March	0.04	0.21	0.29	0.41
April	0.07	0.16	0.26	0.27
May	0.15	0.23	0.24	0.20
June	0.31	0.22	0.14	0.10
July	0.26	0.34	0.18	0.47
August	0.23	0.33	0.15	0.48
September	0.10	0.26	0.16	0.39
October	0.06	0.34	0.29	0.25
November	0.04	0.21	0.31	0.34
December	0.03	0.15	0.32	0.27

The advent of remote sensing technology, new international field studies, and numerical computer models promise new insights into the old question of how forests influence the weather and climate. An introduction to recent advances in this field can be obtained by consulting the following references: A. Henderson-Sellers et al. [3928, 3930] J. C. Andre et al. [3901], and J. L. Kinter and J. Shukla [3934].

Although there are substantial experimental design problems associated with establishing a clear statistical relationship between forest cover growth (removal) and precipitation enhancement (reduction) at the microscale, the enormous amount of experimental data on the subject seems to indicate that forests may affect precipitation depth perhaps by as much as 10%. Efforts to establish a causal mechanism have generally focused upon two areas. First, since forests are generally located in local uplands they are thought to help strengthen the local orographic effect. Second, the aerodynamically rough forests are very effective at extracting momentum from the overlying atmosphere. This results in reduced surface wind which may lead to an increase in precipitation and/or precipitation catch by surface rain gauges. The precipitation may be increased as a result of speed convergence and resulting rising air as the forest slows the wind speed. This may be particularly noticeable at the edge of a forest. As the forest decreases the wind speed, precipitation will fall more vertically and instead of blowing over, a higher percentage will fall into rain gauges resulting in a greater catch. For further information concerning the effects of deforestation on the climate and water balance, the reader is referred to R. L. Desjardins et al. [4805]

All rain gauges produce an obstruction to the airflow near their openings, which influences the efficiency of the precipitation catch. The magnitude of this aerodynamic effect varies with the rain gauge type, positioning height, presence or absence of a momentum shield, wind speed and raindrop diameter (or precipitation intensity). Results obtained by C. C.

Mueller and E. H. Kidder [3941] based upon wind tunnel models and aerodynamic drag characteristics of raindrops indicate that significant precipitation catch errors can occur with raindrops of less than 2.0 mm in diameter.

Toward the end of the 19th century, a new technique for tracing forest influences was brought into use in Europe. Twin forest observation stations were set up, consisting of meteorological stations in pairs, one inside and one outside the forest, close to each other, so as to provide a good opportunity to study the differences between the climate of the trunk area and the open country. This technique provided our first knowledge of precipitation conditions within and outside the forest, giving us information about the amount of interception (Section 36), and allowing us to numerically express the moderation of temperature variation in the forest. Figure 39-2 shows examples of so-called "characteristic curves for tree type" observations made over 15 years by A. Müttrich [3943] with five paired stations in a fir stand, four in pine, and six in beech. These annual curves show by how many degrees the mean diurnal temperature variation is lower in the trunk area than at the same level outside the stand. The difference is naturally greatest in the summer. The annual variation is more uniform in the coniferous forests, in comparison with beech forests where breaking into leaf in the spring and disappearance of foliage in the fall cause noticeable discontinuities. It is known that the presence of forests has a moderating influence on climatic extremes at the microscale. Forested areas have a milder temperature regime more similar to a marine climate than nearby areas without forests.

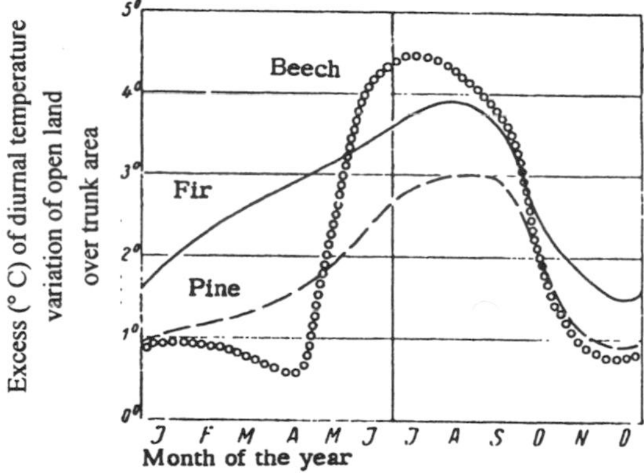

Figure 39-2. Temperature variations between the trunk area and open land for different types of trees. (After A. Müttrich [3943])

A different approach for studying the climatic influence of forests is to examine the microclimatic effects at the outer effective surface above nearby forested and cleared sites. H. G. Bastable, et al. [3904] undertook a comparative study of the microclimate above a 35 m tall forest canopy in an undisturbed primary forest in Amazonia (3°S, 80 m), and within a nearby 10 km^2 clearing of pasture grass. Observations were made over a 60 day period from 12 October-10 December 1990, which spanned the end of a dry season and the beginning of a wet season. Average results for the whole study period are given in Table 39-2. Although the mean values of the climate variables are not that different, substantially greater daily variation was found over the cleared site, especially during dry periods. Wind flow above the canopy was characterized by a weak diurnal flow pattern, with winds ranging between 1.5-2.2 m sec^{-1}. Daytime surface winds in the clearing were comparable to those above the canopy, however, nighttime winds were much lower over the clearing. The lower winds of the cleared site as compared to the forested canopy and its lower nighttime grass surface temperatures led to the creation of a strong nocturnal inversion, which was associated with the decoupling of the near-surface wind from the airflow above the canopy. The cleared site average temperature range was nearly twice that of the forest. Lower nighttime temperatures were found in the cleared site because of the reduced vertical mixing, and higher daytime maximum temperatures occurred, especially during dry periods, because of greater water stress, reduced evaporation, and increased convection. The daily range in specific humidity (g kg^{-1}) was also nearly twice as great over the cleared site. Reduced nighttime winds and lower temperatures often produced fog and dew over the cleared site, while higher daytime temperatures and reduced evaporation resulted in higher average daytime specific humidity deficits. The cleared site allowed greater penetration of solar radiation to the ground surface, which produced both a greater amount and longer duration (2 hours) of heat conduction into the soil. Lower nighttime ground temperatures in the clearing, however, led to greater night-time conduction toward the surface, and therefore a greater diurnal range in the amount of heat conduction.

Table 39-2. Mean hourly minimum, maximum and daily temperature, specific humidity deficit, wind speed, and albedo for the period 12 October-10 December 1990 at the outer effective surface above a forest and nearby clearing in Amazonia. (After H. G. Bastable et al. [3904])

	Clearing			Above Forest		
Variable	Minimum	Maximum	Mean	Minimum	Maximum	Mean
Air temperature (°C)	21.8	31.4	25.8	23.9	29.6	26.5
Specific humidity deficit (g kg^{-1})	0.0	12.7	4.3	2.0	9.8	5.3
Wind speed (m sec^{-1})	0.3	2.2	0.9	1.5	2.2	1.8
Albedo (percent)	–	–	16.3	–	–	13.1

The radiation terms of the energy balance also varied between the two sites. Daytime net radiation totals were approximately 10 percent greater for the forest canopy due to the lower canopy albedo (Table 39-2), and the higher clearing surface temperature, which produced greater outgoing radiation. These effects were best developed during the dry season.

I. R. Wright et al. [3959] investigated the changes in the surface energy budget that result from large-scale conversion of tropical rainforest into grasslands of hardy pastures and grass-es near Manaus, Brazil (2°19'S). When soil moisture was readily available, such as in the aftermath of a storm, the two surfaces both lost about 3.8 mm d^{-1}. Under such condition, Q_E averaged 0.70 of available net radiation Q^* for both surfaces while the average Bowen ratio β was 0.43. This suggests that the larger canopy resistance for the grassland was compensated by its higher surface temperature and humidity deficit. Prolonged absence of rainfall, however, eventually led to a rapid divergence in evapotranspiration and energy partitioning. Q_E for the forest, with its deeper root system, continued to account for 0.60-0.75 of Q^*. For the more shallow-rooted grassland surface evapotranspiration fell to an average of 2.1 mm d^{-1}, Q_E averaged only 0.50 of Q^*, and the average Bowen ratio increased to 0.67.

The influence of the forest is greater in areas less hospitable to the growth of trees. In the boundary zones between arid and humid climates, at the tree line in mountains and polar regions, and in regions of frequent storms or constant strong winds, the beneficial effects of forests may become very great.

In Section 37, the ability of forest edges to filter out particulates was introduced. Figure 39-3 shows the distribution of particulates measured by M. Rötschke [3726] during an automobile journey near Leipzig in the afternoon of 10 April 1935, using a Zeiss conimeter. Not only the edge, but also the whole area of the forest, shows a low particulate content. H. Zenker [3960] found a similar reduction in the smaller and more numerous condensation nuclei. For further information dealing with the forest atmospheric pollutants the reader is referred to D. Fowler et al. [3924].

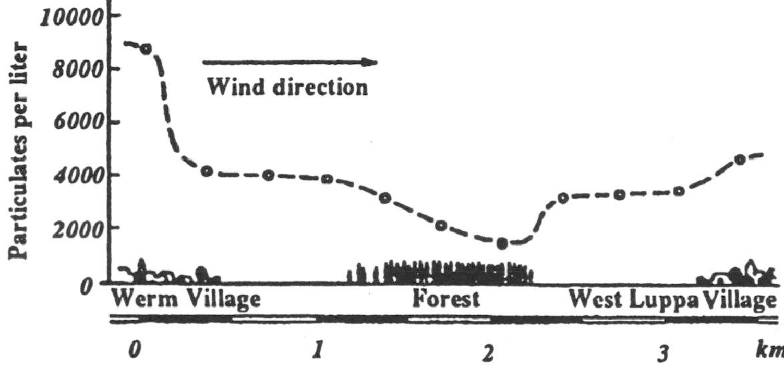

Figure 39-3. Particulate concentrations in open land and forests. (After M. Rötschke [3726])

It has long been suggested that shrubs and trees can reduce noise within a city, around factories, airports, and highways. In parks or along highways, they are often planted not only to provide a visual but also a noise screen. Plants change the propagation of sound by scattering, reflection, and absorption. There has been some controversy as to how much vegetation reduces noise. G. Reethoff and G. Heisler [3947] suggest that " . . . even the thickest barks on large trunks are very poor acoustic absorbers. On ther other hand, typical forest floors of decaying leaves or needles are excellent acoustic absorbers over a fairly wide range of frequencies. Generally speaking, the denser the forest, the better the absorption." D. E. Carlson [3914] suggests that trees, while absorbing a small amount of noise, act primarily as scatterers of sound. The multiple scattering effect which occurs within an area of trees provides the propagating sound waves with more opportunities to strike a more absorptive surface, like the ground, thus increasing absorption. J. Borthwick et al. [3908] and Federal Highway Administration [3923] suggest that sound attenuation of narrow belts of vegetation is negligible. Only dense belts at least 30 m wide and 15 m tall have a noticeable impact. D. I. Cook and D. F. Van Haverbeke [3916] for wide shelterbelts reported noise attenuations on the order of 5 to 8 decibels were common and attenuations of 10 decibels (approximately half as loud) were not unusual. B. E. Mulligan, et al. [3942] has suggested that the visual and acoustic aspects may interact to alter the perception of noise. A reduction in noise levels from relatively narrow belts of vegetation have been reported when it actually has little effect on sound transmission. "Removal of the vegetation invariably produces numerous complaints from nearby residents who perceived an immediate increase in ambient noise levels" (R. A. Harris et al. [3926]). While most (J. Borthwick, et al. [3908]; Federal Highway Administration [3923]; R. A. Harris, et al. [3926]) agree that wide (at least 30 m) high (15 m) belts of vegetation can attenuate noise, R. A. Harris, et al. [3926] found a limited reduction (2-3 dB) even by a dense narrow strip of vegetation. Others have also reported some noise reduction of 3-6 dB by relatively narrow (3-6 m) shelterbelts (D. I. Cook and D. F. Van Haverbeke [3916, 3917]; B. K. Huang [3927]). R. A. Mecklenburg, et al. [3937] found that the greatest noise attenuation by plants occurred in the lower and higher hearing frequency regions. S. W. Tromp [4830] suggests that "noise reflected by a large concrete wall or buildings can be reduced threefold by planting trees between the concrete surfaces." D. I. Cook and D. F. Van Haverbeke [3916, 3917] evaluated sound attenuation properties of different types, heights and densities of trees, shrubs and landforms, and made planting recommendations for optimum noise reduction for a number of situations. D. Aylor [3903], F. Fricke [3925], M. J. M. Martens [3936] and T. F. W. Embleton [3920] evaluated noise attenuation for a variety of plant types at various frequencies. They found that noise attenuation while modest at lower frequencies are substantial in the higher auditory range. F. Fricke [3925] suggests that in a dense pine forest excess attenuation at high frequencies is on the order of 0.1 dB/m. G. Beck [3905] and C. E. Whitcomb and J. F. Stowers [3958] found that differences in sound attenuation among plants was due primarily to size, shape, mutual position, and density of their leaves and needles. He also found bushes at the edge of a stand have a strong influence on the sound attenuation properties of the stand (from S. Kellomäki et al. [3933]). S. Kellomäki et al. [3933] developed an equation, using regression, for estimating sound attenuation provided by different types and ages of plants at various distances from the sound source. They found that the total amount of needles or leaves and branches is the most important factor in estimating sound attenuation. However, density, height and the age of a stand are also important. Earlier successional stages of a stand give better attenuation than mature stands.

The decibel (dB) unit used for measuring the intensity of sound is a logarithmic scale. For each ten dB increase the actual sound pressure on the ear increases ten times, but the actual sound is only perceived as having approximately doubled. Thus, the 2-6 dB sound attenuation attributed to relatively narrow vegetation widths, while only reducing the perceived sound by a small (but potentially noticeable) amount, will result in a 38-75% decrease in the pressure to which the ear is exposed.

For further information about atmospheric acoustics, sound absorption by tree bark and the forest floor, propagation of noise over and through a forest stand and land use planning in noise control besides the articles referenced above the set of articles found in the Conference on Metropolitan Physical Environment [3944].

G. Flemming's [3922] book *Wald Wetter Klima-Einführung in Die Forestmeteorologie* is a good source for further information on the variation of climatic elements in the forest.

CHAPTER VII

THE INFLUENCE OF TOPOGRAPHY ON THE MICROCLIMATE

In Chapter II, the microclimate of a level surface without vegetation was considered. Then in Chapters V and VI the interactions of vegetation and microclimate were studied. Now the assumption that the ground is level will be abandoned, and the influences exerted on microclimate by the topographic variations will be investigated.

When the sun is shining, slopes of different inclination and orientation receive different amounts of radiant energy, and the currents of air thus set in motion (such as upslope or anabatic winds) determine the microclimate. The orientation of the slope is of decisive importance. At night, when the temperature distribution is regulated by the downward flow of air that has been cooled in contact with the surface, the orientation of the slope is of no importance, but differences of height become significant.

When only level ground was considered, the microclimate was restricted to the lowest few meters of the atmosphere, and we ventured out of this shallow layer only occasionally to obtain a better understanding of the processes involved. When the land has strong relief, however, the microclimate must be extended farther. There is a continuous transition from the microclimate of a furrow in a plowed field to the climate of a long mountain valley, a scale more suitable for the field of general climatology. Since this is a microclimatology textbook, the exposition will become briefer as problems at a scale dealt with in textbooks on general climatology are approached.

40 Insolation on Various Slopes

Sloping ground is described by its slope angle, or gradient, measured from the horizontal and the slope azimuth, or direction toward which it faces, measured in a clockwise lateral direction from true north. These two quantities determine the slope situation or exposure. The connection between slope angle and gradient is as follows:

Slope angle:	0.1°	0.5°	1°	3°	7°	11°	27°
Gradient:	1:573	1:115	1:57	1:19	1:8	1:5	1:2

A west slope means a slope facing toward the west; a valley running N-S therefore has a west slope on its eastern side and an east slope on its western side.

Slope climate (sometimes called terrain climate or exposure climate) is determined by the different amounts of shortwave and longwave radiation received by an inclined surface as compared with a horizontal surface. This difference can have great significance. For example, a surface inclined at 20° facing toward the south, even allowing for the high degree of cloudiness in Germany, receives roughly twice as much radiation in January as a horizontal surface. The amount of radiation it receives is equivalent to a substantial displacement toward the equator. Although extreme slope angles of 90° can be found in natural environments, such as along cliffs and open rock faces, the slope angles in natural terrain are usually not that great. G. A. Olyphant [4020], for example, reported slope angles between 7° and 11° for the floors of alpine cirque basins along the Front Range of Colorado. The solar radiation received on a slope is thus most affected by the direction the slope is facing.

The climate of a slope has always been of significance. In agriculture and horticulture, it decides the quality of arable land and whether it will be possible to cultivate certain plant species. The early crop of strawberries that are on sale in Tokyo two months in advance of the main crop are grown on the steep terraces of Shizuoka, the exposure climate of which has been described by S. Suzuki [4031]. J. E. Radcliffe and K. R. Lefever [4026] suggested that within New Zealand the difference in the microclimate of a 25° north and south-facing slope is equivalent to a latitudinal displacement of about 9°. Also when looking for sites for hospitals and sanatoriums, an attempt is made to find sunny slopes. Vertical slopes of 90°, that is, walls facing various directions, are of great importance in the architecture of dwellings, town planning, technology, and the cultivation of fruit on trellises. We are already familiar with some of the effects from the microclimates of stand borders (Section 37) and vineyards (Section 32).

The receipt of solar radiation on an inclined surface consists of three separate components: direct-beam radiation from the sun, diffuse radiation from the sky, and reflected solar radiation from the surrounding ground surface. A review of methods to assess these three components is given in J. E. Hay and D. C. McKay [4010] and J. E. Hay [4009]. The receipt of longwave radiation on an inclined surface has two components: atmospheric radiation and emitted radiation from the surrounding terrain. D. S. Munro and G. J. Young [4017] and A. L. Flint and S. W. Childs [4005] present models for the determination of the individual shortwave components in mountainous terrain, while G. A. Olyphant [4021] developed a model that examines all incoming radiation sources. Each of these terms will be discussed in order.

The direct-beam solar radiation incident upon an inclined surface under clear-sky conditions is a function of the slope-solar geometry, and depends on five factors: latitude, solar declination (time of year), solar altitude (time of day), slope angle, and slope azimuth. K. Schütte [4029] devised an early simple procedure for calculating direct-beam solar radiation on a slope. The fundamental approach used in the evaluation of slope-solar geometry is described in K. Ya Kondratyev [212]. Both the Smithsonian Meteorological Tables prepared by R. J. List [4015] and K. Ya Kondratyev [212] present listings of the amount of incoming solar radiation for slope angles at different latitudes, times of the day, seasons of the year and degrees of cloudiness (or atmospheric turbidity).

Direct-beam solar radiation on an inclined slope also varies with atmospheric conditions, including cloudiness, atmospheric turbidity, ozone and water vapor concentrations, and atmospheric pressure. G. A. Olyphant [4019] found that clear-sky conditions occurred on only

9% of the days in the Front Range of Colorado, and that atmospheric transmission normally varied significantly during the day due to the afternoon development of cumulus clouds. Since all methods of estimating direct-beam solar radiation on an inclined surface for an hour or less share a common approach to slope-solar geometry, estimation differences result from different approaches to modeling atmospheric transmission. We will examine the simpler but less representative clear-sky case first. Examples of modeling efforts to determine direct-beam radiation on an inclined surface can be found in M. A. Atwater and J. T. Ball [4000] and J. A. Davies and D. C. McKay [4001]. Extensions of such modeling efforts to rugged terrain, such as in the work of D. S Munro and G. J. Young [4017], A. L. Flint and S. W. Childs [4005], and G. A. Olyphant [4021], must also consider the shadowing effects of topography, and the spatial variation of surface temperature and albedo of the surrounding ground.

The basic principles of slope irradiation for clear-sky conditions are illustrated by Figure 40-1. It is based on measurements of direct-beam solar radiation in clear weather during the years 1930 to 1933 at Trier (49°45'N) by W. Kaempfert [4012]. The abscissa is the slope angle in degrees and the ordinate is local time. The figure consists of nine diagrams, three slope azimuths for three selected days; the isopleths give the quantity of direct-beam solar radiation received for clear skies with a normal amount of atmospheric turbidity.

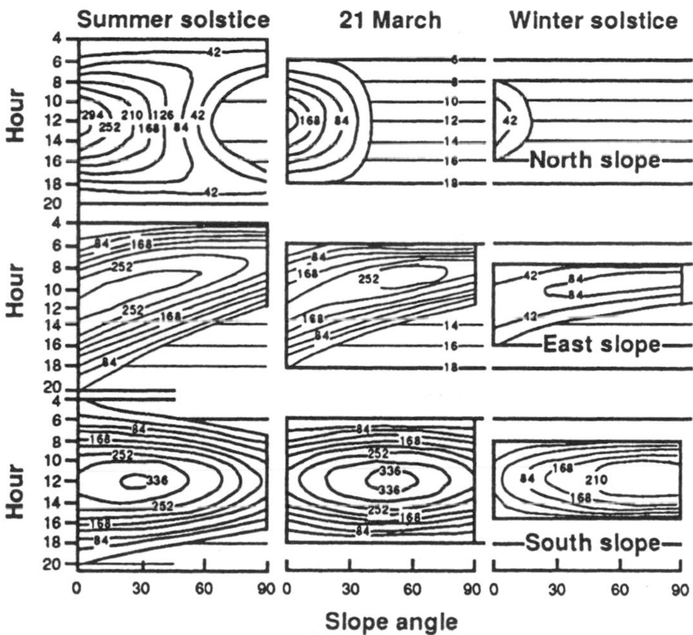

Figure 40-1. Direct-beam solar radiation (MJ m^{-2} hr^{-1}) on clear days on N, E, and S slopes (49°45'N).

Since the slope scale of each diagram for each season begins with a slope angle of 0°, the left margin of each shows the radiation received on the horizontal, which is identical for each column. The right margin of each diagram indicates the radiation received on a vertical wall, the upper limit being the time of sunrise and the lower limit the time of sunset. The north and south-facing slopes naturally show a symmetric distribution of isopleths about the noon line, while the east slope gives an asymmetric distribution.

Considering first the lower row, for the south slope, on 21 March (center diagram), the sun rises exactly in the east and sets exactly in the west. The right diagram shows that throughout the winter the sun strikes southern slopes of all inclinations at the same instant, namely, at sunrise. The situation is different in the summer. At that time the sun rises to the NE and some time elapses after it appears over the horizon, and its rays strike a slope inclined toward the south; the steeper the slope, the longer this will take. The upper and lower boundaries in the left diagram are therefore curved lines, compared with the straighter boundaries in the center and right diagrams. The intensity, which is always greatest at noon, is higher the more perpendicular the slope is to the sun's rays. The maximum, therefore, moves from a relatively flat slope in summer (left) to a steep slope in winter (right). On 21 December, a southern wall receives at noon about the same intensity of radiation as that received on the flat surface by 09:00 at the summer solstice.

For the north-facing slope in the top row in Figure 40-1 at midsummer (left), the times of sunrise and sunset are the same for all the slope angles. If the slope is very steep, the sun is no longer able to shine on it at noon, hence the "neck" cut out of the right margin. Steep north slopes receive direct sunshine only in early morning and late evening. Similar to south slopes, the greatest intensity on north slopes with respect to time is at noon, but the maximum with respect to slope is always on the horizontal surface, in contrast to south slopes.

East slopes (center row) differ from both north and south slopes in that the location of greatest radiation intensity moves according to slope angle and time of year. The maximum radiation is on flatter slopes in summer and steeper slopes in winter. Sunrise is always at the same time, while sunset occurs earlier on the steeper slopes.

East and west slopes receive about the same quantity of solar energy, the former mostly in the forenoon, the latter in the afternoon. Thus, an estimate can be made of the direct-beam solar radiation falling on a west facing slope by taking the inverse of the east facing slope. At noon, both east and west facing slopes are identical. The radiation received on a west facing slope at 13:00 would be the same as an east facing slope at 11:00, west at 14:00, east at 10:00, etc.

Figure 40-1 is designed to allow differences in slope angle and position to be studied in detail. Table 40-1 gives the amount of energy received per square meter in direct-beam solar radiation for each month of the year. The calculations were made by A. Schedler [4028] and are based on radiation measurements for Vienna.

For practical applications in topoclimatology, diffuse solar radiation also needs to be taken into account. In general, the relative importance of diffuse solar radiation to global solar radiation increases with increasing cloudiness, atmospheric turbidity, and surface albedo. Figure 40-2 shows an approximate relationship between the diffuse and direct-beam components of global solar radiation as a function of atmospheric transmission for a horizontal surface obtained at Oxfordshire, UK (52°N) by H. R. Oliver [4018]. In middle latitudes, where cloudiness is common, more solar radiation is typically received as diffuse than direct-beam solar radiation (Figure 25-1). According to radiation measurements made at the Hamburg Meteorological Observatory by W. Collmann [204], for example, a horizontal surface annually receives around 1420 MJ m^{-2} from direct sunshine, but 1800 MJ m^{-2} in diffuse radiation.

Table 40-1. Monthly totals (MJ m^{-2}) of direct-beam solar radiation, from radiation measurements at Vienna, 1930-1932. (After A. Schedler [4028])

	January	February	March	April	May	June	July	August	September	October	November	December	Total
Level Ground	44.8	75.0	159.1	245.0	310.0	343.0	373.0	330.0	238.3	123.1	44.8	34.3	2320.4
10° slope N	21.8	48.6	124.0	213.6	288.9	324.1	346.3	294.4	193.5	85.8	24.3	13.0	1978.3
10° slope NE (NW)	27.6	55.7	133.6	221.1	291.5	328.3	354.7	304.9	203.5	94.6	28.9	18.8	2063.2
10° slope E (W)	43.6	74.5	157.0	241.6	306.1	340.0	367.3	325.8	232.4	117.7	42.7	31.0	2279.7
10° slope SE (SW)	59.5	91.3	178.4	259.6	319.9	350.1	380.2	346.7	256.3	142.8	57.4	44.4	2486.6
10° slope S	66.2	99.7	186.8	266.3	323.7	352.2	387.4	356.8	272.2	155.8	64.9	52.3	2584.3
20° slope N	1.7	21.8	85.4	174.6	255.9	294.0	311.6	249.6	143.2	47.7	2.9	0	1588.4
20° slope NE (NW)	15.1	38.5	107.2	193.5	265.1	303.6	324.1	269.7	168.3	69.1	17.2	8.8	1780.2
20° slope E (W)	43.6	73.7	153.3	234.1	295.2	325.8	352.2	314.9	226.1	116.8	42.7	31.0	2209.4
20° slope SE (SW)	73.3	105.9	195.1	270.5	319.9	345.5	380.7	354.3	271.8	160.0	69.1	56.1	2602.2
20° slope S	85.0	121.0	209.8	283.5	330.8	354.7	388.2	370.2	302.8	185.1	80.8	67.0	2778.9

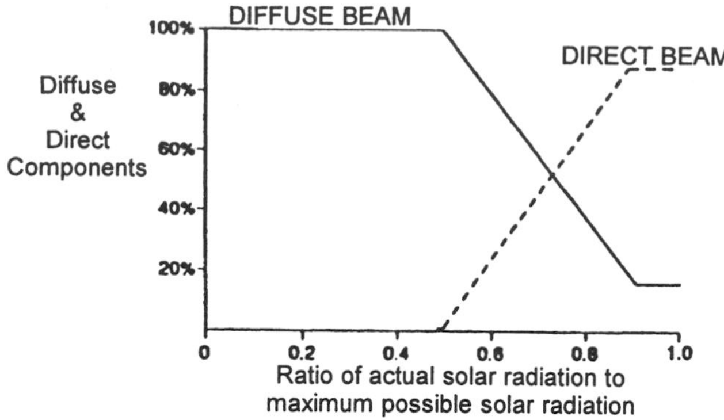

Figure 40-2. Relationship between direct-beam and diffuse solar radiation with the ratio of the measured to the maximum possible solar radiation at Oxfordshire, UK. (After H. R. Oliver [4018])

Diffuse solar radiation received by slopes is proportional to the sky view factor of the slope, and is dependent upon the angular direction of the slope with respect to the solar disk, and the atmospheric transmission. Diffuse solar radiation (and atmospheric longwave emittance) increases with increasing cloudiness, water vapor content, atmospheric turbidity, and atmospheric pressure, while global solar radiation and direct-beam solar radiation decrease with increases in the same factors. Because diffuse solar radiation is more evenly distributed over slopes of different slope angle and slope azimuth than is direct-beam solar radiation, smaller spatial variations in global solar radiation are found in rugged terrain with decreased atmospheric transmission. An impressive example of how diffuse radiation may even out differences due to slope orientation was given in 1952 by J. Grunow [4007]. He set up recorders on the N and S slopes of the Hohenpeissenberg, with their receiving surfaces parallel to the slopes, which were about 30°. According to Kaempfert's [4012] tables, the N slope receives about 2 percent in December and about 73 percent in June of the direct-beam radiation of the S slope. Measurements which included diffuse radiation, however, showed values that varied between 32 percent (December) and 94 percent (July). Similar results were reported by W. R. Rouse and R. G. Wilson [4027] based upon measurements made over north and south-facing slopes at Lake Hill near Mont St. Hilaire, Quebec during May-August of 1964 and 1967. When direct-beam radiation for a horizontal surface is compared to that received on north and south slopes, the ratio of direct-beam radiation for south:horizontal:north surfaces was found to be 120:100:72, while similar ratios for global solar radiation were only 107:100:81. Topographic control of incident global solar radiation is thus enhanced under clear-sky conditions. The distribution of global solar radiation becomes more uniform in complex terrain, however, with increasing cloudiness and reduced atmospheric transmission.

Climatologists have often assumed that the sky scatters diffuse solar radiation equally in all directions. Such a diffuse radiation distribution, showing no directional bias, is referred to as an isotropic distribution, and is a suitable assumption for most microclimatic studies. For applications in topoclimatology and solar energy resource evaluation, however, the spatial distribution of the diffuse radiance over the sky hemisphere needs to be taken into account. Using an actinometer, M. D. Steven [4030] produced standard normalized clear-sky radiance distributions for various zenith angles at Sutton Bonington, UK [53°N). L. J. B. McArthur and J. E. Hay [4016] combined actinometer measurements and all-sky photographs to create a more detailed mapping of the diffuse sky radiance. An example of their results for clear-sky conditions at 13:40 on 10 February 1978 is shown in Figure 40-3. The isoline patterns for incoming diffuse radiation reveal three basic trends. First, there is a brightening of the sky along the horizon due to the increased scattering mass, which occurs as the solar zenith angle increases. Second, increased scattering is found in the circumsolar region near the solar disk because of preferred forward scattering by aerosols. Finally, a region of minimum sky brightness is found in a general region approximately 90° from the solar disk. In Figure 40-3, for example, when the sun is in the south-southwest minimum brightness is centered in the northeast quadrant. Clear-sky radiance distributions were found to be between 7 and 118.6 W m^{-2} sr^{-1}. Since radiation to a point on the surface comes from the entire sky hemisphere, it is often useful to measure how much radiation comes from different parts of the sky. A steradian (sr) is the unit for measuring incoming radiation from a portion of the sky and is equal to the solid angle subtended at the center of a sphere of a given radius by an area equal to the radius squared on the surface of the sphere. Diffuse radiation can also be affected by screening and reflection from the surrounding topography (J. A. Olseth and A. Skartviet [4019]).

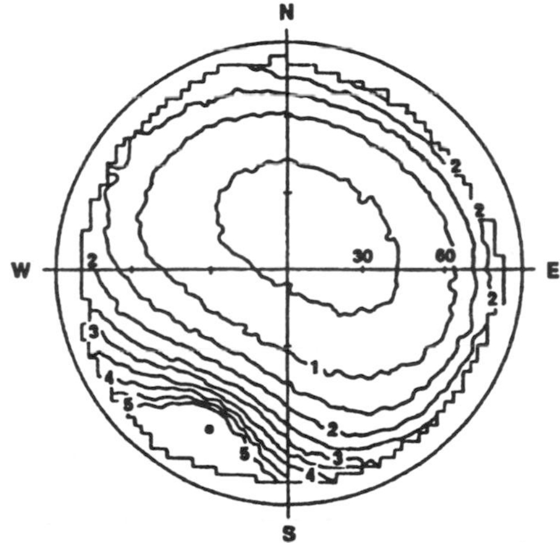

Figure 40-3. Clear-sky normalized sky radiance distribution (per steridian) for 13:40 on 10 February 1978 at Vancouver, BC. The solar disk is shown by a black dot in the southwest quandrant. (After L. J. B. McArthur and J. E. Hay [4016])

Cloud cover introduces considerable complexity into the sky radiance distribution due to the greater scattering effect of cloud droplets. L. J. B. McArthur and J. E. Hay [4016] found that sky radiances for partly cloudy conditions ranged between 20.9 and 132.6 W m^{-2} sr^{-1}, and displayed some spatial similarities to the clear-sky pattern. The most important differences were an increased minimum radiance due to the water droplets, and large radiance values associated with reflectance of solar radiation from cloud elements near the solar disk. Under overcast conditions the clear-sky radiance distribution was nearly completely obscured and, although the sky radiances did range between 14 and 28 W m^{-2} sr^{-1}, the assumption of an isotropic distribution was found to be reasonably valid for the central portion of the sky. J. E. Hay [4008] and R. Perez, et al. [4024] provide examples of efforts to model the effect of the anisotropic distribution of sky radiance upon the receipt of diffuse solar radiation on inclined surfaces.

Many studies of slope exposure have assumed clear-sky conditions as a simplifying condition. G. A. Olyphant [4020] found that global solar radiation transmission during clear-sky summer days averaged 0.80 at the Indian Peaks section of the Colorado Front Range, and was symmetrical about solar noon. Such days, however, were not typical, because orographic and convectional lifting normally produced cumulus clouds in the afternoon. On more typical days, global transmission averaged 0.75 before noon and 0.55 in the mid-afternoon. These diurnal changes in atmospheric transmission affected the incident global solar radiation, with daily $K\downarrow$ totals being greatest on south-facing slopes on clear-sky days, but greatest on east-facing slopes on days with partly cloudy skies.

The third source of solar radiation on an inclined surface, the ground-reflected radiation, depends upon the amount of global solar radiation incident upon the ground, the surface albedo, and the view factor of the inclined surface to the ground. Measurements of ground-reflected radiation on inclined surfaces by P. Ineichen, et al. [4011] indicate mean errors of less than 9.8 W m^{-2} when an isotropic distribution of ground-reflected solar radiation is assumed, and the surface albedo of the site in question is known. I. Dirmhirn and F. D. Eaton [4004], however, found an anisotropic distribution to ground-reflected solar radiation from snow-covered surfaces in Cache Valley, Utah. Estimation of ground-reflected solar radiation is further complicated by the large spatial variation in surface albedo found in many environments. Estimation of net solar radiation K^* is especially sensitive to the determination of surface albedo. D. S. Munro and G. J. Young [4017] found the following average albedo values in a glacial basin in the Canadian Rocky Mountains: 0.24 ice, 0.25 bare soil, 0.50 firn, 0.61 old snow, and 0.74 new snow. The migration of the snow line with time, the presence of new falling snow, and snow aging combine to create extremely variable and dynamic surface albedo conditions in mountainous regions that exert a strong effect upon net solar radiation K^*.

The global solar radiation incident on walls facing N, E, S, and W, is given by the observations made at the Hamburg Meteorological Observatory over a period of years. Figure 40-4 from K. Gräfe [4006] shows the annual variation of global solar radiation received at a north-facing wall. This shows the effect of diffuse radiation at its strongest. The lower diagrams (A and B) have had added to them the direct-beam solar radiation S, from W. Kaempfert and A. Morgen [4013], which is restricted to the half year between the spring and autumn equinoxes.

The substantial ground reflection from snow cover and backscattering from the atmosphere in A raises the winter observations far above the normal maximum shown by the dashed line. Diagram B shows the 3 yr average value of global solar radiation on the north slope, and C the ratio of this value to that on a horizontal surface. In autumn, the measurement points lie closer to the dashed curve (A) than in spring. This is because of greater turbidity at that time of the year, which increases diffuse radiation at the expense of direct-beam solar radiation. Figure 40-5 gives the 3 yr average values of global solar radiation, as measured in Hamburg, for a S wall, a N wall, an E (W) wall, and a south-facing 45° slope. Here

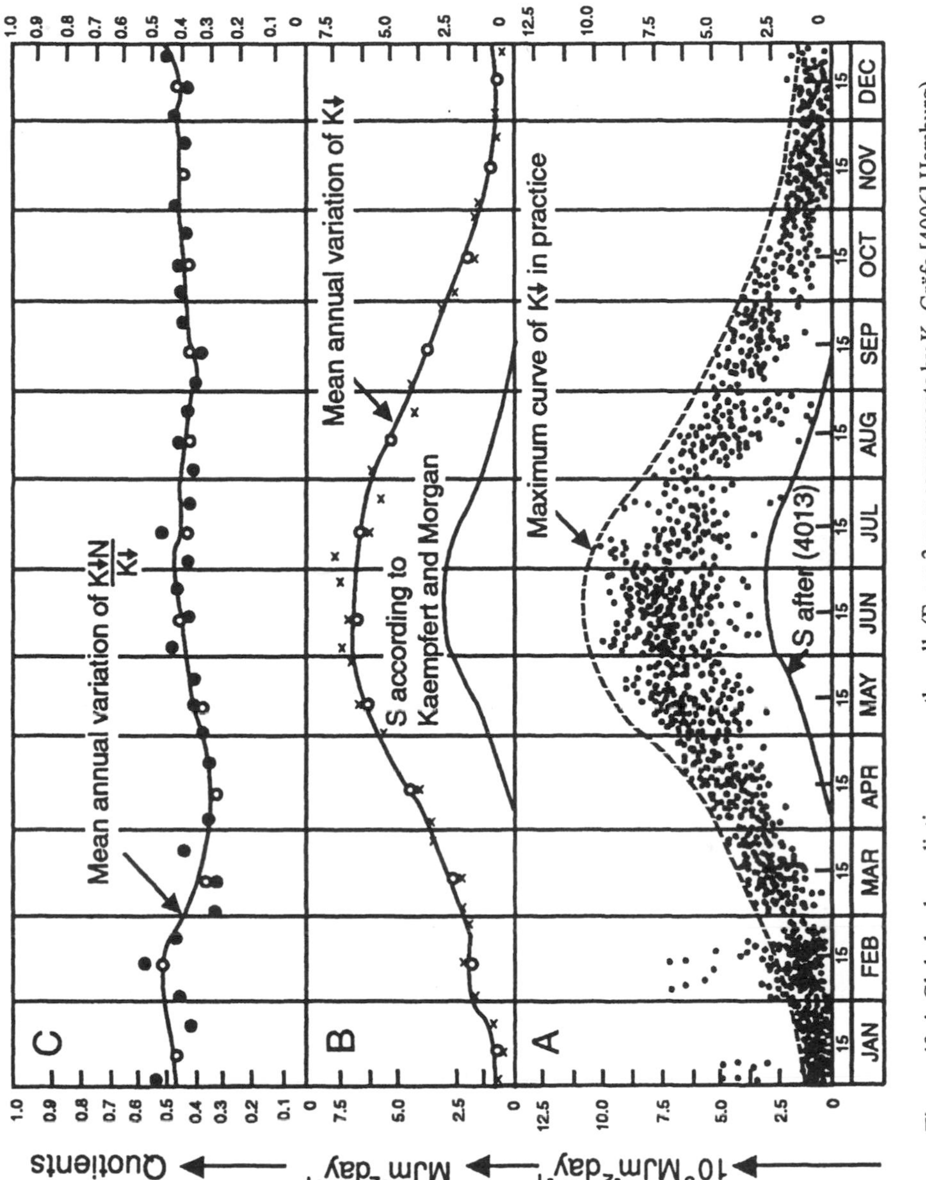

Figure 40-4. Global solar radiation on a north wall. (From 3 yr measurements by K. Gräfe [4006] Hamburg)

the east-facing wall receives more $K\downarrow$ in mid-summer than the south-facing wall due to the increased afternoon cloudiness characteristic of Germany.

The radiation on walls facing the eight principal directions was calculated for each month, for both clear and cloudy weather, by M. Decoster, et al. [4003] for Leopoldville at 4°S and by M. Decoster and W. Schüepp [4002] for Stanleyville at 1°N. Figure 40-6 shows global solar radiation ($S + D$) on a horizontal surface and on walls in clear weather at the equator (Stanleyville). While a south-facing wall in the winter half year receives about the

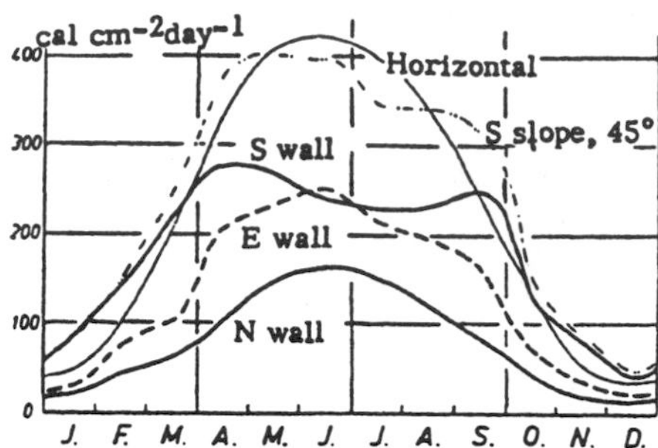

Figure 40-5. Global solar radiation measured on a horizontal surface, a south-facing 45° slope, and on three walls. (After K. Gräfe [4006])

same amount of energy as a horizontal surface in temperate latitudes (Figure 40-5), the amount received by a wall in the tropics remains far below that falling on the horizontal surface. The E wall receives about the same quantity of energy throughout the year, and depicts the double zenith position of the sun (as does the horizontal surface). The N and S walls exhibit a marked seasonal variation and reach their maximum when the sun stands farthest north or south, respectively.

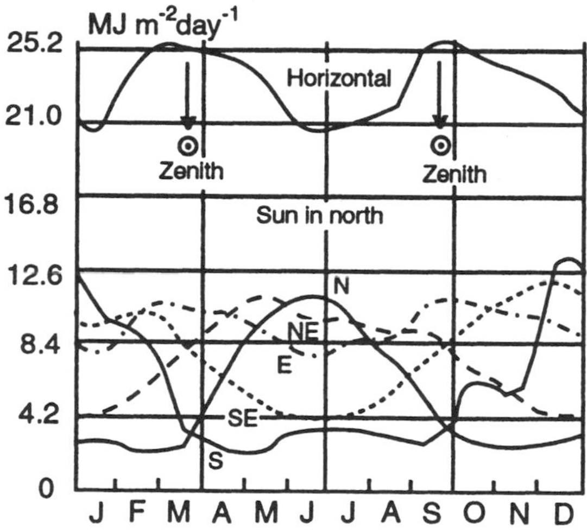

Figure 40-6. Global radiation on a horizontal surface and on vertical walls near the equator. (After H. Decoster and W. Schüepp [4002])

A comparison of Figures 40-5 and 40-6 illustrates the difference between temperate latitudes and the tropics. In regions where the sun comes near the zenith, it is clear that the orientation of a slope does not have the practical importance it acquires in mid-latitudes. In the arctic regions, the low altitude and increased optical path length of the sun results in increased scattering such that global solar radiation is predominantly comprised of diffuse solar radiation. This somewhat balances out the difference due to varying slopes. On the other hand, it is so cold that even a small gain of heat by virtue of a slope's direction can be very significant.

The global solar radiation results shown in Figures 40-4 through 40-6 are for vertical surfaces in which the effects of slope angle and slope orientation are most extreme. Naturally occurring surfaces generally experience much smaller variations in the cumulative receipt of global solar radiation because they normally have a much smaller range of slope angles. Frequent cloud cover causes increased diffuse and decreased direct-beam solar radiation. H. R. Oliver [4018] measured global solar radiation over a 17° north-facing and a 10° south-facing slope at Oxfordshire, UK (52°N) from December 1989 to July 1990. The cumulative results over the eight month period, which covered a complete high-sun/low-sun solar season, differed by only 25%, with clear days accounting for essentially all of the difference.

The principles governing atmospheric longwave radiation input $L\downarrow$ to individual surface facets have already been discussed (Section 5). It should be mentioned, however, that atmospheric longwave radiation also follows an anisotropic distribution, so that the surface receipt of $L\downarrow$ will vary with the orientation of the surface to the sky and with atmospheric conditions. Longwave radiation emitted from the surrounding terrain $L\uparrow_t$ also exhibits significant spatial variations resulting from changes in surface emissivity and surface temperature.

Table 40-2: Individual energy term totals for modeled net radiation over a snowcover in the upper Green Lakes Valley of Colorado. (From G. A. Olyphant [4021])

Term	Mean MJ m^{-2} d^{-1}	St. Deviation MJ m^{-2} d^{-1}	Range MJ m^{-2} d^{-1}
S	15.28	2.15	8.37 – 17.79
D	7.83	0.65	4.94 – 8.67
$K\uparrow_t$	1.55	0.57	0.38 – 3.68
$L\downarrow$	19.55	1.54	13.23 – 22.23
$L\uparrow_t$	5.23	2.12	1.63 – 13. 90

The combined effect of these individual terms can be summarized from the results in Table 40-2. This table presents daily totals from the modeled energy terms over the upper Green Lakes Valley of the Colorado Front Range (G.A. Olyphant [4021]). The direct-beam total of 15.28 MJ m^{-2} d^{-1} was only 0.29 MJ m^{-2} d^{-1} less than the measured unobstructed ridge top total. A large range in S totals was found due to the effects of variable slope angle, slope azimuth, and solar obstructions. Diffuse solar radiation from the sky D and terrain $K\uparrow_t$ were more than a third of total $K\downarrow$, and less variable than S. Radiation input to the basin was dominated by the longwave terms that were also quite variable. Heterogeneous terrain is thus subject to considerable spatial variation in the individual energy terms. Simple characterization of the all-wave radiation input to mountainous terrain surfaces by the direct-beam radiation input alone are likely to produce large errors. G. A. Olyphant, however [4021], did find that the individual energy components, though individually spatially variable, combine in such a way as to decrease the inter-site variability in radiation input.

Whether some particular type of slope is favorable or unfavorable for plants, animals, or man, depends on the regional climate. Where rainfall is plentiful but solar energy is somewhat lacking, as in Germany, southern slopes provide a better habitat for most plants. However, where there is plenty of solar energy but a shortage of water, the shady northern slopes support more luxuriant vegetation. R. Geiger writes that he was very impressed during his first flight over the Sinai Peninsula in October 1956 to observe the green shimmer of a thin vegetation cover only on the north side of the mountains. R. H. Aron had a similar experience in flying over California in summer. On east-west oriented mountain ranges, he observed that the north-facing slopes supported scrub oaks and appeared green, while south-facing slopes contained yellow or brown grass. The ridge formed distinct lines between the two types and colors of vegetation. In another example, the saguaro cactus grows on north-facing slopes in the southernmost part of its range and in the northernmost part on south-facing slopes and in protected areas (G. Krulik [2918]). J. Walther wrote in his book on the laws governing the formation of deserts (Berlin, 1900): "While in polar lands most horizontal surfaces are empty of vegetation, because only the shrubs on the mountains can receive enough radiation to meet the requirements of growth, a mountain slope in the desert becomes more bare the steeper it is and the more it is baked in the blistering sun. Only the northern slopes (in the Northern Hemisphere) are able to support a richer vegetation, and shady valleys and sheltered hollows favor the settlement of plant communities." As a graduate student P. E. Todhunter remembers studying a fern that actually grew in the extreme desert environment near Palm Springs, California (34°N). The fern was only found beneath narrow overhanging rock crevices that faced directly north. This location enabled the fern to avoid essentially all direct-beam solar radiation, and most diffuse solar radiation, while obtaining additional moisture input from runoff off the adjacent rocks. N. Polunin [4025] noted that farther poleward vegetation on north-facing slopes (northern hemisphere) becomes more and more limited due to the late or non-melting of snow on these slopes. In Lapland and Labrador, however, he observed that because of the drying out of the south-facing slopes the north-facing slopes had more luxuriant vegetation. The south slopes were vegetated primarily by lichens and xeric undershrubs while the north-facing slopes had herbs, ferns and mosses. With respect to erosion, K. L. Pierce and S. L. Colman [4022 and 4023] found less vegetation and more degradation on south than north-facing scarps in Idaho. On the 2 m high scarps the degradation rate was 2 times and on 10 m high scarps 5 times more on south than on north-facing scarps.

41 The Effect of Differing Amounts of Sunshine on the Microenvironment

We will now turn our attention to the consequences of varying amounts of radiation on small scale features.

A lone tree trunk standing erect is circled by the sun during the day. One half of the surface of the trunk is exposed to the sun at any time. K. Krenn [410] made use of measurements of direct-beam solar radiation intensity from Vienna at 202 m above sea level and from the top of the Kanzel Mountain in Carinthia, Austria, at 1474 m, to calculate the solar radiation received by the sides of a tree trunk on a clear day. The trunk was considered circular in cross section, 1 m in diameter, and divided into 16 sectors corresponding to points of the compass, each 1 cm in height. The quantity of direct-beam solar radiation received by each of these areas was calculated for each hour during the day. The values arrived at for 1 April on the summit of the Kanzel are plotted by K. Krenn [4107], as shown in Figure 41-1.

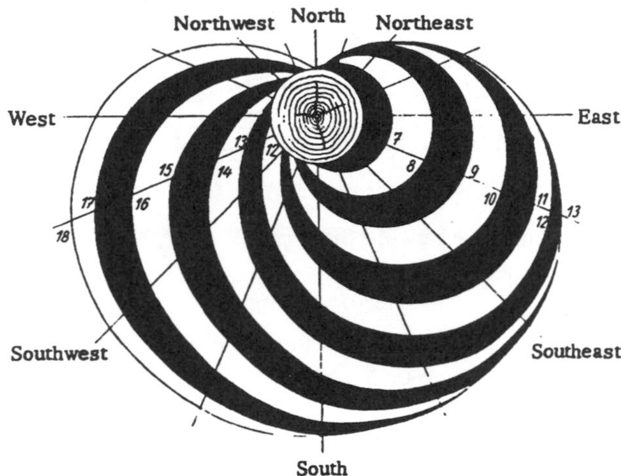

Figure 41-1. Amount of direct-beam solar radiation per hour on a standing tree trunk. (After K. Krenn [4107])

The amount of direct-beam solar radiation received by each sector during each hour is represented by a vector length, laid out along the appropriate radius, the end of the vector being marked by the hour in question (for the scale see Figure 41-3). Lines are then drawn connecting the points marked for the same hours, and the spaces between are alternately black and white to make the distribution more apparent and show the effect of the sun's movement from east to west. The outer curve represents the daily total of solar radiation received. Since clear weather has been presupposed, it is symmetric about the N-S axis.

The temperature variation in the bark of the tree, in the living cambium below, and in the woody parts of the trunk corresponds to the pattern of solar radiation received by the trunk. The energy absorbed by the bark penetrates inward as if into a soil consisting of three layers. The density, thermal conductivity, and thermal capacity of these layers depend on the type of tree and its moisture content at the time. Numerical values for these quantities may be found in B. Koljo [4106]. Temperature measurements have frequently been made in the trunks of standing trees. Figure 41-2 gives a simplified version of the daily temperature in a 50 yr old stand of Sitka spruce on the Nystrup plantation in West Jutland, Denmark, on 20 July 1951, as measured by N. Haarløv and B. B. Petersen [4104] at a height of 1.3 m above the ground. The full lines give the temperatures below the bark for the exposures marked on them, the dotted curve is for the temperature at a depth of 5 cm in the wood on the SW side, and the dashed line is the air temperature.

The temperature maximum follows the sun from S to SW. The parts which receive solar radiation later in the day are drier and have already been warmed to some extent. Therefore, the maxima increases as the day continues, in spite of the fact that solar radiation is symmet-

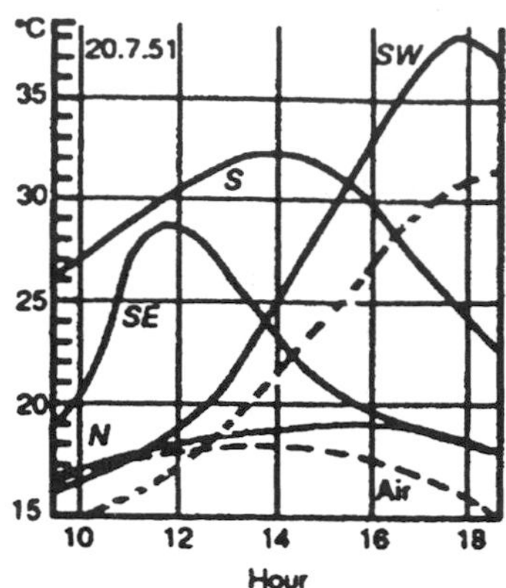

Figure 41-2. Daily temperature in the bark of a Sitka spruce. (After N. Haarløv and B. B. Petersen [4104])

ric about the noon line. The greatest maximum in the SW is 9°C higher than the SE maximum (Figure 41-2). Bark temperatures on the N side are a little above air temperature at first because of absorption of diffuse radiation, but it is only toward evening that the increasing difference with respect to air temperature shows how the whole trunk has warmed up. The temperature of the N side increases as a result of heat conducted into the tree.

The differences are reduced when shading occurs from branches of neighboring trees, or with increased cloudiness. The highest bark temperatures measured by N. Haarløv and B. B. Petersen when the air temperature was 23°C were: 26°C in the N, 33°C in the SE, 38°C in the S, and 42°C in the SW. Further data on the temperature of trees can be found in B. Primault [4111], B. Koljo [4106], E. Gerlach [4103], and H. Aichele [4100].

In Figure 41-3, the daily direct-beam solar radiation of the tree trunk on the Kanzel on 1 April from Figure 41-1 is shown (dashed line), but only its right half because of the symmetry about the noon line. Figure 41-3 shows the seasonal effect by thc inclusion of radiation curves for 1 January (dotted line), 1 April (dashed line), and 1 July (solid line). The influence of elevation is also shown by similar curves for a tree in Vienna on the left of the diagram. The seasonal effect is least on the north side. On the south side at the higher elevation site, the low winter sun provides the vertical trunk with more energy than at any other time of the year. On 1 January for the Kanzel, this is more than double the value for Vienna. Even in Vienna, the S side (but no other) receives more direct-beam solar radiation in January than in July.

Great temperature contrasts are, therefore, experienced in winter between the cold air, which may be well below the freezing point, and the surface of the trunk on its southern side. The layer of cambium cells below the bark, which at the end of winter contain considerable amounts of water in preparation for the approaching rise of sap, may be unable to withstand the violent alternation of midday heat and night frost. It often splits longitudinally with an audible report. The crack in the bark so formed may be hair-fine at first, widening later to form fissures in the bark. Figure 41-4 shows the damage caused to a red beech in a photograph taken by M. Seeholzer [411]. H. Schulz [4113] demonstrated a relation between the widening of frost cracks and continuing frost action. He was able to establish that the width of a crack varied with the rhythm of daily temperature fluctuations, and it was greatest in

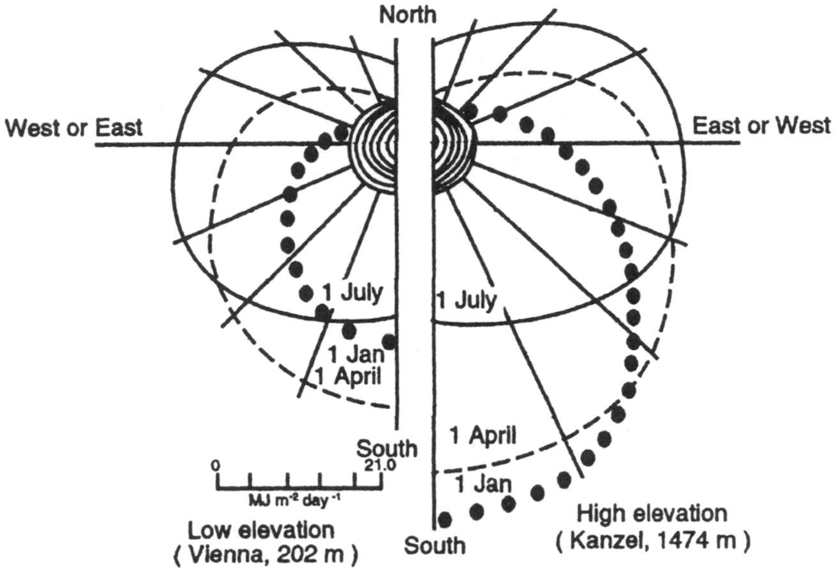

Figure 41-3. Direct solar radiation on a tree standing on level ground (left), and on a mountain (right), on three cloudless days. (After K. Krenn [4107])

poplar and least in oak. Trees suddenly exposed by felling are particularly susceptible. Fruit trees are also subject to this kind of damage.

The risk can be reduced by taking measures to reduce the amount of radiation absorbed by the surface of the trunk. B. Primault [4111] found that a straw matting cover reduced the diurnal temperature variation on the SW side of trees by 29 percent, and white paint by 35 percent. H. Aichele [4101] painted a ring of whitewash on fruit trees and compared cambium temperatures under the white ring and under untreated bark. The temperature differences between the N and S sides of the tree were reduced from 8°C (not whitewashed) to 4°C (whitewashed) in January, and from 20°C to 14°C in February. Tree bark can assist in controlling temperatures. V. Nicolai [3916] states that tree species with white bark avoid overheating of their surface by reflection of solar radiation. Trees with fissured and scaly bark shade the inner parts of the bark and lose heat from the outer surface through convection and radiation. Bark can also insulate trees to avoid overheating (cork). V. Nicolai [3916] further suggests that trees with thin smooth bark form closed stands to avoid overheating.

A trunk's cross section is usually oval (not circular). In stands on level ground, the maximum diameter is usually in the direction of the prevailing wind, and the smallest diameter perpendicular to this direction, as shown by G. Müller [4110]. The cross sections of tree trunks standing on slopes are no longer controlled by wind direction but by the light received, which is least in the upslope direction and greatest in the opposite direction. This results in the greatest diameter following the line of steepest slopes. Figure 41-5, for example, shows the results obtained in measuring 331 oaks in two test areas in the Palatinate Forest. The sketch on the left shows the area with contour lines drawn in; on the right the

Figure 41-4. Cracks and peeling of bark of a red beech as a result of intense midday solar radiation at low air temperatures. (After M. Seeholzer [4114])

frequency distribution of the greatest (solid lines) and smallest (dashed lines) diameters are arranged on a compass rose.

Hitherto only the trunk of the tree has been discussed. Now the whole tree will be considered. G. Eisenhut [4102] examined 5276 separate blossoms uniformly spread over the crown of an isolated 90 yr old lime tree *(Tilia euchlora)* in Jesenwang. His objective was to establish how many blossoms were in full bloom (or withered) according to their position on the crown and exposure. G. Eisenhut [4102] found the following percentages:

Exposure	S	W	E	N
Upper crown	38.7	-	-	-
Middle crown	30.2	22.1	16.9	2.9
Lower crown	14.3	19.5	21.9	5.7

Blossoming appears to depend on orientation. The tip of the crown, which receives sunlight all day, blossoms first. In the middle crown the directions follow in an order corresponding to that of maximum temperatures in Figure 41-2. In the lower part of the crown, the contrasts are more balanced. The sun at low solar altitude gives preference to the E in the morning and the W in the evening over the S and N. The number of blooms present also

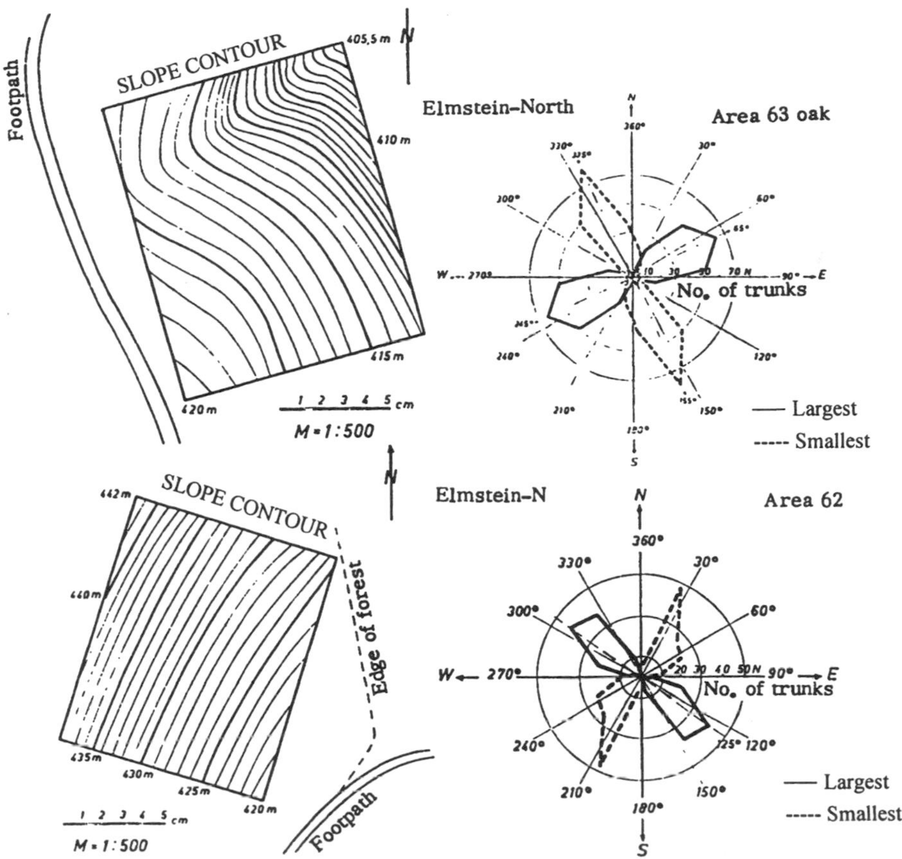

Figure 41-5. The greatest trunk diameter of oaks, which need much light, in the Palatinate Forest, is in the direction of the slope. (After G. Müller [4110])

depended on exposure. In every 100 g of freshly weighed petals there were 296 blooms in the tips of the crowns, 327 in the middle crowns, and 251 in the lower crowns. South facing exposures contained 130 blooms and north-facing sides 115.

A. Scamoni [4112] studied the daily sequence of blossoming of 181 male blooms on a 15 yr old isolated pine. The flowers were on parts of branches at a height of 1.1 m above the ground, and the sequence in which they opened is shown graphically in Figure 41-6.

Although there are many other influences at work, such as differences of shadow, density of flowers, supply of sap, and so forth, the effect of orientation is clear. Prof. H. Walter of Stuttgart-Hohenheim wrote to R. Geiger about this diagram: "I can give you a good example from Arizona. The enormous saguaro cactus always blooms first on the SW side. No buds develop on the NE side. The concentration of sap is also greatest in the cells on the SW side, and lowest on the NE side. The growth of *Echinocactus wislizeni is* restricted on its SW side by the greater dryness there, so that it gradually curves over toward the SW and finally topples."

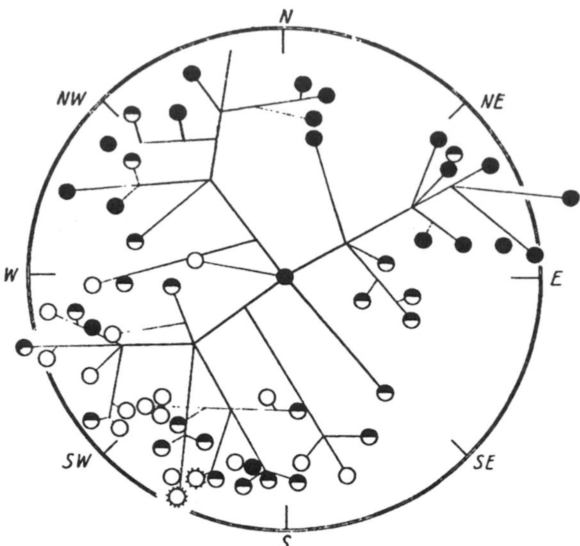

Figure 41-6. Time of blossoming of a pine tree, standing by itself, near Eberswalde. (After A. Scamoni [4112])

In agriculture, level fields are often modified by planting crops in rows oriented in various directions. Although the total amount of radiation received on horizontal surfaces may be the same, it is nevertheless distributed differently over the small ridges. Before the young plants had time to alter the microclimate, N. Weger [4115] in Geisenheim and H. Lessmann [4109] in Freiburg near Emmandingen investigated the temperature at depths of 5 cm and 10 cm respectively in various experimental plots. Figure 41-7 shows H. Lessmann's [4109] results in the form of 2°C isotherms for rows running N-S, 13 to 15 cm high, 26 to 30 cm apart, with sides sloping at 45°, at three times on a clear day.

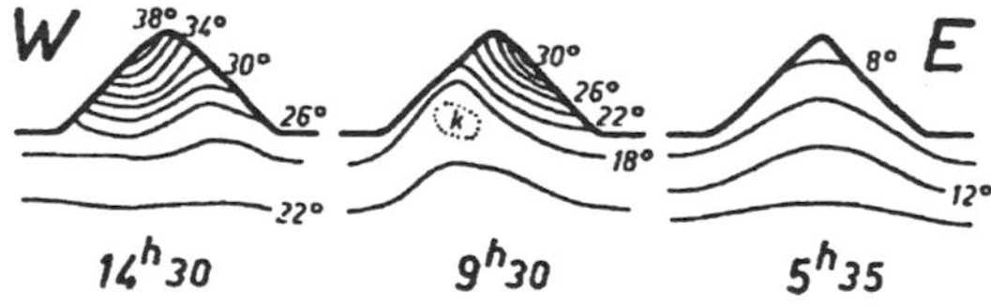

Figure 41-7. Temperatures in unplanted ridges, running N-S, on a sunny day in August. (After H. Lessmann [4109])

Twenty minutes after sunrise (right diagram) the top of the ridge was the coldest part, and it is colder than the level ground. It may be colder in part because it was farther from the heat coming up from the ground and in part because the ridge top has the highest sky view factor, resulting in increased loss of energy by net longwave exchange. By 09:30 the ridge has been heated to 30°C at the surface of the sunny eastern slope (center diagram), while the west side was still cold from the night. Heating of the western slope to over 39°C occurs in the early afternoon (left diagram). Similarly during the day plants in N-S rows receive substantially more heat. Both investigations agree on this point; for rows running E-W, N. Weger [4115] was able to establish a gain, though a small one, over level ground. H. Lessmann [4109] did not observe this.

N. Weger [4115] extended his temperature measurements at a depth of 5 cm to the ridge and furrow of the rows, and to flat dams 100 cm wide at the base and 50 cm at the top, which ran E to W, and to circular mounds which were 18 cm high and 50 cm in diameter at the surface. He was able to establish a strong seasonal influence. In four periods of clear weather, the average daily gain of heat of each feature was compared with level ground. All elevated features had temperatures that were equal to, or in excess of, the level ground. Differences in computed degree-hours for 2 hr intervals are shown in Table 41-1.

Table 41-1. Average daily gain of heat (degree-hour) of various terrain configurations in excess of that of a level surface.

Periods (number of days of clear weather)	Conical mound	Dam E-W	Rows: Side N-S	Rows: Side E-W	Rows: Furrow N-S	Rows: Furrow E-W
21 July to 1 August (9)	50	30	16	14	0	0
29 August to 10 September (6)	39	32	14	10	1	2
19 to 25 September (4)	34	34	12	7	6	10
3 to 10 October (6)	26	30	6	5	4	2

The conical mounds achieve the greatest gain. It is worth noting that the gain at the dam increased until autumn. In Table 41-1, the temperature deficit in the valleys is neglected because nothing is growing there. The young plants on the ridges respond to the extra heat by sprouting sooner and growing more quickly.

Some of the advantages of planting on mounds to improve the microclimate are described by D. L. Spittlehouse et al. [4420]. "There are often significant regeneration problems with white spruce on relatively high yielding sites in the Sub-Boreal Spruce and Englemann Spruce-Subalpine Fir zones in the interior of British Columbia." In an attempt to ameliorate these problems they planted the spruce seedlings on mounds about 0.3 m high. This resulted in more favorable growing conditions including increased light, temperature, doubling of growing degree days at 0.1 m, earlier warming in spring, and drier soils. After five years the spruce seedlings on the mounds had a higher survival rate, were on the average over 70 percent taller, had straighter trunks and more extensive root systems than those not planted on mounds.

When the plants have grown, the radiation picture becomes quite different. W. Kaempfert [4105] has studied the influence of the direction in which the rows of plants lie in relation to the ratio (E) of the plant's height to the spacing between rows. Figure 41-8 considers the plants as vertical walls and portrays the annual variation of sunshine duration. The time of day is read on the ordinate. The sunrise and sunset curves delimit the possible sunshine duration. The first diagram shows the conditions prevailing in rows running N-S. The second diagram is for rows turned at an angle of 22.5°, and so on, until the E-W rows are reached. The diagrams can then be continued by inverting the time scale.

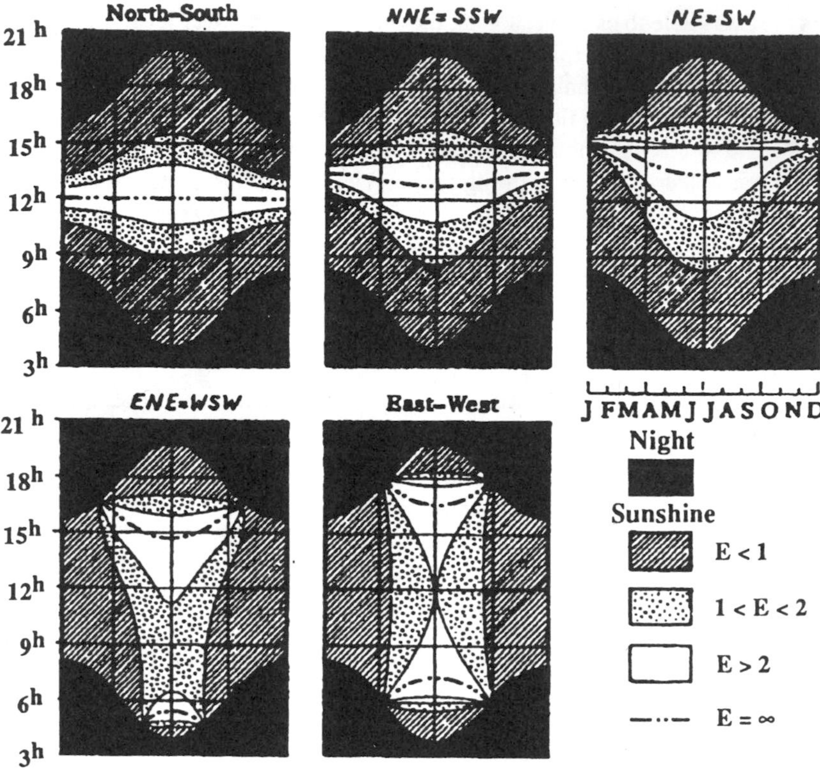

Figure 41-8. Sunshine duration in rows of plants, for various row directions, and various widths between plants. (After W. Kaempfert [4105])

The boundary case $E = \infty$ (dot-dash line) means that the rows of plants are in contact with each other (distance apart = 0). With rows running N-S, the sun is able to penetrate to the ground only for an instant at noon through the opening between the rows. This instant is independent of season. With NE-SW rows, however, this instant will be earlier in summer and later in winter. If the rows run ENE-WSW, the sun can no longer penetrate the opening in winter and in spring and autumn only in the late afternoon from the WSW; in midsummer, when the sun rises far in the NE, it can even penetrate a second time in the early morning. If $E = 2$, that is, if the plants are twice as high as their perpendicular distance apart, the sun will shine into the rows only at the times covered by the white area (Figure 41-8). If $E = 1$, then it will also shine in during the period covered by the dotted area.

In a drought-threatened tree nursery, young firs were set in hollows. The weed-covered topsoil was turned over, and divots placed around the plant hollows (Figure 41-9). Only the plants marked B, with the divots on the eastern side of the hollow, came through well since they were placed on a slight W slope and were able to make use of the night dew and humidity for a few extra hours in the morning before the heat of the day was felt. The plants marked *A* on the E slope were soon dried by the morning sun, and the afternoon shade was no substitute for the loss of moisture.

Figure 41-9. In a new planting, microclimate is determined by the position on which the divot is placed.

The magnitude of the spatial variation in total solar radiation input in complex terrain due to slope angle and slope aspect effects is normally greatest in the middle latitudes. Here, topography can create small-scale microclimatic variations that are comparable to those found over level terrain over a large latitude range. The receipt of solar radiation on south-facing slopes (northern hemisphere) may be similar to that received on level terrain at much lower latitudes. Surface solar radiation input will be correlated with moisture availability, air and soil temperature, humidity, and light intensity. These variations affect soil formation and characteristics that influence vegetation patterns. On north-facing slopes (northern hemisphere) the reduced input of solar radiation will decrease surface evapotranspiration that may affect surface soil moisture content as well.

In natural environments the surface soil moisture regime is not only determined by the input of solar radiation as determined by slope angle and slope aspect, but also to the position along the slope and slope shape as well. These last two variables affect the receipt of precipitation, the collection of surface waters, the movement of surface runoff, and subsurface drainage patterns. Soil moisture influences soil development, soil depth, erosion rates and nutrient levels along the slope. Thus all four slope factors combine to produce a diversity of microclimatic environments that have profound effects upon the composition, diversity, and dominance of plant species. Within such a diverse setting it is difficult to separate the effects of climatic factors from non-climatic effects on individual plants and plant communities.

V. J. Lieffers and P. A. Larkin-Lieffers [4108] provide an example of how slope angle, slope aspect, slope position, and slope shape affect the distribution of plants and soil chemical properties along a coulee in a semi-arid grassland community near Lethbridge, Alberta.

42 Small-Scale Topographic Influences at Night (Cold Air Currents, Frost Hollows)

There is a transition from ridge tops, where topographic effects are minimal and L* and regional air flow dominate, and down-valley locations where the sky is progressively shielded a bit, and topographic effects become more important. There is a two-fold transition, one of declining influence of the regional airflow, and one of increasing influence of topography. At lower local elevations large lapse rates between the surface and upper-air temperature are associated with stable atmospheric conditions, calm winds, and clear sky conditions. Smaller or even negative lapse rates are associated with unstable atmospheric conditions, overcast skies, and strong winds. When these latter conditions are found, the lapse rate will approach the dry adiabatic lapse rate and terrain effects on surface air temperature will approach zero.

Minimum surface air temperatures are also influenced by the presence of cold air drainage and the stagnation of surface air. R. Geiger and G. Fritzsche [2008] investigated night temperature minima at a height of 10 cm above the ground during the spring and summer of 1939 in order to find the causes for the varied severity of frost damage to a stand of pines near Eberswalde. The effect of height was evident both on individual nights with late frost and for the average of all observations, although the ground appeared to be level, and surveying disclosed only a gentle slope. Table 42-1 gives the temperature minima observed at five points within 100 m of each other. Differences of a few tenths of a meter give rise to substantial temperature differences. The lowest observation point always had the lowest temperature at night, and it fell below the freezing point on 17 nights during the growing period. It was warmest at the highest measurement point, where there were only 12 nights with frosts.

Table 42-1. Night minimum temperatures (°C) over nearly level ground near Eberswalde.

Nights with frost, 1939	Elevation (m)				
	36.1	36.2	36.3	36.6	37.7
23-24 May	-7.6	-6.9	-5.4	-5.1	-3.7
2-3 June	-9.4	-7.9	-8.2	-6.7	-5.0
2-3 July	-2.1	-1.3	-1.1	0.0	0.1
11-12 July	-2.5	-1.4	0.0	1.6	1.9
Mean of 30 coldest nights	-0.6	-0.4	0.1	0.7	1.7

In calm conditions air that is cooled in contact with the ground or the upper surface of plants will increase in specific gravity and start to flow down to lower levels. It is, therefore, customary to speak of a nocturnal cold air current by analogy to water, which also flows in a density current toward lower levels. The outflowing air has to be replaced by a compensating airflow aloft. The downslope flow is frequently very shallow and can be impeded by the presence of minor barriers, such as stone fences, bridges, and hedge rows, which can lead to the stagnation and accumulation of cold air. As the slope of the terrain increases, the density flow becomes stronger, and the pooling of cold air by such barriers becomes less common.

M. Reiher [4322] calculated the rate of airflow from given density differences and arrived at values that were, for the most part, less than 1 m sec^{-1}. Small-scale air currents are, therefore, usually weak, hardly detectable by an observer, and easily upset by obstructions or winds of even moderate strength. They move as part of a closed circulation, and can develop only after night cooling has progressed for a time. This movement of air will continue throughout the night as long as a negative net radiation regime is maintained at the outer effective surface.

Figure 42-1 shows a fir plantation damaged by frost in the great Anzing-Ebersberg forest near Munich. The left part of the diagram shows the distribution of the different age classes of trees, and the right half shows the topography by height contours drawn at 10 cm intervals. To the eye, it appears to be a level surface. Table 42-2 gives the night minima at 5 cm above the ground, measured by R. Geiger [4201] at the nine observation sites during the spring of 1925.

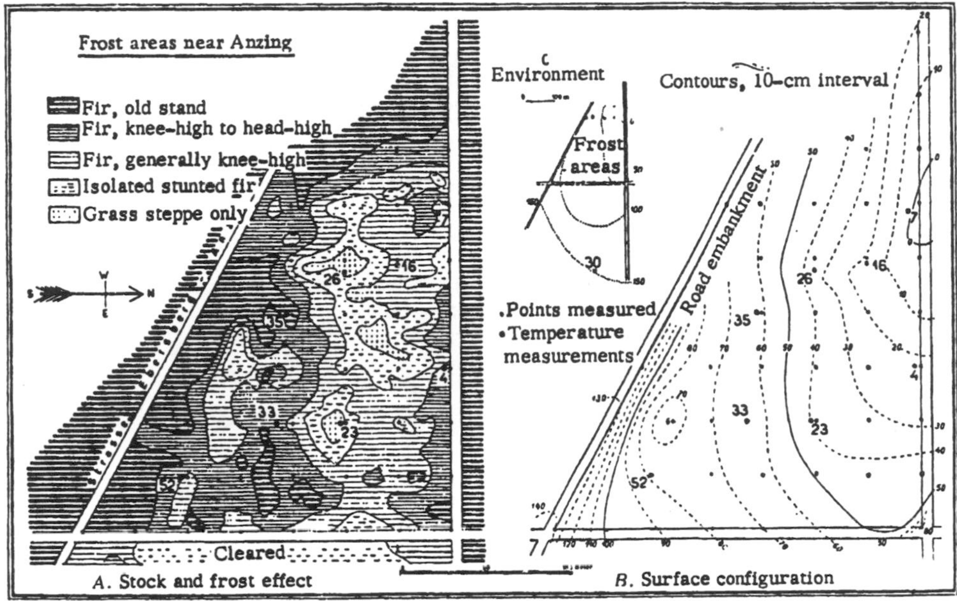

Figure 42-1. Plant growth, terrain, and observation points in the frost-prone areas of the Anzing Forest near Munich. (After R. Geiger [4201])

The most striking thing is the extremely high frequency of nights with frost. Data from two climatic stations of the Weather Service in Munich, which are included in Table 42-2, give an erroneous perception of the frost hazard for the vegetation of the area, and illustrates the problems associated with the extrapolation of climate data in regions possessing even minor topographic variation. These figures explain why young firs were unable to grow over the previous 30 years in this particular frosty area.

More detailed analysis shows that the temperature was not determined only by a location's relative height. The cold air current at night was directed toward the border of old trees to the north, but the protective effect of the high trees (Section 37) reduces the severity of the frost. It is also difficult for the cold air to penetrate into the old stand, which causes a cold air dam to develop.

An example from agriculture, taken from the fruit growing area of southwest Germany, is illustrated in Figure 42-2, by F. Winter [4212]. The areas shaded in black show the extent of frost damage in May 1957 to the fruit on individual trees in a peach orchard. The figures on which the diagram is based were obtained by estimation and by random sampling. The difference in height amounted to only 3 m, but the boundary at which frost damage occurred went through the middle of the orchard.

A hollow is an enclosed basin without a drainage outlet. A lake of cold air will build up in these low lying areas as a result of downslope flow. They are called frost hollows if the

Table 42-2. Night minimum temperatures (°C) in Munich and in the Anzing Forest.

Station	Height of obser-vation	May			June		
		Mean tempera-ture	Coldest night, 3-4 May	Number of nights with frost	Mean tempera-ture	Coldest night, 7-8 June	Number of nights with frost
		A. For comparison (general climate)					
Munich City	8.4 m	8.8	2.1	0	10.6	8.2	0
Outside Munich	1.4 m	6.5	-1.8	1	9.0	4.2	0
		B. Anzing Forest near Munich					
Anzing Sauschütte	5 cm	1.6	-8.4	12	4.5	-3.9	4
At the frost area:							
Point No.30		0.1	-10.7	17	1.2	-5.2	9
52		-0.3	-11.0	17	0.4	-7.9	12
35		-0.3	-10.8	19	1.4	-7.1	8
33	5 cm	-0.6	-12.4	20	0.4	-7.0	12
4		-0.7	-11.9	20	-0.1	-8.0	14
16		-0.8	-12.8	20	0.3	-7.2	13
7		-1.1	-13.5	22	0.1	-7.1	13
23		-1.5	-13.5	22	-0.2	-7.1	15
26		-2.0	-14.4	23	-0.7	-8.8	15

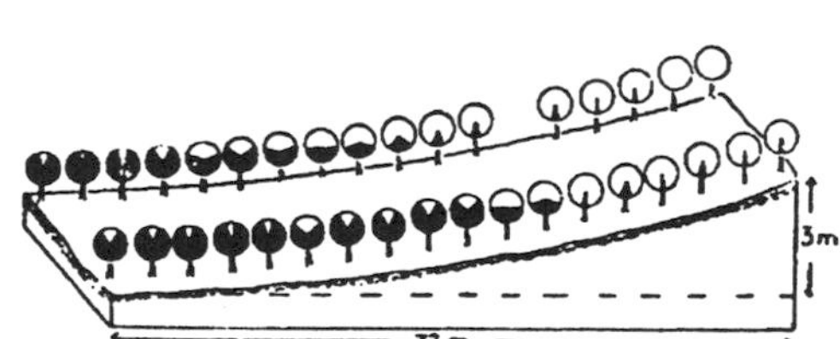

Figure 42-2. Proportion of young peach fruits (black) frozen on a slope in May 1957. (After F. Winter [4212])

temperature reaches below 0°C, and cold hollows if the temperature is above 0°C. This phenomenon has a variety of other names including frost pocket, cold island, frost hole, cold air pool, frost hollow, etc. The old fashioned rule still applies: at night, concave land surfaces are cold and convex surfaces warm. The strength of these drainage winds will depend upon the relative density of the cold air, the height of the nocturnal inversion, the slope of the local terrain, the magnitude of the surface frictional resistance, and the direction and strength of upper air flows arising from the regional pressure gradient.

Analysis of the numerous aspects of the cooling phenomena has shown that the cold air current just described is only one of five factors that affect the microclimate in hollows. The other four factors are: (2) the increase in counterradiation which results from the screening of the horizon in the hollow, leading to a change in the longwave radiation balance that reduces the frost hazard; (3) the reduction in turbulent exchange produced by the dish shape, which keeps the temperature low at night; (4) the heat supply from the soil of the slopes when the hollows are deep and narrow; and finally, (5) the shortening of the duration of daylight as the sun rises later and sets earlier. As valleys become steeper and deeper, the sun will rise later and set earlier.

The interplay of these five factors may vary according to the size and form of the depression in the ground. H. M. Bolz [3807] divided depressions into three types, shown in the sketches in Figure 42-3. The flat hollow (I) derives its low night temperature primarily from the flow of cold air; it is merely a cold air pocket. The low temperatures in deeper hollows (II) are mainly due to a reduction in turbulence, while a narrow depression (III) remains comparatively warmer in spite of the absence of turbulence because of the heat supplied from the soil (side walls) and the strong screening of the cold sky. The minimum temperatures given by Bolz [3807] illustrate the temperature characteristics of the three types of depressions.

Figure 42-3. The three types of small hollows. (After H. M. Bolz [3807])

A few experiments made in open country give some insight into the combined influence of the five factors. K. Brocks [4200] dug six 20 cm deep trenches in Eberswalde, with the same base width but sides sloping at different angles. He recorded the night minimum temperatures at a height of 10 cm above the base and compared them with temperatures above level ground. His results were:

Slope angle	0°	15°	30°	45°	60°	75°	90°
24 May 1937 (°C):	6.3	6.6	7.0	7.3	7.5	7.5	8.1
Average of 138 nights (°C):	6.23	6.23	6.27	6.34	6.44	6.59	6.67
With snow cover (°C):	-2.5	--	-4.4	-3.5	-2.7	-2.4	-2.4

When a shallow fresh snow cover cuts off the supply of heat from the soil, the flow of cold air led to lower temperatures in the flat trenches than over level ground (the last line of the table, 24 March). Apart from this, the trenches were warmer than the air above level ground because of screening against longwave radiation loss and conduction of heat from the soil.

In addition to detailed field assessments of terrain and land cover to evaluate frost hazard potential in undulating terrain, use has been made in recent years of thermal scanner data from low-flying aircraft and thermal sensor data from satellites. The regional coverage provided by these methods enables microclimatologists to distinguish between the effects of broad topographic controls and local surface characteristics on nighttime minimum temperatures. Examples of these approaches can be found in L. Mahrt and R. C. Heald [4205], and J. D. Kalma et al. [4203, 4204].

Attention will now turn from small-scale depressions to features of larger dimensions. W. Schmidt [4210] studied the Gstettneralm Sink hole at an elevation of 1270 m near Lunz in Austria. The sink hole is a funnel shaped depression in a limestone mountain, formed by faulting.

Figure 42-4 shows a cross section of the sink hole with the height slightly exaggerated. The figures show the temperatures measured before sunrise on 21 January 1930 at the points indicated on the slopes. On the upper slope, temperatures were only 1 to 2°C below zero. But below the edge of the WSW slope, over which the cold air can flow through the saddle and down into the Lechner Valley, temperature decreased very quickly in the deep cold air basin, and at the bottom of the valley they reached -28.8°C on that particular day. This sink hole records the lowest temperatures for central Europe. The minimum thermometers, which were read 73 times between 1928 and 1942 in this isolated and barren landscape, gave temperatures lower than -40°C 27 times, 8 times below -50°C, and an absolute extreme of -52.6°C. Night frosts occur even in the middle of summer. Whereas vegetation elsewhere becomes hardier at higher altitudes in mountain regions, the reverse is true in the sink hole. There are tall forests at the upper edge, followed lower down by stately firs, mixed with Alpine roses. Then come the dwarf firs, and, at the bottom of the sink hole, only vegetation consisting of a few hardy grasses and shrubs that are able to survive the winter under the snow. J. Horvat [4202] described a similar reversal of the vegetation pattern for the karst sink holes in the former Republic of Yugoslavia. For more information dealing with the relation of microclimates and vegetation in depressions the reader is referred to J. Rikkinen [4207] and A. Rajakorpi [4206].

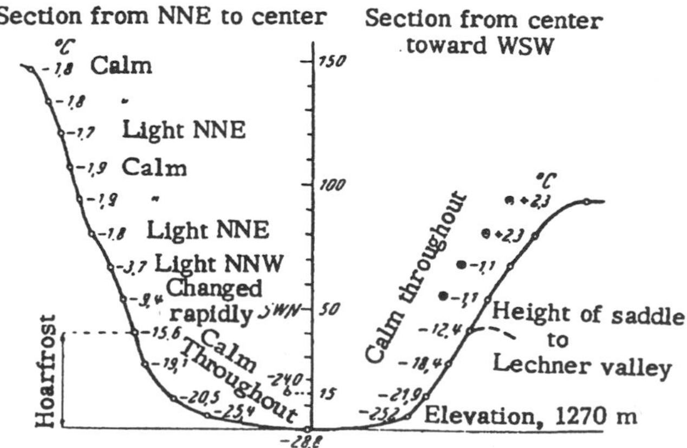

Figure 42-4. Night temperatures in the Gstettneralm Sink hole on 21 January 1930. (After W. Schmidt [4210])

F. Sauberer and I. Dirmhirn [4208, 4209] have analyzed conditions prevailing on the Gstettner hill pastures. For record low temperatures to be reached, a low initial temperature is necessary; this is obtained when the season, elevation, topographic features, and weather situation are favorable. Winter snow cover also plays an important part by insulating the air layer from stored heat in the ground (Section 24). This is shown by the fact that the lowest temperatures appear with a fresh snow cover. In all observations calm conditions prevailed below the rim of the 70-80 m deep sink hole. The decrease of temperature from the top to the bottom of the sink hole averages 10°C in summer and 16°C in winter, with 25°C as an extreme value. The regional wind frequently flows over the top of this stagnating pool of cold air and causes only a regular undulating motion in the boundary layer.

Many things take place in a different way in these large depressions compared with what happens in smaller features. At the time of the winter solstice, the north-facing slopes remain in shadow throughout the day, and these slopes produce cold air masses (even during the day). The large-scale flow of cold air acquires new significance since counterradiation from the atmosphere comes from a remarkably shallow atmospheric layer (72 percent from the lowest 87 m). It follows that in such deep sink holes, counterradiation will be greatly reduced when they are filled with cold air. The essential condition for the formation of extreme low temperatures in sink holes is that air cooled in contact with the ground should be able to flow in from a sufficiently wide area, that is, the "frost catchment area" must be sufficiently large, and the surrounding slope must be sufficiently great to initiate and sustain drainage flow. If these conditions were not fulfilled on the Gstettneralm, it would not produce such record low temperatures.

Finally, something must be said about water vapor. The cold bottom of the sink hole is often filled with fog. This means that the temperature has fallen to the dew point. R. Wagner [4211] has described (with photographs) the fluctuating fogs of the sink holes in the Bükk Mountains of Hungary in 1953-54 and compared them with temperature measurements. When a snow cover lay in the Gstettneralm, however, the excess water vapor sublimated directly onto the snow without forming fog and hence dried the air considerably. Quantitatively, the latent heat of sublimation released was insignificant in comparison with the increased amount of longwave radiation lost through the dehydrated air; the energy gain was estimated to be only 6 percent of the loss (4.2 compared with 75.4 10^4 J m^{-2}).

Figure 42-5 illustrates a few of the night temperature recordings made on the slope (1), and at the bottom (2) of the sink hole, with a height difference of 56 m. Curves *(a)* at the top left are for a normal, clear night; (*b*) shows intervention by a foehn wind about 23:00, which initiated a slight heating on the slope about 20:00, but reached the bottom only at 23:00. By 02:00 the foehn effect had stopped, and cooling began again. At (*c*) the foehn is unable to break through but moderates the temperature drop somewhat at the bottom. Curves (*d*) show a fresh influx of cold air at 05:00. This cools the slope, but the blowing wind warms up the bottom of the sink hole. Curves *(e)* illustrate a cold air outbreak with a homogeneous air mass, overcast skies, and a strong wind that effectively eliminates the microclimatic differences between the sinkhole and slope.

Because of the large number of factors, it is difficult to accurately forecast the temperature distribution in broken terrain. The effects of the ground surface, described in Chapter III, and of the plant cover, also have to be taken into account.

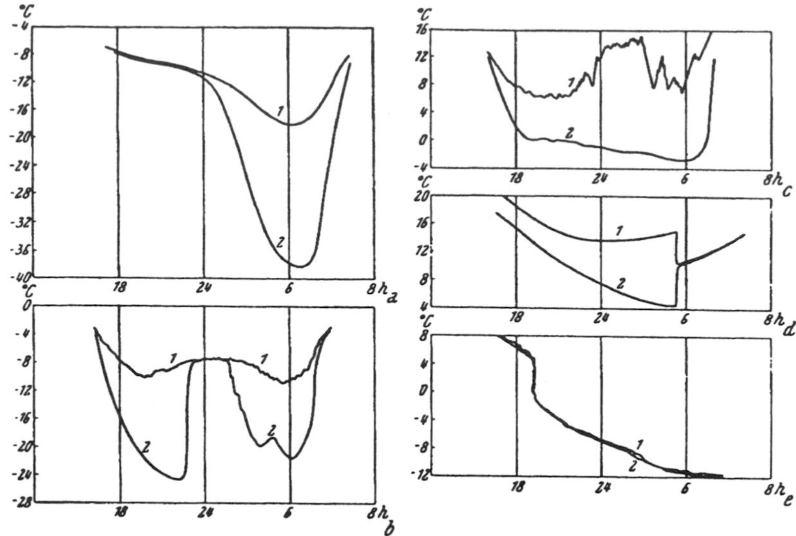

Figure 42-5. Types of night temperature variation on a slope (l) and at the bottom (2) of the Gstetteneralm Sink hole. (After F. Sauberer and I. Dirmhirn [4208, 4209])

43 Local Winds in Hilly and Mountainous Terrain

In Section 37 a distinction was made between active and passive influences of the forest on the wind field. A similar distinction can be made of the effect of the land's configuration upon the air flowing over it. We can, therefore, speak of an active topographic effect when differences of temperature and air pressure, caused by topographic variations, give rise to air currents. The most important of these are the mountain winds, which vary periodically during the day. The influence is described as a passive topographic effect when topographic features, mountains, valleys, and slopes affect and modify the existing wind field through dynamic influences. This section will be confined to discussion of active topographic effects. Dynamic influences of hilly and mountainous terrain upon the overlying air flow are reviewed in the several micrometeorology texts cited in Section 1, and in an American Meteorological Society publication [4302].

A. Wagner [4334] subdivided mountain winds with a diurnal periodic fluctuation into three types: compensating winds, mountain and valley winds, and slope winds. Differential heating of two adjacent areas creates compensating winds. Greater heating causes the isobaric surfaces over one area to bulge upward, thus creating a horizontal pressure gradient from one area to the other. A return flow can occur at night if there is a reversal of the temperature difference. In a snow-covered mountain, the valley below heats up, increasing both the temperature contrasts between the mountain and valley below and the strength of the winds from below. These air movements linking mountains and plains can increase to gale force in narrow passes. C. Troll [4333] has studied their development in the tropical highlands of South America and Africa, and has described their influence on cloud formation, precipitation, and vegetation. He has also indicated that they are often inseparably associated with valley winds (to be described later), or with land and sea breezes, trade winds, and monsoons.

Let us now consider the slope winds, which usually blow upslope during the day and downslope at night. Nighttime downslope winds are a result of cold airflow described in Section 42. In mountainous terrain, downslope winds develop on slopes at night and are replaced by upslope winds during the day.

F. Defant [4304] determined the wind components parallel to the slope as a function of height above the surface of the slope. He found the following vertical profile from measurements made on five undisturbed nights with downslope winds (Table 43-1):

Table 43-1. Variation of slope winds by height and time of day. (After F. Defant [4304])

Height above slope (m)	5	10	20	30	50	100
Downslope wind at night (m sec^{-1})	1.0	1.5	2.3	2.4	1.9	0.2
Upslope wind at midday (m sec^{-1})	2.3	2.9	3.7	3.9	3.4	2.4

The greatest speed was found to be at a height of 20 to 40 m, close to the surface of the slope, although at some distance from it due to friction from the surface. Over steeper slopes, air flows in gusts; according to A. Defant [4303], this begins when gradients reach about 1:100. M. Reiher [4322] measured air temperature on a steep 18° slope at heights of 10, 30, and 50 cm above the ground. The lower half of Figure 43-1 shows the temperature variation at the three measurement points, and the upper half gives the temperature stratification by the distribution of isotherms in the lowest half meter of the air over a period of only 5 1/2 minutes. The increase of temperature with height is typical of the nocturnal negative radiation balance. At 19:04, a typical cold air temperature drop can be recognized from the upward arch of the isotherms. Such movements took place every 4 to 5 min; with an average wind speed of 1.4 m sec^{-1}, the horizontal extent of the cold air temperature drop was 300 to 400 m.

F. W. Nitze [4320] observed a similar process and gave the following vivid explanation: "It seems that the cold air dams up on the plateau of the slope. As long as its vertical dimensions remain small, it only flows slowly; friction in contact with the ground is very great. When the cold air has collected in sufficient quantity, however, it begins to slide down. In doing so it entrains the air behind it so that a strong cold air current develops which persists until there is no longer enough cold air on the plateau to support it." Continued cooling of the air by radiative cooling is required to sustain the cold air current.

W. R. Porch, et al. [4321] suggest that the timing of the flow of cold air pulses down a valley can be strongly affected by the flow out of its tributaries. In addition, much of the cold air flowing down a valley may have entered through tributaries.

In mountain areas, these cold air temperature drops may assume the form of "air avalanches," to use the term of A. Schmauss [4325]. He has described such avalanches (katabatic winds) in the Bavarian Alps. H. Scaëtta [4324] encountered them in such intensity in central Africa, northeast of Lake Kiwu, that on two nights the katabatic winds almost swept away his tent. J. Kütter [4318] observed that these winds follow a rhythmic pulsation. He has described it as follows: "I was able to follow the microstructure of cold air drops during two nights that I had to spend in the upper part of the wall, when making a tour of the north wall of the inner Höllental peak (Wetterstein Mountains) on 22 to 24 July 1947. The point of

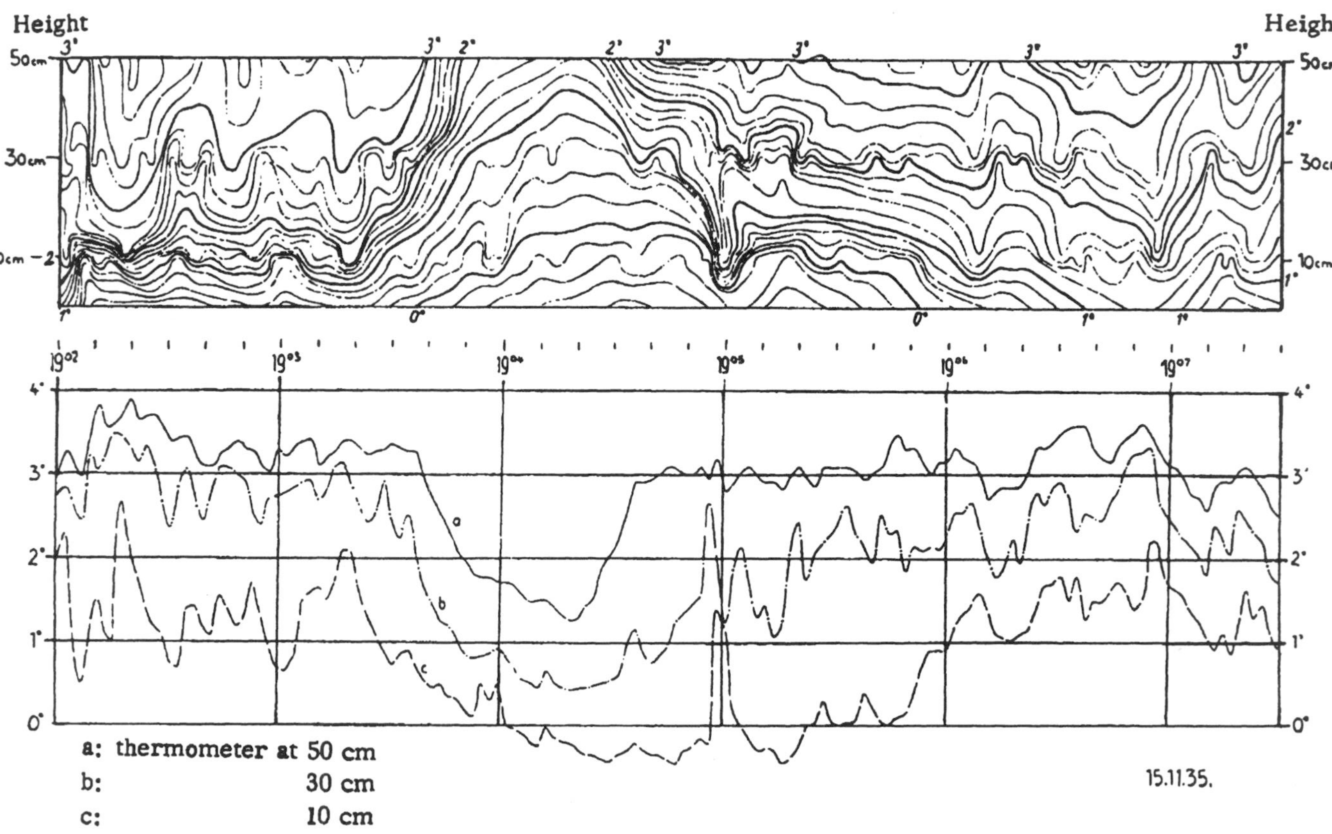

Figure 43-1. Movement of cold air on a steep slope near Göttingen. (After M. Reiher [4322])

observation was at 2700 m, about 100 m below the summit. A high pressure weather situation with light winds prevailed. After sunset at 21:01 central European time, a weak air current developed which died down again within a minute. At 21:06 there was another air movement, and another at 21:11, followed by gusts at regular intervals of 5 minutes, which persisted throughout the night, disturbing our sleep in a most unpleasant way. An occasional gust would be missing, but the next one would appear punctually on time. The greatest deviation was ± 1 min. The rhythm continued until 6 o'clock in the morning with an unbelievable regularity, always bringing a shower of cold air at h+1, h+6, etc. Soon after sunrise these periodic winds stopped, although the north wall still stood in unchanging shadow." The same 5 minute period continued the following night.

A. Schmauss [4326] proved, during a forest fire that lasted from 4-6 October 1942 in the dwarf fir region of the Karwendel Mountains (2000 m), that the fire had been carried by the downslope wind at night into separated stands of dwarf fir a few hundred meters lower. The nocturnal downslope circulation was established over the whole area in spite of the rising convection currents in the area of the fire. Figure 43-2 shows the nocturnal distribution of wind speeds in the Finkenbach Valley prior to and after complete deforestation. Prior to deforestation downslope winds are elevated above the canopy and winds speeds are considerably reduced due to the frictional drag.

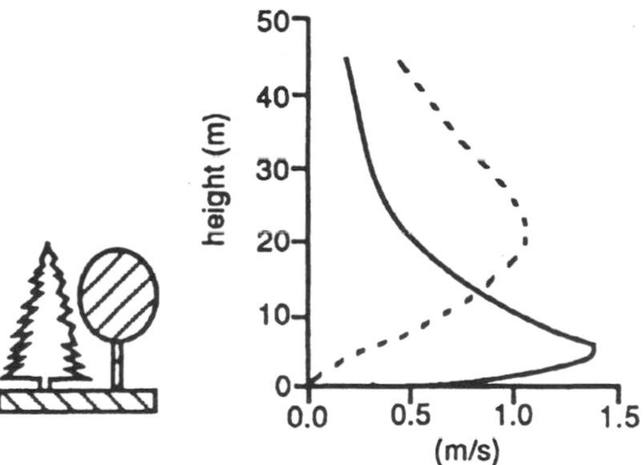

Figure 43-2. Vertical profiles of downslope winds (------ with canopy, _____ after deforestation). (From G. Gross [4311])

Contrasting with the downslope winds at night are the winds blowing up the slope during the day. These are usually stronger because the radiation exchange is greater and also because the lighter upslope winds have less friction with the surface. Table 43-1 indicates the profile of the upslope winds. Here too, the wind is strongest at 20 to 40 m above the ground. This means that glider pilots, who wish to use these winds as thermals, must fly with their wings very close to the slope. R. Maletzke [4319] reported that flying corps students from the Technical College at Munich, touring the Alps by glider in 1958, were able to cover a distance of 30 km using the upslope wind from a single hillside, without losing altitude. Development of local slope winds occurs more frequently with an unstable temperature stratification of the atmosphere. H. G. Koch [4316] was able to show on the stormy day of 28 June 1952, from the 14:00 synoptic observations of the meteorological network, that winds were blowing up the slopes on both sides of the ridge of the Thuringian Forest.

J. Grunow [4312] measured upslope winds for the SE slope on which a ski jump had been built for the international championships held at Oberstdorf in the Allgäu from 28 January to 6 February 1950. They were found to be 2 to 3 m sec^{-1}, with a maximum of 3.6 m sec^{-1}. They were most strongly developed between 11:30 and 13:00, and it could be proved that they increased the length of jumps. The ten best skiers made jumps of between 92 and 117 m with upslope winds of 1.8 m sec^{-1}, and 106 to 128 m with upslope winds of 3.2 m sec^{-1}.

Figure 43-3 is a cross section of a mountain valley around noon. The winds flowing up both slopes, which are in sunshine, increase in vertical extent farther up the slope, in proportion to the areas shaded with fine dots. The rising air must be continuously replaced. This is achieved not only by the cross valley flow shown in the sketch, but also by the wind flowing up the valley bottom, which is indicated by the larger dots, intended to show a wind blowing perpendicular to the plane of the paper.

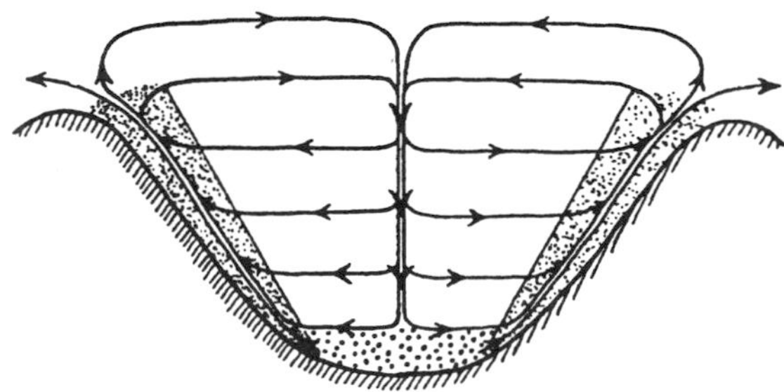

Figure 43-3. The system of upslope and upvalley winds. (After A. Wagner [4334])

If the directions of the arrows in Figure 43-3 are reversed, we obtain a picture of the system of downslope and downvalley winds at night. In this case, however, the finely dotted areas have to be reversed so as to be narrow at the top, broadening out toward the bottom of the slope.

There is, therefore, a double system of periodic mountain winds; during the day there are upslope and upvalley winds, and at night downslope and downvalley winds. These are sometimes called valley and mountain breezes because of their source. F. Defant [4305] has suggested that the slope winds are always initiated first, then the valley winds follow, and hence during the course of the day there is a change taking place with a phase difference, which is illustrated in Figure 43-4 for a clear summer day with light winds.

Diagram *A* shows the situation shortly after sunrise. The sun has come up and the surface is heated. The density of the air over the slope decreased relative to that of the air at the same level near the center of the valley. The resulting pressure difference causes the air to flow up the slope as an upslope wind. While upslope winds have set in (light arrows), the air in the valley is still colder from the night than the air on the plain outside. The downvalley wind is still blowing (black arrow) and is fed by the return flow from the upslope circulation. As the day progresses, the downvalley wind dies out with further heating. For a time (*B*), the upslope winds occupy the picture alone. This is the time in the morning when the air in the valley heats most quickly. Toward midday (*C*), the wind up the valley begins and enhances the upslope winds, and is supported by the return flow descending from greater heights in the

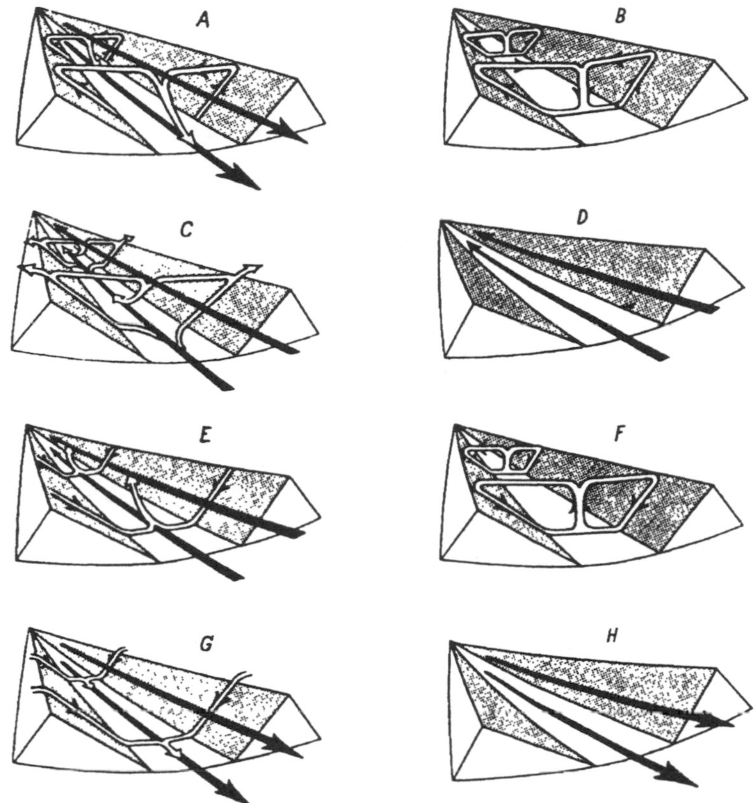

Figure 43-4. The interplay of slope and valley winds during the course of a day. (After F. Defant [4305])

center of the valley (Figure 43-3). This is the hour at which chains of cumulus clouds sometimes appear above the slopes while the sky remains clear where the air is sinking over the center of the valley. In the late afternoon (*D*), the upslope winds cease to blow; however, the upvalley wind continues to blow. H. Berg [4301] has shown that in valleys running N-S in the Allgäu Alps, an upslope wind continues to blow up the west slope when a downslope wind has set in over the east slope which is in shadow. Such situations with different degrees of solar heating must give rise to a cross valley wind (called overturning), according to the theory of T. A. Gleeson [4309].

As evening draws on, the downslope wind sets in (*E*), but the upvalley winds continue for a while. Diagrams *F* and *G* in Figure 43-4 corresponds to *B* and *C* in the morning. Diagram *H* represents the period shortly after sunrise when the downslope winds have ceased due to surface heating, but the stronger downvalley winds continue for a time. Clouds reduce both the loss of energy by terrestrial radiation from the surface and the gain of energy from solar radiation so that both downslope and upslope winds are lighter with cloudy skies. The speed attained by slope winds is related to the degree of the slope of the surface over which they occur and its roughness; over average slopes they are a few meters per second but may be much stronger over unobstructed steep slopes. Over a tree-covered slope they may be light or absent. There is a seasonal difference as well; in winter the downslope winds are stronger and upslope winds are light or absent, especially on snow-covered or north-facing slopes.

As with slope winds, the behavior of valley winds depends on the time of day, amount of clouds present, slope and roughness of the valley floor, season of the year, and strength of the prevailing wind. The valley winds, which blow in both directions, are on a larger scale and, as a rule, are more strongly developed than slope winds. As in the case of slope winds, cloudy skies will reduce the gain of energy by the ground during the day and the loss of energy at night, so that valley winds become lighter. The strength of valley winds tend to vary directly with the slope of the valley floor and inversely with the vegetative cover (E. W. Hewson [711]).

Salzburg airport lies at the exit of the Salzach Valley. E. Ekhart [4307] analyzed surface winds recorded at a height of 18 m for the period 1949 to 1951 to determine their daily variation. Figure 43-5 shows the results after smoothing. Wind components in the direction of the valley show a marked change of direction from downvalley (hatched) to upvalley wind (white). Between 10:00 and 13:00, depending on the season, the upvalley wind sets in and reaches a speed of 2.5 m sec^{-1} in summer. The downvalley wind at night sets in at different times in different seasons and reaches its maximum during the late hours of the night. The speeds are low because these values have been averaged for all kinds of weather situations. Nevertheless, the periodicity of the wind system is so great that it clearly shows up. A. S. Devito and D. R. Miller [3403], using smoke release tracers to determine wind direction, observed that the stronger winds aloft often dissipated or diminished the density-induced flows. However, when winds were light, density flows developed near the surface beneath the overlying winds.

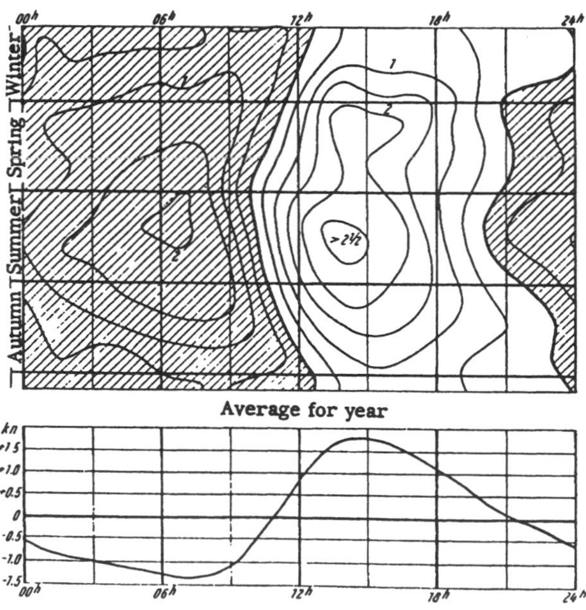

Figure 43-5. Wind components (m sec^{-1}) in the direction of the valley at Salzburg airport. (After E. Ekhart [4307])

The downvalley wind has been given special attention since it is partly responsible for the night frost danger. One such wind ("Wisperwind") blows at night out of the Wisper Valley from the east into the warm valley of the Rhine with a speed of 3 to 4 m sec^{-1}. H. Schultz [4330] showed that the speed increased linearly with the magnitude of the temperature reversal in the valley of the Wisper. W. Schüepp [4329] investigated valley winds both in vertical extent and in their relation to temperature and vapor pressure in the Davos Valley in Switzerland. He made measurements close to the ground on 20 winter mornings, using observations from a meteorological station, the Davos Observatory, and the top of a 70 m church steeple. The temperature and wind profile is shown in the upper part of Figure 43-6; the water vapor distribution is shown in the lower part.

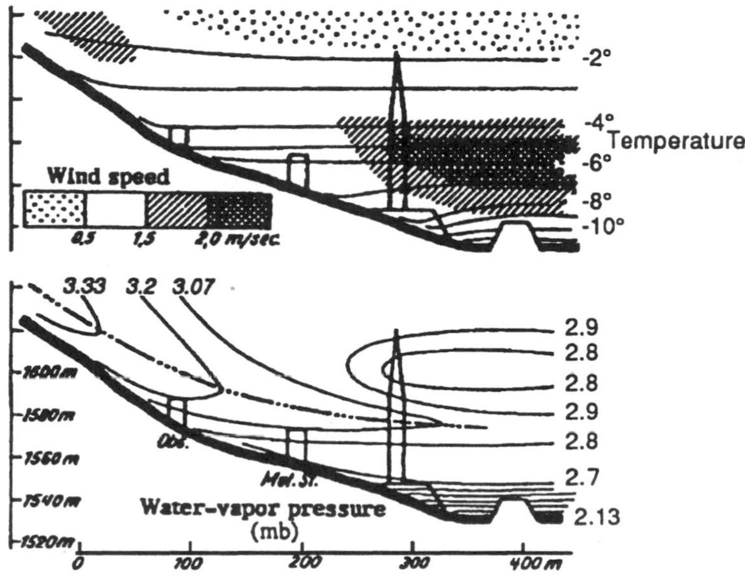

Figure 43-6. Vertical cross section of the nocturnal downvalley wind in Davos Valley. (After W. Schüepp [4329]) Top diagram shows wind speed and temperature, bottom vapor pressure.

The maximum speed of the downvalley wind is 2.5 m sec^{-1} at a height of 40 m above the bottom of the valley. It is reduced to a speed of 1 m sec^{-1} by friction at the slopes where the wind flows perpendicular to the height contours as a downslope wind; while at higher levels, it follows the general course of the valley as a downvalley wind. The temperature inversion amounted to about 9°C. The maximum vapor pressure of 3.33 mb was found at the top of the slope, and, while the air at the cold valley floor is dry from the point of view of absolute humidity, its relative humidity was 70 to 90 percent. The dot-dash line joining the vapor pressure maxima in the vertical section shows a downward drifting of moister air from the upper slope, along with the downslope wind.

R. H. Aron and I-M. Aron [4300] investigated the slope and valley winds in the Chien Hwa Valley in the Republic of China. The site location, valley contours, and instrument location are shown in Figure 43-7. This valley is oriented north-south and is about 200 m in width and 1,200 m in length, with nearly parallel hills on either side about 100 m high.

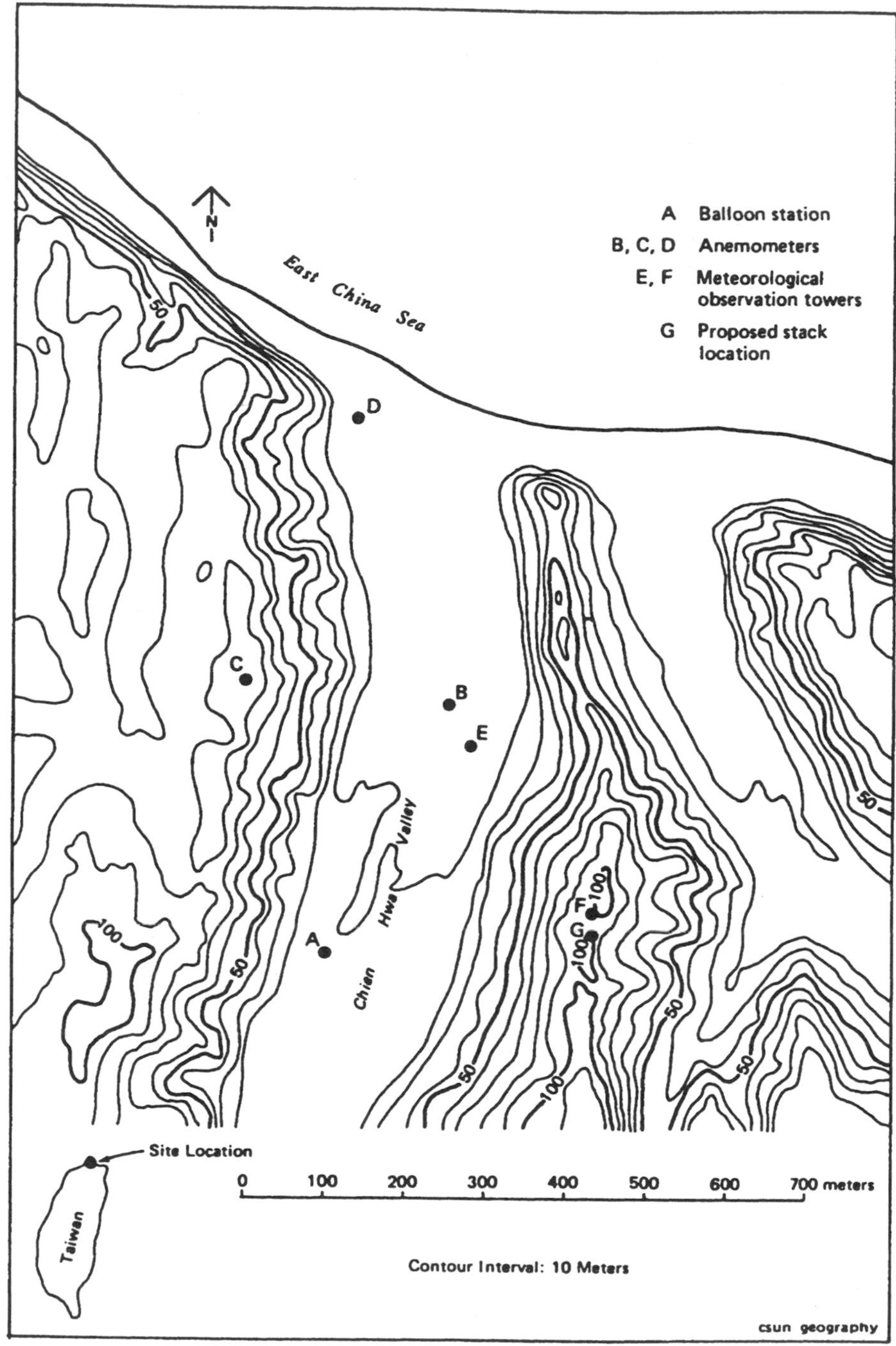

Figure 43-7. Meteorological observation stations in the Chien Hwa Valley. (From R. H. Aron and I-M. Aron [4300])

In this valley, the winds most frequently come from the north or the south (Figure 43-8). The great predominance of the north and south direction of the wind is due to the mountain and valley wind circulation and the land and sea breeze. In addition, when the general surface winds flow at an angle to the valley, they tend to turn parallel to the valley and flow through it. This effect of topography on the wind direction, known as channeling, often sub-

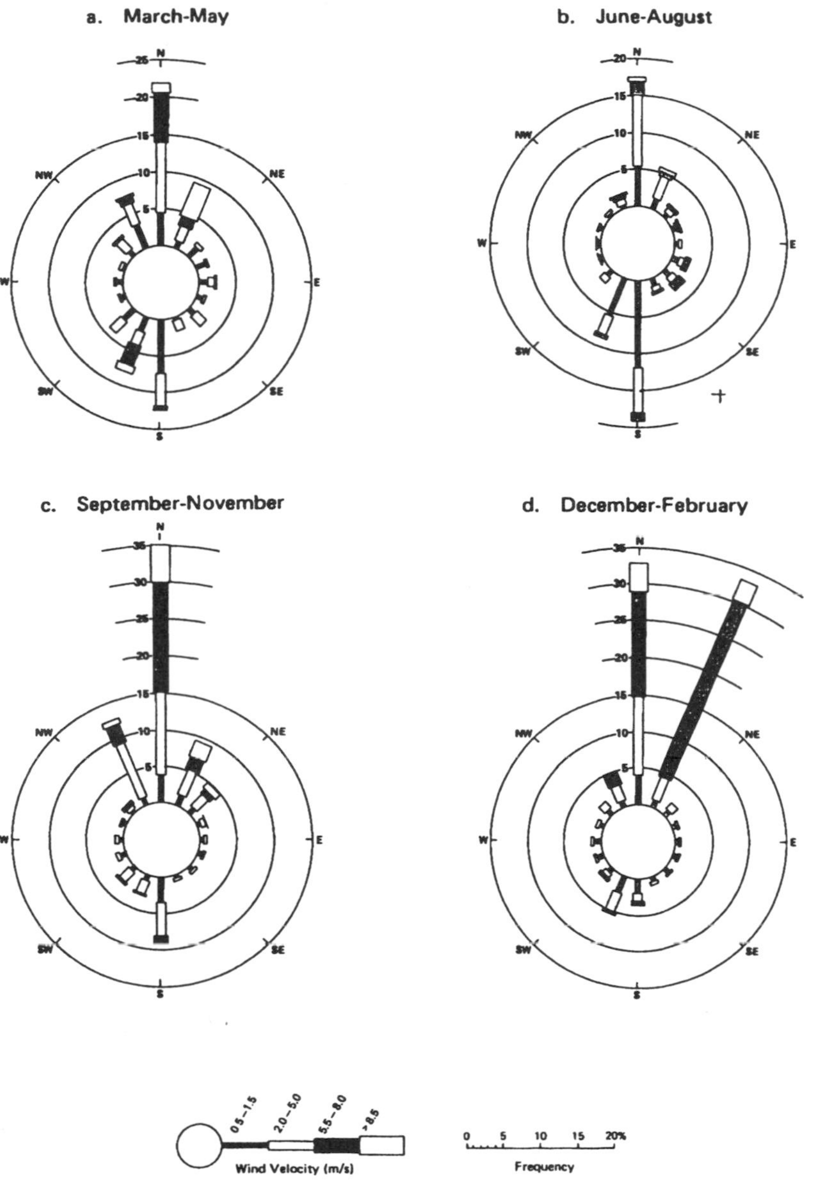

csun geography

Figure 43-8. Valley wind roses from the Chien Hwa Valley. (Site B in Figure 43-7) (From R. H. Aron and I-M. Aron [4300])

stantially modifies the thermally induced winds. In the Chien Hwa Valley, north winds occur 27 percent of the time (Figure 43-8), while on the hilltops winds from the north occur only 2.4 percent of the time (Figure 43-9). South winds occur 10.8 percent of the time in the valley (Figure 43-8) but only 1.6 percent of the time on the hill tops (Figure 43-9). Table 43-2 also shows the change in wind direction with increased height. Since the general wind

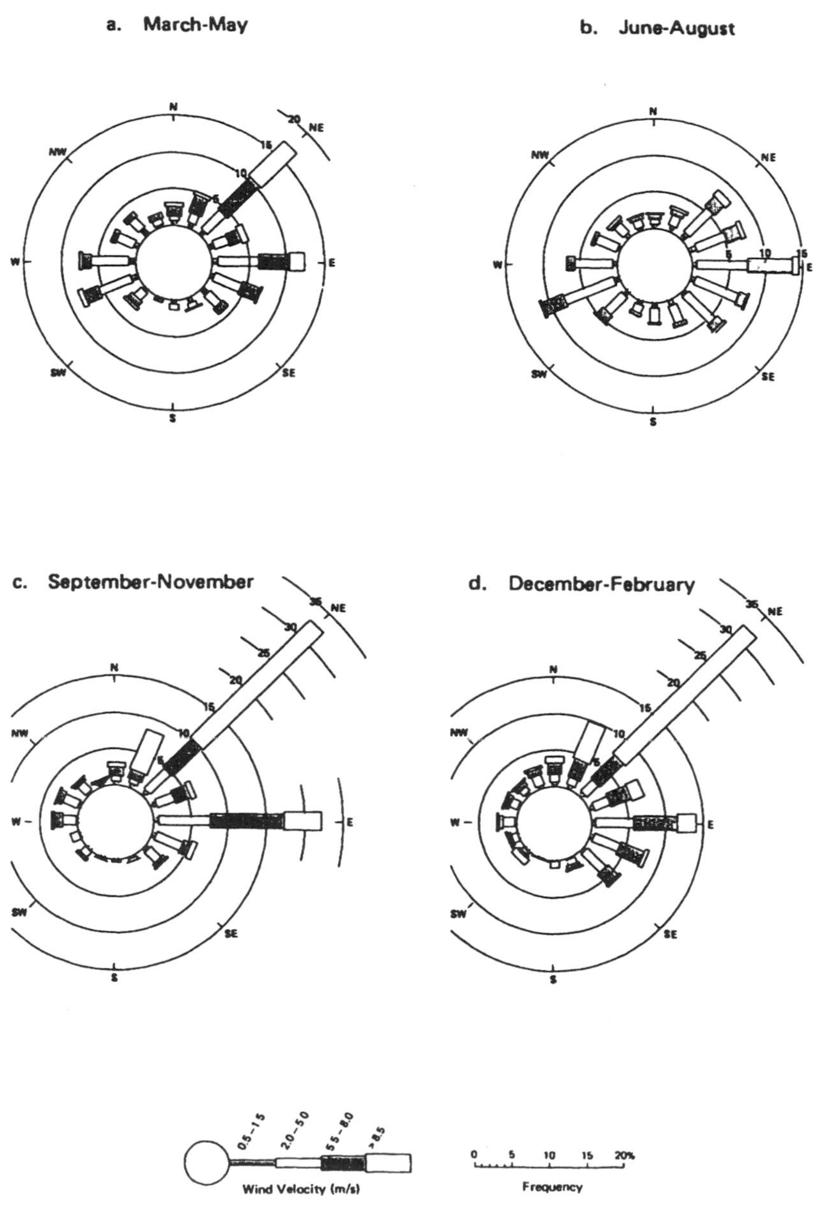

Figure 43-9. Hill wind roses from the Chien Hwa Valley. (Site F in Figure 43-7) (From R. H. Aron and I-M. Aron [4300])

circulation and the sea breezes generally have a greater depth than the tops of these hills, the influence of the valley orientation on the wind direction (channeling) can easily be seen. At night in April, the northeast monsoon dominates at 300 m while the local circulation dominates at 100 m (Table 43-2).

Table 43-2. Frequency distribution of the North and South winds (percent) at 02:00 and 14:00, Chien Hwa Valley.

Month	Hour	Elevation (m)					
		100		200		300	
		N	S	N	S	N	S
April	02:00	15.4	40.3	15.4	15.4	16.7	0
	14:00	42.9	0	19.1	0	15.8	0
May	02:00	5.0	42.0	5.0	26.0	6.0	12.0
	14:00	24.0	12.0	8.0	4.0	4.0	4.0
June	02:00	15.0	15.0	5.0	35.0	10.0	20.5
	14:00	36.3	9.1	22.6	9.2	13.7	0
Sept.	02:00	9.0	22.7	9.0	18.1	4.5	18.1
	14:00	29.2	4.2	12.5	4.2	8.3	8.3

The mean daily wind distributions for the north-south and east-west wind components are shown in Figures 43-10 and 43-11, and were obtained by using captive and free balloons. These are more representative of relatively calm conditions since these balloons are not suitable at high wind speeds, and use discontinued when the surface wind speed was greater than 5.0 m sec^{-1}. Figure 43-10 shows the distribution of the thermally induced mountain and valley wind and land and sea breeze. At night, the south wind dominates and reaches a maximum speed in excess of 3.5 m sec^{-1} around 150 m (125 m above the floor of the valley). During the day, winds from the north prevail and reach a maximum wind speed in excess of 5 m sec^{-1} around 75 m (50 m above the floor of the valley at the observation site).

When the prevailing wind speed is light, the wind pattern is often influenced by thermally induced slope winds. This results in a high frequency of east and west winds on the hill-sides on these days (Figure 43-11). On clear nights, air drains down the slopes of the hills (Figure 43-11(a)). On sunny days when winds are light, the wind flows in the opposite direction. In the morning, R. H. Aron and I-M. Aron [4300] also observed a weaker, secondary circulation above the primary slope winds, probably initiated by the frictional drag of the primary slope wind (Figure 43-11(b)). Since the balloon station (Figure 43-7, site A) was located on the west side of the valley, the east and west components reflect the slope winds associated with the western slopes. A similar pattern would be expected on the eastern slopes. In the early morning and late afternoon, when opposite facing slopes receive unequal amounts of solar radiation, overturning frequently occurs as illustrated in Figure 43-11(c). Overturning is most pronounced on days with clear skies and light winds.

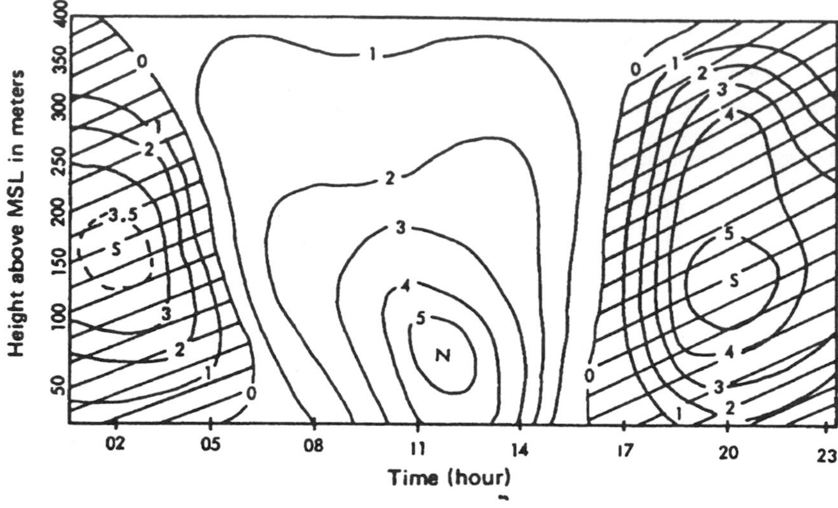

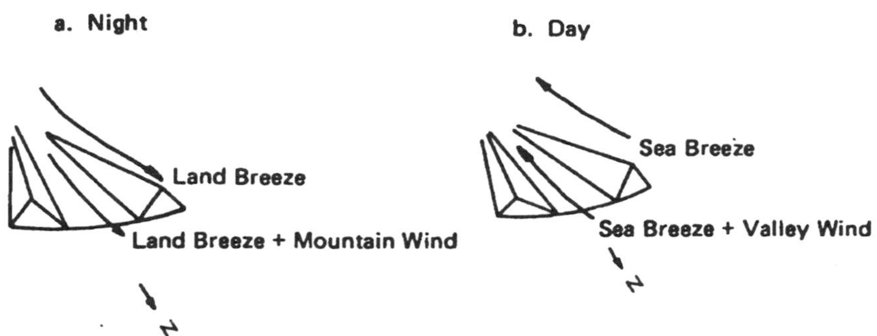

Figure 43-10. The diurnal variation of the North-South wind components (m sec^{-1}) in Chien Hwa Valley (data recorded at balloon station shown on Figure 43-7); south winds (shaded) and north winds (unshaded). (From R. H. Aron and I-M. Aron [4300])

When winds blow across an obstacle, a downwash phenomenon may result in which the air descends on the lee side of the obstacle. Smoke release tests with NE winds were made in the late afternoon from the meteorological observation tower (Site F, Figure 43-7). It was observed that the smoke was carried downwind and came in contact with the slope of the hills west of the valley and then dispersed in the valley. This was probably due to a combination of the downwash phenomenon and overturning.

As part of the extensive study of the meteorology conditions at Brush Creek Valley in western Colorado, P. H. Gudiksen and D. L. Shearer [4313] observed strong down-valley winds at night (Figure 43-12(a)). On the east-facing slope, they observed moderate upslope winds which were initiated within minutes after sunrise (Figure 43-12(b)). Using a PDCH tracer to detect circulation, they found that the upslope and crossvalley wind circulation in

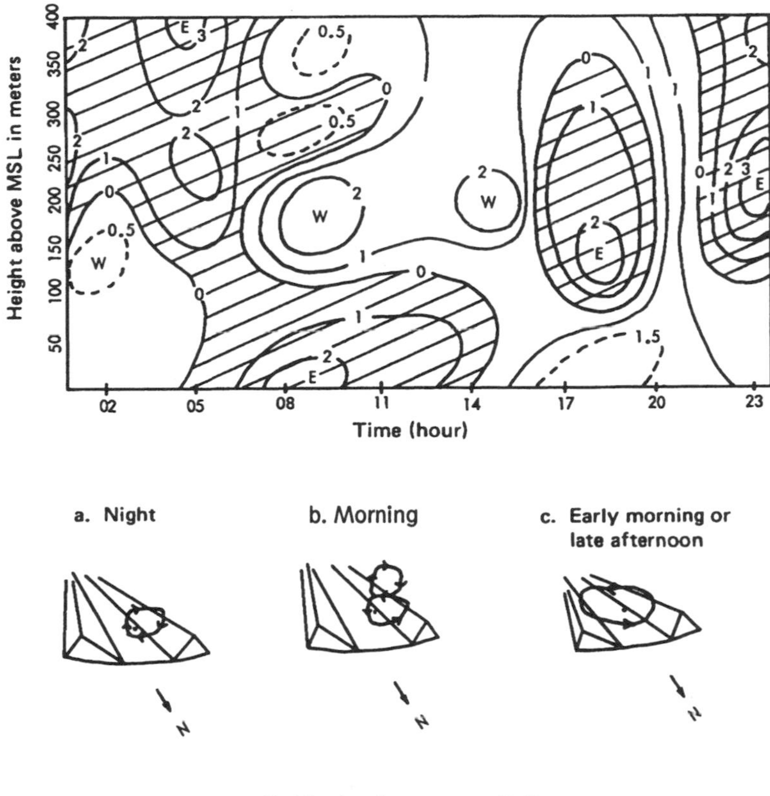

Figure 43-11. The diurnal variation of the East-West (shaded) wind components (m sec^{-1}) in Chien Hwa Valley (data recorded at balloon station shown on Figure 1). (From R. H. Aron and I-M. Aron [4300])

the morning remained within the nocturnal drainage flow. Essentially no tracers were observed on the eastern side of Brush Creek until the daytime circulation pattern was established (Figure 43-12(c)).

In a well-illustrated study in Israel, Y. Goldreich [4310] analyzed the interactions of the valley/mountain and sea/land breezes. He was able to identify three vertically separated wind regimes: the mountain/valley winds dominated the flow within the valley walls; the regional gradient flow was discernible 400 m above the local topography; and the land/sea breeze was sandwiched between the two.

M. H. McCutchan and D. G. Fox [4630] analyzed the wind patterns at San Antonio, an isolated conical mountain in New Mexico. When wind speeds at the peak were less than 5 m sec^{-1}, the daytime flow was upslope and the nighttime flow downslope (Figure 43-13). The upslope flow usually started within minutes after the sun first struck the slope and continued to strengthen during the day. A transition to a downslope flow began soon after the slope goes into the afternoon shadow and continued through the night. When the wind

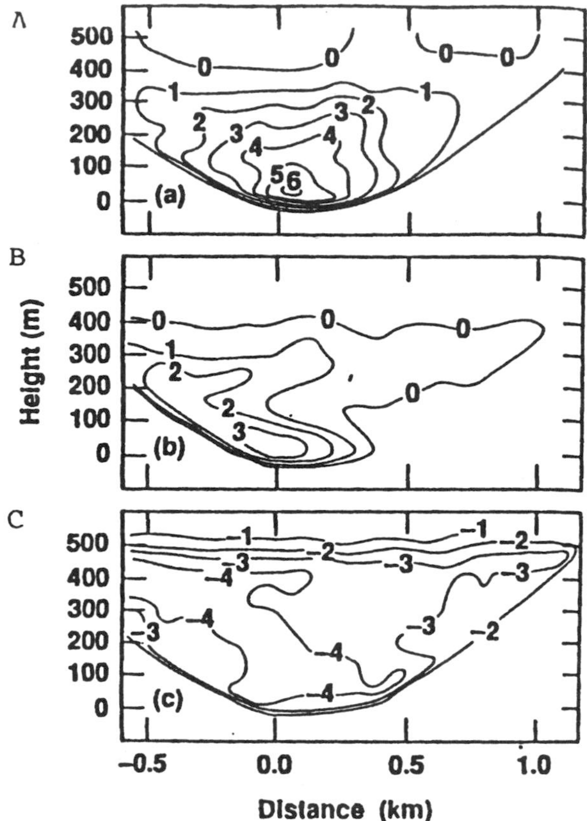

Figure 43-12. Crossvalley views of the flows observed during Experiment 4 by the Doppler lidar within the Brush Creek Valley. The data were obtained during (a) nighttime, (b) morning transition-period, and (c) daytime. Positive values denote downvalley flows while negative values indicate up-valley flows in units of m sec^{-1}. (After P. H. Gudiksen and D. L. Shearer [4313])

speeds at the peak were strong ($\geq$ 5 m sec^{-1}), the slope winds were overpowered by the general wind circulation.

After World War II, expansion of the Odenwald fruit-growing industry was considered in Germany. However, a narrow limit was set to this extension by the risk of late frosts. The German Weather Service, the Agricultural College, and the Fruit Industry Advisory Center of the district joined their resources to carry out a microclimatologic survey, under F. Schnelle [4327, 4328]. The upper limit of the frost danger area for the slopes to be brought under cultivation was mapped on a scale of 1:10,000. These maps were based on carefully collected information from fruit farmers, on night temperature measurements over a period of several years, on observations of the frequent valley fogs (especially their upper limits), and on a detailed study of the topography of the area.

It was found (Figure 43-14) that wherever a valley narrows, the flow of cold air becomes dammed up. Such restrictions to the flow of cold air current can occur due to frictional effects caused by obstructions, topographic narrowings, and vegetation. In the Mossau

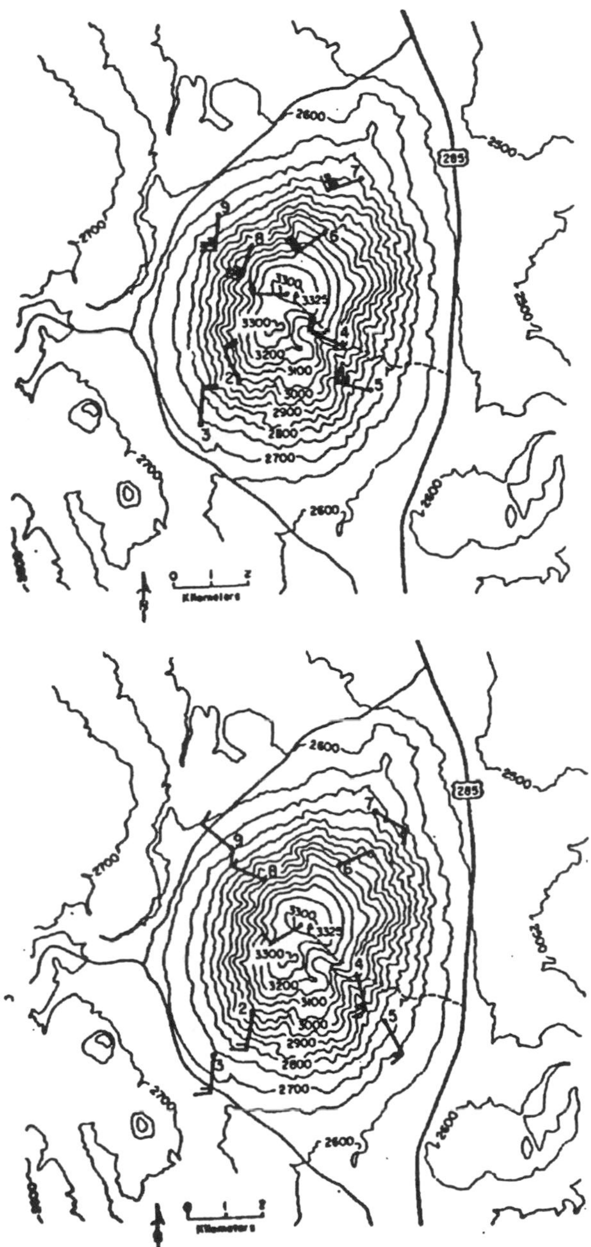

Figure 43-13. Resultant wind directions and windspeed for the mean *u* and *v* wind components when windspeeds at the peak were less than 5 m sec^{-1} in September and October 1981 and 1982 at nine stations on San Antonio Mountain for 28 days at 0000 MDT (top diagram), and 27 days at 1200 MDT (bottom diagram). Pennant 5 m sec^{-1}, full barb 1 m sec^{-1}, and half barb 0.5 m sec^{-1}. Terrain contours in meters; contour interval 50 meters. (After M. H. McCutchan and D. G. Fox [4630])

Valley, which is uniform in shape and about 8 km long, the upper limit of the cold air sinks very slowly, becomes horizontal, and is dammed up at all junctions of 10° to 20° slopes that are covered with tall trees. More detailed analysis of 20 such valley narrows indicated that only an opening of 400 to 500 m or more in the valley would allow the cold air to flow through in sufficient amounts to avoid a lake of cold air building up behind the narrows. "Field observations showed repeatedly that cold air does not flow like water, but more like porridge or thick syrup." The cold air flowing down both slopes forms a flat surface in the upper part of each section of the valley, extending up to 55 m vertically above the narrow.

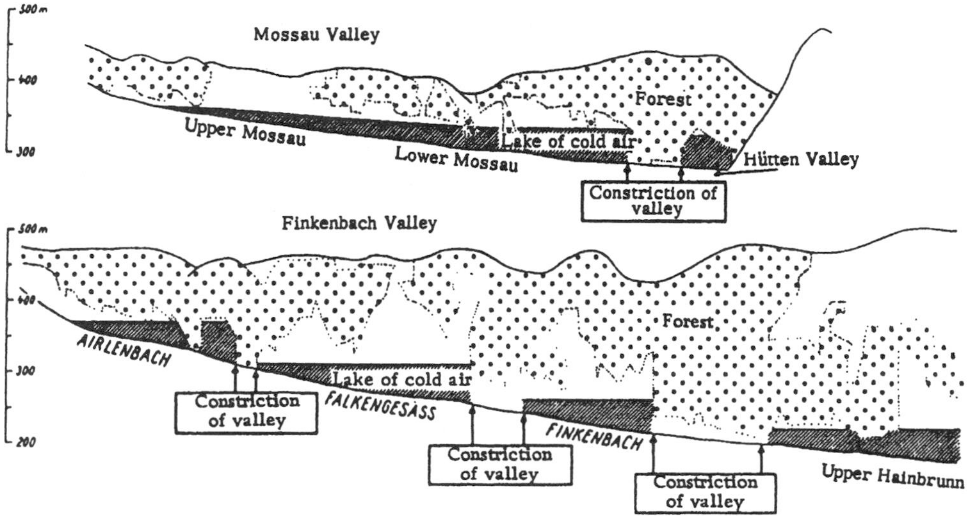

Figure 43-14. Damming of cold air flowing down the valley, in two Odenwald valleys. (From F. Schnelle [4327, 4328])

Whether topography exerts an active or passive effect on the overlying air flow can vary with the local vertical temperature gradient. R. S. Scorer [107], for example, showed (Figure 43-15) that the gradient wind will frequently follow the terrain when the surface is cold but can be separated from the surface when it is warm.

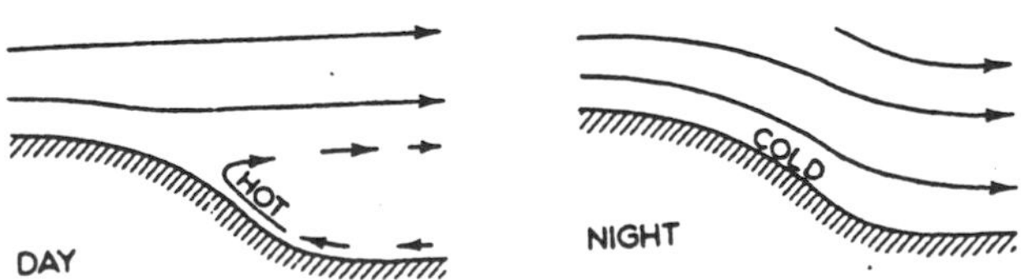

Figure 43-15. Effect of surface temperature on the flow over irregular terrain. (After R. S. Scorer [107]).

The glacier wind is included in the list of active mountain winds since it owes its origin to the contrast between the temperature of the glacier ice and the sunny ground in the surrounding areas and thus tends to be strongly developed on warm days. Glacial winds were first described by H. Tollner [4332] and later investigated by E. Ekhart [4306] and H. Hoinkes [4314].

Figure 43-16 shows how the glacier wind fits into the daily wind system. In clear weather, both an upvalley and glacier wind blowing in opposite directions can develop below the gradient wind. The glacier wind, which is only a few tens of meters deep and has a maximum speed of 3 m sec^{-1} (which is reached at 2.5 meters above the surface of the glacier), slides down underneath the upvalley wind.

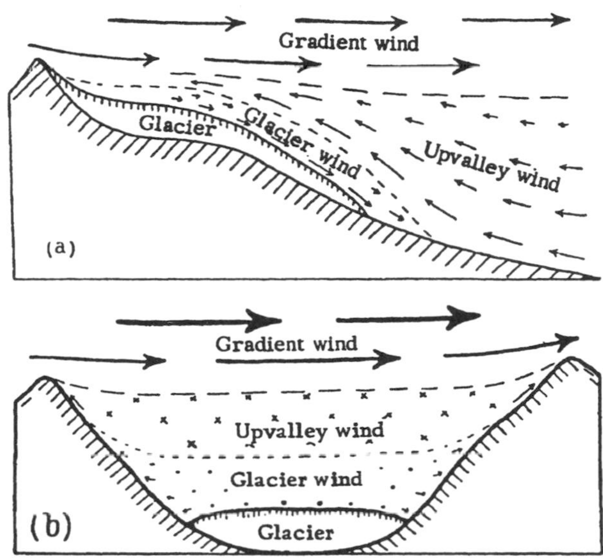

Figure 43-16. Incorporation of a glacier wind into the wind system on a fair weather day: (a) profile, (b) cross section. (After E. Ekhart [4306])

The glacier wind blows night and day in summer with approximately constant strength. There is a first maximum before sunrise, corresponding roughly to the maximum of cold downvalley airflow, and a second maximum before sunset when the temperature contrast between the heated neighborhood and the glacier ice has reached its peak. In discussing the temperature of the air close to the ice, we follow H. Hoinkes [4314] in making a distinction between air that belongs to the glacier and air that has come from the outside. During the day, air which has flowed in has transferred only a little heat to the glacier and exhibits only a small increase of temperature with height (in contrast to the temperature decrease above the neighboring rock surface). The glacier air, on the other hand, has become highly stabilized by long contact with the ice, has a steep temperature gradient (6°C in 2 m), and short-term temperature fluctuations joined with downvalley wind pulses. The steep temperature gradient means that there is a strong flow of heat from the surrounding air to the ice surface. Water vapor from the surrounding air frequently deposits onto the ice surface (Table 24-6 and Figure 26-5). This assists the process of ablation, as was proved by H. Hoinkes [4315]. The often-expressed opinion that the cold glacier wind has a "preserving" effect on the ice, therefore, is incorrect.

The glacier wind is quickly depleted downvalley. Nevertheless, it has a refreshing effect on the mountaineer who approaches the glacier from below and has an important influence on plant life in the neighborhood. According to H. Friedel [4308], the Pasterze Glacier on the Grossglockner causes various types of wind damage such as wind tracks in grass, damaged mosses, and stunted trees, and it reduces the elevation at which plants can grow (*Elynetum,* for example, was reduced by about 500 m in this vicinity). During the summer of 1931, G. Schreckenthal-Schimitschek [4636] measured air temperature at a height of 1 m and ground temperature at depths of 5 and 20 cm at distances of 3, 30, and 500 m in the front of the Mittelberg Glacier in Pitztal. At a distance of 30 m from the glacier, ground temperatures in August and September no longer fell below the freezing point and were 2-3°C higher at the 5 cm depth, and 2-4°C higher at the 20 cm depth, than they were at 3 m from the glacier terminus. The first sparse vegetation was encountered at 30 m from the glacier. Vegetation is often stunted or deformed for considerable distances in front of the glacial terminus because of the cool glacial wind. The glacier wind usually dies out about one-half kilometer in front of the glacier.

The manner in which trees in hilly and mountainous terrain grow is often influenced by the prevailing wind direction. For more information on this topic, consult W. Weischet [4335], F. Runge [4323], and M. M. Yoshino [2714]. H. R. Scultetus [4331] described the effect of topography, wind direction, and lapse rate on the distribution of wind speed and direction in a forest village near Cleves. The influence of overturning has been studied in Japan by M. M. Yoshino [4336 to 4338] in several valleys, using various wind speed, direction, and gustiness recorders.

44 The Climate of Various Slopes (Exposure Climate)

The different amounts of radiation received during the day on different slopes was the subject of Section 40. Section 41 dealt with the small-scale effects of these differences. This discussion will now be broadened to include larger features. We will begin by looking at a regular cone-shaped hill, sloping uniformly on all sides. It is possible to construct such a shape on a small scale, but it is seldom found in nature on a large scale.

The solar radiation received by the slope, described in Section 40, is only one of a number of factors in the energy balance (Section 25), and the actual character of the exposure climate is influenced by many other factors. Air movements created by topographic features were discussed in Sections 42 and 43. On a uniform cone-shaped hill situated in an air stream, the greatest speeds are found on the hillsides. The windward side will favor the formation of upslope winds, and the wind in the lee will be gusty in both strength and direction.

R. Geiger [4404] investigated the microclimate on the Hohenkarpfen in 1926 to evaluate the climate near the ground on hill slopes, and to determine whether temperature contrasts created by topographic effects might be eliminated by the orographic winds generated. The Hohenkarpfen is a conical hill of almost ideal shape, situated on the edge of the hilly area of Württemberg. It has a relative height of 100 m with a slope in the upper part of 30° and lower down a gentle slope of 11°. Figure 44-1 shows the profile of the hill. The air layer close to the slope has been enlarged 50 times to show the observation points at 25 and 100 cm. The temperatures are averages for 8 points on the steep upper slopes (25 cm high) arranged in the 8 principal directions, 16 stations at the point of inflection of the slope, and 8 stations on the lower slope below (25 cm high). The diagram shows that the highest and lowest diurnal temperatures during summer both occur at the summit.

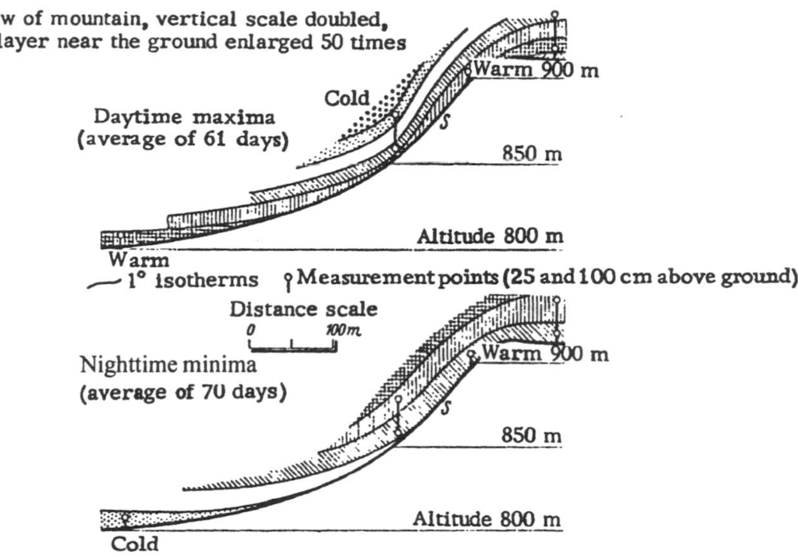

Figure 44-1. Microclimate near the ground of an isolated hill. (After R. Geiger [4404])

Temperature stratification close to the ground can be recognized even on these steep slopes, but it is different from that of the adjoining plain. About midday (Figure 44-1, top) in summer, the strong upslope wind at the foot of the steeper slope draws the cooler air farther away from the slope, with the result that the lapse rate becomes large at this position. Therefore, the highest temperatures near the ground are not found at the steep slope but at the bottom of the valley and on the flat plateau, which is protected from the wind by a growth of shrubs. At night (Figure 44-1, bottom), the stratification of the 100 m layer and the drop of temperature with decreasing height can be recognized.

The climate of slopes facing in different directions is affected to a large extent by moisture conditions as well as radiation and wind. Mountains influence the distribution of precipitation at two different spatial scales. Large scale mountains like the Harz Mountains receive more rainfall on the west (windward) side due to lifting of the air by the mountains, while the east (leeward) side is dry due to foehn-like drying on wind-protected slopes. With smaller scale topographic relief, however, the local precipitation distribution will be determined by the wind field and local topography. Topographically-induced local wind flow patterns often result in a redistribution of falling raindrops, which can be opposite that of the large-scale pattern. More precipitation is found in the lee side than on the windward side, especially with brisk winds. The ordinary type of rain gauges on the lee side of the Hohenkarpfen picked up 5 to 10 percent more precipitation than on the windward side. Snow may often be observed to pile up behind a fence or an undulation in the ground (Figure 52-

11). Some rain and snow falling on the windward side is carried over the top to the leeward side. This is called spillover and plays a very significant role in the distribution of precipitation in some mountainous areas (R. D. Fletcher [4403]).

The slope of the topography also affects precipitation, because the angle of incidence of falling raindrops on sloping surfaces varies with slope angle, slope orientation, wind speed, wind direction, and rainfall intensity. By general agreement, precipitation is measured by means of gauges set up so that the receiving surface is horizontal. If a gauge parallel to the slope is used, of such a size that the horizontal projection of its receiving surface is equal to that of a standard gauge, the amount collected is usually different and will give a measure of the rainfall actually received by the slope, which of course is what interests us. The difference depends on the direction the slope faces, its gradient, and the angle at which the rain is falling (which depends upon the wind speed). M. M. Yoshino [2714] states that "...it is thought that trees in a mountain area receive more raindrops than those coming into a rain gauge with a level receiver."

Figure 44-2 (J. Grunow [4405]) shows the amount of precipitation received in a gauge parallel to a 20° slope on the Hohenpeissenberg, compared with a horizontal gauge, as a function of wind speed. When the wind is blowing at the slope (windward), it always catches more rain than either a horizontal surface or the lee slope. For snow, the difference increases with increasing wind speed (solid line) which is more easily borne on the wind than rain (broken line). When the slope is in the lee, and winds are moderate, the "rain shadow" of the mountain shows up clearly. Only when wind speeds rise above 8 m sec^{-1}, does more rain fall on the lee slope than on the horizontal surface. The relation illustrated depends on the angle at which the precipitation is falling. This angle is shown on a scale at the right edge of the diagram. On the average, more rain falls on the windward slope than a horizontal surface when the angle of falling precipitation was 20° for rain and 31° for snow. It was 8° for both on the lee slope. With winds of more than 7 m sec^{-1}, the angle of the falling precipita-

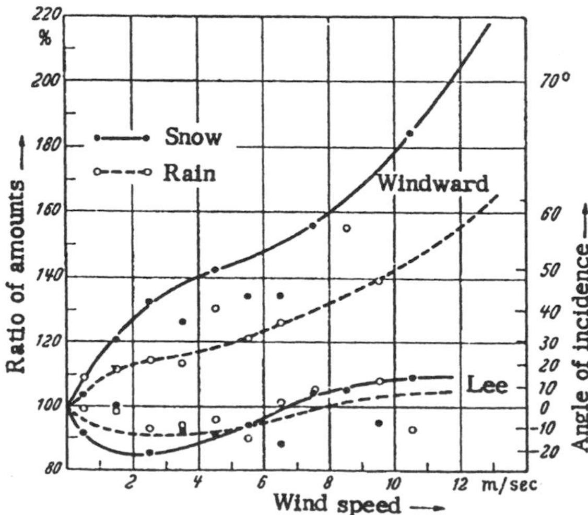

Figure 44-2. Proportion of precipitation falling on a receiving surface parallel to the slope, compared to a horizontal surface on the Hohenpeissenberg. (After J. Grunow [4405])

tion begins to change, turning gradually against the direction of the general wind, as a lee eddy forms. On other slopes, different values for the proportions of rainfall collected and for the angle of rainfall are found, but the basic principles remain the same. J. L. M. P. de Lima [4402] developed a nomograph to estimate rain gauge correction factors by which standard horizontal rain gauge observations can be multiplied to obtain precipitation estimates for inclined surfaces in rugged terrain. The correction factors are dependent upon wind speed at 10 m, wind direction, rainfall type (intensity), and the slope angle and orientation of the inclined surface. D. Sharon [4419] evaluated the use of simple trigonometric adjustments to horizontal rain gauge measurements to estimate effective precipitation on sloping surface in hilly terrain based upon a dense network of stations in the Negev, Israel. Such models worked well only if the slope geometry and storm vector could be specified within narrow limits. Under more general conditions, when such information could not be accurately prescribed, the use of direct measurements from inclined rain gauges was recommended.

When we come to consider soil moisture, both the amount of precipitation and the rate at which the soil dries, which is influenced by the soil type, air temperature, wind field and radiation climate, are of importance. We have already seen from Figure 41-2 that radiation has an asymmetric effect. Although it is approximately evenly distributed before and after noon, a significant amount of the energy received before noon is used to dry the surfaces on which it falls, while in the afternoon, most of it is used for heating the air and soil. The results found for the bark temperatures of a tree trunk (Section 41) are similar to the surface temperatures on differently oriented slopes.

It is for these reasons (in the northern hemisphere) that generally the warmest slopes are not those facing south, but southwest. This was demonstrated as early as 1878 by E. Wollny [4423] using boxes filled with sifted soil facing eight directions and tilted at 15°. Figure 44-3 shows the results of an experiment by A. Kerner [4411] on the Judenbühel near Innsbruck. This hill projects like a peak on the south slope of the Hungerberg plateau and today is called the "Spitzbühel" because of its shape. Soil temperatures were measured at depths of 70 to 80 cm. The temperature differences are small, but they are sufficient to indicate the essential features of the different slope exposures.

The circular shape of Figure 44-3 illustrates the slope azimuths, while months are shown as concentric circles. The average soil temperature has been evaluated for all directions, for each month, and deviations from this mean are shown. Hatched areas are relatively cold, and dotted areas are warm. The greatest temperature differences are found in summer (center of the circular band). The north slope is coldest. The warmest zone varies in position, however, with season. The southwestern position is warmest, as expected, from autumn to spring. The warmest area moves around to the southeast in summer. This is caused by the diurnal variation in cloudiness discussed in Section 40. In the Alps, strongly developed cumulus clouds build up by early afternoon and occasionally produce thunderstorms. The ground, therefore, receives more radiation before noon when the cloud cover is less.

G. H. Schwabe [4417], using artificial soil mounds in south Chile, studied the effect of the contrast between the comparatively low air temperatures caused by the Humboldt Current and the large amount of incoming radiation, found at this latitude, on the plants on various slopes. He used circular mounds 1.2 to 1.5 m high with side slopes of about 35° (as estimated from the diagram). The surfaces of the mounds and flat surrounding areas were sown uniformly and later harvested by sectors.

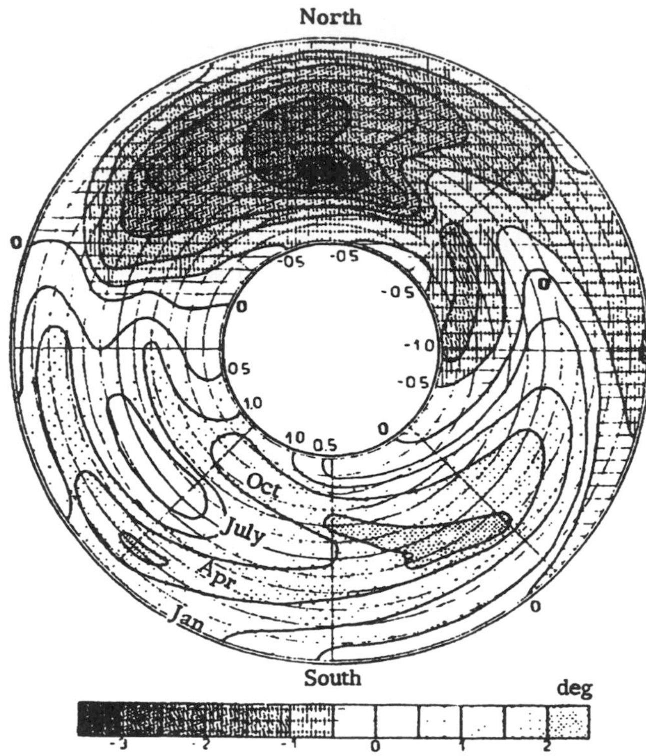

Figure 44-3. Departure from the mean ground temperature at a depth of 70 cm according to slope azimuth and time of year. (From observations by A. Kerner [4411] near Innsbruck)

Two examples of G. H. Schwabe's [4417] results are given in Figure 44-4. In order to facilitate a comparison between these measurements from the southern hemisphere and the other diagrams, the sunny side, which in Chile is the north side, has been put at the bottom of the diagram to accommodate those who live north of the Equator. The isopleths in Figure 44-4(A) give the percentages of beans planted on 3 October 1955 that survived a night with late frost on 28 to 29 October on two mounds near Valdivia (40°S). The assessment was made for eight sectors of the mounds at positions shown by the small circles. The differences were between 10.8 and 30.3 percent and were naturally greatest on the slopes. The soil acted as a heat reservoir, with those areas gaining the most energy during the day being warmer at night.

Figure 44-4(B) shows the yield of grass (kg m^{-2}) on eight mounds in a natural meadow near Valdivia. It had been unusually dry in spring and early summer, and the drying west winds had an unfavorable effect on the western slopes of the mounds, so that the region of highest yield was determined by both humidity and temperature, the highest yields being to the north and east.

W. Haude [4407] made a series of temperature measurements on the slopes of a sand dune in the arid Gobi Desert. He selected a dune 23 m high, in a position some 1400 m above sea level, near the winter camp of the Sven Hedin Expedition to NW China in 1931-

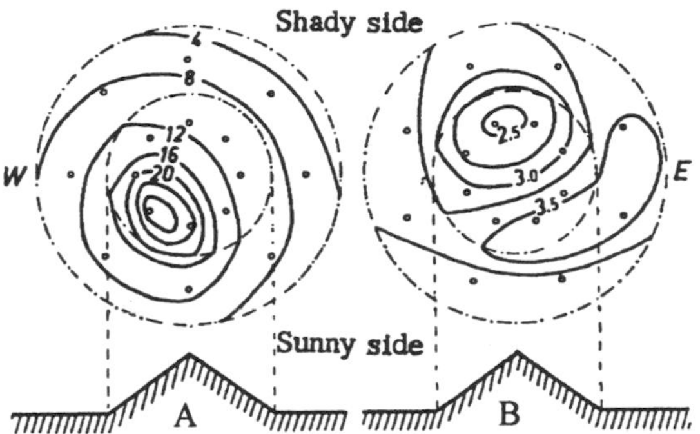

Figure 44-4. A: Beans (percent) surviving a night frost, and B: Yields of pasture grass (kg m^{-2}) on artificial mounds 1.5 m high. (After G. H. Schwabe [4417])

32, and took regular soil temperature readings on the E, S, and W slopes and on the top of the dune. Figure 44-5 shows the daily temperature variation from the averages of 12 nearly clear days. The air temperature was read at the same time in a shelter on the sand dune. While the air temperatures stayed below freezing, the temperature of the soil on a south-facing dune at a depth of 2 mm reached values of 22°C and on one occasion, 18 February 1932, rose as high as 32.8°C. The E slope was warmer than the W because there was a flat pebbly area to the east, while there were other dunes to the west. Soil temperatures on the

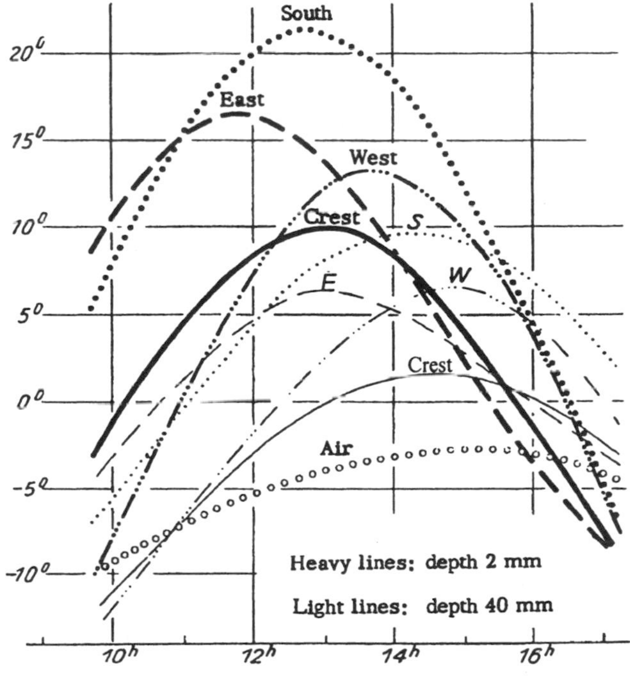

Figure 44-5. Winter soil temperatures (°C) in sand dunes in the Gobi Desert. (After W. Haude [4407])

more exposed crest had lower maximum values than for the three directions shown. Soil temperatures at 40 mm are lower than at 2 mm during the morning and early afternoon, when solar radiation is strong, but exceed the 2 mm temperatures in the late afternoon when surface radiative cooling produces surface cooling.

A remarkable microclimatologic phenomenon in the eskers of southern and central Finland was investigated by V. Okko [4415] and discussed by C. Troll [4421]. The debris carried off by melting glacier waters is piled up to a relative height of 25 m. It consists of coarse blocks and loose gravel with only a thin covering of fine sand on the surface where plants find enough nourishment and moisture. In some places on the crests of such eskers, there are snow-free areas throughout winter (V. Okko [4415] located 28 places from 60° to 64°N). In mid-winter, warm air having a mean temperature of about 3°C streams out from these places. This occurs independently of the temperatures of the surrounding air, which are as low as -30°C. The air current coming out of the ground is strong enough to blow out a match. At temperatures below -20°C, a cloud of condensed water vapor 2 to 3 m high is produced. The circulation is fed from the foot of the esker where there are coarse blocks, and air can enter freely from the outside. This winter circulation begins about the end of September and ends with gradually diminishing speeds in March and April. The explanation is found in the seasonal variation of the supply of heat in the ground.

During the summer, the southwest slope of the eskers undergo considerable heating (temperatures up to 62°C were measured). At this time, cold air flows out from the cold interior, and an inward flow of air at the top may be observed. The esker thus stores more heat during the summer than a neighboring level soil surface. As soon as the first snow falls on the thin sand cover, the stored heat within the mass of rubble is protected against substantial energy loss. A rough computation suffices to show that this heat supply is able to provide the energy for the outward flow of air in winter, which keeps surrounding areas free of snow. The ground water level is usually a few meters down and the water has a temperature of 5° to 6°C, thus also providing heat. It is not surprising that the warm air flowing outward is always saturated, in contrast to the outside air. Vegetation near the snow-free areas is often loaded with a heavy deposit of rime.

Fewer systematic investigations have been made covering all the sides of a hill, mountain, or mountain peak. The forest often, however, provides us with a picture of the varying climate of different slopes. For example, T. Künkele [4413] describes the Palatinate forest as follows: "Anyone who looks out from a single peak over this area, which from a geologic point of view is an apparently uniform terrain, deeply dissected by narrow valleys and steep mountain slopes, and directs his gaze in a NNE direction (that is, toward the slopes facing the sun and the wind), will have the impression of an almost uninterrupted dark bluish sea of firs, interspersed with very few deciduous trees. Looking toward the SSW, however, it is surprising even to a forester to observe how completely the picture of the forest changes into the soft green shimmer of extensive areas of deciduous trees with a decreasing admixture of conifers on the winter sides. This difference shows up on the maps of the Forestry Department and of hiking clubs." J. Parker [4416] has also described forest distributions on the N and S slopes of the Palouse Range in Idaho and has shown the climatic differences by temperature and soil moisture records.

F. K. Hartmann, et al. [4406] studied the slope climate on a completely wooded hill in the Harz in 1954. One of the three experimental areas used (the Grosse Staufenberg near Zorge in the South Harz) was completely covered with a high beech forest on a deep and well aerated soil. The stands, however, showed markedly different characteristics on different slopes. During a summer and an autumn period, measurements of radiation, illumination, temperature, humidity, and evaporation were made near the ground at 20 points around this hill with slopes inclined from 15° to 40° and a relative height of 200 m. Clear and unequivocal effects of the orientation of the slope were found even in the strongly subdued microclimate in the interior of the stands where only 5 to 15 percent of the external light penetrated.

Figure 44-6 shows the distribution around a hill of the daily maximum, and Figure 44-7 that of the daily minimum air temperature for four clear June days, measured at a height of 40 cm above the forest floor. The temperature maxima have, to a considerable extent, the same distribution as global radiation values. The southwest slope is warmest and the northeast slope is coldest. Evaporation is also greatest in the middle of the SSW slope, with a maximum of 18 $cm^3\ d^{-1}$. It is least in the NE, with a maximum of only 9 $cm^3\ d^{-1}$, and the relative humidity is correspondingly lowest in the SSW. The north slope appears to be relatively warm in Figure 44-6 because when the sun is high in summer, it also shines on this slope in the late afternoon. This is not the case during the autumn.

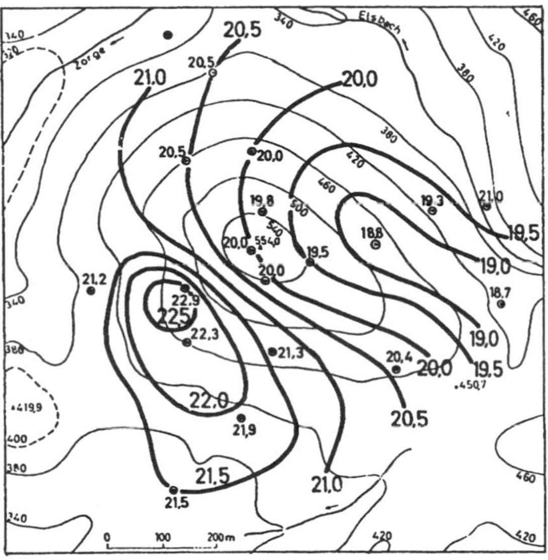

Figure 44-6. Maximum temperatures near the forest floor on the Staufenberg on four clear June days in 1954. (From F. K. Hartmann, et al. [4406])

Night minimum temperatures during the summer period (Figure 44-7) are in part determined by the flow of cold air down the slopes and from the north out of the Zorge Valley. This air is dammed up by the Staufenberg and produces the lowest temperatures in the Elsbach Valley to the NE. The cold air is able to penetrate into the closed stand around the bottom of the hillside. This cold air encroachment around the hillside makes the lower part of the southern slope the warmest part in other seasons as well. The night minimum pattern is also partly explained by the nighttime release of heat stored during the day.

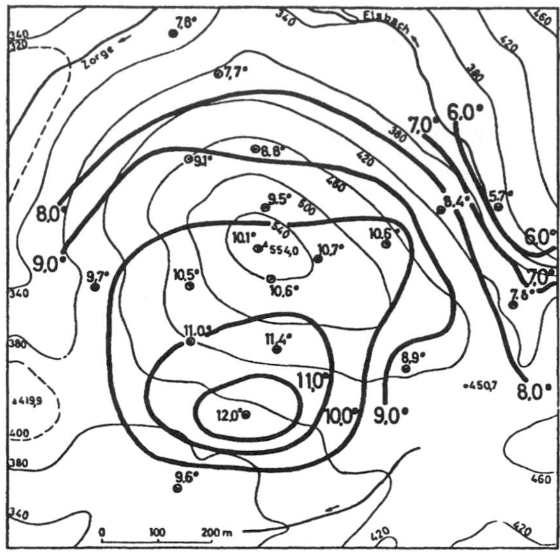

Figure 44-7. Minimum temperatures for the same location as in Figure 44-6. (From F. K. Hartmann, et al. [4406])

Figure 44-8 shows observed values of daytime net radiation (S), temperature (T), and vapor pressure (F) divided into five gradations, where 1 indicates the highest and 5 the lowest values. The distribution arrived at was closely related to the type of stand. As might be expected, the warmest temperatures are generally associated with higher net radiation and vapor pressure.

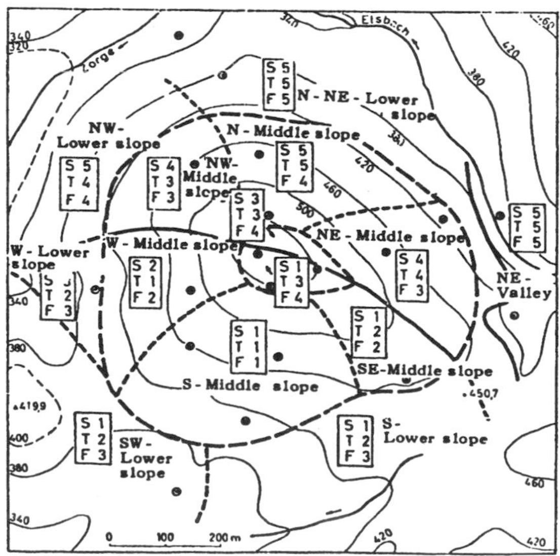

Figure 44-8. Distribution of microclimate on the Staufenberg. (From F. K. Hartmann, et al. [4406])

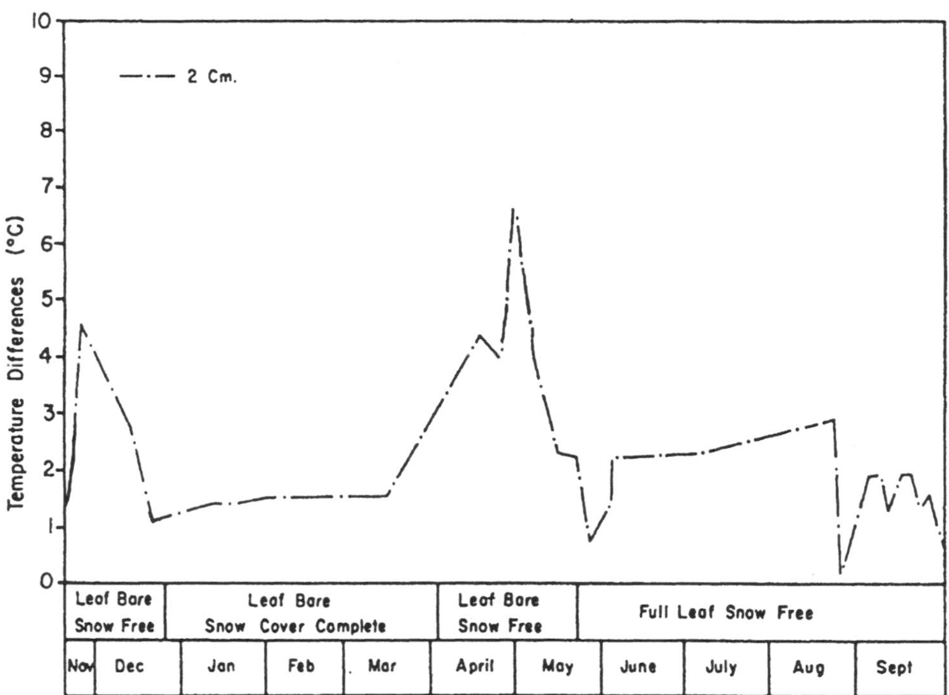

Figure 44-9. Temperature difference between the north and south-facing slopes at a depth of 2 cm in Quebec. (After R. G. Wilson [4422])

Figure 44-9 shows the temperature differences throughout the year between the north and south slopes at a depth of 2 m. The differences are largest during the fall after the leaves have fallen and in spring prior to bud break. During these periods the sun is relatively high in the sky and a large percent of the solar energy is absorbed at the soil surface. At the end of April the surface temperature on the south-facing slopes was 5°C while those on the north slope were still frozen. Soil temperatures at 2 cm were 6°C warmer on the southern slopes in the spring, and were 2°C warmer in the late summer.

J. B. Kirkpatrick and M. Nuñez [4412] examined variations in surface solar radiation input and species distribution along a north-south transect in the hills near Risdon, Tasmania. The largest variations in solar radiation input were found during the spring and fall seasons, and the smallest variations during the summer. Radiation totals along the two ridges were similar throughout the year, while north-facing slopes experienced increased solar input in all seasons for this southern hemisphere site. A yearly dryness index calculated from estimated precipitation and surface evaporation varied by a factor of more than two along the transect. Vegetation responses to these microclimatic variations were numerous. The most xeric vegetation was generally found on the northwest-facing slopes, while the most mesic vegetation was found along the southeast-facing slopes. Herbaceous species diversity was lowest on the north-facing slopes and highest on the south-facing slopes. The three plant communities in the hills generally changed gradually in height and relative dominance along the transect except in areas of sharp changes in slope and aspect, where more abrupt plant changes occurred.

J. E. Cantlon [4401] investigated the air temperature and humidity at four heights (5, 20, 100, and 200 cm) and the maximum and minimum soil temperature at 4 cm depth at two stations on the N and S slopes of Cucketunk Mountain in New Jersey. During the day, on the south slope under heavy shade, he observed nearly isothermal conditions, whereas under light shade, there was a marked decrease of temperature with height. On the north slope, however, there was always an increase of temperature with height regardless of the density of the vegetation. He also found daytime vapor pressure deficits increasing toward the surface, particularly on the north slope. R. E. Shanks and F. H. Norris [4418] investigated the danger of late frosts on the N and S slopes of a valley running E-W near Knoxville, Tennessee by recording extreme temperatures at a height of 30 cm at 14 stations. The S slope had the higher maxima, the N slope the lower minima. The daily average was 1.7°C warmer on the S slope, while the minima on the south slope was only 0.6 to 1.1°C higher. The duration during which temperatures were below the freezing point showed a better correlation with observed frost damage than did minimum temperatures.

Large differences in soil moisture can also be found on different slopes. Figure 44-10 shows the change in soil moisture from May through October 1953 on the 20° S and N slopes of the Hohenpeissenberg, measured at a depth of 5 cm below the grass-covered surface (after K. Heigel [4409]). The two scales show soil moisture in percent weight (with an inverted scale). The N slope is moister throughout the period. In the period of clear weather in the autumn, the north slope's soil moisture fell to 20 percent at the time of strongest drying, whereas it decreased to 10 percent on the S slope.

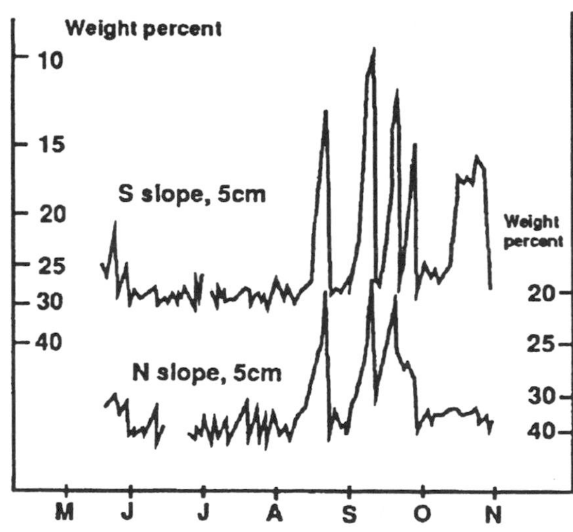

Figure 44-10. Soil moisture variation from May through October 1953 on the S and N slopes of the Hohenpeissenberg. (After K. Heigel [4409])

It is often assumed, however, that in water-limiting environments the reduced input of solar radiation on north-facing slopes (northern hemisphere) will result in increased levels of soil moisture on those slopes. The greater biomass produced in that more mesic environment, however, can complicate this general pattern. E. Ng and P. C. Miller [4414], for example, found greater levels of soil moisture on south-facing slopes among southern California chaparral communities (32°54'N, 925 m), even though they had higher soil evaporation rates. The increased vegetation cover, larger leaf area indices and deeper rooting systems found on north-facing slopes resulted in higher transpiration rates which depleted soil moisture levels to a greater degree than on south-facing slopes.

K. Heigel [4408] investigated the dependence of phenologic processes on exposure and height on the slopes of the Hohenpeissenberg. Blossoms of the sweet cherry were delayed 2 days for every 100 m of height in 1951; the difference between N and S slopes amounted to 5 to 7 days. The harvest of winter rye, on the other hand, responded primarily to height and was delayed on the average by 7 days per 100 m. The opening of dandelion *(Taraxacum officinale)* flowers was found to be extremely sensitive to sunshine duration on the slope. Comparing plants on only the N and S slopes with equal gradients of 18 to 20°, the following dates for flowering were found in 1954:

Elevation (m):	820	860	900	940
South slope:	25 April	27 April	2 May	11 May
North slope:	9 May	17 May	(no vegetation)	

45 The Thermal Belt

Section 44 was primarily concerned with the influence of slope orientation on the local climate; the effect of elevation or relative height will now be considered.

If the rules for the nocturnal movement of cold air given in Section 42 are applied, valleys would contain an enormous lake of cold air as shown in the top sketches of Figure 45-1. The effect of the greater dimensions here is that individual circulations are built up between the air that is cooling on the slope and the reservoir of warmer air above the valley floor, as shown in the lower sketches in Figure 45-1. A lake of cold air only develops near the bottom of the valley. Since a layer of cold air near the ground remains over the plateau above, an intermediate zone, known as the thermal belt, develops on the slope, where temperatures are higher at night. This corresponds to the nocturnal inversion above level ground.

This vertical division into three zones can be recognized from temperatures recorded by F. D. Young [4512] at five different heights, during the night of 27-28 December 1918 on the slopes of the San Jose Mountains in the Pomona Valley, California. The lowest curves for 0 and 8 m in Figure 45-2 show the freezing temperatures in the lake of cold air at the bottom of the valley; they run almost horizontal in the undisturbed air before sunrise. At 15 m, temperatures increase because the warm thermal belt is approached; the temperature pattern is much more irregular. The warmest zone is around 68 m, above which (84 m) temperatures decrease again.

During the day, a division into three zones is maintained, but the temperature pattern is different. Figure 45-3 shows the arrangement of observation points in the investigation carried out on the Grosse Arber by R. Geiger, et al. [4504]. Stations with meteorological shelters were set up in the valley in the SW (Bodenmais, 665 m), in the E (Seebachschleife, 645

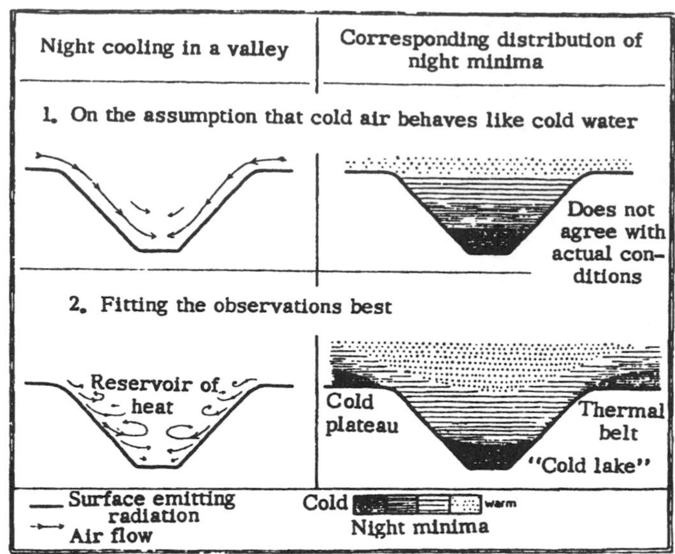

Figure 45-1. Development of the thermal belt.

m), on the hillsides at Kopfhäng (1008 m) in the SW, on a level area in the N at Mooshutten (946 m), and also on the peak (1447 m). Along the line of crosses, 99 measuring points were set up to record night temperatures near the ground. Figures 45-4 and 45-5 give the averages of the shelter readings for 25 clear days in May and June in the form of curves showing the diurnal air temperature and relative humidity.

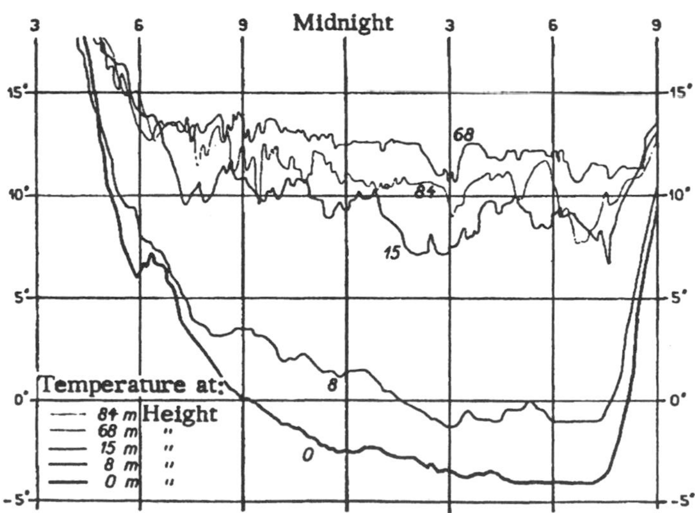

Figure 45-2. Night temperature variation at five different heights in the Pomona Valley, California. (After F. D. Young [4512])

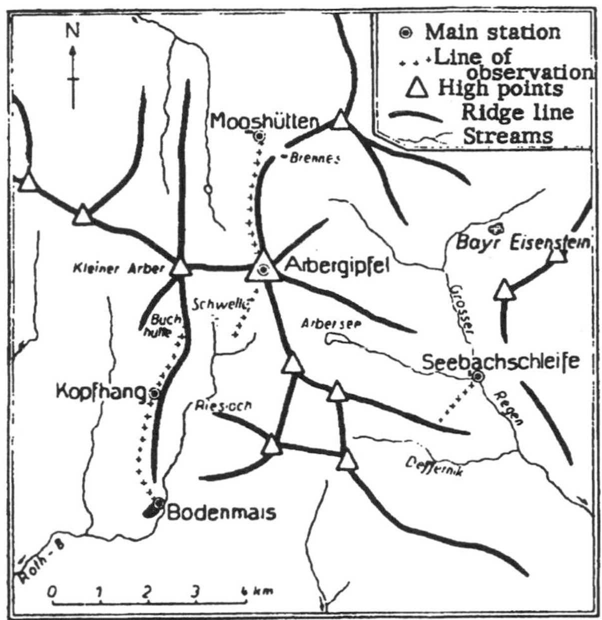

Figure 45-3. Experimental setup on the Grosse Arber, 1931-1932. (After R. Geiger, et al. [4504])

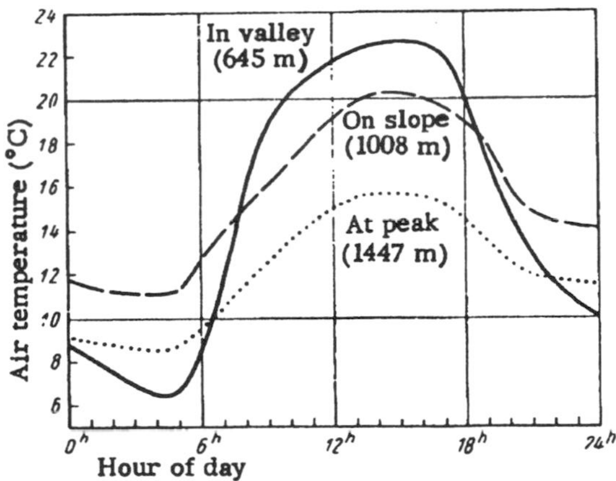

Figure 45-4. Diurnal air temperature variation on clear spring days on the Grosse Arber. (After R. Geiger, et al. [4504])

The values in Figure 45-4 are averages. The temperatures at 24:00 are warmer than at 00:00 because the season was progressing. The valley site has the lowest air temperature and highest relative humidity at night, and the highest air temperature and lowest relative humidity during the day (Figures 45-4, 45-5). The hillside station (thermal belt) is the warmest location at night. Its daytime temperatures depend on its relative height, and its relative humidity is intermediate throughout the day. The peak has the smallest diurnal variation. It is coldest during the day, but the valley is cooler at night. Above the thermal belt, temperatures decrease with elevation. If the peak is high enough, its temperature at night will also be lower than the valley. The peak has the smallest diurnal variation in relative humidity (Figure 45-5). It was driest at night and moistest during the day. However, on the days selected for observation on the Grosse Arber, there was none of the condensation frequently found on mountaintops.

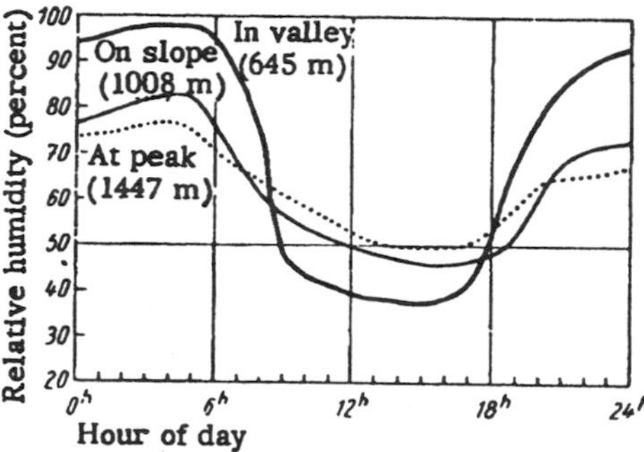

Figure 45-5. Diurnal relative humidity for the same time and place as Figure 45-4. (After R. Geiger, et al. [4504])

A contour map of the investigation on the Grosse Falkenstein study site in the Bavarian Forest is shown in Figure 45-6. The line of measurement followed a straight, clear path on the WSW slope, which was open enough for the instruments to be set up outside the stand and yet narrow enough to inhibit slope winds. Figure 45-7 shows the arrangement of instruments. There were eight stations with shelters, near which were small experimental phenologic plots, arranged down the hillside. Between these were points at which the night minimum temperatures at two heights, precipitation from clouds and fog, and the depth of snow were measured. In the stand nearby, semiconductor thermistors were set up in shelters at six levels: on the forest floor, in the lower trunk area, in the crown, and at the ends of long poles several meters above the treetops (W_1 to W_6).

Table 45-1 gives the diurnal variation of air temperature, relative humidity, and mean vapor pressure (from the 07:00, 14:00, and 21:00 observations) for May 1955 for six selected shelter stations, indicated in Figure 45-7 by the numbers 2 (1307 m at the peak) to 14 (622 m in the valley). These results, published by A. Baumgartner [4501], illustrate the principal features of the transition of slope climate from the bottom of the valley to the peak.

The three part division of slope climate previously described is most conspicuous in situations with relatively calm conditions and clear skies. When there are clouds, precipitation or strong winds present, a normal decrease of temperature with an increase of height is observed. In Figure 45-8, the observations made on the Grosse Arber have been arranged according to the three most frequent air masses, for the warmest midday hours and the coldest hours at night. At midday, the temperature always decreases with height. At night, temperatures only decrease with height with maritime polar air (*mP*), that is, when the temperature is comparatively low, precipitation and clouds are frequent, and winds are brisk. With a

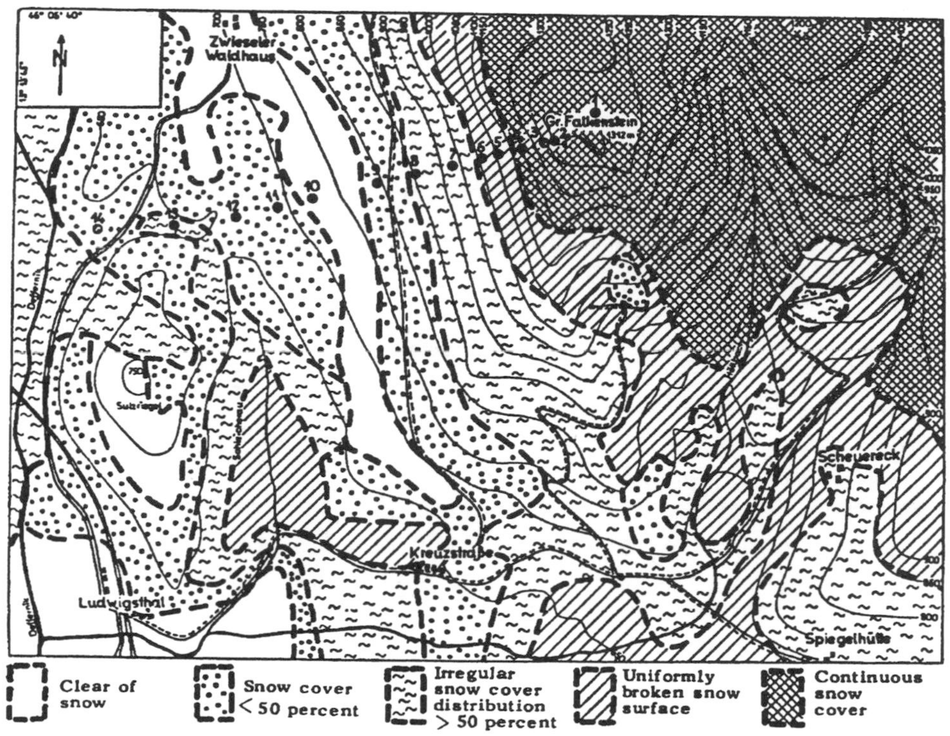

Figure 45-6. Map of snow cover on the Grosse Falkenstein on 19 April 1955, showing early melting in the thermal belt. (After G. Waldmann [4513])

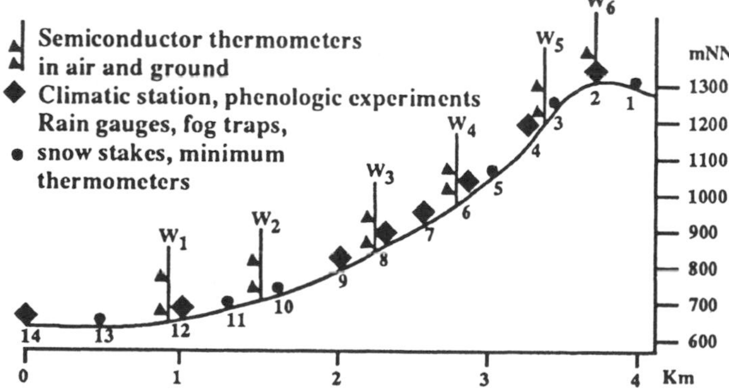

Figure 45-7. Experimental setup on the WSW slope of the Grosse Falkenstein in 1955. (After A. Baumgartner and G. Hofmann [3402])

continental air mass (*c*), however (always associated in spring and summer with high temperatures and light winds), an inversion and a warm thermal belt develop. While temperatures are somewhat different on the W and E slopes of the Arber, the pattern of

Table 45-1. Daily variation in meteorological shelters on the west slope of the Grosse Falkenstein, in May 1955. Asterisks mark the warmest and relatively driest zones of the vertical profile; plus signs indicate the coldest and most humid zones. (After A. Baumgartner [4501])

Station No. (Fig 45-7)	Elevation (m)	Hour of the day 2	4	6	8	10	12	14	16	18	20	22	24	Daily average
Air temperature (°C)														
2	1,307	2.9	2.7	2.7	3.6+	5.2+	6.6+	7.0+	6.5+	5.2+	4.0+	3.4+	3.0	4.4
4	1,157	4.2	3.8	3.6	4.2	5.8	7.4	8.1	8.1	7.3	5.6	4.7	4.3	5.6
7	925	5.8	5.5	5.1	5.9	8.1	9.7	10.5	10.4	9.3	7.3	6.3	5.7	7.4
9	796	6.4*	6.0*	5.9*	6.9	9.9	11.4	12.0	12.0	11.0*	8.8*	7.2*	6.5*	8.6
12	658	3.8	3.4	3.6	6.6	10.2	11.9	12.3	12.0	10.6	8.4	5.8	4.4	7.7
14	622	1.9+	1.5+	2.4+	8.0*	11.0*	12.6*	13.2*	12.9*	10.9	6.8	3.8	2.4+	7.8
Relative humidity (percent)														Vapor pressure (mb)
2	1,307	89	89	90	86	82+	75+	75+	77+	82+	86	88	89	7.1
4	1,157	88	88	89	86	81	73	72	73	75	82	86	88	7.5
7	925	86*	86*	88*	83	70	60	60	62	68	79*	84*	86*	8.0
9	796	91	92	92	80	66	56	58	60	66	80	88	91	8.7
12	658	97+	97	96	88+	68	58	59	63	72	85	95	97+	8.8
14	622	97+	98+	97+	81*	64*	55*	56*	58*	66*	88+	95+	97+	8.3

temperature variation with height is similar. Although the temperature at the peak is 8°C higher, the valley below the W slope is colder at night with *mP* than with *c* air. The maritime air mass (*m*), shown between these two, also is associated with the formation of inversions.

In Japan, H. Mano [4508] found, on the slopes (1819 m) above Lake Inawashiro (514 m) in the autumn of 1954, that there were regular night temperature fluctuations at a station set up at 830 m, which is normally the area of the thermal belt. These variations were of the order of 1°C, and there was a time interval of 2 hr from one extreme to the other. By measuring the temperature and wind field above the mountain side up to a height of 1800 m, the cause of the fluctuation was established to be a displacement of the boundary layer between the lower wind close to the hillside and the upper gradient wind. These regular displacements were initiated partly by thermal and partly by dynamic processes, and they had the effect that the station at 830 m was sometimes inside and sometimes above the inversion layer.

In regions where there are persistent subsidence inversions important changes can be observed in the thermal belt. T. W. Giambelluca and D. Nullet [4505] examined daily,

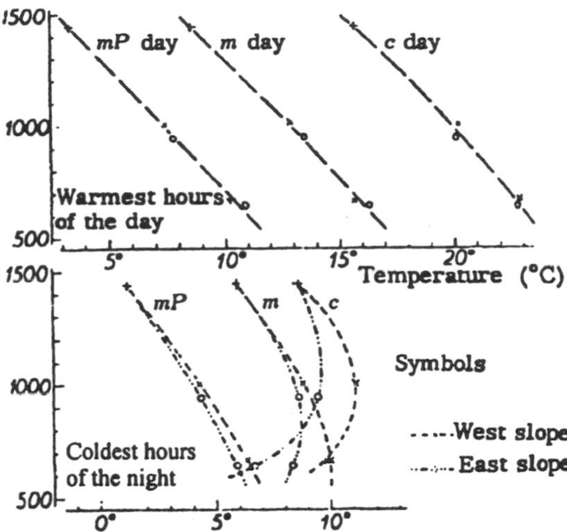

Figure 45-8. Vertical temperature profiles on the Grosse Arber in spring as a function of time of day and air mass. (After R. Geiger, et al. [4504])

monthly, and annual variations in global solar radiation, net radiation, air temperature, relative humidity, vapor pressure, and wind speed at five elevations along the leeward slope of Haleakala volcano on Maui (20°45') in the Hawaiian Islands. In this subtropical island a subsidence inversion associated with the descending limb of the Hadley cell circulation is found at an elevation of between 1200-2400 m on 70 percent of all days. The subsidence inversion strongly affects vertical mixing that, in turn, influences cloudiness, precipitation, solar transmission, net longwave radiation, air temperature, humidity and wind speed. The climate, natural vegetation, and wildlife in the middle and upper-elevations on the lee side of the volcanic peak are greatly affected by the presence and movement of the inversion throughout the day and year.

Four climatic zones are evident from their analyses. A marine zone is found below 1200 m that is almost always below the inversion layer. This zone consists of well-mixed air in contact with the ocean. Temperature and humidity in this marine layer are high and decrease linearly with elevation. A cloud (fog) zone is found between the mean cloud base at 1200 m and the most frequent lower limit of the inversion at 1800 m. Here the cloud layer is often in contact with the surface, global and net radiation are often low, and climate conditions can change suddenly in response to changes in the height of the inversion. A transition zone occurs between 1800-2400 m where the inversion base is most frequently found. This zone experiences the highest degree of climatic variability. Finally, an arid zone is found above 2400 m above the inversion base that is isolated from the marine influence. Here the climate is sunny, relatively warm, dry, and winds are strong. Global solar radiation and net radiation increase in this layer due to the high solar transmission resulting from the absence of clouds and lower water vapor concentration. The local climate in each zone also shows the effects of the diurnal cycle of upslope and downslope winds commonly developed in mountains (Section 43).

In spite of the many influences affecting the location of the thermal belt, its position on any hillside is remarkably constant when established by statistical averaging. Figure 45-9 shows a side view of the section marked on Figure 45-3 by a line of crosses running SW from Seebachschleife. On it, the positions of the 23 measurement points are shown by short vertical lines. Night minimum temperatures observed in the springs of 1931 and 1932 were used to establish the height of the highest temperature for each individual night. The resulting frequency distribution is given on the right of Figure 45-9, showing that the warmest temperatures occur most frequently at elevations slightly higher than 800 m. The second weak maximum at the bottom of the valley results from windy or cloudy weather situations in which temperature decreases with height.

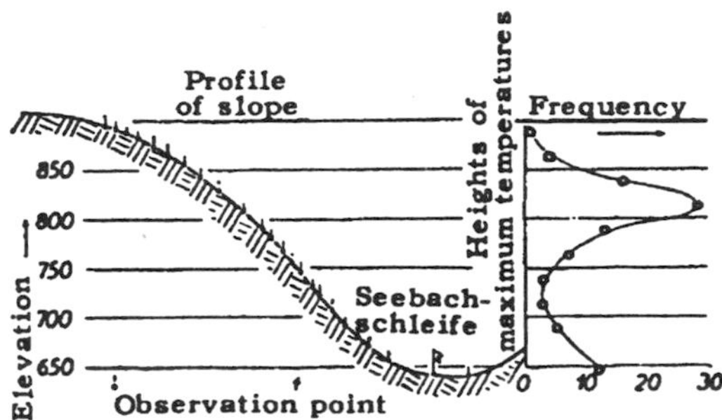

Figure 45-9. Location of the warmest temperatures (thermal belt) on the slopes of the Grosse Arber in spring. (After R. Geiger, et al. [4504])

It is not surprising that the existence of the thermal belt was known long before there was any scientific knowledge of climatology. In Germany, this area was preferred for the earliest villages, monasteries, and country houses. E. Bylund and A. Sundborg [4503] have given some good examples from Swedish Lapland, where places 2 to 3 km apart horizontally and with a height difference of less than 100 m can show a temperature difference of as much as 8 to 9°C.

On a small conical hill in China, S. B. Huang [4410] reported the distribution of frost damage to citrus trees shown in Figure 45-10. The increased frost damage in the lower areas to the east and north are a result of low minimum temperatures in this region (Figure 44-8). The decreased frost damage on the north and east slopes is a result of the thermal belt.

The effects of relative height are also apparent in the distribution of snow cover in mountainous terrain. Figure 45-6, taken from the Falkenstein experiment, shows the distribution of melting snow on 19 April 1955. Snow depth measurement locations were selected to be comparable and unaffected by the forests in order to clarify the effect of slope position, slope orientation, and terrain configuration. By examining the height contours, it can be seen that the warm thermal zone at about 700 m is already clear of snow. Below this level, snow depth increases, and above, it increases even more since winter accumulation was somewhat greater there. This also provides an example of the close relation between the microclimate of an area and the snowmelt pattern mentioned in Section 24. M. M. Yoshino [2714] stated that the height of the center of the thermal belt is 100-400 m (usually 200-300 m) above the valley floor or base of the mountain, is usually higher in winter than in summer, and is higher and more distinct on calm clear nights. B. Obrebska-Starkel [4509], however, in her summation of previous studies found the average height of the center of thermal belts to be

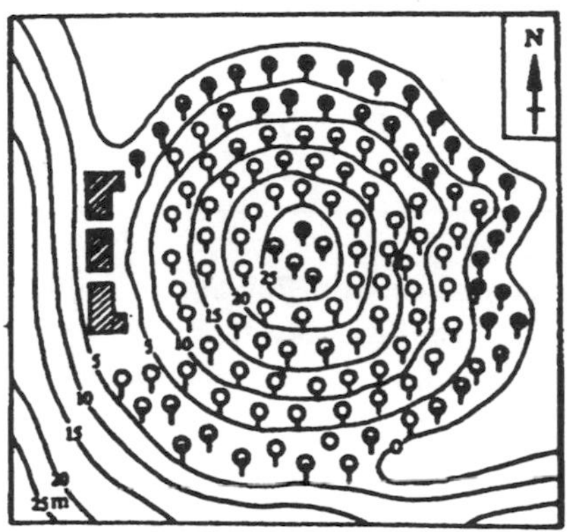

Figure 45-10. Ponkan citrus trees on a circular hill. Shading shows amount of frost damage in March 1977. (After S. B. Huang [4410])

150-200 m above the base level. A. Baumgartner [4501] also found that the thermal belt was usually higher in winter.

The development of plants also reflects the location of the thermal belt. F. Schnelle [106] made use of German and British observations to draw attention to the fact that vegetation first makes its appearance up to 20 days in advance, and trees blossom a few days earlier in the thermal belt than in the valley 200 m below. From observations also made by F. Schnell [4510] in Germany from 1936 to 1939, thc cars of winter rye first appeared as follows:

height (m):	150	200	300	500	700	1000
date:	17	16	15	22	28 May	7 June

The position of earliest winter rye appearance in this case was, therefore, at 300 m.

The curve on the left of Figure 45-11 gives the mean night minima over a period of 68 nights in May and June 1931 and 1932, for the area above the Seebachschleife (shown in Figures 45-3 and 45-9). Although the period covers all kinds of weather, the warm thermal belt is easily recognized. On the right are three phenological graphs that show the starting date of selected phenological stages at different elevations. The time scale runs from right to left to make comparison easier. The phenologic and temperature curves are very similar.

Figure 45-12 shows the development of phenologic phase over a period of time on the Falkenstein (compare Figure 45-6). The first signs of green on the red beech (*Fagus silvatica*) appeared in the thermal belt before 1 May 1955. From that position the green area spread upward and downward and did not reach positions in the valley until 9 days later. A series of similar phenologic charts were published by A. Baumgartner, et al. [4502], in which four species of trees and ground flora were observed continuously. The duration of the growing period in 1955, measured from the time when more than half of all twigs bore open leaves until the time when one tree had lost more than half its leaves, was shown to depend on elevation as follows:

Elevation (m):	620	700	800	900	1000	1100	1200
beech (days):	139	157	163	161	135	127	121
maple (days):	137	146	146	140	134	129	126

The longer growing season duration between 700-900 m is clearly evident. J. J. Higgins [4507] determined the state of phenologic development experimentally on garden peas (*Pisum sitavum*) by counting the nodes and measuring the length of the internodes. This method was successfully used on the phenologic experimental plot on the Falkenstein. Five sowings, between 12 May and 30 August, gave a maximum development on hillside positions from 800 to 850 m (thermal belt).

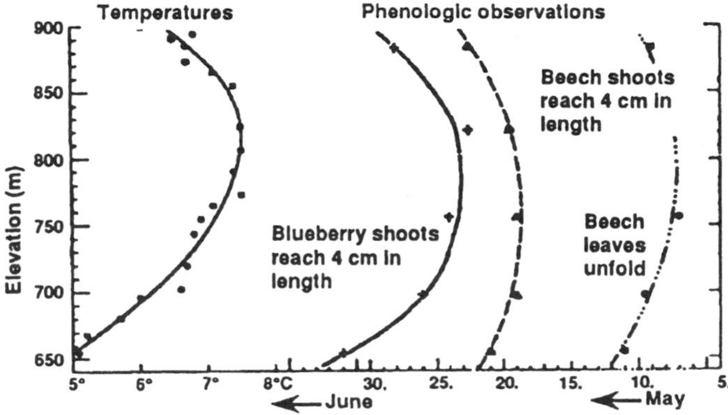

Figure 45-11. The close relation between average minimum night temperatures (left) in May and June and plant growth on the Grosse Arber. (After R. Geiger, et al. [4504])

In contrast to the warm belt on the hillside, valley sites are cold at night, and, therefore, exposed to frost danger; the air there is also still, and fogs are frequent. Fog is therefore a characteristic feature of valleys. For example, it was established by J. Grunow [4506] that in the period 1950-1956 the Ammer basin had 200 hr less sunshine per year than the Hohenpeissenberg, and in December alone the difference was 32 hr. Above the warm thermal belt, however, fog is often the result of drifting clouds. Figure 45-13, after A. Baumgartner [4500], shows the change in the frequency of fog type with height on the western slopes of the Grosse Falkenstein. Three groups of days are differentiated in the period from 16 April to 15 November 1955; their totals add up to 100 percent for each height. The days with fog in which no rain falls (dotted curve) are rare and increase slightly with height. When it is raining in the valley, the rain is nearly always associated with good visibility (rain without fog). The higher one goes up the slope, the more frequently rain is associated with fog. Measurements of precipitation from fog, using J. Grunow's [3609] catchment nets, indicated supplemental precipitation from about 1000 m upward in comparison with readings of ordinary rain gauges. These supplemental readings exceeded 100 percent at some places on the peak. This caused substantial differences in local conditions and provided an explanation for the moorland vegetation on the peak, which requires a high relative humidity. Because of

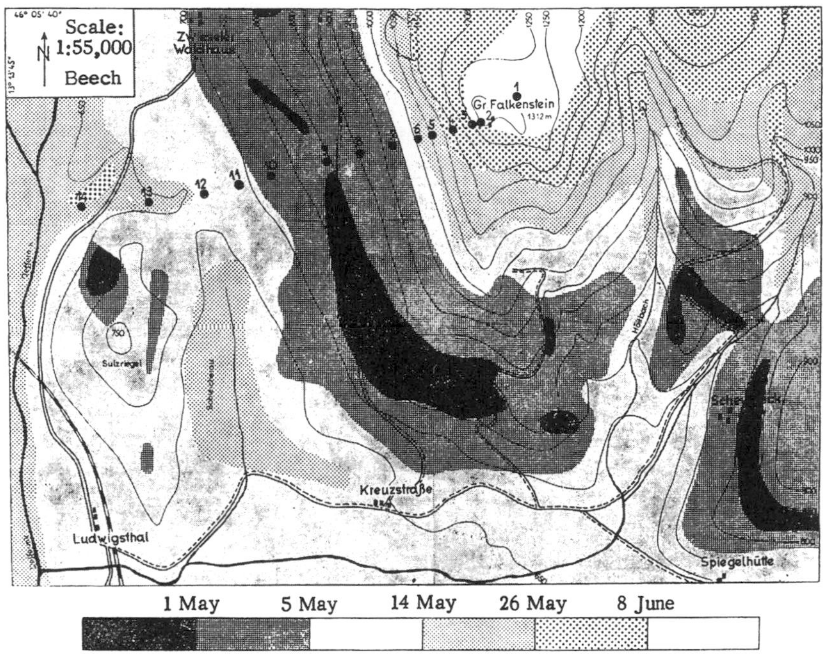

Figure 45-12. Appearance of the first green beech leaves on the slopes of the Grosse Falkenstein. (From A. Baumgartner, et al. [4502])

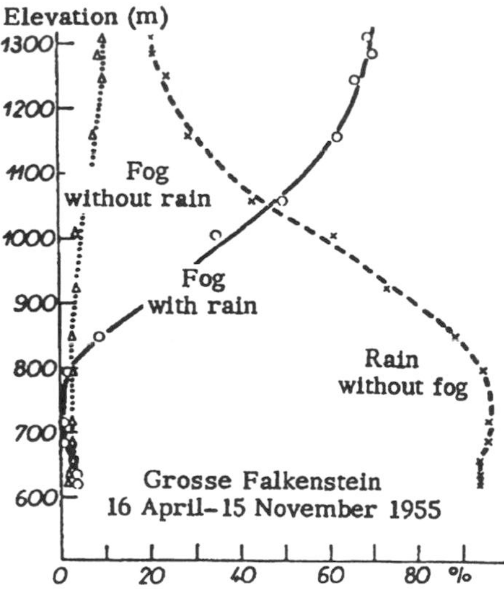

Figure 45-13. Fog types on the west slope of the Falkenstein with changing altitude. (After A. Baumgartner [4500])

the limited areal extent of the peak region, the fog supplement is of limited importance to the regional water budget.

46 Microclimate in the High Mountains

Mountain and high plateau regions occupy nearly 20 percent of the land surface of Earth. With increasing elevation the climate near the ground acquires new characteristics that become increasingly pronounced at higher elevations.

Increasing population pressure and more intensive economic exploitation have caused far-reaching and serious damage in the high mountains. Trees subject to human abuse are particularly susceptible at the tree line where they grow slowly and are subjected to more severe and frequent damage by weather. The construction of new roads is being progressively extended to higher regions to meet the demands of tourist traffic. The construction, use and maintenance of such roads is often influenced by the climate near the ground.

No other mountain range in Europe has been so thoroughly investigated from the plains to the summits as the Eastern Alps. The figures given in Table 46-1 are of average conditions, and may at times lack precision because of interpolation. Therefore, these macroclimatic values should not be taken as valid for any particular place at a given elevation. Nevertheless, they provide a general picture of mountain climate.

Direct-beam solar radiation increases and diffuse solar radiation decreases with height above sea level since atmospheric mass, water vapor content and turbidity, with their scattering and absorbing properties, decrease with elevation. Table 46-1 gives the daily global radiation on a horizontal surface from F. Sauberer and I. Dirmhirn [4634] and I. Dirmhirn [4620] for clear and cloudy days in June and December. From 200 to 3000 m above sea level, global solar radiation increases by 21 percent with clear sky, and by 160 percent with cloudy skies, or by 1 and 4 percent, respectively, per 100 m.

The longwave radiation budget, on the other hand, is virtually independent of height; this is because decreasing surface temperature reduces the amount of outgoing radiation from the ground, and the counterradiation is also reduced with elevation because of the decreasing temperature, density, thickness, and moisture content of the atmosphere.

The irregular cloud cover pattern with increasing elevation, however, can complicate the relationship between global radiation and elevation. Under partly cloudy conditions, the surface radiation is the sum of direct-beam solar radiation that shines through gaps in the clouds and diffuse radiation reflected from the sides of clouds. H. Turner [4644] made a detailed study of radiation at the Obergurgl station (1940 m), and measured instantaneous values of as much as 1570 W m^{-2}, which is 115 percent of the solar constant. Figure 29-7 gave a similar example from a height of 2720 m in the mountains of Japan. In the presence of drifting clouds, therefore, solar radiation at the surface is subject to very great temporal fluctuations. H. Turner [4645] was able to record changes of surface temperature of 10°C within a few seconds. Vegetation at high elevations must be able to cope with these sharp variations in temperature and solar radiation.

At high elevations, the horizon is often screened by mountains, which reflect strongly when covered with snow or glaciers. H. Turner [4644] calculated for Obergurgl that there was a 10 percent loss of global solar radiation in cloudless weather, through obstruction of the horizon, with the sun at its highest elevation in June. This value increased to 60 percent

Table 46-1. Changes in climatic conditions with height above sea level in the Eastern Alps.

Elevation (m)	Mean daily global-radiation totals MJ $m^{-2}d^{-1}$				Mean air temperature (°C)				Annual number of:			
	Clear		Overcast									
	June	December	June	December	January	July	Year	Annual variation	Summer days	Days without frost	Days of frost change	Days with frost
1	2	3	4	5	6	7	8	9	10	11	12	13
200	28.9	5.4	6.5	1.3	-1.4	19.5	9.0	20.9	48	272	67	93
400	29.7	5.7	7.0	1.3	-2.5	18.3	8.0	20.8	42	267	97	98
600	30.3	5.9	7.5	1.4	-3.5	17.1	7.1	20.6	37	250	78	115
800	30.8	6.1	8.0	1.5	-3.9	16.0	6.4	19.9	31	234	91	131
1,000	31.2	6.3	8.6	1.6	-3.9	14.8	5.7	18.7	15	226	86	139
1,200	31.8	6.5	9.2	1.7	-3.9	13.6	4.9	17.5	11	218	84	147
1,400	32.3	6.6	9.9	1.8	-4.1	12.4	4.0	16.5	7	211	81	154
1,600	32.8	6.7	10.6	2.0	-4.9	11.2	2.8	16.1	4	203	78	162
1,800	33.1	6.8	11.4	2.1	-6.1	9.9	1.6	16.0	2	190	76	175
2,000	33.5	7.0	12.3	2.3	-7.1	8.7	0.4	15.8	0	178	73	187
2,200	33.8	7.0	13.1	2.4	-8.2	7.2	-0.8	15.4	0	163	71	202
2,400	34.1	7.1	14.0	2.6	-9.2	5.9	-2.0	15.1	0	146	68	219
2,600	34.4	7.1	15.0	2.8	-10.3	4.6	-3.3	14.9	0	125	66	240
2,800	34.7	7.2	15.9	2.9	-11.3	3.2	-4.5	14.5	0	101	64	264
3,000	34.9	7.2	16.9	3.1	-12.4	1.8	-5.7	14.2	0	71	62	294

Table 46-1. Changes in climatic conditions with height above sea level in the Eastern Alps.

Ele-vation (m)	Annual number of days with:		Relative humidity (percent)	Annual precipi-tation (mm)	Relative snow frequency (percent)		Number of days with snowfall	Average quantity (cm d^{-1})	Maximum depth of snow	
	Dry ground	Snow cover			Summer	Winter			Depth (cm)	Beginning on:
1	14	15	16	17	18	19	20	21	22	23
200	187	38	71	615	0	49	27	4.6	20	18 January
400	173	55	74	750	0	61	32	5.2	31	23 January
600	160	81	77	885	0	70	38	5.8	51	28 January
800	147	109	78	1,025	0	79	45	6.4	73	3 February
1,000	133	127	76	1,160	0	85	53	7.0	93	11 February
1,200	120	138	74	1,295	1	90	62	7.6	100	14 February
1,400	107	152	73	1,430	2	93	73	8.2	120	21 February
1,600	93	169	73	1,570	5	96	85	8.8	142	3 March
1,800	80	189	74	1,700	10	97	98	9.4	168	14 March
2,000	67	212	74	1,835	16	98	113	10.0	199	26 March
2,200	53	239	75	1,970	24	99	128	---	242	8 April
2,400	40	270	78	---	34	100	143	---	296	20 April
2,600	27	301	80	---	44	100	158	---	366	3 May
2,800	13	332	82	---	55	100	173	---	446	15 May
3,000	0	354	84	---	67	100	188	---	545	29 May

when the sun was at its lowest in December. This loss caused by the screening of the horizon offsets the gain resulting from decreased atmospheric turbidity at higher elevations in cloudless weather. Only in the months of May and June was there a gain in solar radiation in the mountains as compared to lower elevations.

The spectral composition of global solar radiation also changes at high altitudes. The most important effect of this is an increase of ultraviolet radiation with increasing elevation. This arises from three separate causes. First, in Section 5 it was shown that the sea-level optical path length m_o was a direct function of the solar zenith angle. At higher elevations the optical path length m_z is reduced due to the decreased station air pressure:

$$m_z = m_o^{Pz/Po}$$

where P_z is the station air pressure, and P_o is the sea-level air pressure. This elevation adjustment reduces the ozone optical depth and increases the transmission of ultraviolet radiation. Second, the snow and ice common at high altitudes selectively reflect ultraviolet radiation, resulting in increased backscattering of ultraviolet radiation (Section 24). Finally, selective Rayleigh scattering of shorter wavelength solar radiation enhances the receipt of ultraviolet radiation.

These effects are illustrated from measurements of global solar radiation $K\downarrow$, UV-A radiation ($0.315 \leq \lambda \leq 0.400$ μ), and UV-B radiation ($0.280 \leq \lambda \leq 0.315$ μ) at Jungfraujoch (3576 m) and Innsbruck (577 m) in Austria by M. Blumthaler et al. [4606]. The seasonal variation of daily totals of $K\downarrow$, UV-A and UV-B are shown in Figure 46-1 for the two stations. Maximum daily totals for the two stations, taken from curves drawn along the upper limit of the seasonal values in Figure 46-1, are presented in Table 46-2C, along with annual totals of the three radiation terms. UV-B radiation is presented in Sunburn Units SU, where one SU is equal to the threshold dose of erythemal reaction (SU = 200 J m^{-2}). The ratio UV-A/$K\downarrow$ was found to consistently vary between 4-6% throughout the year at both stations. The ratio UV-B/$K\downarrow$, however, showed a pronounced summer maximum, increasing by a factor of more than four from winter to summer. The high altitude station at Jungfraujoch received 28% more global solar radiation than the lower altitude station at Innsbruck on an annual basis, but 58% more UV-B radiation. The increased transmission of UV-B radiation in summer compared to winter is striking. In summer, the high altitude station could potentially receive 54% more UV-B radiation than a lower elevation site under comparable meteorological conditions.

Table 46-2. Daily and annual totals of global solar radiation, UV-A and UV-B radiation fluxes at Jungfraujoch (3576 m) and Innsbruck (577 m), Austria. (After M. Blumthaler et al. [4605A-1])

	Summer	Winter	Annual
Jungfraujoch			
$K\downarrow$	36.50 MJ m^{-2}d^{-1}	7.39 MJ m^{-2}d^{-1}	8167 MJ m^{-2} y^{-1}
UV-A	1.89 MJ m^{-2}d^{-1}	0.32 MJ m^{-2}d^{-1}	412 MJ m^{-2} y^{-1}
UV-B	28.70 SU h^{-1}	1.61 SU h^{-1}	5154 SU y^{-1}
Innsbruck			
$K\downarrow$	29.45 MJ m^{-2}d^{-1}	5.09 MJ m^{-2}d^{-1}	6344 MJ m^{-2} y^{-1}
UV-A	1.48 MJ m^{-2}d^{-1}	0.24 MJ m^{-2}d^{-1}	307 MJ m^{-2}y^{-1}
UV-B	18.65 SU h^{-1}	0.96 SU h^{-1}	3265 SU y^{-1}

Although elevation plays the dominant role in controlling surface air temperature and vertical temperature lapse rates in mountain environments, differences due to slope exposure, described in Sections 40 and 44, also become more important with elevation. The surface

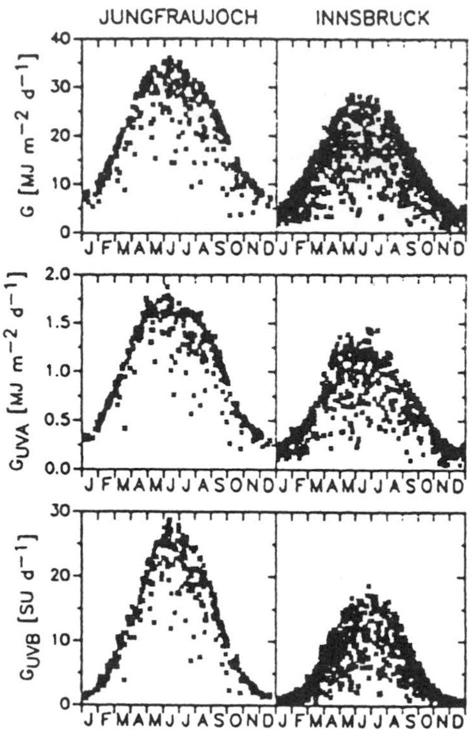

Figure 46-1. Seasonal variation of daily totals of global, UV-A, and UV-B solar radiation fluxes at Jungfraujoch (3576 m) and Innsbruck (577 m) Austria. (After M. Blumthaler et al. [4606])

climate becomes more extreme in character due to the more extreme radiation conditions. Stronger solar radiation produces higher surface temperatures. As long ago as 1900, J. Schubert [3953] showed by means of paired observation points in forests that the excess of soil temperature at a depth of 60 cm compared with air temperature at a height of 2 m was 0.75°C on the plain, and 2°C for stations at 1000 m. J. Maurer [4629] found that the excess of soil temperature at a depth of 1.2 m over air temperature in Switzerland was:

Elevation (m):	600	1200	1800	2400	3000
Excess (°C):	0.5	1.3	2.0	2.5	2.9

At a height of 2070 m in the Ötztal, at the tree line, H. Turner [4645] was able to measure absolute surface temperatures that were substantially in excess of the highest known values for the European plain, in spite of the initially lower air temperature at this height (Section 20). The 80°C (highest estimated value 84°C) in the hot July of 1957 can be understood if we consider the more intense solar radiation regime, the SW exposure and slope of 35°, a low albedo of 9 percent, and the very low thermal conductivity of a dark raw humus soil without vegetation (barren from excessive heating). In comparison with air temperature at 2 m, this is an excess of 50°C; at the same time, the surface temperature on the NE slope was 57°C lower. Only plants that have an unusually high resistance to heat can repopulate these hot, bare patches of ground. Leaf temperatures up to 44°C have been measured.

In September 1951, I. Dirmhirn [4611] measured the surface temperatures of a gneiss slab 0.25 m^2 in area and 6 cm thick with an albedo of 33 percent, lying horizontally at 3050 m on the Hohen Sonnblick. It reached 29°C during the day and -4°C at night. The diurnal temperature variation in the surface of the stone, compared with that of the air, on days of clear weather showed the following excesses arranged according to wind speed:

Wind speed (m sec^{-1}):	3	6	8	11	14
Excess (°C):	24.1	20.6	18.5	17.2	16.4

On clear September days (0-0.2 cloud cover), the average temperatures during the day were:

Hour of day	2	4	6	8	10	12	14	16	18	20	22
In the air (shelter) (°C)	2.4	2.7	3.2	4.3	4.8	5.2	5.7	5.4	4.7	4.0	3.6
On the gneiss surface (°C)	-0.8	-1.0	0.0	9.0	19.6	24.0	22.7	13.0	6.6	0.0	-0.6

The neighboring rock was heated to a smaller extent because some of the energy was conducted away to lower layers. Very thin laminae of rock are, therefore, subjected to a high degree of physical weathering.

Values of air temperatures from F. Lauscher [4624, 4625] are given in columns 6 to 13 of Table 46-1. Mean annual temperature decreases with elevation (Table 46-1, Column 16) due to the reduced Greenhouse effect associated with decreasing atmospheric density, water vapor content, and turbidity. The large vertical temperature gradients found in mountainous environments produces a distinct climatic zonation that is apparent in the changes in natural vegetation and agricultural practices with elevation. Although absolute maximum air temperatures decrease with elevation (not shown), absolute minimum air temperatures created by the effects of relative relief and cold air drainage (not shown) normally show no relationship with elevation. Daily and annual air temperature ranges (column 9) decrease with height. Nighttime air temperature inversions are common throughout the year due to cold air drainage. These usually burn off during the day in summer, but can persist throughout the day under particularly intense conditions in winter. The frequent winter temperature inversion (thermal belt of Section 45) can be recognized from the fact that the temperature ceases to decrease between 800 and 1200 m in elevation in January in Table 46-1. Column 16 from F. Lauscher [4626] contains the average daily relative humidity for the year.

While in the midlatitudes there is often a great difference in temperature and vegetation between north and south-facing slopes, in the tropics these differences are less (J. M. B. Smith [4638]). However, on Mt. Wilhelm in Papua New Guinea, J. M. B. Smith [4638] found differences between east- and west-facing slopes. Maximum temperatures were consistently higher on east-facing slopes. The distribution of species of grass tended to occur substantially higher on the east than on west-facing slopes. He suggested that this was a result of insolation on the west-facing slopes being significantly reduced by a build-up of clouds during the late morning and afternoon.

As a rule, wind speed increases with height. The surface layer in which extreme conditions develop, therefore, on the average, has a smaller vertical extent. Ventilation is also quite varied since it depends on local topography and wind direction. Wind recordings made by H. Aulitzky [4601] from May to November 1953 over a steep slope in the Ötztal, facing WNW, gave only moderate speeds of 1 to 3 m sec^{-1} at 10 m above the slope, and 0 to 1.5 m sec^{-1} at 40 cm above Alpine rose bushes. During the 5-1/2 months the instantaneous value at 10 m never exceeded 10.2 m sec^{-1}, and normal daily maxima were between 2 and 9 m sec^{-1}. Analysis of wind directions showed that in 70 percent of all cases these were local slope and valley winds (Section 43). In 30 percent of the cases, the wind was a gradient wind, corresponding to the pressure field. The slope winds were steered by the thickly branched trees at the forest edges. The predominantly N and W daytime winds left the southern edges in a wind shadow where the air was heated by the sun to temperatures fatal to young plants.

Precipitation is normally increased on the windward side of a mountain, giving rise to what is called orographic precipitation, or an enhancement of precipitation on the side of a mountain facing a storm track. Areas on the lee side of a mountain, in turn, often receive decreased precipitation. When storm tracks come from a single prevailing direction the difference in annual precipitation totals between the windward and leeward sides can be pronounced. The windward side of a mountain may be a single side, such as when storms result during a single season from a single meteorological cause, or on different slopes during different seasons due to different meteorological causes. M. M. Yoshino [2714], in describing the variation of precipitation with increasing elevation, states that the maximum amount of precipitation usually comes on the slopes of a high mountain. While there are rather large regional variations in the height of maximum precipitation, it is usually lower in the tropics and higher in the midlatitudes. In the tropics, the height of maximum precipitation is usually around 1000 m, and in the midlatitudes it is often a few hundred meters higher. The reason maximum precipitation is found on the slopes of high mountains is that as the air rises up the slopes, it reaches its lifting condensation level, resulting in cloud formation and possible rain fall. Rain may fall from this level on up, with lower elevations receiving rainfall less frequently. Thus, we would expect rainfall to increase with elevation; however, with increased elevation, temperatures decrease, and the resulting decrease in the saturation vapor pressure of the air contributes to a decrease in the rainfall. Thus, the zone of maximum precipitation is typically on the slopes of high mountains. The zone of maximum precipitation is usually higher in summer than winter due to the higher summer air temperatures.

Some mountains do not have their maximum precipitation on the slopes. Table 46-1, column 17, and R. G. Barry [4603], for example, report precipitation continuously increasing with increasing height.

In addition, to fog precipitation discussed in Section 37, significant moisture input on mountains can also result from cloud interception. J. L. Collett et al. [4608] used a passive cloudwater monitoring system to examine cloud interception and precipitation along an elevation gradient in the Sierra Nevada Mountains of California. Rates of cloudwater deposition to the collectors were frequently in excess of 1.0 mm h^{-1} at several sites, and were as much as 25-33 percent of precipitation. Cloudwater deposition rates were found to vary by a factor of four among their 14 measurement sites, and were controlled by four factors: the liquid water content of the cloud, the droplet size distribution of the cloud, the forest canopy structure, and the ambient wind speed. Maximum rates occurred during frontal storm passages along ridge-top sites that experienced strong ambient wind speeds and a slope exposure facing the direction of the frontal passage and prevailing surface winds. Cloudwater interception was found to always increase with elevation as well since in this case frontal lifting was the principal precipitation mechanism.

In addition, with increased height and decreasing temperatures, precipitation falls more frequently in the form of snow. Columns 15 and 18-22 in Table 46-1, from E. Ekhart [4613] contain data on snow cover and snowfall. Columns 18 and 19 give the percentage of precipitation days with snowfall; column 21 gives the average snow depth in centimeters per day of snowfall; and column 22 gives the maximum depth of snow to be expected on the average throughout the year from the values of F. Steinhauser [4641] for locations below 1200 m. The time of the greatest snow depths are shown in column 24 (from H. Steinhäusser [4640]).

Fewer studies on surface energy budgets in mountainous terrain exist. Considerable small-scale variation in surface energy exchanges is expected, however, due to small-scale spatial and temporal variations in the physical factors that control the surface energy budget. These include spatial variations in slope orientation, slope angle, surface albedo, soil depth and soil texture, diurnal variations in cloud cover and wind speed, and elevation gradients of wind speed and precipitation amount and type.

S. A. Isard [4619] investigated the surface energy budget along an alpine ridge crest and alpine slopes on 3-10 July 1985 during a summer drying event at Niwot Ridge, Colorado (40°3'N). As discussed in Section 40, cloud cover due to convective lifting normally increased systematically during the day leading to increased global solar radiation on east-facing slopes in comparison to west-facing slopes. In fact, cloud cover was found to be the primary factor controlling the energy available ($Q^* - Q_G$) for the sensible and latent heat fluxes ($Q_H + Q_E$), with topographic controls (slope and aspect) having a secondary influence on available energy. The role of topography was influential, however, in determining the turbulent transfer coefficient for momentum, and hence the transfer of sensible and latent heat. Wind speed normally increases with elevation due to reduced surface friction, leaf size frequently decreases with elevation in response to the need to reduce moisture loss, and plants normally lie closer to the ground surface at higher elevations. These factors lead to an enhancement of sensible heat flux, a reduction of latent heat flux, and increased Bowen ratios at higher elevation. S.A. Isard [4619], for example, found minimum Bowen ratios of only 1.0 following rain events when soils were moist, but values as high as 7.5 following extensive drying events. Wind direction also has a strong control over wind speed. West-facing slopes experienced stronger wind speeds, reduced leaf-air temperature differentials, weaker global solar radiation, and reduced available energy for sensible and latent heat flux ($Q^* - Q_G$) in comparison to east-facing slopes. The thin sandy and gravely soils common at high elevations resulted in limited soil moisture content that exerted a strong negative feedback on evapotranspiration, and was a primary control on latent heat flux Q_E.

As long as a plant is under the snow cover, it is protected from destructive weather influences. Parts projecting above the snow, however, are subjected to the extreme climatic conditions near the snow, which were described in Section 24. If the ground is bare of snow, as is often the case on sunny slopes in winter, plants are stimulated to increase transpiration and may suffer from the lack of moisture, especially when the ground is frozen. W. Larcher [4623] has reported the investigations of this phenomenon made in an alpine garden.

In tropical mountainous environments the presence of the trade-wind inversion (Section 45) produces changes in the microclimatic patterns obtained in mid-latitude high mountains, and the prevailing trade winds interact with the diurnal circulation of sea and land breezes as well. These effects were examined on the island of Hawaii (19°30'N) by J. O. Juvik and D. Nullet [4622]. On the windward side of the island the trade winds dominated, and were overlain by a weaker diurnal cycle. Maximum annual precipitation of nearly 7000 mm yr^{-1} occurred at 1000 m, and the absence of convective cloud development led to a symmetrical diurnal global solar radiation pattern in which $K\downarrow$ increased with elevation. On the leeward side of the mountains annual precipitation was as low as 250 mm yr^{-1} in places, and the diurnal sea and land breeze circulation dominated. This resulted in frequent afternoon cumulus clouds and a strong morning maximum of $K\downarrow$. The clear skies, low relative humidity, and increased wind speeds above the trade-wind inversion caused evaporation to increase with elevation above the base of the inversion, in contrast to mountains in middle latitudes where Q_E normally continuously decreases with elevation. The lapse rates for the maximum, mean, and minimum air temperatures were also constant with elevation at a value of about 0.64°C/100 m.

W. Tranquillini [4643] published the first series of microclimatic observations at the mountain station near Obergurgl. Results for the snow melting period in April and May 1955 are shown in Figure 46-2. In addition to meteorological data on global solar radiation, the air temperature at the shelter, and the rapidly increasing temperature of the ground as the snow melts, there are also records of the temperature variation in the needles of young *Pinus cembra* 1.5 m high. Within the snow cover, the needle temperature is usually near 0°C; above the snow, it may increase to 30°C or fall to -12°C. The greatest diurnal temperature variation measured in the pine needles amounted to 34°C on an April day.

Photosynthesis in alpine or arctic areas may be more than expected due to this elevation of plant temperatures. J. L. Hadley and W. K. Smith [4617] found daytime temperatures within an alpine krummholz mat were frequently 10°C and as much as 23.7°C above ambient air temperatures. P. Mølgaard [4631] also found plant temperatures in the arctic often more than 20°C above that of the ambient air. Elevated needle temperatures are also evident in Figure 46-2.

The way snow melts in mountain areas is closely related to topography. Even if the snow melts at different times in different years, the spatial pattern of melting is often similar. H. Friedel [4615] made a photogrammetric survey of melting snow for an area of 32 km^2 in the eastern Alps near the Pasterze Glacier, and the analysis was supplemented by direct measurements on the ground. Division of the area into 620 sections enabled the effect of soil type and slope exposure on the melt pattern to be evaluated as a function of elevation. Figure 46-3 is an extract from the results.

Observations in the top left of the diagram are for snow on the ground, while those on the right are for snow on the glacier ice. The solid line gives the percentage of the ground free of snow at the height in question on the 1st of each month, and the broken line is for the 15th. It may be seen, for example, that at 2800 m on 1 August, 60 percent of the ground is free of snow, at 2500 m 80 percent, and at 2000 m 100 percent. The date lines converge with the 0 and 100 percent lines because the very first bare patches occur early in certain high positions, and snow continues to lie for a long time in some hollows. The firn line lies where the upward extension of the melt pattern in September comes in contact with the first of the new snow falling in autumn (surfaces above this border are crosshatched).

The effect of slope exposure is shown in the lower part of Figure 46-3. This applies to snow on the ground between 2400 and 3000 m. It follows from the surprising configuration of the date lines that the steep southern slopes thaw first. The next to thaw are the more

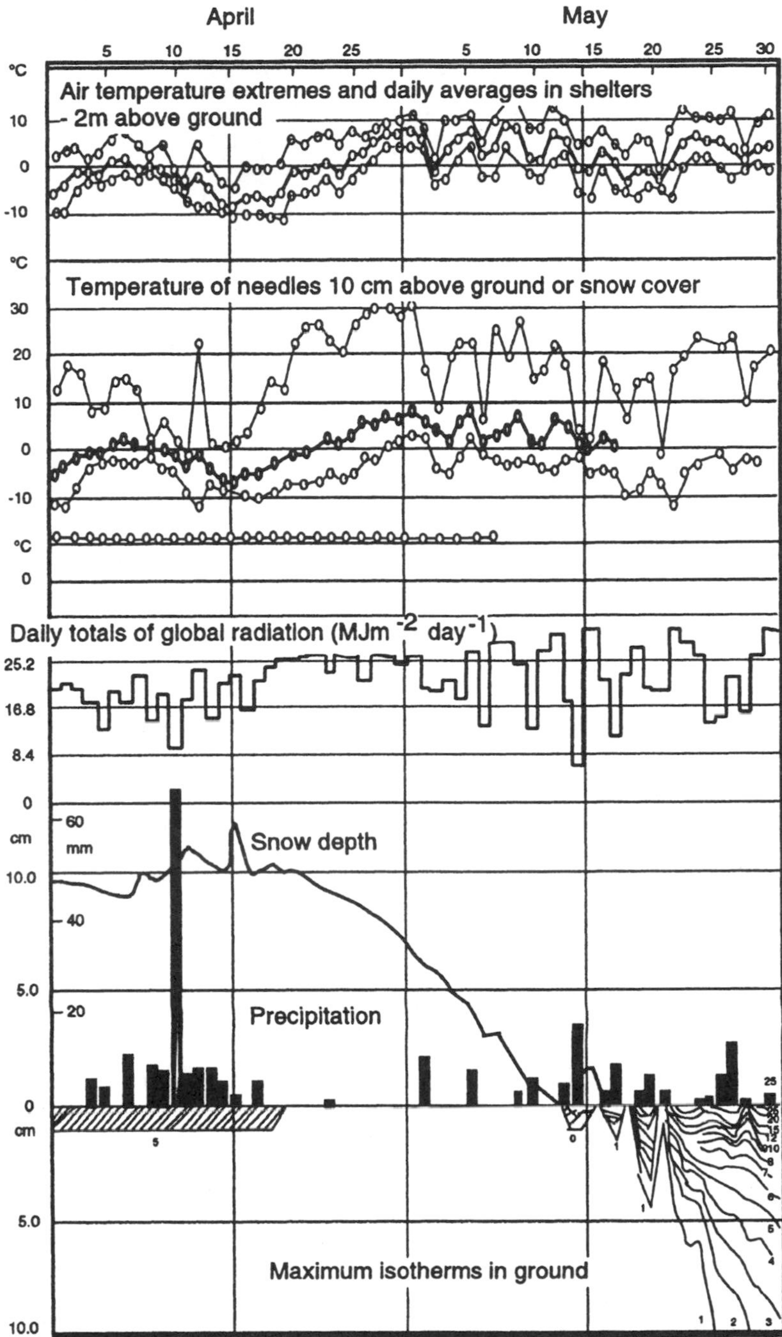

Figure 46-2. Measurement of microclimate at the Alpine tree line near Obergurgl, during snowmelt. (After W. Tranquillini [4643])

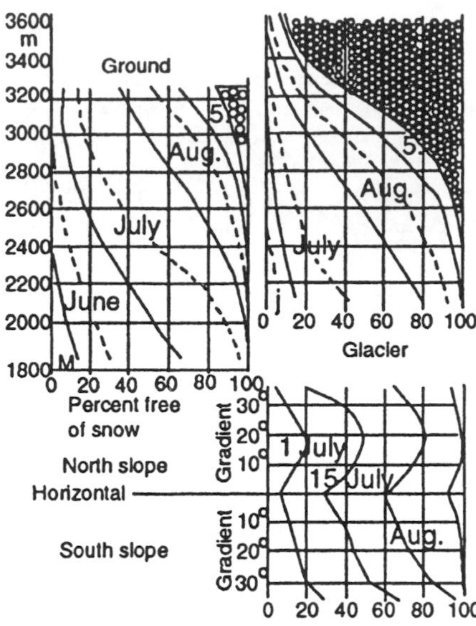

Figure 46-3. The dependence of the percentage of the ground free of snow in high mountains upon elevation (left), underlying surface (right) and slope exposure (below). (After H. Friedel [4615])

gentle southern and northern slopes. These are followed, but with a marked delay, by level surfaces. This is because the snow depth is always greater in level areas from which the snow is unable to slide downward. The steep northern slopes sheltered from the sun are the last surfaces to begin to thaw. For a more detailed examination and discussion of the surface climate and energy exchange of a melting snow surface in an alpine basin during a complete water year the reader is referred to D. Marks and J. Dozier [4628].

Since the influence of relief is so strong, the high mountain microclimate is a mosaic of vastly different conditions. H. Aulitzky [4602] published a map of the patterns of melting snow in the experimental area near Obergurgl for the spring of 1955, and a map of the vegetation of the same area was published by H. Friedel [4615]. These two maps give an impressive picture of the way in which plant life is entirely dependent on these rapidly changing conditions.

The number of days without frost decreases with increasing elevation (Table 46-1 column 11). The frost-free growing season is defined as the time between the last frost in spring until the first frost in autumn. The frost-free growing season declines with increasing elevation. D. N. Jordan and W. K. Smith [4620] in their study of a subalpine environment in the Medicine Bow Mountains of Wyoming found that by the above definition the growing season was only five days in 1993. They found many microclimatological variations, however, including a greater frequency and duration of frosts in the center of a clearing least within the forest (Table 46-3). They also found that the broadleaf *E. Peregrinus*, by creating a larger boundary layer (and reduced circulation over its leaf), experienced more nights with frost conditions (41 nights) than the needle leaf *A. Lasiocarpa* (25 nights). Radiative losses to the night sky were strongly dependent on microsite sky exposure. "Shelter by the forest overstory dramatically reduced both the frequency and duration of radiative frost events for both broadleaves and conifers."

Table 46-3: The number of frost events and total hours of frost at 8 m during the summer of 1993. (After D. N. Jordan and W. K. Smith [4610])

Sensor	Site	Number of Frost Events	Number of Frost Hours
E. Peregrinus	clearing center	41	226
leaf	clearing edge	33	151
	forest understory	16	82
air	center clear	26	138
	clearing edge	25	125
	forest understory	13	82

Vegetation also has a reciprocal influence on microclimate. H. Zöttl [4647] carried out an investigation on a west slope of 20° at a height of 1830 m in the Wetterstein Mountains. Three equally exposed stations were used to measure temperature in a pebble field, a belt of grass, and dwarf fir bushes. The plants were arranged in up-and-down strips so that differences in microclimate were due only to the influence of vegetation at the site. On the pebble field, pioneer types of grass had taken root over 30 percent of the surface. The grass itself *(Caricetum firmae)* covered 85 percent of the ground. The dwarf firs were an association of *Pinus montana prostrata* and *Erica carnea.* These three types of vegetation are characteristic of subalpine elevations. As an example, temperatures on 2 July 1949 (rounded to whole degrees) were:

Height above the ground (cm)	Afternoon at 13:00			Early morning 04:30		
	Grass	Pebbles	Dwarf fir	Grass	Pebbles	Dwarf fir
20	15	13	15	2	2	2
0	41	23	18	0	1	2
-20	10	8	7	11	8	8

The grass-covered surface had the most extreme microclimate. This was due to the low air mobility and thermal conductivity between the ground and the top of the grass, while incoming radiation was practically unhindered. The dwarf firs reduced the diurnal variation to a great extent. With increased height vegetation becomes more hardy and changes from trees, to dwarf trees (Krummholz), sedges, and and tundra grasses. D. J. Grace [4616] states that "trees do not generally grow in places where the mean temperature of the warmest month is less than about 10°C. At their limit, trees are often short and crooked (Krummholz) ... the transition from tall forests to dwarf shrubby vegetation is often abrupt, forming a distinct tree line."

While the word treeline generally evokes images of alpine or arctic ecotones, treelines appear under a much wider variety of circumstances. Wet treelines occur along margins of bogs or swamps, dry treelines mark the transition of forest and grasslands, and cold treelines include arctic, alpine or sink hole conditions (Section 42) (G. C. Stevens and J. F. Fox [4642]).

At the highest elevations vegetation is only found in favorable microclimates. Some of the highest altitude observations of vegetation were made by S. Halloy [4618]. At fumaroles (warm spots) in the Andes at 5760-6060 m he found algae, lichens, mosses and liverworts (36 taxa of bryophytes and lichens). Close to the fumaroles the vegetation was dense and varied. As distance from the fumaroles increased, the vegetation became drier, less varied and appeared wilted at the edge of the patches. A few isolated patches of mosses and lichens were also found on the rocks not related to the fumaroles. Except for the vegetation around the fumaroles the area was bare of vegetation. The highest lichens have been found at 6600 m (M. S. Mani [4627]), liverworts at 6060 m (S. Halloy [4618]), mosses at 6035 m (H. N. Dixon [4612]), ferns at 5500 m (M. S. Mani [4627]) and flowering plants at 6350 m (A. Zimmermann [4646]).

At their cold, hot, or dry margins some plant seedlings primarily become established in the protection of other plants. The protecting plants are often called nurse or cushion plants and protect the seedlings from the harsh environment (B. Å. Carlsson and T. V. Callaghan [4607], A. C. Franco and P. S. Nobel [4614], K. L. Bell and L. C. Bliss [4608], P. S. Nobel [4633], and P. W. Jorden and P. S. Nobel [4621]).

The influence of the forest on climate, described in Chapter VI, is also apparent on high mountains. This has been shown by H. Desing's [4609] observations at the stations in the fir stand. H. Aulitzky [4600] has pointed out that as the tree line is approached and the forest becomes thinner and lighter, the protection it offers to the next generation of trees gradually decreases. M. M. Yoshino [2714], R. G. Barry [4604] T. Niedzwiedz [4632], and R. B. Smith [4639] are good sources for additional information on the variation of climatic elements in mountainous areas. I. R. Saunders and W. G. Bailey [4635] provide an excellent long-term study of the radiation budget and surface energy balance characteristics of a high mountain alpine tundra site.

47 The Microclimate of Caves

Speleology is the study of natural caves and the phenomena pertaining to them. The climatic conditions of caves correspond, in many respects, to what has been discussed in Sections 10 and 19-21 about the climate of soils. Openings, such as entrances and air shafts, also affect the climate in caves. The most striking feature observed upon first entering a mountain cave on a summer day is the rapid decrease in light intensity, and the way the air becomes cooler and more humid. The way in which animal and plant life have adapted to this environment are unique.

The cave atmosphere in an open system is strongly influenced by the air exchange with the outside atmosphere. According to P. A. Smithson [4719] the airflow in a cave is controlled by three factors, gravity, density-induced convection and forced advection. The magnitude of the airflow and the resulting temperature fluctuations within a cave are then affected by cave morphology such as the size and shape of the tunnels and chambers, height of an observation site above the cave entrance, distance from the entrance, temperature difference between inside and outside of the cave, and external wind speed and direction relative to the cave entrance. Normally the stronger the airflow the greater the influence of outside temperatures on those in the cave. The magnitude of these effects, however, can be reduced by cave passages with a highly variable radius, high surface roughness, sinuous cave passages, and a sloping cave floor. As a result, P. A. Smithson [4719] emphasizes that each cave has a unique element to its microclimate.

R. Oedl [4715] distinguished between static caves, which have a single opening in which there is little air circulation, and transit caves (also called windpipes or dynamic caves), which have multiple openings and greater air circulation. In static caves, turbulent mixing at the entrance may penetrate a short distance inside, or there may be an inward or outward flow of air as a result of differences in air density due to variations in temperature and/or vapor pressure. In transit caves, on the other hand, there is a circulation of air, perhaps reaching considerable speeds in some cases; especially where there are narrow areas in the passages. It is not necessary to have "entrances" that can be used by people; wind shafts or cracks of sufficient width in rocks are enough to generate a throughflow circulation. The circulation is enhanced if the openings differ substantially in height from one another.

Static caves will be discussed first. Figure 47-1 shows a longitudinal static cave at Jenin in Palestine. Measurements made by P. A. Buxton [4702] at midday on 7 June 1931 show the typical temperature distribution in a cave with only one opening. At point *A*, where daylight still penetrates and a human can stand upright, it is hot and relatively dry. The cave narrows after 7 m (*B*) and it was possible to penetrate farther only by crawling. At this point frogs and the larvae of water insects were found in puddles. Farther inside, the dry-bulb temperature decreased rapidly, and about 25 m from the entrance it became equal to the wet-bulb temperature, which was constant. From that point onward the air was saturated and the temperature constant.

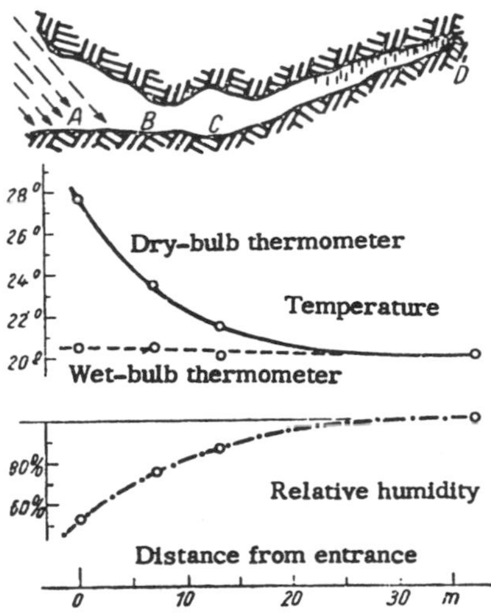

Figure 47-1. Temperature and relative humidity measurements in a static cave with one opening. (After P. A. Buxton [4702])

The change from open air to cave conditions, illustrated here by a single example, is best supported by the series of measurements made over a year in the stalactite cave of Baradla in Hungary by E. Dudich [4703]. In this cave, which has its entrance on a southern slope, a tunnel 45 m long leads downward in steps. The values given in Table 47-1 were recorded over this stretch in 1928-1929. The minimum relative humidity given shows the extent of its variation, which changes just as quickly as temperature with distance from the entrance. Throughout the cave the relative humidity was 100 percent.

Table 47-1. Temperature and relative humidity in a cave. (After E. Dudich [4703])

Location entrance	Entrance to cave		Steps leading downward			45 m from
	In doorway	Behind door	10 steps	40 steps	68 steps	
Annual temperature (°C)						
Maximum	17.3	14.6	11.8	10.6	10.2	10.4
Minimum	1.8	4.8	6.6	7.1	7.8	8.8
Difference (°C)	15.5	9.8	5.2	3.5	2.4	1.6
Relative humidity, minimum (%)	19	66	77	91	95	96

It might be expected that the same kind of mean annual temperature is found within a cave as is observed with deep mines and borings. The configuration of the cave, however, exerts considerable influence. If the cave slopes downward from the entrance, cold air flows downward and is only marginally affected by the warmer and lighter air above. Caves of this type are called ice, cold storage, or sock caves, and act as cold reservoirs. If the entrance itself lies at the cold floor of a valley, this effect is enhanced. In extreme cases the shaft of the cave goes vertically downward. The "Eisbinge," at an elevation of 900 m in the Erzgebirge, explored by H. Mrose [4714], is a cleft in the rock 1 m wide and 20 m deep. In winter, 1.5 m of snow falls into the cave. The mean annual temperature of the rock is 4°C, which is conducive to melting. In summer, the cold air is trapped in the cave, and there is often a thin fog layer about 5-10 m deep in which there is only a small amount of mixing with the warm external air. Although the process of melting continues from the spring onward by means of heat conduction through the rock, it is not altogether completed when the first new snow falls in the autumn. Despite a mean annual temperature of 8.4°C ice remains in the downward sloping Fuji ice caves throughout the year. Cold air enters and cools the cave from November through February. In March and April water from melting snow and rain seeps into the cave and freezes on contact with the cool rocks. From July through October some melting occurs (T. Ohaia, et al. [4716, 4717]). The absence of any significant circulation makes the microclimate in the above two caves substantially colder than might be expected.

Meteorological conditions in a vertical shaft were investigated by A. Baumgartner [4701] in the "Holloch" (hellhole) in the Allgau, in 1949. This cave may be described as having a single entrance since connections with the open air through narrow cracks and joints in the rock have only a small effect. An almost circular vertical shaft, 71 m deep and about 8 m in diameter, led into the cave. On 3 September 1949, light intensity decreased with depth in the following way:

Depth in shaft (m):	0	5	10	15	20
Light intensity (%):	100	90	30	10	4

From 30 m down, there was only scattered light with an intensity of about 0.5 percent of the external light, which, after a long period of adaptation, allowed objects to be recognized even at the bottom of the shaft.

The air temperature variation with depth on a clear summer day in a sock cave is shown in Figure 47-2. The characteristics of this type of cave are easy to distinguish. Below 15 m, temperatures remain between 4.5 and 5.5°C; midday temperatures in the air outside are able to penetrate downward by mixing for only a few meters and rapidly decrease in intensity. The relative humidity behaves inversely, reaching the saturation point at a depth of 15 m. When the transport of air through cracks in the rock and the joints where water percolated was sufficient, the cold air in the cave flowed over the upper rim of the shaft. This intruding air was up to 20°C colder than the environment and flowed down the mountainside at a speed of 0.5 m sec^{-1}, giving prospective visitors to the cave a foretaste of its microclimate.

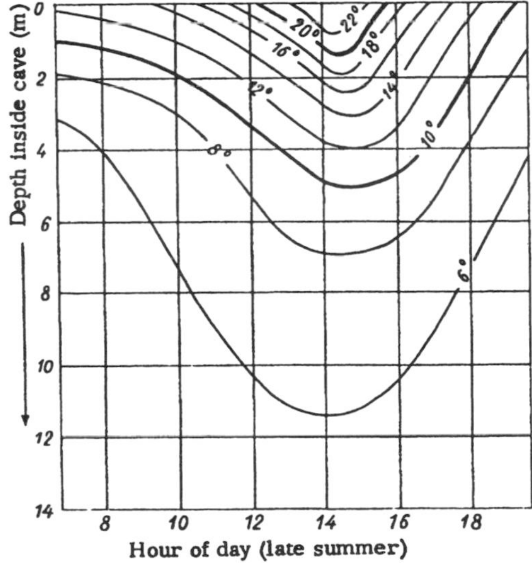

Figure 47-2. Penetration of diurnal temperature fluctuation into a sock cave. (After A. Baumgartner [4701])

The opposite thermal effect is obtained when a static cave slopes upward from its single entrance. Cold air is able to flow out from the cave if it is cooler than the environment, and warm air from the outside is able to penetrate within. While there is often a sharp decrease in temperature at the entrance to a sock cave, especially in summer, the temperature may suddenly increase at the entrance to a rising cave, particularly in winter. Such caves are, therefore, preferred by insects and bats for hibernation.

In the transit cave the degree of air exchange or "breathing" between the outside and cave atmospheres varies on a daily, synoptic, and annual scale. The surface daily heating and cooling cycle can create a density-induced reversal of wind movement whose strength depends upon the relative temperature and, to a lesser degree, vapor pressure of the cave and outside air. Pressure changes associated with the synoptic-scale movement of high and low pressure systems can also increase air exchange. P. A. Smithson [4719] observed considerable seasonal change in air exchange within Poole's Cavern, a rising static cave near Buxton, UK. In summer and early fall, when outside temperatures were much greater than inside the cave, the cave's thermal environment was very stable and had little interaction with the outside air. A strong air current drained out from the cave entrance, and was capped by a much weaker warm inflow aloft. By fall, temperature differences between the inside and outside air decreased, leading to reduced air exchange. During winter, when outside temperatures fell below those in the cave, a reversal of flow occurred, with the more distant and higher elevation cave locations being the last to experience a temperature drop due to the cold air drainage. The cold winter air, however, quickly reached equilibrium with the cave walls. Such cold air intrusions were more common in winter, and became less frequent in spring.

J. S. McLean [4713] in his study of Carlsbad Caverns also reported a flow of cool air out of the caverns in summer and an in flow during winter. The cool air flowing into the cave in winter is warmed by the cavern. As the air is warmed the saturation vapor pressure increases and relative humidity drops. This explains why evaporation is at a maximum in these caverns during the winter. J. S. McLean believes that the addition and loss of water vapor through an elevator shaft may be increasing winter evaporation and is an important explanation of why many of the lakes in the caverns are decreasing in size.

Conditions are quite different in caves with a number of openings. These caves are to a certain extent included in the atmospheric circulation. The main entrance is usually at a low level, and the links with upper levels are made through airshafts from which the flowing water that hollows out the caves is able to penetrate. The circulation pattern inside the cave will be determined by the number, type and relative positioning of the cave openings, the altitude of the openings, the temperature difference between the cave and outside atmospheres, and the dynamic circulation of the free atmosphere. These factors are then modified by the cave morphology, including the radius, sinuosity, slope and roughness of the cave.

A good example is the "Eisriesenwelt" in the Dachstein Mountains, where the climatic conditions have been described in some detail by R. Oedl [4715], and R. Saar [4718]. The main entrance to this cave is at 1458 m on the steep downward slope from the Dachstein plateau, while the open wind ducts end on the plateau at heights of from 1600 to 1900 m. The average speed of the air current in the tubular entrance gallery of the Eisriesenwelt, 13 m^2 in area, was found to be 4.0 m sec^{-1}. The maximum recorded was 10 m sec^{-1}. This upper limit apparently is due to the effect of friction on the walls. It was estimated that an average of 1.6×10^9 m^3 of air flows through in a year.

The direction of the air current is decided mainly by the difference in temperature between the outside and inside air. The movement of air has an effect on the temperature of the cave. In winter, from about December to March, cold air streaming in through the lower entrance is warmed by the rocks of the cave and escapes upward. T. M. L. Wigley and M. C. Brown [4724] refer to this as the chimney effect. This air is comparatively dry; relative humidities as low as 40 percent have been measured. When cave temperatures fall considerably below 0°C, ice can form, particularly in spring as warmer air entering the cave is cooled and sublimates.

Warm summer periods are scarcely felt within the cave. From May to the middle of November, its temperature ranges between -1°C and 1°C. The circulation from July to September is the reverse of that in winter, with cold air flowing out the lower entrance of the cave. In summer, air is sucked in from above it and warms the rock mass in the higher levels of the cave system. As the air enters and cools, the relative humidity of the air increases, and is close to the saturation level.

When outside and inside temperatures do not differ by much, such as during the transitional seasons of spring and autumn, or during changing synoptic circulation patterns there can be a short-term reversal of the direction of air circulation in a cave. W. Gressel [4706] has stated that dynamic processes in the atmosphere, the pressure gradient at the several openings of the cave, the enhancement or inhibition of local mountain winds, and frontal pressure differences can affect the air circulation in a cave. The change in pressure in a cave often lags behind the changes outside, resulting in air entering or leaving the cave for several hours after a change has occurred. If the cave entrance is properly exposed external winds may also be funneled through it.

Another source of air movement described by T. M. L. Wigley and M. C. Brown [4724] occurs in caves with streams. The stream can pull or drag a current of air along with it. In a static cave this stream-driven air current near the surface may result in a balancing counter flow in the opposite direction above it.

Carbon dioxide concentrations in the soil are typically higher than the atmosphere as a result of plant and animal respiration. As rain water passes through the soil it absorbs carbon dioxide. If it infiltrates into a cave it will release some of the carbon dioxide. As a result carbon dioxide levels in caves are generally higher than in the atmosphere. (C. Hill and P. Forti [4709]).

The microclimate of caves is of scientific interest, but it is also of practical importance when the caves are used for dwellings, tourism, cave exploration, or food storage. Tourism can have an effect on the microclimate of caves. For example, in a study of Mammoth Cave in Kentucky, L. M. Trapasso and K. Kaletsky [4721] found that both food preparation and the presence of tourists had an effect on humidity and temperatures within the cave. In a study of the Altamira cave in Spain, E. Villar et al. [4723] measured increases in temperature as different size tourist groups passed through. They found that every additional five people in a group resulted in the air temperatures increasing by approximately 0.2°C. Within an hour after the visit temperatures had returned to their original level. J. S. McLean [4713], however, suggests that while the body heat given off by visitors in Carlsbad Caverns was minimal (25 W), the heat released by lighting, food preparation, and other activities (43,970 W) was nearly six times that of the natural energy exchange through openings. Bat colonies in Cuban caves (which contain from 3-8 million bats each) were also found to affect the temperature. Heat is released both directly from the bats and from their fermenting excrement (A. Stoev and P. Maglova [4720]).

M. Hell [4707] proved that in the famous beer cellars of Kaltenhausen near Hallein, Salzburg that galleries 6 m long had been tunneled into the mountain leading to natural air vents.

When the temperature outside was 20°C, there was a wind blowing inside with a temperature of 4°C. The climate of the cave gave the name to the whole district of Kaltenhausen. H. Lautensach [4712] observed that people in Korea moved silkworm eggs into cold air shafts, similar to caves, so that when it suddenly became warm in spring, the caterpillars would not hatch out before the mulberry trees, their food source, were in leaf.

Caves have long been used in the treatment of respiratory diseases. Even today many researchers report beneficial results from placing patients in caves. T. Horvath [4710], for example, in a study of 4,000 patients found significant improvement in those who stayed in a cave four hours a day for three weeks. T. Horvath [4711] further suggests aspects of cave microclimate that may be beneficial to health. L. G. Varcha [4722] summarizes the results of speoleotherapeutic treatment in Hungary and suggested this may be a source of future tourism.

To conclude, it is worth mentioning two rare and special types of cave formations. The first is the thermal cave where the microclimate is entirely controlled by the flow of thermal water through it. Anyone who has been in the galleries of the famous hot springs at Gastein will remember the extremely damp and warm climate. The second type is the ice cave and the tunnels hewn out in glaciers. The measurements made by W. Paulcke [2440a], H. Hess [4708], and W. R. B. Battle and W. V. Lewis [4700] should be referred to for details. Further information on the circulation, temperature and moisture regimes of cave climates can be found in C. R. de Freitas, et al., [4704] and C. R. de Freitas and R. N. Littlejohn [4705] and T. M. L. Wigley and M. C. Brown [4724].

CHAPTER VIII

INTERRELATIONS OF ANIMALS AND HUMANS TO THE MICROCLIMATE

Bioclimatology, in the more restricted sense of the term, investigates the connections between climate and life. It determines, for example, how humans adapt their style of house building to the various climate types on earth, or the way in which sunshine duration on grassland affects the milk yield of cattle.

In its broader sense, bioclimatology has developed into a special science of such wide scope that it is not possible here to treat even its micrometeorological and microclimatologic aspects in detail. Instead, this chapter will provide a few selected problems of microclimatological interest, to illustrate principles discussed in earlier chapters.

The survey will be restricted to the living conditions of healthy organisms. It is well known that an ailing plant reacts more sensitively to atmospheric influences, just like a sick animal or man, than it does in a healthy state. Those aspects of botany, veterinary science, and medicine, which are interested in the relations between the sick organism and weather and climate, must also be concerned with microclimate. Unfortunately, these important practical problems are outside the scope of this book.

Phytopathology, or the study of plant illnesses and animal pests, will, therefore, be omitted in our discussion of plants and animals. From the point of view of microclimate, phytopathology investigates the temperature and humidity conditions that favor the development of fungi and other sources of disease, or the type of climate which will stimulate damage by, for example, aphids. A comprehensive survey of these problems from the meteorological point of view is to be found in H. Schrödter [4826], K. Unger and H. J. Müller [4831], H. J. Müller, et al. [4820], and C. V. Smith and M. C. Gough [4827].

In the following sections (48-50), this book will give only a basic introduction to some aspects of animal and human bioclimatology. For more complete and detailed coverage, the reader is referred to S. W. Tromp's [4830] book, *Biometeorology: The Impact of the Weather and Climate on Humans and Their Environment (Animals and Plants).* This book's particular strength is its coverage of weather and climatic effects on humans and human disease including allergies, asthma, cancer, cardiovascular disease, physiological processes, mental development, sex ratio, sport performances, suicides, building microclimate, architectural applications, etc.

T. R. Oke's [3332] book, *Boundary Layer Climates*, contains an excellent review of the physical principals governing the interrelationship of animals and the climate, as do the general texts by G. Campbell [4802] and D. M. Gates [2909]. Two books dealing with the effect of the weather and climate on insects are M. J. Tauber's, et al. [4829] *Seasonal Adaptions of Insects*, and J. L. Hatfield and I. Thomason's [4811] *Biometeorology and Integrated Pest Management.* C. Huffaker and R. Rabb's [4812] *Ecological Entomology* also has several sections dealing directly or indirectly with the effects of weather and climate on insects.

48 Animal Behavior

Plants are unable to escape from the microclimate of their environment, no matter how adverse the conditions may be. Animals, on the other hand, have freedom of movement, are able to avoid an unfavorable microclimate, and can search for suitable conditions. The present section will discuss some basic principles of animal thermoregulation and give a few examples of the behavior of animals in response to the microclimatic environment.

Every animal has its own, usually rather restricted, sphere of movement, and is bound to a locality. This especially applies to lower and less mobile forms such as larvae, worms, and caterpillars. But even the swiftest and freest of creatures, like birds, are restricted in a sense by having nests, in which they spend the night and bring up their young, and, therefore, are subject to local conditions. The microclimate of animal habitat will be treated in Section 49.

The physical environment, in particular solar radiation, air temperature, humidity, and wind speed, has a profound effect upon the energy and water budgets of all warm and cold-blooded organisms. We will first examine selected microclimatic adaptations in endothermic (warm-blooded) organisms.

Endothermic organisms generate their own metabolic heat, exhibit distinct physiological adaptations to heat and cold, and make optima use of the microclimate in order to maintain a constant core body temperature. Small warm-blooded organisms with small mass and high surface-area-to-volume ratios face particular challenges in this respect. Their small mass and limited thermal inertia limits their energy and water storage capacity, demands a high rate of energy production per unit of body mass, and requires a rapid response to changes in the thermal environment in order to maintain a constant core body temperature. Their high surface-area-to-volume ratio, however, results in rapid convective cooling or heating under extreme cold or warm environments. In such small organisms, the energy expended for basal metabolic and thermoregulatory needs may be a large percent of the total daily energy expenditure. Organism behavior that reduces thermoregulatory energy costs will reduce the total energy requirement of the organism, allow reallocation of energy to other important needs, conserve time required for obtaining energy, and free time for other functions. Thus, the efficiency of microclimatic selection can have an important bearing upon survival rates and competitiveness, particularly in extreme environments.

The Verdin *Auriparus flaviceps* of the southwestern United States and northern Mexico desert illustrate these principles. B. O. Wolf and G. E. Walsberg [4834] found that this small songbird had a basal metabolic rate of about 54.0 W m^{-2} at 36°C. Daytime resting metabolic rates ranged from a high of 182.9 W m^{-2} at 10°C to a low of 59.8 W m^{-2} at 36°C, and then increased to 87.5 W m^{-2} at 48°C. Evaporative water loss was between 6-10 mg g^{-1} hr^{-1} at temperatures below 37°C, but increased by more than an order of magnitude to 71.6 mg g^{-1} hr^{-1} at 48°C. Through behavioral adjustments Verdins seek to avoid cold microclimatic sites that increase resting metabolic rates, and hot microclimatic conditions that result in excessive water loss. Under calm wind conditions (0.4 m s^{-1}) and at air temperatures (15°C) and solar irradiances (1000 W m^{-2}) typical of the desert environment, B. O. Wolf and G. E.

Walsberg [4834] found that a Verdin could supply 46% of its metabolic heat requirements by selecting microclimatic sites that optimized the absorption of solar radiation. This value was reduced to only 3% of metabolic needs, however, at higher wind speeds (3.0 m sec^{-1}). Small organisms like the Verdin will respond rapidly and frequently to changing microclimatic conditions in an effort to avoid unnecessary expenditures of energy and water. During the winter, when air temperatures are low and food is less available, Verdins spend 75-95% of their active day foraging for food. Careful microsite selection through solar positioning and the avoidance of windy locations can significantly reduce their rate of energy expenditure. During the summer, the severe environment challenges the ability of the Verdins to dissipate heat and maintain a favorable water balance. During summer, Verdins seek shaded environments, avoid windy conditions, and reduce their foraging activity to as little as 9-21% of their active day.

Microclimate also affects the location of fixed site activities, such as nesting site selection (Section 49), and the timing and location of free-ranging activities, such as foraging. In cold environments, roosting and foraging sites with warmer air temperatures, increased receipt of solar energy, and protection from the wind will reduce free and forced convective heat loss and ameliorate radiative heat loss. In extreme cold environments the energy savings achieved through the careful selection of such activity space can be critical for survival.

D. G. Wachob [4832] compared the microclimate of foraging and non-foraging sites used by the mountain chickadee (*Parus gambeli*) during winter in Wyoming, and found that the foraging sites were significantly warmer and less windy than non-foraging sites. Although many factors such as risk of predation, food availability, and competition for food affect foraging site selection, he found that the metabolic rate could be reduced by 10-12% by selecting favorable forage sites. As air temperatures increased during the late winter and spring the chickadees could also tolerate windier environments and therefore were able to forage in a wider range of sites without experiencing increased convective or radiative heat heat loss.

Many species also exhibit physiological adaptations to extreme environments. The Mountain chickadee in Wyoming, for example, experiences nocturnal hypothermia in which nighttime body temperatures are depressed by as much as 10°C in winter (B. O. Wolf and G. E. Walsberg [4834]). Like most desert birds the rock pigeon (*Columba livia*) of Israel exhibits many highly specialized physiological adaptations that enable it to survive in an extreme ambient environment where air temperatures reach as high as 50-60°C, relative humidity is normally less than 30%, and direct solar radiation is quite strong. The rock pigeon experiences controlled hypothermia in which increased core body temperatures can be sustained for finite periods of time, as well as panting and gular fluttering that promotes evaporative cooling. Both physiological adaptations, however, result in increased metabolic heat production that requires additional behavioral mechanisms to reduce heat stress and excessive water loss. J. Marder et al. [4817] reviews how the rock pigeon, like many birds that inhabit extreme cold or warm environments, utilizes fleather fluffing (piloerection) to reduce heat stress and water loss. In piloerection, the feathers are fluffed to create an air cavity between the inner skin surface and the outer feather shell. This enables extreme temperatures to be experienced on the outer feather shell, while the insulating air layer protects the inner skin surface of the pigeon from heat gain by convection ($-Q_H$). The effectiveness of this adaptation, which works well in either cold or warm environments, depends upon the thickness of the insulating air layer, how intact the feather shell is kept as it is expanded, and how fully air movement is restricted within the air layer. For the rock pigeon, most of the absorbed solar radiation is removed as net longwave radiation (L^*) or convective heat loss (Q_H). A small amount of evaporative cooling from the inner skin layer is able to offset the heat gain from metabolic heat production and sensible heat flux ($-Q_H$). The pigeon is thus able to maintain a normal metabolic rate at air temperatures of 50-60°C in a manner that is energetically efficient and not water-demanding.

In extremely cold environments a significant amount of energy can be consumed while simply resting. This energy loss is especially acute in small organisms with a large surface area-to-volume ratio that are unable to significantly increase their bodily insulation through physiological means. Consequently, social aggregation is one of the many methods endotherms use to reduce heat loss in cold environments. In the prairies of Canada, the muskrat (*Ondatra zibethicus*) is gregarious in winter, with as many as six muskrats occupying a single winter lodge. R. C. Bazin and R. A. MacArthur [4800] found that by huddling within their lodges, muskrats were able to increase their effective surface area-to-volume ratio and thereby reduce their total heat loss. The animals reshuffled themselves at an average periodicity of 12 minutes, and by doing so were able to maintain air temperatures between adjacent animals between 24-28°C when the air temperature was between 0-10°C. By constructing well-insulated winter lodges (Section 49) and generating significant metabolic heat through group occupation, muskrats can usually maintain air temperatures above 0°C in their lodges. When air temperatures were between –10 and 0°C a group of four muskrats were found to consume 11-14% less oxygen than four single muskrats under the same temperatures. Since muskrats may spend 14-15 hours per day during winter resting in lodges this can result in a large daily energy savings, and reduced winter foraging needs.

Ectothermic organisms are unable to maintain a constant core body temperature. Instead they display a range of temperatures over which they can survive, and a narrower preferred range of body temperatures over which they function best. Temperatures above and below the thresholds of their range are lethal, so ectotherms must carefully regulate their body temperature through behavioral adaptations to the surrounding microclimate.

The land iguana (*Conolophus pallidus*) of Isla Santa Fé in the Galapagos Islands illustrates how an ectotherm adjusts behaviorally to maintain preferred body temperatures. These islands experience two distinct seasons, a cool (Garua) season that lasts from late June through September and is marked by cool and foggy conditions, and a sunny hot season that lasts from January through May. Typical air temperature at 10 cm, substrate temperature, and global solar radiation for these two seasons are shown in Figure 48-1 (upper). The diurnal variation in measured body temperature for a female land iguana during a typical day for each season is shown in Figure 48-2. Observed body temperatures were always cooler during the cool season than during the hot season, with midday body temperatures averaging 4.4°C less during the cool season. Modeling results showed that without behavioral adjustments the land iguanas could experience lethal body temperatures (> 40°C) between 1200-1600 hours (cool season) and 0900-1800 hours (hot season), respectively.

The land iguana also inhabits two distinct habitats with significant microclimatic differences. A north-facing cliff face habitat is found along steep east-west trending ridges. These cliff faces occur within a wind shadow, and receive strong global solar radiation and greatly reduced wind speeds. As a result high air and substrate temperatures are found (Figure 48-1 lower). Their plateau habitat is found on flat regions above the cliff faces and is

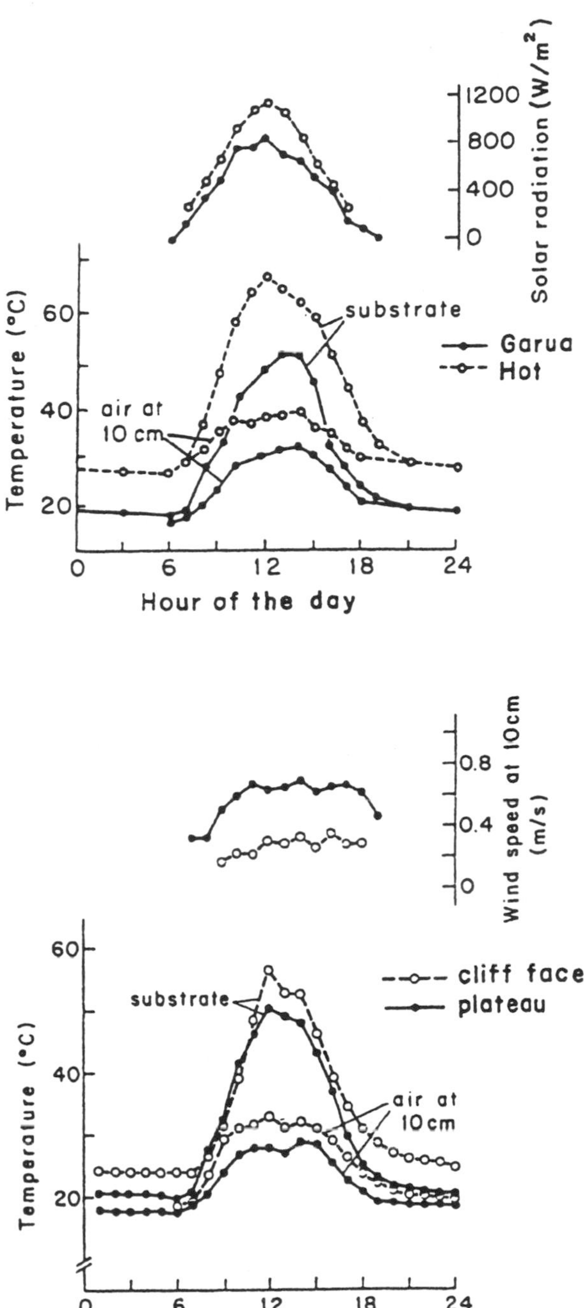

Figure 48-1. Upper: Microclimatic data on a typical day during the Garua (solid circle) and hot seasons (open circles) on Isla Santa Fé, Galapagos Islands. Lower: Microclimatic data on a typical day during the Garua season for a plateau (solid circles) and cliff face site (open circles). (After K. Christian et al., [4803])

characterized by much stronger wind conditions, and significantly lower air and substrate temperatures.

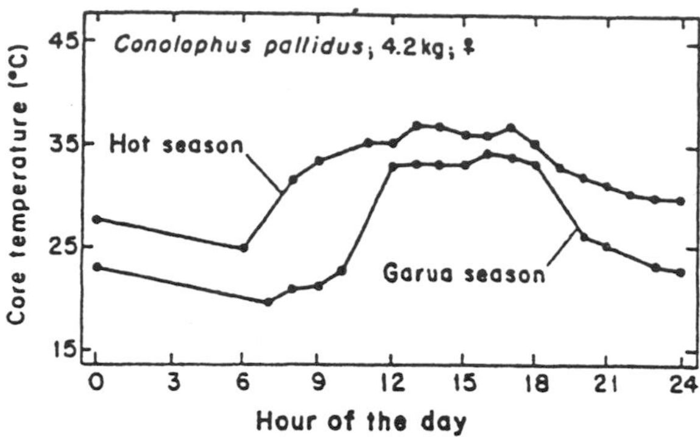

Figure 48-2. Daily pattern of measured body temperature (°C) for a female land iguana during a typical cool (Garua) and hot season day. (After K. Christian et al., [4803])

Through thermoregulatory behavior the land iguana utilizes these two habitats to maintain the longest possible period of constant body temperature while also avoiding lethally high body temperatures. On a daily basis, the land iguana's preferred behavior during both seasons is to bask in the early morning sun in order to quickly raise its body temperature to a preferred level. On a seasonal basis, land iguanas were found on the cooler plateau sites 39% of the time during the cool season, but 78% of the time during the hot season. During the cool season, land iguanas also tended to seek out plateau locations more frequently on clear days and cliff face sites more often on cloudy days. Thus land iguanas display daily, seasonal, and weather-dependent changes in thermoregulatory behavior in order to optimize utilization of microclimatic space.

A mobile animal will quickly adjust itself to the microclimate of its immediate environment by reacting to local weather conditions. Dancing swarms of midges avoid the changing directions of a freshening wind, as observed at times by F. Lauscher [4816], by taking shelter in the lee of a hedge. These behavioral adaptations to extreme microclimatic conditions are especially evident in the movement and habitat selection of desert organisms. According to W. Mosauer [4819], the sand lizard (*Uma notata*) can withstand the extreme temperatures of desert sand by raising its body a little above the heated ground when running fast. The African cricket, according to F. S. Bodenheimer [4801], turns its body in the direction of the sun's rays about noon, so as to present the smallest possible cross-sectional area to it, but lies broadside to the sun in the cool of the morning. N. F. Hadley [4810] found that by controlling the depth of burrow penetration, varying its movement within a burrow, and adopting a nocturnal feeding pattern, adult scorpions in the Arizona desert were able to experience environmental conditions that ranged between 32-40°C and 55-70 percent relative humidity, when surface conditions were between 20-65°C and the relative humidity was between 5-100 percent.

E. T. Nielsen [4822] observed in Denmark that one type of cricket (*Tettigonia viridissima*) begins its song in the afternoon on low forms of growth, sitting on low plants or reeds. At night, it is heard singing higher up, in the trees. In order to find out whether different types of crickets were involved, or if it was a single kind that sang below, and climbed higher as darkness fell, he attached some test specimens to a few meters of thread. The crickets did climb spirally up the tree when the air layer near the ground began to cool. Nielsen assumes that they seek the more comfortable temperatures of higher layers.

R. Geiger was always impressed while making observations in stands of tall pines by the regular accumulation of hordes of insects above the tree crowns in the first hours of the morning. Thousands upon thousands of hovering insects, gnats, and butterflies would collect there, such a mass of living creatures that it was hardly possible to believe they could exist in such numbers. This living cloud was cut off so sharply at the boundary surface of the microclimate (Figure 35-1) that the observer climbing up the observation ladder seemed to be pushing his head through a barrier.

Sea gulls take the wind field and lapse rate in the layer close to the water into account when flying over the sea. A. H. Woodcock made systematic observations; the results have been illustrated by G. Neumann [4821] in Figure 48-3. When the water is warmer than the air, and wind speeds are light, a cellular circulation develops over the sea. The gulls make use of the updrafts in these cells and climb by gliding upward in a circle. Such cases are indicated by small triangles in Figure 48-3. When the water is colder than the air, on the other hand, or when the wind is stronger than Beaufort force 6 with the water warmer, there are no updrafts and the gulls climb in a straight line by flapping their wings (crosses in Figure 48-3). Between these two possibilities, there were two intermediate forms: a straight line glide upward (small circles), and an upward glide partly in a straight line and partly in a circle (large circles with dots). Each type of flight corresponds to a state of microclimate over the sea with wellmarked boundaries.

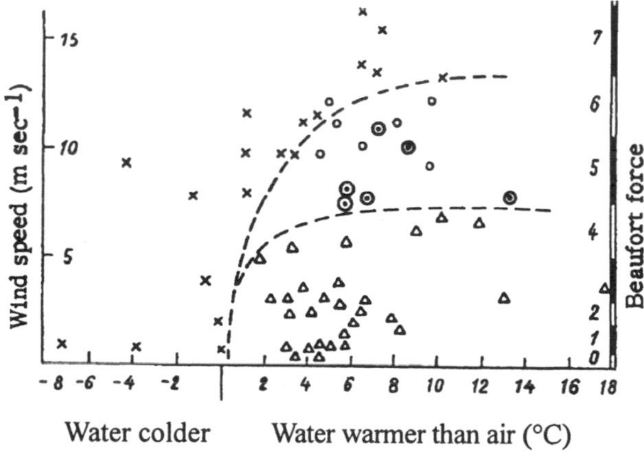

Figure 48-3. The flight tactics of sea gulls are designed to take advantage of the wind and temperature structure. (After A. H. Woodcock. From G. Neumann [4821])

The Japanese lark (*Alauda arvensis japonica*) usually flies in a horizontal circle about 60 m in diameter at a definite height. S. Suzuki and his colleagues [4828] established from flight path measurements that this height depended on the surface temperature. It was 80 to 100 m with temperatures of 24° to 28°C, and averaged only 40 to 70 m at 16°C.

The activity space of organisms also varies with changes in humidity and wind speed. The surface area of an organism of a given body shape will vary as a function of the two-thirds power of its mass. Therefore smaller organisms of the same body shape will tend to desiccate faster than larger organisms of similar body shape. In the tropical forests of Costa Rica, M. Kaspari [4815] found that larger ants tended to forage beneath both closed and open forest canopies that experienced a wider range of air temperatures and vapor pressure deficits. Smaller ants, however, foraged only within the more uniform forest floor beneath closed forest canopies. Thus, body size increases the breadth of microclimates utilized by ants also increases. In the Namib Desert of southwest Africa, B. A. Curtis [4804] found that the activity of the Emery ant (*Camponotus detritus*) between foraging and nesting sites was unaffected by wind speeds below 11 km hr^{-1} somewhat affected by wind speeds up to 16 km hr^{-1}, markedly affected with wind speeds up to 25 km hr^{-1}, and completely inhibited at wind speeds above 25 km hr^{-1}.

A consequence of the way animals adjust themselves to the circumstances in which they live is that even in a small area, microclimatic differences will influence the species composition, density, and spatial distribution of organisms. An example of this kind of adjustment is taken from E. Schimitschek [4825] on the development of bark beetles (*Ips typographus*) in a tree trunk lying on the ground. Figure 48-4 shows a cross section of a fir log lying in a NW to SE direction. In sector 1 on the SW side, where the effect of the sun is greatest, bark temperatures of 50°C were recorded and no beetle eggs were found to have been laid. Eggs laid in both of the sectors marked 2 died. Larvae developed from eggs laid in sectors 3, but these dried up later. It was only in sector 4, which is narrow on the sunny side and broad on the shady side, that conditions suitable for normal development were found. Where the trunk was in contact with the ground (5) and became very damp in wet weather, the mortality rate rose again to 75 to 92 percent.

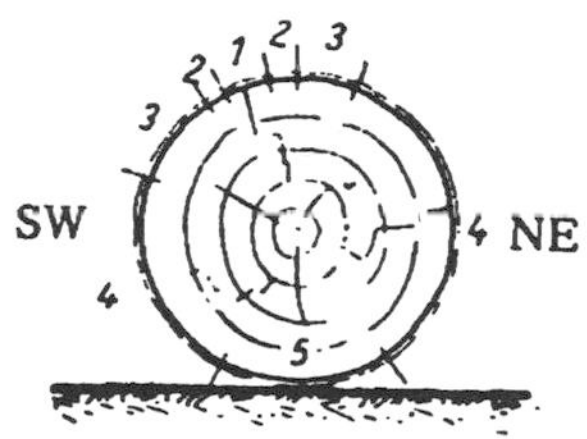

Figure 48-4. Development of bark beetles around a tree trunk. (From E. Schimitschek [4825])

More topographically complex environments allow for significant changes in local habitat quality for organisms capable of moving over short distances. S. B. Weiss, et al. [4833] found that the phenology of larval and pupal development of the butterfly *Euphydryas editha* was strongly influenced by slope exposure along the Coast Range of central California. Later senescence of host plants on the cooler north-facing slopes offered high quality habitat for prediapause larvae, while the warmer south-facing slopes provided a preferred environment for the growing pupae. Dispersing larvae were able to partially ameliorate their living conditions by moving only short distances across ridges and v-shaped gullies.

Some animals have been shown to be very sensitive to slight variations in temperature. For example, the plant-parasitic nematode *Ditylenchus dipsaci* and *Pratylenchus penetrans* have been shown to migrate in response to air temperature variations of as little as 0.03°C cm^{-1} (M. El-Sherif and W. F. Mai [4808]) and the root-knot nematode *Meloidogyne incognita* with soil temperature variations of as little as 0.001°C cm^{-1} (M. Pline et al. [4823], J. A. Diez and D. B. Dusenbery [4806] and D. B. Dusenbery [4807]).

The distribution of animal life, therefore, depends not only on the general large-scale climatic conditions, but also on microclimate. The limit of the area in which any particular species is found is a boundary zone in which the creatures are found to exist in pockets. In much the same way as plants, animals are able to exist in an unfavorable general climate only if they can find places where the microclimate is favorable. The egg and the larva of the hookworm, for example, which are adapted to tropical temperatures, find suitable environments in tunnels and mine shafts where they create risks for miners. In underground heating systems in Paris, the rat flea that carries plague is able to thrive, although it is a visitor from warmer countries. E. Martini and E. Teubner [4818] proved by laboratory studies and field observations that the true malaria mosquito (*Anopheles*) needs microclimatic conditions different from those favorable to other mosquitoes. This has an immediate effect on the malaria risk to which humans are exposed in the tropics. The microclimate of tropical dwellings determines which kinds of insects are able to live alongside human beings. Many additional examples can be found to illustrate this point.

49 Animal Dwellings

Most of the higher animals have dwellings in which they can raise their young. Such animals exhibit an inherent instinct in creating microclimatic conditions which are optimal in regards to survival. The modern zoologist must be familiar with microclimatic principles in order to study the ecology of breeding places. The examples that follow will provide some indication of the extent to which the surface climate influences animal nesting and roosting site selection.

As was the case with foraging and roosting site selection (Section 48), the microclimate also influences the selection of nesting sites. Thermal advantage can be gained by placing nests along a specific side of a tree or other obstruction, at a particular vertical location within a canopy or along an obstruction, or with a particular entrance orientation.

Mountain chickadees (*Parus gambeli*) in Wyoming (41°N) were found by D. G. Wachob [4919] to select nesting sites near trees with a more open southeastern exposure in order to maximize solar access for rapid warming during the early morning. Sites with a more restricted exposure to the northwest were also selected to minimize nest exposure to rain and snow that normally came from that direction. Air temperature, substrate temperature, and global solar radiation were all significantly higher at successful mountain chickadee nesting sites than at unused or unsuccessful nesting sites (D. G. Wachob [4919]).

In addition to providing a thermoregulatory savings for the adult mountain chickadees due to reduced convective cooling around the nesting sites, careful nesting site selection also contributes to reproductive success. Selection of nesting sites in warmer microclimates provides a thermal benefit to the incubating eggs and developing nestlings, as well as to the incubating females. These energy savings may decrease the number of interruptions in egg laying, reduce the energetic costs of incubation, and increase the length of time allowable for adult off-nest foraging searches, all of which increase the likelihood of successful reproduction.

Nest site selection is especially important to small birds because of their limited mass and large surface area-to-volume ratio. Furthermore, because small birds are ectothermic during the early stages of their development any thermoregulatory advantage that can be gained by careful nest selection will improve their competitiveness and chances of survival.

Along the coastal plain of Israel, Y. Sidis et al. [4916] found that the orange-tufted sunbird (*Nectarinia osea*) selected nest sites near sheltering structures that helped them to avoid direct-beam solar radiation. This selection, in a microclimate where summer maximum air temperatures often reach 40°C, helped to reduce the risk of overheating. Along the coastal plain of Israel, where mean monthly minimum air temperatures can be as low as 5°C, wind also plays a major factor in the selection of nest sites for the sunbird. Placement of nests and nest entrances away from the prevailing wind can reduce the wind speeds near nest sites by nearly a factor of ten, which significantly reduces the rate of cooling for the birds (Y. Sidis et al. [4916]).

Larger animals, on the other hand, with their greater body mass and smaller surface area-to-volume ratios have significant physiological limitations on their ability to dissipate heat. C. W. Barrows [4900] documented this in his study of the spotted owl (*Strix occidentalis*) in North America. The relative intolerance of this bird to heat stress plays an important factor in its selection of roosting sites. During the summer the spotted owl selects roosting sites beneath dense multi-layered canopies, and along north-facing slopes in deep ravines. The reduced solar radiation input of these microclimates consistently resulted in roosting sites that were 5-6°C cooler than open sites. Depending upon the air temperature, global solar radiation, and wind speed, spotted owls begin gular fluttering at between 27-31°C. C. B. Barrows [4900] estimated that the preferred summer roosting sites could result in a 40-90% decrease in the number of days per year in which continuous gular fluttering (with its associated increased rate of oxygen consumption) was required. During the winter, in contrast, the spotted owl avoided the north-facing roosting sites altogether, and consistently selected roosting sites on the south side of trees with open exposure to the sun.

R. Geiger was amazed at the way rabbits on the North Sea island of Sylt arranged their burrows. One had its entrance, for example, on the upper slope of a sand dune, so that when there was heavy rain it collected below, and none came through the opening. An overhanging growth of heather roofed over the entrance and protected it from rain and dripping water. It had a southerly exposure so that the entrance was sunny and shielded from north winds. A very large bush to the west provided additional protection from the stormy west winds.

The high dome shaped nests built by ants in the European climate apparently are intended to exploit to full advantage the small amount of solar radiation in forests. Since the materials used (pine needles, straw) and the loose methods of construction make thermal conductivity poor, differences of exposure within short distances are not evened out. G. Wellenstein [4921] investigated in September 1927 a nest of red forest ants, 80 cm high and more than 12 m in circumference, in a stand of young firs on a steep slope not far from Trier. The nest, 25 cm under the surface, was 3 to 4°C warmer on the shady side than the air in the environment, and below the sunny side it was 5 to 9°C warmer than the shady side.

A. Steiner [4917] describes the construction and operation of a nest in the following way: "A dome shaped nest with numerous air spaces in the interior is built of earth and parts of plants in a place with a southern exposure, sheltered from the wind. The shape varies from that of a hemisphere to a cone, depending on the prevailing radiation and precipitation conditions, and serves to capture heat. The relatively greatest degree of efficiency in this respect is obtained when the elevation of the sun is low; it has been calculated for a latitude of 47°N that a hemispherical dome receives twice the incident solar radiation on a horizontal surface at noon on 21 December, 1.25 times as much at the equinoxes, and 1.05 times the quantity of 21 June. The increased quantity of heat received is protected from loss by various means, principally by the thick roof of the dome, constructed of plant materials with poor conductivity, by the insulated air chambers in the interior, and by closing the openings at night. By these means the average temperature in the interior of the nest, at a depth of 30 cm, often varies between 23° and 29°C, and is 10°C higher than the corresponding soil temperature. The varying temperatures throughout the upper parts of the dome are utilized to the optimum extent by tireless shifting of the eggs, which are also protected from overheating by being transported down into deeper parts of the nest."

Although extreme environments, such as a desert, are good locations for identifying microclimatic adjustments among animals, considerable small-scale variation in the adjustments occur that, upon closer inspection, each reveal a separate adjustment. The Cactus Wren (*Campylorhynchus brunneicapillus*) of the southwestern desert of the United States builds a closed nest of twigs, grass, fur and feathers with a long passageway entrance. Interior air temperatures in the closed nest can be as much as 6°C warmer than the outdoor air when the nest is fully illuminated by the sun. Nest entrances oriented toward the north reduce the input of direct-beam solar radiation to the nest interior through the passageway and help to maintain lower interior air temperatures. Nest entrances that bisect the prevailing wind direction at a 90 angle, however, also have a microclimatic advantage since a vacuum effect is created that draws air out from from the interior through the nest entrance. This produces a ventilation effect in which outside air comes into the nest through the top, bottom, and sides of the closed nest and helps to limit the accumulation of sensible heat in the interior. The orientation of the nest entrances were found by C. F. Facemire et al. [4904] to vary depending upon microclimatic conditions. Nests located on the eastern side of a mountain where a well-developed valley breeze circulation developed during the day with winds normally from the southwest had nest entrances oriented to the north-northeast. This entrance orientation provided both ventilation and shading benefits. Nests located on the western slopes of mountains where winds were randomly oriented and highly variable had nest entrances oriented to the north. Although no ventilation benefit was gained due to the variable wind directions, the north entrances maximized the shading benefits. Finally, nests located on the valley floor where daytime winds were generally from the southeast and nighttime winds from the northwest had nest entrances oriented to the southwest to provide ventilation benefits.

The metabolic heat released by animals themselves can also be utilized to their advantage by proper construction of their dwellings. Figure 49-1 after R. Hesse [4909] shows photographs of a termite hill in Arnhem Land in northern Australia. Its asymmetric shape was developed as a protection against strong solar radiation. The flat nests, similar to those built by ants in Europe, of one type of termite (*Eutermes exitiosus*) were investigated by F. G. Holdaway and F. J. Gay [4910]. Figure 49-2 shows the temperatures measured on a winter day, 9 August 1933, in a circular nest 38 cm high and 80 to 90 cm in diameter, near the Australian capital of Canberra (35°S). When air temperatures were below 9°C, the wall of the nest was heated extensively by radiation, up to 25°C on the N side (Southern Hemisphere). The termites were, however, always able to maintain temperatures above 25°C in the interior of the nest through the expenditure of metabolic heat. The authors were able to establish that inhabited termite nests were on the average 8.1 to 10.3°C warmer than those that had been abandoned.

Figure 49-1. Termite nest orientated in compass direction in northern Australia. (After R. Hesse [4909])

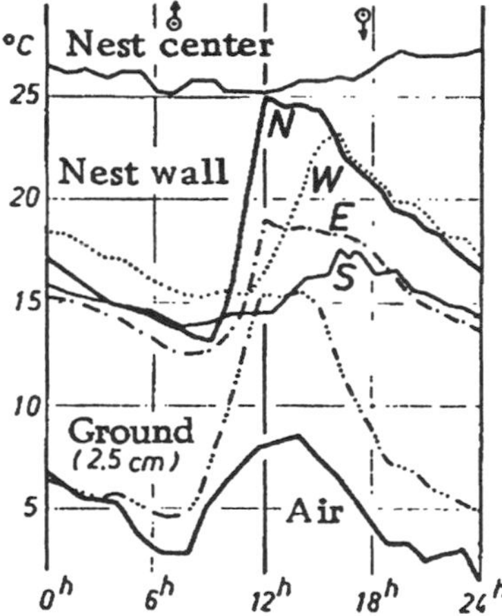

Figure 49-2. Temperature variation on a winter day in a flat termite nest near Canberra, Australia. (After F. G. Holdaway and F. J. Gay [4910])

The beaver (*Castor canadensis*) of the boreal forests of Manitoba, Canada (50°10'N) construct lodges of branches and mud that provide a year-round thermoneutral environment while also allowing adequate gas exchange. A. P. Dyck and R. A. MacArthur [4903] found that even though the ambient air temperature varied by 73.8°C over the year

(-41.4 to 32.4°C), the interior wall temperatures of the lodges varied by only 34.2°C (1.4 to 35.6°C). Even when exterior air temperatures reached as low as -40°C, the interior walls were normally between 3-5°C, well within the 0-28°C thermoneutral zone of the beaver. Thermoregulatory control was achieved by varying the lodge wall thickness and the amount of mud used in construction, and also by the metabolic heat generated by the beavers that live in communal groups of a male-female pair and their offspring. Some carbon dioxide accumulation and oxygen depletion was observed within the lodges as a result of the construction methods. However, the levels of each gas were well within the tolerable limits of the beaver since adequate ventilation was achieved through vents in the lodge roofs, thinner roof walls, and less extensive application of mud on the roofs.

In addition to their own metabolic heat, some animals will take advantage of additional energy sources to warm their nests. An extraordinary degree of microclimate regulation has been achieved by the Australian megapods *Leipoa ocellata* (mound bird, jungle fowl), which has been called the "thermometer bird" as a consequence. H. J. Frith [4905] has shown that the temperature of its nest is maintained at 34.5°C, with a fluctuation of only 1 to 2°C when surface air temperatures vary between 16° and 49°C during the year. Nests are roughly circular in shape, approximately 5 m in diameter. About 20 eggs are laid over a period of a week and hatch in about 6 months. During the cold spring, the birds collect large quantities of leaves, which are then covered with about 50 cm of sand and left to ferment and produce heat, thus raising the temperature of the nest. During the summer, depending on the amount of solar radiation and the air temperature, adjustments are made to the sand thickness (involving 2 to 3 hours of work).

Bird nestlings show two different patterns of thermoregulation following birth. Precocial birds are able to regulate their body temperature from birth or at a very early stage of development. Altricial birds initially behave like ectotherms and only regulate their body temperatures (endothermy) at a later date. Thermoregulation during the nestling stage can be critical for these altricial species. The White stork (*Ciconia ciconia*) shows several behavioral adjustments that enable its nestlings to achieve effective endothermy at an earlier date than they achieve physiological endothermy (F. S. Tortosa and R. Villafuerte [4918]). The White Stork builds a large nest 1-2 m in diameter on the top of trees and structures that are not protected from the wind. The nests have a sunken center 30-40 cm deep that they line with straw and other fine materials. The nest bottoms are then covered with moist cow dung gathered from surrounding agricultural fields. The nestlings have an initial body temperature of 33°C. After about 30 days they achieve physiological endothermy and can maintain a constant body temperature of 40°C. One parent is always present with the nestlings during the first five days after birth when the young birds have a low body temperature, and both adult birds are present during cold or wet weather. As the body temperature of the nestlings begins to slowly increase the adults will leave the nestling alone for short periods to forage and look for dung. The length of these periods increases until at 30 days the young birds can be left unattended for longer periods. The moist cow dung, with its high moisture content, plays an important role in regulating the temperature of the nest. The high thermal inertia of the moisture in the fresh cow dung helps to maintain an elevated and more stable thermal environment within the nest by acting as a source of sensible heat. As the nestlings age and achieve a higher body temperature the adults collect fresh dung less frequently until they stop collecting it altogether when their young attain physiological endothermy.

Nesting site selection and construction can also improve the water balance conditions of organisms in dry environments. In the desert near Salt Lake City, there lives a type of rabbit that obtains most of its water supply from its scanty nourishment. To investigate these unusual living conditions H. Burckhardt [4901] investigated the humidity conditions of its environment and burrow. To do this a rabbit was caught in a trap in front of its burrow, and a recording device was tied to its tail. This device was smaller in cross section than the rabbit and could record temperature and humidity for a few hours. On being freed, the animal rushed back into the protection of its burrow, and thus the conditions in the interiors, and usually also in the environment, were recorded. By this method, it was shown that the animal was able to maintain its fluid balance without ever drinking a drop of water because of the surprisingly high relative humidity of its burrow where it remained for long periods.

Although the five-lined skink (*Eumeces fasciatus*) of the eastern deciduous forests of Ontario, Canada (42°10N) also selects cooler locations for their brooding sites, moisture conditions play a more important role in nest site selection (S. J. Hecnar [4908]). The egg is the most vulnerable stage in the life cycle of the skink, so nest sites with adequate temperature, moisture, and gas exchange conditions are crucial for their survival. The females select large moderately decayed logs that provide a relatively constant microclimate for egg development. Skink eggs allow for the transfer of moisture through their shell as they lie on a substrate, and can thus suffer from desiccation due to drying as well as excessive swelling due to moisture gain. Mortality rates are lowest at moderate ranges of substrate moisture levels. Female skinks adjust the horizontal and vertical position of their brooding sites to regulate the moisture levels of the egg substrate.

Animals will also seek shelter for extended periods of time from extreme thermal environments. A good example of such long-term roosting behavior is seen in the willow grouse (*Lagopus lagopus*) that seeks shelter from extreme cold and wind in open or closed snow burrows. K. Korhonen [4911] studied the closed snow burrows of willow grouses in Simo, Finland (65°37'N) and found that after 60 minutes of roosting the burrow wall temperature was only -3°C when the outside air temperature was as low as -24°C. Increased snow temperatures extended only 5-7 cm from the burrow walls as a result of the low conductivity of snow. Because the relative humidity within the snow burrows was always 100%, expired water vapor immediately deposited onto the burrow walls, which increased the density of the snow along the burrow walls, and resulted in the release of latent heat. Water contained in bird excretions within the burrow also froze and released its latent heat. These two forms of latent heat were estimated to be nearly 35% of the total heat loss for an inactive roosting grouse. This heat source, combined with the well insulated willow grouse plummage, enabled the snow burrows to provide a thermoneutral environment in which the burrow wall temperatures remained above the grouse's lower critical temperature of -6°C.

Urban microclimates (Section 51) are also exploited by organisms for warmth and protection. H. Löhrl [4913] showed that one type of swallow (*Nyctalus noctula* Schreb.) exhibited considerable skill in selecting, out of the wide area in which it lived, a place to nest in winter with the warmest microclimate. Those observed in Munich first selected a location in town where it is warmer in winter than in the country. The favored position was the corner of a house pointing inward toward the warmest part in the center of the town with an exposure to the southeast. About 12 m above the street, and, therefore, above the cold layer of surface air, the birds hollowed out holes about 50 cm deep in the walls behind the gutter of the roof which protected them from rain. The interior of the house was heated, and in addition, one of the main pipes of the steam central heating ran close to the nest. The following temperatures were measured simultaneously on one occasion: -14°C at the meteorological station, -5°C on the roof of the house, and almost 0°C in the nesting area.

Finally, cave dwelling bats (*Miniopterus schreibersii blepotis*) studied by L. S. Hall [4907] in New South Wales, Australia (35°45'S) exhibit a unique hibernation pattern that enables them to avoid periods of food shortages and to reduce thermoregulatory energy loss during winter. This involves a combination of physiological and behavioral adjustments. These bats can behave as both endotherms and ectotherms. During hibernation, or periods of winter torpor, their body temperatures fall to within 1°C of the cave air temperature, which lowers their rate of metabolic heat production. If the cave microclimate changes, the bats can be aroused and will change their location to a more suitable microclimate. Air temperatures ranged between 9 and 19.5°C in Thermocline Cave, a small 60 m long and 2-5 m wide cave with a west-facing entrance. In fall, when food was abundant, the bats selected the warmest area in the cave for daytime roosting. These were the highest locations in the cave where warm air was trapped. As the bats would put on body weight in the fall they began to move to cooler parts of the cave (T_A = 11°C) to reduce their metabolic heat loss while roosting. When maximum weights were reached in the winter the bats relocated again to the coolest areas in the cave (T_A = 9.5°C) to further reduce metabolic heat loss while they underwent more extended periods of torpor. In spring, they returned to warmer daytime roosting sites. Through this combination of physiological and behavioral adjustments the bats were able to minimize their body temperatures and metabolic rates during torpor, while building up their fat reserves prior to hibernation.

50 Bioclimatology

In its broadest sense, bioclimatology is the study of the relationship between climate and living organisms (R. E. Munn, [5029). Several of the subdivisions of bioclimatology, therefore, have already been covered in earlier sections on animal behavior (Section 48), animal dwellings (Section 49), agricultural fields (Sections 29-32), forests (Sections 33-39), and mountain environments (Section 46). Our discussion in this section will concentrate on the fields of human bioclimatology, animal bioclimatology, and aerobiology.

The assessment of human thermal comfort can best be approached by means of the human energy balance equation. P. Höppe [5018] developed one such model that illustrates the basic principles. The human energy balance equation is given by:

$$Q_M + Q^* + Q_G + Q_H + Q_{ED} + Q_{ES} + Q_{ER} + Q_{HR} = 0$$

where energy inputs (W m^{-2}) are positive and energy outputs are negative. In this model, Q_M = internal heat produced by human metabolism, Q^* = net radiation, Q_G = conduction, Q_H = sensible heat flux, Q_{ED} = latent heat flux by diffusion of water vapor, Q_{ES} = latent heat flux by evaporation of sweat, Q_{ER} = latent heat flux by respiration, and Q_{HR} = sensible heat flux due to breathing.

As warm-blooded creatures humans must keep their inner core body temperature within a narrow range (around 37°C) for survival. Humans gain heat energy from the environment by shortwave and longwave radiation, conduction, convection, and by producing internal metabolic heat (Q_M). Metabolic heat production consists of two basic types, basal metabolic heat that is produced when the human body is at rest (or asleep), and muscular metabolic heat that is released during periods of work, exercise, or shivering. The flux of metabolic heat from the body core to the skin (Q_{CS}) is given by (P. Höppe [5018]):

$$Q_{CS} = A_S V_B r_B c_B (T_C - T_{SK})$$

where A_S = the surface area of the skin (m^2), V_B = the blood flow density from the core to the skin ($l\ s^{-1}\ m^{-2}$), r_B = the density of blood ($kg\ l^{-1}$), c_B = the specific heat of blood ($J\ kg^{-1}\ K^{-1}$), T_C = the core body temperature (°C), and T_{SK} = the mean skin temperature (°C).

Humans lose heat energy to the environment by radiation, conduction, sensible heat flux (convection and breathing), and latent heat flux (sweating and breathing). The flux of heat from the skin, through the clothing, and to the outer surface of the clothing (Q_{SC}) can be modeled as (P. Höppe [5018]):

$$Q_{SC} = A_C / R_{CL} (T_{SK} - T_{CL})$$

where A_C = the clothed surface area (m^2), R_{CL} = the resistance of the clothing to the transfer of heat ($m^2\ K\ W^{-1}$), and T_{CL} = the surface temperature of the clothing (°C).

By taking observations of atmospheric conditions (solar radiation, air temperature, wind speed, relative humidity), establishing the human characteristics (age, sex, weight, height, metabolic rate, body position, and clothing characteristics), and assuming that the complex human body can be approximated by simpler shapes it is possible to study human comfort using the energy balance approach. J. E. Burt et al. [5004, 5005] provide another example of a human energy balance model developed for an urban environment setting, while H. Mayer and P. Höppe [5028] review some of the simpler biometeorological indices that have also been used in assessing human comfort.

Through involuntary physiological mechanisms the human body will work to maintain a constant inner core body temperature. Humans can also undertake voluntary bchavioral adjustments to help regulate body temperature (R. E. Munn [5029]).

Heat stress is caused by a prolonged positive imbalance in the human energy balance equation, and can raise the human core body temperature above optimum levels. Environmental conditions that promote heat stress include high solar radiation inputs, high longwave radiation inputs, air temperatures above the core body temperature (37°C) that result in heat gain through convection, and high atmospheric humidity that reduces the loss of latent heat from the body. Strong winds enhance heat stress when combined with air temperatures above the core body temperature, but the effects of high atmospheric humidity on heat stress are exacerbated by calm wind conditions. Human behavior that promotes heat stress includes continued strenuous physical activity and insufficient fluid intake. The latter factor is particularly critical for persons engaged in strenuous physical activity in strong sunlight and warm, humid, and calm conditions.

When the core body temperature begins to increase the human body will involuntarily initiate a series of measures designed to dissipate heat from the body core (R. E. Munn [5029]). These include the dilation of blood vessels near the skin surface (to increase blood flow to the skin so body heat can be more easily lost via longwave emission, convection or evaporation), increased sweating (to increase evaporational cooling), and an increased respiration rate (to increase sensible and latent heat loss via breathing). Voluntary behavioral adjustments that reduce the threat of heat stress include the avoidance of radiational heat sources, seeking out shaded microclimates, seeking out well-ventilated microclimates, increased fluid intake to promote evaporative cooling, reducing the rate of physical activity, and changing clothes to increase heat loss by sensible and latent heat flux.

Prolonged exposure to a negative imbalance in the human energy balance equation can similarly result in a lowering of the human core body temperature below optimum levels that can result in cold stress. Cold stress is enhanced by exposure to low temperatures that increase the loss of energy by net longwave radiation and convection, and thin or wet clothing that offers insufficient resistance to the transfer of heat from the skin to the clothing surface. Under such conditions strong wind always acts to increase cold stress. Involuntary physiological mechanisms that the body initiates to reduce heat loss include the constriction of blood vessels near the skin (to reduce the flow of blood to the skin surface), and the initiation of shivering (to increase the rate of metabolic heat production). Behavioral adjustments that reduce the threat of cold stress include the movement to locations sheltered from cold, wet and windy environments, an increased rate of physical activity to increase metabolic heat production, and the addition of dry clothing layers to increase the resistance to heat transfer from the skin surface.

Through the careful selection of clothing humans are able to inhabit an extraordinarily wide range of microclimates. The key principles are to select materials with a high resistance to heat transfer, dress in multiple layers to trap still air between the clothing layers, to select materials that allow the movement of water vapor, and to avoid wetting the clothes. The interplay of human metabolism, clothing, and ambient conditions are illustrated in the field measurements of core body temperature, skin temperature, inner clothing temperature, and air temperature reported by K. Natsume et al. [5030]. Core body temperature (rectal temperature) was lowest during sleep when human metabolism was minimum but increased by 1-2°C during wakefulness as internal metabolic heat production increased. In spite of the core body temperature being lowest during sleep, however, chest skin temperature and inner clothing temperature were highest during sleep and lowest during wakefulness. This resulted from the increased air circulation across the human body during wakefulness that resulted from human activity. This activity increased the convective heat transfer coefficient and reduced the resistance to the transfer of heat from the skin through the clothing to the air. The inner clothing temperature was also more variable during wakefulness due to the increased range of human movement (and thus more variable clothing resistance to the transfer of heat) and more variable ambient environment.

Humans also have a direct interest in making dwelling conditions suitable for themselves. This leads up into the realm of cryptoclimate, the climate of completely or partially enclosed spaces. C. E. P. Brooks and G. J. Evans [5003] published an early annotated bibliography describing the climate of dwelling houses, schools, theaters, hospitals, libraries, museums, factories, warehouses, offices, laboratories, cellars, ships, tunnels, tents, and many other places. The relations between climate (and microclimate) and architecture are discussed in books by J. E. Aronin [5001] and B. Givoni [5015]. Only a few of the basic ideas and some explanatory examples from this range of bioclimatic topics will be presented.

With respect to the microclimate of buildings H. Landsberg [5027] showed that people feel most comfortable with a temperature between 18° and 32°C but are able to extend their habitation into parts of Earth where temperature extremes range from -76°C to 63°C. The climate of a room depends, in the first instance, on the situation of the house and on the position of the room inside the house. The direction in which it faces, the arrangement of windows, and the immediate neighborhood determine the atmospheric conditions to which the dwelling is exposed. The way in which a room is able to cope with these conditions depends on the building material, the strength of the walls, the size of the room, and the size and number of its doors and windows.

The exchange of air between a room and its environment takes place mainly through gaps in doors and windows and also occurs through pores in the material of the walls. H. W. Georgii [5013] was able to separate these two ventilation factors by enriching the air in a sealed room with carbon dioxide. That gases penetrate through walls has been known since 1858 from the work of Max von Pettenkofer who carried out the same experiment with an artificial aerosol (a sprayed $CaCl_2$ solution). In both cases, the quantity measured was the rate of decrease of concentration with time. This was expressed as a function of wind speed and temperature difference between the outside and inside, with both doors and windows closed.

The results refer to a second floor laboratory, in the Frankfurt Meteorological Institute, with brick walls, two doors, and one window. Results showed that there was no discernible connection between the temperature of the exterior and interior, and the amount of ventilation. By contrast, the influence of the wind blowing outside was strong as may be seen from Figure 50-1. In calm conditions, the CO_2 readings gave an exchange of air amounting to 2.3 l sec^{-1} which increased very quickly as wind speeds rose. The carbon dioxide was involved in both methods of self-ventilation, whereas the aerosol could escape only through cracks. This indicated a much lower rate of air exchange. The aerosol undergoes a certain amount of alteration with time through sedimentation, coagulation, and diffusion at the walls, which results in the corrected curve (dashed line) shown. The average of 35 series of meas-

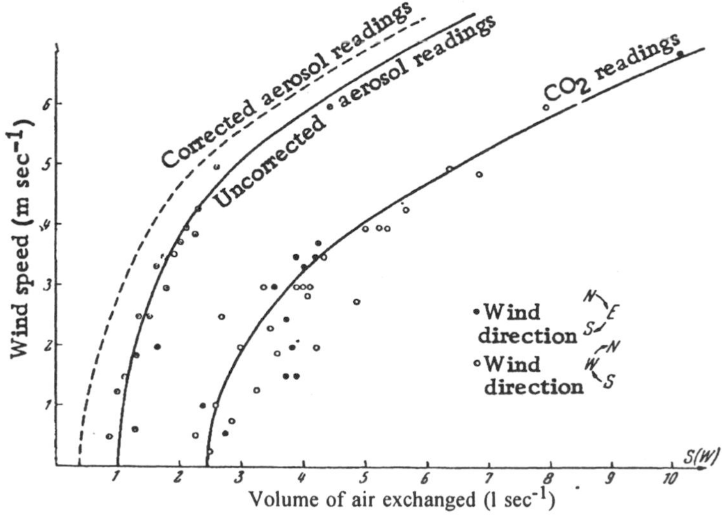

Figure 50-1. Self-ventilation of a laboratory, in relation to the wind outside. (After H. W. Georgii [5013])

urements showed that 46 percent of the air volume in the room was exchanged every hour by autoventilation (the ventilation coefficient). An air exchange value of 1, an exchange of air equivalent to the total volume of the laboratory in an hour, was reached with winds of 6 to 7 m sec^{-1}.

It is worth noting that the ventilation coefficient of a cellar in the same building was more than double this value (1.15). This was found to result from the suction effect of the warm staircase. As a consequence, there was a marked dependence of the effect, in this case, on the temperature difference between the exterior and interior. R. H. Aron, et al. [5000], in a study of the factors affecting the winter heating and summer cooling energy requirements at Central Michigan University, had somewhat different results. They found that the primary factor affecting energy requirements was the level of university activity (vacations, weekends, evenings, etc.). The second most important factor was the outdoor temperature. Contrary to H. W. Georgii [5013] and what might normally be expected, wind speed was found to be only marginally significant in explaining energy requirements. Clearly as building technology has improved with time the microclimate of modern interior living and working spaces has become independent of the exterior climate.

Measurements made in Innsbruck by R. Giner and V. F. Hess [5014], over a whole year, established that the particulate content of a closed room was, on the average, only 58 percent, and the content of condensation nuclei was only 31 percent of the values on the balcony. After ventilation, this state became reestablished in 3 to 4 hours. The situation, with regard to particulates, is rather different where people are active (98 percent in schoolrooms) or some form of work is being done (the rag-pulping room of a paper mill has 800 to 900 percent). The "dead inside air" (C. Dorno [2003]) in a room, according to measurements made by F. Dessauer, et al. [5009] in Frankfurt, has a much higher ion count than the "live outside air," amounting roughly to the quantity found only with summer thunderstorms.

Adding insulation reduces heat flow into or out of a building. The primary goal of most types of insulation is to reduce air movement. Still air has a very low thermal conductivity (Table 6-2). In thermal pane windows, for example, there are two or more panes of glass. Their distance apart is determined by two conflicting factors. On the one hand, the greater the distance, the more air there will be between the panes. On the other hand, as the space between the panes increases, so too will the air circulation. This increased circulation increases the heat flow through the air space and thus needs to be minimized. The optimum gap between the panes is around one-half inch, but is dependent on the emissivity of the glass coatings. For more information, the reader is referred to Cardinal I. G. [5007].

Vegetation can affect residential winter heating and summer cooling energy requirements. D. R. DeWalle et al. [5041] found that shading by trees can significantly reduce solar heating and by lowering wind speeds can lessen infiltration of outside air. They found that in mobile homes in one deciduous stand in Pennsylvania summer energy needs were 75 percent less than in the open. J. H. Parker [5032, 5033, 5034] found landscaping around a house reduced energy needs in excess of 50 percent and on hot sunny days reduced surface temperatures by 13.5 – 15.5°C. Steen et al. [5040] estimated that for a typical air-conditioned house in Florida the total heat gained via infiltration exceeded the heat gained by radiation and conduction through the windows and walls. In winter, D. R. DeWalle et al. [5041] stated that shading is counterproductive, offsetting savings from reduced infiltration. In a deciduous stand in winter they reported energy savings of 8 percent while in a dense pine forest heating energy needs rose 12 percent.

For additional information dealing with architectural design to maximize human comfort with the minimal amount of energy expended consult B. Givoni [5015, 5016]. Wind stresses and protection against wind have been studied by M. Jensen [5020] and by H. Blenk and H. Trienes [5201]. Questions such as these have an added importance for hospitals and convalescent homes (H. Landsberg [5027]).

The microclimate of transportation provides a special form of human dwelling space. I. Dirmhirn [5010] investigated the conditions to which travelers were exposed in 13 types of unheated streetcars in Vienna, in winter and summer. The excess temperature in winter of the interior above the air outside depended on radiation, wind, inflow of fresh air through windows and doors, and the number of passengers. It varied from 2.9°C in trailers with open platforms to 10.3°C in automobiles. The excess was 3.5°C in empty cars, 6.0°C in those half filled, 7.7°C in fully occupied cars, and 12.7°C when they were overcrowded. There was a marked temperature stratification within the car; on one occasion, 0°C was measured at the floor, 6.0°C at 60 cm, and 8.5°C at a height of 1.7 m. The relative humidity fell, as was to be expected, as the temperature excess increased.

Early forms of sheds were designed mainly as a protection against theft and severe weather, but modern breeding requirements have given great significance to the controlled climate of stalls. P. Lehmann [4912] and W. Zorn and G. Freidt [4922] drew attention to the importance of this factor some time ago. Modern selective breeding is designed for specific purposes so that the genetic factors can become fully effective only in an environment that has a comfortable microclimate.

Sweating and excretion by animals make it necessary to have good ventilation. In the construction of animal sheds, there are two competing goals. The first is to have good ventilation. The second is to protect the animals from the physical environment, which is most easily accomplished, particularly in winter, by restricting the airflow. When constructing stalls one must try to strike a balance between these two opposing requirements.

Even an open shed offers some protection against weather. A. Raeuber [4915] showed that in a stall for calves that was open toward the south, the temperature and humidity were approximately the same as in the open. However, there was protection against rain and wind.

A closed shed results in still air and a uniform microclimate. Heating depends on the body heat of the animals. By making comparative measurements in a shed when empty and fully occupied, A. Mehner and A. Linz [4914] showed that the temperature fluctuation in the former case was one-half and in the latter one-eighth that of the open air. Figure 50-2 gives an example of the balanced type of climate prevailing in a shed. H. Wächtershäuser [4920] made measurements in a barn measuring 27 x 11 x 4 m in which there were sheds for 35 to 40 milk cows. The average diurnal temperature variation for the cold period from October to April for the years 1949-1952 was 3 to 4°C. It was hardly ever more than 6°C inside (top diagram), while outside the values were widely scattered, following the general weather conditions. The lower part of Figure 50-2 shows the frequency of various mean temperatures over the same period. In the shed, the daily average never fell below 8°C even when outside temperatures were -8°C.

The skin temperature of the animals and the longwave radiation exchange with the walls are of importance. The ultimate aim of modern stabling is stalls with controlled artificial climate similar to heated and air conditioned houses. H. Dahmen [4902] has reported one such attempt. In the tropics, it is important that animal houses have good ventilation so as to

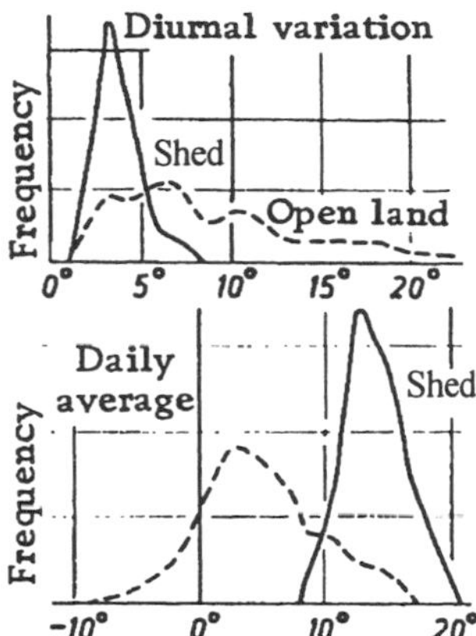

Figure 50-2. Average daily temperature variations (top) and temperatures(bottom) for four winters in a shed for 40 milk cows, compared with open land. (After H. Wächtershäuser [4920])

avoid a buildup of heat. R. M. Gatenby, et al. [4906], describes the design of animal houses in Indonesia. It is beyond the scope of this book to investigate this in more detail.

Food storage offers an example of an intentional modification to the non-growing or harvested plant environment. Winter supplies of potatoes, sugar beets, turnips, and fruits are stored in the open in clamps (produce heaped in piles covered with straw or earth to prevent freezing). The most favorable storage temperature within the hill shaped clamp is 2° to 4°C. W. Kreutz [5024] published the first temperature measurements in the study of the climate of clamps in 1948. Covering the clamp with straw heaped up with loose earth, which has poorer thermal conduction than the soil itself eliminates diurnal variations and small irregularities of temperature. A layer of winter snow will also provide added insulation. Melting snow, particularly on the southside, may increase the risk of frost by increasing the thermal conductivity of the snow. Details of clamp microclimate are to be found in F. Hummel [5019], H. Kern [5021], and W. Kreutz [5024-5026] for Germany, and in E. M. Crook and D. J. Watson [5008] for Britain. In a modification of the clamp storage system, carrots and some other root crops are sometimes left in the ground unharvested, with leaves, wood chips, or other insulating items piled on top. The root crops are then harvested through winter.

Modern agricultural meteorology now extends energy balance investigations to include the heat budget of the animal's body. In Australia, C. H. B. Priestley [5036] calculated the amount of direct-beam, diffuse solar radiation, and reflected shortwave radiation from the ground which was incident on a sheep standing under the tropical sun, as well as its longwave radiation budget, and its energy exchange with the surrounding air by means of conduction and eddy diffusion. All these factors were calculated for an idealized theoretical sheep in the form of a horizontal circular cylinder. Two layers were distinguished in the calculations, the fleece with its poor thermal conductivity, and the actual body of the sheep. The investigation made it possible to evaluate the external heat exchanges in terms of measured meteorological conditions, and hence to distinguish them from internal physiologic processes.

Microclimatological principles have also been extended to the field of biology, leading to the development of biophysical ecology, which uses the energy balance approach to examine the thermoregulatory behavior animals employ to maintain their body temperatures within an optimal range. P. N. Bartlett and D. M. Gates [5002] developed a model to determine where a lizard (*Sceloporus occidentalis*) might position itself at different times of the day on a valley oak in California in order to maintain its characteristic body temperature within an acceptable range. Good correspondence was found between the observed and predicted behavior. J. R. Scott, et al. [5038] and B. W. Grant and A. E. Dunham [5017] provide further examples of thermoregulatory studies of reptiles. D. M. Gates, a pioneer in the field, has provided a thorough review of basic principles and applications in his text [5012].

The quality and quantity of meat and dairy production, as well as the work output of animals is closely related to their physiological comfort. D. E. Ray et al. [5037] found that summer heat stress often occurred among dairy cattle in the dry subtropical climate of Phoenix, Arizona. Sustained air temperatures above the zone of thermoneutrality for the dairy cattle led to reduced reproductive performance of dairy cattle as measured by several measures of reproductive success. These negative effects were shown to be reduced by the use of evaporative sprayers in combination with shading. A positive carryover effect was also observed into the fall and spring seasons by the use of the cooling systems.

High air temperature and humidity has also been associated with reduced milk yields among dairy cattle in temperate and tropical regions. J. D. Kabuga [5022] found that Holstein-Friesian dairy cows imported from Canada to Ghana in the humid tropics experienced an average 20% decrease in milk yield, while their Ghana-born progeny experienced a 50% reduction in milk yield. He found that variations in climate only explained about 2-9% of the daily variations in milk production, however, with the majority of the reduction due to other factors such as nutrition, disease, and stunted growth. In natural settings it thus is very difficult to separate the small effects of climate alone, significant thought they may be, from the effects of other more important variables.

The transportation of living organisms is also dependent upon the climate, and has already been mentioned in the discussion of bird flight in Section 48. Other forms of movements, such as those of insects, pollen, spores, and rust are also considered in the field of bioclimatology.

Small, light insects, easily borne on the wind, exhibit a sensitivity of distribution in the atmosphere similar in many respects to that of the inorganic particles and suspended matter described in Section 17. C. G. Johnson [4814] investigated the number of aphids in the air near Bedford, England, up to heights of 600 m, by towing nets. From 151 measurements at six different heights, the density was found to be between 0.0008 and 1.8 aphids per cubic meter. The decrease with height h was found to follow an approximate index law $D = h^{-b}$, in which D is the density relative to that at a height of 1 m, h (m) is the height, and the index b varies to a marked degree with the lapse rate. With an adiabatic structure, b was approximately 1, and the density was, therefore, inversely proportional to height. With a smaller lapse rate, the density of aphids aloft was always very low. The greater the decrease of temperature with height, the higher the aphids were carried. After conversion of units, the value of the index b was found to decrease from 1.04 to 0.80 to 0.55 for a lapse rate of 1.0, 1.2 to 1.4°C/100 m, respectively. Using these figures, there is a decrease in aphid density with height as follows:

Height (m)	1	10	50	100	500	1000
for 1.0°C/100 m	1	0.091	0.017	0.008	0.002	0.001
for 1.2°C/100 m	1	0.158	0.044	0.025	0.007	0.004
for 1.4°C/100 m	1	0.282	0.116	0.079	0.033	0.022

For these reasons the decrease in aphids with height was slower in spring and summer (mean value from May to August, $b = 0.78 \pm 0.21$) than in autumn (mean of September and October, 1.25 ± 0.31). Two maxima were found in the diurnal distribution, one in the late forenoon, and one toward evening.

S. A. Isard, et al., [4813] utilized a helicopter mounted aphid collector to examine the vertical distribution of aphids between 7.5 and 1100 m during a summer in Illinois. The initiation of aphid flight only occurred when temperatures were between 12.5-25.0°C and wind speeds were below a critical threshold. The extent of aphid ascent was controlled by the thermal stability of the atmosphere, with inversions inhibiting upward movement and producing a stratification of aphid numbers.

Locusts fly in close swarms, but here too the density within the swarm is a function of the lapse rate and varies, according to the observations of R. C. Rainey [4824], between 0.001 and 10 locusts per cubic meter in Kenya and Somaliland. The swarms have a vertical spread between a few meters and several kilometers. Rainey's excellent air photographs show swarms sometimes looking like a thick fog lying on the ground, and sometimes like a veil of cumulus shape. The text by R. L. Edmonds, [5011] should be consulted for a more formal introduction to the field of aerobiology.

51 Urban Climate

Urbanization results in the transformation of natural landscapes through the clearing of natural surface covers and their replacement with single-family houses, multiple-family residential complexes, transportation systems, business, commercial, educational and governmental centers, and industrial parks. These surface alterations produce significant changes in the surface energy and water budgets, leading to the creation of distinct urban microclimates. The majority of the population in developed countries already live in cities, and are continuing to concentrate in urban-suburban complexes. At present rates of urbanization, the majority of the population of developing countries will also soon live in cities.

The pioneering work in the study of urban climates was L. Howard's [5126] examination of the climate of London, first published in the early 18th century. This was followed by A. Kratzer's [5131] monograph on urban climate in 1956. A. Sundborg [5159] published a study of the climate of Uppsala in 1951. Comprehensive studies of Vienna were later published by F. Steinhauser, et al. [5158] and in Lunz by F. Lauscher, et al. [5133]. The monograph by H. E. Landsberg [5132] provides a comprehensive review of earlier work on urban climates, while the series of reviews and bibliographies by T. R. Oke [5140-5144] provide a summary of more recent research.

The primary reason for the microclimatological changes associated with urban climates is the alteration of the surface energy and water budgets, which is caused by three basic factors. First, direct emission of waste heat, gases, and particulates into the urban atmosphere modifies the nature of the urban boundary layer and alters the surface receipt of solar and atmospheric radiation. Second, modification of the land surface cover changes the surface thermal and radiative properties, and, because the natural ground cover is largely replaced by impervious materials from which precipitation is quickly lost, changes the partitioning of surface net radiation between sensible heat and latent heat. Lastly, alteration of the surface geometry by the presence of buildings impacts surface radiative exchanges and increases the surface roughness. Secondary effects created by the modified surface radiative and free and forced convection regimes produced by these three primary factors modify every aspect of urban weather and climate (H. Landsberg [5132]).

A striking impression of the extent of atmospheric pollution in a city is obtained when it is approached on a day characterized by anticyclonic weather conditions, especially from a higher level such as an aircraft. A dark gray, sometimes colored, dust dome often lies over the city.

Figure 51-1 shows measurements made in the lowest atmospheric layer in Leipzig by A. Löbner [5135] when an ENE wind was blowing. Air blowing over areas not built up is comparatively clean, but the particulate content increases immediately upon entering Leipzig.

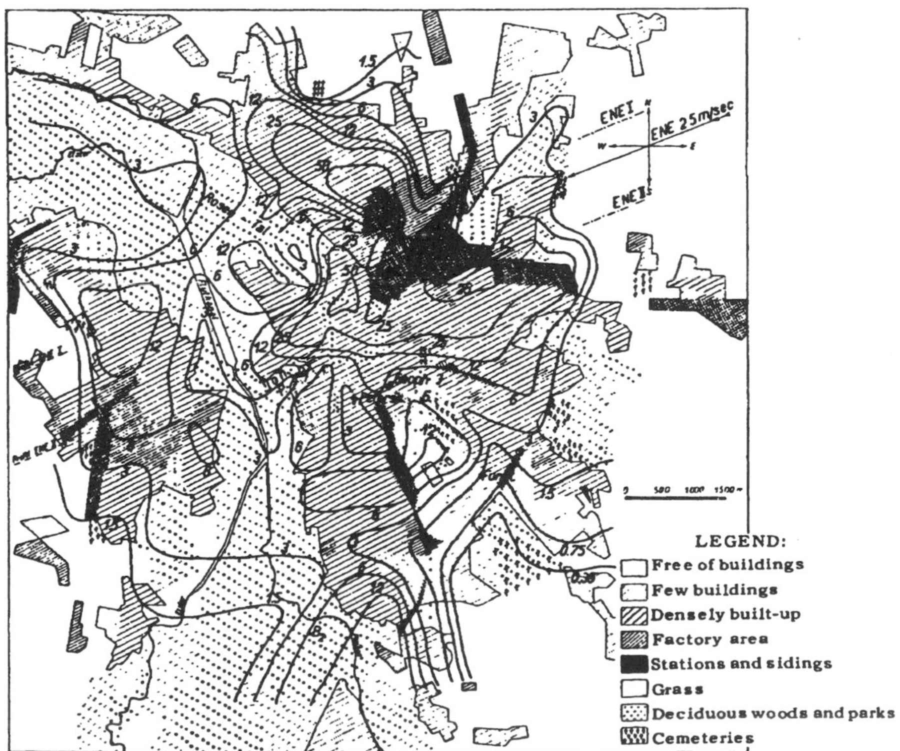

Fig. 51-1. Distribution of particles in Leipzig with an ENE wind ((number of particles x 100)/liter). (After A. Löbner [5135])

The increase is large and occurs suddenly in the area of the railroad station in the northeast part of the city. The vegetated areas of Rosental (upper left of diagram) filter the particulates out again almost as quickly. In passing through the industrial area to the west, the increase in particulates is insignificant probably because the high industrial smokestacks feed the uppermost air layer in which these measurements were made.

A later study of the spatial distribution of particulates in the Greater Cincinnati area by W. Bach [5107] revealed that the dust dome is not a monolithic feature, but can be described as a series of coalescing minidomes. A distinct spatial pattern occurs with local particulate maximums found near industrial sites, the city center, and transportation arteries. Local minimums occurring near residential, suburban, park environments, and hilly terrain with less urban development. The horizontal and vertical extent of the dust dome depends on the rate of vertical mixing, the strength of surface ventilation by the wind, and the magnitude and distribution of surface particulate sources (smokestacks, roads, and chimneys) and sinks (vegetative cover).

Figure 51-2 shows horizontal scattering coefficient profiles (and mass concentration profiles derived from them) obtained by W. Bach and A. Daniels [5109] across a 50 km automobile transect in Greater Cincinnati during two contrasting weather conditions. On 12 July 1970 (top) anticyclonic weather, characterized by calm, foggy conditions, and an elevated inversion, led to high levels of air pollution. A well-developed dust dome developed with dust loading in the city center approximately twice as high as those on the city outskirts.

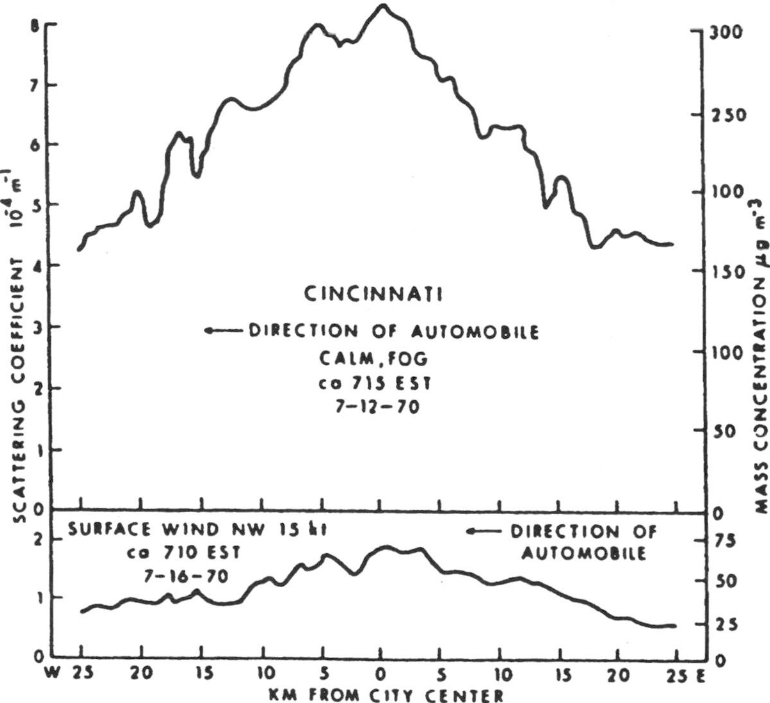

Fig. 51-2. Horizontal scattering profiles under different weather conditions obtained from automobile traverses in Greater Cincinnati. (After W. Bach and A. Daniels [5109])

During cyclonic weather on 16 July 1970 along the same transect, which featured winds from the NW at 15 knots, dust loading was only one-fourth as high as those on the previous four days. This was due primarily to the greater ventilation on 16 July. Wind erodes the spatial detail in the dust dome and reduces the level of particulates, but it does not affect the ratio of aerosol concentration between the city center and city outskirts. The dips and peaks in both sets of transects result from the effect of variable surface cover.

W. Bach and A. Daniels [5109] found that the vertical extent of the dust dome was best developed during anticyclonic conditions, but even then was only 1000-1500 m above ground level. A complicated and dynamic pattern of aerosol stratification occurs in the urban atmosphere that is strongly linked to the stability structure of the atmosphere. Alternating polluted and clean layers can occur due to stratified inversion and lapse condition layers.

More dangerous than particulates are chemically active solid and gaseous products. Their composition depends on the type of industry present. High concentration of sulfur oxides (SO_X) may be injurious to plants and human health. Figure 51-3 shows the mean distribution of sulfur dioxide in and surrounding Linz in summer (left) and in winter (right), from E. Weiss and J. W. Frenzel [5163]. The figures in small circles give the points at which measurements were made. In summer, the maximum sulfur content in Lunz is found in the industrial area in the southeast of the town. In winter, when consumption of domestic coal is higher, there is a second maximum in the heart of the city, and concentrations are higher on the whole.

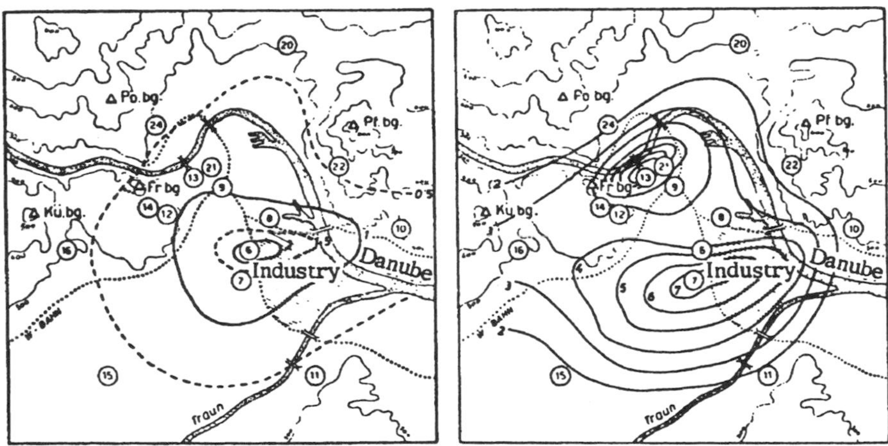

Fig. 51-3. Distribution of sulfur dioxide in Linz in summer (left) and winter (right). (After E. Weiss and J. W. Frenzel [5163])

In Taipei in the Republic of China, R. H. Aron [5102] found that sulfur dioxides were typically lower from July through November than for the rest of the year. For unexplained reasons concentrations were significantly higher on Thursdays than during the rest of the week. Carbon monoxide concentrations in Los Angeles were found to be higher during the week and lower on weekends (R. H. Aron and I-M. Aron [5105]). Carbon monoxide concentrations in Taipei, however, were found to be higher on Monday, Tuesday, Friday and Saturday and lower during the middle of the week and on Sunday (R. H. Aron [5103]).

These variations in the level of pollution concentrations are related to variations in human activity. Another example of this occurs with oxidant concentrations in the Los Angeles Basin. However, in this case, the relation is not as obvious. The precursors necessary for high oxidant levels build up for a number of days. Thus one of the best measures of a given day's oxidant concentrations is the preceding day's level (R. H. Aron and I-M. Aron [902] and R. H. Aron [5104]). In Los Angeles, automobile and other activities are less on Sunday, and fewer oxidant precursors are created. Thus, oxidant levels decrease from a peak on Friday and Saturday and reach a minimum on Monday. Concentrations then begin to build up again through the week (R. H. Aron [5104]).

The increased aerosol concentration in urban atmospheres includes an especially large number of giant nuclei with diameters between 1 and 10 μ. These suspended particulates are active as condensation nuclei, and, because many are hygroscopic, can effectively attract water for condensation at relative humidities less than 100 percent. R. Bornstein and T. R. Oke [5110] report an increased frequency of fog in urban areas as a result.

Air quality trends over the past few decades have changed considerably due to a combination of economic, technological, demographic, and public policy factors. In general, developed countries have shown a trend toward improved air quality, while developing nations have experienced declining air quality. H. W. de Koning, et al. [5130] and S. Eggleston, et al. [5118] provide a review of such trends on a global and national scale. Individual case studies of air quality trends for cities in a developed and developing country can be found in C. I. Davidson [5116] and C. O. Collins and S. L. Scott [5115], respectively.

Artificial heat is also released to the urban atmosphere from the domestic, service, industrial, and transportation sectors. Figure 51-4 shows the spatial distribution of average annual artificial heat release within Greater London for the period 1971-76. Artificial heat released ranged from less than 4.9 W m^{-2} in the outer fringes to greater than 50.3 W m^{-2} in the city center. Superimposed upon this spatial heterogeneity are temporal variations created largely by seasonal demands for space heating and diurnal variations in transport requirements. In Greater London, for example, the two daily peaks in artificial heat release at the work-related rush hours of 08:00 and 17:00 are nearly 2.5 times larger than the daily minimum at 05:00. Artificial heat generation from the urban surface across a metropolitan environment thus shows diurnal, synoptic, and seasonal temporal variations as well as spatially varying production patterns. J. D. Kalma and K. J. Newcombe [5129] describe similar energy use patterns for the two very different climate regimes provided by Hong Kong and Sydney.

The absorption of solar radiation by aerosols emitted into the atmosphere and by ozone photochemically produced in cities both lead to reduced atmospheric transmission of global solar radiation, and especially direct-beam solar radiation (S). Measurements taken in Los Angeles by J. T. Peterson and E. C. Flowers [5153] during 47 cloud-free days revealed an average 11 percent reduction in global solar radiation (0.3-3.0 μ) ($K\downarrow$) for five urban sites compared to one rural site. Daily average depletion values were between 4 and 20 percent. J. T. Peterson and E. C. Flowers [5153] report average ultraviolet (0.3-0.385 μ) depletions of 8 percent for St. Louis, and 29 percent for Los Angeles, in comparison to rural environments. Scattering of solar radiation by aerosols also leads to an increase in the diffuse solar radiation (D), as well as an increase in the diffuse fraction ($D/K\downarrow$). Since both aerosols and ozone preferentially attenuate shorter wavelengths, the spectral composition of solar radiation in urban areas also changes.

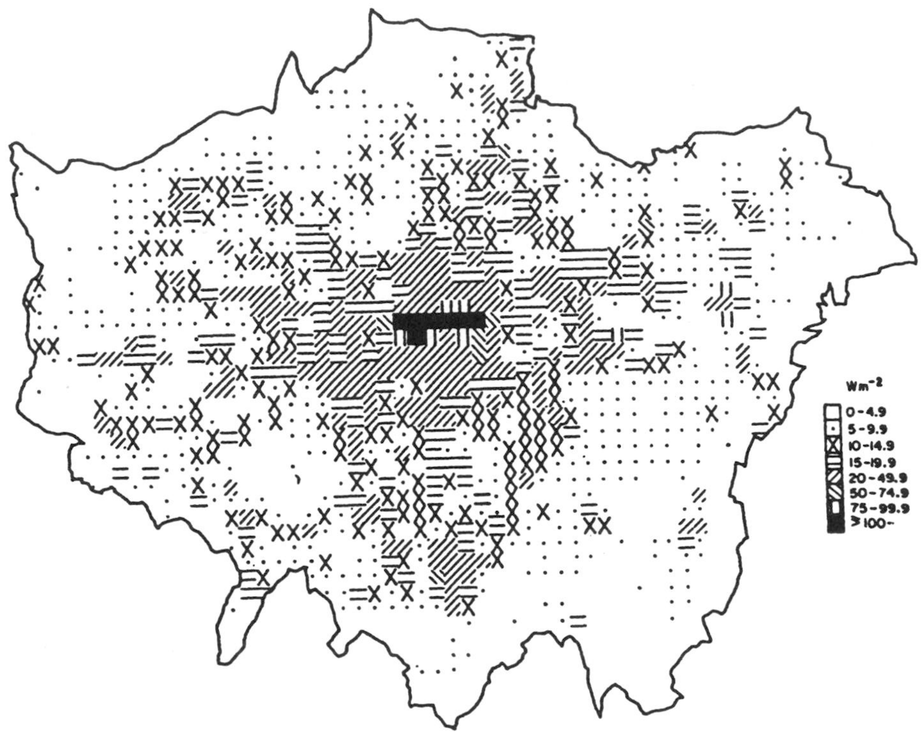

Fig. 51-4. Distribution of annual average artificial heat release from Greater London, 1971-76. (After R. Harrison, et al. [5125])

Observations of the spectral composition of global solar radiation taken during cloud-free days at Barcelona, Spain between 1989-1992 illustrate these principles. Four atmospheric turbidity classes were examined: D1 – very clear air, D2 – relatively clear air, D3 – moderately turbid air, and D4 – extremely turbid air. The results are summarized in Table 51-1 for optimal air mass (m) values between 1.1 and 1.2 corresponding to noon values in summer. Global solar radiation ($K\downarrow$) was reduced by 10% during extremely turbid conditions. Urban aerosols affected the spectral composition of the solar radiation, with greater attenuation of $K\downarrow$ occurring in the ultraviolet (UV) band as compared to the visible (VIS) or near infrared (NIR) bands. Particle scattering due to atmospheric aerosols led to sizeable increases in diffuse radiation across all spectral bands, particularly in the longer wavelengths. Though not measured in their study, direct-beam solar radiation (S) must experience sharp reductions if global solar radiation decreases and diffuse radiation increases with increasing urban aerosol. Under the clear-sky conditions (D1) shown in Table 51-1, diffuse ultraviolet radiation comprised nearly 50 percent of global ultraviolet radiation. This percentage increased and approached 100 percent with increasing optical air mass and atmospheric turbidity. Diffuse radiation thus makes a significant contribution to sunburn under conditions of higher turbidity.

Table 51-1. Global solar radiation ($K\downarrow$) and diffuse radiation (D) for selected spectral bands (nm) as a percentage of clear-sky conditions at Barcelona, Spain during cloud-free days in 1989-1992. The optimal air mass (m) range is $1.1 < m < 1.2$. (After J. Lorente et al., [5136])

Turbidity Class	UVB 300-320	UVA 320-400	UV 300-400	VIS 400-700	NIR 700-1100	Total 300-1100
$K\downarrow$						
D1	100	100	100	100	100	100
D2	96	94	92	96	94	95
D3	91	86	87	94	90	92
D4	86	73	74	91	91	90
D						
D1	100	100	100	100	100	100
D2	127	120	120	171	192	164
D3	127	138	137	224	242	209
D4	127	153	151	303	336	278

W. Bach [5108] has shown that precise determination of the magnitude of urban-rural contrasts in solar radiation is complicated by several factors. The dust dome often spreads over the surrounding countryside and can be stretched into a dust plume by the prevailing wind. Solar transmission in the urban atmosphere is a dynamic feature that varies with the optical air mass, surface production of hydrocarbons and nitrous oxides, surface emissions of particulates, degree of surface heating, wind speed and direction, cloud amount, vertical height above the ground, and atmospheric stability.

The urban albedo is also spatially variable and subject to human modification. The urban albedo depends upon the materials comprising the surface cover and the geometric arrangement of the structures making up the urban fabric. A distinction must be made between the albedo of urban materials and the effective albedo of the urban surface complex. The materials, which make up the urban complex generally have lower albedos compared to natural surfaces.

M. Aida [5101] conducted a field experiment, which measured the albedo of a flat bed of concrete blocks and compared them to the albedo of the same materials arranged in block-canyon form to simulate a series of urban canyons. The irregular arrangement of the concrete blocks resulted in multiple scattering of solar radiation between the concrete walls and, therefore, increased absorption. The effective albedo of the black-canyon system was 10 percent lower on an annual average than for the same materials arranged in slab form, and 20 percent lower in winter.

It is very difficult to determine precise urban-rural albedo differences because of the complex mosaic nature of the urban surface cover, the highly irregular urban geometry, spatial sampling considerations, and measurement errors introduced by the urban aerosol layer. T. Takamura [5161] obtained measurements of solar reflectance along a 5 km helicopter flight path above Tokyo during clear days in winter, and compared them to similar measurements taken above agricultural land near Tsukuba. Although significant differences were found among the primary land use types present in Tokyo, a general decrease in solar reflectance was noted above the urbanized areas.

The approximate 10 percent decrease in global solar radiation and effective albedo experienced in urban areas generally offset one another, resulting in only very minor urban-rural differences in absorbed solar radiation (T. R. Oke [5146]).

Nighttime counterradiation ($L\downarrow$) above cities is generally 5-10 percent greater than for rural areas, with somewhat larger differences for daytime values. T. R. Oke and R. F. Fuggle [5149] measured air temperature and counterradiation along an automobile traverse route on the Island of Montreal on clear nights during a variety of seasons in 1969-70. They found that the increased nighttime counterradiation found over the city, which averaged 6-12 percent and was as much as 25 percent in extreme cases, could be entirely explained by the higher urban air temperatures, since the majority of counterradiation received at the surface is emitted by the lowest 100 m of the atmosphere. During the day, when solar absorption by the aerosol layer is important, the presence of the dust dome may also help to enhance counterradiation.

The general waterproofing of the city and presence of materials which evaporate only immediately following wetting results in much warmer urban surface temperatures than in rural areas. Adequate determination of this parameter was not available, due to spatial sampling considerations, until the advent of airborne and satellite thermal imagery. Recent studies, however, have shown that urban surface longwave emissions ($L\uparrow$) are 5-12 percent greater at night, and as much as 20 percent greater during the day. M. Roth, et al. [5154] analyzed mid-day thermal imagery from Vancouver, Seattle, and Los Angeles and observed a close relationship between land use cover and surface temperature patterns. Average daytime urban-rural surface temperature differences were between 5.5-7.5°C, while similar cold season values were between 1.0-2.7°C. K. P. Gallo, et al. [5120] have shown that there is an inverse linear relationship in urban regions between minimum surface temperatures and the degree of vegetative cover. Urban-rural surface temperature differences are greatest during the middle of the day, while urban-rural air temperature differences are maximum at night.

Interestingly, each of the four radiative terms ($K\downarrow$, $K\uparrow$, $L\downarrow$, $L\uparrow$) show significant urban-rural differences. As with solar radiation above, however, the effects largely offset one another, so that urban-rural net radiation (Q^*) differences are negligible (T. R. Oke [5146]).

Although the receipt of net radiation varies little between urban and rural environments, the partitioning of that energy into evaporation (Q_E), convection (Q_H), and conduction (Q_G) shows marked urban-rural contrasts. In general, evaporation is lower in urban areas due to the reduced vegetative cover and drier surfaces, convection is higher because of the drier surfaces and increased aerodynamic roughness, and conduction is higher due to the increased building mass and higher thermal conductivities of the building materials. These are generalities which only apply to average values. In reality an enormous variety of urban surface energy budgets can be obtained. This arises from several factors. First, the urban landscape is a complex mosaic of numerous land uses, each of which is characterized by great heterogeneity of surface physical properties, including thermal conductivity, thermal capacity, albedo, emissivity, leaf area index, stomatal resistance, surface roughness length, and zero-plane displacement. Second, the partitioning of urban net radiation is extremely dynamic, not only because of synoptic weather changes, but also because of anthropogenic irrigation of the urban and suburban landscapes. Finally, urban surface energy budgets are characterized by continual local advection of heat and moisture from dry surfaces onto wetted or vegetated surfaces. Two reviews of the urban energy balance by T. R. Oke [5140, 5141] should be consulted for a more detailed discussion of recent research.

Two case studies illustrate the great contrast of urban energy budgets. M. Nuñez and T. R. Oke [5139] examined energy balance measurements from an urban canyon comprised of two opposing walls and a canyon floor in Vancouver. Average totals for 9-11 September 1973 revealed that 60 percent of net radiation (Q^*) was lost via convection, 30 percent by heat storage within the canyon, and only 10 percent by evaporation, resulting in an average Bowen ratio of 6.0. In contrast, H. A. Cleugh and T. R. Oke [5114] conducted simultaneous energy balance measurements from suburban (rural) sites in Vancouver from 18 July to 22 September 1983, and found that average Q_H, Q_G, and Q_E totals were 44 (30), 22 (4), and 34 (66) percent of radiation. The average suburban Bowen ratio was 1.28, in comparison to a rural average of 0.46. These average values, however, hide a considerable degree of day-to-day variation. H. A. Cleugh and T. R. Oke [5114], for example, found that during this period, average daily Bowen ratios for the rural site ranged between 0.35 and 0.55, while the corresponding suburban values varied from 0.65 to 1.80.

Information concerning seasonal changes in the urban energy budget can be found in C. S. B. Grimmond's [5124] study of the suburban energy budget for Vancouver in winter and spring, while H. P. Schmid, et al. [5155] provide a discussion of the spatial variability of the suburban energy budget in Vancouver.

The most thoroughly documented aspect of urban climate is the air temperature anomaly known as the urban heat island. Taken over the whole year, a city is warmer than the surrounding countryside by 1-2°C. Under a high pressure system with light winds, this difference may become considerably greater. An observant city dweller may be able to observe this temperature difference from the behavior of plants, such as an earlier date of spring budding or a later date for leaf fall in autumn, from the faster rate of snowmelt in cities, or the greater likelihood of precipitation in cities in the form of rain than snow.

Figure 51-5 shows the essential features of the urban heat island from a study of Karlsrude by A. Peppler (cited in A. Kratzer [5131]). Map A shows the arrangement of isotherms on the midsummer evening of 23 July 1929 from air temperatures measured along the automobile traverse routes shown. Graphs B and C show the west-east and north-south profiles, respectively, during mid-day (b) and the late evening (a). On 23 July it was up to 7°C warmer in the city center than in the outskirts of town. The magnitude of the urban heat island varies with time and is closely linked to comparative urban and rural cooling rates. In their study of the relationship between urban heat island intensity and urban and rural cooling rates in Montreal, Quebec, and Vancouver, T. R. Oke and G. B. Maxwell [5151] found that rural areas generally cool rapidly following sunset, while urban locations cool at a lesser and more gradual rate. Consequently, the maximum urban heat island intensity is normally reached 3-5 hours following sunset (in all seasons). Evidence of this can be seen by comparing the temperature profiles obtained for Karlsrude from 1315-1445 and 1900-2130 in Figure 51-5. The urban heat island displays significant variations over the diurnal, seasonal, and annual time scales, as shown for London by T. J. Chandler [5111].

Possible causes of the urban heat island were speculated upon long before the energy and water balances of urban environments were adequately understood. T. R. Oke [5145] has provided a review of the current understanding of causal mechanisms. These include reduced longwave radiative cooling from within urban canyons because of smaller sky view factors, increased solar absorption due to a lower effective albedo caused by multiple reflections in urban canyons, greater daytime heat storage and subsequent nocturnal heat release due to the thermal properties of urban materials, increased Bowen ratios resulting from the

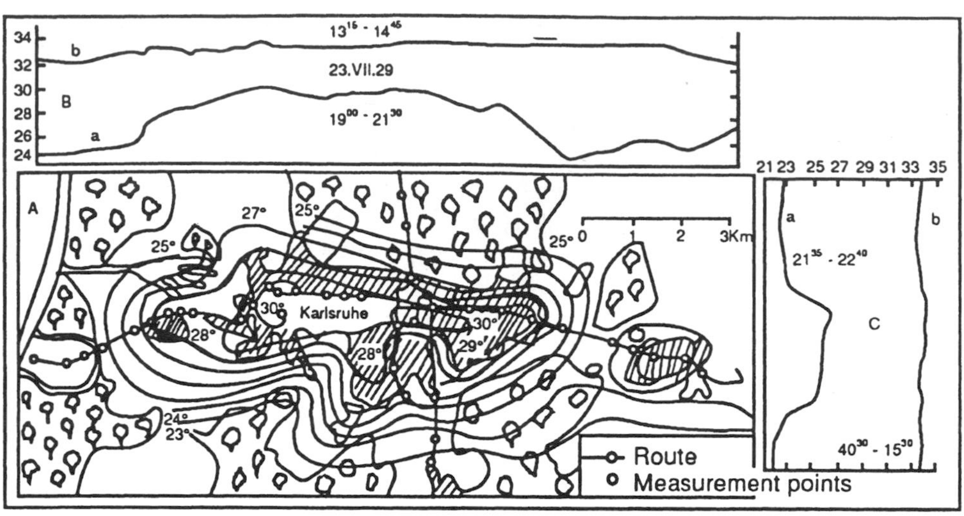

Fig. 51-5. Distribution of air temperature in Karlsrude on a hot summer day. (After A. Peppler [5131])

removal of vegetative cover and general surface waterproofing, release of anthropogenic heat directly and indirectly into urban canyons, and reduced atmospheric mixing due to poorer ventilation caused by the greater surface roughness. LANDSAT Thematic Mapper data can be used to gather information about the magnitude and extent of the urban heat island. J. E. Nichol [5138] analyzed Singapore's heat island using this technique.

T. R. Oke, et al. [5150] conducted a modeling study of an urban canyon on calm cloudless nights in an attempt to quantify the significance of each factor on the nocturnal urban heat island. Their results indicated that the effect of urban geometry (sky view factor) in reducing longwave cooling and of increased daytime heat storage and subsequent nighttime release were the primary, and nearly equal, causative factors. Winter release of anthropogenic heat can also be an important factor in selected high latitude cities during winter.

Decreased frictional mixing caused by the greater urban surface roughness, and increased convective mixing resulting from the urban heat island also modify urban wind speed, wind direction, and atmospheric stability patterns. From his analysis of hourly wind speeds around London during 1961-70, D. O. Lee [5134] found that average urban wind speeds were reduced by 20 percent at night, and 30 percent during the day. With strong gradient winds, urban winds were decelerated because of greater frictional drag, while with light gradient winds, urban winds were often accelerated as a result of increased vertical mixing due to greater convective and frictional mixing. Urban acceleration of winds were most common at night under anticyclonic weather conditions when gradient winds were light, the urban heat island was strongly developed, and urban-rural atmospheric stability differences were greatest. D. O. Lee [5134] found that the critical wind speed separating urban acceleration/deceleration was greater at night than during the day (2.0-2.5 m sec^{-1} at night, as compared to 1.1-2.1 m sec^{-1} during the day), and varied with the seasons. D. Pearlmutter et al. [5152] analyzed the attenuation of wind speed in an urban area with respect to the relative wind direction and the axis of the streets. He found that when the streets

were nearly parallel to the wind direction the wind was about 66% of the free air flow. However, when the streets were perpendicular to the wind the speed was attenuated more severely to about 33% of the free flow.

Wind direction can also be modified over cities due to changes in the Coriolis effect since deceleration of winds produces cyclonic turning, and acceleration of winds leads to anticyclonic turning. W. Bach [5106] describes how an urban circulation regime can occur under anticyclonic weather conditions with calm winds and clear skies, and a strong urban heat island. The urban heat island can initiate thermal updrafts over the city which then descend over the countryside. A small thermally-induced pressure gradient results which leads to the formation of a country breeze. Unlike the other local-scale shallow closed circulation systems described in Section 43, the urban circulation does not exhibit a nighttime reversal of direction.

Increased urban convective and frictional mixing is also responsible for the precipitation anomalies that have been observed in many urban areas. Results from the Metropolitan Meteorological Experiment (METROMEX) study of St. Louis have helped clarify the nature of the urban precipitation changes and their physical causes. S. A. Changnon, et al. [5112, 5113] identified maximum precipitation enhancements during winter, spring, summer, and fall of 0, 4, 25 and 17 percent, respectively. These effects are not centered over the city as, for example, is the urban heat island, but are displaced toward the downwind half of the city and the neighboring countryside. Urban precipitation anomalies are greatest during summer, and are most noticeable during the afternoon period of maximum surface heating and greatest atmospheric instability. The effects are most evident in convective processes, such as severe thunderstorms, squall lines, cold fronts, and air mass thunderstorms, and are least or not evident, for stratiform events. An increase in severe thunderstorm related activity, such as heavy precipitation, hail, and strong gusts have also been noted.

The most probable cause of the increased urban precipitation are the greater daytime surface heating, which leads to heightened convective activity and a deeper mixing layer and the mechanical effects of the urban structures which produce convergence and uplift above the city. Changes in the number and size distribution of cloud condensation and ice nuclei, as well as the reduced rate of frontal passage over cities due to greater frictional drag, may also be contributing factors.

In a study of precipitation in Detroit, R. C. Snider [5157] noted that while there was increased precipitation in Detroit as compared to its surrounding rural environment, the number of days with precipitation were less. He suggested that on days when precipitation was light, the higher urban temperatures and lower relative humidities resulted in the precipitation evaporating before it reached the surface.

In general, the topography in and around a city (Chapter VII) will be a deciding factor in determining the extent to which the influences described above will be effective. It is only natural that a city in a sheltered location such as a valley, where winds are light, will have a greater difference in climate between its center and its outskirts than a city situated on a plateau exposed to winds. Locations on mountain slopes or on coasts will also favor certain aspects of urban climate. The interactions between urban climates and topoclimates are discussed more fully in Y. Goldreich [5123].

B. Givoni [5119B] states that "Many features of the physical structure of the city can affect the urban climate. ... it is possible to modify the urban climate through appropriate design so as to improve the comfort of the inhabitants, both outdoors and indoors, and to reduce the energy demand of the buildings for cooling in summer, as well as for heating in winter." He further suggests that natural ventilation is very important for controlling summer cooling requirements, and thermal stress. In locating a city windward slopes are thus preferable to leeward. With respect to the street layout he states

> "The best ventilation within the streets and the sidewalks is achieved when the streets are parallel to the direction of the prevailing winds during the afternoon hours (when the urban temperature reaches its maximum). However, when streets are parallel to the wind direction and the buildings along them are very close together and face the street, the ventilation potential of the buildings is compromised. The reason is that with this orientation, all the walls of the building are in "suction" zones. Effective indoor cross-ventilation can occur in a building only when at least one of its walls (and windows) are in a "pressure" zone."

Streets perpendicular to the wind direction may block the wind and retard urban ventilation. Thus B. Givoni [5119B] states that

> "A good street layout from the urban ventilation aspect in a hot-humid region [or time of the year] is when wide main avenues are oriented at an oblique angle to the prevailing winds (e.g. between 30 and 60 degrees). This orientation enables penetration of the wind into the heart of the town. The buildings along such avenues are exposed to different air pressures on their front and back facades. The upwind wall is at a pressure zone while the downward wall is at a suction zone thus providing the potential for natural ventilation."

He further suggests that protection for pedestrians against the sun can be provided by buildings with overhanging roofs, colonnades in which the ground floor is set back from the edge of the road, with upper stories jutting out or covering walks with awnings, or eaves. White colored walls will reduce summer indoor heat load (while dark colored walls will reduce winter heating requirements) but will increase outside glare. Urban vegetation also affects urban temperatures and environmental conditions. For more than a century, popular garden books have been reminding homeowners of the advantages of strategic placement of trees and shrubs (particularly deciduous) to give summer shade, allow access to the winter sun, and break winter winds.

Trees in urban areas produce shade, add moisture to the air, reduce surface nocturnal radiation losses, filter particulates and other pollutants, intercept rain and snow, reduce noise (Section 39), and through evapotranspiration cool the air and raise the relative humidity. In his paper on urban forests T. R. Oke [5147] pointed out that the thermal pattern produced by urban parks is usually most pronounced under calm clear nights. The measurements in Figure 51-6 were taken under these conditions and show that the urban park temperatures are between those of urban and rural areas. The urban, urban park, and rural temperature differentials were established shortly after sunset after which the three areas cool at somewhat similar rates (T. R. Oke [5147]). Typically, the urban park nocturnal temperatures are not more than 1-2°C cooler than surrounding areas. Figure 51-7 shows that the effect of parks on nocturnal temperatures in Vancouver, British Columbia (a) and in Montreal, Quebec (b) can extend for a considerable distance beyond the park itself. T. R. Oke [5147] hypothesized that this might generate a cool air outflow from the park (park breeze) with warm air from the city returning aloft.

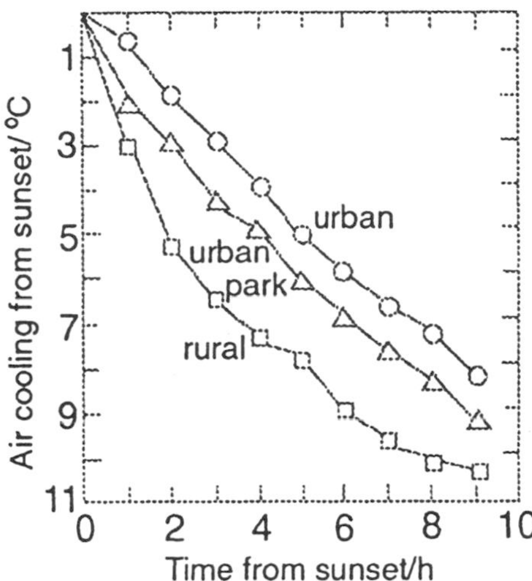

Figure 51-6. Nocturnal cooling of urban (city centre), urban park and rural environments in Vancouver, British Columbia. Values are derived from automobile traverses on three near calm, cloudless nights in August 1971. Data for each environment are normalized to their respective sunset temperatures. The parks were 1.1°C, and the rural area more than 5°C, cooler than the urban center at sunrise. (After T. R. Oke [5147])

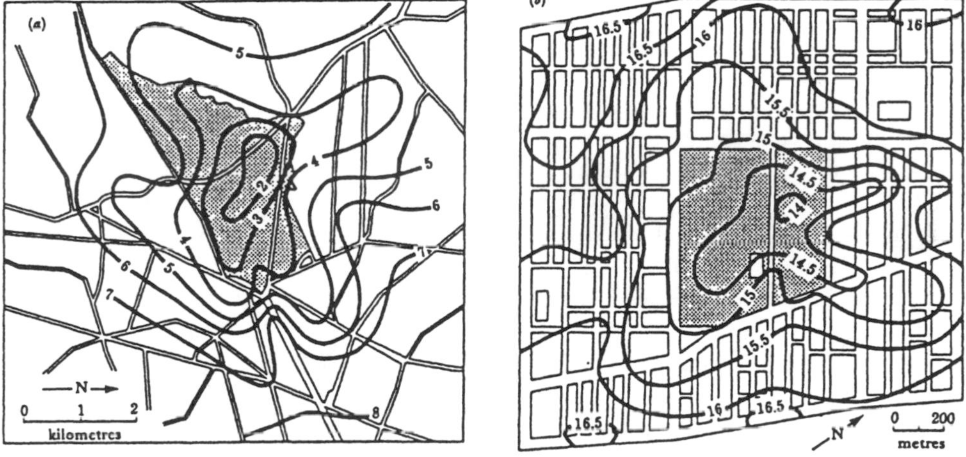

Figure 51-7. Distribution of near-surface air temperature (°C) in the vicinity of urban parks. *(a)* Chapultepec Park, Mexico City (500 ha) with calm cloudless conditions at 05:28-06:48 on 3 December 1970 (After E. Jauregui [5128]). *(b)* Parc LaFontaine, Montreal (38 ha) with cloudless skies and winds of 2 m sec^{-1} from the southwest at 20:15-21:15 on 28 May 1970 (From T. R. Oke [5147]). (Park area shaded.)

H. Upmanis et al. [5283] found that both the magnitude of the park urban temperature differential and the extension of the cool park nocturnal climate into the built-up area surrounding the park depended upon the size of the park. The cooling effect of the park reached a maximum about one park width from the park's edge, but this nocturnal cooling influence decreased with increased distance from the park. The distance of the park's cooling effect into surrounding areas may be influenced by a number of factors including topography and the pattern of built-up areas surrounding the park, which may impede park air from intruding into the built-up areas.

The importance of evapotranspiration is shown in Figure 51-8 in which Golden Gate Park stands out as 8°C cooler than the surrounding city. H. Upmanis and D. Chen [5162] in their study of nocturnal park-urban temperature differences found that the maximum temperature difference typically occurs two to three hours after sunset. The only geographical factor affecting this difference was the distance from the park border. The meteorological factors that influenced this difference were wind speed and cloud cover. Wind direction had no effect.

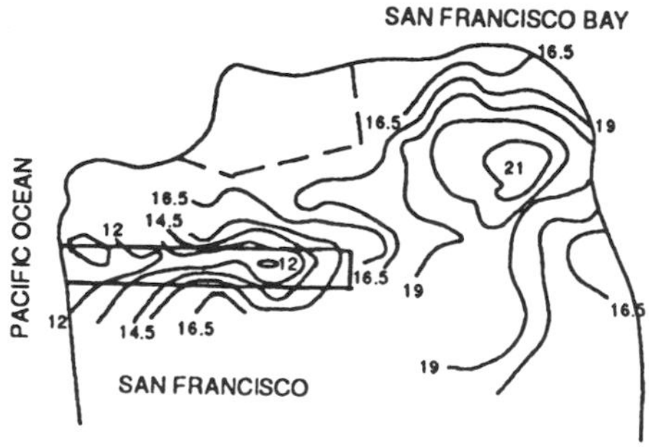

Figure 51-8. Temperatures in San Francisco at 23:00 on March 26, 1952. (After F. S. Duckworth and J. S. Sandberg [5117])

Figure 51-9 compares the temperatures in three streets in Vienna on a warm calm summer day. C. A. Federer [5119] suggests that "For narrow streets orientation also plays an important role. North-south streets will not be shaded from intense noontime solar radiation. They will have higher temperatures than narrow east-west streets." D. Pearlmutter et al. [5152] pointed out that during the day shading by buildings can actually create "cool islands" in a desert environment. We agree with C. A. Fender [5119] conclusion that "... more research is needed before the effect of parks, greenbelts, rooftop gardens and isolated trees on the temperatures can be predicted quantitatively." The studies by Y. J. Huang et al. [5127] and E. G. McPherson et al. [5137] provide good examples of recent work on these problems.

Plants can also have an affect on urban air. First, vegetation is an important sink for particulates, ozone, sulfur dioxide, nitrogen dioxide, carbon monoxide and heavy metals (J. A. Schmid [5156]). With respect to particulates, conifers generally are more effective filters than broadleaf deciduous trees. Besides filtering air pollution, large expanses of vegetation, as in a major park, represent an area of lower surface runoff. These areas can also reduce surrounding air pollution by dilution (J. A. Schmid [5156]). For further understanding of climate in an urban setting the following articles are recommended (H. Swaid [5160], T. R. Oke [5142, 5147, 5148] and B. Givoni [5121, 5122]).

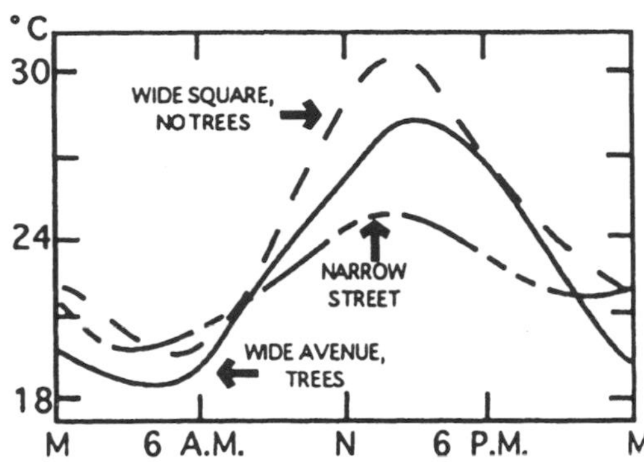

Figure 51-9. Diurnal temperature in Vienna on 4 and 5 August, 1931. (From A. Kratzer [5131])

52 Artificial Protection against Wind

Shelterbelts (rows of trees or plants) and windbreaks (any nonliving fence or structure) can have significant effects on reducing wind speed. These protections from the wind may also have significant effects on the microclimate. As suggested by C. J. Stigter [5278] the primary direct factor of wind causing physical stress and damage to the plant is the asymmetrical air pressure acting on the plant. Damage to the whole plant is caused by swaying, shaking, bending and lodging, while damage to plant parts is by premature fruit, leaf, and flower shedding, breakage, bruises lesions, and abrasion. Secondary direct wind-related injury is a result of soil particles carried by the wind abrading the plant, rubbing of adjacent leaves, the carrying of pathogens and insects, evaporation and soil moisture depletion, and soil erosion.

We shall begin our discussion with the modification of the wind field, assuming that the ground is level. Shelter from wind is generally provided by long narrow strips set up perpendicular to the direction of the prevailing wind. The length, breadth, height, and material used may vary considerably. Trees with and without undergrowth, hedges, and rows of bushes, stone walls, sunflowers or other tall plants, lath trellises, straw matting, fences of woven reeds, wire netting with a small mesh, and board fences or walls are used. Figure 52-1 (W. Nägeli [5249]) shows the average braking effect on the wind within a shelterbelt 75 m wide consisting of a mixed stand 20 m in height for three wind speed groups. The two-fold effect of any loose shelterbelt can be recognized; the trunks, branches, leaves, or needles reduce the wind speed by friction, while at the same time as the paths open for air flow are reduced or narrowed the wind speed increases. It, therefore, appears, particularly with stronger winds, that the braking effect is delayed, and there are even cases in which an increase of wind has been measured in the entry zone. Rows of single trees sometimes produce a kind of "funnel" effect. In the investigations on Mount Aso by K. Sato, et al. [5266], the greatest wind speed was recorded inside the 16 rows of trees of the shelterbelt because the first branches of the cypress trees were several meters above the ground. The greatest shelter slowing of the wind was produced 50 m behind the belt.

The extent to which a single shelterbelt is able to reduce the wind speed depends on wind direction, and the height, breadth, and nature of the shelterbelt. In all cases, the protec-

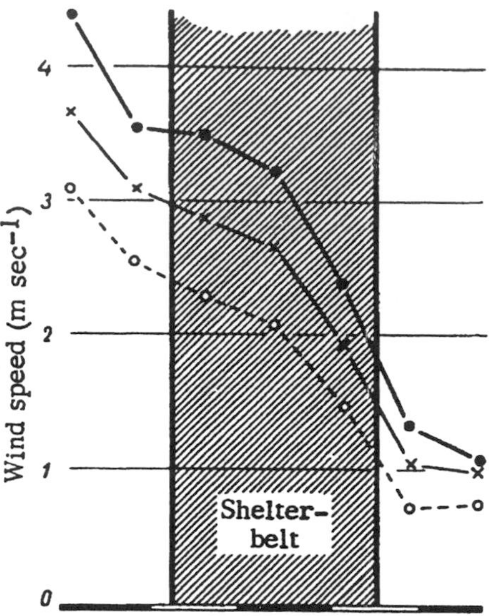

Figure 52-1. Slowing of wind in a shelterbelt of trees 20 m high. (After W. Nägeli [5249])

tive effect is a function of distance from the shelterbelt. It is substantially greater downwind than upwind. All investigations have shown that the effects downwind are proportional to the height h of the shelterbelt and to the wind speed u. Distances x in front of the shelterbelt (-) or behind it (+) are expressed in dimensionless units of h; that is, the quantity used is $\pm x/h$ (multiples of the height of the belt). Wind speeds are expressed as percentages of the wind speed in the open. Figure 52-2 illustrates the shelter effect of various types of shelterbelts for winds blowing at right angles to them. They are based on measurements made in the open by W. Nägeli [5249] at 12 different shelterbelts in Switzerland at a height of 1.4 m.

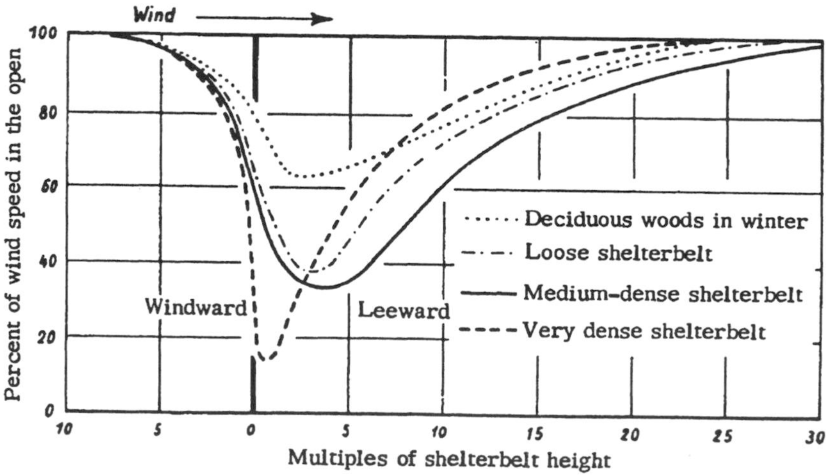

Figure 52-2. The effect of a shelterbelt as a function of its penetrability. (After W. Nägeli [5249])

The denser the obstruction, the greater the wind reduction immediately behind it, but the downward effect diminishes more rapidly (Figure 52-2). Thus a very dense shelterbelt reduces the wind speed by the greatest amount just behind the shelterbelt. On the other hand, the medium dense shelterbelt, while less effective just behind, is more effective downwind (Figure 52-2). According to M. Jensen [5227], the most favorable practical results are obtained with a penetrability of 35 to 40 percent; H. Blenk and H. Trienes [5201] found it to be from 40 to 50 percent. N. J. Rosenberg, et al. [5264], however, suggested "...for best wind reduction and greatest downwind influence the wind break should be more porous near the ground where the wind speed is lowest. Ideally, the density of the barrier should increase logarithmically with height in accordance with the wind speed profile." It follows from the above that the shelter effect of a dense forest does not extend as far as that of a narrow penetrable strip of trees (W. Nägeli [5251]). G. M. Heisler and D. R. DeWalle [5224] suggest, however, that there is a larger difference in turbulence generated between solid barriers and slightly porous barriers than between slightly porous and very porous barriers. The shape of the openings has not been found to have any effect on the reduction of the wind speed. Shelterbelts without foliage have, according to M. Jensen [5227], 60 percent of the effect they have in summer when in leaf. J. van Eimern [5215] made detailed comparative measurements in summer and winter in a double row of maples. He also found that a change in wind direction of $\pm$ 45° from the perpendicular did not produce any significant change in the shelter effect. The wind structure and turbulence characteristics behind a dense shelterbelt differ from those for a porous barrier in that the wind forced over the solid barrier is pulled back down in the lee of the barrier (Figure 52-3), resulting in a flow reverse to the general wind flow. R. C. Schwartz et al. [5269] pointed out that shelterbelt porosities less than 20 percent resulted in a reverse flow and that decreasing barrier porosity below 20% did not have any effect on wind speeds. The porous barrier, in allowing air to penetrate, breaks up the larger eddies and prevents the wind from being drawn back down to form a lee eddy.

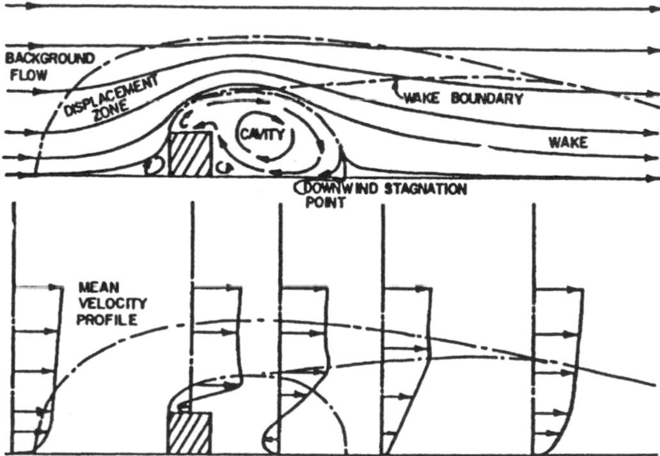

Figure 52-3. Wind flow in the vicinity of a building or impenetrable barrier. (After J. Halitsky [5223])

All the results mentioned up to now have been based on wind measurements made with instruments placed 1 to 1.5 m above the ground, high enough to be free of most effects due to the ground surface, but low enough to measure the conditions in which lower growing crops would be exposed. Figure 52-4 gives an example of the measurements made by W. Nägeli [5250] at nine levels up to 8.8 m above the ground, on both sides of a reed screen 2.2 m in height in the Fur Valley near Zurich. H. Wang and E. S. Takle [5286] use wind tunnels to investigate the wind flow through and over shelterbelts with varying porosities.

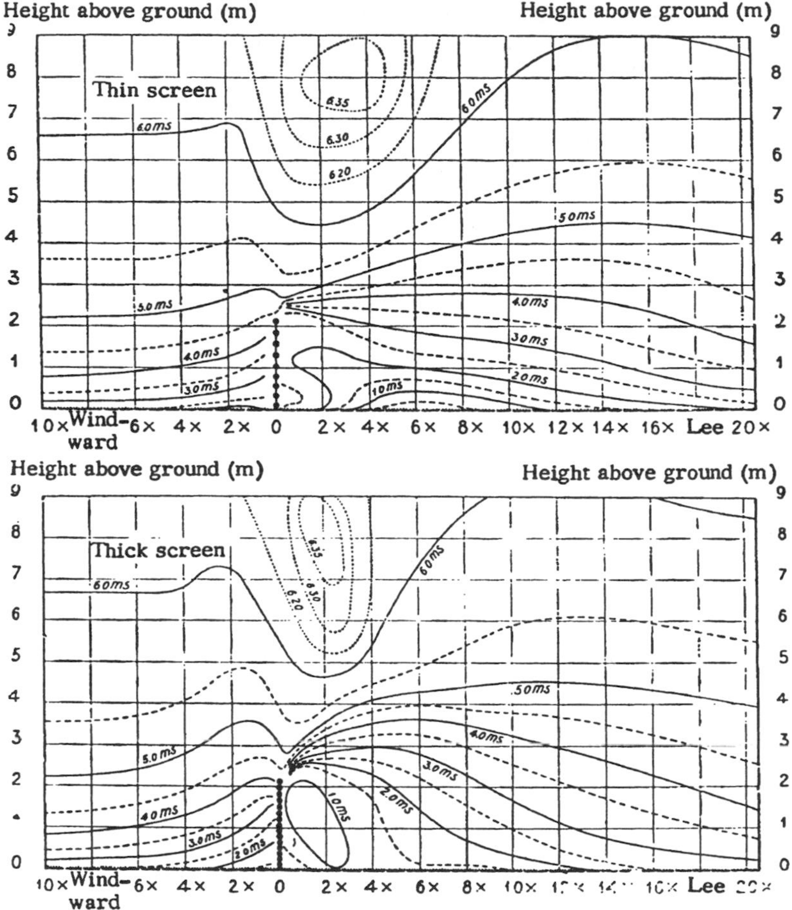

Figure 52-4. Wind field around two reed screens of different density.(After W. Nägeli [5249])

The lines of equal wind speed (isotachs) in the upper diagram are for a reed screen with a penetrability of 45 to 55 percent, and the lower diagram for 15 to 20 percent penetrability. The zone of increased wind speeds above the screen can be followed up to more than 9 m and is displaced slowly toward the lee as height increases. This effect higher up in the diagram is accentuated because intermediate isotachs have been inserted.

The lower diagram shows the stronger shelter effect in the neighborhood of the denser reed screen. The isotachs rise more steeply, are crowded more tightly above the obstruction, and there is a closed isotach of 1 m sec^{-1} in the lee eddy. However, in conformity with the basic rule, the isotachs bend down again toward the ground more quickly, showing that the denser screen has a lesser effect at a distance.

The length of time observations have to be continued in order to obtain a representative mean was the subject of experiments made by J. van Eimern [5215] with the double row of trees. The results illustrated in Figure 52-5 are for measurements made at a height of 3 m. They are summaries of the summer conditions at distances of 15, 48, and 100 m with winds perpendicular to the orientation of the shelter. The abscissa is the wind speed in the open, and the ordinate is the wind speed in the shelter, expressed as a percentage of the former. Each dot corresponds to an hourly average, and the symbols used indicate the time of day. The points shown below the hatched line on the left lie below the limits of the accuracy of measurement.

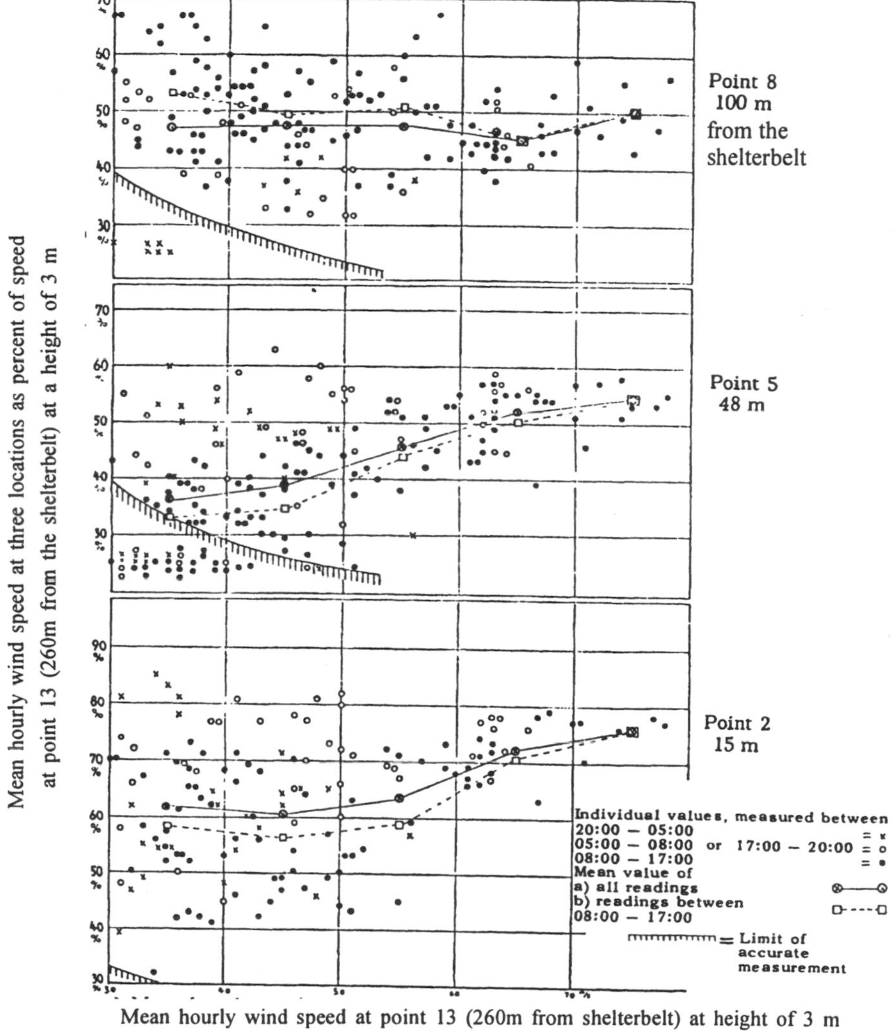

Figure 52-5. Wind measurements at three points behind a double row of trees in foliage. (After J. van Eimern [5215])

The scatter of hourly averages is greater with weak winds. Similar measurements made in winter gave only half the amount of scatter. Analysis of the results in this case showed that, for a given wind direction, at least 20 hours of recording were required to collect sufficient data for evaluation. The broken line, joining daylight values only, show a relatively lower speed at a distance of 15 m and a relatively higher speed at 100 m than the average of all readings. Therefore, the row of trees appear to behave as if it were denser during the day and less dense by night. The reason for this is the more unstable structure of the air during the day. H. R. Scultetus [5270, 5271] confirms this effect of temperature stratification and shows that the shelter effect will also depend on the type of air mass.

Wider tracts of land are often shielded by a number of shelterbelts in succession. Figure 52-6, which is based on W. Nägeli's measurements [5252], illustrates the protective influence of additional belts of trees. H. Kaiser [5230] called this the "back coupling" of shelterbelts. It is difficult to determine the most effective distance between two shelterbelts. There are many factors at work, such as the nature of the shelterbelt, the frequency distribution of wind directions, local features, and the topography of the area. In a later article W. Nägeli [5252] looked at the effects of back coupling with shelterbelts at different spacing. As might be expected with closer spacing the wind remains lower. However, the maximum reduction of each successive belt is somewhat less (Figure 52-6). When the wind direction is not perpendicular to the shelterbelt the reduction of the wind speed is less.

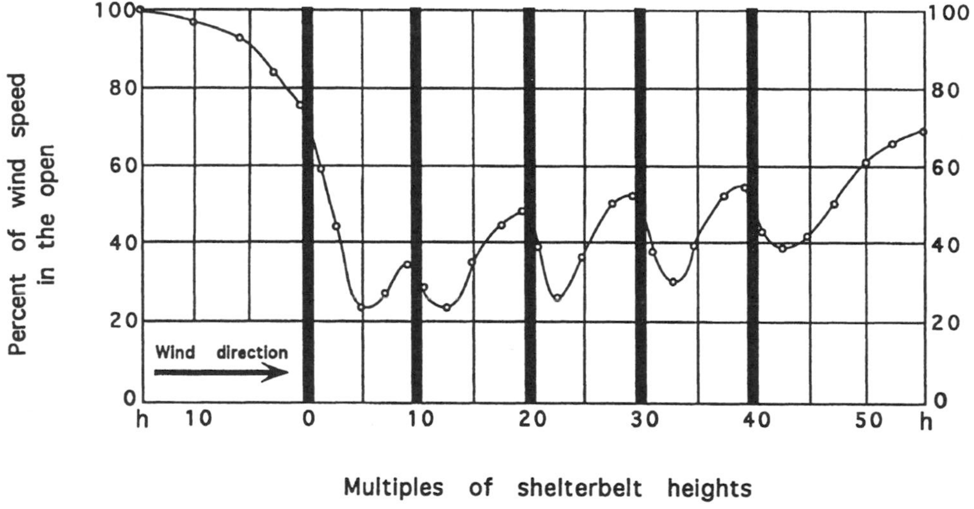

Figure 52-6. The effect of shelterbelts arranged behind one another. (After W. Nägeli [5252])

Experience shows that in passing from the sea where friction is small, the wind field becomes adjusted to the greater friction over land after crossing a coastal transition zone. A similar kind of modification takes place where there is a change from unprotected to protected areas. Shelterbelts established one behind the other increase the aerodynamic roughness of the land, and have been shown by experience to have a substantially more favorable effect than might be expected from the individual experiments described above.

A more favorable arrangement is that of a network of shelterbelts perpendicular to each other, in which the orientation of the belts is of fundamental importance only when the ground is not level. In this case, the distance between the belts can be increased greatly: the shelter effect is more uniform and is generally least near the center of the network. The corners provide positions where it is particularly suitable to make the openings necessary for access. The hope that it might be possible to use rows of high trees as "wind steering lines," to guide the wind into paths that would be convenient for the protective hedges, ignores the turbulent character of the wind field.

P. Todhunter has observed that both back coupling and perpendicular orientation of shelterbelts are employed in the Red River Valley of North Dakota to minimize wind erosion. The Red River Valley, a former glacial lakebed, is intensively farmed and experienced severe wind erosion during the Dust Bowl Era of the 1930s. Following that era, extensive planting of shelterbelts was undertaken to reduce soil loss. The original settlers in the region were given a quartersection of land (0.25 square miles or 160 acres) as an inducement to move to the Dakota Territory. Although modern farms in the region are now much larger than 160 acres, the quartersection remains the basic field unit. By planting shelterbelts of 90° angles along the northwest corners of fields, farmers are able to shelter their fields from oblique winds. The density of shelterbelts (P. E. Todhunter and L. J. Cihacek [5281]) in this once near-treeless grassland is now among the highest in North America, and wind erosion has been greatly reduced.

Wind tunnel shelterbelt simulations (H. Wang and E. S. Takle [5286, 5288 and 5289] demonstrate that the effect of shelterbelts in reducing wind speed varies little with changes in their width, internal structure or external shape. Their results show that wind speed decreased by only 15-18 percent as the width increased by a factor of 100. Differing internal structures and shape had even less effect on wind speed. However, width greatly affects the location of the minimum wind speed, pressure perturbations and permeability of the shelterbelt. The location of the minimum wind speed moves closer to the shelter as the width increases (H. Wang and E. S. Takle [5286 and 5288]). H. Wang and E. S. Takle [5287] also studied the effect of shelterbelts on wind speed reduction when the wind is blowing at an angle oblique to the shelterbelt (not 90°). They found:

1. Wind reductions for oblique flow resemble those for 90°. However, the channeling of the wind with oblique flow reduces the sheltering effect especially for dense shelterbelts in the middle lee. In this situation winds may exceed undisturbed wind speed values.
2. The location of the minimum wind speed (maximum reduction) moves closer toward the shelterbelt with increasing obliqueness (increasing deviation from 90°).
3. The sheltered distance in oblique flow decreases with increasing obliqueness.
4. The rate of decrease of the sheltered distance increases with increasing shelterbelt density and decreases with shelterbelt height.

The indirect effects of the modification of the wind field are worth noting. Since a limit is placed on the peak values of wind speed, the danger of soil erosion is reduced sharply, as was proved statistically by M. Jensen [5227] and J. van Eimern [5215]. W. S. Chepil, et al. [5210] established, based upon field studies at Garden City, Kansas, that the wind erodibility of soil particles varies directly with the cube of wind velocity at 30 feet, and inversely by approximately the square of effective surface moisture. Both parameters are directly affected by the spacing and density of shelterbelts. W. S. Chepil [5209] also demonstrated that the maximum unprotected distance governs the erodibility of soils in the field across a farm field along the prevailing wind direction.

Since wind energy increases to the cube of the wind speed, in slowing the wind a shelterbelt will help reduce wind erosion. In many parts of the world wind erosion is the primary cause for establishing shelterbelts, and removing trees has been blamed for the spread of deserts (J. Kort [5234]). E. L. Skidmore and L. J. Hagen [5275] evaluated the effect of a wind barrier on the erosional force of the wind at Great Falls, Montana, where the prevailing wind direction was from the southwest (Figure 52-7). Their study showed the importance of the barrier's orientation to the prevailing wind. Shelterbelts reduce wind damage to crops including sandblasting (which is the effect of soil particles abrading plant tissue), and lodging (which is flattening of crops in the field by wind or rain) (J. Kort [5234]). K. A. Ticknor [5280] has recommended various shelterbelt designs to limit soil erosion.

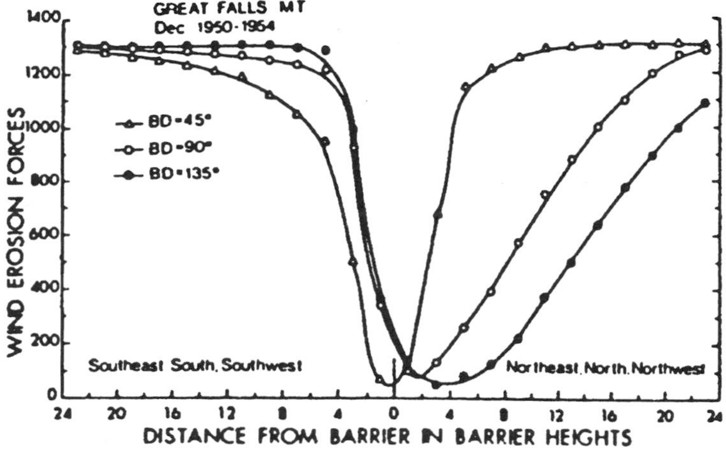

Figure 52-7. Wind erosion forces at indicated distances perpendicular from 40 percent porous barrier when the barrier direction [BD] is 45°, 90° (east-west), and 135°, respectively. Wind data are for Great Falls, Montana. (After E. L. Skidmore and L. J. Hagen [5275])

An incidental result of employing a single windbreak is the undesirable increase in wind speed, which always occurs around the ends of the obstruction. Figure 52-8 gives one example for the shelterbelt shown in Figure 52-6. The fields at the edge of the belt experienced winds over 20 percent stronger. The same applies to gaps (funnel effect); thus, wind damage may be greater near or through these gaps.

Establishment of shelterbelts requires a thorough preliminary study of the area. The natural distribution of wind speeds and directions should be analyzed, and a survey should be made of the microclimatic features. W. Kreutz [5236] gives two examples of this kind of planning, and another example is given in J. van Eimern's investigation [5216]. K. Illner [5226] and H. Schrödter [5268] have investigated the effect of wind protection on the dissemination of weed seeds and spores that might cause a risk of infection.

The next discussion deals with the effect of shelterbelts on the microclimate. The first simplified rule established for the wind field was that low shelterbelts, short distances apart,

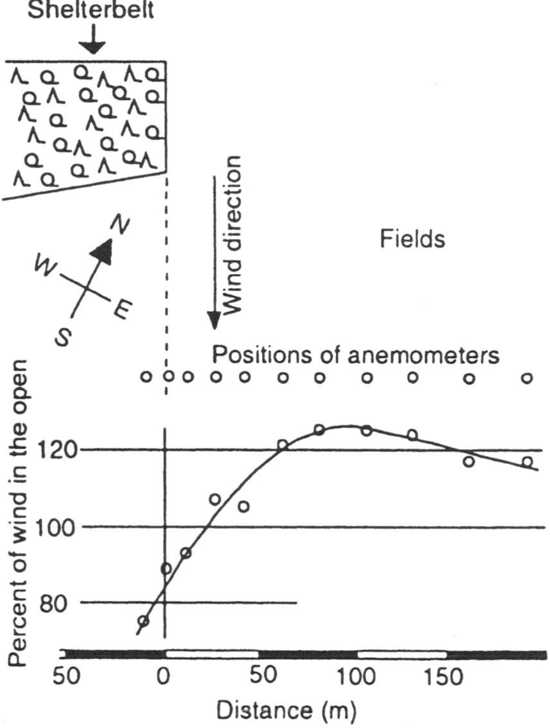

Figure 52-8. Side effects of a shelterbelt. (After W. Nägeli [5249])

have the same effect as high shelterbelts at correspondingly greater distances. This is not necessarily valid for the other elements of the microclimate.

The radiation balance is modified by the shadows in the shelterbelt. More can be learned about this from Section 41 and in R. J. van der Linde and J. P. M. Woudenberg [3717], I. Dirmhirn [5212] and S. Sato [5267]. I. Dirmhirn [5212] computed, for Vienna, the amount of global solar radiation ($K\downarrow$), taking into account the observed duration of sunshine received by the ground between distances of 0 and $10h$ from shelterbelts, for a number of directions, over the four seasons. The differences are less in summer than in spring and autumn because of the high altitude of the sun. For example, the total daily global solar radiation in spring, expressed as a percentage of the undisturbed global solar radiation, is shown in Table 52-1.

The reduction of shortwave radiation is restricted to the proximity of the shelterbelt. The shading effect is much less on N-S shelterbelts than on E-W ones. Areas to the north of an E-W shelterbelt, particularly during the winter, may be in the shade for long periods of time. At night, outgoing longwave radiation is reduced near the shelterbelt because of the reduced sky view factor (Section 5, and Section 37). The sky view factor would be 0.50 adjacent to the shelterbelt and would increase with distance from it. Its value is determined by the height, length, and the ground position relative to the shelterbelt (G. T. Johnson and I. D. Watson [5228]).

Table 52-1. Daily global solar radiation in spring (percent of undisturbed radiation) near shelterbelts at Vienna.

Direction of Shelterbelts	Side of belt	Distance in units of tree height (h)				
		0	0.2	0.5	1	2
West-East	N	27	33	39	48	97
	S	81	85	90	95	98
Southwest-	NW	37	45	60	79	92
Northeast	SE	71	77	81	92	97
South-North	Both	53	60	72	84	94

During the day, especially in clear and dry weather, soil and air temperatures are substantially higher in the shelter areas. On the average, the difference in the uppermost soil layer is about 2°C and 0.5 to 1°C in the air near the ground. G. Casperson [5208] observed that excess temperatures of 10°C were found on days of strong radiation in the shelter of a 3 m high hedge of hawthorn near Potsdam. An extreme example of this diurnal temperature excess (°C) for a day in May is given in Table 52-2 and were primarily a result of increased daytime maxima. This gain in heat, in conjunction with a reduction in evaporation, often results in an improvement in growth and increased crop yield in areas protected from the wind. The gain of heat around a shelterbelt is due both to a reduction in mixing and a reduction in evaporation.

Table 52-2. Air and ground temperature excesses (°C) near a shelterbelt in May, near Potsdam.

	In the ground			In the air		
Height (cm)	-10	-5	-2	5	75	140
In the open	4.1	5.9	8.7	26.9	22.9	22.0
8 m behind hedge	7.2	11.4	15.5	32.9	26.0	24.2
Excess	3.1	5.5	6.8	6.0	3.1	2.2

J. E. Ujah and K. B. Adeoye [5282] also observed higher maximum temperatures on the leeward side of a shelterbelt than in an open field (Table 52-3). Minimum temperatures were about the same at both locations. The maximum soil temperatures 2-4*h* distant on the leeward side at a depth of 5 cm were from 0.5-1°C higher than in the open. Increased crop yields at 2*h* were observed behind the shelterbelt (Table 52-4), which they ascribe to the competition for moisture and nutrients between the roots of the trees of the belt and the crop. Crops also differ in their responsiveness to shelter. Winter wheat, barley, rye, millet, alfalfa, and hay are highly responsive, while spring wheat, oats, and corn are less affected (Table 52.5) (J. Kort [5234]). R. L. Norton [5252] summarizes the effects of shelter on vineyards and orchards. In most cases production and/or fruit quality were improved.

Table 52-3. Mean monthly maximum air temperature (°C) in the open and in sheltered farmlands at Dambatta, Nigeria. (From J. E. Ujah and K. B. Adeoye [5282])

Period	Open farmland	Protected farmland Distance from shelterbelt (multiples of tree height, *h*)				
		2*h*	4*h*	10*h*	15*h*	20*h*
July	32.1	33.5	33.3	32.9	32.2	32.1
Aug.	32.1	33.9	33.4	32.8	32.0	32.1
Sept.	33.2	34.1	33.8	33.0	33.2	33.1
Oct.	34.3	35.0	34.7	34.4	34.4	34.2
Nov.	34.3	35.2	34.9	34.5	34.6	34.3
Dec.	32.7	33.5	33.3	33.3	33.0	33.0
Jan.	30.0	30.9	30.7	30.0	29.6	30.1
Feb.	31.0	31.8	31.6	31.1	31.0	31.2
March	37.1	38.0	37.5	37.5	37.2	37.2
April	39.8	40.9	40.2	39.7	39.7	39.5
May	38.2	39.0	38.9	38.4	38.1	38.2
June	34.3	35.1	34.8	34.3	34.3	34.1

Table 52-4. Grain yield (percent of open farmland) at various distances from the belt. (From J. E. Ujah and K. B. Adeoye [5282])

Distance from belt (multiples of tree height, *h*)				
2*h*	4*h*	10*h*	15*h*	20*h*
115	121	115	110	107

Table 52-5. Relative responsiveness of various crops to shelter. (From J. Kort [5234])

Crop	No. of field years	Weighted mean yield increase (%)
Spring wheat	190	8
Winter wheat	131	23
Barley	30	25
Oats	48	6
Rye	39	19
Millet	18	44
Corn	209	12
Alfalfa	3	99
Hay (mixed grasses and legumes)	14	20

Animals increase their food intake and energy expenditure when subject to temperatures below their comfort zone. Wind chill can add to animal energy expenditures. Windbreaks can thus reduce animal feed requirements as much as thirty percent (B. C. Wright and L. R. Townstead [5291]).

K. J. McAneny, et al. [5243] observed higher maximum air and soil temperatures but no clear pattern of minimum temperatures in their six year study of temperature changes associated with a growing shelterbelt. The higher daytime temperatures associated with a shelterbelt results in a faster development of plants. Sheltered rapeseed and safflower flowered earlier (PFRA [5257]). Corn also flowered up to two weeks (Y. Zohar and J. R. Brandle [5292]) and cotton 4-5 days earlier (R. A. Read [5259]).

J. K. Radke's [5258] observations of soybeans are shown in the schematic diagram (Figure 52-9) of a sheltered crop's response to porous wind barriers with an east-west row orientation. Two vertical lines represent the barriers, and the numbers represent relative increases of plant growth or yields. The solid and dashed curves in the lower part of the figure represent plant response to southerly and northerly winds, respectively. The light shaded areas represent negative plant response to competition for soil nutrients and water, and the hatched areas represent additional decreases due to shading. The upper curve is the sum of the lower curves and represents the expected yield. Note that the expected yields are lower near the shelterbelt due to shading and competition for water and nutrients from shelterbelt roots. E. L. Skidmore [5272] found that "vegetative growth and dry matter production of winter wheat are usually higher in the area sheltered by the wind barrier than in the open field. Sometimes grain yield also is increased. However, because growth and yield next to the barrier are reduced and the land occupied by the barrier is unavailable for crop production, the net effect on grain yield is often negligible." Most researchers, however, have reported that shelterbelts increased yields due to reduced wind erosion, improved microclimate, increased soil moisture and reduced crop damage. In the drier Sahel A. J. Brenner et al. [5203], however, found a decrease in crop production which they suggested was primarily due to the daytime increase in temperature behind the shelterbelt. For further information dealing with which crop type yields benefit most, both overall and in the zone near

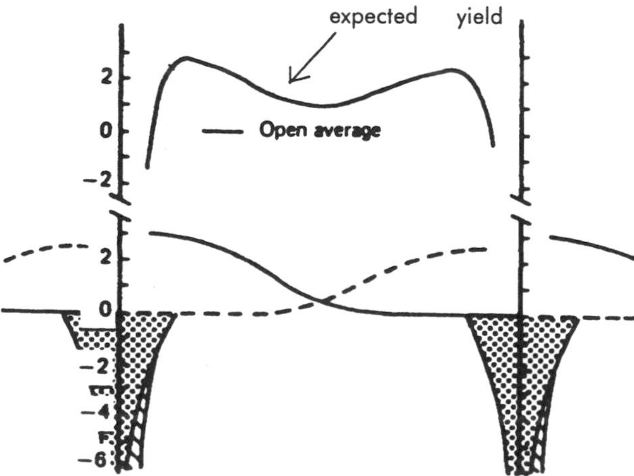

Figure 52-9. Effect of a shelterbelt and wind direction on plant growth. (After J. K. Radke [5258])

the shelterbelt, and other problems and benefits associated with shelterbelts the reader is referred to J. Kort [5234] and R. L. Norton [5253].

A much debated indirect effect of the reduction in wind is the increase in the risk of frosts. It is clear that in sloping areas the cold air dammed (Section 42) behind the upper side of the shelterbelt will produce lower temperatures. This has been proved by Y. Daigo and E. Maruyama [5211] for a 6° slope with a hedge of Japanese cedars, 1.2 m in height. Care must therefore be taken to allow spaces for the cold air to flow downward when planning wind protection in such situations. Studies by R. J. van der Linde and J. P. M. Woudenberg [5240] and M. Jensen [5227] in flat areas found that when it is calm at night, temperatures are higher in the shelterbelt because outgoing radiation is reduced and soil temperatures have been higher during the day. With brisk winds the temperature is the same inside and outside. M. Jensen [5227] suggests that only with weak winds (0.5 to 2 m sec^{-1}) is there increased danger of night frost in the areas of the shelterbelt, particularly in the first weeks of spring, when the soils are still cool from winter.

A. Kaminski [5233] observed an increase in temperatures near a shelterbelt and a decrease 4-16*h* distant. M. Nuñez and D. Sander [5254] observed a 1-2°C nocturnal temperature excess within a eucalyptus shelterbelt as compared with a cleared area. In the shelterbelt, they observed a 21-31 percent reduction in net longwave radiation loss and felt this was the primary reason for the observed temperature excesses. The Controller General of the United States [5221] reported temperatures near the shelterbelt up to 2.2°C warmer on cool days and during the night. R. H. Aron (unpublished) has also observed decreased frosts near vegetation.

A shelterbelt has three influences on nocturnal cooling. First, the shelterbelt reduces the net loss of longwave radiation. This is most effective close to the shelterbelt and probably is the primary factor accounting for the higher temperatures (reduced frosts) observed at this location. Second, since the soil (and air) becomes warmer during the day (Table 52-3), more energy will be liberated from the soil at night. This is particularly the case in early autumn when the soil is still relatively warm. Cool soils in spring may explain the increased risk of frost observed by M. Jensen [5227]. Third, the reduction of wind speed and resulting mixing means less warm air will be brought down from aloft, resulting in lower nocturnal temperatures (Table 19-1). The reduction in mixing, together with decreased soil temperatures and reduced longwave emissions from trees, probably explain the lower nocturnal temperatures found at a distance from the trees. J. M. Caborn [5206] found slightly greater risk of frosts behind a shelterbelt. N. J. Rosenberg [5262, 5263, 5264], G. Guyot and B. Segiun [5222] and J. K. Marshall [5242] also report that nocturnal temperatures are generally lower near a shelterbelt; this is shown in Figure 52-10. Additional research may be necessary to determine in exactly what conditions nocturnal temperatures are lower near a shelterbelt.

Greater quantities of dew deposition may occur in places sheltered from the wind, as show by measurements made by L. Steubing [5277]. The difference was greatest on nights with strong winds. The greater amount of dew results from the stillness of the air behind the obstruction and also to some extent from the higher air humidity there. L. Steubing found a maximum increase of up to three times the value in the open at a distance of 2 to 3*h* behind the obstruction. J. van Eimern [5214] also established that there was a delay in the disappearance of dew after sunrise, and found the greatest amount of dew midway between the hedges where night temperatures were lowest (Figure 38-2). N. J. Rosenberg [5262] also

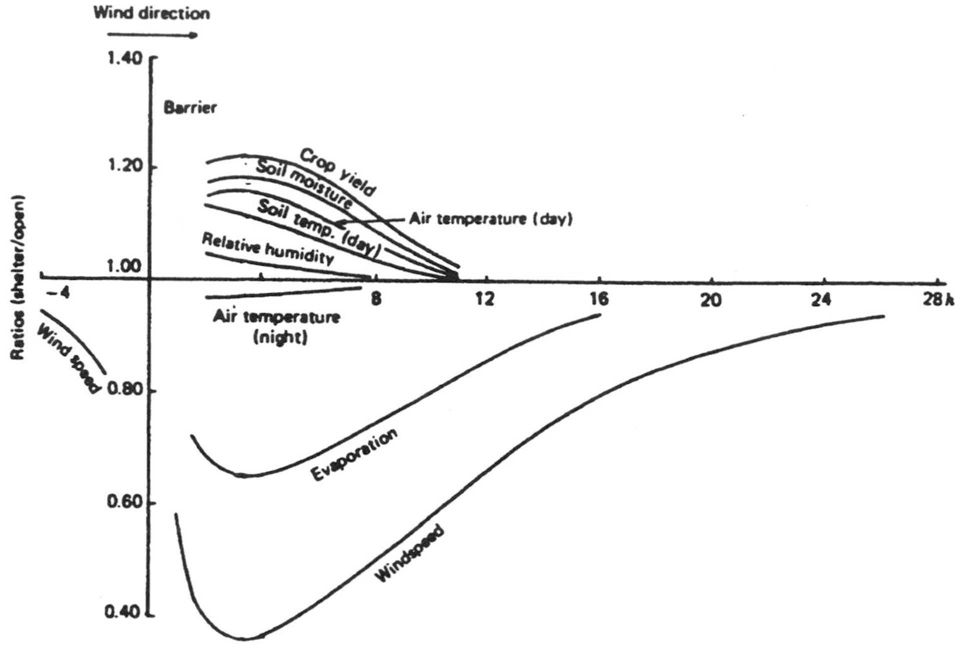

Figure 52-10. The effect of barriers on micrometeorological factors. The arrows indicate the directions in which values of different factors have been found to vary relative to the control values measured in unsheltered areas. (After J. K. Marshall [5242])

reported greater dew deposition and duration, and he states that near a shelterbelt absolute humidity is usually higher throughout the day and night. He also reported that relative humidity is generally higher during the night and is also usually higher during the day despite higher daytime temperatures. He suggests that the higher daytime relative humidity is a result of reduced water vapor transport associated with the reduction of vertical mixing.

Thus, the reduction in mixing associated with a shelterbelt not only results in higher daytime temperatures but also higher relative humidity, reduced evaporation, and higher soil moisture (Figure 52-10). E. L. Skidmore and L. J. Hagen [5274] have suggested that in many agricultural regions, especially where evaporation is high, windbreaks are used to reduce evaporation and provide a more favorable environment for plants. While the reduction of evaporation is often considered to be the most valuable result of artificial protection against the wind, studies indicate that the actual reduction is sometimes small and occasionally evaporation even increases as a result of increased leaf area due to greater plant growth (E. J. George [5220] and M. Dixon and J. Grace [5213]). M. Jensen [5227] pointed out that the increased heat during the day and increased growth will increase evaporation. This productive evaporation has the effect that the water budget, as measured by M. Jensen [5227], was not substantially different inside and outside the area protected by the hedges. A. J. Brenner et al. [5203] found higher transpiration for plants behind a shelterbelt and ascribed this to greater leaf surface area and higher saturation vapor pressure deficits, both a result of higher temperatures. K. W. Brown and N. J. Rosenberg [5204] found that with light winds a shelterbelt did not have a marked effect on daily water usage of sugar beets. However, J. K.

Marshall [5242] found reduced evaporation up to about 16*h* downwind, while R. J. Bouchet, et al. [5202] found a reduction in evaporation as far as 30*h* downwind. These are consistent with most reports, which suggest a decrease in water usage (e.g., J. van Eimern [5217]).

E. L. Skidmore and L. J. Hagen [5273] evaluated the effect of a shelterbelt's porosity on evaporation (Figure 52-11). They found that in a less porous shelterbelt minimum leeward evaporation occurred closer to the windbreak, and after reaching a minimum, tended to increase more quickly than with more porous shelterbelts. At 4*h* leeward of the solid shelterbelt, evaporation had recovered to 92 percent of open-field evaporation, whereas at 4*h* leeward of 40 and 60 percent porous shelterbelts, evaporation rates were 65 and 75 percent of open-field evaporation, respectively. Shelterbelts were found to reduce evaporation in proportion to the reduction of the wind speed. D. R. Miller et al. [5247] also found that windbreaks reduce evaporation of sheltered crops. They attributed this to a reduction in the advection of sensible heat. They also found that the shelterbelt, by reducing evaporation, decreased the plant stress, resulting in the stomata remaining open longer and thereby increasing the potential photosynthesis. In summation, it appears that despite the higher temperatures in the vicinity of a shelterbelt, evaporation is reduced for most crops in most situations. This reduction is primarily caused by the increased relative humidity and reduced transpiration. N. J. Rosenberg [5262] reported photosynthesis is usually increased for plants protected by a shelterbelt, due in part to the reduced evapotranspiration and lower nocturnal respiration associated with lower nocturnal temperatures.

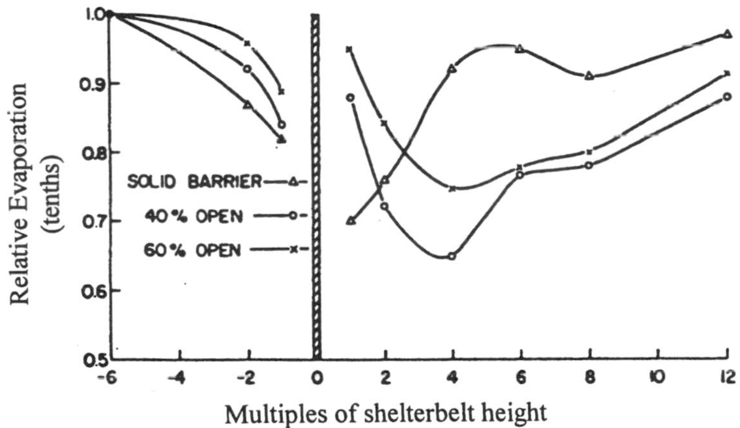

Figure 52-11. Evaporation as a function of shelterbelt porosity and distance from the barrier. (After E. L. Skidmore and L. J. Hagen [5273])

J. D. Rüsch [5265] found that the air in the sphere of influence of the shelterbelt contains less carbon dioxide than it does outside. E. Lemon [5237] also found lower carbon dioxide concentrations around a shelterbelt, particularly with low wind speeds. The increased photosynthesis may explain the reduced carbon dioxide concentrations reported by J. D. Rüsch [5265] and E. Lemon [5237]. K. W. Brown and N. J. Rosenberg [5205] found that daytime carbon dioxide concentrations in a shelterbelt deviated by 9 to 12 ppm and nighttime concentrations were about 3.5 ppm greater than that found in the open. The higher nighttime concentration is attributed to plant respiration and lower atmospheric mixing.

Research by T. E. Kowalchuk and E. de Jong [5235] may explain some of the reported variability in crop yields associated with a shelterbelt. In investigating crop yields protected by a shelterbelt bordered by an approximately five meter wide grass border they found that in dry years yields were reduced for about ten meters (2 heights) into the field. This is consistent with Figure 38-6 which shows lower soil moisture at the edge of a clearing. T. E. Kowalchuk and E. de Jong also reported that during dry years yields were slightly above the field mean from 10 - 20 m (2-4 heights) from the sheltering edge. This would be due to either reduced or more productive evaporation. In a wet year when water was abundant, however, the competition effect near the shelterbelt was smaller and the zone of improved crop growth further out was absent. This suggests that the competition for moisture or reduction of evaporation may be the primary factor explaining improved crop yields associated with a shelterbelt and, particularly in dry years, a major factor explaining reduced crop yields near the shelterbelt.

Vapor pressure deficits (E_0-e) are affected both by the vapor pressure and by the air temperature. The reduction of mixing associated with a shelterbelt slows the transfer of water vapor from the foliage and together with an increase in transpiration should result in a decrease in the vapor pressure deficits (K. W. Brown and N. J. Rosenberg [5205] and S. P. Long and N. Persuad [5241]). Others, however, have reported an increase in the vapor pressure deficits which they ascribe to higher air temperatures (A. J. Brenner et al. [5203], M. K. U. Carr [5207] and H. C. Aslying [5200]).

Shelterbelts also provide a habitat in which many types of wildlife can flourish (G. McClure [5244] and R. J. Johnson and M. M. Beck [5229]). Wingless and very small insects generally depend on air currents to carry them to new sites, but tend to settle in areas of low wind speeds where they have greater flight control. Insects, therefore, accumulate leeward of windbreaks (J. E. Pasek [5255] and T. Lewis [5238]). Both beneficial and harmful insects may be harbored in shelterbelts. T. Lewis and B. D. Smith [5239] found greater numbers of pollinating insects in sheltered fruit orchards. J. E. Slosser et al. [5276], however, found shelterbelts harboring the cotton boll weevil.

The problem of protection against wind is associated with many problems, which cannot be dealt with here. For example, in areas exposed to strong winds, the best plants to be used for the establishment of shelterbelts, the width of the belt, the distances between the rows and plant stands, and the suitable mixture of trees must be properly selected. H. H. Hilf [5225] published a guide to these practical problems in 1959. Despite all the beneficial effects of shelterbelts, a report by the Controller General of the United States [5221] indicates that increasing land values and incompatibility of wind breaks with large scale farming operations have resulted in many being removed in that country.

Using wind tunnels some researchers analyzing the effect of shelterbelts have observed different patterns of wind alternation, evaporation, and shelterbelt density effects on wind retardation than those observed in field (K. G. McNaughton [5245 and 5246]). They observed a reduction in wind speed up to about eight shelterbelt heights down wind which they refer to as a quiet zone, and an increase in wind speed beyond eight heights which is referred to as a wake zone. For information on numerical simulation on the affect of shelterbelts on the wind refer to H. Wang and E. S. Takle [5284, 5285, 5286] and the classic work by J. D. Wilson [5290]. For additional information on shelterbelts' effect on crop yields, plant physiological responses and water usage, the reader is referred to N. J. Rosenberg, et al. [5264], and the series of articles on windbreaks in the journal Agriculture, Ecosystems and Environment (1988, vol. 22-23).

We now turn our attention to windbreaks. The snow fence is a familiar type of windbreak seen on country roads. Snow fences are also used for controlling snowdrifts on roadways. "Plowing snowdrifts costs about 100 times more than snow fences" R. D. Tabler [5279]. There are three primary types of snow fences: the collecting fence (which decelerates the wind speed and thus causes deposition), the guide fence (which deflects the snow for deposition elsewhere) and the blower fence (which accelerates the wind speed locally for deposition elsewhere). By reducing wind speeds, snow fences lower the kinetic energy of the overlying air, which causes snow to be removed from suspension, thus keeping the snow off the roadway. Observations and measurements of this process are to be found in J. van Eimern [5215], A. Kaminski [5233], H. Kaiser [5232], W. Kreutz [5236], and W. Nägeli [5249]. Snow fences are also used to retain snow cover to enhance groundwater recharge for water supply purposes.

T. Müller [5248] mapped the "funnel effect" by observing the way in which snow was swept away in and behind a gap in a hedge. Similar processes can be observed at work in drifting sand, where it is possible to follow the effects for a considerably longer time than with snow. Figure 52-12 shows the results obtained by H. Kaiser [5231] with piles of sand. The wind speed profile was measured in the lee of the obstruction at a height of 1.4 m in (*a*), 38 cm in (*b*), and 45 cm in (*c*). The sand accumulation has been exaggerated four times. The figure shows accumulation behind: (*a*) an impenetrable brushwood barrier after 7 years; (*b*) a penetrable brushwood fence after 2 years; and (*c*) a snow fence with a clearance at the ground, after 6 months.

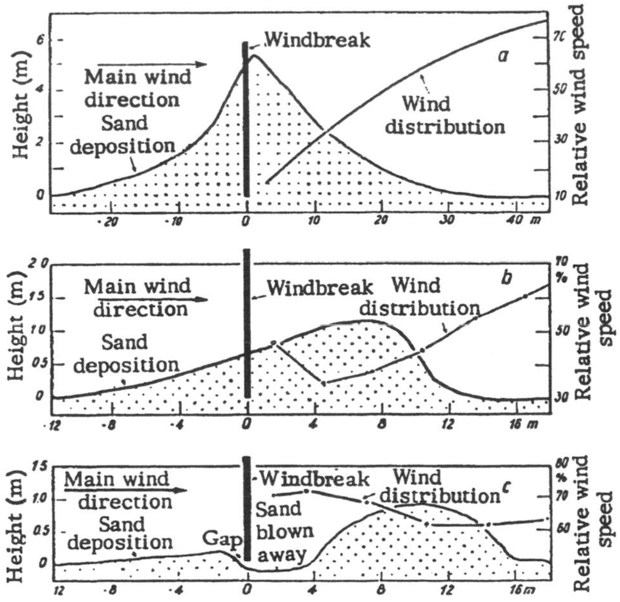

Figure 52-12. Sand accumulation at windbreaks of different densities. (After H. Kaiser [5231])

The effects of porous (Figure 52-13) and nonporous (Figure 52-14) windbreaks on snow accumulation can be substantial. A solid snow fence will result in higher accumulation near the fence, while a porous fence accumulation spreads a greater distance both windward and leeward of the fence (Figures 52-13 and 52-14). For maximum snowdrift accumulations S. L. Ring [5261] suggests a bottom gap of about 30 cm, with slats of fifty percent porosity at a 15° downward inclination from vertical. A downward slope in excess of 10% also increases storage capacity. As long ago as the 1930's E. A. Finney [5218] showed that the configurations of highways and driveways would maximize or minimize snow accumulations (Figures 52-15 and 52-16). While major freeways frequently are built to provide aerodynamically drift-free roadways, city streets and driveways frequently totally disregard his advice, requiring added snow plowing and removal efforts. To minimize the effort of keeping roadways free of snow E. A. Finney [5219] further advises that:

1. The street or driveway should be raised above the adjacent ground at least to the average depth of snow accumulation. Slopes should be rounded at the top and bottom.
2. Road cuts or sunken driveways should be avoided (Figure 52-17). S. L. Ring et al. [5260] states that a solid barrier or a road cut will generate drifts up 7-10 times its height downwind. Thus to help ensure a drift-free roadway the distance from the barrier or edge of the cut should exceed 7-10 heights downwind.
3. Slopes with a grade of 4:1 or greater (Figure 52-15) with wide shoulders and shallow wide ditches, and rounded-slope intersections are advised. There is little or no accumulation downwind of a slope of 6:1 (Figure 52-16).
4. Guard rails, curbs and other features that act as a snow fence should be minimized.

For readers desiring more knowledge on the effects and remedies of various meteorological factors on road conditions and highway construction, the books on these topics by S. L. Ring et al. [5260] and A. H. Perry and L. J. Symons [5256] are recommended. For information on the relationship between vegetation barriers and snow accumulation S. L. Ring et al. [5260] is recommended. For more information dealing with snow fences S. L. Ring et al. [5260] or R. D. Tabler [5279] are recommended.

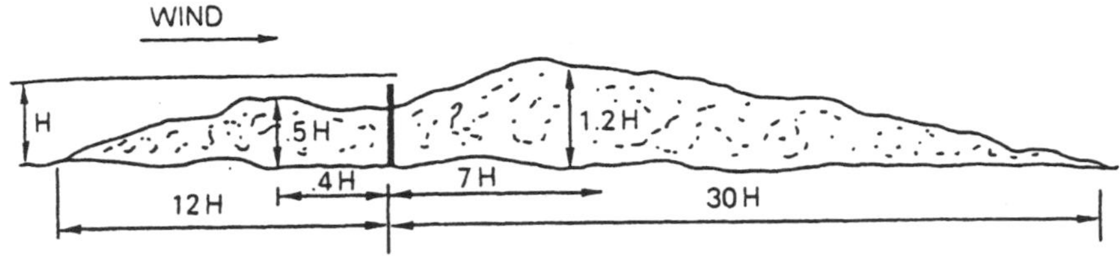

Figure 52-13. Snow drift accumulation around a porous windbreak. (After S. L. Ring [5261])

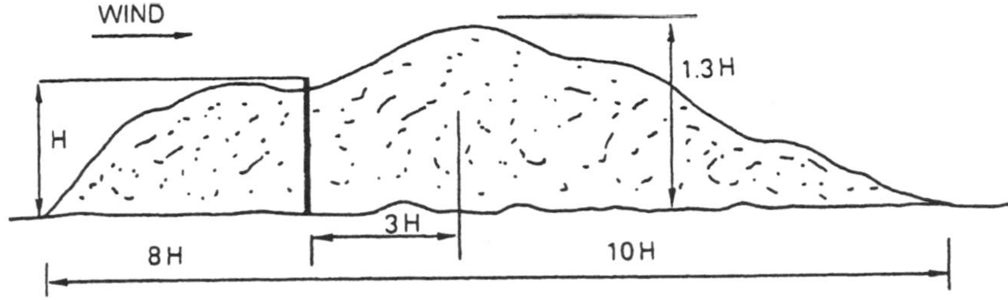

Figure 52-14. Snow drift accumulation around a nonporous windbreak. (After S. L. Ring [5261])

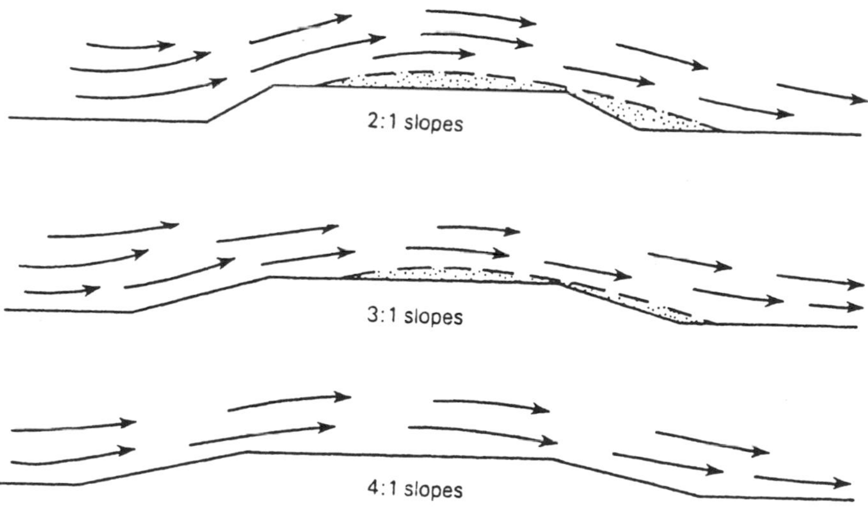

Figure 52-15. Effect of highway slopes on air flow and snow accumulation. (After E. A. Finney [5218])

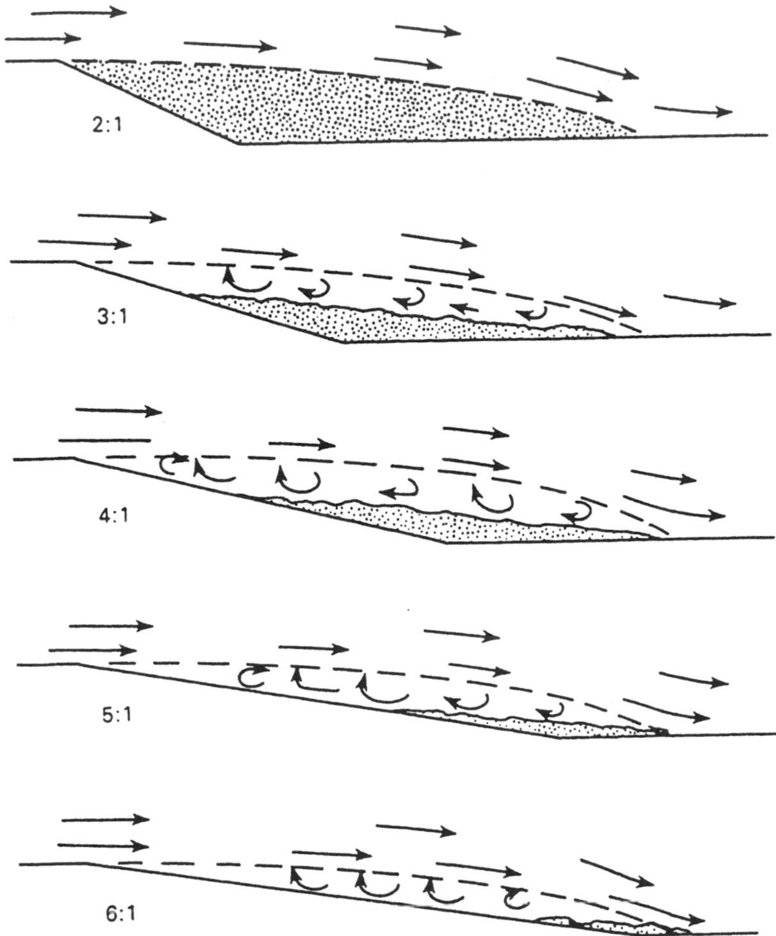

Figure 52-16. Effect of variations in highway slope on snowdrift accumulation. (After E. A. Finney [5218])

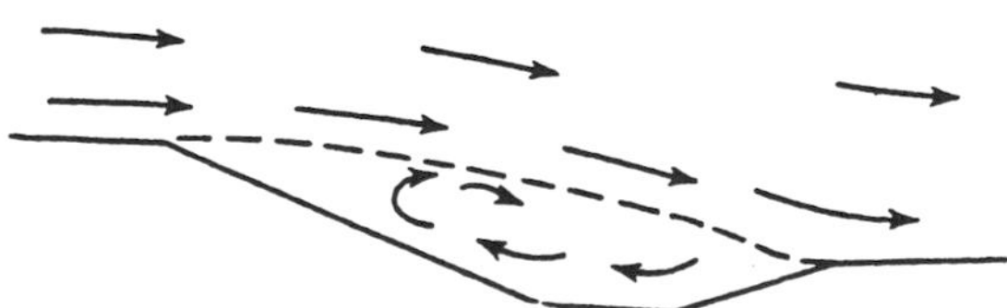

Figure 52-17. Drift forming eddy in a highway cut. (After A. E. Finney [5218])

53 Artificial Protection Against Low Temperatures

Low temperatures can cause injury in plants. While temperatures at or below 0°C are referred to as frosts, it should be pointed out that in some plants, temperatures substantially above freezing can cause injury, while other plants can tolerate temperatures substantially below freezing. A plant's tolerance to cold temperatures depends to a large extent upon its stage of development. For example, while dormant, some varieties of pear can withstand temperatures as low as -196°C. These same plants may suffer frost damage with temperatures only a few degrees below freezing immediately before, after, or during full bloom (M. N. Westwood [5369]). Both M. N. Westwood [5369] and N. J. Rosenberg, et al. [5264] present lists of critical temperatures at which plants suffer damage during various stages of development. In addition, M. N. Westwood [5369] points out that at any given phase in the plant's development, there is a range of temperatures at which damage will occur. For example, at the first sign of pink in apricot flowers, ten percent were killed when the temperature fell to -3.9°C. However, the temperature had to drop to -7.2°C for 90 percent of the flowers to be killed.

It is useful to make a distinction between advection frosts and radiation frosts. Advection frosts have their origin in the advection of cold air, are controlled by the general synoptic weather patterns, and are usually associated with stronger winds. Radiation frosts are caused by radiative cooling at night and form best with low wind speeds, clear skies, and low relative humidity. Strong radiative cooling often intensifies advection frosts.

The most effective weapon against frost damage is preventive action. "The best time to protect an orchard against frost is when the orchard is being established," as W. J. Humphreys said in 1914. It is incomprehensible that even today, the most fundamental laws of microclimatology are disregarded repeatedly when new orchards are laid out at great cost in areas subject to repeated frosts.

Information should be collected on frost danger as a function of topography when protective measures are prepared. The best method of doing this is to map the danger areas; good examples are to be found in W. Baier's [5307] report for the wine country in the north of Württemberg and Baden, in A. Vaupel [5368] for the wine industry of the Rhineland-Palatinate, and in J. Lomas', et al. [5340] *Frost Atlas of Israel*. J. D. Kalma et al. [5331] discuss methods of mapping frosts based upon topography. Satellites are providing a new methodology both for the mapping and forecasting of areas that may be subject to frosts. For more information dealing with this topic the reader is referred to E. Y. Chen [5313], E. Y. Chen et al. [5314, 5315], J. D. Martsolf [5343], T. W. Oswalt [5348], and J. D. Kalma et al. [5331]. One problem with the use of satellites for forecasting frost has been that with the advection of cold air, counterradiation ($L\downarrow$) is reduced and the soil and air cool faster than expected. J. D. Martsolf et al. [5345, 5346] discuss this problem.

Of equal importance to careful site selection is the selection of an appropriate crop or crop variety, as some plant varieties are more resistant to frost damage. Deciduous orchards, for example, are quite resistant to frosts (cold hardy) as long as the trees are in rest (winter dormancy). While in the rest phase of its growth cycle, plant growth inhibitors (S-ABA) dominate the plant's hormonal balance and warm temperatures will not promote renewed growth. Exposure to cool temperatures through the winter (below approximately 7.2°C (chilling temperatures)) results in a change in the plant's hormonal balance so that growth promoters (GA) begin to dominate. When the plant's chilling requirement is satisfied, growth promoters will dominate the hormonal balance. However, plants will remain cold hardy (frost resistant) and dormant (imposed dormancy) until the warm temperatures of spring promote renewed growth. Thus, if the area is subject to frequent spring frosts, it may be advisable to plant a crop variety with a somewhat larger chilling requirement than may be expected to have occurred in early spring. The plant would remain dormant and relatively frost resistant later into the spring avoiding some frost problems. However, care must be taken not to plant a variety whose chilling requirement is too long. If the plant is exposed to an inadequate number of hours of chilling temperatures, a whole host of problems may arise which collectively are referred to as prolonged dormancy. For more information about chilling, factors that affect chilling, prolonged dormancy, and a listing of the chilling requirement of numerous deciduous fruit varieties, the reader is referred to R. H. Aron [5301]. For information dealing with estimating chilling hours and environmental factors that may affect a plant's chilling requirement, the reader is referred to R. H. Aron [5302, 5303] and R. H. Aron and Z. Gat [5304]. Growth regulators have also been used in spring to keep the plants cold hardy and frost resistant by retarding growth (R. Ryugo et al. [5358]).

If sufficient warning has been given of an impending frost when a crop is nearing maturity, one method of evading or minimizing the problem is to harvest as much of the crop as possible. For annuals one of the best methods of frost protection is to plant them late enough to avoid spring frosts. In cases where an earlier harvest has substantially greater value, early planting may require protection from frosts.

Combating frost damage effectively can require a substantial amount of money. This expenditure is economically justifiable only when the crop has a high value such as vines or fruit. If there are a number of late night frosts in a single spring, as is often the case, the effectiveness of frost protection over the entire frost episode will depend on the night when protection has been least successful. All types of frost protection will be more effective if protective measures are started before critical temperatures are reached. It is easier to maintain rather than to increase the temperature.

Frost protection may be achieved either by reducing the amount of radiative cooling (protection against outgoing radiation), which will be discussed first, and/or by directly or indirectly supplying heat.

For low plants, coverings have been used for decades to mitigate the effects of frost. Straw matting has proven effective in protecting garden plots from frost. K. Baker and R.H. Aron [5300] investigated the value of polyethylene terephthalate (plastic) milk jugs or polyurethane (Styrofoam like) rose cones in providing frost protection in spring. While temperatures in both coverings were warmer on clear days and in the milk jugs on cloudy days, nocturnal temperatures averaged 0.9°C and 0.6°C cooler under clear and cloudy skies than the ambient air. Rose cones on the other hand offered some protection with temperatures averaging 0.6°C above that of the ambient air. The three factors that influenced nocturnal temperatures within these containers were the reduction of the mixing of environmental air, heat flow from the soil, and the nocturnal radiation balance. The temperatures inside the rose cones were warmer than that of the ambient air because the loss of heat from the soil was slowed, which more than offset the reduction of the heat gain from the atmosphere. The nocturnal temperatures in the milk jugs were cooler than the ambient air because they reduced the heat gain from the air while not appreciably slowing the heat loss from the soil (see greenhouses in Section 20).

As shown in Table 31-2 taller plants are less subject to frost. M. L. Blanc et al. [5309] has suggested that increasing the height to which plants are grown can increase the minimum temperature to which the plants are exposed by as much as 1-2°C.

It has long been recognized that on cold nights, frost damage will be considerably less or perhaps eliminated if fog or low clouds (high clouds tend to be less effective (Section 5)) cover an area. Fogs and clouds assist in minimizing frost damage by absorbing outgoing longwave radiation and emitting much of it back toward the surface, thus slowing the net loss of longwave radiation. For many years, farmers have tried to duplicate the effects of natural fogs by covering areas threatened by frosts with dense smoke. This is referred to as smudging. In creating smoke clouds, many types of things have been used including leaves, wet straw, old tires, wood, oil-soaked rags, and oxygen-starved heaters. For the most part, these efforts were unsuccessful because most of the smoke particles were too small and thus transparent to longwave radiation. Frost damage tends to be less if the plants are allowed to thaw slowly. While smoke is transparent to longwave radiation, it does reflect incoming shortwave radiation and thus has some benefit in reducing crop damage by slowing thawing in the morning. Smudging has fallen out of favor, in part because it is not very effective at slowing nocturnal cooling, and in part because it is often unpopular with nearby communities. Smudging has even been outlawed in some areas.

Creation of artificial fogs has been shown to be effective in retarding nocturnal cooling in some situations. It has the additional advantage of being economical. It is most effective on nights with high relative humidity and light winds. If the air is dry some of the fog will evaporate and cool the air. Intermittent increases in wind speed may cause problems because they not only may blow the fog away but also result in increased evaporation. If the leaves of a plant are wet, the infusion of dry air may result in the leaf being cooled to the wet-bulb temperature. To minimize the problem of evaporation, the drops can be coated with evaporation suppressants. Although this method has been shown to be effective either alone or together with heaters, it is not in widespread use. For more information on man-made fogs, the reader is referred to T. R. Mee [5347] or B. Itier, et al. [5330].

Another method of frost protection that is designed to retard the loss of radiation from the soil or the plant is the application of foam. The air bubbles in the foam have a low thermal conductivity thus insulating the crop. While this method has been tried on citrus, it is most effective for protecting low growing crops, having increased the temperature under the foam by as much as 10°C. Lower temperatures may occur at the top of the foam because of the reduction of the heat flow from the soil. Foams have been designed to last in excess of two days, but the benefit is greatest the first night and decreases with time due to the normal deterioration of the foam. The use of foam for frost protection has not gained widespread popularity because of high costs and the slow rate of application. For more information regarding foam insulation for frost protection, the reader is referred to J. F. Bartholic [5308] and R. L. Desjardins and D. Siminovitch [5317].

A number of chemicals have been applied to plants to increase their resistance to cold temperatures. None of these have proven to be effective and cheap enough to be commercially acceptable (J. D. Kalma et al. [5331]).

Plants can be protected completely from cooling if they are covered with water. This is done in the United States with cranberries, which are grown in fields surrounded by earthen walls. These fields are flooded when a damaging frost has been forecast. The water is then drained when the threat of frost has passed. Cool nocturnal temperatures cause sterility in rice. M. L. Peterson et al. [5352] reported that due to warmer nocturnal temperatures raising water levels in rice fields substantially reduced this problem.

The protective influence of small bodies of water on plants close to them or at a higher level is often spoken of, but it is too small quantitatively to be of practical importance. J. van Eimern and E. L. Loewel [5322] have been able to prove the existence of a "tube of slightly warmer air" over the ditches in the marshy land near Hamburg. J. Hanyu and K. Tsugawa [5329] showed that the watering of Japanese rice fields for frost protection gave a warming effect up to 60 or 70 cm above the surface.

Snow cover, by insulating the ground, can provide critical protection against frost damage and winterkill of crops and plants. I. Saudie, et al. [5359] pointed out that the reliability of early winter snow cover appears more critical in minimizing winterkill of winter wheat than the total amount. They also suggest that the method of trapping snow by leaving the stubble of the preceding crop has permitted the expansion of the North American crop northeastward to include most of western Canada's agricultural area.

There are four different methods of providing frost protection by supplying heat. This can be accomplished by using the heat stored in the soil, by ventilation (mixing the cold air near the ground with warmer air from aloft), by heating with burners, or by utilizing the latent heat released during the freezing of water.

Figure 53-1 shows a propeller with three blades, 6 m in diameter, mounted on a tubular steel pole 9 m high above apple trees in a fruit growing experimental station at Buxtehude. The axis of the propeller can be adjusted to any inclination, and the propeller can be set automatically to traverse horizontally through any required angle. When a strong radiation inversion is established the propeller blows the air aloft, which is considerably warmer, down into the cold layer. Helicopters flying just above the treetops have also been used with success as described by R. T. Small [5364] and J. D. Martsolf [5344].

Figure 53-1. Propeller used for frost protection near Buxtehude. (Photo by J. van Eimern [5321])

J. van Eimern [5321], in experiments among apple trees, found that the propeller shown in Figure 53-1 had a continuous effect of raising the temperature 0.5°C over 1.2 ha and 1°C over 0.8 ha when inclined at 45° and traversing around 180°. N. J. Doesken, et al. [5320] suggest that a growing interest in wind machines can be directly attributable to the cost of fuel. Their analysis gave strong indications that wind machines would assist in minimizing frost damage on 93 percent of the springtime nights with damaging freezes and 100 percent of the nights with the most severe freezes in western Colorado. Many small propellers are always more effective than a few big ones. J. F. Gerber [5323] states:

> "The combination of wind machines and heaters is much more energy efficient than heaters alone and reduces the risks from depending on wind machines alone. The investment for heaters and wind machines in combination is less than their combined costs because only about half the number of heaters are needed."

Wind machines are most effective on clear calm nights with strong inversions. Because buds frequently have temperatures lower than that of the ambient air (Figure 29-5), some have suggested that wind machines might even be effective for frost protection during nights with no, or very weak inversions. However, A. R. Renquist [5356] has shown that this would seldom be the case. He gives two explanations for this. First, nocturnal air/bud temperature differences are quite small and would only rarely span critical temperatures. Second, larger air/bud temperature differentials will only occur under conditions conducive to stronger inversions (clear skies and light winds).

The advantages of wind machines are that a minimum of personnel is needed (for turning the propellers on) and full effectiveness is reached within 1 or 2 minutes. There is little need to be afraid of the cooling caused by evaporation as all the dew present is blown away in 5 to 10 minutes. For further information on wind machines, the reader is referred to J. F. Gerber [5323].

As pointed out in Section 20, loosening the soil will reduce its thermal conductivity and thermal capacity and will result in lower nocturnal air temperatures (R. H. A. van Duin [1903]). Dry soils, due to their lower thermal conductivity and capacity, result in a larger nocturnal temperature drop (R. H. A. van Duin [1903] and J. D. Kalma et al. [5331]). In California, citrus is frequently irrigated prior to a threatened frost. This increases both the thermal conductivity and capacity of the soil. J. A. Businger [5312] suggests that irrigation prior to a frost may result in warming by as much as 4°C. This method is particularly effective with the first threatened frost in the fall because the soil has a relatively large heat reservoir. Increased surface evaporation and crop susceptibility to frost may offset some of the gains. Compaction (rolling) of the soil, either alone or in conjunction with irrigation to increase its conductivity, may also assist in releasing some of the soil's stored heat. S. F. Bridley, et al. [5311] found that rolling and watering resulted in a 0.6°C increase in nocturnal temperatures at about one meter. This was more than was observed with either rolling or watering alone. A ground cover in an orchard acts as an insulating barrier both making the orchard slightly warmer during the day and slowing the heat loss from the soil at night, resulting in slightly cooler nocturnal temperatures. While the air above the ground cover is cooler, the soil remains warmer (Sections 20 and 31). B. S. Sharratt and D. M. Glenn [5360, 5361, 5362] in a series of articles have suggested that the soil is a heat reservoir that could be used to assist in frost protection. Increasing the heat flow into the soil during the day should result in a greater heat flow out of the soil at night. By adding coal dust to the surface of the soil in an orchard, they observed a 0.5°C increase in nocturnal temperatures. The effect of soil covers such as coal dust is discussed in more detail in Section 20. R. L. Snyder and J. H. Connell [5365] have shown that removing ground cover under an orchard increases the nocturnal temperatures. In addition to the insulating effect of the ground cover it also has a higher albedo than the soil. Thus the soil without the ground cover absorbs more energy during the day which is released at night.

Frost protection by heating has been described in detail by O. W. Kessler and W. Kaempfert [5332] in their fundamental work published in 1940. Heat has been provided by numerous types of sources including burning wood, tires, coal, carbon filament lamps with aluminum reflectors, oil or kerosene heaters or by briquettes which are either arranged in small heaps on the ground or burned in stoves. Figure 53-2 shows a vineyard in the Saar where seven rows of briquette stoves, arranged 4.5 m apart in the lowest row, can be seen burning during the frosty night of 2-3 May 1935. Farther up the slope, the distance between burners can be increased because the heat and smoke are blowing upward. In Japan, as reported by Y. Tsuboi, et al. [5367], the oil stoves are sometimes not placed on the ground in vineyards, but at a height of 90 cm, so that a cold layer of air, which is not a problem, is left near the ground, and more of the heat goes to the benefit of the grapes.

Figure 53-2. Stoves for burning briquettes are lit in a vineyard when night frost sets in. (After O. W. Kessler and W. Kaempfert [5332])

O. Dinkelacker and H. Aichele [5318] in 1958 produced an increase in temperature of 2 to 3°C by using two to three briquette stoves per 100 m^2. With oil stoves, which are much easier to light, and which can be adjusted to suit the temperature variation during the night, an increase of about 3°C is obtained with two to five stoves per 100 m^2.

Frost protection by heating is dependent on terrain and the vagaries of the weather. Figure 53-3 illustrates the results of measurements by O. W. Kessler with 28 thermometers inside and 17 outside a heated experimental field near Oppenheim on the evening of 26 April 1929. The initial temperatures of the thermometers, which were set up at 50 cm above the ground, were between 6.0 and 6.5°C before heating was started (diagram at top left). After the first stoves were lit at 22:15, the temperature rose to 8°C in the middle of the field, while at the southern edge, where the stoves had not yet been lit, there was a decrease to 5°C (cold air drawn in by the first fires?). A quarter of an hour later, when all the stoves had been lit (below left), the warm zones were well distributed and showed an increase of 2 to 3°C in the center.

The temperature increase was kept up in this stand, as shown by the distribution of the isotherms an hour later (below right). The only difference is that the warm center has been displaced a little to the side. As mentioned earlier, heaters are generally more effective when used in combination with wind machines, sprinklers, or fog generators. For more information on heating for frost protection, the reader is referred to J. D. Martsolf [5342].

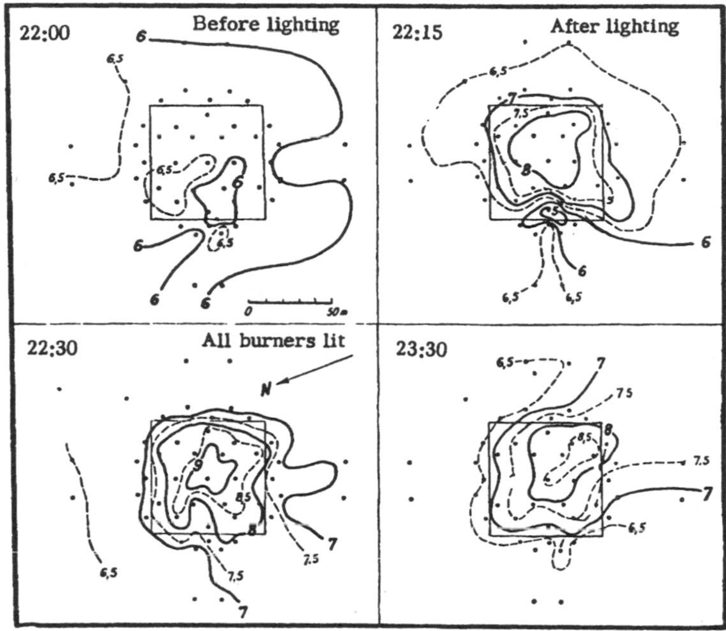

Figure 53-3. Frost protection using oil burners (isotherms at 50 cm above the ground). (After O. W. Kessler)

Sprinklers that spray water on plants also provide considerable frost protection. Heat is gained both as the water cools to freezing (4.19×10^3 J kg^{-1} K^{-1}) and as the water begins to freeze (334×10^3 J kg^{-1}). When temperatures are below 0°C, the leaves become covered with a layer of ice inside, but of spray water on top. The upper surface of the leaves give off heat by radiation, convective heat exchange with the surroundings, and evaporation. At the boundary between ice and water, however, the freezing process causes latent heat to be liberated, maintaining the boundary at a temperature of 0°C. This temperature can be tolerated by many plants. If no further water is supplied, and the ice is allowed to dry, its upper surface would soon fall below 0°C, and the plants might be killed since ice is a good conductor of heat (Table 6-2 and Section 24). Figure 53-4, from O. Dinkelacker and H. Aichele [5319], shows a Moselle vineyard being protected by spraying.

The stream of water from a rotating sprinkler moves around in a circle, while that from a swinging sprinkler moves to and fro. Interruption of the water supply to a single plant must never last long enough for the film of water to freeze completely. This may be avoided by making the time of interruption small or by having a heavy fall of spray. Table 53-1 gives the relation between the minimum density of spray necessary for complete frost protection, the period of interruption with a rotating sprinkler, and the observed temperatures found by H.

Figure 53-4. Sprinkler arrangement for frost protection in a Moselle vineyard. (After O. Dinkelacker and H. Aichele [5319])

von Pogrell [5353] in 20 open air experiments in late winter nights with frost and primarily weak winds.

Table 53-1: Minimum temperatures (°C) for different spray densities and interruption times

Spray density	Interruption time (min)			
(mm hr^{-1})	3	2	1	0.5
<1.5	-2.5 to -3	-3	--	--
1.8-3.0	-4.5	-4 to-5	-5 to -6	--
3.4-4.1	--	--	-6.5	-7.5
11.3	--	--	--	9

Naturally, the temperature that can be counteracted will depend on the spray density and the period of interruption, ranging from a low density with long interruptions to high density and short interruptions. The greater the quantity of water, the more latent heat will be released, and the lower the temperatures that can be coped with successfully. P. J. C. Hamer [5327] has evaluated the amount of water that needs to be applied to protect apple buds and blossoms at varying temperatures.

Damage often occurs in the border areas where the spray is less consistent. This is not only because the quantity of spray is insufficient, or because the interruptions are too long, but also because evaporation from the wetted plants cool them (this is particularly a problem at the edge where the relative humidity and temperature are lower), making them more liable to suffer frost damage. In some cases damage was observed in these areas when plants within and outside the area were untouched. In addition, P. J. C. Hamer [5327] has shown that the distribution of water from sprinklers is very uneven resulting in varying degrees of protection through the crop.

It is important to start and cease watering at the right time. The first drops cool the plants immediately to the wet-bulb temperature. When the air is relatively dry and motionless, cooling can be substantially below the air temperature at the moment. If spraying is begun too late, after the freezing point is reached, the plants may be killed before the latent heat released becomes effective. Many of the ill effects in early days are attributable to this mistake. The opinion is often heard that sprinkling should be stopped only after sunrise when all ice has melted. If sprinkling is discontinued too soon, heat may be drawn from the plant to melt the ice causing damage to the plant.

One of the main problems with sprinklers for certain crops is ice loading. According to J. F. Gerber and J. D. Martsolf [5324]:

> "Low growing crops are rarely broken from the ice even though they may be bent to the ground. Often ice pillars are formed which help support such crops. Foliated trees with weak scaffold branches such as some of the prunus (peaches, plums, nectarines, almonds) cultivars are usually so severely damaged as to preclude the use of sprinkling for cold protection. Aside from pruning and bracing there is little which can be done with mature trees."

To minimize the problem of the breaking of branches associated with overhead sprinkling for frost protection, intermittent applications and microsprinkler irrigation have been proposed. Both these methods minimize ice buildup by reducing the quantity of water used. With intermittent applications instead of the water being applied continuously, it is turned off for brief periods as discussed by K. B. Perry, et al. [5351]. Microsprinkler irrigation reduces water usage by employing a finer spray. They are usually positioned closer to the ground than overhead sprinklers. Microsprinkler irrigation is the most commonly used method of cold protection in the Florida citrus industry (L. R. Parsons, et al. [5349]). While microsprinkler irrigation and intermittent application do not give as good protection as overhead sprinklers, they do minimize the problem of ice buildup. With both intermittent applications and microsprinkler irrigation, there is concern that under windy or dry conditions the crops might actually be cooled by evaporation below ambient air temperatures. K. B. Perry, et al. [5351] investigated this problem with intermittent applications and L. R. Parsons and T. A. Wheaton [5350] investigated it with microsprinkler irrigation. In both situations, the problem appears to be manageable. For more information on microsprinklers, the reader is referred to W. J. Bourgeois, et al. [5310], L. R. Parsons, et al. [5349], and L. R. Parsons and T. A. Wheaton [5350]. F. S. Davies et al. [5316] found that sprinkler irrigation combined with wrapping the tree with insulating material provided greater protection for young citrus trees than either sprinklers or wraps used separately. M. Rieger et al. [5357] also found that while wrapping young trees kept them up to 2-3°C warmer than the air temperature the use of sprinklers offered an additional 1-5°C of protection.

An additional use of sprinklers in spring for frost protection with deciduous fruit trees (and some other crops) is to sprinkle on warm days in order to keep the developing buds cooler. Once started in spring, the rate of bud development is related to its exposure to warm temperatures. As the bud develops, it becomes increasingly frost sensitive (less cold hardy). The object of this type of sprinkling is to prolong the period of environmentally imposed dormancy and thus avoid spring frosts. For additional information dealing with the use of sprinklers to retard spring bud development as a method of frost protection, the reader is referred to P. J. C. Hamer [5326, 5328], J. D. Strang, et al. [5366], J. D. Kalma et al. [5331] and R. E. Griffin [5325].

Another method of delaying bud break in spring and keeping a plant cold hardy longer is to whitewash it, thereby increasing the albedo. This will keep the plant in imposed dormancy longer. H. Aichele [4101] found that white-washing reduced temperatures substantially. G. B. Sibbert and M. Bailey [5363] reported that by painting walnut trees they delayed bud break by as much as two or three weeks. One of the primary advantages in planting fruit trees or perennial frost sensitive crops downwind of large lakes is that as the lake warms slowly in spring it keeps the air temperatures cooler and thus delays spring bud break.

Spraying with water as a means of frost protection is gaining popularity. It is effective to much lower temperatures than other methods, and does not lose its effectiveness with light winds that would blow away artificial fogs or heated air. The topography of the area to be protected, which often causes great difficulties with other methods, is of less importance provided the spray is able to reach the plants. The great expense of the installations is offset to some extent by low operational costs (as compared to heaters). In addition, sprinklers can also be used for irrigation, pest control, and for heat suppression and increasing the relative humidity for crops planted in areas where they may occasionally be exposed to temperatures or humidity conditions that may injure the crop or the plant itself. J. Lomas and M. Mandel [5341], for example, found that overhead sprinklers reduced the maximum temperature while increasing the relative humidity in an avocado plantation. They found that the amount of the temperature reduction was directly related to the actual air temperature. The reduction was 6°C when the air temperature was 24°C, 10°C at 44°C, and up to 13°C on exceptionally hot days. Sprinkling also increased the relative humidity as much as 33 percent. J. R. Recasens et al. [5355] found that sprinkling apple trees in summer to reduce their exposure to stress associated with high temperatures resulted in larger fruit, increased yield and soluble solids (primarily sugars) and equal firmness and storage life.

Considerable attention has been given to the formation of ice as a plant freezes. In the 1970's, it was discovered that certain agents (primarily bacteria) referred to as ice nucleation-active (INA) bacteria, commonly inhabit plant surfaces, and significantly limit the ability of water in herbaceous plant tissues to supercool below -2°C (E. L. Proebsting, Jr. and D. C. Gross [5354]). Plants with low concentrations of INA bacteria seem to lack other intrinsic ice nucleating agents active above -8 to -10°C (E. N. Ashworth, et al. [5306]). This finding suggested that by applying disinfectants or inhibitors to INA bacteria, plants might be able to supercool to a greater degree. Many researchers have shown positive correlations between the presence of INA bacteria and higher temperatures at which frost damage occurs. However, some studies, particularly with fruit trees, have shown no correlation (E. N. Ashworth and G. A. Davis [5305], E. L. Proebsting, Jr. and D. C. Gross [5354]). Thus, while the results to date appear promising in some situations for some crops, we are unable to conclude if this is a new viable method that will assist in protecting plants from frost.

P. springae is the most common of the five known INA bacteria (S. E. Lindow [5338]). Using recombinant genetics, S. E. Lindow [5339] modified it so that it would have no ice nucleating activity at temperatures warmer than -15°C (the original bacteria will be called ice$^+$ and the mutant ice *P. springae*.)

S. E. Lindow [5337] has suggested that the number of sites on leaves or fruit with sufficient nutrients or suitable sites for bacterial adhesion may be limited, and this may limit the population size of epiphytic bacteria. Bacteria competing for these limited sites may exhibit considerable antagonism against strains requiring similar resources. S. E. Lindow and N. J. Panopoulos [5339] have also shown that by inoculation and direct application of ice$^-$ *P. syringae*, thus increasing its populations, that there was a large reduction in the population of ice$^+$ *P. syringae*. This was effective only when ice$^-$ *P. syringae* was applied before the colonization by the ice$^+$ bacteria. In field test, frost injury to treated plants were reduced from 20 to 95% compared with untreated plants (S. E. Lindow [5333, 5334] and S. E. Lindow, et al. [5335, 5336]). Despite these positive results, using ice$^-$ *P. syringae* for frost protection is still in an experimental stage. For additional information dealing with frost, the reader is referred to J. D. Kalma's, et al. [5331] book dealing with the occurrence, impact and protection against frost.

REFERENCES

SECTION 1

[100] *Arya, S. P.*, Introduction to Micrometeorology, Academic Press, 1988.

[101] *Geiger, R.*, Das Stationsnetz z. Untersuchung d. bodennahen Luftschicht. Deutsch. Met, Jahrb. f. Bayern, 1923-1927.

[102] *Jones, H. G.*, Plants and the Microclimate – A Quantitative Approach to the Environmental Plant Physiology, Cambridge University Press, 1992.

[103] *Kraus, G.*, Boden u. Klima auf kleinstem Raum. Fischer, Jena, 1911.

[104] *Lettau, H. H.* and *Davidson, B.*, Exploring the Atmosphere's First Mile, Vol. I and II. Pergamon Press, London, 1957.

[104a] *Linacre, E. T.*, Climate Data and Resources: A Reference and Guide, Routledge, 1992.

[105] *Munn, R. E.*, Descriptive Micrometeorology, Academic Press, 1966.

[106] *Schnelle, F.*, Pflanzen-Phanologie, Probleme der Bioklimatologie 3, Akad. Verl. Ges. Leipzig, 1955.

[107] *Scorer, R. S.*, Environmental Aerodynamics, Ellis Horwood Limited, 1978.

[108] *Stull, R. B.*, An Introduction to Boundary Layer Meteorology, Kluwer Academic Publishers, 1988.

[109] *Sutton, O. G.*, The development of meteorology as an exact science. Quart. J. 80, 328-338, 1954.

[110] *Yoshino, M. M.*, Local Climatology. In: J. E. Oliver and R. W. Fairbridge (Eds.) The Encyclopedia of Climatology, 551-558, Van Nostrand and Reinhold Co., 1987.

SECTIONS 2-4

[200] *Ahmad, B. S.*, and *Lockwood, J. G.*, Albedo. Prog. Phy. Geog. 3, 510-543, 1979.

[201] *Ångström, A.*, The albedo of various surfaces of ground. Geograf. Ann. 7, 323-342, 1925.

[202] *Baumgartner, A.*, Das Eindringen des Lichtes in den Boden. Forstw. C. 72, 172-184, 1953.

[203] *Büttner, K.*, and *Sutter, E.*, Die Abkühlungsgrösse in den Dünen etc. Strahlentherapie 54, 156-173, 1935.

[204] *Collmann, W.*, Diagramme zum Strahlungsklima Europas. Ber. DWD 6, Nr. 42, 1958.

[205] *Coulson, K. L., and Reynolds, D. W.*, The spectral reflectance of natural surfaces, J. App. Met. 10, 1285-1295, 1971.

[206] *Dirmhirn, I.*, Einiges über die Reflexion der Sonnen- u. Himmelsstrahlung an verschied. Oberflächen. Wetter u. Leben 5, 86-94, 1953.

[207] *Falckenberg, G.*, Absorptionskonstanten einiger met. wichtiger Körper für infrarote Wellen. Met. Z. 45, 334-337, 1928.

[208] *Fleischer, R.*, and *Gräfe, K.*, Die Ultrarot-Strahlungsströme aus Registrierungen des Strahlungsbilanzmessers nach Schulze. Ann. d. Met. 7, 87-95, 1955/56.

[209] *Gates, D. M.*, and *Tantraporn, W.*, The reflectivity of deciduous trees and herbaceous plants in the infrared to 25 microns. Science 95, 613-616, 1952.

[210] *Gayevsky, U. L.*, Surface temperature of large territories. Proc. Main Geophys. Obs., No. 26, 1951.

[211] *Köhn, M.*, Zur Kenntnis des Lichthaushaltes dünner Pulverschichten, insbesondere v. Böden. Naturw. 34, 89-90, 1947.

[212] *Kondratyev, K. Ya*, Radiation in the Atmosphere, Academic Press, 1969.

[213] *Lauscher, F.*, Strahlungs- u. Wärmehaushalt. Ber. DWD 4, Nr. 22, 21-29, 1956.

[214] *Sauberer, F.*, Das Licht im Boden. Wetter u. Leben 3, 40-44, 1951.

[215] —, Beiträge zur Kenntnis des Strahlungsklimas von Wien. Wetter u. Leben 4,187-192, 1952.

[216] —, and *Dirmhirn, I.*, Untersuchungen über die Strahlungsverhältnisse auf den Alpengletschern. Arch. f. Met. (B) 3, 256-269, 1951.

[217] *Van de Griend, A. A., Owe, M., Groen, M., and Stoll, M. P.*, Measurement and spatial variation of thermal infrared surface emissivity in a savanna environment. Wat. Resourc. Res. 27, 371-379, 1991.

SECTION 5

[500] *Angell, J. K.* and *Korshover, J.*, Update of ozone variations through 1979. In J. London (ed.), Proceedings of the Quadrennial International Ozone Symposium, Boulder, Colorado, 393-6, 1980.

[501] *Ångström, A.*, Über die Gegenstrahlung der Atmosphäre. Met. Z. 33, 529-538, 1916.

[502] *Arnfield, A. J.*, Evaluation of empirical expressions for the estimation of hourly and daily totals of atmospheric longwave emission under all sky conditions. Q. J. R. Meteorol. Soc. 105, 1041-1052, 1979.

[503] *Berdahl, P. and Martin, M.*, Emissivity of clear skies. Sol. Ener. 32, 663-664, 1984.

[504] *Boden, T. A., Kaiser, D. P., Spanski, R. J.*, and *Stoss, F. W. Eds.*, Trends 93: A Compendium of Data on Global Change, Carbon Dioxide Information Analysis Center, World Data Center – A For Atmospheric Trace Gases, Center for Global Environmental Studies, Oak Ridge National Laboratory, 1993.

[505] *Bolz, H. M.*, Über die Wirkung der Temperaturstrahlung des atmosphärischen Ozons am Erdboden. Z. f. Met. 2, 225-228, 1948.

[506] —, Die Abhängigkeit der infraroten Gegenstrahlung von der Bewölkung. Z. f. Met. 3, 201-203, 1949.

[507] —, and *Falckenberg, G.*, Neubestimmung der Konstanten der Angströmschen Strahlungsformel. Z. f. Met. 3, 97-100, 1949.

[508] *Brunt, D.*, Notes on radiation in the atmosphere. Q. J. R. Meteorol. Soc. 58, 389-418, 1932.

[509] *Brutsaert, W.*, On a derivable formula for long-wave radiation from clear skies. Wat. Resourc. Res. 11, 742-744, 1975.

[510] *Czepa, O.*, and *Reuter, H.*, Über den Betrag der effektiven Ausstrahlung in Bodennähe bei klarem Himmel. Arch. f. Met. (B) 2, 250-258, 1950.

[510a] *Davies, J. A. and McKay, D.C.*, Estimating solar irradiance and components. Sol. Ener. 29, 55-64, 1982.

[511] *Dubois, P.*, Nächtliche effektive Ausstrahlung. Gerl. B. 22, 41-99, 1929.

[512] *Falckenberg, G.*, Die Konstanten der Angströmschen Formel zur Berechnung der infraroten Eigenstrahlung d. Atmosph. aus dem Zenit. Z. f. Met. 8, 216-222, 1954.

[513] *Fishman, J., Ramanathan, V., Crutzen, P. J.* and *Liu, S. C.*, Tropospheric ozone and climate. Nature, 282, 818-20, 1979.

[514] *Goody, R. M., and Robinson, G. D.*, Radiation in the troposphere and lower stratosphere. Quart. J. 77, 151-187, 1951.

[515] *Hatfield, J. L., Reginato, R. J. and Idso, S. B.*, Comparison of longwave radiation calculation methods over the United States. Wat. Resourc. Res. 19, 285-288, 1983.

[516] *Haude, W.*, Ergebnisse der allgemeinen met. Beobachtungen etc. * Rep. Scient. Exped. to the NW China under the leadership Dr. Sven Hedin IX Met. 1, Stockholm 1940.

[517] *Hinzpeter, H.*, Die effektive Ausstrahlung u. ihre Abhängigkeit v. d. Absorptionseigenschaften im Fenster der Wasserdampfbanden. Z. f. Met. 11, 321-329, 1957.

[518] *Hov, O.*,: Ozone in the troposphere: high level pollution. Ambio, 13, 73-9, 1984.

[519] *Howard, J. H.*, The transmission of the atmosphere in the infrared. Proc. I. R. E. 47, 1451-1457, 1959.

[520] *Idso, S. B.*, A set of equations for full spectrum and 8- to 14-μm thermal and 10.5- to 12.5 μm thermal radiation from cloudless skies. Wat. Resourc. Res. 17, 295-304, 1981.

[521] *—, and Jackson, R. D.*, Thermal radiation from the atmosphere. J. Geophys. Res. 74, 5397-5403, 1969.

[522] *Kley, D., Crutzen, P. J., Smit, H. G. J., Vömel, H., Oltmans, S. J., Grassl, H., and Ramanathan, V.*, Observations of near-zero ozone concentrations over the convective Pacific: Effects on air chemistry. Science 274, 230-233, 1996.

[523] *Lal, M., Dube, S. K., Sinha, P. C.* and *Jain, A. K.*, Potential climatic consequences of increasing anthropogenic constituents in the atmosphere. Atm. Env., 20, 639-42, 1986.

[524] *Lauscher, F.*, Bericht uber Messungen der nächtl. Ausstrahlung auf der Stolzalpe. Met. Z. 45, 371-375, 1928.

[525] —, Wärmeausstrahlung u. Horizonteinengung. Sitz-B. Wien. Akad. 143, 503-519, 1934.

[526] *Liou, K-N.*, An Introduction to Atmospheric Radiation, Academic Press, 1980.

[527] *Morgan, D. L., Pruitt, W. O. and Lourence, F. J.*, Estimation of atmospheric radiation. J. App. Met. 10, 463-469, 1971.

[528] *Sauberer, F.*, Registrierungen der nächtlichen Ausstrahlung. Arch. f. Met. (B) 2, 347-359, 1951.

[529] *Schnaidt, F., Zur* Absorption infraroter Strahlung in dünnen Luftschichten. Met. Z. 54, 234-242, 1937.

[530] *Sugita, M. and Brutsaert, W.*, Cloud effect in the estimation of instantaneous downward longwave radiation. Wat. Resourc. Res. 29, 599-605, 1993.

[531] *Swinbank, W. C.*, Long-wave radiation from clear skies. Q. J. R. Meteorol. Soc. 89, 339-348, 1963.

[532] *Unsworth, M. H. and Monteith, J. L.*, Longwave radiation at the Ground I Angular distribution of incoming radiation. Q. J. R. Meteorol. Soc. 101, 13-24, 1975.

[533] *Volz, A.* and *Kley, D.*, Evaluation of the Montsouris series of ozone measurements made in the nineteenth century. Nature, 332, 240-2, 1988.

SECTION 6

[600] *Albrecht, F.*, Messgeräte des Wärmehaushalts an der Erdoberfläche als Mittel d. bioklimat. Forschung. Met. Z. 54, 471-475, 1937.

[601] *Becker, F.*, Die Erdbodentemperaturen als Indikator der Versickerung. Met. Z. 54, 372-377, 1937.

[602] *Diem, M.*, Bodenatmung. Gerl. B. 51, 146-166,1937.

[603] *Hofmann, G.*, Die Thermodynamik der Taubildung. Ber. DWD 3, Nr. 18, 1955.

SECTIONS 7-8

[700] *Defant, A.*, Über die Abkühlung der untersten staubbeladenen Luftschichen. Ann. d. Hydr. 47, 93-105, 1919.

[701] *Lettau, H.*, Über die Zeit- u. Höhenabhängigkeit d. Austauschkoeff. im Tagesgang innerhalb d. Bodenschicht. Gerl. B. 57, 171-192, 1941.

[702] *Monin, A. S. and Obukov, A. M.*, Basic turbulent mixing laws in the atmospheric surface layer. Tr. Geofiz. Inst. Akad. Nauk. SSSR 24(151), 163-187, 1954.

[703] *Priestley, C. H. B.*, Free and forced convection in the atmosphere near the ground. Quart. J. 81, 139-143, 1955; 82, 242-244, 1956.

[704] *Ramdas, L. A.*, and *Malurkar, S. L.*, Surface convection and variation of temperature near a hot surface. Indian J. Physics 7, 1-13, 1932.

[705] *Schmauss, A.*, Die nächtliche Abkühlung der untersten Luftschichten. Ann. d. Hydr. 47, 235-236, 1919.

[706] *Schmidt, W.*, Der Massenaustausch in freier Luft u. verwandte Erscheinungen. H. Grand, Hamburg, 1925.

SECTION 9

[900] Aron, R. H., Mixing Height - An Inconsistent Indicator of Potential Air Pollution Concentrations. Atm. Env. 17, 2193-2197, 1983.

[901] —, Author Reply to Hawke and Heggie. Atm. Env. 19, 1732-1733, 1985.

[902] —, *and Aron, I-M.*, Statistical forecasting models: II. Oxidant concentrations in the Los Angeles Basin. J. Air Poll. Con. Assoc. 28, 686-688, 1978.

[903] *Berger-Landefeldt, U.*, *Kiendl, J.*, and *Danneberg, H.*, Betrachtungen zur Temp.-u. Dampfdruckunruhe über Pflanzenbeständen. Met. Rundsch. 10, 11-20, 1957.

[904] *Boubel, R. W.*, *Fox, D. L.*, *Turner, D. B.* and *Stern, A. C.*, Fundamentals of Air Pollution, Academic Press, 1994.

[905] *Church, P. E.*, Dilution of waste stack gases in the atmosphere. Ind. Eng. Chm. 41, 2753-2756, 1949.

[906] *Deng, J.P. and Aron, R. H.*, Further investigations into mixing height. Atm. Env. 19, 1563-1564, 1985.

[907] *Firbas, F.*, and *Rempe, H.*, Über die Bedeutung der Sinkgeschwindigkeit für d.Verbreit. des Blütenstaubes durch d. Wind. Biokl. B. 3, 49-53, 1936.

[908] *Haude, W.*, Temperatur u. Austausch der bodennahen Luft über einer Wüste. Beitr. Phys. d. fr. Atm. 21, 129-142, 1934.

[909] *Hewson, E. W.*, The meteorological control of atmospheric pollution by heavy industry. Quart J. 71, 266-282, 1945.

[910] —, Industrial air pollution meteorology, Meteorological Laboratories of the College of Engineering, University of Michigan, Ann Arbor, 1964.

[911] —, and *Gill, G. C.*, Meteorological Investigations in Columbia River Valley near Trail, B. C., In: report submitted to the Trail Smelter Arbitral Tribunal, U. S. Bureau of Mines Bull. 453, 23-228, 1944.

[912] *Holzworth, G. C.*, A study of air pollution potential for the Western United States. J. App. Met. 1, 366-382, 1962.

[913] *Kohlermann, L.*, Untersuchungen über d. Windverbreitung der Früchte u. Samen d. mitteleurop. Waldbäume. Forstw. C. 69, 606-624, 1950.

[914] *Lyons, W. A. and Cole, H. S.*, Fumigation and plume trapping on the shores of Lake Michigan during stable onshore flow. J. App. Met. 12, 494-510, 1973.

[915] *Scorer, R. S.*, The behavior of chimney plumes. Int. J. Air Poll. 1, 198, 1959.

[916] U. S. Weather Bureau, Meteorology and Atomic Energy, Washington, 1968.

[917] W.M.O., Guide to International Meteorological Instruments and Methods of Observation, World Meteorological Organization, Geneva, No. 8, TP. 3, 1983.

SECTION 10

[1001] *Albrecht, F.*, Ergebnisse von Dr. Haudes Beobachtungen etc. Rep. Scient. Exped. to the NW Prov. China under the leadership Dr, Sven Hedin 9, Met. 2, Stockholm, 1941.

[1002] *Ameyan, O., and Alabi, O.*, Soil temperatures in Nigeria. Phys. Geog. 8, 275-286, 1987.

[1003] *Batta, E.*, Tägliche Normalwerte der Bodentemp. u. die Bestimmung einer Aussaattemperatur. Idöjárás 59, 351-358, 1955.
[1004] *Chang, Jen-hu*, World patterns of monthly soil temperature distribution. Ann. Assoc. Amer. Geograph. 47, 241-249, 1957.
[1005] —, Global distribution of the annual range in soil temperature. Tran. Amer. Geophys. Union 38, 718-723, 1957.
[1006] —, Ground temperature (2 Bde.). Blue Hill Met. Observ., Milton Mass., 1958.
[1007] *Dirmhirn, I.*, Tagesschwankung d. Bodentemp., Sonnenscheindauer u. Bewölkung. Wetter u. Leben 3, 216-219, 1951.
[1008] *Hausmann G.*, Unperiodische Schwankungen der Erdbodentemp. in 1 m bis 12 m Tiefe. Z. f. Met. 4, 363-372, 1950.
[1009] *Herr, L.*, Bodentemperaturen unter besonderer Berücksichtigung der ausseren met. Faktoren. *Diss.* Leipzig, 1936.
[1010] *Homén*, T., Der tägliche Wärmeumsatz im Boden u. die Wärmestrahlung zwischen Himmel u. Erde. Leipzig, 1897.
[1011] *Katić , P.*, The soil temperatures at Novi Sad. Ann. Scient. Work Fac. of Agriculture Novi Sad 1, 1957.
[1012] *Kullenberg, B.*, Biological observations during the solar eclipse in southern Sweden etc. Oikos 6, 51-60, 1955.
[1013] *Leyst, E.*, Über die Bodentemperaturen in Pawlowsk. Rep. f. Met. 13, Nr. 7,1-311, Petersburg, 1890.
[1014] —, Untersuchungen über die Bodentemp. in Königsberg. Schr. d. phys.ökonom. Ges. Königsberg 33, 1-67, 1892.
[1015] *McCulloch, J. S. G.*, Soil temperatures near Nairobi 1954-55. Quart. J. 85, 51-56, 1959.
[1016] *Paulsen, H. S.*, On radiation, illumination and meteorological conditions in S. Norway during the total solar eclipse of June 30,1954. Arbok Univ. Bergen Nr. 7, 1955.
[1017] *Schmidt*, A., Theoretische Verwertung der Königsberger Bodentemperaturbeobachtungen. Schr. d. phys.-ökonom. Ges. Königsberg 32, 97-168, 1891.
[1018] *Unger, K.*, Bearbeitung der Bodentemperaturen von Quedlinburg. Angew. Met. 1, 85-90, 1951.

SECTIONS 11-12

[1101] *Best, A. C., Knighting, E., Pedlow, R. H.*, and *Stormonth, K.*, Temperature and humidity gradients in the first 100 m over SE-England. Geophys. Mem. 89, London, 1952.
[1102] *Brocks, K.*, Über den tägl. u. jährl. Gang der Höhenabhängigkeit der Temp. in den untersten 300 m d. Atmosphäre u. ihren Zusammenhang mit d. Konvektion. Ber. DWD-US Zone 1, Nr. 5, 1948.
[1103] —, Die Höhenabhängigkeit der Lufttemperatur in der nächtlichen Inversion. Met. Rundsch. 2, 159-167, 1949.
[1104] —, Temperatur u. Austausch in der untersten Atmosphäre. Ber. DWD-US Zone 2, Nr. 12, 166-170, 1950.
[1105] *Flower, W. D.*, An investigation into the variation of the lapse rate of temperature in the atmosphere near the ground at Ismailia, Egypt. Geophys. Mem. 71, London, 1937.
[1106] *Henning, H.*, Pico-aerologische Untersuchungen über Temperatur- and Windverhältnisse d. bodennahen Luftschicht bis 10 m Höhe in Lindenberg. Abh. Met. D. DDR 6, Nr. 42, 1-66, 1957.
[1107] *Johnson, N. K.*, A study of the vertical gradient of temperature in the atmosphere near the ground. Geophys. Mem. 46, London, 1929.
[1108] —, and *Heywood, G. S. P.*, An investigation of the lapse rate of temperature in the lowest 100 m of the atmosphere. Geophys. Mem. 77, London, 1938.

SECTION 13

[1300] *Aron, R. H., Francek, M. A., Nelson, B. D., and Bisard, W. J.*, The persistence of atmospheric misconceptions: How they cloud our judgment. Sci. Teach. 61, 30-33, 1994.
[1301] *Baum, W.* A., Note on the theory of super-autoconvective lapse rates near the ground. J. Met. 8, 196-198, 1951.
[1302] *Best, A. C.*, Transfer of heat and momentum in the lowest layers of the atmosphere. Geophys. Mem. 65, London, 1935.
[1303] *Czepa, O.*, Über die Energieleitung durch langwellige Strahlung in der bodennahen Luftschicht. Z. f. Met. 5, 292-300, 1951.
[1304] *De Mastus, H. L.*, Pressure disturbances in the vicinity of dust devils. Bull. Am. Met. Soc. 35, 497-498, 1954.
[1305] *Flower, W. D.*, Sand devils. Met. Office London Profess. Notes 71, 1936.
[1306] *Franssila, M.*, Mikroklimatische Temperaturmessungen in Sodankylä. Mitt. Met. Z. Anst. Helsinki 26, 1-29, 1945.
[1307] *Graetz, R. D.* and *Cowan, I.*, Microclimate and evaporation. In: Arid Land Ecosystems: Structure, Functioning and Management. Cambridge University Press, International Biological Programme, 409-433, 1979.
[1308] *Ives, R. L.*, Behaviour of dust devils. Bull. Am. Met. Soc. 28, 168-174, 1947.
[1309] *Klauser, L.*, Beobachtung einiger Kleintromben bei Potsdam. Z. f. Met. 4, 187-188, 1950.
[1310] *Kyriazopoulos, B. D., Micrometeorological* phenomenon in Byzantine decoration. Publ. Met. Inst. Univ. Thessaloniki 4, 1955.
[1311] *Möller, F.*, Strahlungsvorgänge in Bodennähe. Z. f. Met. 9, 47-53, 1955.
[1312] *Nelson, B. D., Aron, R. H. and Francek, M. A.*, Clarification of selected misconceptions in physical geography. J. Geog. 91, 76-80, 1992.
[1313] *Schlichting, H.*, Kleintrombe. Ann. d. Hydr. 62, 347-348, 1934.
[1314] *Thornthwaite, C. W.*, Micrometeorology of the surface layer of the atmosphere. Publ. Climat. 1-5, 1948-52.

SECTION 14

[1400] *Bottsma, A.*, Estimating grass minimum temperatures from screen minimum values and other climatological parameters. Ag. For. Met. 16, 103-113, 1976.
[1401] *De Quervain, F.*, and *Gschwind, M.*, Die nutzbaren Gesteine der Schweiz. H. Huber, Bern, 1934.
[1402] *Dimitz, L.*, Untersuchungen über die Frostdauer in 2 m and 5 cm über dem Erdboden. Wetter u. Leben 2, 58-61, 1950.
[1403] *Eimern, J. van*, and *Kaps, E.*, Lokalklimatische Untersuchungen im *Raum* der Harzburger Berge u. d. benachbart. Elbniederung. Landwirtsch.-Verl. Hiltrup, 1954.
[1404] *Falckenberg,* G., Der Einfluss der Wellenlängentransformation auf das Klima der bodenn. Luftschichten u. d. Temp. der freien Atmosphäre. Met. Z. 48, 341-346, 1931.
[1405] —, Der nächtliche Wärmehaushalt bodennaher Luftschichten. Met. Z. 49, 369-371, 1932.
[1406] —, Experimentelles zur Absorption dünner Luftschichten für infrarote Strahlung. Met. Z. 53, 172-175, 1936.
[1407] —, and *Stoecker,* E., Bodeninversion u. atmosphärische Energieleitung durch Strahlung. Beitr. Phys. d. fr. Atm. 13, 246-269, 1927.
[1408] *Fleagle, R.* G., A theory of fog formation. J. Marine Res. 12, 43-50, 1953.
[1409] *Hader, F.*, Kann der Erdbodenabstand der Thermometerhütte verkleinert werden? Wetter u. Leben 6, 27-31, 1954.
[1410] *Heyer, E.*, Über Frostwechselzahlen in Luft u. Boden. Gerl. B. 52, 68-122, 1938.

[1411] *Murthy, A. S. V., Srinivasan, J.,* and *Harasimha, R.* A theory of the lifted temperature minimum on calm clear nights. Phil. Trans. R. Soc. Lond. A. 344, 183-206, 1993.
[1412] *Narasimha, R.,* The dynamics of the Ramdas layer. Current Science. 66, 16-23, 1994.
[1413] *Narasimha, R.* and *Murthy, A. S. V.,* The energy balance in the Ramdas layer. Boundary-Layer Met. 76, 307-321, 1995.
[1414] *Oke, T. R.,* The temperature profile near the ground on calm clear nights. Quart. J. R. Met. Soc. 96, 14-23, 1970.
[1415] *Ramanathan, K. R.,* and *Ramdas, L. A.,* Derivation of Angströms formula for atmospheric radiation and some general considerations regarding nocturnal cooling of air-layers near the ground. Proc. Ind. Acad. Sciences 1, 822-829, 1935.
[1416] *Ramdas, L. A.,* and *Atmanathan, S.,* The vertical distribution of air temperature near the ground during night. Gerl B. 3, 49-53, 1936.
[1417] *Raschke, K.,* Über das nächtliche Temperaturminimum über nacktem Boden in Poona. Met. Rundsch. 10, 1-11, 1957.
[1418] *Schwalbe, G.,* Über die Temperaturminima in 5 cm über dem Erdboden Met. Z. 39, 41-46, 1922.
[1419] *Siegel, S.,* Messungen des nächtlichen Gefüges in der bodennahen Luftschicht. Gerl. B. 47, 369-399, 1936.
[1420] *Sverdrup, H. U.,* Austausch u. Stabilität in der untersten Luftschicht. Met. Z. 53, 10-15, 1936.
[1421] *Troll, C.,* Die Frostwechselhäufigkeit in den Luft- u. Bodenklimaten der Erde. Met. Z. 60, 161-171, 1943.
[1422] *Witterstein, F.,* Die Differenz zwischen Hütten- u. Erdbodenminimumtemp. nach heiteren and trüben Nächten in Geisenheim. Met. Rundsch. 2, 172-174, 1949.

SECTION 15

[1500] *Lehmann, P.,* and *Schanderl, H.,* Tau and Reif. R. f. W. Wiss. Abh. 9, Nr. 4, 1942.
[1501] *Monteith, J. L.,* Dew. Q. J. R. Meteorol. Soc. 83, 322-341, 1957.
[1502] *Ramdas, L. A.,* The variation with height of the water vapour content of the air layers near the ground at Poona. Biokl. B. 5, 30-34, 1938.
[1503] *Raman, C. R. V., Venkataraman, S. and Krishnamurthy, U.,* Dew over India and its contribution to winter-crop water balance. Ag. Met. 11, 17-35, 1973.
[1504] *Subramaniam, A. R., and Kesava Rao, A. V. R.,* Dew fall in sand dune areas of India. Int. J. Biomet. 27, 271-280, 1983.
[1505] *Tuller, S. E., and Chilton, R.,* The role of dew in the seasonal moisture balance of a summer-dry climate. Ag. Met. 11, 135-142, 1973.
[1506] *Vieser, W.,* Temperatur- u. Feuchtigkeitsverhältnisse in bodennahen Luftschichten. Beitr. z. naturkundl. Forsch. in Südwestdeutschland 10, 3-34, 1951.
[1507] *Vowinckel, E.,* Temperatur u. Feuchtigkeit der bodennahen Luftschicht in Pretoria. Met. Rundsch. 4, 22-23, 1951.

SECTION 16

[1600] *Blackadar, A. K.,* Boundary layer wind maxima and their significance for the growth of nocturnal inversions. Bull. Amer. Met. Soc. 38, 283-290, 1957.
[1601] *Bonner, W. D.,* Climatology of the low level jet. Mon. Wea. Rev. 96, 833-850, 1968.
[1602] *Chen, Y.-L., Chen, X. A.,* and *Zhang, Y.-X.,* A diagnostic study of the low-level jet during TAMEX 10P 5. Mon. Wea. Rev. 122, 2257-2284, 1994.
[1603] *Deacon, E. L.,* Vertical profiles of mean wind in the surface layers of the atmosphere. Geophys. Mem. 11, Nr. 91, London, 1953.

[1604] *Hellmann, G.*, Über die Bewegung der Luft in den untersten Schichten der Atmosphäre. Met. Z. 32, 1-16, 1915, and Sitz-B. Berlin Akad. 404-416, 1919.

[1605] *Heywood, G. S. P.*, Wind structure near the ground and its relation to temperature gradient. Quart. J. 57, 433-455, 1931.

[1606] *McAdie, A. G.*, Studies in frost protection-effect of mixing the air. Mon. Weath. Rev. 40, 122-123, 779, 1912.

[1607] *Paeschke, W.*, Experimentelle Untersuchungen zum Rauhigkeits- and Stabilitätsproblem in der bodennahen Luftschicht. Beitr. Phys. d. fr. Atm. 24, 163-189, 1937.

[1608] *Prandtl, L.*, Führer durch die Strömungslehre, 5. Aufl. Fr. Vieweg, Braunschweig, 1957.

[1609] *Sutton, O. G.*, Note on the variation of the wind with height. Quart. J. 58, 74-76, 1932.

[1610] *Walters, C.*, A synoptic climatology of warm-season low level wind maxima in the Great Plains and their relationship to convection. Unpublished Ph.D. Thesis, Michigan State University, 1997.

[1611] *Wexler, H.*, A boundary layer interpretation of the low-level jet. Tellus 13, 368-378, 1961.

SECTION 17

[1701] *Ashwell, I. Y.*, Meteorology and duststorms in Central Iceland. Arc. Alp. Res. 18, 223-234, 1986.

[1702] *Attmannspacher, W. and Hartmannsgruber, R.*, On extremely high values of ozone near the ground. Pure Appl. Geophys. 106-108, 1091-1096, 1973.

[1703] *Brazel, A. J., and Nickling, W. G.*, Dust storms and their relation to moisture in the Sonoran-Mojave Desert region of the South-western United States. J. Env. Man. 24, 279-291, 1987.

[1704] *Becker, F.*, Messung des Emanationsgehalts der Luft in Frankfurt a. M. and am Taunusobservatorium. Gerl. B. 42, 365-384, 1934.

[1705] *Buch, K.*, Kohlensäure in Atmosphäre u. Meer. Ann. d. Hydr. 70, 193-205, 1942.

[1706] *Davis, D. R.*, Influence of thunderstorms on environmental ozone. Proc. Ann. Tall Timber Fire Ecology Conf., 505-516, 1973.

[1707] *Chatfield, R., and Harrison, H.*, Ozone in the remote troposphere: Higher levels?. Proc. Conf. on Ozone-Oxidants Interaction with the Total Environment, APCA, 77-83, 1976.

[1708] *Chung, Y. S., and Dann, T.*, Observations of stratospheric ozone at the ground level in Regina, Canada. Atm. Env. 19, 157-162, 1985.

[1709] *Demon, L., DeFelice, P., Gondet, H., Pontier, L. and Kast, Y.*, Recherches effectuées par la section de physique du centre de recherches sahariennes en 1954,1955 et 1956. J. des Recherches du Centre Nat. de la Rech. Scientif., Lab. de Bellevue 38, 30-63, 1957.

[1710] *Dougherty, P.M.*, Net carbon exchange characteristics of a dominant white oak tree. Unpublished Ph.D. Dissertation. School of Forestry, Fisheries and Wildlife, University of Missouri, Columbia, 1977.

[1711] *Effenberger, E. F.*, Kern- u. Staubuntersuchungen am Collmberg. Veröffentlicht. Geoph. I. Leipzig 12, 305-359, 1940.

[1712] *Garrett, H. E., Cox, G. S.*, and *Roberts, J. E.*, Spatial and temporal variations in carbon dioxide concentrations in an oak-hickory forest ravine. Forest Sci. 24, 180-190, 1978.

[1713] v. *Gehren, R.*, Die Bodenverwehungen in Niedersachsen 1947-51. Veröff. Niedersächs. Amt f. Landesplanung u. Statistik, Reihe G, Bd. 6, Hannover, 1954.

[1714] *Goldschmidt, H.*, Messung der atmosphärischen Trübung mit einem Scheinwerfer. Met. Z. 55, 170-174, 1938.

[1715] *Huber, B.*, Über die vertikale Reichweite vegetationsbedingter Tagesschwankungen im CO_2-Gehalt der Atmosphäre. Forstw. C. 71, 372-380, 1952.

[1716] —, Der Einfluss, der Vegetation auf die Schwankungen des CO_2-Gehalts der Atmosphäre. Arch. f. Met. (B) 4, 154-167, 1952.

[1717] *Iizuka, I.*, On carbon dioxide in the peach orchard. J. Agr. Met. Japan 11, 84-86, 1955.
[1718] *Jauregui, E.*, The dust storms of Mexico City. J. Climatol. 8, 1-12, 1988.
[1719] *Lamb, R. G.*, A case of stratospheric ozone affecting ground-level oxidant concentrations. J. App. Met. 16, 780-794, 1977.
[1720] *Manabe, S., and Wetherald, R. J.*, Thermal equilibrium of the atmosphere with a given distribution of relative humidity. J. Atm. Sci. 24, 241-259, 1967.
[1721] *Middleton, N. J.*, A geography of dust storms in South-west Asia. J. Climatol. 6, 183-196, 1986.
[1722] *Moses, H., Stehney, A. F., and Lucas, Jr., H. F.*, The effect of meteorological variables upon the vertical and temporal distributions of atmospheric radon. J. Geophys. Res. 65, 1223-1238, 1960.
[1723] *Mühleisen, R.*, Atmosphärische Elektrizität. Hand. d. Physik (herausg. v. S. Flügge) 48 (Geophysik 11) 541-607, Springer, 1957.
[1724] *Nickling, W. G.*, Eolian sediment transport during dust storms: Slims River Valley, Yukon Territory. Can. J. Earth Sci. 15, 1069-1084, 1978.
[1725] *Ohtaki, E., and Oikawa, T.*, Fluxes of carbon dioxide and water vapor above paddy fields. Biomet. 35, 187-194, 1991.
[1726] *Peterson, J. T., Komhyr, W. D., Waterman, L. S., and Gammon, R. H.*, Atmospheric CO_2 variations at Barrow, Alaska, 1973-1982. J. Atm. Chem. 4, 491-510, 1986.
[1727] *Priebsch, J.*, Die Höhenverteilung radioaktiver Stoffe in der freien Luft. Met. Z. 49, 80-81, 1932.
[1728] *Schaedle, M.*, Tree photosynthesis. Ann. Rev. Plant Physiol. 26, 101-115, 1975.
[1729] *Siedentopf, H.*, Zur Optik des atmosphärischen Dunstes. Z. f. Met. 1, 417-422, 1947.
[1730] *Sindowski, K. H.*, Korngrössen- and Konformen-Auslese beim Sandtransport durch Wind. Geolog. Jahrb. Hannover 71, 517-525, 1956.
[1731] *Spittlehouse, D. L., and Ripley, E. A.*, Carbon dioxide concentrations over a native grassland in Saskatchewan. Tellus 29, 54-65, 1977.
[1732] *Teichert, F.*, Vergleichende Messung des Ozongehaltes der Luft am Erdboden and in 80 m Höhe. Z. f. Met. 9, 21-27, 1955.
[1733] *Wakamatsu, S., Uno, I., Ueda, H., Uehara, K., and Tateishi, H.*, Observational study of stratospheric ozone intrusions into the lower troposhere. Atm. Env. 23, 1815-1826, 1989.
[1734] *Wilkening, M. H.*, Daily and annual courses of natural atmospheric radioactivity. J. Geophys. Res. 64, 521-526, 1959.

SECTION 18

[1800] *Albert, D. G.*, Acoustic pulse propagation over a seasonal snow cover. Proc. East. Snow Conf. 47-53, 1992.
[1801] *Beranek, L. L.*, Acoustic properties of gases. In: American Institute of Physics Handbook. D. W. Gray (ed.) 3-57-3-66, 1957.
[1802] *Bohn, D. A.*, Environmental effects on the speed of sound. J. Audio Eng. Soc. 36, 223-231, 1988.
[1803] *Brocks, K.*, Die terrestrische Refraktion, ein Grenzgebiet der Meteorologie and Geodäsie. Ann. d. Met. 1, 329-336, 1948.
[1804] *Delasso, L. P.* and *Knudsen, V. O.*, Propogation of sound in the atmosphere. J. Acoust. Soc. Am. 12, 417, 1941.
[1805] *Fényi, J.*, Über Luftspiegelungen in Ungarn. Met. Z. 19, 507-509, 1902.
[1806] *Fraser, A. B. and Mach, W. H.*, Mirages. Sci. Amer., 102-111, Jan., 1976.
[1807] *Greenler, R.*, Rainbows, Halos, Glories, Cambridge University Press, 1990.
[1808] *Heuer, K.*, Rainbows, Halos and other Wonders, Dodd, Mead, and Co., 1978.
[1809] *Kaye, G. W. C.* and *Evans, E. J.*, Sound absorption by snow. Nature 143, 80, 1939.

[1810] *Klug, H.*, Sound-speed profiles determined from outdoor sound propagation measurements. J. Acoust. Soc. Am. 90, 475-481, 1991.

[1811] *Können, G. P.*, Polarized light in nature, Cambridge University Press, 1985.

[1812] *Kurze, V.* and *Beranek, L. L.*, Sound propogation outdoors. In: Noise and Vibration Control. Institute of Noise Control Engineering, 164-193, 1988.

[1813] *Lehm, W. H. and Schroeder, I.*, The horse merman as an optical phenomenon. Nature 289, 362-366, 1981.

[1814] *Lynch, D. K.*, and *Livingston, W.*, Color and Light in Nature. Cambridge University Press, New York, 1995.

[1815] *McCartney, E. J.*, Optics in the Atmosphere, Wiley Interscience, 1976.

[1816] *Meisser, O.*: "Luftseismik," Handb. d. Experimentalphysik (edited by W. Wien and F. Harms), Vol. 25, Angew. Geophys., part 3, Chapter 1, 211-251, Akad. Verl., Leipzig, 1930.

[1817] *Minnaert, M.* The Nature of Light and Colour in the Open Air, Dover, 1954.

[1818] *Neuberger, H.*, Introduction to Physical Meteorology. The Pennsylvania State University, 1957.

[1819] *Rasmussen, K. B.*, Outdoor sound propogation under the influence of wind and temperature gradients. J. Sound Vib. 104, 321-335, 1986.

[1820] *Pierce, A. D.*, Acoustics – An Introduction to its Physical Principles and Applications. McGraw-Hill, New York, 1981.

[1821] *Sobel, M. I.*, Light. The University of Chicago Press, 1987.

[1822] *Tape, W.*, The topology of mirages. Sci. Amer. June, 120-129, 1985.

[1823] *Thomson, D. W.*, Personal communication, 1986.

[1824] *Tricker, R. A. R.*, Introduction to Meteorological Optics, American Elsevier Pub. Co., Inc., 1970.

[1825] *Whipple, F. J. W.*, The high temperature of the upper atmosphere as an explanation of zones of audibility. Nature 111, 187, 1923.

[1826] *Williamson, S. J. and Cummins, H. Z.*, Light and Color in Nature and Art, John Wiley and Sons, © New York, 1983.

SECTION 19

[1901] *Baden, W.*, and *Eggelsmann, R.*, Ein Beitrag zur Hydrologie der Moore. Moor u. Torf 4, Beilage 3, 1952.

[1902] *Bender, K.*, Die Frühjahrsfröste an der Unterelbe u. ihre Bekämpfung. Z. f. angew. Met. 56, 273-289, 1939.

[1903] *Duin, R. H.* A. *van*, Influence of tilth on soil- and air temperature. Netherlands J. Agric. Sci. 2, 229-241, 1954.

[1904] *Booysen, P. de V.* and *Tainton, N. M.*, Ecological Effects of Fire in South African Ecosystems. Ecological Studies, V. 48. Springer-Verlag, Berlin, 1984.

[1905] *Homén, T.*, Der tägliche Wärmeumsatz im Boden u. die Wärmestrahhlng zwischen Himmel u. Erde. Leipzig, 1897.

[1906] *Kern, H.*, Die Temperaturverhältnisse in Niedermoorboden im Gegensatz zu Mineralboden. Landw. Jahrb. f. Bayern 29, 587-602, 1952.

[1907] *Morgen, A.*, Zur künstl. Wärmesteuerung im Wurzelraum der Pflanze. Met. Rundsch. 10, 135-139, 1957.

[1908] *Norton, B. E.*, and *McGarity, J. W.*, The effect of burning of native pasture on soil temperature in northern New South Wales. J. Brit. Grassland So. 20, 101-105, 1965.

[1909] *Olsson, A.*, Undersökning över vältuingens iuverkan på marktemp. och på lufttemp. närmast markytan. Lantbruksakad. Tidskr. Stockholm 92, 220-241, 1953.

[1910] *Pessi, Y.*, Studies on the effect of the admixture of mineral soil upon the thermal conditions of cultivated peat land. Publ. Finnish State Agric. Res. Board 147, 1956 (89 S.).

[1911] —, On the effect of rolling upon the barley and oat crop yield and upon the thermal conditions of cultivated peat land. Publ. Finnish State Agric. Res. Board 151, 1956 (23 S.).

[1912] *Philipps, H.*, *Zur* Theorie der Wärmestrahlung in Bodennahe. Gerl. B. 56, 229-319, 1940.

[1913] *Reuter, H.*, Zur Theorie der nächtlichen Abkühlung der bodennahen Schicht u. Ausbildung der Bodeninversion. Sitz-B. Wien. Akad. 155, 333-358, 1947.

[1914] *Sauberer, F.*, Messungen des nächtlichen Strahlungshaushaltes der Erdoberfläche. Met. Z. 53, 296-302, 1936.

[1915] *Schmidt, W.*, Über kleinklimatische Forschungen. Met. Z. 48, 487-491, 1931.

[1916] *Vries, D. A. de*, Het Warmtegeleidingsvermogen van grond. Med. Landbouwhogeschool Wageningen 52, 1-73, 1952.

[1917] *Winter, G.*, Antibiose u. Symbiose als Elemente der Mikrobenentwicklung im Boden u. Wurzelbereich. Naturw. Rundsch., 116-123, 1951.

[1918] *Yakuwa, R.*, On the effect of soil dressing upon the temperature of peat soil. J. Agr. Met. Japan 8, 92-96, 1953.

SECTION 20

[2001] *Aderikhin, P. G.*, Ob uteplenii pochv putem izmeneniia ikh tsveta. Met. i. Gidrologiia 8, 28-30, 1952.

[2002] *Andó, M.*, Beitrag zur Bodentemperatur des Flugsandes. Acta Geographica 1, 1-7, Szeged, 1955.

[2003] *Dorno, C.*, Über die Erwärmung von Holz unter verschiedenen Arlstrichen. Gerl. B. 32, 15-24, 1931.

[2004] *Dufton, A. F.*, and *Beckett, H. E.*, Terrestrial temperatures. Met. Mag. 67, 252-253, 1932.

[2005] *Ehrenberg, W. W.*, Künstliche Geländefärbung als Beispiel für physikalische Katalyse. Arch. f. Met. (B) 4, 470-482, 1953.

[2006] *Fleagle, R.*, and *Businger, J.*, An Introduction to Atmospheric Physics, International Geophysical Series, Vol. 25, Academic Press, 1980.

[2007] *Firbas, F.*, Über die Bedeutung des thermischen Verhaltens der Laubstreu für die Frühjahrsvegetation des sommergrünen Laubwaldes. Beih. z. Botan. Centralbl. 44, Abt. II, 179-198, 1927.

[2008] *Geiger, R.*, and *Fritzsche, G.*, Spätfrost u. Vollumbruch. Forstarchiv *16*, 141-156, 1940.

[2009] *Gurnah, A. M.*, and *Mutea, J.*, Effects of mulches on soil temperatures under arabila coffee at Kabete, Kenya. Agr. For. Met. 25, 237-244, 1982.

[2010] *Hanson, K. J.*, The radiative effectiveness of plastic films for greenhouses. J. App. Met. 2, 793-797, 1963.

[2011] *Hatfield, J. L.* and *Prueger, J. H.*, Microclimate effects of crop residues on biological processes. Theor. Appl. Climatol. 54, 47-59, 1996.

[2012] *Helgerson, O. T.*, Heat damage in tree seedlings and its prevention. New Forests 3, 333-358, 1990.

[2013] *Horton, R.*, *Bristow, K. L.*, *Kluitenberg, G. J.* and *Sauer, T. J.*, Crop residue effects on surface radiation and energy balance – review. Theor. Appl. Climatol. 54, 27-37, 1996.

[2014] *Huber, B.*, Der Wärmehaushalt der Pflanzen. Datterer, Freising-München, 1935.

[2015] *Keil, K.*, Frostbekämpfung im hohen Norden. Met. Rundsch. 1, 40-41, 1947.

[2016] *Kiss, A.*, Temperaturextreme auf dem Sande von Üllés. Acta Geographica 1, 9-13, Szeged, 1955.

[2016a] *Kubecka, P.*, A possible world record maximum natural grounds surface temperature. Weather 56, 218-221, July 2001.

[2017] *Ludlow, M.* and *Fisher, M.*, Influence of soil surface litter on frost damage. J. Aust. Inst. Agric. Sci. 42, 134-136, 1976.

[2018] *Manera, C.*, *Picuno, P.*, and *Mognozza, G. S.*, Analysis of nocturnal microclimate in single skin cold greenhouses in mediterranean countries. Acta Hort. 281, 47-56, 1990.

[2019] *Mason, B.* In: *Blanc, M. L., Geslin, H., Holzberg, I. A.* and *Mason, B.*, Protection against frost damage, WMO Tech. Note 51, 1963.

[2020] *Martsolf, J. D., Wiltbank, W. J., Hannah, H. E., Bucklin, R. A.*, and *Harrison, D. S.*, Modification of temperature and wind by an orchard cover and heaters for freeze protection. Proc. Fla. State Hort. Sci. 101, 44-48, 1988.

[2021] *Neubauer, H. F.*, Notizen über die Temperatur der Bodenoberfläche in Afghanistan. Wetter u. Leben 4, 165-168, 1952.

[2022] *Neumann, J.*, Some microclimatological measurements in a potato field. Israel Met. Service, Ser. C, Misc. Pap. 6, 1953.

[2023] *Novak, M., Chen, W., Orchansky, A. L*, and *Ketler, R.*, Turbulent exchange processes within and above a straw mulch. Part 1: mean wind speed turbulent statistics. Agr. For. Met. 102, 139-154, 2000.

[2024] *Nijskens, J., Deltour, J., Coutisse, S.* and *Nisen, A.*, Heat transfer through covering materials of greenhouses. Agr. For. Met. 33, 193-214, 1984.

[2025] *Oke, T. R.*, and *Hannell, F. G.*, Variations of temperatures within a soil. Weather 21, 21-28, 1966.

[2026] *Ramdas, L. A.*, and *Dravid, R. K.*, Soil temperatures. Current Science 3, 266-267, 1934.

[2027] *Ramin, v.*, Massnahmen gegen das Verwehen der jungen Zuckerrüben. Zuckerrübenbau 17, 66, 1935.

[2028] *Regula, H.*, Die Wetterverhältnisse während der Expedition and die Ergebnisse der met. Messungen. Ergeb. d. Antarkt. Exped. 1938/39 2, 16-40, Hamburg, 1954.

[2029] *Sato, S.*, Studies on the methods to lower the high temp. of paddy field in the warm districts in Japan-mulching with rice-straw and grass. J. Agr. Met. Japan 11, 39-40, 1955.

[2030] —, Heat economy in clear-water-plot and in carbon-black powdered plot. J. Agr. Met. Japan 13, 30-32, 1957.

[2031] —, and *Funahashi, Y.*, Studies on the methods of lowering the field-temp. in the common-cultivation of rice in warm districts of Japan. J. Agr. Met. Japan 13, 89-92, 1958.

[2032] *Sauberer, F.*, Über die Strahlungsbilanz verschiedener Oberflächen and deren Messung. Wetter u. Leben 8, 12-26, 1956.

[2033] *Schanderl, H.*, and *Weger, N.*, Studien über das Mikroklima vor verschiedenfarbigen Mauerflächen and der Einfluss auf Wachstum and Ertrag der Tomaten. Biokl. B. 7, 134-142, 1940.

[2034] *Schmeidl, H.*, Oberflächentem peraturen in Hochmooren, Wetter u.Leben 17, 87-97, 1965.

[2035] *Schropp, K.*, Die Temperaturen technischer Oberflächen unter dem Einfluss der Sonnenbestrahlung and der nächtl. Ausstrahlung. Gesundheits-Ing., 729-736, 1931.

[2036] *Stanhill, G.*, Observations on the reduction of soil temperatures. Ag. Met. 2, 197-203, 1965.

[2037] *Stoutjesdijk, P.*, High temperatures of trees and pine litter in winter and their biological importance. Int. J. Biomet. 21, 325-331, 1977.

[2038] *Takakura, T.*, Predicting air temperatures in glasshouses (I). J. Met. Soc. Japan, Ser. II 45, 40-52, 1967.

[2039] —, Temperature gradients in the greenhouse. J. App. Met. 6, 956-957, 1967.

[2040] *Vaartaja, O.*, High surface soil temperatures. Oikos 1, 6-28, 1949.

[2041] *Van Wijk, R. R., Larson, W. E.*, and *Burrows, W. L.*, Soil temperature and early growth of corn from mulched and unmulched soils. Proc. Soil Sci. Soc. Am. 23, 428-434, 1959.

[2042] Vaupel, A., Mikroklima and Pflanzentemperaturen auf trocken-heissen Standorten. Flora 145, 497-541, 1958.

[2043] *Vries, D. A. de*, and *Wit, C. T. de*, Die thermischen Eigenschaften der Moorböden and die Beeinflussung der Nachtfrostgefahr dieser Böden durch eine Sanddecke. Met. Rundsch. 7, 41-45, 1954.

[2044] *Waggoner, P. E., Miller, P.. M.*, and *De Roo, H. C.*, Plastic Mulching-Principles and Benefits, Conn. Agric. Exp. St. Bull. No. 634, New Haven, Conn., 1960.

[2045] *Weger, N.*, Beiträge zur Frage der Beeinflussung des Bestandsklimas, des Bodenklimas and der Pflanzenentwicklung durch Spaliermauern and Bodenbedeckung. Ber. DWD-US Zone 4, Nr. 28, 1951.

[2046] —, Höhere Tomaten- u. Gurkenerträge durch Abdecken des Bodens mit Glas. Z. f. Acker- u. Pflanzenbau 97, 115-128, 1953.

[2047] *Whittle, R. M.* and *Lawrence, W. J. C.*, The climatology of glasshouses. III Air temperature. J. Agr. Eng. Res. 5, 165-178, 1960.

[2048] *Wood, R. W.*, Note on the theory of the greenhouse. Phil. Mag. 17, 319-320, 1909.

[2049] *Yakuwa, R.*, and *Yamabuki, F.*, Studies on the raising of the temperature of irrigation water. Bull. Experim. Farm Fac. of Agric., Hokkaido Univ. 10, 28-35, 1952.

[2050] *Zhang, Y., Gauthier. L., de Halleux, D., Dansereau, B.*, and *Gosselin, A.*, Effect of covering materials on energy consumption and greenhouse microclimate. Agr. For. Met. 82, 227-244, 1996.

SECTION 21

[2101] *Berndtsson, R. Nodomi, K., Yasuda, H., Persson, T., Chen, H.* and *Jinno, K.*, Soil water and temperature patterns in an arid desert dune sand. J. Hydrol. 185, 221-240, 1996.

[2102] *Bracht, J.*, Über die Wärmeleitfähigkeit des Erdbodens u. des Schnees u. den Wärmeumsatz im Erdboden. Veröff. Geoph. 1. Leipzig 14, 145-225, 1949.

[2103] *Brooks, F. A.*, and *Rhoades, D. G.*, Daytime partition of irradation and the evaporation chilling of the ground. Trans. Amer. Geophys. Union 35, 145-152, 1954.

[2104] *Brown, R. J. E.*, The distribution of permafrost and its relation to air temperaure in Canada and the U.S.S.R. Arctic 13, 163-177, 1960.

[2105] —, Factors influencing discontinuous permafrost in Canada, In: The Periglacial Environment — Past and Present. T. L. Péwé (Ed.). McGill – Queen's University Press, Montreal, 11-53, 1969.

[2106] *Dücker, A.*, Der Bodenfrost im Strassenbau. E. Schmidt, Berlin and Detmold, 1947.

[2107] *Fleischmann, R.*, Vom Auffrieren des Bodens. Biokl. B. 2, 88-90, 1935.

[2108] *Franssila, M.*, Mikroklimatische Untersuchungen des Wärmehaushalts. Mitt. Met. Zentralanstalt Helsinki 20, 1-103, 1936.

[2109] *Fukuda, H.*, Über Eisfilamente im Boden. J. College of Agric. Tokyo 13, 453-481, 1936.

[2110] *Hayhoe, H., and Tarnocai, C.*, Effect of site disturbance on the soil thermal regime near Fort Simpson, Northwest Territories, Canada. Arc. Alp. Res. 25, 37-44, 1993.

[2111] *Herdmenger, J.*, Flugzeug u. Vorgeschichte. Orion 4, 474-475, 1949.

[2112] *Keränen, J.*, On frost formation in soil. Fennica 73, Nr. 1, 1-14, 1951.

[2113] *Kretschmer, G.*, Messungen der vertikalen Volumänderung von Ackerböden. Wiss. Z. Fr. Schiller Univ. Jena 4, 639-645, 1954/55.

[2114] *Kreutz, W.*, Das Eindringen des Frostes in Böden unter gleichen and verschiedenen Witterungsbedingungen während des sehr kalten Winters 1939/40. R. f. W. Wiss. Abh. 9, Nr. 2, 1942.

[2115] —, Bodenfrost. Umschau in Wiss. u. Techn. 1950, Heft 2.

[2116] *Liddle, M. J.*, and *Moore, K. G.*, The microclimate of sand dune tracks: The relative contribution of vegetation removal and soil compression. J. Appl. Ecol. 11, 1057-1068, 1974.

[2117] *Lowry, W. P. and Lowry, P. P.*, Fundamentals of Biometeorology: Interactions of Organisms and the Atmosphere, Vol. 1, The Physical Environment, Peavine Publications, 1990.

[2118] *Mayer, H.*, Beobachtungen über die Wärmeleitfähigkeit. Synopt. Bearbeit. d. Frankfurter Wetterd., Linke-Sonderheft 1933, S. 67.

[2119] *Monteith, J. L.*, Visible microclimate. Weather 13, 121-124, 1958.

[2120] *Nakshabandi, G. A., and Kohnke, H.*, Thermal conductivity and diffusivity of soil as related to moisture tension and other physical properties. Ag. Met. 2, 271-279, 1965.

[2121] *Passerat de Silans, A. M. B.*, Apparent soil thermal diffusivity, a case study: HAPEX-Sahel experiment, Ag. For. Met. 81, 201-216, 1996.

[2122] *Price, A. G., and Bauer, B. O.*, Small-scale heterogeneity and soil-moisture variability in the unsaturated zone. J. Hydrol. 70, 277-293, 1984.

[2123] *Rettig, H.*, Beitrag zum Problem der Wasserbewegung im Boden. Met. Rundsch. 9, 182-184, 1956.

[2124] *Ruckli, R.*, Der Frost im Baugrund. Springer, Wien, 1950.

[2125] *Schaible, L.*, Frost- and Tauschäden an Verkehrswegen and deren Be kämpfung. W. Ernst, Berlin, 1957.

[2126] *Schmid, J.*, Der Bodenfrost als morphologischer Faktor. A. Hüthig, Heidelberg, 1955.

[2127] *Thompson, R. D.*, Climate and permafrost in Canada. Weather 29, 298-305, 1974.

[2128] *Tuller, S.E.*, Energy balance microclimate variations on a coastal beach. Tellus 24, 260-270, 1972.

[2129] *Uhlig, S.*, Die Untersuchung and Darstellung der Bodenfeuchte. Ber. DWD-US Zone 4, Nr. 30, 1951.

[2130] *Unger, K.*, Bodenfeuchtigkeitsbestimmungen unter einer Standardfläche and unter versch. Pflanzenbeständen zur Charakterisierung der Dürrewirkung. Abh. Met. D. DDR 3, Nr. 19, 23-27, 1953.

[2131] *Unglaube, E.*, Ergebnisse der Bodenfeuchtigkeitsmessungen in Geisenheim. Ber. DWD-US Zone 6, Nr. 38, 1952.

[2132] *Vries, D. A. de*, Some results of field determinations of the moisture content of soil from thermal conductivity measurements. Netherl. J. Agr. Sci. 1, 115-121, 1953.

[2133] *Weger, N.*, Die Wasserbewegung u. die Wassergehaltsbestimmung in gefrorenem Boden. Met. Rundsch. 7, 45-47, 1954.

[2134] *Zhang, T., and Berndtsson, R.*, Temporal patterns and spatial scale of soil water variability in a small humid catchment. J. Hydrol. 104, 111-128, 1988.

SECTION 22

[2201] *Czepa, O., Über die* spektrale Lichtdurchlässigkeit von Binnengewässern. Wetter u. Leben 6, 122-128, 1954.

[2202] *Dirmhirn, I.*, Neuere Strahlungsmessungen in den Lunzer Seen. Wetter u. Leben 3, 258-260, 1951.

[2203] *Herzog, J.*, Thermische Untersuchungen in Waldteichen. Veröff. Geoph. I. Leipzig 8, Nr. 2, 1936.

[2204] *Höhne, W.*, Experimentelle u. mikroklimatische Untersuchungen an Kleingewässern. Abh. Met. D. DDR 4, Nr. 26, 1954.

[2205] *Keil, K.*, Verfahren zur Verhütung des Zufrierens von Häfen. Met. Rundsch. 1, 312, 1948.

[2206] *Nuñez, M., Davies, J. A., and Robinson, P. J.*, Surface albedo of tower site in Lake Ontario. Bound. Lay. Met. 3, 77-86, 1972.

[2207] *Pesta, O.*, Alpine Hochgebirgstümpel u. ihre Tierwelt. Naturwiss. Rundsch. 8, 65-68, 1955.

[2208] *Rathschüler, E.*, Der Einfluiss eines Wasserfalles auf die Luftfeuchtigkeit der Umgebung. Arch. f. Met. (B) 1, 108-114, 1948.

[2209] *Sato, S.*, Macro and microclimates in rice culture and artificial control of microclimates of paddy fields in warm region of Japan. Bull. Kyushu Agric. Exp. Stn. 6, 259-364, 1960.

[2210] *Sauberer, F.*, Über das Licht im Neusiedlersee. Wetter u. Leben 4, 12-15, 1952.

[2211] —, and *Ruttner, F.*, Die Strahlungsverhältnisse der Binnengewässer. Probl. d. kosm. Physik 21, Akad. Verl. Ges. Leipzig, 1941.

[2212] *Schanderl, H.*, Studien über die Körpertemperatur submerser Wasserpflanzen. Ber. D. Bot. G. 68, 28-34, 1955.

[2213] *Schmidt, W.*, Absorption der Sonnenstrahlung im Wasser. Sitz-B. *Wien.* Akad. 117, 237-253, 1908.

[2214] —, Über Boden- u. Wassertemperaturen. Met. Z. 44, 406-411, 1927.

[2215] *Sverdrup, H. U., Johnson M. W.*, and *Fleming, R. H.*, The Oceans, Their Physics, Chemistry and General Biology, Prentice Hall, 1946.

[2216] *Uchijima, Z.*, Microclimate of the rice crop. Intern. Proc. of the Symposium on Climate and Rice, 115-140, 1976.

[2217] *Volk, O. H.*, Ein neuer für botanische Zwecke geeigneter Lichtmesser. Ber. D. Bot. G. 52, 195-202, 1934.

SECTION 23

[2301] *Brocks, K.*, Atmosphärische Temperaturschichtung and Austauschprobleme über dem Meer. Ber. DWD 4, Nr. 22, 10-15, 1956.

[2302] *Bruch, H.*, Die vertikale Verteilung von Windgeschwindigkeit u. Temperatur in den untersten Metern über der Wasseroberfläche. Veröff. Inst. f. Meereskde. Berlin, Heft 38, 1940.

[2303] *Conrad, V.*, Oberflächentemperaturen in Alpenseen. Gerl. B. 46, 44-61, 1935.

[2304] —, Zum Wasserklima einiger alpiner Seen Österreichs. Beih. z. Jahrb. d. Zentralanst. f. Met. u. Geodyn. Wien 1930, Wien, 1936.

[2305] *Eckel, O.*, Über die interdiurne Veränderlichkeit der Fluss- and Seeoberflächentemperaturen. Wetter u. Leben 3, 203-212, 1951.

[2306] —, Mittel and Extremtemperaturen des Hallstättersees. Wetter u. Leben 4, 87-93, 1952.

[2307] —, Zur Thermik der Fliessgewässer: Über die Änderung der Wassertemperatur entlang des Flusslaufs. Wetter u. Leben Sonderheft 2, 41-47, 1953.

[2308] —, and *Reuter, H.*, Zur Berechnung des sommerlichen Wärmeumsatzes in Flussläufen. Geograf. Ann. 32, 188-209, 1950.

[2309] *Kagan, B. A.*, Ocean-Atmosphere Interaction and Climate Modelling, Cambridge University Press, Cambridge, 1995.

[2310] *Kraus, E. B.*, and *Businger, J. A.*, Atmosphere-Ocean Interaction, Oxford University Press, New York, 1994.

[2311] *Kuhlbrodt, E.*, and *Reger, J.*, Wissenschaftl. Ergebnisse der Deutsch. Atlantischen Expedition auf dem Meteor 1925-27, Bd. 14, Berlin 1938.

[2312] *Peppler, W.*, Beitrag zur Kenntnis der Oberflächentemperatur des Bodensees. Z. f. angew. Met. 44, 250-256, 1927; 45, 14-20, 99-105, 1928.

[2313] *Pickard, G. L. and Emery, W. J.*, Descriptive Physical Oceanography, Pergamon Press, 1990.

[2314] *Reiter, E. R.*, Der mitführende Einfluss einer Flussoberfläche auf die darüberliegenden Luftschichten. Arch. f. Met. (A) 8, 384-396, 1955.

[2315] *Roll, H. U.*, Zur Frage des tägl. Temperaturgangs u. des Wärmeaustauschs in den unteren Luftschichten über dem Meere. Aus d. Arch. d. Seewarte 59, Nr. 9, 1939.

[2316] —, Wassernahes Windprofil u. Wellen auf dem Wattenmeer. Ann. d. Met. 1, 139-151, 1948.

[2317] —, Temperaturmessungen nahe der Wasseroberfläche. D. Hydrograph. Z. 5, 141-143, 1952.

[2318] *Sinokrot, B. A., and Stefan, H. G.*, Stream temperature dynamics: Measurements and modeling. Wat. Resourc. Res. 29, 2299-2312, 1993.

[2319] *Wahl, E.*, Temperaturmessungen in der Nordsee im Sommer 1948. Ann. d. Met. 2, 65-71, 1949.

[2320] —, Strahlungseinflüsse bei der Wassertemperaturmessung an Bord von Schiffen. Ann. d. Met. 3, 92-102, 1950.

[2321] *Webb, B.W.*, and *Walling, D. E.*, Long term water temperature behavior and trends in a Devon, UK, river system. Hydrol. Sci. Jour. 37, 567-580, 1992.

[2322] ___., and *Zhang, Y.*, Spatial and seasonal variability in the components of the river heat budget. Hydrol. Proc. 11, 79-101, 1997.

[2323] *Wegner, K. O.*, Windprofilmessungen über Flussoberflächen bei schwachem Wind. Arb. d. Met. Inst. Univ. Köln, 1956.

SECTION 24

[2401] *Abels, H.*, Beobachtungen der tägl. Periode der Temp. im Schnee u. Bestimmung des Wärmeleitungsvermogens des Schnees als Funktion seiner Dichtigkeit. Rep. f. Met. 16, Nr. 1, 1892.

[2402] *Ambach, W.*, Über den nächtlichen Wärmeumsatz der gefrorenen Gletscheroberfläche. Arch. f. Met. (A) 8, 411-426, 1955.

[2403] —, Über die Strahlungsdurchlässigkeit des Gletschereises. Sitz-B. Wien. Akad. 164, 483-494, 1955.

[2404] *Ångström, A.*, Der einfluss der bodenoberfläche auf das lichtklima. Gerlands Beitr. Geoph. 34, 123-130, 1931.

[2405] *Backhouse, S. L.*, and *Pegg, R. K.*, The effects of the prevailing wind on trees in a small area of south-west Hampshire. J. Biogeog. 11, 401-411, 1984.

[2406] *Baker, D. G., Ruschy, D. L., Skaggs, R. H., and Wall, D. B.*, Air temperature and radiation depression associated with a snow cover. J. App. Met. 31, 247-254, 1992.

[2407] —,—, and *Wall, D. B.*, The albedo decay of prairie snows. J. App. Met. 29, 179-187, 1990.

[2408] —, *Skaggs, R. H., and Ruschy, D. L.*, Snow depth required to mask the underlying surface. J. App. Met. 30, 387-392, 1991.

[2409] *Band, G.*, Strahlung u. Temperaturmessung über Schnee. La Mét., 363-369, 1957.

[2410] *Bauer, K. G., and Dutton, J. A.*, Flight investigations of surface albedo. University of Wisconsin, Depart. of Met. Tech. Rep. 2, 1960.

[2411] *Bührer, W.*, Über den Einfluss der Schneedecke auf die Temperatur der Erdoberfläche. Met. Z. 19, 205-2ll, 1902.

[2412] *Colbeck, S. C.*, An overview of seasonal snow metamorphism. Rev. Geophys. 20, 45-61, 1982.

[2413] *Dewey, K. F.*, Daily maximum and minimum temperature forecasts and the influence of snow cover. Mon. Weath. Rev. 105, 399-401, 1977.

[2414] *Dirmhirn, I.*, Über neuere Strahlungsmessungen auf den Ostalpengletschern. La Mét., 345-351, 1957.

[2415] *Doesken, N. J*, and *Judson, A.*, The Snow Booklet: A Guide to the Science, Climatology, and Measurement of Snow in the United States. Colorado Climate Center, Colorado State University, Fort Collins, 1997.

[2416] *Eckel, O.*, and *Thams, C.*, Untersuchungen über Dichte, Temperatur and Strahlungsverhältnisse der Schneedecke in Davos. Geologie d. Schweiz, Hydrol. 3, 275-340, 1939.

[2417] *Friedrich, W.*, Schneerollen. Wetter u. Leben 5, 82-83, 1953.

[2418] *Fukutomi, T.*, Effect of ground temperature upon the thickness of snow cover. Low Temp. Sc. Sapporo Japan 9, 145-148, 1953.

[2419] *Gressel, W.*, Über das Auftreten von Schneerollen u. Schneewalzen in Niederösterreich. Met. Rundsch. 6, 94-96, 1953 (s. auch Universum Wien 13, l39-141, 1958).

[2420] *Grunow, J.*, Zum Wasserhaushalt einer Schneedecke etc. Ber. DWD-US Zone 6, Nr. 38, 385-393, 1952.

[2421] *Hennessey, J. P.*, A critique of "Trees as a local climatic wind-indicator deformed vegetation as an indicator," *Pseudotsuga taxifolia*. J. App. Met, 19, 1020-1023, 1980.

[2421a] *Hoinkes, H.*, Zur Mikrometerologie der eisnahen Luftschicht. Arch. f. Met (B) 4, 451-458, 1953.

[2422] ____, Über Messungen der Ablation and des Wärmeumsatzes auf Alpengletschern etc. Publ. Assoc. Internat. d'Hydrolog. 39, 442-448, 1954.

[2423] *Keränen, J.*, Über die Temperaturen des Bodens and der Schneedecke in Sodankylä. Helsinki, 1920.

[2424] *Kierkus, W. T., and Colborne, W. G.*, Diffuse solar radiation - daily and monthly values as affected by snow cover. Sol. Ener. 42, 143-147, 1989.

[2425] *Köhn, M.*, Über den Einfluss einer Schneedecke auf die Bodentemperaturen. Wetter u. Klima 1, 303-306, 1948.

[2426] *Kreeb, K.*, Die Schneeschmelze als phänologischer Faktor. Met. Rundsch. 7, 48-49, 1954.

[2427] *Langham, E. J.*, Physics and properties of snowcover. In: D. M. Gray and D. H. Male, Handbook of Snow. Pergamon Press, Oxford, 1981.

[2428] *Liljequist, G. H.*, Radiation and wind and temperature profiles over an antarctic snowfield. Proc. Toronto Met. Conf. 1953, Am. Met. Soc. & R. Met. Soc. 78-87, 1954.

[2429] *Löhle, F.*, Absorptionsmessungen an Neuschnee u. Firnschnee. Gerl. B. 59, 283-298, 1943.

[2430] *Loewe, F.*, Etudes de glaciologie en terre Adélie 1951- 1952. Expéd. polaires francaises (P. E. Victor) IX, Paris, 1956.

[2431] *McClung, D,* and *Schaerer, P.*, The Avalanche Handbook, Mountaineers, Seattle, 1993.

[2432] *Michaelis,* P., Ökologische Studien an der alpinen Baumgrenze V. Jahrb. f. wiss. Botanik 80, 337-362, 1934.

[2433] *Möller, F.*, On the backscattering of global radiation by the sky. Tellus 17, 350-355, 1965.

[2434] *Monteith, J. L.*, The effect of grass-length on snow melting. Weather 11, 8-9, 1956.

[2435] *Müller, H. G.*, Zur Wärmebilanz der Schneedecke. Met. Rundsch. 6, 140-143, 1953.

[2436] *Musselman, R. C.*, Using wind-deformed confers to measure wind patterns in alpine transition at Glees. U.S.D.A. Forest Service, Technical Report Int-270, 80-84, 1990.

[2437] *Niemann, A.*, Die Schutzwirkung einer Schneedecke, dargestellt am farbigen Frostschadenbild. Photograph. u. Wissensch. 5, 27-28, 1956.

[2438] *Nkemdirim, L. C.*, A comparison of radiative and actual nocturnal cooling rates over grass and snow. J. App. Met. 17, 1643-1646, 1978.

[2439] *Nyberg, A.*, Temperature measurements in an air layer very close to a snow surface. Geograf. Ann. 20, 234-275, 1938.

[2440] *Paterson, W. S. B.*, The Physics of Glaciers, Pergamon, 1994.

[2440a] *Paulcke, W.*, Praktische Schnee- u. Lawinenkunde (Verständl. Wissensch. 38). Springer, 1938.

[2441] *Pichler, F.*, Über den Kohlensäure- u. Sauerstoffgehalt der Luft unter einer Schneedecke. Wetter u. Leben 1, 15, 1948.

[2442] *Reuter, H.*, Über die Theorie des Wärmehaushalts einer Schneedecke. Arch. f. Met. (A) 1, 62-92, 1948.

[2443] —, Zur Theorie des Wärmehaushalts strahlungsdurchlässiger Medien. Tellus 1, 6-14, 1949.

[2444] *Rossmann, F.*, Beobachtungen über Schneerauchen u. Seerauchen. Z. f. angew. Met. 51, 309-317, 1934.

[2445] *Sauberer, F.*, Versuche über spektrale Messungen der Strahlungseigenschaften von Schnee u. Eis mit Photoelementen. Met. Z. 55, 250-255, 1938.

[2446] —, Die spektrale Durchlässigkeit des Eises. Wetter u. Leben 2, 193-197, 1950.

[2447] *Schlatter, T. W.*, The Local Surface Energy Balance and Subsurface Temperature Regime in Antarctica. J. App. Met. 11, 1048-1062, 1972.

[2448] *Sharratt, B. S., Benoit, G. R.*, and *Voorhees, W. B.*, Winter soil microclimate altered by corn residue management in the northern corn belt of the USA. Soil Till. Res. 49, 243-248, 1998.

[2449] *Slanar, H.*, Schneeabschmelzung im bewachsenen Gelände. Met. Z. 59, 413-416, 1942.

[2450] *Takahashi, Y., Soma, S., and Nemoto, S.*, Observations and a theory of temperature profile in a surface of snow cooling through nocturnal radiation. Seppyo, Tokyo 18, 43-47, 1956.

[2451] *White, D.*, Nature's powdered doughnuts? Country Extra 2, 63, Jan. 1992.

SECTIONS 25-26

[2501] *Albrecht, F.*, Untersuchungen über den Warmehaushalt der Erdoberfläche in verschiedenen Klimagebieten. R. f. W. Wiss. Abh. 8, Nr. 2, 1940.

[2502] *Baumgartner, A.*, Untersuchungen über den Wärme- u. Wasserhaushalt eines jungen Waldes. Ber. DWD 5, Nr. 28, 1956.

[2503] *Budyko, M. I.*, Atlas teplovogo balansa. Leningrad 1955 (s. Ref. H. Flohn, Erdk. 12, 233-237, 1958).

[2504] —, The Earth's Climate: Past and Future, Academic Press, 1982.

[2504a] *Fleischer, R.*, Der Jahresgang der Strahlungsbilanz u. ihrer Komponenten. Ann. d. Met. 6, 387-364, 1953/54.

[2505] —, Registrierung der Infrarotstrahlungsströme der Atmosphäre u. des Erdbodens. Ann. d. Met. 8, ll5-123, 1957.

[2506] *Frankenberger, E.*, Über vertikale Temperatur-, Feuchte- and Windgradienten in den untersten 7 Dekametern der Atmosphäre, den Vertikalaustausch u. den Wärmehaushalt an Wiesenboden bei Quickborn/Holstein 1953/54. Ber. DWD 3, Nr. 20, 1955.

[2507] *Gjessing, Y.*, and *Øvstedal, D. O.*, Microclimates and water budgets of algae, lichens and a moss on some nunataks in Queen Maud Land. Int. J. Biomet. 33, 272-281, 1989.

[2508] *Hoinkes, H.*, Wärmeumsatz u. Ablation auf Alpengletschern II. Geograf. Ann. 35, 116-140, 1953.

[2509] —, and *Untersteiner, N.*, Wärmeumsatz u. Ablation auf Alpengletschern 1. Geograf. Ann. 34, 99-158, 1953; 35, 116-140, 1953.

[2510] *Kraus, H.*, Untersuchungen über den nächtlichen Energietransport and Energiehaushalt in der bodennahen Luftschicht bei der Bildung von Strahlungsnebeln. Ber. DWD 7, Nr. 48, 1958.

[2511] *Miller, D. H.*, The influence of snow cover on local climate in Greenland. J. Met. 13, 112-120, 1956.

[2512] *Mitchell, J. M., Jr.*, Theoretical Paleoclimatology, In: The Quaternary of the United States, H. E. Wright and D. G. Frey (Editors), Princeton University Press, 883, 1965.

[2513] *Munro, D. S.*, Boundary layer climatology. In: J. E. Oliver and R. W. Fairbridge (Eds.) The Encyclopedia of Climatology, 172-183, Van Nostrand and Reinhold Co., 1987.

[2514] *Niederdorfer, E.*, Messungen des Wärmeumsatzes über schneebedecktem Boden. Met. Z. 50, 201-208, 1933.

[2515] *Rider, N. E.*, and *Robinson, G. D.*, A study of the transfer of heat and water vapour above a surface of short grass. Quart. J. 77, 375-401, 1951.

[2516] *Sverdrup, H. U.*, The eddy conductivity of the air over a smooth snow field. Geofysiske Publ. 11, Nr. 7, 1-69, 1936.

[2517] *Untersteiner, N.*, Glazial-meteorologische Untersuchungen im Karakorum. Arch. f. Met. (B) 8, 1-30, 137-171, 1957.

SECTION 27

[2701] *Berg, H.*, Mikroklimatische Beobachtungen am Rande einer Wasserfläche. Ann. d. Met. 5, 227-235, 1952.

[2702] *Craig, R. A.*, Measurements of temperature and humidity in the lowest 1000 feet of the atmosphere over Massachusetts bay. Mass. Inst. Technology, Pap. Phys., Ocean. and Met. 10, Nr. 1, 1946.

[2703] *Helmis, C. G.*, *Asimakopoulos, D. N.*, *Deligiorgi, D.*, and *Lalas P. D.*, Observations of sea breeze fronts near the shoreline. Bound. Lay. Met. 38, 395-410, 1987.

[2704] *Knochenhauer, W.*, Inwieweit sind die Temperatur- u. Feuchtigkeitsmessungen unserer Flughäfen repräsentativ? Erfahr. Ber. d. D. Flugwetterd., 9. Folge Nr. 2, 1934.

[2705] *Krawezyk, B.*, The structure of the heat balance of the human body at the Polish coast of the Baltic Sea. Z. Meteorol. 34, 175-183, 1984.
[2706] *Landeck, J.*, and *Uhlig, S.*, Kondensationserscheinungen über Brachland. Met. Rundsch. 5, 107, 1952.
[2707] *Mäde, A.*, Über die Methodik der meteorologischen Geländevermessung. Sitz-B. Deutsche Akad. d. Landwirtsch. Wiss. Berlin 5, Nr. 8, 1-25, 1956.
[2708] *Nyberg, A. and Raab, L.*, An experimental study of the field variation of the eddy conductivity. Tellus 8, 472-479, 1956.
[2709] *Runge, H.*, Entstehung von Bodennebel durch Auspuffgase. Z. f. angew. Met. 2, 289-300, 1956.
[2710] *Simpson, J. E.*, Sea Breeze and Local Winds, Cambridge University Press, New York, 1994.
[2711] *Tuller, S. E.*, Onshore flow in an urban area: Microclimatic effects. Int. J. Climatol., 15, 1387-1396, 1995.
[2712] *Verber, L. J.*, The climates of South Bass Island, Western Lake Erie. Ecol. 36, 388-400, 1955.
[2713] *Visher, S. S.*, Some climatic influences of the Great Lakes, latitude and mountains; an analysis of climatic charts in climate and man (1941). Bull. Am. Met. Soc. 24, 205-210, 1943.
[2714] *Yoshino, M. M.*, Climate in a Small Area, University of Tokyo Press, 1975.
[2715] *Zhong, S.* and *Takle, E. S.*, An observational study of sea- and land-breeze circulation in an area of complex coastal heating. J. App. Met. 31, 1526-1538, 1992.

SECTION 28

[2801] *Avissar, R., Avissar, P., Mahrer, Y.* and *Ami Bravdo, B.*, A model to simulate response of plant stomata to environmental conditions. Agr. For. Met. 34, 21-29, 1985.
[2802] *Brutsaert, W.*, Evaporation into the Atmosphere: Theory, History and Application, D. Reidel, 1992.
[2803] *Beven, K.*, A Sensitivity analysis of the Penman-Monteith actual evapotranspiration estimates. J. Hydrol. 44, 169-190, 1979.
[2804] *Dolman, A. J., Gash, J. H. C., Roberts, J.* and *Shuttleworth, W. J.*, Stomatal and surface conductance of tropical rainforest. Agr. For. Met. 54, 303-318, 1991.
[2805] *Eisenlohr, W. S.*, Water loss from a natural pond through transpiration by hydrophytes. Wat. Resourc. Res. 2, 443-453, 1966.
[2806] *Hofmann, G.*, Die Thermodynamik der Taubildung. Ber. DWD. 3, Nr. 18, 1955.
[2807] —, Verdunstung u. Tau als Glieder des Wärmehaushalts. Planta 47, 303-322, 1956.
[2808] *Linacre, E. T., Hicks, B. B., Sainty, G. R., and Grause, G.*, The evaporation from a swamp. Ag. Met. 7, 375-386, 1970.
[2809] *Mansfield, W. W.*, Reduction of evaporation of stored water. Proc. Canberra Symposium 1956, 61-64, UNESCO Paris, 1958.
[2810] *McIlroy, I. C.*, The measurement of natural evaporation. J. Aust. Inst. Agr. Sci. 23, 4-17, 1957.
[2811] *Monteith, J. L.*, Editorial note. Weather 12, 225, 1957 (vgl. S. 203-210).
[2812] *National Research Council*, Global Change in the Geosphere-Biosphere, National Academy Press, 1986.
[2813] *Penman, H. L.*, Evaporation: an introductory survey. Netherlands J. Agr. Science 4, 9-29, 1956.
[2814] *Rijks, D. A.*, Evaporation from a papyrus swamp. Q. J. R. Met. Soc. 96, 643-649, 1970.
[2815] *Saugier, B.* and *Katerji, N.*, Some plant factors controlling evapotranspiration. Agr. For. Met. 54, 263-277, 1991.
[2816] *Shuttleworth, W. J.*, Evaporation from Amazonian rainforest. Proc. R. Soc. Lond. B233, 321-346, 1988.

[2817] *Speidel, D. H., and Agnew, A. F.*, The world water budget. In: D. H. Speidel, L. C. Ruedisili, and A. F. Agnew (Eds.), Perspectives on Water: Uses and Abuses. Oxford University Press, 1988.

[2818] *Van der Leeden, F., Troise, F., and Todd, D.*, Water Encyclopedia, 2nd ed., Lewis Pub. Co., 1991.

SECTION 29

[2901] *Angerer, E. V.*, Landschaftsphotographien in ultrarotem u. ultraviolettem Licht. Naturw. 18, 361-364, 1930.

[2902] *Brown, D. S.*, A comparison of the temperature of the flower buds of royal apricot with standard and blackbulb thermograph records during winter, Proc. Amer. Soc. Hort. Sci. 72, 113-122, 1958.

[2903] *Büdel, A.*, Das Mikroklima der Blüten in Bodennähe. Z. f. Bienenforschung 4, 131-140, 1958.

[2904] *Clark, J. A., and Wigley, G.*, Heat and mass transfer from real and model leaves. In: deVries, D. A., and Afgan, N. H., Heat and Mass Transfer in the Biosphere, Part 1 Transfer Processes in the Plant Environment, 413-422, 1975.

[2905] *Egle, K.*, Zur Kenntnis des Lichtfeldes u. der Blattfarbstoffe. Planta 26, 546-583, 1937.

[2906] *Ehleringer, J.*, Comparative microclimatology and plant responses in *encelia* species from contrasting habitats. J. Arid Env. 8, 45-56, 1985.

[2907] *Forseth, I. N.*, and *Teramura, A. H.*, Field photosynthesis, microclimate and water relations of an exotic temperate liana, *Pueraria lobata*, kudzu. Oecologia 71, 262-267, 1987.

[2908] *Gates, D. M.*, Energy Exchange in the Biosphere, Harper and Row, 1962.

[2909] ——, Biophysical Ecology. Springer-Verlag, Berlin, 1980.

[2910] ——, Radiant energy, its receipt and disposal. In: Agricultural Meteorology, P. E. Waggoner (Ed.), Boston, Am. Met. Soc., 1965.

[2911] *Guyot, G.*, Measurement of plant canopy fluoresence. In: C. Varlet-Grancher, R. Bonhomme and H. Sinoquet (Eds.), Crop Structure and Light Microclimate: Characterization and Applications. Institut Nationale de La Recherche Agronomique, 77-91, 1993.

[2912] *Huber, B.*, Der Wärmehaushalt der Pflanzen. Verl. Datterer, FreisingMünchen, 1935.

[2913] *Hummel, K.*, Über Temperaturen in Winterknospen bei Frostwitterung. Met, Rundsch. 1, 147-150, 1947.

[2914] *Jordan, D. N.*, and *Smith, W. K.*, Energy balance analysis of nighttime leaf temperatures and frost formation in a subalpine environment. Ag. For. Met. 71, 359-372, 1994.

[2915] ——, and ——, Microclimate factors influencing the frequency and duration of growth season frost for subalpine plants. Ag. For. Met. 77, 17-30, 1995.

[2916] *Kessler, O. W.*, and *Schanderl, H.*, Pflanzen unter dem Einduss verschiedener Strahlungsintensitäten. Strahlentherapie 39, 283-302, 1931.

[2917] *Knutson, R. M.*, Heat production and temperature regulation in eastern skunk cabbage. Science 186, 746-747, 1974.

[2918] *Kruilk, G.*, The warm-blooded cactus. The National Cactus and Succulent Journal, 37, 100-103, 1982.

[2919] *Kunii, K.*, The tree temperature of Yoshino-Zakura. J. Met. Res. Tokyo 4, 1035-1038, 1953.

[2920] *Lange, O. L.*, Hitze- u. Trockenresistenz der Flechten in Beziehung zu hrer Verbreitung. Flora 140, 39-97, 1953.

[2921] —, Einige Messungen zum Wärmehaushalt poikilohydrer Flechten and Moose. Arch. f. Met. (B) 5, 182-190, 1953.

[2922] *Leuning, R.*, and *Cremer, K. W.*, Leaf temperatures during radiation frosts. Part I. Observations. Ag. For. Met. 42, 121-133, 1988.

[2923] ——, Leaf energy balances: Developments and applications. Phil. Trans. R. Soc. Lond. 3248, 191-204, 1989.

[2924] *Lu, S., Reiger, M., and Duemmel, M. J.,* Flower orientation influences ovary temperatures during frost in peach. Ag. For. Met. 60, 181-191, 1992.

[2925] *Michaelis, G.* and *Michaelis, P.*, Über die winterlichen Temperaturen der flanzlichen Organe, insbesondere der Fichte. Beih. z. Bot. Centralbl. 52, 333-377, 1934.

[2926] *Miller, D. H.*, Water at the Surface of the Earth – An Introduction to Ecosystem Hydrodynamics. Academic Press, New York, 1977.

[2927] *Nagy, K. A., Odell, D. K., and Seymour, R. S.*, Temperature regulation by the inflorescence of philodendron. Science 178, 1195-1197, 1972.

[2928] *Raschke, K.*, Mikrometeorologisch gemessene Energieumsätze eines Alocasiablattes. Arch. f. Met. (B) 7, 240-268, 1956.

[2929] —, Über die physikalischen Beziehungen zwischen Wärmeubergangszahl, trahlungsaustausch, Temperatur u. Transpiration eines Blattes. Planta 48, 200-238, 1956.

[2930] —, Über den Einfluss der Diffusionswiderstände auf die Transpiration and die Temperatur eines Blattes. Flora 146, 546-578, 1958.

[2931] *Rohmeder, E.*, and *Eisenhut, G.*, Untersuchungen über das Mikroklima in estäubungsschutzbeuteln. Silvae Genetica 8, 1-36, 1959.

[2932] *Salisbury, F. B.*, and *Ross, C. W.*, Plant-Physiology. Wadsworth Publishing Co., 1992.

[2933] *Sauberer, F.*, Zur Kenntnis der Strahlungsverhältnisse in Pflanzenbeständen. Biokl. B. 4, 145-155, 1937.

[2934] —, Über die Strahlungseigenschaften der Pflanzen im Infrarot. Wetter u. Leben 1, 231-234, 1948.

[2935] *Seybold,* A., Über den Lichtfaktor photophysiologischer Prozesse. Jahrb. f. wiss. Bot. 82, 741-795, 1936.

[2936] Suzuki, S., The nocturnal cooling of plant leaves and hoarfrost deposited thereon. Geophys. Mag. Tokyo 25, 219-235, 1954.

[2937] *Takasu, K.*, Leaf temperatures under natural environments. Mem. College of Science Kyoto (B) 20, 179-187, 1953.

[2938] *Tranquillini, W.*, Über den Einfluss von Übertemperaturen der Blätter bei Dauereinschluss in Küvetten auf die ökologische CO_2-Assimilationsmessung. Ber. D. Bot. G. 67, 191-204, 1954.

[2939] *Ullrich, H.*, and *Mäde, A.*, Untersuchungen über den Temperaturverlauf beim Gefrieren von Blättern u. Vergleichsobjekten. Planta 31, 251-262, 1940.

[2940] *Varlet-Grancher, C., Moulia, B., Sinoquet, H.* and *G. Russell, G.* Spectral modification of light within plant canopies: How to quantify its effects on the architecture of the plant stand. In: C. Varlet-Grancher, R., Bonhomme and H. Sinoquet (Eds.), Crop Structure and Light Microclimate: Characterization and Applications. Institut Nationale de La Recherche Agronomique, 427-451, 1993.

[2941] *Weger, N.*, Über Tutentemperaturen. Biokl. B. 5, 16-19, 1938.

[2942] —, *Herbst, W.*, and *Rudloff, C. F.*, Witterung u. Phänologie der Blühphase des Birnbaums. R. f. W. Wiss. Abh. 7, Nr. 1, 1940.

SECTION 30

[3001] Ångström, A., The albedo of various surfaces of ground. Geograf. Ann. 7, 323-342, 1925.

[3002] *Bartels, J.*, Verdunstung, Bodenfeuchtigkeit u. Sickerwasser. Z. f. F. u. Jagdw. 65, 204-219, 1933.

[3003] *Filzer, P.*, Untersurhungen über den Wasserumsatz künstlicher Pflanzenbestände. Planta 30, 205-223, 1939.

[3004] *Freidrich, W.*, Über die Verdunstung vom Erdboden. Gas- and Wasserfaich 91, Heft 24, 1950.

[3005] *Göhre, K.*, Der Wasserhaushalt im Boden. Z. f. Met. 3, 13-19, 1949.
[3006] *Kanitscheider, R.*, Temperaturmessungen in einem Bestand von Legföhren. Biokl. B. 4, 22-25, 1937.
[3007] *Köstler, J., N.*, Waldbau. Verlag P. Parey, Berlin, 1950.
[3008] *Lang, A. R. G.*, Cauchy's theorems and estimation of surface areas of leaves, needles and branches. In: C. Varlet-Grancher, R. Bonhomme and H. Sinoquet (Eds.), Crop Structure and Light Microclimate: Characterization and Applications. Institut Nationale de la Recherche Agronomique, 175-182, 1993.
[3009] *Malek, E.*, Night-time evapotranspiration vs. daytime and 24 h evapotranspiration. J. Hydrol. 138, 119-129, 1992.
[3010] *Paeschke, W.*, Mikroklimatische Untersuchungen innerhalb and dicht über erschiedenartigem Bestand. Biokl. B. 4, 155-163, 1937.
[3011] *Palmer, J. W.*, Canopy manipulation for optimum utilization of light. In: Manipulation of Fruiting, C. J. Wright. Butterworths, London, 1989.
[3012] *Unger, K.*, Agrarmeteorologische Studien 1. Abh. Met. D. DDR 3, Nr. 19, 1-22, 1953.
[3013] *Walter, H.*, Grundlagen des Pflanzenlebens, 2. Aufl. E. Ulmer, Stuttgart 1947.

SECTION 31

[3101] *Ball, M. C., Egerton, J. J. G., Cunningham, R. B.*, and *Dunne, P.*, Microclimate above grass adversely affects spring growth of seedling snow gum (*Eucalyptus pauciflora*). Plant Cell Env. 20, 155-166, 1997.
[3102] *Gadre, K. M.*, Microclimatic survey of a sugarcane field. Ind. J. Met. Geophys. 2, 142-150, 1951.
[3103] *Grover, P. E., Grover, J.*, and *Gwynne, M. D.*, Light rainfall and plant survival. In: E. Africa II. Dry Grassland Vegetation. J. Ecol. 50, 199-206, 1962.
[3104] *Khera, K. L.*, and *Sandhu, B. S.*, Canopy temperature of sugarcane as influenced by irrigation regime. Ag. For. Met. 37, 245-258, 1986.
[3105] *Kim, J.* and *Verma, S. B.*, Components of surface energy balance in a temperate grassland eco-system. Bound. Lay. Met. 51, 401-417, 1990.
[3106] *Knapp, R., Lieth, H.*, and *Wolf, F.*, Untersuchungen über die Bodenfeuchtigkeit in verschiedenen Pflanzengesellschaften nach neueren Methoden. Ber. D. Bot. G. 65, 113-132, 1952.
[3107] *Mäde*, A., Die Agrarmeteorologie in der Pflanzenzüchtung. R. f. W. Wiss. Abh. 9, Nr. 6, 1-48, 1942.
[3108] *Monteith, J. L.*, The heat balance of soil beneath crops. Proc. Canberra Symposium 1956, 123-128, UNESCO Paris, 1958.
[3109] *Müller-Stoll, W. R.* and *Freitag, H.*, Beiträge Zur Bestandsklima Analtsisvon Wiesengesellschaften, Angew. Met. 3, 16-30, 1957.
[3110] *Norman, J. T., Kemp, A. W.*, and *Tayler, J. E.*, Winter temperatures in long and short grass. Met. Mag. 86, 148-152, 1957.
[3111] *Ramdas, L. A., Kalamkar, R. J.*, and *Gadre, K. M.*, Agricultural studies in microclimatology. Ind. J. Agric. Sci. 4, 451-467, 1934 and 5, 1-ll, 1935.
[3112] *Reicosky, D. C., Deaton, D. E., and Parsons, J. E.*, Canopy air temperatures and evapotranspiration from irrigated and stressed soybeans, Ag. For. Met. 21, 21-35, 1980.
[3113] *Sandhu, B. S., and Morton, M. L.*, Temperature response of oats to water stress in the field. Ag. For. Met. 19, 329-336, 1978.
[3114] *Specht, R. L.*, Micro-environment (soil) of a natural plant community. Proc. Canberra Symposium 1956, 152-155, UNESCO Paris 1958.[563] *Szász, G.*, Das Bestandsklima der Wintergerste. Debrecen Met. Univ. Inst. Nr. 13, 1956.
[3115] *Tamm, E.*, and *Funke, H.*, Pflanzenklimatische Temperaturmessungen in einem aisbestand. Z. f. Acker- u. Pdanzenbau 100, l99-210, 1955.

[3116] *Waterhouse, F. L.*, Microclimatological profiles in grass cover in relation to biological problems. Quart. J. 81, 63-71, 1955.

[3117] *Wright, J. R., Pierson, F. B., Hanson, C. L.*, and *Flerchinger, G. N.*, Influence of sagebrush on the soil temperatures. In: Proceedings – Symposium on Ecology and Management of Riparian Shrub Communities, Intermountain Research Station, U.S. Dept. of Ag. Forest Service, General Technical Report Int–289, 181-185, 1992.

SECTION 32

[3201] *Aron, R. H.*, Oregon's climatic suitability for premium wine grapes, Calif. Geog. 16, 53-61, 1976.

[3202] *Broadbent, L.*, The microclimate of the potato crop. Quart. J. 76, 439- 454, 1950.

[3203] *Heilman, J. L., McInnes, K. J., Savage, M. J., Gesch, R. W.*, and *Lascano, R. J.*, Soil and canopy energy balances in a west Texas vineyard. Ag. For. Met. 71, 99-114, 1994.

[3204] *Hirst, J. M., Long, I. F.*, and *Penman, H. L.*, Micrometeorology in the Potato crop. Proc. Toronto Met. Conf. 1953, 233-237, 1954.

[3205] *Kliewer, M. N.*, and *Wolpert, J. A.*, Integrated canopy management practices for optimizing vine microclimate, crop yield and quality of table and wine grapes. Bard, Bet Degan Israel, 1991.

[3206] *Linck, O.*, Der Weinberg als Lebenstraum. Hohenlohesche Buchh. Öhringen, 1954.

[3207] *Smart, R. E.*, Principles of grapevine canopy microclimate manipulation with implications for yield and quality. Am. J. Enol. Vitic. 36, 230-239, 1985.

[3208] —, Principles of grapevine canopy microclimate manipulation with implications for yield and quality. Am. J. Enol. Vitil. 36, 230-239,1985.

[3209] *Sonntag, K.*, Bericht über die Arbeiten des Kalmit-Observatoriums. D. et. Jahrb. f. Bayern 1934, Anhang D.

[3210] *Reynolds, A. G., Wardle, D. A.*, and *Maylor, A. P.*, Impact of training system, vine spacing and basal leaf removal on riesling vine performance, berry composition, canopy microclimate and vineyard labor requirements. Am. J. Enol. Vitic. 47, 63-76, 1996.

[3211] *Weise, R.*, Wettkundliches bei Rebenerziehungsversuchen. Weinberg u. Keller 1, 85-90, 1954.

[3212] —, Wie beeinflusst die Erziehungsform die Temperaturen im Rebinnern? Weinberg u. Keller 3, 332-338, 383-390, 1956.

[3213] *Winkler, A. J. Cook, J. A., Kliewer, W. M.*, and *Lider, L. A.*, General Viticulture, University of California Press, Berkeley, 1974.

SECTION 33

[3301] *Baumgartner, A.*, Die Strahlungsbilanz in einer Fichtendickung. Forstw. C. 71, 337-349, 1952.

[3302] *Barradas, V. L., Jones, H. G.*, and *Clark, J. A.* Sunfleck dynamics and canopy structure in a Phaeolus Vulgarls L. Canopy. Int. J. Biomet. 42, 34-43, 1995.

[3303] —, Beobachtungswerte u. weitere Studien zum Wärme- u. Wasserhaushalt ines jungen Waldes. Wiss. Mitt. Met. Inst. d. Univ. München 4, 1957.

[3304] *Brocks, K.*, Die räumliche Verteilung der Beleuchtungsstärke im Walde. Z. f. F. u. Jagdw. 71, 47-53, 1939.

[3305] *Chazdon, R. L. and Fetcher, N.*, Photosynthetic light environments in a lowland rain forest in Costa Rica. J. Ecol. 72, 553-564, 1984.

[3306] *Deinhofer, J.*, and *Lauscher, F.*, Dämmerungshelligkeit. Met. Z. 56, 53-159, 1939.

[3307] *Egle, K.*, Zur Kenntnis des Lichtfeldes u. der Blattfarbstoffe. Planta 26, 546-583, 1937.

[3308] *Eidmann, H.*, Meine Forschungsreise nach Spanisch-Guinea. D. Biologe 10, 1-13, 1941.

[3309] *Ellenberg, H.*, Über Zusammensetzung, Standort and Stoffproduktion odenfeuchter Eichen- and Buchen-Mischwaldgesellsch. NW-Deutschlands. itt. flor.-soz. Arb. Gem. Niedersachsen 5, Hannover 1939.

[3310] *Evans, G. C.*, An area survey method of investigating the distribution of light intensity in woodland, with particular reference to sun flecks. J. Ecol. 44, 391-428, 1956.

[3311] *Fliervoet, L. M.*, and *Werger, M. J. A.*, Canopy structure and microclimate of two wet grassland communities. New Phytol. 96, 115-130, 1984.

[3312] *Fritschen, L. J., Walker, R. B., and Hsia, J.*, Energy balance of an isolated Scots Pine. Int. J. Biomet. 24, 293-300, 1980.

[3313] *Gardner, B. R., Blad, B. L.*, and *Watts, D. G.*, Plant and air temperatures in differentially irrigated corn. Ag. For. Met. 25, 207-217, 1981.

[3314] *Granberg, H. B., Ottosson-Löfvenius, M.*, and *Odin, H.*, Radiative and aerodynamic effects of an open pine shelterwood on calm nights. Ag. For. Met. 63, 171-188, 1993.

[3316] *Grubb, P. J. and Whitmore, T. C.*, A comparison of montaine and lowland rainforest in Ecuador. II. The climate and its effects on the distribution and physiology of the forests. J. Ecol. 54, 303-333, 1966.

[3317] *Gusinde, M.*, and *Lauscher, F.*, Meteorologische Beobachtungen im KongoUrwald. Sitz-B. Wien. Akad. 150, 281-347, 1941.

[3318] *Harada, Y.*, A study of sunlight forests on foggy days. Bull. Forest Exp. Science Tokyo 64, 170-181, 1953.

[3319] *Hutchison, B. A., and Matt, D. R.*, The annual cycle of solar radiation in a deciduous forest. Ag. For. Met. 18, 255-265, 1977.

[3320] *Jarvis, P. G., James, G. B., and Landsberg, J. J.*, Coniferous Forests. In: Monteith, J. L. (ed.), Vegetation and the Atmosphere, Vol. 2, Case Studies, Academic Press, 171-240, 1976.

[3321] *Lauscher, F.*, and *Schwabl, W.*, Untersuchung über die Helligkeit im Wald and am Waldrand. Biokl. B. 1, 60-65, 1934.

[3322] *Löfvenius, M. O.*, Observed and simulated global radiation in pine shelterwood. In: Temperature and Radiation Regimes in Pine Shelterwood and Clear-cut Area. Swedish Univ. of Ag. Sci. Dept. of For. Ecol., by *M. O. Löfvenius*, 1993.

[3323] *March, W. J., and Skeen, J. H.*, Global radiation beneath the canopy and in a clearing of a suburban hardwood forest. Ag. For. Met. 16, 321-327, 1976.

[3324] *Mauerer, K.*, Der Versuchsbestand. Forstw. C. 71, 324-330, 1952.

[3325] *McCaughey, J. H.*, The albedo of a mature mixed forest and a clear-cut site at Petawawa, Ontario. Ag. For. Met. 40, 251-263, 1987.

[3326] *Miller, D. H.*, Snow cover and climate in the Sierra Nevada California. Univ. Calif. Publ. Geography 11, 1-218, Berkeley and Los Angeles, 1955.

[3327] —, Transmission of insolation through pine forest canopy, as it affects the melting of snow. Mitt. Schweiz. Vers. Anst. f. d. Forstl. Versuchswesen 35, 59-79, 1959.

[3328] *Mitscherlich, G.*, Das Forstamt Dietzhausen. Z. f. F. u. Jagdw. 72, 149-188, 1940.

[3329] *Monteith, J. L. and Szeicz G.*, The radiation balance of bare soil and vegetation. Quart. J. 87, 159-170, 1961.

[3330] *Moro, M. J., Pugnaire, F. I., Haase, P.* and *Puigde Fábregas. J.*, Mechanisms of interaction between a leguminous shrub and its understory in a semi-arid environment. Ecography. 20, 175-184, 1997.

[3331] *Nägeli, W.*, Lichtmessungen im Freiland and im geschlossenen Altholzbestand. Mitt. Schweiz. Vers. Anst f. d. Forstl. Versuchswesen 21, 50-306, 1940.

[3332] *Oke, T. R.*, Boundary Layer Climates, Methuen, 1987.

[3333] *Rauner, J. V. L.*, Deciduous forests. In: Montieth, J. L. (ed.), Vegetation and the Atmosphere, Vol. 2, Case Studies, Academic Press, 1976.

[3334] *Richards, P. W.*, The Tropical Rain Forest (hier: Kap. 7: microclimates 58-190). Cambridge Univ. Press, 1952.

[3335] *Sacharow, M. I.*, Influence of wind upon illumination in a forest. Akad Nauk SSSR 67, 913-916, 1949 (MAB 8, 528, 1957).
[3336] *Sauberer, F.*, and *Trapp, E.*, Helligkeitsmessungen in einem Flaumeichenbuschwald. Biokl. B. 4, 28-32, 1937.
[3337] *Scheer, G.*, Über Änderungen der Globalbeleuchtungsstärke durch Belaubung and Horizonteinengung. Wetter u. Leben 5, 65-71, 1953.
[3338] *Shuttleworth, W. J.*, Micrometeorology of temperate and tropical forest. Phil. Trans. R. Soc. Lond. B 324, 299-334, 1989.
[3339] *Shuttleworth, W. J., et al.*, Observation of radiation exchange above and below Amazonian forest. Quart. J. R. Met. Soc. 110, 1163-1169, 1984.
[3340] *Sirén, G.*, The development of spruce forest on raw humus sites in Northern Finland and its ecology (408 S.). Acta Forest. Fennica 62, Helsinki, 1955.
[3341] *Slanar, H.*, Das Klima des östlichen Kongo-Urwalds. Mitt. Geograph. es. Wien, 1945.
[3342] *Stanhill, G.*, Some results of helicopter measurements of the albedo of different land surfaces. Sol. Ener. 13, 59-66, 1970.
[3343] *Stewart, J. B.*, The albedo of a pine forest. Quart. J. 97, 561-564, 1971.
[3344] *Tong, H.*, and *Hipps, L. E.*, The effect of turbulence on the light environment of alfalfa. Ag. For. Met. 80, 249-261, 1996.
[3345] *Trapp, E.*, Untersuchung über die Verteilung der Helligkeit in einem uchenbestand. Biokl. B. 5, 153-158, 1938.
[3346] *Whitmore, T. C. and Wong, Y. K.*, Patterns of sunfleck and shade in tropical rain forest. Malays. For. 22, 50-62, 1959.

SECTIONS 34-35

[3401] *Aston, A. R.*, Heat storage in a young eucalypt forest. Ag. For. Met. 35, 281-297, 1985.
[3402] *Baldocchi, D. D.*, *Verma, S. B.*, and *Anderson, D. E.*, Canopy photosynthesis and water-use efficiency in a deciduous forest. Jour. Appl. Ecol. 24, 251-260, 1987.
[3403] *Baumgartner, A.*, and *Hofmann, G.*, Elektrische Fernmessung der Luftund Bodentemperatur in einem Bergwald. Arch. f. Met. (B) 8, 215-230, 1957.
[3404] *Devito, A. S.*, and *Miller, D. R.*, Some effects of corn and oak forest canopies on cold air drainage. Ag. For. Met. 29, 39-55, 1983.
[3405] *Evans G. C.*, Ecological studies on the rain forest of S. Nigeria II. J. Ecol. 27, 436-482, 1939.
[3406] *Geiger, R.*, Untersuchungen über das Bestandsklima. Forstw. C. 47, 29-614, 848-854, 1925; 48, 337-349, 495-505, 523-532, 749-758, 1926.
[3407] —, and *Amann H.*, Forstmeteorologische Messungen in einem Eichenbestand. Forstw. C. 53, 237-250, 341-351, 705-714, 809-819, 1931; 54, 371-383, 1932.
[3408] *Göhre, K.*, and *Lützke, R.*, Der Einfluss von Bestandsdichte and *-struktur* auf das Kleinklima im Walde. Arch. f. Forstwesen 5, 487-572, 1956.
[3409] *Inoue, E.*, An aerodynamic measurement of photosynthesis over a paddy field. Proc. 7. Jap. Nation. Congress f. Appl. Mech., 211-214, 1957.
[3410] —, Energy budget over fields of waving plants. J. Agr. Met. Japan 14, 6-8, 1958.
[3411] —, *Tani, N.*, *Imai, K.*, and *Isobe, S.*, The aerodynamic measurement of photosynthesis over a nurserey of rice plants. J. Agr. Met. Japan 14, 45-53, 1958.
[3412] *Jaeger, L.* and *Kessler, A.*, Twenty years of heat and water balance climatology at the Hartheim pine forest, Germany. Agr. For. Meteor. 84, 25-36, 1997.
[3413] *Lindroth, A.*, Seasonal and diurnal variation of energy budget components in coniferous forests. Jour. Hydrol. 82, 1-15, 1985.
[3414] *Lindroth, A.*, Canopy conductance of coniferous forests related to climate. Wat. Resourc. Res. 21, 297-304, 1985.

[3415] *Jacobs, A. F. G., Van Boxel, J. H.,* and *Nieveen, J.,* Nighttime exchange processes near soil surface of a maize canopy. Agr. For. Met. 82, 155-169, 1996.

[3416] *Kratochvílová, E. P., E. I., Janouš, D., Marek, M.,* and *Masarovičová,* E., Stand microclimate and physiological activity of tree leaves in an oak-hornbeam forest. I. Stand microclimate In: Trees: Structure and Function, 4, 227-233, 1989.

[3417] *McCaughy, J. H.,* and *Saxton, W. L.,* Energy storage terms in a mixed forest. Agr. For. Met. 44, 1-18, 1988.

[3418] *Marek, M., Masarovičová, E., Kratochvílová, I., Eliáš, P.,* and *Janouš, D.,* Stand microclimate and physiological activity of tree leaves in an oak-hornbeam forest. II. Leaf photosynthetic activity. In: Trees: Structure and Function. 4, 234-240, 1989.

[3419] *Moore, C. J.,* and *Fisch, G.,* Estimating heat storage in Amazonian tropical forest. Agr. For. Met. 38, 147-169, 1986.

[3420] *Oliver, S. A, Oliver, H. R., Wallace, J. S.,* and *Roberts, A. M.,* Soil heat flux and temperature variations with vegetation, soil type and climate. Agr. For. Met. 39, 257-269, 1987.

[3421] *Schimitschek, E.,* Bioklimatische Beoabachtungen and Studien bei Borkenkäfcrauftreten. Wetter u. Leben 1, 97-104, 1948.

[3422] Shuttleworth, W. J., Micrometeorology of temperate and tropical forests. Phil. Trans. R. Soc. Lond. B 324, 229-334, 1989.

[3423] *Spittlehouse, D. L.,* and *Stathers, R. J.,* Seedling microclimate. Land management report number 65. B. C. Ministry of Forests, 1990.

[3424] *Tan, C. S.,* and *Black, T.A.,* Factors affecting the canopy resistance of a Douglas-fir forest. Bound. Lay. Met. 10, 475-488, 1976.

[3425] *Ungeheller, H.,* Mikroklima in einem Buchenhochwald am Hang. Biokl. B. 1, 75-88, 1934.

[3426] *Verma, S. B., Baldocchi, D. D., Anderson, D. E., Matt, D. R.,* and *Clement, R. J.,* Eddy fluxes of CO_2, water vapor, and sensible heat over a deciduous forest. Bound. Lay. Met. 36, 71-91.

SECTION 36

[3601] *Brandt, J.,* The transformation of rainfall energy by a tropical rainforest canopy in relation to soil erosion. J. Biogeog. 15, 41-48, 1988.

[3602] *Delfs, J.,* Die Niederschlagszurückhaltung im Walde. Mitt. d. Arbeitskreises, Wald u. Wasser 2, Koblenz 1955.

[3603] —, Friedrich, W., *Kiesekamp, H.,* and *Wagenhoff, A.,* Der Einfluss des aldes u. des Kahlschlages auf den Abflussvorgang, den Wasserhaushalt u. den Bodenabtrag. Mitt. Niedersächs. Landesforstverwaltung 3, mit Tab.band, Hannover 1958.

[3604] *Calder, I. R., Wright, I. R.,* and *Murdiyarso, D.,* A study of evaporation from tropical rain forest - West Java. J. Hydrol. 89, 13-31, 1986.

[3605] *Eidmann, F. E.,* Die Interception in Buchen- u. Fichtenbeständen. UGGI, Symp. of Hannov.-Münden 1, 5-25, 1959.

[3606] *Eitingen, G. R.,* Interception of precipitation by the canopy of forests. Les i Step, Moskau 8, 7-16, 1951 (MAB 8, 530, 1957).

[3607] *Freise, F.,* Das Binnenklima von Urwäldern im subtropischen Brasilien. Peterm. Mitt. 82, 301-304, 346-348, 1936.

[3608] *Godske, C. L.,* and *Paulsen, H. S.,* Investigations carried through at the station of forest met. at Os II. Univ. Bergen Årb. 1949, Naturw. Rekke 8, 1-39, Bergen, 1950.

[3609] *Grunow, J.,* Der Niederschlag im Bergwald. Forstw. C. 74, 21-36, 1955.

[3610] *Haworth, K.* and *McPherson, G. R.,* Effects of *Quercus Emory* trees on precipitation distribution and microclimate in a semi-arid savanna. J. Arid Env. 31, 153-170, 1995.

[3611] *Hesselman, H.*, Einige Beobachtungen über die Beziehung zwischen der amenverbreitung von Fichte u. Kiefer u. die Besamung der Kahlhiebe. eddel. Fran. Stat. Skogförsöksanst. Stockholm 27, 145-182, 1934; 31, 1-64, 1938.

[3612] *Höppe, E.*, Regenmessungen unter Baumkronen. Mitt. a. d. Forstl. Versuchswesen Österreichs 21, Wien, 1896.

[3613] *Linskens, H. F.*, Niederschlagsmessungen unter verschiedenen Baumronentypen im belaubten u. unbelaubten Zustand. Ber. D. Bot. G. 64, 15-221, 1951.

[3614] —, Niederschlagsverteilung unter einem Apfelbaum im Laufe einer egetationsperiode, Ann. d. Met. 5, 30-34, 1952.

[3615] *Lloyd, C. R., and de O. Marques F°, A.*, Spatial variability of throughfall and stemflow measurements in Amazonian rainforest. Agr. For. Met. 42, 63-73, 1988.

[3616] *Löfvenius, M. O.*, Observations of the nocturnal temperature regime in pine shelterwoods and a nearby clear-cut area. In: Temperature and Radiation Regimes in Pine Shelterwood and Clear-cut Area. Part III, by *M. O. Löfvenius*, 1993.

[3617] *Mäde*, A., Zur Methodik der Taumessung. Wiss. Z. d. Martin-Luther-Univ. Halle-Wittenberg 5, 483-512, 1956.

[3618] *Moličová, H., Hubert, T. P.*, Canopy influence on rainfall fields' microscale structure in tropical forests, J. App. Met. 33, 1464-1467, 1994.

[3619] *Nicolai, V.*, The bark of trees: Thermal properties, microclimate fauna. Oecologia 69, 148-160, 1986.

[3620] *Ovington, J. D.*, A comparison of rainfall in different woodlands. Forestry London 27, 41-53, 1954.

[3621] *Pressland, A. J.*, Rainfall partitioning by an arid woodland (*Acacia aneura*—F. Muell.) in south-western Queensland. Aust. J. Bot., 21, 235-245, 1973.

[3622] ——, Soil moisture redistribution as affected by throughfall and stemflow in an arid zone shrub community. Aust. J. Bot. 24, 641-649, 1976.

[3623] *Priehäusser, G.*, Bodenfrost, Bodenentwicklung u. Flachwurzeligkeit der ichte. Forstw. C. 61, 329-342, 381-389, 1939.

[3624] *Rosenfeld, W.*, Erforschung der Bruchkatastrophen in den ostschlesischen Beskiden in der Zeit v. 1875-1942. Forstw. C., 1-31, 1944.

[3625] *Rossmann, F.*, Das Rauchen der Wälder nach Regen and die Unterscheidung on Wasserdampf and Wasserrauch. Wetter u. Leben 4, 56-57, 1952.

[3626] *Rowe, P. B.*, and *Hendrix, T. M.*, Interception of rain and snow by second growth ponderosa pine. Trans. Amer. Geophys. Union 32, 903-908, 1951.

[3627] *Rutter, A. J., Kershaw, K. A., Robins, P. C., and Morton, A. J.*, A predictive model of rainfall interception in forests, I. Derivation of the model from obersations in a plantation of Corsican Pine. Ag. Met. 9, 367-384, 1971.

[3628] —, *and Morton, A. J.*, A predictive model of rainfall interception in forests, III. Sensitivity of the model to stand parameters and meteorological parameters. J. App. Ecol. 14, 567-588, 1977.

[3629] —, —, *and Robins, P. C.*, A predictive model of rainfall interception in forests, II. Generalization of the model and comparison with observation in some coniferous and hardwood stands. J. App. Ecol. 12, 367-380, 1975.

[3630] *Schubert, J.*, Niederschlag, Verdunstung, Bodenfeuchtigkeit, Schneedecke in Waldbeständen and im Freien. Met. Z. 34, 145-153, 1917.

[3631] *Slatyer, R. O.*, Measurements of precipitation interception by an arid zone plant community – (*Acacia aneura* F. Muell). Arid Zone Res. 25, 181-192, 1965.

[3632] *Stoutjesdijk, P.*, The open shade – an interesting microclimate. Acta. Bot. Neerl. 23, 125-130, 1974.

[3633] *Vis, M.*, Interception, drop size distribution and rainfall kinetic energy in four Colombian forest ecosystems. Earth Surf. Proc. Land. 11, 591-603, 1986.

SECTION 37

[3701] *Azevedo, J.*, and *Morgan, D. L.*, Fog precipitation in coastal California forests. Ecol. 55, 1135-1141, 1974.

[3702] *Archibold, O. W., Ripley, E. A.*, and *Bretell, D. L.*, Comparison of the microclimates of a small aspen grove and adjacent prairie in Saskatchewan. Am. Mid. Nat. 136, 248-261, 1996.

[3703] *Baumgartner, A.*, Licht und Naturverjüngung am Nordrand eines Waldbestandes. Forstw. C. 74, 59-64, 1955.

[3704] *Cavelier, J.*, and *Goldstein, C.*, Mist and fog interception in elfin cloud forests in Columbia and Venezuela. J. Trop. Ecol. 5, 309-322, 1989.

[3705] *Chen, J., Franklin, J. F., and Spies, T. A.*, Contrasting microclimates among clearcut, edge and interior of old growth Douglas-fir forest. Agr. For. Met. 63, 219-237, 1993.

[3706] *Didham, R. K.* and *Lawton, J. H.*., Edge structure determines the magnitude of change in microclimate and vegetation structure in tropical rain forest fragments. Biotropica 31, 17-30, 1999.

[3707] *Diem, M.*, Höchstlasten der Nebelfrostablagerungen an Hochspannungsleitungen im Gebirge. Arch. f. Met. (B) 7, 84-95, 1955.

[3708] *Dörffel, K.*, Die physikalische Arbeitsweise des Gallenkampschen Verdunstungsmessers etc. Veröff. Geoph. 1. Leipzig 6, Nr. 9, 1935.

[3709] *Geiger, R.*, Die Beschattung am Bestandsrand. Forstw. C. 57, 789-794, 1935; 58, 262-266, 1936.

[3710] *Grunow, J.*, Nebelniederschlag. Ber. DWD-US Zone 7, Nr. 42, 30-34, 1952.

[3711] —, Kritische Nebelfroststudien. Arch. f. Met. (B) 4, 389-419, 1953.

[3712] *Herr, L.*, Bodentemperaturen unter besonderer Berücksichtigung der äusseren met. Faktoren. Diss, Leipzig, 1936.

[3713] *Hori, T.*, Studies on fog (399 S.). Tanne Trading Co. Sapporo, Hokkaido, Japan 1953.

[3714] *Kerfoot, O.*, Mist precipitation on vegetation. For. Abstr. 29, 8-20, 1968.

[3715] *Koch, H. G.*, Der Waldwind. Forstw. C. 64, 97-111, 1942.

[3716] *Lauscher, F.*, and *Schwabl, W.*, Untersuchungen über die Helligkeit im Wald and am aldrand. Biokl. B. 1, 60-65, 1934.

[3717] *Linde, R. J. van der*, and *Woudenberg, J. P. M.*, A method for determining the daily variations in width of a shadow in connection with the time of the year and the orientation of the overshadowing object. Med. en Verh. (A) Nr. 52, Kon. Nederland. Met. Inst. Nr. 102, 1946.

[3718] *Linke, F.*, Niederschlagsmessung unter Bäumen. Met. Z. 33, 140-141, 1916; 38, 277, 1921.

[3719] *Lüdi, W.*, and *Zoller, H.*, Über den Einfluss der Waldnähe auf das Lokalklima. Ber, Geobotan. Forsch. Inst. Rübel Zürich f. d. Jahr 1948, 85-108, 1949.

[3720] *Marloth, R.*, Über die Wassermengen, welche Sträucher u. Bäume aus treibendem Nebel u. Wolken auffangen. Met. Z. 23, 547-553, 1906.

[3721] *Miller, D. R.*, The two-dimensional energy budget of a forest edge with field measurements at a forest-parking lot interface. Agr. For. Met. 22, 53-78, 1980.

[3722] *Nagel, J. F.*, Fog precipitation on Table Mountain. Quart. J. 82, 52-460, 1956.

[3723] *Ooura, H.*, The capture of fog particles by the forest. J. Met. Res. Tokyo 4, Suppl., 239-259, 1952.

[3724] *Pfeiffer, H.*, Kleinaerologische Untersuchungen am Collmberg. Veröff. Geoph. I. Leipzig 11, Nr. 5, 1938.

[3725] *Rink, J., Die* Schmelzwassermengen der Nebelfrostablagerungen R. f. W. Wiss. Abh. 5, Nr. 7, 1938.

[3726] *Rötschke, M.*, Untersuchungen über die Meteorologie der Staub atmosphäre. Veröff. Geoph. I. Leipzig 11, 1-78, 1937.

[3727] *Schemenauer, R. S.* and *Cereceda, P.*, Fog-water collection in arid coastal locations. Ambio. 20, 303-308, 1991.
[3728] —, —. Water from fog covered mountains. Waterlines. 10, 10-13, April 1992.
[3729] —, —. The role of wind in rainwater catchment and fog collection. Water International 19, 70-76, 1994.
[3730] *Schmauss,* A., Seewinde ohne See. Met. Z. 37, 154-155, 1920.
[3731] *Schubert, J.*, Die Sonnenstrahlung im mittleren Norddeutschland nach den Messungen in Potsdam. Met. Z. 45, 1-16, 1928.
[3732] *Scott, K. I, Simpson, J. R.*, and *McPherson, E. G.*, Effects of tree cover on parking lot microclimate and vehicle emission. J. Arboricul. 25, 129-142, 1999.
[3733] *Vogel, J., and Huff, F.*, Atmospheric effects of cooling lakes. Illinois Institute of Natural Resources Project 578, Electrical Power Research Institute, 1981.
[3734] *Waibel,K., Die* meteorologischen Bedingungen für Nebelfrostablagerungen n Hochspannungsleitungen im Gebirge. Arch. f. Met. (B) 7, 74-83, 1955.
[3735] *Woelfle, M.*, Windverhältnisse im *Walde.* Forstw. C. 61, 65-75, 461-475, 1939; 64, 69-182, 1942.
[3736] —, Bemerkungen zu "Der Waldwind" von *H. G. Koch.* Forstw. *C.* 1944, 131-136.
[3737] *Young, A.*, and *Mitchell, N.*, Microclimate and vegetation edge effects in a fragmented podocarp-broadleaf forest in New Zealand. Biological Conservation 67, 63-72, 1994.
[3738] *Ziegler, O.*, Die Bedeutung des Windes u. der Thermik für die Verbreitung der Insekten etc. Z. f. Pflanzenbau u. -schutz 1, 241-266, 1903.

SECTION 38

[3801] *Aichele, H.*, Beitrag zum Mikroklima eines Käferkahlschlages. Arch. d. Wiss. Ges. f. Land- u..Forstw. Freiburg 1, 43-49, 1949.
[3802] *Ångström, A.*, Jordtemperatur i bestand av olika täthet. Medd. Stat. Met. Hydr. Anst. Stockholm 29, 187-218, 1936.
[3803] *Ashton, P. M. S.*, Some measurements of the microclimate within a Sri Lankan tropical rainforest, Agr. For. Met. 59, 217-235, 1992.
[3804] *Barden, L. S.*, A comparison of growth efficiency of plants on the east and west sides of a forest canopy gap. Bull. Torrey Bot. Club 123, 240-242, 1996.
[3805] *Berry, G. L.*, and *Rothwell, R. L.*, Snow ablation in small forest openings in southwest Alberta. Can. J. For. Res. 22, 1326-1331, 1992.
[3806] *Black, P. E.*, Watershed Hydrology, Prentice Hall, 1990.
[3807] *Bolz, H. M.*, Der Einfluss der infraroten Strahlung auf das Mikroklima. Abh. Met. D. DDR 1, Nr. 7, 1-59, 1951.
[3808] *Camargo, J. L. C.*, and *Kapos, V.*, Complex edge effects on soil moisture and microclimate in Central Amazonian Forest. J. Trop. Ecol. 11, 205-221, 1995.
[3809] *Danckelmann, B.*, Spätfrostbeschädigungen im märkischen Wald. Z. f. F. u. Jagdw. 30, 389-411, 1898.
[3810] *Geiger, R.*, Das Standortklima in Altholznähe. Mitt. H. Göring Akad. d. Forstwiss. 1, 148-172, 1941.
[3811] *Ghuman, B. S.*, and *Lal, R.*, Effects of partial clearing on microclimate in a humid tropical forest. Agr. For. Met. 40, 17-29, 1987.
[3812] *Göhre, K.*, Kleinklimatische Untersuchungen auf einer Kiefernkultur unter Birkenvorwald. Arch. f. Forstwesen 3, 441-474, 1954.
[3813] *Koch, H. G.*, Temperaturverhältnisse u. Windsystem eines geschlossenen aldgebietes. Veröff. Geoph. I. Leipzig 3, Nr. 3, 1934.
[3814] *Lal, R.*, and *Cummings, D. J.*, Clearing a tropical forest 1. Effects on soil and micro-climate. Field Crops Res. 2, 91-107, 1979.

[3815] *Nuñez, M.*, and *Bowman, D. M. J. S.*, Nocturnal cooling in a high altitude stand of *Eucalyptus delegatensis* as related to stand density. Aust. For. Res. 16, 185-197, 1986.

[3816] *Satterlund, D. R.*, Wildland Watershed Management, John Wiley, 1972.

[3817] *Slavík, B.*, *Slaviková, J.*, and *Jeník, J.*, Ökologie der gruppenweisen Ver jüngung eines Mischbestandes. Rozpravy Tschechoslow. Akad. 67, 2, 1957.

[3818] *Windsor, D. M.*, Climate and moisture variability in a tropical forest: Long-term records from Barro Colorado Island, Panamá, Smithsonian Inst. Press, Washington D.C., 1990.

[3819] *Wrede, C. V.*, *Die* Bestandsklimate u. ihr Einfluss auf die Biologie der Verjüngung unter Schirm u. in der Gruppe. Forstw. C. 47, 441-451, 91-505, 570-582, 1925.

SECTION 39

[3901] *Andre, J. C.*, *Goutorbe, J. P.*, and *Perrier, A.*, HAPEX-MOBILHY: A hydrologic atmospheric experiment for the study of water budget and evaporation flux at the climatic scale. Bull. Amer. Met. Soc. 67, 138-144, 1986.

[3902] *Anon*, Johore Tengahand Tanjong Penggerand regional master plan--Hunting Technical Services, Ltd., 1971.

[3903] *Aylor, D.*, Noise reduction by vegetation and ground. J. Acoust. Soc. of Am. 51, 197-205, 1972.

[3904] *Bastable, H. G.*, *Shuttleworth, W. J.*, *Dallarosa, R. L. G.*, *Fisch, G.*, and *Nobre, C. A.*, Observations of climate, albedo, and surface radiation over cleared and undisturbed Amazonian forest. Int. J. Climatol. 13, 783-796, 1993.

[3905] *Beck, G.*, Untersuchung über Planungsgrundlagen für eine Lärmbekämpfung im Freiraum mit Experimenten zum artspezifischen Lärmminderungsvermögen verschidener Bau- und Straucharten. Arbeit aus dem Institut für Gartenkunst und Landschaftsgestaltung der Technischen Universität Berlin, Nr. 178, 1965.

[3906] *Bell, R. W.*, *Schofield, N. J.*, *Loh, I. C.*, and *Bari, M. A.*, Groundwater response to reforestation in the Darling Range of Western Australia. J. Hydrol. 119, 179-200, 1990.

[3907] *Blanford, H. F.*, On the influence of Indian forests on the rainfall. J. Asiat. Soc. of Bengal 56, II, 1, 1887.

[3908] *Borthwick, J.*, *Halverson, H.*, *Heisler, G. M.*, *McDaniel, O. H.*, and *Reethof, G.* Attenuation of highway noise by narrow forest belts. *Report No. FHWA-RD-77-140.* Federal Highway Administration. Washington, D.C., 1978.

[3909] *Bosch, J. M.*, and *Hewlett, J. D.*, A review of catchment experiments to determine the effect of vegetation changes on water yield and evapotranspiration. J. Hydrol. 55, 3-23, 1982.

[3910] *Brubaker, K. L.*, *Entakhabi, D.*, and *Eagleson, P. S.*, Estimation of continental precipitation recycling. J. Climatol. 6, 1077-1089, 1993.

[3911] *Burch, G. J.*, *Bath, R. K.*, *Moore, I. D.*, and *O'Loughlin, E. M.*, Comparative hydrological behavior of forested and cleared catchments in Southeastern Australia. J. Hydrol. 90, 19-42, 1987.

[3912] *Burckhardt, H.*, Lokale Klimaänderungen auf einem Berggipfel durch ahlhieb. Angew. Met. 1, 150-154, 1952.

[3913] *Burger, H.*, Einfluss des Waldes auf den Stand der Gewässer II-V. Mitt. d. Schweiz. Anst. f. d. forstl. Vers. w. 18, 311-416, 1934; 23, 167-222, 1943; 24, 133-218, 1944; 31, 7-58, 1954/55.

[3914] *Carlson, D. E.*, Theoretical and experimental analysis of the acoustical characteristics of forests. Unpublished Ph.D. Thesis. The Pennsylvania State University, 1979.

[3915] *Charney, J.*, Dynamics of deserts and droughts in the Sahel. Q. J. R. Met. Soc. 101, 193-202, 1975.

[3916] *Cook, D. I.*, and *Haverbeke, Van D. F.*, Trees and shrubs for noise abatementment. Rep. 246 Neb. Agr. Exp. Sta. Lincoln, 1971.

[3917] —, —, Trees, shrubs and landforms for noise control. J. Soil Wat. Cons. 27, 259-261, 1972.

[3918] *Daniel, J. G.*, and *Kulasingham, A.*, Problems arising from large scale forest clearing for agricultural use--the Malaysian experience. Malayan Forester 37, 152-160, 1974.

[3919] *Eagleson, P. S.*, The emergence of global-scale hydrology. Wat. Resourc. Res. 22, 6-14, 1986.

[3920] *Embleton. T. F. W.*, Sound propagation in homogenous deciduous and evergreen woods. J. Acoust. Soc. of Am. 35, 1119-1125, 1963.

[3921] *Entekhabi, D.*, and *Eagleson, P. S.*, Land surface hydrology para-meterization for atmospheric general circulation models including subgrid scale spatial variability. J. Climatol. 2, 816-831, 1989.

[3922] *Flemming, G.*, Wald Wetter Klima-Einführung in dic Forestmeteorologie. Deutscher Landuirtschaftsuerlag. Berlin, 1995.

[3923] *Federal Highway Administration.* Fundamentals and Abatement of Highway Traffic Noise. U.S. Department of Transportation, Washington, D.C., 1980.

[3924] *Fowler, D.*, *Cape, J. N.*, and *Unsworth, M. H.*, Deposition of atmospheric pollutants on forests. Phil. Trans. R. Soc. Lond. 324B, 247-265, 1989.

[3925] *Fricke, F.*, Sound attenuation in forests. J. of Sound and Vib. 92, 149-158, 1984.

[3926] *Harris, R. A.*, *Asce, A. M.*, *Cohn, L. F.*, and *Asce, M.*, Use of vegetation for abatement of highway traffic noise. J. Urban Plan. Dev. 111, 34-48, 1985.

[3927] *Huang, B. K.* An Ecological Systems Approach to Community Noise Abatement–Phase I, *Report No. DOT-05-30102*. U.S. Department of Transportation. Washington, D.C., 1974.

[3928] *Henderson-Sellers, A.*, The "coming of age" of land surface climatology. Glob. Plan. Change 82, 291-319, 1990.

[3929] —, *Yang, Z. L.*, and *Dickinson, R. E.*, The project for intercomparison of land-surface parameterization schemes. Bull. Amer. Met. Soc. 74, 1335-1349, 1993.

[3930] —, and *Robinson, P. J.*, Contemporary Climatology, Longman Scientific & Technical, 1986.

[3931] *Hibbert, A. R.*, Water yield changes after converting a forested catchment to grass. Wat. Resourc. Res. 5, 634-640, 1969.

[3932] *Kaminsky, A.*, Beitrag zur Frage über den Einfluss der Aufforstung der aldlichtungen in Indien auf die Niederschläge. Nachr. d. Geophys. Centr. I. Leningrad 4.

[3933] *Kellomäki, S.*, *Haapanen, A.*, and *Salohen, H.*, Tree stands in urban noise abatement. Silva Fennica 10, 237-256, 1976.

[3934] *Kinter, J. L.*, and *Shukla, J.*, The global hydrologic and energy cycles: Suggestions for studies in the pre-global energy and water cycle experiment (GEWEX) period. Bull. Amer. Met. Soc. 71, 181-189, 1990.

[3935] *Mägdefrau, K.*, and *Wutz, A.*, Die Wasserkapazität der Moos- u. Flechtendecke des Waldes. Forstw. C. 70, 103-117, 1951.

[3936] *Martens, M. J. M.*, Noise abatement in plant monocultures and plant communities. Appl. Acoust., 14, 167-189, 1981.

[3937] *Mecklenburg, R. A.*, *Rintelmann, W. F.*, *Schumaier, D. R.*, *Van den Brink, C.*, and *Flores, L.*, The effect of plants on microclimate and noise reduction in the urban environment. Hortsci. 7, 37-39, 1972.

[3938] *Micklin, P. P.*, An inquiry into the Caspian Sea problem and proposals for its alleviation, Doctoral Dissertation, University of Washington (unpublished), 1971.

[3939] *Mintz, Y.*, The sensitivity of numerically simulated climates to land surface boundary conditions. In: The Global Climate. J. T. Houghton (Ed.), Cambridge University Press, 79-105, 1984.

[3940] *Mrose, H.*, Der Einfluss des Waldes auf die Luftfeuchtigkeit. Angew. Met. 2, 281-286, 1956.

[3941] *Mueller, C. C.*, and *Kidder, E. H.*, Rain gage catch variation due to airflow disturbances around a standard rain gage. Wat. Resourc. Res. 8, 1077-1082, 1972.

[3942] *Mulligan, B. E.,Goodman, L.S., Faupel, M., Lewis, S.,* and *Anderson. L.M.* , Interactive effects of outdoor noise and visible aspects of vegetation on behaviour. In: Proceedings. Southeastern Recreation Researchers Conference, Asheville, N.C., 265-279, 1981.

[3943] *Müttrich, A.*, Über den Einfluss des Waldes auf die periodischen Veränderungen der Lufttemperatur. Z. f. F. u. Jagdw. 22. 385-400, 449-458, 513-526, 1890.

[3944] *Northeastern Forest Experiment Station.* The conference on metropolitan physical environment. USDA Forest Service General Technical Report NE-25, U.S. Dept. of Ag. 1977.

[3945] *Peck, A. J.,* and *Williamson, D. R.,* Effects of forest clearing on groundwater. J. Hydrol. 94, 47-65, 1987.

[3946] *Pinker, R. T., Thompson, O. E.,* and *Eck, T. F.,* The albedo of a tropical evergreen forest. Quart. J. R. Met. Soc. 106, 551-558, 1980.

[3947] *Reethof, G.* and *Heisler, G.*, Trees and forests for noise abatement and visual screening. USDA For. Ser. Gen. Tech. Rep. NE For. Exp. Sta. 39-48, 1976.

[3948] *Roundtree, P. R.,* Review of general circulation models as a basic for predicting the effects of vegetation change on climate. Proceedings of the United Nations Univerity Workshop on Forests, Climate and Hydrology, Oxford, 26-30 March 1984, 1985.

[3949] *Rutter, A. J.,* The hydrological cycle in vegetation. In: J. L. Monteith, (ed.), Vegetation and the Atmosphere, Vol. I. Principles, London, Academic Press, 1975.

[3950] *Salati, E.,* and *Vose, P.*, The Amazon Basin: A system in equilibrium. Science 225, 129-138, 1984.

[3951] —, —, and *Lovejoy, T. E.,* Amazon rainfall potential effects of deforestation and plans for future research. In: Tropical Rain Forests and the World Atmosphere, G. T. Prance (Ed.), Westview Press, Inc. 1986.

[3952] *Sellers, P. J., Mintz, Y., Sud, Y. C.,* and *Dalcher, A.,* A simple biosphere model (SiB) for use within general circulation models. J. Atm. Sci. 43, 505-531, 1986.

[3953] *Schubert, J.,* Der jährliche Gang der Luft- u. Bodentemperatur im Freien and in Waldungen. J. Springer, Berlin, 1900.

[3954] —, Über den Einfluss des Waldes auf die Niederschläge im Gebiet der Letzlinger Heide. Z. f. F. u. Jagdw. 69, 604-615, 1937.

[3955] *Schultze, J. H.,* Neuere theoretische u. praktische Ergebnisse der Bodenerosionsforschung in Deutschland. Forsch. u. Fortschr. 27, 12-18, 1953.

[3956] *Sud, Y. C., Shukla, J.,* and *Mintz, Y.,* Influence of land surface roughness on atmospheric circulation and precipitation: A sensitivity study with a general circulation model. J. App. Mct. 27, 1036-1054, 1988.

[3957] *Swift, L. W., Jr.,* and *Swank, W. T.,* Long term responses of streamflow following clearcutting and regrowth. Hydro. Sci. Bull. 26, 245-255, 1981.

[3958] *Whitcombe, C. E.* and *Stowers, J. F.,* Sound abatement with hedges. Hortsci. 8, 128-129, 1973.

[3959] *Wright, I. R., Gash, J. H. C., DaRocha, H. R., Shuttleworth, W. J., Nobre, C. A., Maitelli, G. T., Zamparoni, C. A. G. P.,* and *Carvalho, P. R. A.,* Dry season micrometeorology of central Amazonian ranchland. Quart. J. R. Met. Soc. 118, 1083-1099, 1992.

[3960] *Zenker, H.,* Waldeinfluss auf Kondensationskerne u. Lufthygiene. Z. f. Met. 8, 150-159, 1954.

SECTION 40

[4000] *Atwater, M. A.,* and *Ball, J. T.,* A numerical solar radiation model based on standard meteorological observations. Sol. Ener. 21, 163-170, 1978.

[4001] *Davies, J. A.,* and *McKay, D. C.,* Estimating solar irradiance and components. Sol. Ener. 29, 55-64, 1982.

[4002] *Decoster, M.,* and *Schüepp, W.,* Le rayonnement sur des plans verticaux à Stanleyville. Serv. Mét. Congo Belge 15, Léopoldville, 1956.

[4003] —, —, and *Elst, N. van der,* Le rayonnement sur des plans verticaux à Léopoldville. Serv. Mét. Congo Belge 7, 1955.

[4004] *Dirmhirn, I.,* and *Eaton, F. D.,* Some characteristics of the albedo of snow. J. App. Met. 14, 375-379, 1975.

[4005] *Flint, A. L.* and *Childs, S. W.,* Calculation of solar radiation in mountainous terrain. Agric. For. Meteor. 40, 233-249, 1987.

[4006] *Gräfe, K.,* Strahlungsempfang vertikaler ebener Flächen; Globalstrahlung von Hamburg. Ber. DWD 5, Nr. 29, 1-15, 1956.

[4007] *Grunow, J.,* Beiträge zum Hangklima. Ber. DWD-US Zone 5, Nr. 35, 293-298, 1952.

[4008] *Hay, J. E.,* A revised method for determining the direct and diffuse components of the total shortwave radiation. Atmos. 14, 278-287, 1976.

[4009] —, Calculation of solar irradiances for inclined surfaces: validation of selected hourly and daily models. Atmos.-Ocean 24, 16-44, 1986.

[4010] —, and *McKay, D. C.,* Estimating solar irradiance on inclined surfaces: a review and assessment of methodologies. Int. J. Sol. Ener. 3, 203-240, 1985.

[4011] *Ineichen, P., Guison, O.,* and *Perez, R.,* Ground-reflected radiation and albedo. Sol. Ener. 44, 207-214, 1990.

[4012] Kaempfert, W., Sonnenstrahlung auf Ebene, Wand u. Hang. R. f. W. Wiss. Abh. 9, Nr. 3, 1942.

[4013] —, and *Morgen, A.,* Die Besonnung. Z. f. Met. 6, 138-146, 1952.

[4014] *Kienle, J. V.,* Die tatsächliche and die astronomisch mögliche Sonnenscheindauer auf verschieden exponierte Flächen. D. Met. Jahrb. f. Baden 1933, Anhang.

[4015] *List, R. J.,* Smithsonian Meteorological Tables, Smithsonian Inst. Press, 1949.

[4016] *McArthur, L. J. B.,* and *Hay, J. E.,* A technique for mapping the distribution of diffuse solar radiation over the sky hemisphere. J. App. Met. 20, 421-429, 1981.

[4017] *Munro, D. S.* and *Young, G. J.,* An operational net shortwave radiation model for glacier basins. Wat. Resourc. Res. 18, 220-230, 1982.

[4018] *Oliver, H. R.,* Studies of surface energy balance of sloping terrain. Int. J. Climatol. 12, 55-68, 1992.

[4019] *Olseth, J. A.,* and *Skartveit, A.,* Spatial distribution of photosynthetically active radiation over complex topography. Agr. For. Met. 86, 205-214, 1997.

[4020] *Olyphant, G. A.,* Insolation topoclimates and potential ablation in alpine snow accumulation basins: Front Range, Colorado. Wat. Resourc. Res. 20, 491-498, 1984.

[4021] *Olyphant, G. A.,* The components of incoming radiation within a mid-latitude watershed during the snowmelt season. Arc. Alp. Res. 18, 163-169, 1986.

[4022] *Pierce, K. L.,* and *Colman, S. M.,* Effect of height and orientation (microclimate) on geomorphic degration rates and processes, late glacial terrace scarps in central Idaho. Geol. Soc. Amer. Bull. 97, 869-885, 1986.

[4023] –, –, Effect of height and orientation (microclimate) on degradation rates of Idaho terrace scarps. Directions in Paleoseismology Proceedings Conference XXXIX 22-25, 1987.

[4024] *Perez, R., Seals, R.,* and *Michalsky, J.,* All-weather model for sky luminance distribution - preliminary configuration and validation. Sol. Ener. 50, 235-245, 1993.

[4025] *Polunin, M.,* Plant succession in Norwegian Lapland. J. Ecol. 24, 372-391, 1946.

[4026] *Radcliffe, J. E.* and *Lefever, K. R.,* Aspect influences on pasture microclimate at Coopers Creek, North Canterbury. N. Z. J. Ag. Res. 24, 55-66, 1981.

[4027] *Rouse, W. R.,* and *Wilson, R. G.,* Time and space variations in the radiant energy fluxes over sloping forested terrain and their influence on seasonal heat and water balances at a middle latitude site. Geograf. Ann. 51A, 160-175, 1969.

[4028] *Schedler, A.,* Die Bestrahlung geneigter Flächen durch die Sonne. Jahrb. Zentralanst. f. Met. u. Geodyn. Wien, N. F. 87, D 51-64, 1950, Wien 1951.

[4029] *Schütte, K.,* Die Berechnung der Sonnenhöhen für beliebig geneigte Ebenen. Ann. d. Hydr. 71, 325-328, 1943.

[4030] *Steven, M. D.*, Standard distribution of clear sky radiance. Q. J. R. Met. Soc. 103, 457-465, 1977.

[4031] *Suzuke, S.*, Early strawberries and their cultivation on slopes. Agric. Hort. Japan 16, 1185-1188, 1941.

SECTION 41

[4100] *Aichele, H.*, Der Temperaturgang rings um eine Esche. Allg. Forst- u. J. Zeitung 121, 119-121, 1950.

[4101] —, Untersuchunger über die Frostschutzwirkung eines Kalkanstrichs an Obstbäumen. Ber DWD-US Zone 5, Nr. 32, 70-73, 1952.

[4102] *Eisenhut, G.*, Blühen, Fruchten u. Keimen in der Gattung Tilia. Diss. Univ. München, 1957.

[4103] *Gerlach, E.*, Untersuchung über die Wärmeverhältnisse der Bäume. Diss. Univ. Leipzig, 1929.

[4104] *Haarløv, N.*, and *Petersen, B. B.*, Measurement of temperature in bark and wood of Sitka spruce. Kopenhagen, 1952.

[4105] *Kaempfert, W.I*, Ein Phasendiagramm der Besonnung. Met. Rundsch. 4, 141-144, 1951.

[4106] *Koljo, B.*, Einiges über Wärmephänomene der Hölzer u. Bäume. Forstw. C. 69, 538-551, 1950.

[4107] *Krenn, K.*, Die Bestrahlungsverhältnisse stehender u. liegender Stämme. Wien. Allg. F. u. Jagdz. 51, 50-51, 53-54, 1933.

[4108] *Lieffers, V. J.* and *Larkin-Lieffers, P. A.*, Slope, aspect, and slope position as factors controlling grassland communities in the coulees of the Oldman River, Alberta. Can. J. Bot. 65, 1371-1378, 1987.

[4109] *Lessmann, H.*, Temperaturverhältnisse in Häufelreihen. Jahr. Ber. d. Bad. Landeswetterd., 35-45, 1950.

[4110] *Müller, G.*, Unterschungen über die Querschnittsformen der Baumschäfte. Forstw. C. 77, 41-59, 1958.

[4111] *Primault, B.*, L'influence de l'insolution sur la température du cambium des arbres frutiers. Rev. Romande Agric., Vitc. Arboric. 10, 26-28, 1954.

[4112] *Scamoni, A.*, Über Eintritt u. Verlauf der männlichen Kiefernblüte. Z. f. F. u. Jagdw. 70, 289-315, 1938.

[4113] *Schulz, H.*, Untersuchungen an Frostrissen im Frühjahr 1956. Forstw. C. 76, 14-24, 1957.

[4114] *Seeholzer, M.*, Rindenschäle u. Rindenriss an Rotbuche im Winter 1928/29. Forstw. C. 57, 237-246, 1935.

[4115] *Weger, N.*, Bodentemperaturen in Beeten verschiedener Form u. Richtung. Met. Rundsch. 2, 291-295, 1949.

SECTION 42

[4200] *Brocks, K.*, Nächtliche Temperaturminima in Furchen mit verschiedenem Böschungswinkel. Met. Z. 56, 378-383, 1939.

[4201] *Geiger, R.*, Spätfröste auf den Frostflächen bei München. Forstw. C. 48, 279-293, 1926.

[4202] *Horvat, J.*, Die Vegetation der Karstdolinen. Geografski Glasnik 14-15, 1-25, Zagreb, 1953.

[4203] *Kalma, J. D., Byrne, G. F., Johnson, M. E.,* and *Laughlin, G. P.*, Frost mapping in Southern Victoria: an assessment of HCMM thermal imagery. J. Climatol. 3, 1-19, 1983.

[4204] —, *Laughlin, G. P., Green, A. A., and O'Brien, M. T.*, Minimum temperature surveys based on near-surface temperature measurements and airborne thermal scanner data. J. Climatol. 6, 413-430, 1986.

[4205] *Mahrt, L.*, and *Heald, R. C.*, Nocturnal surface temperature distribution as remotely sensed from low-flying aircraft. Agr. Met. 28, 99-107, 1983.
[4206] *Rajakoppi, A.*, Topographic, microclimatic and edaphic control of the vegetation in the central part of the Hämeenkahgas esker complex, western Finland. Acta Bot. Fennica. 134, 1-70, 1987.
[4207] *Rikkinen, J.*, Relations between topography, microclimates and vegetation on the Kalmari-Saarijärvi esker chain, central Finland. Fennia. 167, 87-150, 1989.
[4208] *Sauberer, F.*, and *Dirmhirn, I.*, Über die Entstehung der extremen Temperaturminima in der Doline Gstettner Alm. Arch. f. Met. (B) 5, 307-326, 1953.
[4209] —, —, Weitere Untersuchungen über die Kaltluftansammlungen in der Doline etc. Wetter u. Leben 8, 187-196, 1956.
[4210] *Schmidt, W.*, Die tiefsten Minimumtemperaturen in Mitteleuropa. Naturw. 18, 367-369, 1930.
[4211] *Wagner, R.*, Fluktuierende Dolinen-Nebel. Idöjárás 58, 289-298, 1954.
[4212] *Winter, F.*, Das Spätfrostproblem im Rahmen der Neuordnung des südwestdeutschen Obstbaus. Gartenbauwissensch. 23, 342-362, 1958.

SECTION 43

[4300] *Aron, R. H.*, and *Aron, I-M.*, A case study of the influence of topographic features on dispersion climatology in a small valley. Gt. Plains-Rocky Mt. Geo. J. 11, 32-41, 1983.
[4301] *Berg, H.*, Beobachtungen des Berg- u. Talwindes in den Allgäuer Alpen. Ber. DWD-US Zone 6, Nr. 38, 105-109, 1952.
[4302] *Blumen, W.* (Ed.), Atmospheric Processes over Complex Terrain. Am. Met. Soc., 1990.
[4303] *Defant, A.*, Der Abfluss schwerer Luftmassen auf geneigtem Boden. Sitz-B. Berlin. Akad. 18, 624-635, 1933.
[4304] *Defant, F.*, Zur Theorie der Hangwinde, nebst Bemerkungen zur Theorie der Berg- u. Talwinde. Arch. f. Met. (A) 1, 421-450, 1949.
[4305] —, Local winds. Compend. of Met. (Am. Met. Soc.) 655-672, Boston, 1951.
[4306] *Ekhart, E.*, Neuere Untersuchungen zur Aerologie der Talwinde. Beitr. Phys. d. fr. Atm. 21, 245-268, 1934.
[4307] —, Über den täglichen Gang des Windes im Gebirge. Arch. f. Met. (B) 4, 431-450, 1953.
[4308] *Friedel, H.*, Wirkungen der Gletscherwindc auf die Ufervegetation der Pasterze. Biokl. B. 3, 21-25, 1936.
[4309] *Gleeson, T. A.*, On the theory of cross-valley winds arising from differential heating of the slopes. J. Met. 8, 398-405, 1951.
[4310] *Goldreich, Y.*, *Druyan, L. M.*, and *Berger, H.*, The interaction of valley/mountain winds with a diurnally veering sea/land breeze. J. Climatol. 6, 551-561, 1985.
[4311] *Gross, G.*, Some effects of deforestation on nocturnal drainage flow and local climate – a numerical study. Bound. Lay. Met. 38, 315-337, 1987.
[4312] *Grunow, J.*, Der Wetterdienst bei der internationalen Skiflugwoche vom 8. 2. bis 6. 3. 1950 in Oberstdorf. Met. Rundsch. 4, 62-64, 1951.
[4313] *Gudiksen, P. H.*, and *Shearer, D. L.*, The dispersion of atmospheric tracers in nocturnal drainage flows. J. App. Met. 28, 602-608, 1989.
[4314] *Hoinkes, H.*, Beiträge zur Kenntnis des Gletscherwindes. Arch. f. Met. (B) 6, 36-53, 1954.
[4315] —, Der Einfluss des Gletscherwindes auf die Ablation. Z. f. letscherkde. u. Glazialgeologie 3, 18-23, 1954.
[4316] *Koch, H. G.*, Ein lokaler Tagesfallwind in Mittelsardinien. Veröff. Geophys. I. Leipzig 15, 40-60, 1949.
[4318] *Küttner, J.*, Periodische Luftlawinen. Met. Rundsch. 2, 183-184, 1949.
[4319] *Maletzke, R.*, siehe Alpenwandersegelflug 1958, Mitt.heft I/1958 der Akaflieg München.

[4320] *Nitze, F. W.*, Untersuchung der nächtlichen Zirkulationsströmung am Berghang durch stereophotogrammetrisch vermessene Ballonbahnen. Biokl. B. 3, 125-127, 1936.
[4321] *Porch, W. R., Clements, W. E.*, and *Coulter, R. L.*, Nighttime valley waves. J. App. Met. 30, 145-156, 1991.
[4322] *Reiher, M.*, Nächtlicher Kaltluftfluss an Hindernissen. Biokl. B. 3, 152-163, 1936.
[4323] *Runge, F.*, Windgeformte Bäume in den Tälern der Zillertaler Alpen bzw. Allgäuer Alpen). Met. Rundsch. 11, 28-30, 1958 (bzw. 12, 98-99, 1959).
[4324] *Scaëtta, H.*, Les avalanches d'air dans les Alpes et dans les hautes montagnes de l'Afrique centrale. Ciel et Terre 51, 79-80, 1935.
[4325] *Schmauss, A.*, Luftlawinen in Alpentälern. D. Met. Jahrb. f. Bayern 1926, Anhang F.
[4326] —, Absinken einer Inversion. Z. f. angew. Met. 59, 260-263, 1942.
[4327] *Schnelle, F.*, Kleinklimatische Geländeaufnahme am Beispiel der Frostschäden im Obstbau. Ber. DWD-US Zone 2, Nr. 12, 99-104, 1950.
[4328] —, Ein Hilfsmittel zur Feststellung der Höhe von Frostlagen in Mittelgebirgstälern. Met. Rundsch. 9, 180-182, 1956.
[4329] *Schüepp, W.*, Untersuchungen über den winterlichen Kaltluftsee in Davos. Verh. Schweiz. Naturf. Ges., 127-128, 1945.
[4330] *Schultz, H.*, Über Klimaeigentümlichkeiten im unteren Rheingau, unter besonderer Berücksichtigung des Wisperwindes. Frankfurter Geogr. Hefte 7, Nr. 1, 1933.
[4331] *Scultetus, H. R.*, Geländeausformung u. Bewindung in Abhängigkeit von der Austauschgrösse. Met. Rundsch. 12, 73-80, 1959.
[4332] *Tollner, H.*, Gletscherwinde in den Ostalpen. Met. Z. 48, 414-421, 1931.
[4333] *Troll, C.*, Die Lokalwinde der Tropengebirge and ihr Einfluss auf Niederschlag and Vegetation. Bonner Geogr. Abh. 9, 124-182, 1952.
[4334] *Wagner, A.*, Theorie and Beobachtungen der periodischen Gebirgswinde. Gerl. B. 52, 408-449, 1938.
[4335] *Weischet,* W., Die Baumneigung als Hilfsmittel zur geographischen Bestimmung der klimatischen Windverhältnisse. Erdk. 5, 221-227, 1951.
[4336] *Yoshino, M.*, The structure of surface winds crossing over a small valley. J. Met. Soc. Japan 35, 184-195, 1957.
[4338] —, Winds in a V-shaped small valley. J. Agr. Met. Japan 13, 129-134, 1958.

SECTION 44

[4401] *Cantlon, J. E.*, Vegetation and microclimates on north and south slopes of Cucketunk Mountain, New Jersey. Ecolog. Monogr. 23, 41-270, 1953.
[4402] *de Lima, J. L. M. P.*, The effect of oblique rain on inclined surfaces: a nomograph for the rain-gauge correction factor. J. Hydrol. 115, 407-412, 1990.
[4403] *Fletcher, R. D.*, Hydrometeorology in the United States, Comp. Met. 1033-1049, 1951.
[4404] *Geiger, R.*, Messung des Expositionsklimas. Forstw. C. 49, 665-675, 853-859, 914-923, 1927; 50, 73-85, 437-448, 633-644, 1928; 51, 37-51, 305-315, 637-656, 1929.
[4405] *Grunow, J.*, Niederschlagsmessungen am Hang. Met. Rundsch. 6, 85-91, 1953.
[4406] *Hartmann, F. K., Eimern, J. van,* and *Jahn, G.*, Untersuchungen reliefbedingter kleinklimatischer Fragen in Geländequerschnitten der hochmontanen Stufe des Mittel- u. Südwestharzes. Ber. DWD 7, Nr. 50, 1959.
[4407] *Haude, W.*, Ergebnisse der allgemeinen met. Beobachtungen etc. Rep. cient. Exp. to the NW Prov. China (Sven Hedin) IX, Met, 1, Stockholm, 1940.
[4408] *Heigel, K.*, Exposition u. Höhenlage in ihrer Wirkung auf die Pflanzenentwicklung. Met. Rundsch. 8, 146-148, 1955.
[4409] —, Ergebnisse von Bodenfeuchtemessungen mit Gipsscheibenelektroden. Met. Rundsch. 11, 92-96, 1958.

[4410] *Huang, S. B.*, Protecting citrus trees from freezing injury by use of topoclimate in China. Ag. For. Met. 55, 95-108, 1991.

[4411] *Kerner, A.*, Die Änderung der Bodentemperatur mit der Exposition. Sitz-B. Wien. Akad. 100, 704-729, 1891.

[4112] *Kirkpatrick, J. B.* and *Nuñez, M.*, Vegetation-radiation relationships in mountainous terrain: eucalypt-dominated vegetation in the Risdon Hills, Tasmania. Jour. Biogeog. 7, 197-208, 1980.

[4413] *Künkele, T.*, and *Geiger, R.*, Hangrichtung (Exposition) u. Pdanzenklima. Forstw. C. 47, 597-606, 1925.

[4414] *Ng, E.*, and *Miller, P. C.*, Soil moisture relations in the southern California chaparral. Ecol. 61, 98-107, 1980.

[4415] *Okko, V.*, On the thermal behaviour of some Finnish eskers. Fennia 81, Nr. 5, 1957.

[4416] *Parker, J.*, Environment and forest distribution of the Palouse Range in northern Idaho. Ecol. 33, 451-461, 1952.

[4417] *Schwabe, G. H.*, Der künstliche Erdkegel als Gegenstand der experimentellen Ökologie. Arch. f. Met. (B) 8, 108-127, 1957.

[4418] *Shanks, R. E.*, and *Norris, F. H.*, Microclimatic variation in a small valley in eastern Tennessee. Ecol. 31, 532-539, 1950.

[4419] *Sharon, D.*, The distribution of hydrologically effective rainfall incident on sloping ground. J. Hydrol. 46, 165-188, 1980.

[4420] *Spittlehouse, D. L.*, *Daper, D. A.*, and *Binder, W. D.*, Microclimate of mounds and seedling response. Can. For. Ser. British Columbia. 109, 73-76, 1990.

[4421] *Troll, C.*, Unterirdische Jahreszeitenwinde in finnischen Äsern. Erdk. 13, 150-152, 1959.

[4422] *Wilson, R. G.*, Topographic influences on a forest microclimate, Climatological Research Series No. 5. McGill University, Montreal, 1970.

[4423] *Wollny, E.*, Untersuchungen über den Einfluss der Exposition auf die Erwärmung des Bodens. Forsch. a. d. Geb. d. Agrik. Physik 1, 263-294, 1878.

SECTION 45

[4500] *Baumgartner, A.*, Nebel u. Nebelniederschlag als Standortsfaktoren etc. Forstw. C. 77, 57-272, 1958.

[4501] —, Die Lufttemperatur als Standortsfaktor C. am Grossen Falkenstein (Bayer. Wald). Forstwissenschaftliches Centralblatt, Part 1 (79), 362-373, 1960, Part 2 (80), 107-120, 1961, Part 3 (81), 17-47, 1962.

[4502] —, *Kleinlein, G.*, and *Waldmann, G.*, Forstlich-phänologische Beobachtungen u. Experimente am Grossen Falkenstein. Forstw. C. 75, 290-303, 1956.

[4503] *Bylund,* E., and *Sundborg, A.*, Lokalklimatische Einflüsse auf die Platzwald der Siedlungen etc. Ymer Stockholm 1, 1-30, 1952.

[4504] *Geiger, R.*, *Woelfle, M.*, and *Seip, L. P.*, Höhenlage u. Spätfrostgefährdung. Forstw. C. 55, 579-592, 737-746, 1933; 56, 141-151, 221-230, 253-260, 57-364, 465-484, 1934.

[4505] *Giambelluca, T. W.* and *Nullet, D.*, Influence of the trade-wind inversion on the climate of a leeward mountain slope in Hawaii. Clim. Res. 1, 207-216, 1991.

[4506] *Grunow, J.*, Die Abschirmung des Sonnenscheins durch Talnebel im Alpenvorland. Wetter u. Leben 9, 99-104, 1957.

[4507] *Higgins, J. J.*, Instructions for making phenological observations of garden peas. Publ. Climat. 5, Nr. 2, 1952.

[4508] *Mano, H.*, A study of the sudden nocturnal temperature rises in the valley and on the basin. Geophys. Mag. Tokyo 27, 169-204, 1956.

[4509] *Obrebska-Starkel, B.*, Über die thermische themperaturschichtung in Berg Talern. Alta. Clima. 9, 33-47, 1970.

[4510] *Schnelle, F.*, Studien zur Phänologie Mitteleuropas. Ber. DWD-US Zone 1, Nr. 2, 1948.

[4511] *Waldmann, G.*, Schnee- and Bodenfrost als Standortsfaktoren am Grossen Falkenstein. Forstw. C. 78, 98-108, 1959.

[4512] *Young, F. D.*, Nocturnal temperature inversion in Oregon and California. Mon. Weath. Rev. 49, 138-148, 1921.

SECTION 46

[4600] *Aulitzky, H.*, Waldkrone, Kleinklima u. Aufforstung. Centralbl. f. d. ges. Forstw. 73, 7-12, 1954.

[4601] —, Über die lokalen Windverhältnisse einer zentralalpinen HochgebirgsHangstation. Arch. f. Met. (B) 6, 353-373, 1955.

[4602] —, Waldbaulich-ökologische Fragen an der Waldgrenze. Centralbl. f. d. ges. Forstw. 75, 18-33, 1958.

[4603] *Barry, R. G.*, Climatic environment of the east slope of the Colorado Front Range. Inst. Arctic and Alpine Res. Occas. Paper 3, 1-206, 1972.

[4604] —, Mountain Weather and Climate, Routledge, 1992.

[4605] *Bell, K. L.* and *Bliss, L. C.*, Plant reproduction in a high arctic environment. Arc. Alp. Res. 12, 1-10, 1980.

[4606] *Blumthaler, M.*, *Ambach, W.* and *Rehwald, W.*, Solar UV-A and UV-B radiation fluxes at two alpine stations at different latitudes. Theor. Appl. Climatol. 46, 39-44, 1992.

[4607] *Carlsson, B. Å.* and *Callaghan, T. U.*, Positive plant interactions in tundra vegetation and the importance of shelter. J. Ecol. 79, 973-983, 1991.

[4608] *Collett, Jr., J. L.*, *Daub, Jr., B. C.*, and *Hoffmann, M. R.*, Spatial and temporal variations in precipitation and cloud interception in the Sierra Nevada of central California. Tellus 43B, 390-340, 1991.

[4609] *Desing, H.*, Klimatische Untersuchungen auf einer grossen Blaike. Wetter u. Leben 5, 46-82, 1953.

[4610] *Dirmhirn, I.*, Untersuchungen der Himmelsstrahlung in den Ostalpen mit esonderer Berücksichtigung ihrer Höhenabhängigkeit. Arch. f. Met. (B) 2, 301-346, 1951.

[4611] —, Oberflächentemperaturen der Gesteine im Hochgebirge. Arch. f. Met. 4, 43-50, 1952.

[4612] *Dixon, H. N.*, Miscellanea Broyologia-IX, on some mosses from high altitudes, J. Bot. British and Foreign, 62, 228-231, 1924.

[4613] *Ekhart, E.*, Zur Kenntnis der Schneeverhältnisse der Ostalpen. Gerl. B. 56, 321-358, 1940.

[4614] *Franco. A. C.* and *Nobel, P. S.*, Effect of nurse plants on the microhabitat and growth of cacti. J. Ecol. 77, 870-886, 1989.

[4615] *Friedel, H.*, Gesetze der Niederschlagsverteilung im Hochgebirge. Wetter u. Leben 4, 73-86, 1952.

[4616] *Grace J.*, Tree lines. Phil. Trans. R. Soc. London, 324B, 233-245, 1989.

[4617] *Hadley, J. L.* and *Smith, W. K.*, Influence of Krummholz mat microclimate on needle physiology and survival. Oecologia 73, 82-90, 1987.

[4618] *Halloy, S.*, Islands of life at 6000m altitude: The environment of the highest autrophic communities on earth (Socompa Volcano, Andies). Arc. Alp. Res. 23, 247-262, 1991.

[4619] *Isard, S. A.*, Topoclimatic controls in an alpine fellfield and their ecological significance. Phys. Geog. 10, 13-31, 1989.

[4620] *Jordan, D. N.* and *Smith, W. K.*, Microclimate factors influencing the frequency and duration of growth season frost subalpine plants. Agr. For. Met. 77, 17-30, 1995.

[4621] *Jordan, P. W.*, and *Nobel, P. S.*, Infrequent establishment of seedlings of agave deserti *(agavaceae)* in the northwestern Sonoran Desert. Amer. J. Bot. 9, 1079-1084, 1979.

[4622] *Juvik, J. O.*, and *Nullet, D.*, A climate transect through tropical montane rain forest in Hawaii. Jour. Appl. Meteor. 33, 1304-1312, 1994.

[4623] *Larcher, W.*, Frosttrocknis an der Waldgrenze and in der alpinen Zwergstrauchheide. Veröff. Ferdinandeum Innsbruck 37, 49-81, 1957.

[4624] *Lauscher, F.*, Neue klimatische Normalwerte für Österreich. Beih. z. ahrb. d. Zentralanst. f. Met. u. Geodyn. 1932, 1-13, Wien, 1938.

[4625] —, Langjährige Durchschnittswerte für Frost u. Frostwechsel in Österreich. Beih. z. ahrb. d. Zentralanst. f. Met. u. Geodyn. 1946, D 18-30, Wien, 1947.

[4626] —, Normalwerte der relativen Feuchtigkeit in Österreich. Wetter u. Leben 1, 289-297, 1949.

[4627] *Mani, M. S.*, Ecology and Phytogeography of High Altitude Plants of the Northwest Himalaya. Chapman and Hall, New York, 1978.

[4628] *Marks, D.*, and *Dozier, J.*, Climate and energy exchange at the snow surface in the alpine region of the Sierra Nevada 2. Snow cover energy balance. Wat. Resourc. Res. 28, 3043-3054, 1992.

[4629] *Maurer, J.*, Bodentemperatur u. Sonnenstrahlung in den Schweizer Alpen. Met.-Z. 33, 193-199, 1916.

[4630] *McCutchan, M. H., and Fox, D. G.*, Effect of elevation on wind, temperature and humidity. J. Appl. Met. 25, 1996-2013, 1986.

[4631] *Mølgaard, P.*, Temperature observations in high arctic plants in relation to microclimate in the vegetation of Peary Land, North Greenland. Arc. Alp. Res. 14, 105-115, 1982.

[4632] *Niedzwiedz, T.*, Climate of the Tatra Mountains. Mount. Res. Develop. 12, 131-146, 1992.

[4633] *Nobel, P. S.*, Morphology, nurse plants, and minimum apical temperatures for young *carnegiea gigantea*. Bot. Gaz. 141, 181-191, 1980.

[4634] *Sauberer, F.*, and *Dirmhirn, I.*, Das Strahlungsklima. Klimatographie von Österreich 13-102, Wien, 1958.

[4635] *Saunders, I. R.*, and *Bailey, W. G.*, The physical climatology of alpine tundra, Scott Mountain, British Columbia, Canada, Mount. Res. Develop. 16, 51-64, 1996.

[4636] *Schreckenthal-Schimitschek, G.*, Klima, Boden u. Holzarten an der Waldu. Baumgrenze in einzelnen Gebieten Tirols. Univ. Verl. Wagner, Innsbruck, 1934.

[4638] *Smith, J. M. B.*, Vegetation and microclimate of east- and west-facing slopes in the grasslands of Mt. Wilhelm, Papua New Guinea. J. Ecol. 65, 39-53, 1977.

[4639] *Smith, R. B.*, The influence of mountains on the atmosphere. Adv. Geophys. 21, 87-230, 1979.

[4640] *Steinhäusser, H.*, Normalhöhen zur Kennzeichnung der Schneedeckenverhältnisse. Met. Rundsch. 3, 32-34, 1950.

[4641] *Steinhauser, F.*, Die Schneehöhen in den Ostalpen and die Bedeutung der winterlichen Temperaturinversion. Arch. f. Met. (B) 1, 63-74, 1949.

[4642] *Stevens, G. C.* and *Fox, J. F.*, The causes of treelines. Ann. Rev. Ecol. System. 22, 177-191, 1991.

[4643] *Tranquillini, W.*, Standortsklima, Wasserbilanz u. $C0_2$-Gaswechsel junger Zirben an der alpinen Waldgrenze. Planta 49, 612-661, 1957.

[4644] *Turner, H.*, Über das Licht- u. Strahlungsklima einer Hanglage der Ötztaler Alpen etc. Arch. f. Met. (B) 8, 273-325, 1958.

[4645] —, Maximaltemperaturen oberflächennaher Bodenschichten an der alpinen aldgrenze. Wetter u. Leben 10, 1-12, 1958.

[4646] *Zimmermann, A.*, The highest plants in the world: In the mountain world. Harper, 130-136, 1953.

[4647] *Zöttl, H.*, Untersuchungen über das Mikroklima subalpiner Pflanzengesellschaften. Ber. Geobot. Forsch. Inst. Rübel für 1952, 79-103, Zürich, 1953.

SECTION 47

[4700] *Battle, W. R. B.*, and *Lewis, W. V.*, Temperature observations in Bergschrunds and their relationship to cirque erosion. J. Geol. 59, 537- 545, 1951.

[4701] *Baumgartner, A.*, Meteorologische Beobachtungen am Hölloch. Wissenschaftliche Alpenvereinshefte, Nr. 18, 61-84, Innsbruck 1961.

[4702] *Buxton, P. A.*, Climate in caves and similar places in Palestine. J. Animal Ecol. 1, 152-159, 1932.

[4703] *Dudich, E.*, Biologie der Aggteleker Tropfsteinhöhle "Baradla" in ngarn. Speläolog. Monogr. 13, Wien, 1932.

[4704] *Freitas, C. R. de, Littlejohn, R. N., Clarkson, T. S.*, and *Kristament, I. S.*, Cave climate: assessment of airflow and ventilation. J. Climatol. 2, 383-397, 1982.

[4705] —, and *Littlejohn, R. N.*, Cave climate: assessment of heat and moisture exchange. J. Climatol. 7, 553-569, 1987.

[4706] *Gressel, W.*, Über die Bewetterung der alpinen Höhlen. Met. Rundsch. 11, 54-57, 1958.

[4707] *Hell, M.*, Die kalten Keller von Kaltenhausen in Salzburg. Forsch. u. Fortschr. 10, 336, 1934.

[4708] *Hess, H., L.* Handl's Temperaturmessungen des Eises and der Lutt in den Stollen des Marmolata-Gletschers etc. Z. f. Gletscherkde. 27, 168-171, 1940.

[4709] *Hill, C.* and *Forti, P.*, Cave Microclimate and Speleothems. In: Cave Minerals of the World, Hill, C. and Forti, P. Eds., National Speleological Society, 258-261, 1997.

[4710] *Horvath, T.*, Speleotherapy: A special kind of climatotherapy, its role in respiratory rehabilitation. Int. Rehabil. Med. 8, 90-92, 1986.

[4711] ____., Speleotherapy: A special employment of the cave microclimate In: International Symposium on Physical, Chemical and Hydrological Research of Karst. Kosice Czechoslovakia, 128-131, May 10-15, 1988.

[4712] *Lautensach, H.*, Unterirdischer Kaltluhstau in Korea. Peterm. Mitt. 85, 353-355, 1939.

[4713] *McLean, J. S.*, The microclimate in Carlsbad Caverns, New Mexico. United States Geological Survey, Albuquerque, New Mexico. Open File Report, 67, 1971.

[4714] *Mrose, H.*, Eine seltsame Höhlenvereisung. Z. f. angew. Met. 56, 350- 53, 1939.

[4715] *Oedl, R.*, Über Höhlenmeteorologie mit besonderer Rücksicht auf die grosse Eishöhle im Tennengebirge. Met. Z. 40, 33-37, 1923.

[4716] *Ohata, T., Furukawa, T.*, and *Higuchi, K.*, Glacioclimatological study of perennial ice in the Fuji ice cave, Japan. Part 1. Seasonal variations and mechanism of maintenance. Arc. Alp. Res., 26, 227-237, 1994.

[4717] —, —., and *Osada, K.*, Glacioclimatological study of perennial ice in the Fuji ice cave, Japan. Part 2. Interannual variations and relation to climate Arc. Alp. Res., 26, 238-244, 1994.

[4718] *Saar, R.*, Meteorologisch-physikalische Beobachtungen in den Dachsteinrieseneishöhlen, Oberösterreich. Wetter u. Leben 7, 213-219, 1955.

[4719] *Smithson, P. A.*, Inter-relationships between cave and outside air temperatures. Theor. Appl. Climatol. 44, 65-73, 1991.

[4720] *Stoeu, A.*, and *Maglova, P.*, Influence des facteurs biologiques sur le microclimat en conditions karstiques tropicales. Bull. Société Géographique de Liége. 29, 119-124, 1993.

[4721] *Trapasso, L. M.* and *Kaletsky, K.*, Food preparation activities and the microclimate within Mammoth Cave, Kentucky. NSS Bulletin, 56, 64-69, 1994.

[4722] *Varcha, L. G.*, Carst caves in therapy and the correlation between geography and respiratory diseases in Hungary. Sectio Medico-Geographica Societatis Geographicae Hungaricae. 18, 109-136, 1988.

[4723] *Villar, E., Bonet, A., Diaz-Canja, B., Fernandez, P. L., Gutierrez, I., Quindos, L. S., Solana, J. R.*, and *Soto, J.*, Ambient temperature variations in the hall of paintings of Altimira Cave due to the presence of visitors. Cave Science 11, 99-104, 1984.

[4724] *Wigley, T. M. L.* and *Brown, M. C.*, The physics of caves. In: The Science of Speleology, T. D. Ford and C. H. D. Cullingford (Ed.), Academic Press, 329-358, 1976.

SECTION 48

[4800] *Bazin, R. C.*, and *MacArthur, R. A.*, Thermal benefits of huddling in the muskrat (*Ondatra zibethicus*). J. Mammal. 73, 559-564, 1992.

[4801] *Bodenheimer, F. S.*, Studien zur Epidemiologie etc. der afrikanischen Wanderheuschrecke. Z. f. angew. Entomol. 15, 435-557, 1929.

[4802] *Campbell, G.*, An introduction to environmental biophysics. Springer-Verlag, 1977.

[4803] *Christian, K.*, *Tracy, C. R.*, and *Porter, W. P.*, Seasonal shifts in body temperature and use of microhabitats by Galapagos land iguanas (*Conolophus pallidus*). Ecol. 64, 463-468, 1983.

[4804] *Curtis, B. A.*, Activity of the Namib Desert dune ant (*Camponotus detritus*). S. Afr. Tydskr. Dierk. 20, 41-48, 1985.

[4805] *Desjardins, R. L.*, *Gifford, R. M.*, *Nilson, T.*, and *Greenwood, E. A. N.*, Advances in bioclimatology. I. Springer-Verlag, 1992.

[4806] *Diez, J. A.*, and *Dusenbery, D. B.*, Preferred temperature of *Meloidogyne incognita*. J. Nematology 21, 99-104, 1989.

[4807] *Dusenbery, D. B.*, Behavioral responses of *Meloidogyne incognita* to small temperature changes. J. Nematology 20, 351-355, 1988.

[4808] *El-Sherif, M.*, and *Mai, W. F.*, Thermotactic response of some plant parasitic nematodes. J. Nematol. 1, 43-48, 1969.

[4810] *Hadley, N. F.*, Micrometeorology and energy exchange in two desert arthropods. Ecol. 51, 434-444, 1970.

[4811] *Hatfield, J.* and *Thomason, I.*, Biometeorology and integrated pest management, Academic Press, 1982.

[4812] *Huffaker, C. B.* and *Rabb, R. L.*, Ecological Entomology, John Wiley & Sons, 1984.

[4813] *Isard, S. A.*, *Irwin, M. E.*, and *Hollinger, S. E.*, Vertical distribution aphids (*Homoptera*: *Aphididae*) in the planetary boundary layer. Environ. Entomol. 19, 1473-1484, 1990.

[4814] *Johnson, C. G.*, The vertical distribution of aphids in the air and the temperature lapse rate. Quart. J. 83, 194-201, 1957; 85, 173-174, 1959.

[4815] *Kaspari, M.*, Body size and microclimate use in Neotropical granivorous ants. Oecologia 96, 500-507, 1993.

[4816] *Lauscher, F.*, Mückentanz u. Windschutz. Biokl. B. 6, 186, 1939.

[4817] *Marder, J.*, *Arieli, Y.*, and *Ben-Asher, J.*, Defense strategies against environmental heat stress in birds. Israel J. Zool. 36, 61-75, 1989.

[4818] *Martini, E.*, and *Teubner, E.*, Über das Verhalten von Stechmücken bei verschiedener Temperatur u. Luftfeuchtigkeit. Beih. z. Arch. f. Schiffs- u. Tropenhyg. 37, 1933.

[4819] *Mosauer, W.*, The toleration of solar heat in desert reptiles. Ecol. 17, 56-66, 1936.

[4820] *Müller, H. J.*, *Unger, K.*, *Neitzel, K.*, *Raeuber, A .*, *Moericke, V.*, and *Seemann, J.*, Der Blattlausbefallsflug in Abhängigkeit von Flugpopulation and witterungsbedingter Agilität in Kartoffelabbau- u. Hochzuchtlagen. Biolog. Zentralbl. 78, 341-383, 1959.

[4821] *Neumann, G.*, Bemerkungen zur Zellularkonvektion im Meer and in der Atmosphäre and die Beurteilung des statischen Gleichgewichts. Ann. d. Met. 1, 238-244, 1948.

[4822] *Nielson, E. T.*, Zur Ökologie der Laubheuschrecken. Saertryk af Ent. Medd. 20, 121-164, 1938.

[4823] *Pline, M.*, *Diez, J. A.* and *Dusenbery, D. B.*, Extremely sensitive therotaxis of the nematode *Meloidogyne incognita*. J. Nematology, 20, 605-608, 1988.

[4824] *Rainey, R. C.*, Some observations on flying locusts and atmospheric turbulence in eastern Africa. Quart. J. 84, 334-354, 1958.

[4825] *Schimitschek*, E., Standortsklima u. Kleinklima in ihrer Beziehung zum Entwicklungsablauf and zur Mortalität von Insekten. Z. f. angew. Entomol. 18, 460-491, 1931.

[4826] *Schrödter, H.*, Agrarmeteorologische Beiträge zu phytopathologischen ragen. Abh. Met. D. DDR 2, Nr. 15, 1-83, 1952.

[4827] *Smith, C. V.*, and *Gough, M. C.*, Meteorology and Grain Storage. World Meteorological Organization, 1990.

[4828] *Suzuki, S.*, *Tanioka, K.*, *Uchimura, S.*, and *Marumoto, T.*, The hovering height of skylarks. J. Agr. Met. Japan 7, 149-151, 1952.

[4829] *Tauber, M. J.*, *Tauber, C. A.*, and *Masaki, S.*, Seasonal adaptions of insects, Oxford University Press, 1986.

[4830] *Tromp, S. W.*, Biometeorology, the Impact of the Weather and Climate on Humans and Their Environment (Animals and Plants), Heyden, 1980.

[4831] *Unger, K.*, and *Müller, H. J.*, Über die Wirkung geländeklimatisch unterschiedlicher Standorte auf den Blattlausbefallsflug. Der Züchter 24, 337-345, 1954.

[4832] *Wachob, D. G.*, The effect of thermal microclimate on foraging site selection by wintering mountain chickadees. The Condor 98, 114-122, 1996.

[4833] *Weiss, S. B.*, *Murphy, D. D.*, and *White, R. R.*, Sun, slope, and butterflies: topographic determinants of habitat quality for *Euphydryas editha*. Ecol. 69, 1486-1496, 1988.

[4834] *Wolf, B. O.*, and *Walsberg, G. E.*, Thermal effects of radiation and wind on a small bird and implications for microsite selection. Ecol. 77, 2228-2236, 1996.

SECTION 49

[4900] *Barrows, C. W.*, Roost selection by Spotted Owls: An adaptation to heat stress. Condor 83, 302-309, 1981.

[4901] *Burkhardt, H.*, Zitat verloren and z. Z. nicht zu ermitteln.

[4902] *Dahmen, H.*, Untersuchungen im klimatisierten Stall. Gesundh. Ing. 67, 61-64, 1944.

[4903] *Dyck, A. P.*, and *MacArthur, R. A.*, Seasonal variation in the microclimate and gas composition of beaver lodges in a boreal environment. J. Mamm. 74, 180-188, 1993.

[4904] *Facemire, C. F.*, *Facemire, M. E.*, and *Facemire, M. C.*, Wind as a factor in the orientation of entrances of Cactus Wren nests. The Condor 92, 1073-1075, 1990.

[4905] *Frith, H. J.*, *Wie* regelt dcr Thermometervogel die Temperatur seines Nesthügels? Umschau 56, 238-239, 1956.

[4906] *Gatenby, R. M.*, *Handayani, S. W.*, *Martawidjaja, M.*, and *Waldron, M. C.*, Modification of the environment by animal houses in a hot humid climate. Agr. For. Met. 39, 299-308, 1987.

[4907] *Hall, L. S.*, The effect of cave microclimate on winter roosting behavior in the bat, *Miniopterus schreibersii blepotis*. Aust. J. Ecol. 7, 129-136, 1982.

[4908] *Hecnar, S.J.*, Nest distribution, site selection, and brooding in the five-lined skink (*Eumeces fasciatus*). Can. J. Zool. 72, 1510-1516, 1994.

[4909] *Hesse, R.*, Tiergeographie auf ökologischer Grundlage. G. Fischer, Jena, 1924.

[4910] *Holdaway, F. G.*, and *Gay, F. J.*, Temperature studies of the habitat of Entermes exitiosus etc. Austral. J. Sci. Res. B 1, 464-493, 1948.

[4911] *Korhonen, K.*, Microclimate in the snow burrows of Willow Grouse (*Lagopus lagopus*). Ann. Zool. Fennici, 17, 5-9, 1980.

[4912] *Lehmann, P.*, Das Sonderklima des Stalls. Forsch. d. Landwirtsch. 6, 642-647, 1931.

[4913] *Löhrl, H.*, Der Winterschlaf von Nyctalus noctula Schreb. auf Grund von Beobachtungen am Winterschlafplatz. Z. f. Morph. u. Ökol. d. Tiere 32, 47-66, 1936.

[4914] *Mehner, A.*, and *Linz, A.*, Untersuchung über den Verlauf der Stalltemperatur. Forschungsdienst 8, 525-543, 1939.

[4915] *Raeuber*, A., Meteorologische Vergleichsmessungen zwischen Schuppenstall u. Freiland in Gross-Lüsewitz. Angew. Met. 2, 217-222, 1956.

[4916] Sidis, Y, Zilberman, R., and Amos A., Thermal aspects of nest placement in the orange-tufted sunbird (*Nectarinia osea*). The Auk, 111, 1001-1005, 1994.
[4917] *Steiner, A.*, Neuere Ergebnisse über den sozialen Wärmehaushalt der einheimischen Hautflügler. Naturw. 18, 595-600, 1930.
[4918] *Tortosa, F. S.*, and *Villafuerte, R.*, Effect of nest microclimate on effective endothermy in White Stork *Ciconia ciconia* nestlings. Bird Study 46, 336-341, 1999.
[4919] *Wachob, D. G.*, A microclimatic analysis of nest-site selection by Mountain Chickadees. J. Field Ornothol. 67, 525-533, 1996.
[4920] *Wächtershäuser, H.*, Beitrag zum Stallklima. Ber. DWD-US Zone 7, Nr. 42, 382-384, 1952.
[4921] *Wellenstein, G.*, Beiträge zur Physiologie der roten Waldameise etc. Z. f. angew. Entomol. 14, 1-68, 1929.
[4922] *Zorn, W., and Freidt, G.*, Der Einfluss von Wetter u. Klima auf unsere landwirtschaftlichen Nutztiere, Z. f. Züchtkde. 14, Nr. 1, 1939.

SECTION 50

[5000] *Aron, R. H.*, *Hodgson, S-P.*, and *Aron, I-M.*, Empirical models for measuring winter heating and summer cooling energy requirements, J. Env. Man. 18, 339-344, 1984.
[5001] *Aronin, J. E.*, Climate and Architecture. Reinhold Publ. Corp. New York, 304 S., 1953.
[5002] *Bartlett, P. N.*, and *Gates, D. M.*, The energy budget of a lizard on a tree trunk. Ecol. 48, 315-322, 1967.
[5003] *Brooks, C. E. P.*, and *Evans, G. J.*, Annotated bibliography on the climate of enclosed spaces (cryptoclaimtes). MAB 7, 211-264, 1956.
[5004] *Burt, J. E.*, *O'Rourke, P. A.*, and *Terjung, W. H.*, The Relative Influence of Urban Climates on Outdoor Human Energy Budgets and Skin Temperature. I. Modeling Considerations. Int. J. Biomet. 26, 3-23, 1982.
[5005] *Burt, J. E.*, *O'Rourke, P. A.*, and *Terjung, W. H.*, The Relative Influence of Urban Climates on Outdoor Human Energy Budgets and Skin Temperature. I. Man in an Urban Environment. Int. J. Biomet. 26, 25-35, 1982.
[5006] *Cagliolo, A.*, Marcado gradiente térmico en vagones de ferrocarril. Meteoros 4, 395-398, 1954.
[5007] *Cardinal I. G.*, 1991 Architectural Glass Products Buyline 3148, Cardinal I G Minnetonka, Minn., 1991.
[5008] *Crook, E. M.*, and *Watson, D. J.*, Studies on the storage of potatoes. J. Agr. Sci. Cambridge 40, 199-232, 1950.
[5009] *Dessauer, F.*, *Graffunder, W.* and *Laub, J.*, Beobachtungen über Ionenschwankungen im Fleien and in geschlossenen Räumen. Ann. d. Met. 7, 173-185, 1956.
[5010] *Dirmhirn, I.*, Über das Klima in den Wiener Strassenbahnwagen. Wetter u. Leben 4, 158-162, 1952.
[5011] *Edmonds, R. L.* (Ed.), Aerobiology. Dowden, Hutchinson & Ross, 1979.
[5012] *Gates, D. M.*, Biophysical Ecology, Springer-Verlag, 1980.
[5013] *Georgii, H. W.*, Untersuchung über den Luftaustausch zwischen Wohnräumen u. Aussenluft. Arch f. Met. (B) 5, 191-214, 1953.
[5014] *Giner, R.*, and *Hess, V. F.*, Studie über die Verteilung der Aerosole in der Luft von Innsbruch u. Umgebung, Gerl. B 50, 22-43, 1937.
[5015] *Givoni, B.*, Man, Climate and Architecture, Van Nostrand Reinhold Co., 1981.
[5016] —, Urban design in different climates, WCAP-10, Technical Document. No. 346 WMO, 1989.
[5017] *Grant, B. W.*, and *Dunham, A. E.*, Thermally imposed time constraints on the activity of the desert lizard *Sceloporus merriami*. Ecol. 69, 167-176, 1988.

[5018] *Höppe, P.*, Die Energiebilanz des Menschen. Wiss. Mitt. Meteor. Inst. Univ. München, Nr. 49, 1984.

[5019] *Hummel, F.*, Mietenklima u. Windeinfluss. Ber. DWD-US Zone 5, Nr. 32, 44-47, 1952.

[5020] *Jensen, M.*, The model-law for phenomena in natural wind. Ingeniøren (Dänemark) 2, 121-128, 1958.

[5021] *Kern, H.*, Mietentemperaturmessungen auf Niedermoorboden. Ber. DWD-US Zone 6, Nr. 38, 186-189, 1952.

[5022] *Kabuga, J. D.*, Effect of Weather on Milk Production of Holstein-Friesian Cows in the Humid Tropics. Agric. For. Meteor. 57, 209-219, 1991.

[5023] *King, E.*, Medizin-meteorologische Einflüsse auf den Strassenverkehr. Z. f. Verkehrssicherheit 4, 116-136, 1958 (see also Wetter u. Leben 8, 213-219, 1956).

[5024] *Kreutz, W.*, Beitrag zum Mietenklima. Met. Rundsch. 1, 348-351, 1948.

[5025] —, Merkblatt zur Mietenbehandlung. Mitt. DWD-US Zone 7, 1950.

[5026] —, Der Mietenklimadienst. Mitt. DWD 2, Nr. 14, 168-170, 1955.

[5027] *Landsberg, H.*, Bioclimatology of housing. Met. Monographs 2, 81-98, 1954.

[5028] *Mayer, H.* and *Höppe, P.*, Thermal Comfort of Man in Different Urban Environments. Theor. Appl. Climatol. 38, 43-49, 1987.

[5029] *Munn, R. E.*, Bioclimatology. In: The Encyclopedia of Climatology, 163-169, (Eds.) J. W. Oliver and R. W. Fairbridge, Van Nostrand Reinhold, 1987.

[5030] *Natsume, K.*, *Tokura, H.*, *Isoda, N.*, *Maruta, N.*, and *Kawakami, K.*, Field Studies of Clothing Microclimate Temperatures in Human Subjects During Normal Daily Life. J. Human Ergol. 17, 13-19, 1988.

[5031] *Niemann, A.*, Über die Strahlungsbeeinflussung durch verschmutzte Gewächshausscheiben. Ann. d. Met. 8, 344-352, 1959.

[5032] *Parker, J. H.*, Landscaping to reduce the energy used in cooling buildings. J. Forestry 81, 82-84, 105, 1983.

[5033] —, The use of shrubs in energy conservation plantings. Landscape J. 6, 132-139, 1987.

[5034] —, The impact of vegetation on air conditioning consumption. Proc. Conf. on Controlling the Summer Heat Island. LBL-27872, 46-52, 1989.

[5035] *Porter, W. L.*, Occurence of high temperatures in standing boxcars. Siehe MAB 4, 263, 1954.

[5036] *Priestley, C. H. B.*, The heat balance of sheep standing in the sun. Austral. J. Agr. Res. 8, 271-280, 1957.

[5037] *Ray, D. E.*, *Jassim, A. H.*, *Armstrong, D. V.*, *Wiersma, F.*, and *Schuh, J. D.*, Influence of season and microclimate on fertility of dairy cows in a hot-arid environment. Int. J. Biomet., 36, 141-145, 1992.

[5038] *Scott, J. R.*, *Tracy, C. R.*, and *Pettus, D.*, A biophysical analysis of daily and seasonal utilization of climate space by a montane snake. Ecol. 63, 482-493, 1982.

[5039] *Seemann, J.*, Klima u. Klimasteurerung im Gewächshaus. Bayer. Landwirtsch. Verl., München, 1957.

[5040] *Steen, J.*, *Shrode, W.* and *Stuart, E.*, Basis for development of a viable energy conservation policy for Florida residents. Fla. State Energy Off. 212, 1976.

[5041] *Walle de, D. R.*, *Heisler, G. M.*, and *Jacobs, R. E.*, Forest home sites influence heating and cooling energy. J. Forestry 81, 84-88, 1983.

SECTION 51

[5101] *Aida, M.*, Urban albedo as a function of the urban structure - A model experiment. Bound. Lay. Met. 23, 405-413, 1982.

[5102] *Aron, R. H.*, and *Aron, I-M*, Models for estimating future sulfur dioxide concentrations in Taipei. Bull. Geophys. 25, 47-53, 1984.

[5103] —, Models for estimating and forecasting carbon monoxide concentrations in Taipei. Bull. Geophys. 25, 55-61, 1984.

[5104] —, Forecasting high level oxidant concentrations in the Los Angeles Basin. J. Air Poll. Con. Assoc. 30, 1227-1228, 1980.

[5105] ——, and *Aron, I-M.*, Statistical forecasting models: I. Carbon monoxide concentrations in the Los Angeles Basin. J. Air Poll. Con. Assoc. 28, 681-684, 1978.

[5106] *Bach, W.*, An urban circulation model. Arch. Met. Geoph. Biokl. 18B, 155-168, 1970.

[5107] *Bach, W.*, Atmospheric turbidity and air pollution in Greater Cincinnati. Geogr. Rev. 61, 573-594, 1971.

[5108] ——, Solar irradiation and atmospheric pollution. Arch. Met. Geoph. Biokl. 21B, 67-75, 1973.

[5109] ——, and *Daniels, A.*, Aerometric studies in Greater Cincinnati using an integrating nephelometer. Tellus 25, 499-507, 1973.

[5110] *Bornstein, R. D.*, and *Oke, T. R.*, Influence of pollution on urban climatology. Adv. Envir. Sci. Tech. 2, 171-203, 1981.

[5111] *Chandler, T. J.*, Diurnal, seasonal and annual changes in the intensity of London's heat-island. Meteor. Mag. 91, 146-153, 1962.

[5112] *Changnon, S. A., Jr., Semonin, R. G., Auer, A. H., Braham, R. B., Jr.*, and *Hales, J. M.*, METROMEX: A review and summary. American Meteorological Society Monograph No. 40, 1981.

[5113] —, *Shealy, R. T.*, and *Scott, R. W.*, Precipitation changes in fall, winter, and spring caused by St. Louis. J. App. Met. 30, 126-134, 1991.

[5114] *Cleugh, H. A.*, and *Oke, T. R.*, Suburban-rural energy balance comparisons in summer for Vancouver, B. C. Bound. Lay. Met. 36, 351-369, 1986.

[5115] *Collins, C. O.*, and *Scott, S. L.*, Air pollution in the valley of Mexico. Geogr. Rev. 83, 119-133, 1993.

[5116] *Davidson, C. I.*, Air pollution in Pittsburgh: A historical perspective. J. Air Poll. Cont. Assoc. 29, 1035-1041, 1979.

[5117] *Duckworth, F. S.* and *Sandberg, J. S.*, The effect of cities upon horizontal and vertical temperature gradients. Bull. Am. Met. Soc. 35, 198-207, 1954.

[5118] *Eggleston, S., Hackman, M. P., Heyes, C. A., Irwin, J. G., Timmis, R. J.*, and *Williams, M. L.*, Trends in urban air pollution in the United Kingdom during recent decades. Atm. Env. 26B, 117-129, 1992.

[5119] *Federer, C. A.*, Trees modify the urban microclimate. J. Arboricul. 2, 121-127, 1976.

[5120] *Gallo, K. P., McNab, A. L., Karl, T. R., Brown, J. F., Hood, J. J.*, and *Tarlpey, J. D.*, The use of NOAA AVHRR data for assessment of the urban heat island. J. Appl. Met. 32, 899-908, 1993.

[5121] *Givoni, B.*, Impact of planted areas on urban environmental quality: A review. Atm. Env. 25, 289-299, 1991.

[5122] —, Climatic aspects of urban design in tropical regions. Atm. Env. 26, 397-406, 1992.

[5123] *Goldreich, Y.*, Urban topoclimatology. Prog. Phys. Geog. 8, 336-363, 1984.

[5124] *Grimmond, C. S. B.*, The suburban energy balance: methodological considerations and results for a mid-latitude west coast city under winter and spring conditions. Int. J. Climatol. 12, 481-497, 1992.

[5125] *Harrison, R., McGoldrich, B.*, and *Williams, C. G. B.*, Artificial heat release from Greater London, 1971-1976. Atm. Env. 18, 2291-2304, 1984.

[5126] *Howard, L.*, The Climate of London, Vols. 1-3, Harvey & Darton, 1833.

[5127] *Huang, Y. J., Akbari, H., Taha, H.*, and *Rosenfeld, A. H.*, The potential of vegetation in reducing summer cooling loads in residential buildings. Jour. Clim. Appl. Meteor. 26, 1103-1116, 1987.

[5128] *Jauregui, E.*, Untersuchungen um Stadtklima von Mexico-stadt. Doctoral thesis, Rheinishen Friedrich-Wilhems-Universität, 1973.

[5129] *Kalma, J. D.*, and *Newcombe, K. J.*, Energy use in two large cities: A comparison of Hong Kong and Sydney, Australia. Environ. Stud. 9, 53-64, 1976.

[5130] *Koning, H. W., de, Kretzschmar, J. G., Akland, G. G.*, and *Bennett, B. G.*, Air pollution in different cities around the world. Atm. Env. 20, 101-113, 1986.

[5131] *Kratzer, A.*, Das Stadklima, Verl. Vieweg, Braunschweig, 1956.

[5132] *Landsberg, H. E.*, The Urban Climate, Academic Press, 1981.

[5133] *Lauscher, F., Roller, M., Wacha, G., Grammer, M., Weiss, E.*, and *Frenzel, J. W.*, Witterung u. Klima von Linz. Wetter u. Leben, Sonderh. VI, Wien 1959.

[5134] *Lee, D. O.*, The influence of atmospheric stability and the urban heat island on urban-rural wind speed differences. Atm. Env. 13, 1175-1180, 1979.

[5135] *Löbner, A.*, Horizontale u. vertikale staubverteilung in einer Grossstadt. Veröff, Geoph. I. Leipzig 7, Nr. 2, 1935.

[5136] *Lorente, J., Redaño, A.*, and *De Cabo, X.*, Influence of urban aerosol on spectral solar irradiance. Jour. Appl. Meteor. 33, 406-415, 1994.

[5137] *McPherson, E. G.*, and *Heisler, G. M.*, Impacts of vegetation on residential heating and cooling. Ener. Build. 12, 41-51, 1988.

[5138] *Nichol, J. E.*, Monitoring Singapore's microclimate. Geo. Info. Systems. 3, 51-55, 1993.

[5139] *Nuñez, M.*, and *Oke, T. R.*, The energy balance of an urban canyon. J. Appl. Met. 16, 11-19, 1977.

[5140] *Oke, T. R.*, Review of Urban Climatology, 1968-1973. World Meteorological Organization, 1974.

[5141] —, Review of Urban Climatology, 1973-1976. World Meteorological Organization, 1979.

[5142] *Oke, T. R.*, Canyon geometry and the nocturnal heat island: Comparison of scale model and field observations. J. Climatol. 1, 237-254, 1981.

[5143] —, Bibliography of Urban Climatology, 1977-1980. World Meteorological Organization, 1983.

[5144] —, Bibliography of Urban Climatology, 1968-1973. World Meteorological Organization, 1990.

[5145] —, The energetic basis of the urban heat island. Q. J. R. Meteorol. Soc. 108, 1-24, 1982.

[5146] —, The urban energy balance. Prog. Phys. Geog. 12, 471-508, 1988.

[5147] —, Street design and urban canopy layer climate. Ener. Build. 11, 103-113, 1988.

[5148] —, The micrometeorology of the urban forest. Phil. Trans. R. Soc. Lond. 324B, 335-349, 1989.

[5149] —, and *Fuggle, R. F.*, Comparison of urban/rural counter and net radiation at night. Bound. Lay. Met. 2, 290-308, 1972.

[5150] —, *Johnson, G. T., Steyn, D. G.*, and *Watson, I. D.*, Simulation of surface urban heat islands under 'ideal' conditions at night Part 2: Diagnosis of causation. Bound. Lay. Met. 56, 339-358, 1991.

[5151] —, and *Maxwell, G. B.*, Urban heat island dynamics in Montreal and Vancouver. Atm. Env. 9, 191-200, 1975.

[5152] *Pearlmutter, D., Bitan, A.*, and *Berliner, P.*, Microclimatic analysis of "compact" urban canyons in an arid zone. Atm. Env. 33, 4134-4150, 1999.

[5153] *Peterson, J. T.*, and *Flowers, E. C.*, Interactions between air pollution and solar radiation. Ener. 19, 23-32, 1977.

[5154] *Roth, M., Oke, T. R.*, and *Emery, W. J.*, Satellite-derived urban heat islands from three coastal cities and the utilization of such data in urban climatology. Int. J. Remote Sens. 10, 1699-1720, 1989.

[5155] *Schmid, H. P., Cleugh, H. A., Grimmond, C. S. B.*, and *Oke, T. R.*, Spatial variability of energy fluxes in suburban terrain. Bound. Lay. Met. 54, 249-276, 1991.

[5156] *Schmid, J. A.*, Vegetation types, functions and constraints. In: Metropolitan Environments. Agronomy. Agron. J. 21, 499-528, 1979.

[5157] *Snider, R.*, Personal communication, National Weather Service, Detroit, 1982.

[5158] *Steinhauser. F., Echel, O.,* and *Sauberer, F.,* Klima u. Bioklima von Wien. Wetter u. Leben, Sonderh. III, 1955; V, 1957 u. VII, 1959.

[5159] *Sundborg, A.,* Climatological studies in Uppsala. Geographica 22, Uppsala, 1951.

[5160] *Swaid, H.,* The role of radiative-convective interaction in creating the microclimate of urban street canyons. Bound. Lay. Met. 64, 231-260, 1993.

[5161] *Takamura, T.,* Spectral reflectance in an urban area: A case study for Tokyo. Bound. Lay. Met. 59, 67-82, 1992.

[5162] *Upmanis, H.* and *Chen D.,* Influence of geographical factors and meteorological variables on nocturnal urban-park temperature differences – A case study of summer 1995 in Götenborg, Sweden. Clim. Res. 13, 125-139, 1999.

[5163] *Weiss, E.,* and *Frenzel, J. W.,* Untersuchungen von Luftverunreinigungen durch Rauch- u. Industriegase im Raume von Linz. Wetter u. Leben 8, 131-147, 1956.

SECTION 52

[5200] *Aslying, H. C.,* Shelter and its effect on climate and water balance. Oikos 9, 282-310, 1958.

[5201] *Blenk, H.,* and *Trienes, H.,* Strömungstechnische Beiträge zum Windschutz. Grundlagen d. Landtechn. 8, I u. II, VDI-Verl., Düsseldorf, 1956.

[5202] *Bouchet, R. J., Guyot, G.,* and *de Parcevaux, S.,* Augmentation de l'Efficience de l'eau et Amelioration des Rendements par Reduction de l'Evapotranspiration Potentielle au Moyen de Brise-vent, Proc. UNESCO Symp. on Methods in Agrometeorology, July 23-30, 167-173, 1966.

[5203] *Brenner, A. J., Jarvis, P. G.,* and *van den Belt, R. J.,* Windbreak-crop interactions in the Sahel. 2. Growth response of millet in shelter. Agr. Met. 75, 235-262, 1995.

[5204] *Brown, K. W.,* and *Rosenberg, N. J.,* Turbulent transport and energy balance as affected by a windbreak in an irrigated sugar beet (*Beta vulagaris*) field. Agron. J. 63, 351-355, 1971.

[5205] —, —, Shelter-effects on microclimate, growth and water use by irrigated sugar beets in the Great Plains. Agr. Met. 9, 241-263, 1972.

[5206] *Caborn, J. M.,* The influence of shelterbelts on microclimate. Quart. J. 81, 112-115, 1955.

[5207] *Carr, M. K. V.,* Some effects of shelter on the yield and water use of tea. In: J. Grace (Ed.), Effects of Shelter on the Physiology of Plants and Animals. Swets and Zeitlinger, Lisse. 127-144, 1985.

[5208] *Casperson, G.,* Untersuchungen über den Einfluss von Windschutzanlagen auf den standörtlichen Wärmehaushalt. Angew. Met. 2, 339-351, 1957.

[5209] *Chepil, W. S.,* Wind erodibility of farm fields. J. Soil Wat. Cons. 14, 214-219, 1959.

[5210] —, *Siddoway, F. A., and Armbrust, D. V.,* Climatic factor for estimating wind erodibility of farm fields. J. Soil Wat. Cons. 17, 162-165, 1962.

[5211] *Daigo, Y.,* and *Maruyama, E.,* Experimental studies of windbreaks for protection from frost damage. Mem. Industr. Met. 20, 1-7, 1956.

[5212] *Dirmhirn, I.,* Zur Strahlungsminderung an Windschutzstreifen. Wetter u. Leben 5, 208-213, 1953.

[5213] *Dixon, M.,* and *Grace, J.,* Effect of wind on the transpiration of young trees. *Ann. Bot.* 53, 811-819, 1984.

[5214] *Eimern, J. van,* Beeinflussung meteorologischer Grössen durch ein engmaschiges Heckensystem. Ann. d. Met. 6, 213-219, 1953/54.

[5215] —, Über die Veränderlichkeit der Windschutzwirkung einer Doppelbaumreihi bei verschiedenen meteorologischen Bedingungen. Ber. DWD 5, Nr. 32, 1-21, 1957.

[5216] —, Geländeklimaaufnahmen für landwirtschaftliche Zwecke. Bayer. Landw. Jahrb. 35, 193-210, 1958.

[5217] —, Über, den j Ahrezeitlichen Gang der Gelanderlimatisch Bedingten Differenzen der Nächtlichen Minimum Temperatur, Agr. Met. 1, 149-153, 1964.

[5218] *Finney, E. A.,* Snow Control on the Highway. Unpublished Master's Thesis, Iowa State College, 1934.

[5218] ——, Snowdrift Control by Highway Design. Bull. No. 86, Michigan Engineering Experiment Station, East Lansing, Michigan. 1-58, 1939.

[5220] *George, E. J.*, Effects of tree windbreaks and slat barriers on wind velocity and crop yields. *USDA-ARS Prod. Res. Rep.*, 121-123, 1971.

[5221] General Accounting Office/Controller General of the United States. Report to the Congress: Action needed to discourage removal of trees that shelter cropland in the Great Plains, RED-74-375, 31 pp., 1975.

[5222] *Guyot, G.*, and *Segiun, B.*, Modification of land roughness and resulting microclimatic effects: a field study in Brittany. In: D. A. de Vries and N. H. Afgan (Ed.), Heat and Mass Transfer in the Biosphere. Part 1: Transfer Processes in the Plant Environment. Scripta Book Co., Halsted Press, 467-478, 1975.

[5223] *Halitsky, J.*, Gas Diffusion near Buildings in Meteorology and Atomic Energy (H. D. Slade, Ed.). Tech. Info. Cent. U.S. Dept. Energy, Oak Ridge, Tenn., 221-255, 1968.

[5224] *Heisler, G. M.*, and *Dewalle, D. R.*, Effects of windbreak structure on wind flow. Agr. Ecosys. Env. 22/23, 41-69, 1988.

[5225] *Hilf, H. H.*, Wirksamer Windschutz. Die Holzzucht 13, 33-43, 1959.

[5226] *Illner, K.*, Über den Einfluss von Windschutzpflanzungen auf die Unkrautverbreitung. Angew. Met. 2, 370-373, 1957.

[5227] *Jensen, M.*, Shelter effect (264 S.). The Danish Techn. Press, Kopenhagen, 1954.

[5228] *Johnson, G. T.*, and *Watson, I. D.*, The determination of view-factors in urban canyons. J. Clim. App. Met. 23, 329-335, 1984.

[5229] *Johnson, R. J.* and *Beck, M. M.*, Influences of shelterbelt on wildlife management biology. Agr. Ecosys. Env. 22/23, 301-335, 1988.

[5230] *Kaiser, H.*, Beiträge zum Problem der Luftströmung in Windschutzsystemen. Met. Rundsch. 12, 80-87, 1959.

[5231] —, Die Strömung an Windschutzsteifen. Ber DWD 7, Nr. 53, 1959.

[5232] —, Schneeverwehungen an Windschutzanlagen, eine Gefahr für Felder u. Wege? Umschau 60, 33-36, 1960.

[5233] *Kaminski, A.*, The Effect of a shelterbelt on the distribution and intensity of ground frosts in cultivated fields. Ekol. Polska Ser. A 16, 1-11, 1968.

[5234] *Kort, J.*, Benefits of windbreaks to field and forage crops. Agr. Ecosys. Env. 22/23, 165-190, 1988.

[5235] *Kowalchuk, T. E.* and *de Jong, E.*, Shelterbelts and their effect on crop yield. Can. J. Soil Sci. 75, 543-550, 1995.

[5236] *Kreutz, W.*, Der Windschutz (167 S.). Ardey Verl., Dortmund, 1952.

[5237] *Lemon, E.*, Gaseous exchange in crop stands. In: Physiological Aspects of Crop Yield (F. A. Haskins, C. Y. Sullivan, and C. H. M. van Bavel, Eds.), Am. Soc. Agron., 117-137, 1970.

[5238] *Lewis. T.*, Patterns of distribution of insects near a windbreak of tall trees. Ann. Appl. Biol. 65, 213-220, 1970.

[5239] ——, and *Smith, B. D.*, The insect faunas of pear and apple orchards and the effect of windbreaks on their distribution. Ann. Appl. Biol., 64, 11-20, 1969.

[5240] *Linde, R. J. van der*, and *Woudenberg, J. P. M.*, On the microclimatic properties of sheltered areas (151 S.) K. Nederl. Met. Inst., Nr. 102, 1950.

[5241] *Long, S. P.* and *Persuad, N.*, Influence of neem *(Azardirachta indica)* windbreaks on millet yields, microclimate and water use in Niger West Africa. In: Unger, P. W., Sneed, T. V., Jordan, W. R., and Jensen, R., Challenges in Dryland Agriculture – A Global Perspective. Texas Agricultural Experiment Station, Amarillo, Texas. 313-314, 1988.

[5242] *Marshall, J. K.*, The effects of shelter on the productivity of grasslands and field crops. Field Crops Abstr. 20, 1-14, 1967.

[5243] *McAneny, K. J.*, *Salinger, M. J.*, *Porteous, A. S.*, and *Barber, R. F.*, Modification of an orchard climate with increasing shelter-belt height, Agr. For. Met. 49, 177-189, 1990.

[5244] *McClure, G.*, Shelterbelts-Wooded Islands of Wildlife, Extension Rev., 17-19, 1981.
[5245] *McNaughton, K. G.*, Effects of windbreaks on turbulent transport and microclimate. Agr. Ecosys. Env. 22/23, 17-39, 1988.
[5246] —, Micrometeorology of shelter belts and forest edges. Phil. Trans. R. Soc. Lond. 324B, 351-368, 1989.
[5247] *Miller, D. R, Bagley, W. T.*, and *Rosenberg, N. J.*, Microclimate modification with shelterbelts. J. Soil Wat. Cons. 29, 41-44, 1974.
[5248] *Müller, T.*, Versuche über die Windschutzwirkung von Hecken auf der Schwäbischen Alb. Umschaudienst 6, Nr. 1/2, 1956.
[5249] *Nägeli, W.*, Untersuchungen über die Windverhältnisse im Bereich von Windschutzstreifen. Mitt. d. Schweiz. Anst. f. d. Forstl. Versuchswesen 23, 221-276, 1943; 24, 657-737, 1946.
[5250] —, Untersuchungen über die Windverhältnisse im Bereich von Schilfrohrwänden. Mitt. d. Schweiz. Anst. f. d. Forstl. Versuchswesen 29, 213-266, 1953.
[5251] —, Die Windbremsung durch einen grösseren Waldkomplex. Ber. 11. Kongr. Intern. Verb. Forstl. Forsch. Anst. Rom 1953, 240-246, Florenz, 1954.
[5252] *Nageli, W.*, On the most favorable shelterbelt spacing. Scottish Forestry 18, 4-15, 1964.
[5253] *Norton, R. L.*, Windbreaks: Benefits to orchard and vineyard. Agr. Ecosys. Env. 22/23, 205-213, 1988.
[5254] *Nuñez, M.* and *Sander, D.*, Protection from cold stress in a eucalyptus shelterwood. J. Climatol. 2, 141-146, 1982.
[5255] *Pasek, J. E.*, Influence of wind and windbreaks on local dispersal of insects. Agr. Ecosys. Env. 22/23, 539-554, 1988.
[5256] *Perry, A. H.* and *Symons, L. J.*, Highway Meteorology, E & FN Spon, 1991.
[5257] *PFRA*, 1985 Report of the PFRA Tree Nursery. Ag. Canada-PFRA, 1985.
[5258] *Radke, J. K.*, The use of annual wind barriers for protecting row crops, Proc. Symp. Shelterbelts on the Great Plains, Denver, CO, Great Plains Agric. Council Publ. No. 78, 79-87, 1976.
[5259] *Read, R. A.*, Tree windbreaks for the central Great Plains, Agric. Handbook No. 250, U. S. Dept. of Agric., 1964.
[5260] *Ring, S. L., Iversen, J. D., Sinatra, J. B.*, and *Benson, J. D.*, Wind tunnel analysis of the effects of planting at highway grade separation structures. Iowa Highway Research Board, HR-202, 1979.
[5261] ——, Snow-drift modeling and control. In: Highway Meteorology, Perry, A. H. and Symons, L. J., E & FN Spon, 1991.
[5262] *Rosenberg, N. J.*, Effects of windbreaks on the microclimate, energy balance and water use efficiency of crops growing on the Great Plains. Great Plains Agricultural Council. 78, 49-56, 1976.
[5263] —, Windbreaks for reducing moisture stress, modification of the aerial environment of plants, B. J. Barfield and J. F. Gerber, (Eds.), Am. Soc. Agric. Engin. Monogr., ASAE, 394-408, 1979.
[5264] —, *Blad, B. L.*, and *Verma, S. B.*, Microclimate: the Biological Environment, John Wiley & Sons, Inc., 1983.
[5265] *Rüsch, J. D.*, Der CO_2-Gehalt bodennaher Luftschichten unter Einfluss des Windschutzes. Z. f. Pflanzenernähr, Düngung, Bodenkde. 71, 113-132, 1955.
[5266] *Sato, K., Tamachi, M., Terada, K., Watanabe, Y., Katoh, T., Takata, Y., Sakanoue, T.*, and *Iwasaki, M.*, Studies on windbreaks (201. S.). Nippon-Gakujutsu-Shiukokai, Tokyo, 1952.
[5267] *Sato, S.*, Calculations of the received solar radiation in the shade of windbreak at Miyazaki City. J. Agr. Met. Japan 11, 12-14, 1955.
[5268] *Schrödter, H.*, Untersuchungen über die Wirkung einer Windschutzpflanzung auf den Sporenflug and das Auftreten der Alternaria-Schwärze an Kohlsamenträgern. Angew. Met. 1, 154-158, 1952.

[5269] *Schwartz, R. C., Fryrear, D. W., Harris, B. L., Bilbro, J. D.,* and *Juo, A. R. S.,* Mean flow and shear stress distributions as influenced by vegetative windbreak structure. Agr. For. Met. 75, 1-22, 1995.

[5270] *Scultetus, H. R.,* Windschutz immer noch ein Problem etc. Met. Rundsch. 11, 23-28, 1958.

[5271] —, Bewindung eines Geländes and vertikaler Temperaturgradient. Met. Rundsch. 12, 1-10, 1959.

[5272] *Skidmore, E. L.,* Barrier-induced microclimate and its influence on growth and yield of winter wheat. Symposium on Shelterbelts on the Great Plains, 57-63, 1976.

[5273] —, and *Hagen, L. J.,* Evaporation in sheltered areas as influenced by windbreak porosity. Agr. Met. 7, 363-374, 1970.

[5274] —, —, Potential evaporation as influenced by barrier-induced microclimate. Ecol. Stud. 4, 237-244, 1973.

[5275] —, —, Reducing wind erosion with barriers. Am. Soc. Ag. Eng. 20, 911-915, 1977.

[5276] *Slosser, J. L., Fewin, R. J., Price, J. R., Meinke, L. J.,* and *Bryson, J. R.,* Potential of shelterbelt management for boll weevil (*Coleoptra Curculionidae*) control in the Texas rolling plains. J. Econ. Entomol. 77, 377-385, 1984.

[5277] *Steubing, L.,* Der Tau u. seine Beeinflussung durch Windschutzanlagen. Biolog. Zentralbl. 71, 282-313, 1952.

[5278] *Stigter, C. J.,* Wind protection in traditional microclimate management and manipulation – examples from East Africa. Prog. Biomet. 2, 145-154, 1985.

[5279] *Tabler, R. D.,* Snow Fence Guide, Strategic Highway Research Program, National Research Council, Washington, D.C., 1991.

[5280] *Ticknor, K. A.,* Design and use of field windbreaks in wind erosion control systems. Agr. Ecosys. Env. 22/23, 123-132, 1988.

[5281] *Todhunter, P. E.* and *Cihacek, L. J.,* Historical reduction of airborne dust in the Red River Valley of the North. J. Soil Wat. Cons. 54, 543-551, 1999.

[5282] *Ujah, J. E.* and *Adeoye, K. B.,* Effects of shelterbelts in the Sudan Savanna Zone of Nigeria on microclimate and yield of millet. Agr. For. Met. 33, 99-107, 1984.

[5283] *Upmanis, H., Eliasson I.,* and *Lindqvist, S.,* The influence of green areas on nocturnal temperatures in a high latitude city (Göteborg, Sweden). Int. J. Climatol. 18, 681-700, 1998.

[5284] *Wang, H.* and *Takle, E. S.,* Boundary-layer flow and turbulance near porous obstacles. Bound.-Layer Met. 74, 73-88, 1995.

[5285] —, —, Numerical simulations of shelterbelt effects on wind direction J. Appl. Met. 34, 2206-2219, 1995.

[5286] —, —, A numerical simulation of boundary-layer flows near shelterbelts. Bound.-Layer Met. 75, 141-173, 1995.

[5287] —, —, On shelter efficiency of shelterbelts in oblique wind. Agr. For. Met. 81, 95-117, 1996.

[5288] —, —, On three-dimensionality of shelterbelt structure and its influence on shelter effects. Bound.-Layer Met. 79, 83-105, 1996.

[5289] —, —, Model-simulated influences of shelterbelt shape on wind-sheltering efficiency. J. Appl. Met. 1997.

[5290] *Wilson, J. D.,* Numerical studies of flow through a windbreak. J. Wind Eng. Ind. Aerodyn. 21, 119-154, 1985.

[5291] *Wright, B. C* and *Townsend, L. R.,* Windbreak systems in the western United States. In: Agroforestry and Sustainable Systems: Symposium proceedings, W. J. Rietveld, U.S.D.A. Forest Service, General Technical Report RM-GTR-261, Fort Collins, Aug. 7-10, 1994.

[5292] *Zohar, Y.* and *Brandle, J. R.,* Shelter effects on growth and yield of corn in Nebraska. Layaaran. 28, 11-20, 1978.

SECTION 53

[5300] *Baker, B. M.*, and *Aron, R. H.*, Frost protection from small protective coverings. Geogr. Bull. 39, 29-32, 1997.

[5301] *Aron, R. H.*, Chilling as a factor in crop location with particular reference of deciduous orchards in California, Unpublished Ph.D. Thesis, Oregon State University, 1975.

[5302] —, Climatic chilling and future almond growing in Southern California. Prof. Geog. 23, 341-343, 1971.

[5303] —, Availability of chilling temperatures in California. Agr. Met. 28, 351-363, 1983.

[5304] —, and *Gat, Z.*, Estimating chilling duration from daily temperature extremes and elevation in Israel. Clim. Res. 1, 125-132, 1990.

[5305] *Ashworth, E. N.* and *Davis, G. A.*, Ice nucleation within peach trees, Amer. Soc. Hort. Sci. 109, 198-201, 1984.

[5306] —, *Anderson, J. A.* and *Davis, G. A.*, Properties of ice nuclei associated with peach trees. J. Am. Soc. Hort. Sci. 110, 287-291, 1985.

[5307] *Baier, W.*, Frostbekämpfung im Weinbau. Ber. DWD 2, Nr. 15, 1955.

[5308] *Bartholic, J. F.*, Foam insulation for freeze protection. In: Modification of the Aerial Environment of Plants, B. J. Barfield and J. F. Gerber (Eds.), Am. Soc. Agric. Engin. Monogr., 353-363, 1979.

[5309] *Blanc, M. L.*, *Geslin. H.*, *Holzberg, I. A.*, and *Mason, B.*, Protection Against Frost Damage. WMO Tech. Note 51, 1-62, 1963.

[5310] *Bourgeois, W. J.*, *Adams, A. J.*, and *Oberwortmann, D. H.*, Temperature modification in citrus trees through the use of low-volume irrigation. Hortsci., 22, 398-400, 1987.

[5311] *Bridley, S. F.*, *Taylor, R. J.*, and *Webber, R. T. J.*, The effects of irrigation and rolling on nocturnal air temperatures in vineyards. Agr. Met. 2, 373-383, 1965.

[5312] *Businger, J. A.*, Frost Protection with Irrigation, Agricultural Meteorology, Boston, Am. Meteor. Soc., 1965.

[5313] *Chen, E. Y.*, Estimating nocturnal surface temperature in Florida using thermal data from geostationary satellite data. Ph.D. Dissertation, Univ. Florida, Gainesville, 1979.

[5314] ——, *Allen, L. H., Jr.*, *Bartholic, J. F.*, and *Gerber, J. F.*, Delineation of cold-prone areas using nighttime SMS/GOES thermal data: Effects of soils and water. J. Appl. Met. 21, 1528-1537, 1982.

[5315] ——, ——, ——, and ——, Comparison of winter-nocturnal geostationary satellite infrared-surface temperature with shelter-height temperature in Florida. Rem. Sens. Env. 13, 313-327, 1983.

[5316] *Davies, F. S.*, *Jackson, L. K.*, and *Rippetoe, L. W.*, Low volume irrigation and tree wraps for cold protection of young 'Hamlin' orange trees. Proc. Fla. State Hort. Soc. 97, 25-27, 1984.

[5317] *Desjardins, R. L.*, and *Siminovitch, D.*, Microclimatic study of the effectiveness of foam as protection against frost. Agr. Met. 5, 291-296, 1968.

[5318] *Dinkelacker, O.*, and *Aichele, H.*, Versuche u. Erfahrungen über die Spätfrostbekämpfung in Württemberg. Techn. Mitt. d. Instr.wesens d. DWD, N. F. 4, 9-20, 1958.

[5319] —, —, Frostbekämpfung im Wein- u. Obstbau. Umschau 59, 241-244, 1959.

[5320] *Doesken, N. J.*, *McKee, T. B.*, and *Renquist, A. R.*, A climatological assessment of the utility of wind machines for freeze protection in mountain valleys. J. App. Met. 28, 194-205, 1989.

[5321] *Eimern, J. van,* Frostschutz mittels Propeller. Mitt. DWD 2, Nr. 12, 1955.

[5322] —, and *Loewel, E. L.*, Haben die Wassergräben in der Marsch des Alten Landes einer Bedeutung für den Frostschutz? Mitt. d. Obstbauversuchsrings des Alten Landes 8, Nr. 10, 1953.

[5323] *Gerber, J. F.*, Mixing the bottom of the atmosphere to modify temperatures on cold nights, In: Modification of the Aerial Environment of Plants, B. J. Barfield and J. F. Gerber (Eds.), Am. Soc. Agric. Engin. Monog., 315-326, 1979.

[5324] —, and *Martsolf, J. D.*, Sprinkling for Frost and Cold Protection, In: Modification of the Aerial Environment of Plants B. J. Barfield and J. F. Gerber (Eds.), Am. Soc. Agric. Engin. Monog., 327-333, 1979.

[5325] *Griffin, R. E.*, Micro-climatic control of deciduous fruit production with overhead sprinklers. In: Environmental Aspects of Irrigation and Drainage, University of Ottawa, American Society of Civil Engineers, 278-297. 1976.

[5326] *Hamer, P. J. C.*, A model to evaluate evaporative cooling of apple buds as a frost protective technique. J. Hort. Sci. 55, 157-163, 1980.

[5327] —, An automatic sprinkler system giving variable irrigation rates matched to measured frost protection needs. Agr. For. Met. 21, 281-293, 1980.

[5328] —, The heat balance of apple buds and blossoms, Part III. The water requirements for evaporative cooling by overhead sprinkler irrigation. Agr. For. Met. 37, 175-188, 1986.

[5329] *Hanyu, J.*, and *Tsugawa, K.*, On the effect of frost protection of paddy rice by irrigation. J. Agr. Met. Japan 10, 125-127, 1955.

[5330] *Itier, B.*, *Huber, L.*, and *Brun, O.*, The influence of artificial fog on conditions prevailing during nights of radiative frosts, Report on an experiment over a champagne vineyard. Agr. For. Met. 40, 163-176, 1987.

[5331] *Kalma, J. D.*, *Gregory, P. L.*, *Caprio, J. M.*, and *Hamer, P. J. C.* Advances in Bioclimatology 2, The Bioclimatology of Frost--Its Occurrence, Impact and Protection, Springer-Verlag, 1992.

[5332] *Kessler, O. W.*, and *Kaempfert, W.*, Die Frsotschadenverhütung. R. f. W. Wiss. Abh. 6, Nr. 2, 1940.

[5333] *Lindow, S. E.*, Population dynamics of epiphytic ice nucleation active bacteria on frost sensitive plants and frost control by means of antagonistic bacteria. In: Plant Cold Hardiness and Freezing Stress, Li, P. H. and Sakai, A. (Eds), 394-416. Academic Press, 1992.

[5334] —, Methods of preventing frost injury caused by epiphytic ice-nucleation-active bacteria. Plant Dis. 67, 327-333, 1983.

[5335] —, The role of bacterial ice nucleation in frost injury to plants, Ann. Rev. Phyt. 21, 363-384, 1983.

[5336] —, Integrated control and role of antibiotics in biological control of fireblight and frost injury. In: Biological Control on the Phyloplane, Windels, C. and Lindow, S. F. (Eds), 83-115, American Phytopathological Society Press, 1983.

[5337] —, Competitive exclusion of epiphytic bacteria by ice$^-$ *Pseudomonas syringae* mutants. App. Env. Microbio. 2520-2527, 1987.

[5338] —, Design and results of field tests of recombinant ice$^-$ *Pseudomonas syringae* strains. In: Risk Assessment in Agricultural Biotechnology, Proceedings of the International Conference, 61-68, 1988.

[5339] —, and *Panopoulos, N. J.*, Field tests of recombinant ice$^-$ *Pseudomonas syringae* for biological frost control in potato. In: The Release of Genetically-Engineered Micro-organisms, Sussman, M., Collins, C. H., Skinner, F. A. and Steward-Till (Eds.), Academic Press, 121-138, 1988.

[5340] *Lomas, J.*, *Gat, Z.*, *Borsok, Z.*, and *Raz, Z.*, Frost Atlas of Israel, Israel Meteorological Service, Bet-Degan, 1989.

[5341] —, and *Mandel, M.*, The quantitative effects of two methods of sprinkler irrigation on the microclimate of a mature avocado plantation. Agr. Met. 12, 35-48, 1973.

[5342] *Martsolf, J. D.*, Heating for frost protection, In: Modification of the Aerial Environment of Plants, B. J. Barfield and J. F. Gerber (Eds.), Am. Soc. Agric. Engin. Monog., 291-314, 1979.

[5343] ——, Satellite thermal maps provide detailed views and comparisons of freezes. Proc. Fla. State Hort. Soc. 95, 14-20, 1982.

[5344] ——, Cold protection strategies. Proc. Fla. State Soc. Hort. Sci. 103, 72-78, 1990.

[5345] ——, *Gerber, J. F., Chen, E. Y. , Jackson, J. L.*, and *Rose, A. J.*, What do satellite and other data suggest about past and future Florida freezes? Proc. Fla. State Hort. Soc. 97, 17-21, 1984.

[5346] ——, *Heinemann, P. H.*, and *Jackson, J. L.*, Satellite thermal imagery estimation of air temperature in areas during advective freezes. Proc. Fla. State Hort. Soc. 98, 48-52, 1985.

[5347] *Mee, T. R.*, Man-made fogs, In: Modification of the Aerial Environment of Plants, B. J. Barfield and J. F. Gerber (Eds.), Am. Soc. Agric. Engin. Monogr., 334-352, 1979.

[5348] *Oswalt, T. W.*, Comparison of satellite freeze forecast system thermal maps with conventionally observed temperatures. Proc. Fla. State Hort. Soc. 94, 43-45, 1981.

[5349] *Parsons, L. R., Combs, B. S.*, and *Tucker, D. P. H.*, Citrus freeze protection with microsprinkler irrigation during an advective freeze. Hortsci. 20, 1078-1080, 1985.

[5350] —, and *Wheaton, T. A.*, Microsprinkler irrigation for freeze protection: Evaporative cooling and extent of protection in an advective freeze. J. Am. Soc. Hort. Sci. 112, 897-902, 1987.

[5351] *Perry, K. B., Martsolf, J. D.*, and *Morrow, C. T.*, Conserving water in sprinkling for frost protection by intermittent application. J. Am. Soc. Hort. Sci. 105, 657-660, 1980.

[5352] *Peterson. M. L., Lin, S. S., Jones, D.*, and *Rutger, J. N.*, Cool night temperatures cause sterility in rice. Calif. Ag. 28, 12-14, 1974.

[5353] *Pogrell, H. v.*, Grundlegende Fragen der direkten Frostschutzberegnung. Verl. L. Leopold, Bonn 1958.

[5354] *Probsting, E. L.*, and *Gross, D. C.*, Field evaluations of frost injury to deciduous fruit trees as influenced by ice nucleation-active *Pseudomonas syringae*, J. Am. Soc. Hort. Sci. 113, 498-506, 1988.

[5355] *Recasens, J. R., Recasens, D. I.*, and *Barragan, J.*, Sprinkler irrigation to obtain a refreshing microclimate. J. Acta Hort. 228, 197-204, 1988.

[5356] *Renquist, A. R.*, The extent of fruit bud radiant cooling in relation to freeze protection with fans. Agr. For. Met. 36, 1-6, 1985.

[5357] *Rieger, N., Davies, F. S.*, and *Jackson, L. K.*, Microsprinkler irrigation and microclimate of young orange trees during freeze conditions. Hortscience 21, 1372-1374, 1986.

[5358] *Ryugo, K., Kester, D. E., Rough, D.*, and *Mikuckis, F.*, Effects of alar on almonds. Calif. Ag. 24, 14-15, 1970.

[5359] *Saudie, I., Whitewood, R., Raddatz, R. L.*, and *Fowler, D. B.*, Potential for winter wheat production in western Canada: A CERES model winterkill risk assessment. Can. J. Plant Sci. 71, 21-30, 1991.

[5360] *Sharratt, B. S.* and *Glenn, D. M.*, Orchard microclimatic observations in using soil-applied coal dust for frost protection. Agr. For. Met. 38, 181-192, 1986.

[5361] —, —, Orchard floor management utilizing soil-applied coal dust for frost protection. Part I. Potential microclimate modification on radiation frost nights. Agr. For. Met. 43, 71-82, 1988.

[5362] —, —, Orchard floor management utilizing soil-applied coal dust for frost protection. Part II. Seasonal microclimate effect. Agr. For. Met. 43, 147-154, 1988.

[5363] *Sibbett, G. S.* and *Bailey, M.*, Sunburn protection for newly-grafted Payne walnuts. Calif. Ag. 29, 18, 1975.

[5364] *Small, R. T.*, The use of wind machines and helicopter flights for frost protection. Bull. Am. Met. Soc. 30, 79-85, 1949.

[5365] *Snyder, R. L.*, and *Connell, J. H.*, Ground cover height affects pre-dawn orchard floor temperature. Calif. Ag. 47, 9-12, 1993.

[5366] *Strang, J. D., Lombard, P. B.*, and *Westwood, M. N.*, Effects of tree vigor and bloom delay by evaporative cooling on frost hardiness of 'bartlett' pear buds, flowers, and small fruit. J. Am. Soc. Hort. Sci. 105, 108-110, 1980.

[5367] *Tsuboi, Y., Honda, I., Hatagoshi, K.*, and *Yamato, M.*, On experiments of protection against frost damage by oil-burning in vineyard. J. Agr. Met. Japan 10, 109-112, 1955.

[5368] *Vaupel, A.*, Advektivfrost u. Strahlungsfrost. Mitt. DWD 3, Nr. 17, 1959.

[5369] *Westwood, M. N.*, Temperature Zone Pomology, W. H. Freeman Co., 1978.

SYMBOLS

A Absorption coefficient (fraction or percent)
A Area (m^2)
A Austausch coefficient (kg m^{-1} sec^{-1})
$A\lambda$ Absorptivity (fraction or percent)
A_C Surface area of human clothing (m^2)
A_H Austausch coefficient for transport of sensible heat (kg m^{-1} sec^{-1})
A_E Austausch coefficient for transport of water vapor (kg m^{-1} sec^{-1})
A_S Surface area of the human skin (m^2)
B_S Atmospheric backscatterance (fraction)
°C degrees Celsius
C Canopy water storage (mm)
D Diffuse solar radiation (W m^{-2})
D Transmission coefficient (fraction or percent)
D Transmission coefficient for solar radiation penetration into snow (percent m^{-1})
D Diameter of a forest clearing (m)
$D\lambda$ Transmissivity (fraction or percent)
E Blackbody radiation emission (W m^{-2})
E Evaporation (cm t^{-1}, mm hr^{-1})
E_o Saturation vapor pressure at air temperature (mb)
$\acute{E}$ Saturation vapor pressure at surface temperature (mb)
F_C Carbon dioxide flux above a canopy (mg CO_2 m^{-2} sec^{-1})
G_A Aerodynamic conductance (cm sec^{-1})
G_C Canopy conductance (cm sec^{-1})
G_{ST} Stomatal conductance (cm sec^{-1})
H Mean height of a forest stand (m)
I_o Solar radiation on a snow surface (W m^{-2})
I_z Solar radiation at depth z in snow (W m^{-2})
K Kelvin
$K\downarrow$ Global solar radiation (solar irradiance) (W m^{-2})
$K\uparrow$ Reflected solar radiation (W m^{-2})
K^* Net shortwave radiation (W m^{-2})
$K\downarrow_z$ Global solar radiation at depth z into a canopy (W m^{-2})
LAI Leaf area index of a plant canopy (m^2 m^{-2})
$L\downarrow$ Counterradiation from atmosphere (longwave irradiance) (W m^{-2})
$L\uparrow$ Outgoing longwave radiation (longwave emission, longwave emittance) (W m^{-2})
L^* Effective outgoing radiation (net longwave radiation) (W m^{-2})
$L\downarrow_{zenith}$ Counterradiation from the zenith (W m^{-2})
$L\downarrow_w$ Counterradiation from a cloudy sky (W m^{-2})
M Mean Earth-sun distance (km)

P Atmospheric pressure (mb)
PAR Photosynthetically-active solar radiation (W m^{-2})
Q^* Net radiation or radiation balance (W m^{-2})
Q_A Heat transfer to/from the ground surface due to heat advection (W m^{-2})
Q_A Sensible heat storage within a canopy (W m^{-2})
Q_B Biomass heat storage within a canopy (W m^{-2})
Q_{CS} Heat transfer from the body core to the human skin surface (J s^{-1})
Q_E Heat transfer to/from a surface due to latent heat exchanges (W m^{-2}); the numerical values are approximately equal to those in mm hr^{-1}
Q_{ED} Heat transfer from the human skin due to diffusion of water vapor (W m^{-2})
Q_{ER} Heat transfer by latent heat due to human respiration (W m^{-2})
Q_{ES} Heat transfer from the human skin due to evaporation of sweat (W m^{-2})
Q_{Es} Radiation fraction of evaporation (fraction)
Q_{Ev} Ventilation fraction of evaporation (fraction)
Q_F Heat transfer due to friction (W m^{-2})
Q_G Heat transfer within the soil to/from the ground surface (W m^{-2})
Q_{Gb} Heat transfer to/from a river channel bed (W m^{-2})
Q_H Heat transfer to/from the surface of the ground (solid) and the atmosphere (fluid) (W m^{-2})
Q_{HR} Heat transfer by sensible heat due to human respiration (W m^{-2})
Q_M Heat transfer due to advection from ocean currents (W m^{-2})
Q_{MH} Internal heat production by human metabolism (W m^{-2})
Q_P Photosynthetic energy storage within a canopy (W m^{-2})
Q_R Heat transfer to the ground by precipitation (W m^{-2})
Q_S Energy storage within a canopy (W m^{-2})
Q_{SC} Heat transfer to/from the human skin through clothing to/from the outer clothing surface (J s^{-1})
Q_V Latent heat storage within a canopy (W m^{-2})
Q_W Heat transfer from within a water body to/from its surface (W m^{-2})
R Earth's radius (km)
R Reflection coefficient (albedo) (fraction or percent)
RH Relative humidity of the atmosphere (percent)
R_A Aerodynamic resistance (sec cm^{-1})
R_C Canopy resistance (sec cm^{-1})
R_{CL} Resistance of clothing to the transfer of heat (m^2 K W^{-1})
R_S Surface resistance (sec cm^{-1})
R_{ST} Stomatal resistance (s cm^{-1})
R_T Regional albedo of surrounding terrain (fraction or percent)
$R\lambda$ Reflectivity (fraction or percent)
S Solar constant (W m^{-2})
S Direct-beam solar radiation (W m^{-2})
S Canopy storage parameter (canopy storage parameter) (mm)

T Temperature (°Celsius); data from papers using the °F scale have been converted into Celsius unless specifically stated otherwise.
T Canopy transpiration ($W\ m^{-2}$)
T_A Air temperature (°C)
T_C Human core body temperature (°C)
T_{CL} Surface temperature of clothing (°C)
T_E Mean black body surface temperature of Earth (K)
T_l Leaf surface temperature (°C)
T_R Apparent surface radiative temperature (radiant temperature) (K)
T_S Absolute surface temperature (K)
T_{SK} Human skin temperature (°C)
T_{sky} Sky temperature (K)
VPD Vapor pressure deficit (mb)
V_B Blood flow density from the human body core to the skin ($l\ s^{-1}\ m^{-2}$)
V_S Speed of sound ($m\ sec^{-1}$)
WUE Water use efficiency (mg CO_2 (g H_2O)$^{-1}$)
Z Solar zenith angle (degrees)

a Thermal diffusivity ($m^2\ sec^{-1}$)
a_L Energy transport coefficient ($W\ m^{-2}\ K^{-1}$)
a_p Planetary albedo (fraction or percent)
b Rate of surface soil moisture change ($cm\ t^{-1}$)
c Specific heat ($J\ kg^{-1}\ K^{-1}$)
c_B Specific heat of human blood ($J\ kg^{-1}\ K^{-1}$)
c_p Specific heat of air ($J\ kg^{-1}\ K^{-1}$)
c_s Specific heat of soil organic and inorganic substances ($J\ kg^{-1}\ K^{-1}$)
c_w Specific heat of water ($J\ kg^{-1}\ K^{-1}$)
d Day
d Zero-plane displacement (cm)
e Water vapor pressure of the air (mb)
e_s specific humidity of surface ($g\ kg^{-1}$)
e_a specific humidity of air ($g\ kg^{-1}$)
f Runoff ($cm\ t^{-1}$)
h Screening angle, or angle from the horizon to the top of an obstruction (degrees)
h_c Heat transfer coefficient ($W\ m^{-2}\ K^{-1}$)
hr Hour
k Cloud type factor for counterradiation from a cloudy sky (fraction)
k Eddy diffusivity, A/r ($m^2\ sec^{-1}$)
k von Kármán constant, 0.40
m Optical air mass (fraction)

p	Free throughfall fraction (fraction)
pC_i	picoCurie
q	Specific humidity (g kg^{-1})
r	Precipitation (cm t^{-1})
r_B	Human blood density (kg l^{-1})
r_f	Latent heat of fusion (MJ kg^{-1})
r_s	Latent heat of sublimation (MJ kg^{-1})
r_v	Latent heat of vaporization of water (MJ kg^{-1})
s	Sun's radius (km)
t	Time (unit stated when used)
u	Depth of precipitable water (cm)
u	Wind speed (m sec^{-1})
u_*	Friction velocity (shear velocity) (m sec^{-1})
v	Extinction coefficient (cm^{-1})
v_e	Volume fraction of ice in soil (fraction or percent)
v_l	Volume fraction of air in soil (fraction or percent)
v_s	Volume fraction of organic and inorganic substances in soil (fraction or percent)
v_w	Volume fraction of water in soil (fraction or percent)
w	Fraction of sky covered by clouds (fraction)
z	Height above/below the ground surface (m)
z_0	Roughness length (roughness parameter) (m)

ε	Surface emissivity (fraction)
ε_A	Atmospheric emissivity at shelter-height (fraction)
ε_{sky}	Sky emissivity (fraction)
γ	Vertical temperature gradient (°C/100 m), negative when temperature decreases with height, positive with inversions
ψ_{sky}	Sky view factor (fraction)
Θ	Potential temperature (°C)
λ	Thermal conductivity (W m^{-1} K^{-1})
λ	Wavelength (m)
λ_m	Wavelength of median intensity of radiation for a black body (m)
λ_{max}	Wavelength of maximum intensity of radiation for a blackbody (m)
μ	Micron, 10^{-6} m
ρ	Density of air or homogeneous substance (kg m^{-3})
ρ_m	Density of natural soil (kg m^{-3})
ρ_s	Density of soil organic and inorganic substances (kg m^{-3})
$(\rho c)_m$	Thermal capacity by volume of natural soil (J m^{-3} K^{-1})
σ	Stefan-Boltzmann constant, $5.675 \cdot 10^{-8}$ W m^{-2} K^{-4}
β	Bowen ratio (fraction)
Δ	Slope of the saturation specific humidity curve with air temperature

Conversion Table

Item	Unit Name	Approximate conversions	
Length	meter (m) kilometer (km)	1 m = 39.37 inches = 3.281 feet 1 km = 0.6214 mile	1 inch = 0.0254 m 1 foot = 0.3048 m 1 mile = 1.609 km
Area	square meter (m^2) square centimeter (cm^2) hectare (ha) square kilometer (km^2)	1 m^2 = 10.76 square feet = 1.196 square yards 1 cm^2 = 0.155 sq inch 1 ha = 2.471 acres 1 km^2 = 0.3861 sq mile = 247.1 acres	1 square foot = 9.0929 m^2 1 square yard = 0.8361 m^2 1 square inch = 6.452 cm^2 1 acre = 0.4047 ha 1 square mile = 2.590 km^2
Volume	cubic meter (m^3) cubic centimeter (cm^3)	1 m^3 = 35.31 cubic feet 1 cm^3 = 0.06102 cubic inch	1 cubic foot = 0.02832 m^3 1 cubic inch = 16.39 cm
Liter	liter (l)	1 l = 1.760 pints = 0.220 UK gallon = 0.2642 US gallon	1 pint = 0.5683 l 1 UK gallon = 4.546 l 1 US gallon = 3.785 l
Mass	kilogram (kg) gram (g)	1 kg = 2.205 pounds 1 g = 0.03527 ounce	1 pound = 0.4536 kg 1 ounce avoirdupois = 28.35 g
Density	kilogram per cubic meter (kg/m^3) gram per cubic centimeter (g/cm^3)	1 kg/m^3 = 0.06243 pound ft^{-3} 1 g/cm^3 = 0.03613 pound in^{-3}	1 pound ft^{-3} = 16.02 kg/m^3 1 lb/cu inch = 27.68 g/cm^3 = 27.68 t/m^3
Pressure	Pascal (Pa) millibar (mb)	1 Pa = 0.01 millibar = 0.00001 bar 1 mb = 0.7501 mm mercury	1 millibar = 100 Pa 1 bar = 100 kPa 1 pound force per square inch = 6.895 kPa 1 inch mercury = 3.386 kPa 1 mm mercury = 133.3 Pa = 1.333 hPa
Velocity	meter per second (m sec^{-1}) kilometer per hour (km hr^{-1})	1 m sec^{-1} = 3.281 feet sec^{-1} = 2.237 mph = 1.944 knots = 3.600 km hr^{-1} 1 km hr^{-1} = 0.540 knot	1 foot sec^{-1} = 0.3048 m sec^{-1} 1 mile hr^{-1} = 0.447 m sec^{-1} = 1.609 km hr^{-1} 1 knot = 0.5144 m sec^{-1} 1 km/h = 0.2778 m sec^{-1} 1 foot min^{-1} = 5.080 mm sec^{-1}
Power	watt (W), i.e. J sec^{-1}	1 W = 0.2388 cal sec^{-1}	1 cal sec^{-1} = 4.187 W
Energy, Work	joule (J) megajoule (MJ)	1 J = 0.2388 calorie = 0.0002388 kilocalorie 1 MJ = 0.2778 kWh	1 calorie = 4.186 J 1 kilocalorie = 4.186 kJ 1 kilowatt-hour = 3.6 MJ
Energy, density	watt per square meter (W m^{-2})	1 W m^{-2} = 0.001433 cal cm^{-2} min^{-1} = 2.064 cal cm^{-2} day^{-1}	1cal cm^{-2} min^{-1} = 69.78 mW cm^{-2} 1 calorie cm^{-2} day^{-1} = 0.4845 W m^{-2}
Angle	radian	1 radian = 57°18'	100° = 1.745 radians

AUTHOR INDEX

L

M

Y

Z

SUBJECT INDEX

D

G

H

R

Z